738.134

Stoke-on-Trent Libraries
Approved for Sale

HORACE BARKS
REFERENCE LIBRARY

STOKE-ON-TRENT

ADVANCES IN CERAMICS • VOLUME 12

SCIENCE AND TECHNOLOGY OF ZIRCONIA II

Volume 1   Grain Boundary Phenomena in Electronic Ceramics
Volume 2   Physics of Fiber Optics
Volume 3   Science and Technology of Zirconia
Volume 4   Nucleation and Crystallization in Glasses
Volume 5   Materials Processing in Space
Volume 6   Character of Grain Boundaries
Volume 7   Additives and Interfaces in Electronic Ceramics
Volume 8   Nuclear Waste Management
Volume 9   Forming of Ceramics
Volume 10 Structure and Properties of MgO and Al_2O_3 Ceramics
Volume 11 Processing for Improved Productivity

ADVANCES IN CERAMICS • VOLUME 12

SCIENCE AND TECHNOLOGY OF ZIRCONIA II

Edited by

Nils Claussen,
Manfred Rühle
Max-Planck-Institut für Metallforschung
Institut für Werkstoffwissenschaften
7000 Stuttgart 1
Federal Republic of Germany

Arthur H. Heuer
Case Western Reserve University
Cleveland, Ohio

The American Ceramic Society, Inc.
Columbus, Ohio

Proceedings of the Second International Conference on the Science and Technology of Zirconia (Zirconia '83), held in Stuttgart, Federal Republic of Germany, June 21-23, 1983.

Library of Congress Cataloging in Publication Data

International Conference on the Science and Technology of Zirconia (2nd : 1983 : Stuttgart, Germany)
  Science and technology of zirconia II.

  (Advances in ceramics, ISSN 0730-9546 ; v. 12)
  "Proceedings of the Second International Conference on the Science and Technology of Zirconia (Zirconia '83), held in Stuttgart, Federal Republic of Germany, June 21-23, 1983"—T.p. verso.
  Bibliography: p.
  Includes index.
  1. Zirconium oxide—Congresses. I. Claussen, Nils. II.Rühle, Manfred. III. Heuer, A. H. (Arthur Harold), 1936-   .   IV. American Ceramic Society. V. Title. VI. Series.
TP245.Z8157   1983   666   84-14498
ISBN 0-916094-64-2

Coden: ADCEDE

©1984 by The American Ceramic Society, Inc. All rights reserved.

No part of this book may be reproduced, stored in a retrieval system, or transmitted in any form or by any means, electronic, mechanical, photocopying, microfilming, recording, or otherwise, without written permission from the publisher.

Printed in the United States of America.

Foreword

The Second International Conference on the Science and Technology of Zirconia (Zirconia '83) was held in Stuttgart, Federal Republic of Germany, June 21–23, 1983, under the Chairmanship of Nils Claussen and Manfred Rühle. The Chairmen were assisted by an International Advisory Committee, consisting of A. M. Anthony (France), A. J. Burggraaf (Netherlands), R. C. Garvie (Australia), A. H. Heuer (U.S.A.), V. Longo (Italy), R. Pampuch (Poland), W. Pompe (German Democratic Republic), S. Sōmiya (Japan), B. C. Steele (U.K.), and E. C. Subbarao (India). Zirconia '83 was organized by the Max-Planck-Institut für Metallforschung, Institut für Werkstoffwissenschaften, and was sponsored by the Deutsche Keramische Gesellschaft, the American Ceramic Society, the Ceramic Society of Japan, and the International Union of Pure and Applied Chemistry (IUPAC).

Like its predecessor, Zirconia '80 (published as Volume 3 in this series), Zirconia '83 aimed at bringing together those scientists and engineers interested in developing, exploiting, and understanding the significant promise of ZrO_2-containing ceramics. At least by way of numbers, the conference was spectacularly successful! There were 375 participants from 22 countries; 118 papers were presented in 6 sessions, each of which included keynote lectures, and oral and poster contributions.

We are particularly pleased to note that the attendance of 50 graduate students from Germany, England, and the US was made possible through contributions and exhibitor fees from the following companies: Ceramatec Inc., Salt Lake City, Utah, USA; Ceramica Industriale FER, Seregno, Italy; Coors Porcelain, Golden, Colorado, USA; Corning Glass GmbH, Wiesbaden, Federal Republic of Germany; Daiichi Kigenso Kagaku Kogyo, Osaka, Japan; Daimler-Benz AG, Stuttgart, Federal Republic of Germany; Didier-Werke AG, Wiesbaden, Federal Republic of Germany; Dynamit-Nobel AG, Troisdorf, Federal Republic of Germany; Feldmühle AG, Plochingen, Federal Republic of Germany; Goldschmidt AG, Essen, Federal Republic of Germany; H. C. Starck, Goslar, Federal Republic of Germany; Kureha Chemical Ind. Co. Ltd., Tokyo, Japan; Magnesium Elektron Twickenham, U.K.; Magnesium Elektron, Flemington, New Jersey, USA; NGK Insulators, Ltd., Nagoya, Japan; Nilsen Sintered Products, Northcote, Australia; Norton Co., Worcester, Massachusetts, USA; PCUK-Rubis Synthetique des Alpes, Levallois, France; Robert Bosch GmbH, Stuttgart, Federal Republic of Germany; Smiths Industries, Warwickshire, U.K.; Toshiba Ceramics, Tokyo, Japan; Viking Chemicals Co., Føllenslev, Denmark; Zircar Products, Inc., Florida, New York, USA. We especially thank Daimler-Benz for providing all the transportation at the Conference. Further financial support was provided by the Deutsche Forschungsgemeinschaft, Ministerium für Wissenschaft und Kunst of the State of Baden-Württemberg.

Turning to these Proceedings, they accurately reflect the state of understanding in mid-1983 of advanced ceramics based on ZrO_2. Interest in such advanced ceramics stems from the high ionic conductivity of ZrO_2, and from its ability to toughen ZrO_2-containing ceramics through the martensitic transformation of the tetragonal polymorph to monoclinic symmetry. Both aspects of ZrO_2 are treated in these Proceedings, either from a scientific or a technological viewpoint.

The single most important change in the contents of this volume, compared with the predecessor Zirconia '80, is the burgeoning interest in processing which the technological applications of ZrO_2 have fostered.

Concerning terminology: We have edited all the papers such that the cubic, tetragonal, and monoclinic forms of ZrO_2, whether pure or as solid solutions, are denoted c-ZrO_2, t-ZrO_2, and m-ZrO_2, respectively. Partially stabilized and fully stabilized materials (here of course referring to the content of c-ZrO_2) are denoted Mg-PSZ, Y-CSZ, etc. Composite ceramics, such as Al_2O_3-ZrO_2 are denoted ZTA for ZrO_2-toughened Al_2O_3, ZTS for ZrO_2-toughened $MgAl_2O_4$ spinel, etc. To distinguish the newest type of advanced ceramics, those which contained all (or mostly all) t-ZrO_2, we recommend Y-TZP for Y_2O_3-containing tetragonal ZrO_2 polycrystals, Ce-TZP, etc. We have found these abbreviations and acronyms convenient and useful and recommend them to the scientific community.

The reader should also be appraised of the way these papers were reviewed. Each paper was reviewed by at least one expert in the field, who was asked to recommend outright acceptance, acceptance with recommended changes, acceptance after mandatory modification, or outright rejection, *using the standards of the Journal of the American Ceramic Society*. Surprisingly few papers were recommended for outright acceptance, surprisingly many for outright rejection. One of us then independently reviewed the manuscript; most often, we reached the same conclusions as the reviewer, and then negotiated with the authors the style and content of the final version. Many of our authors were at first surprised at our "hard-nosed" attitude, but understood our position when we explained the reviewing standards we were trying to maintain and the page limitation we were under. We thank them one and all for their forbearance.

Finally, these Proceedings have been published in a reasonably timely period after the Conference. For this, we extend thanks to the staff at the American Ceramic Society, and to our reviewers, who acceded to our requests for speedy reviews.

Nils Claussen and Manfred Rühle
Stuttgart, Republic of Germany

Arthur H. Heuer
Cleveland, Ohio

Contents

SECTION I. PHASE TRANSFORMATIONS AND PHASE STABILITY

Phase Transformations in ZrO_2-Containing Ceramics:
I, The Instability of c-ZrO_2 and the Resulting Diffusion-Controlled Reactions................................. 1
    A. H. Heuer and M. Rühle

Phase Tranformations in ZrO_2-Containing Ceramics:
II, The Martensitic Reaction in t-ZrO_2..................... 14
    M. Rühle and A. H. Heuer

Martensitic Transformations in ZrO_2 and HfO_2—An Assessment of Small-Particle Experiments with Metal and Ceramic Matrices.................................. 33
    I.-W. Chen and Y.-H. Chiao

Theory of Twinning and Transformation Modes in ZrO_2........ 46
    M. A. Choudhry and A. G. Crocker

Acoustic Emission Characterization of the Tetragonal-Monoclinic Phase Transformation in Zirconia............... 54
    D. R. Clarke and A. Arora

The Transformation Mechanism of Spherical Zirconia Particles in Alumina..................................... 64
    W. M. Kriven

Diffusionless Transformations in Zirconia Alloys............. 78
    C. A. Andersson, J. Greggi, Jr., and T. K. Gupta

Short-Range Order Phenomena in ZrO_2 Solid Solutions........ 86
    R. Chaim and D. G. Brandon

Phase Relationships in Some ZrO_2 Systems................... 96
    V. S. Stubican, G. S. Corman, J. R. Hellmann, and G. Senft

Ordered Compounds in the System CaO-ZrO_2................. 107
    J. Hangas, T. E. Mitchell, and A. H. Heuer

Tetragonal Phase in the System ZrO_2-Y_2O_3.................. 118
    V. Lanteri, A. H. Heuer, and T. E. Mitchell

Polydomain Crystals of Single-Phase Tetragonal ZrO_2:
Structure, Microstructure, and Fracture Toughness........... 131
    D. Michel, L. Mazerolles, and M. Perez y Jorba

The Phase $Mg_2Zr_5O_{12}$ in MgO Partially Stabilized Zirconia...... 139
H. J. Rossell and R. H. J. Hannink

Diffusional Decomposition of c-ZrO_2 in Mg-PSZ............... 152
S. C. Farmer, T. E. Mitchell, and A. H. Heuer

High-Resolution Microscopy Investigation of the System
ZrO_2-ZrN... 164
G. van Tendeloo and G. Thomas

Compatibility Relationships of Al_2O_3 and ZrO_2 in the
System ZrO_2-Al_2O_3-SiO_2-CaO.............................. 174
P. Pena and S. De Aza

Phase Relations in the Ternary System ZrO_2-Al_2O_3-SiO_2 by
the Slow-Cooling Float-Zone Method........................ 181
I. Shindo, S. Takekawa, K. Kosuda, T. Suzuki, and Y. Kawata

SECTION II. TRANSFORMATION TOUGHENING—MECHANICAL ASPECTS

Toughening Mechanisms in Zirconia Alloys................. 193
A. G. Evans

A Thermodynamic Approach to Fracture Toughness in PSZ.... 213
R. J. Seyler, S. Lee, and S. J. Burns

R-Curve Behavior in Zirconia Ceramics...................... 225
M. V. Swain and R. H. J. Hannink

Residual Surface Stresses in Al_2O_3-ZrO_2 Composites.......... 240
D. J. Green, F. F. Lange, and M. R. James

Calculations of Strain Distributions in and around ZrO_2
Inclusions... 251
S. Schmauder, W. Mader, and M. Rühle

In-Situ Observations of Stress-Induced Phase Transformations
in ZrO_2-Containing Ceramics................................ 256
M. Rühle, B. Kraus, A. Strecker, and D. Waidelich

In-Situ Straining Experiments of Mg-PSZ Single Crystals...... 275
L. H. Schoenlein, M. Rühle, and A. H. Heuer

Theoretical Approach to Energy-Dissipative Mechanisms
in Zirconia and Other Ceramics............................. 283
W. Pompe and W. Kreher

Microcracking Contributions to the Toughness
of ZrO_2-Based Ceramics.................................... 293
K. T. Faber

Microcrack Extension in Microcracked
Dispersion-Toughened Ceramics.......................... 306
   F. E. Buresch

SECTION III. TRANSFORMATION TOUGHENING— MICROSTRUCTURAL ASPECTS

Microstructural Design of Zirconia-Toughened
Ceramics (ZTC).. 325
   N. Claussen

Microstructural Studies of Y_2O_3-Containing Tetragonal
ZrO_2 Polycrystals (Y-TZP)............................. 352
   M. Rühle, N. Claussen, and A. H. Heuer

Effect of Microstructure on the Strength of Y-TZP
Components... 371
   M. Matsui, T. Soma, and I. Oda

Thermal and Mechanical Properties of
Y_2O_3-Stabilized Tetragonal Zirconia Polycrystals............. 382
   K. Tsukuma, Y. Kubota, and T. Tsukidate

Aging Behavior of Y-TZP................................ 391
   W. Watanabe, S. Iio, and I. Fukuura

Phase Stability of Y-PSZ in Aqueous Solutions............. 399
   K. Nakajima, K. Kobayashi, and Y. Murata

Physical, Mircostructural, and Thermomechanical
Properties of ZrO_2 Single Crystals....................... 408
   R. P. Ingel, D. Lewis, B. A. Bender, and R. W. Rice

Ripening of Inter- and Intragranular ZrO_2 Particles in
ZrO_2-Toughened Al_2O_3............................... 415
   B. W. Kibbel and A. H. Heuer

Anomalous Thermal Expansion in Al_2O_3-15 Vol% $(Zr_{0.5}Hf_{0.5})O_2$.. 425
   W. M. Kriven and E. Bischoff

Improvement in the Toughness of β''-Alumina by
Incorporation of Unstabilized Zirconia Particles.............. 428
   J. G. P. Binner, R. Stevens, and S. R. Tan

Microstructure and Property Development of In Situ-Reacted
Mullite-ZrO_2 Composites................................ 436
   J. S. Wallace, G. Petzow, and N. Claussen

Size Effect on Transformation Temperature of Zirconia
Powders and Inclusions.................................. 443
   I. Müller and W. Müller

Relationship Between Morphology and Structure for
Stabilized Zirconia Crystals.................................. 455
    D. Michel

SECTION IV. STRUCTURAL AND OTHER APPLICATIONS

Structural Applications of ZrO_2-Bearing Materials............ 465
    R. C. Garvie

ZrO_2 Ceramics for Internal Combustion
Engines.. 480
    U. Dworak, H. Olapinski, D. Fingerle, and U. Krohn

Plasma-Sprayed Zirconia Coatings............................. 488
    P. Boch, P. Fauchais, D. Lombard, B. Rogeaux, and M. Vardelle

Microstructure and Durability of Zirconia Thermal
Barrier Coatings... 503
    D. S. Suhr, T. E. Mitchell, and R. J. Keller

Thermal Diffusivity of Zirconia Partially and Fully
Stabilized with Magnesia..................................... 518
    W. J. Buykx and M. V. Swain

Mechanical, Thermal, and Electrical Properties in the
System of Stabilized $ZrO_2(Y_2O_3)/\alpha$-Al_2O_3................ 528
    F. J. Esper, K. H. Friese, and H. Geier

The Reaction-Bonded Zirconia Oxygen Sensor: An Application for Solid-State Metal-Ceramic Reaction-Bonding....... 537
    R. V. Allen, W. E. Borbidge, and P. T. Whelan

Properties of Metal-Modified and Nonstoichiometric ZrO_2..... 544
    R. Ruh

Diffusion Processes and Solid-State Reactions in the
Systems Al_3O_3-ZrO_2(Stabilizing Oxide)(Y_2O_3, CaO, MgO)........ 546
    T. Kosmac, D. Kolar, and M. Trontelj

SECTION V. ELECTROLYTIC PROPERTIES AND APPLICATIONS

Defect Structure and Transport Properties of ZrO_2-Based
Solid Electrolytes... 555
    J. F. Baumard and P. Abelard

Microstructural-Electrical Property Relationships in
High-Conductivity Zirconias............................. 572
    E. P. Butler, R. K. Slotwinski, N. Bonanos, J. Drennan, and
    B. C. H. Steele

Low-Temperature Properties of Samaria-Stabilized
Zirconia... 585
    M. Goge, G. Letisse, and M. Gouet

Influence of Impurities in Solid Electrolytes on the
Voltage Response of Solid Electrolyte Galvanic Cells......... 591
    T. Reetz, H. Näfe, and D. Rettig

Low-Temperature Behavior of ZrO_2 Oxygen Sensors........... 598
    S. P. S. Badwal, M. J. Bannister, and W. G. Garrett

Life and Performance of ZrO_2-Based Oxygen Sensors......... 607
    B. Krafthefer, P. Bohrer, P. Meonkhaus, D. Zook, L. Pertl, and
    U. Bonne

Accurate Monitoring of Low Oxygen Activity in Gases
with Conventional Oxygen Gauges and Pumps................. 618
    J. Fouletier, E. Siebert, and A. Caneiro

Collaborative Study on ZrO_2 Oxygen Gauges................. 627
    A. M. Anthony, J. F. Baumard, and J. Corish

Computer-Controlled Adjustment of Oxygen Partial
Pressure... 631
    F. Vizethum, G. Bauer, and G. Tomandl

Oxygen Sensing in Iron- and Steelmaking................... 636
    D. Janke

Mixed Ionic and Electronic Conduction in Zirconia
and Its Application in Metallurgy........................... 646
    M. Iwase, K. T. Jacob, and I. Ichise

ZrO_2 Oxygen and Hydrogen Sensors: A Geologic
Perspective.. 660
    G. C. Ulmer

Application of Zirconia Membranes as High-Temperature
pH Sensors... 672
    L. W. Neidrach

Preparation and Operation of Zirconia High-Temperature
Electrolysis Cells for Hydrogen Production................. 685
    E. Erdle, A. Koch, W. Schäfer, F. J. Esper, and K. H. Friese

SECTION VI. PROCESSING

Processing Techniques for ZrO$_2$ Ceramics 693
S. Wu and R. J. Brook

Sinterability of ZrO$_2$ and Al$_2$O$_3$ Powders: The Role of Pore Coordination Number Distribution 699
F. F. Lange and B. I. Davis

Sintering Kinetics of ZrO$_2$ Powders 714
A. Roosen and H. Hausner

Sintering of a Freeze-Dried 10 Mol% Y$_2$O$_3$-Stabilized Zirconia ... 727
L. Rakotoson and M. Paulus

ZrO$_2$ Micropowders as Model Systems for the Study of Sintering 733
R. Pampuch

Wet-Chemical Preparation of Zirconia Powders: Their Microstructure and Behavior 744
M. A. C. G. van de Graaf and A. J. Burggraaf

Preparation of Y$_2$O$_3$-Stabilized Tetragonal Polycrystals (Y-TZP) from Different Powders 766
H. Schubert, N. Claussen, and M. Rühle

Preparation of Ca-Stabilized ZrO$_2$ Micropowders by a Hydrothermal Method 774
K. Haberko and W. Pyda

Growth and Coarsening of Pure and Doped ZrO$_2$: A Study of Microstructure by Small-Angle Neutron Scattering 784
A. F. Wright, N. H. Brett, and S. Nunn

Al$_2$O$_3$-ZrO$_2$ Ceramics Prepared from CVD Powders 794
S. Hori, M. Yoshimura, S. Sōmiya, and R. Takahashi

Preparation of Mixed Fine Al$_2$O$_3$-HfO$_2$ Powders by Hydrothermal Oxidation 806
H. Toraya, M. Yoshimura, and S. Sōmiya

Applications of Rapid Solidification Theory and Practice to Al$_2$O$_3$-ZrO$_2$ Ceramics 816
G. Kalonji, J. McKittrick, and L. W. Hobbs

Synthesis and Sintering of ZrO$_2$ Fibers 826
I. N. Yermolenko, T. M. Ulyanova, P. A. Vityaz, and I. L. Fyodorova

Epilogue 833
R. J. Brook

Section I
Phase Transformations and Phase Stability

Phase Transformations in ZrO_2-Containing Ceramics: I, The Instability of c-ZrO_2 and the Resulting Diffusion-Controlled Reactions 1
    A. H. Heuer and M. Rühle

Phase Tranformations in ZrO_2-Containing Ceramics: II, The Martensitic Reaction in t-ZrO_2 14
    M. Rühle and A. H. Heuer

Martensitic Transformations in ZrO_2 and HfO_2—An Assessment of Small-Particle Experiments with Metal and Ceramic Matrices 33
    I.-W. Chen and Y.-H. Chiao

Theory of Twinning and Transformation Modes in ZrO_2 46
    M. A. Choudhry and A. G. Crocker

Acoustic Emission Characterization of the Tetragonal-Monoclinic Phase Transformation in Zirconia 54
    D. R. Clarke and A. Arora

The Transformation Mechanism of Spherical Zirconia Particles in Alumina 64
    W. M. Kriven

Diffusionless Transformations in Zirconia Alloys 78
    C. A. Andersson, J. Greggi, Jr., and T. K. Gupta

Short-Range Order Phenomena in ZrO_2 Solid Solutions 86
    R. Chaim and D. G. Brandon

Phase Relationships in Some ZrO_2 Systems 96
    V. S. Stubican, G. S. Corman, J. R. Hellmann, and G. Senft

Ordered Compounds in the System CaO-ZrO_2 107
    J. Hangas, T. E. Mitchell, and A. H. Heuer

Tetragonal Phase in the System ZrO_2-Y_2O_3 118
    V. Lanteri, A. H. Heuer, and T. E. Mitchell

Polydomain Crystals of Single-Phase Tetragonal ZrO_2: Structure, Microstructure, and Fracture Toughness............ 131
  D. Michel, L. Mazerolles, and M. Perez y Jorba

The Phase $Mg_2Zr_5O_{12}$ in MgO Partially Stabilized Zirconia...... 139
  H. J. Rossell and R. H. J. Hannink

Diffusional Decomposition of c-ZrO_2 in Mg-PSZ.............. 152
  S. C. Farmer, T. E. Mitchell, and A. H. Heuer

High-Resolution Microscopy Investigation of the System ZrO_2-ZrN 164
  G. van Tendeloo and G. Thomas

Compatibility Relationships of Al_2O_3 and ZrO_2 in the System ZrO_2-Al_2O_3-SiO_2-CaO............................. 174
  P. Pena and S. De Aza

Phase Relations in the Ternary System ZrO_2-Al_2O_3-SiO_2 by the Slow-Cooling Float-Zone Method...................... 181
  I. Shindo, S. Takekawa, K. Kosuda, T. Suzuki, and Y. Kawata

Phase Transformations in ZrO_2-Containing Ceramics: I, The Instability of c-ZrO_2 and the Resulting Diffusion-Controlled Reactions

A. H. HEUER

Case Western Reserve University
Department of Metallurgy and Materials Science
Cleveland, OH 44106

M. RÜHLE

Max-Planck-Institut für Metallforschung
Institut für Werkstoffwissenschaften
Stuttgart, Federal Republic of Germany

c-ZrO_2 appears to be intrinsically unstable, and undergoes a variety of diffusion-controlled transformations. The origin of this instability is discussed, as well as the subsolidus phase equilibria in the systems MgO-ZrO_2, CaO-ZrO_2, and Y_2O_3-ZrO_2. The diffusion-controlled reactions resulting from this instability include precipitation of either t-ZrO_2 or one or more ordered defect-fluorite intermediate compounds, or eutectoid decomposition (in the system MgO-ZrO_2) to m-ZrO_2 and MgO. Finally, a displacive $c \rightarrow t$ transformation in the system Y_2O_3-ZrO_2, which we have called "homogeneous massive," is discussed.

Transformation-toughening in ZrO_2-containing ceramics requires the presence of the tetragonal (t) form of ZrO_2 under service conditions, so that the toughening obtainable from the stress-induced martensitic transformation to monoclinic (m) symmetry can be realized; this phenomenon is discussed in detail in the second paper in this series.[1] Thus, the various high-temperature phase transformations which control microstructural evolution, and determine the morphology, size, and location of the t-ZrO_2, become of paramount importance.

Rather than simply discussing all the diffusion-controlled reactions involving t-ZrO_2, we have chosen to focus on the *instability* of the cubic (c) polymorph (c-ZrO_2) from which t-ZrO_2 usually forms. This complements the focus on the instability of t-ZrO_2 in the succeeding paper and emphasizes an aspect of the crystal chemistry of ZrO_2 which has perhaps received insufficient attention in the past. In our treatment, conventional phase equilibria considerations will be of interest, as will the diffusion-controlled reactions that operate in various ZrO_2-containing systems as equilibrium is approached. Because of the sluggish diffusion kinetics which are so common in ceramic systems, equilibrium assemblages cannot always be attained and metastable phases and metastable phase transformations may occur.

In past years, ceramists were greatly concerned with "destabilization" of c-ZrO_2 and the resulting degradation of mechanical properties.[2] In more recent years, the recognition that t-ZrO_2 could precipitate from a c-ZrO_2 matrix and subsequently undergo the stress-induced martensitic transformation[3,4] was one of the crucial discoveries leading to our current interest in transformation-toughening in ceramics. The instability underlying these phenomena needs to be understood.

Instability of c-ZrO$_2$

Pure bulk c-ZrO$_2$ is stable only from $\approx$2370°C to the melting point at $\approx$2680°C.[5] The instability of the c polymorph of ZrO$_2$ at lower temperatures, and that of the c phase in chemically similar HfO$_2$, are unique among the numerous M^{2+}F$_2$ and M^{4+}O$_2$ compounds with the fluorite structure. As is well known, Pauling[6] provided a set of semiempirical "rules" based on radius ratio ($r_{cation} \div r_{anion}$) considerations for predicting crystal structures of inorganic materials. For AX$_2$ compounds, those solids with large cations and radius ratios approaching 1 should have the fluorite structure, whereas those with small cations and radius ratios <0.4 should have tetrahedral coordination and crystallize with one of the silica structures; intermediate-sized cations should form AX$_2$ compounds with the rutile structure. Although exceptions are known, Pauling's rules do derive from fundamental aspects of chemical bonding and are widely applicable.

For ZrO$_2$, however, the compilation of Shannon and Prewitt[7] shows that the radius ratio for ZrO$_2$ in eightfold coordination is 0.59, much lower than for other stable oxides and fluorides with the fluorite structure (CeO$_2$ 0.68, UO$_2$ 0.70, ThO$_2$ 0.75, CaF$_2$ 0.84, HgF$_2$ 0.86, CdF$_2$ 0.88, SrF$_2$ 0.94, etc.) but larger than for compounds with the rutile structile. (In sixfold coordination, ZrO$_2$ would have a radius ratio of 0.51, compared with that for TiO$_2$ itself of 0.43.)

This structural instability does not lead to a phase transformation to a rutile form of ZrO$_2$—the Zr^{4+} ion is possibly too large for this to happen—but rather to the formation of t- and m-ZrO$_2$ at progressively lower temperatures, both polymorphs being distorted versions of the fluorite structure (Fig. 1). To understand the origin of these transformations, a more fundamental approach is required, namely one involving quantum mechanics. Fortunately, the electronic structure of both c- and t-ZrO$_2$ has been recently calculated[8] using quantum mechanical cluster techniques, and has yielded insight into the stability of these two polymorphs. The t-ZrO$_2$ was more covalent than the c-polymorph—28% and 22% covalent character, respectively. This extra degree of covalency was shown to lead to a larger band gap and a lower center of mass energy of the valence band, i.e., a lower energy of the entire valence band. t-ZrO$_2$ is thus more stable than c-ZrO$_2$ from the viewpoint of their respective electronic structures. The modest elevated temperature range above $\approx$2370°C where c-ZrO$_2$ is stable can be attributed to a higher entropic contribution to the free energy, compared to the lower symmetry phases. Unfortunately, similar quantum calculations for m-ZrO$_2$ have not been performed for further comparison.

It is well known[9] that Mg^{2+}, Ca^{2+}, Y^{3+}, and virtually all the rare earth ions "stabilize" the c fluorite structure, i.e., they reduce the temperature of the $c \rightarrow t$ transformation. Similarly, when ZrO$_2$ is heated in reducing environments and loses oxygen, the c phase in the resulting nonstoichiometric ZrO$_{2-x}$ has a larger stability field.[10] The predominantly ionic bonding in ZrO$_2$ ensures that the charge-compensating defects in both doped and oxygen-deficient ZrO$_2$ are oxygen vacancies. The quantum mechanical calculations just cited show that there is a large change in the electronic structure (and hence the energy of the valence band) as anions, in particular, are displaced from their ideal fluorite positions. It is likely[8] that the stabilization of c-ZrO$_2$ in doped or oxygen-deficient ZrO$_2$ is related to this effect—local structural relaxation around these anion vacancies significantly

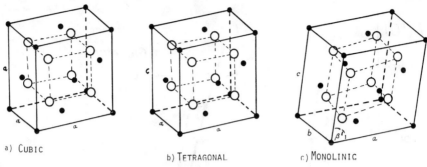

Fig. 1. Three polymorphs of ZrO_2: (a) c phase, (b) t phase $c/a \approx 1.02$, (c) m phase.

affects the electronic structure and energy of the valence band, and thus stabilizes the c polymorph.

From a technological viewpoint, the doped or stabilized ZrO_2's are by far the most important, and the mechanisms by which the instability of c-ZrO_2 is manifested thus have great practical as well as scientific interest. At least four distinctly different reactions involving the instability of c-ZrO_2 have been identified; to set the stage for their discussion, however, we review next the relevant phase diagrams in three important binary systems—MgO-ZrO_2, CaO-ZrO_2, and Y_2O_3-ZrO_2.

Phase Equilibria

MgO-ZrO_2

The most recent diagram is due to Grain[11] (See Fig. 1 of Ref. 38). Subsequent work has confirmed many of the details of the diagram at the high-ZrO_2 end, although the exact solubility of MgO in m- and t-ZrO_2 is not known with certainty, because of the experimental difficulty in determining the absolute (low) MgO concentration with the necessary high spatial resolution. We note for later discussion that an intermediate phase, $Mg_2Zr_5O_{12}$, is known[12] but was not included on the diagram, as it was reported to be unstable below 1850°C. We shall return to this diagram when discussing precipitation of t-ZrO_2 from the c-ZrO_2 solid solution—the formation of optimally aged Mg-PSZ—and when discussing eutectoid decomposition of the c-ZrO_2.

CaO-ZrO_2

The details of phase equilibria at the high-ZrO_2 end are elusive. The last major attempt, that by Stubican and Ray[13] in 1977, has been the subject of four suggested revisions involving either the composition or temperature of the invariant eutectoid near 15 mol% CaO, or the stability fields of the defect-fluorite intermediate compounds, ϕ_1-$CaZr_4O_9$, and ϕ_2-$Ca_6Zr_{19}O_{44}$. As discussed by Stubican et al.[17] (their Fig. 1), the most recent diagram shows that ϕ_1 and ϕ_2 are stable below 1235° and 1355°C, respectively, with the eutectoid for c-ZrO_2 being at 1140°C and 17 mol% CaO; note, however, that there are still some problems, as Hangas et al.[15] suggested that ϕ_1 may be metastable in the system CaO-ZrO_2, as it is in the system CaO-HfO_2.

Y_2O_3-ZrO_2

Discrepancies also exist in the Y_2O_3-ZrO_2 binary system, both at the high- and low-ZrO_2 portions. The most recent determination, by Pascual and Duran,[16] differs

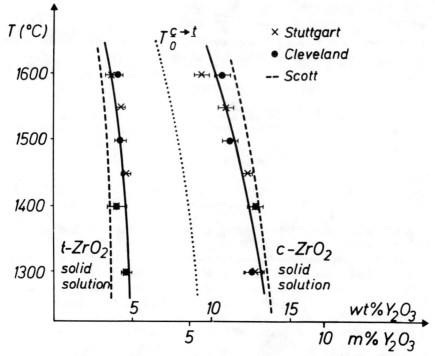

Fig. 2. ZrO$_2$-rich portion of the equilibrium phase diagram in the system ZrO$_2$-Y$_2$O$_3$. The equilibrium compositions were determined by EDS methods (Refs. 18 and 19, "Stuttgart" data and "Cleveland" data, respectively. Previous solvuses (Ref. 42) labeled "Scott" are included for comparison).

in some important details from the previous diagram of Stubican et al.[14] We believe that many of the problems in these phase diagram determinations arise from the use of classical techniques with poor spatial resolution in systems undergoing a martensitic transformation; furthermore, the Gibbs free energy vs composition curves are relatively flat, which can lead to the occurrence of metastable phase transformations.

For technological applications, the high-ZrO$_2$ portion is the more important. The subsolidus equilibria were investigated by Rühle et al.[18] and Lanteri et al.[19]; both groups used analytical electron microscopy to determine the compositions of the coexisting t- and c-ZrO$_2$ with the necessary high spatial resolution, and used calibrated standards to determine absolute concentrations with the necessary accuracy. The data shown in Fig. 2 were determined from the compositions of coexisting c- and t-ZrO$_2$ grains in sintered ceramics fabricated at several temperatures,[18] or from the compositions of coexisting c and t regions in single crystals containing 6 or 8 wt% Y$_2$O$_3$, grown by skull melting, and annealed for long times between 1300° and 1600°C.[19] The data carry an absolute uncertainty of less than or equal to ±0.2%; Fig. 2 probably represents the best determination to date of subsolidus phase equilibria in a ceramic system in which both coexisting phases show appreciable solid solubility.

Diffusion-Controlled Reactions Involving c-ZrO₂
Precipitation of t-ZrO₂

Precipitation of t-ZrO$_2$ following sintering is now recognized as the most critical step in the production of strong, tough, partially stabilized ZrO$_2$ (PSZ). The precipitation reactions in Mg-PSZ,[20–25] Ca-PSZ,[24,26,32] and Y-PSZ[19,24,27,28] have been studied by Heuer and Hannink and their coworkers and, although the precipitate morphologies are different for each solute, certain trends do emerge. Both inter- and intragranular particles are found in Mg-PSZ[20]; the large intergranular particles invariably transform martensitically when cooled to room temperature, as do the larger of the intragranular particles (the "critical" particle thickness for the lens-shaped oblate spheroids in Mg-PSZ is ≈ 0.04 μm,[20,21,24,29] but ≈ 0.1 μm for the more equiaxed particle in Ca-PSZ[24,26]). The intergranular particles are present if samples are densified in the two-phase $t(ss) + c(ss)$ field,[20] or form at high temperatures via a discontinuous precipitation reaction if single-phase c-ZrO$_2$ is slowly cooled through this two-phase field (this discontinuous precipitation reaction has not been studied in detail). The intragranular precipitates in both Mg- and Ca-PSZ appear to form via homogeneous nucleation, followed by conventional diffusional growth and coarsening[20,26,32] (Ostwald ripening). Analysis of the coarsening kinetics in Ca-PSZ yielded an activation energy for the cation interdiffusion in agreement with cation self-diffusion coefficients,[32] as expected, and an estimate of the interfacial energy between c-ZrO$_2$ and the t-ZrO$_2$ precipitates of ≈ 0.2 J/m^2.

The precipitation of t-ZrO$_2$ in the system Y$_2$O$_3$-ZrO$_2$ exhibits some similarities to precipitation of this phase in the systems MgO-ZrO$_2$ and CaO-ZrO$_2$,[24] at least for small precipitates, but there are significant differences. There is a much larger solubility of Y$_2$O$_3$ in t-ZrO$_2$ than is the case for either MgO or CaO in t-ZrO$_2$ (cf. Fig. 2 with Fig. 1 of Ref. 38). This difference in solubility is attributed to the small ionic radius misfit between Zr^{4+} and Y^{3+}, compared to the misfit between Zr^{4+} and either Mg^{2+} or Ca^{2+}. It is interesting in this regard that the molar volume change when t-ZrO$_2$ is precipitated from a c-ZrO$_2$ matrix is positive in Mg-PSZ, negative in Ca-PSZ, and approximately zero in Y-PSZ.[24]

One consequence of this large solubility in Y-PSZ is a greater stability of t-ZrO$_2$ against the martensitic transformation, as discussed in the next paper.[1] This enables the production of strong, tough 100% t-ZrO$_2$ polycrystals, as discussed by Rühle et al.[18] This large solubility implies that the free energy vs composition curves for both t- and c-ZrO$_2$ are rather flat, which leads to the occurrence of a displacive metastable $c \rightarrow t$ transformation, which we discuss below.

The liquidus and solidus curves at the high-ZrO$_2$ end of the Y$_2$O$_3$-ZrO$_2$ binary phase diagram are unusually flat, which also may relate to the similar ionic radii. More significantly, they permit the preparation of large, good-quality single crystals, with compositions up to ≈ 12 mol% Y$_2$O$_3$, which have interesting mechanical properties.[30] Crystals of a range of compositions can be easily quenched through the two-phase field with little precipitation of t-ZrO$_2$; if they are annealed at elevated temperatures (≥ 1300°C), they decompose to t- and c-ZrO$_2$, as predicted from the phase diagram.[19] However, the t-ZrO$_2$ so produced has a unique morphology, different from anything occurring in Mg- or Ca-PSZ, i.e., "colonies" of stacked plates sharing a {110} habit plane, alternate plates being twin-related (their **c** axes are at 90°) (Fig. 3). Similar colonies have been observed in 4.5 mol% Y-PSZ[27] and 6.9 mol%[31] Y-CSZ which have been subjected to ex-

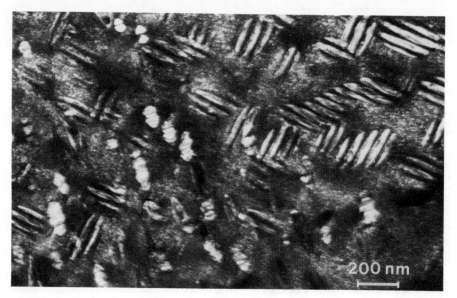

Fig. 3. Colony structure in t-ZrO_2 grains in Y_2O_3-ZrO_2 alloys. Dark-field image (Ref. 31).

tended annealing. The colony structure is maintained on coarsening,[19] and the colonies are very stable against the martensitic transformation.

In fact, it is somewhat surprising that the martensitic transformation occurs so readily in intragranular t-ZrO_2 precipitates in Mg- and Ca-PSZ once they have coarsened past a "critical" size. This particle-size dependence of M_S has been discussed extensively in the literature (e.g., Refs. 23, 24, and 29). It is possible that in PSZ the fine particles are coherent but coarse particles become incoherent, with the microstructural changes that accompany loss of coherency, e.g., the emission of interface dislocations, providing favorable sites to nucleate the martensitic transformation. However, this is difficult to study experimentally, as particles in the electron microscope appear to be either coherent or transformed! This phenomenon is discussed further in Ref. 1.

Precipitation of Ordered Compounds

Hypereutectoid compositions in binary alloys of either CaO-ZrO_2 or Y_2O_3-ZrO_2 can also precipitate one or more ordered defect-fluorite phases. This has not been studied in the system Y_2O_3, where the phase $Zr_3Y_4O_{12}$ is expected from phase equilibria considerations, but two precipitation reactions can occur in the system CaO-ZrO_2.[32-34] Crystals containing between ≈18 and 20 mol% CaO form ϕ_1 (CaZr$_4$O$_9$) on aging,[15,32] whereas crystals containing 20–24 mol% CaO tend to precipitate ϕ_2 (Ca$_6$Zr$_{19}$O$_{44}$).[15] The kinetics of the former precipitation reaction were studied by Marder et al.[32]; the kinetics were similar to those for the precipitation of t-ZrO_2, as expected, but the interfacial energy between ϕ_1 and the c-ZrO_2 matrix was only ≈0.1 J/m^2.

The absence of ϕ_1 in crystals containing 20–24 mol% CaO was tentatively interpreted by Hangas et al.[15] as indicating that ϕ_1 nucleates readily but is metastable, but long-term aging studies are necessary — if ϕ_1 is metastable, it will be replaced by ϕ_2 even in crystals containing ≤20 mol% CaO.

Precipitation of a metastable ordered phase in the system MgO-ZrO$_2$, Mg$_2$Zr$_5$O$_{12}$, was also found by Hannink and Heuer and their coworkers,[25,35–39] although this phase does not appear on the accepted phase diagram. It forms as a precipitate phase within the c grains between the prior t-ZrO$_2$ precipitates on low-temperature ("subeutectoid") aging and was credited by Hannink et al.[25,35,36] with enhancing the thermal shock resistance of such heat-treated PSZ's. On continued aging at these low temperatures, the c-ZrO$_2$ matrix containing both types of precipitates is consumed by a grain-boundary reaction product, which is discussed next.

Eutectoid Decomposition

The c-ZrO$_2$ solid solutions decomposed by proeutectoid precipitation reactions are still unstable with respect to eutectoid decomposition, as can be seen by inspection of the MgO-ZrO$_2$, CaO-ZrO$_2$, and Y$_2$O$_3$-ZrO$_2$ phase diagrams. The low temperature of the eutectoid in the systems Y$_2$O$_3$ and CaO, and the resulting sluggish diffusion kinetics, virtually dictate the absence of eutectoid decomposition in the Y$_2$O$_3$ system and make it improbable in the CaO system.

On the other hand, this reaction occurs readily in the system MgO-ZrO$_2$, the extent of the reaction being important for mechanical properties. The earliest study is due to Viechnicki and Stubican,[40] who found that the decomposition rate of a 20 mol% MgO (hypereutectoid) sample was a maximum at 1200°C. Porter and Heuer[20] reported that decomposition to either (t-ZrO$_2$ plus MgO) or (m-ZrO$_2$ plus MgO) could occur, depending on whether decomposition was carried out above or below 1240°C, a result recently confirmed by Swain et al.[41] The decomposition to (m-ZrO$_2$ plus MgO) occurred in a cellular fashion* (Fig. 4), the nucleation of the cells occurring on grain boundaries of the c matrix. More recently, Hannink and coworkers[25,35,36] studied this reaction because of the improved thermal shock resistance that can be imparted to Mg-PSZ by low-temperature (1100°C), subeutectoid aging. They confirmed the cellular nature of the reaction, but described the MgO as "MgO-rich" pipes. However, Farmer et al.[38] showed by electron diffraction that MgO with the conventional rock-salt structure does indeed form,† and that the "bubblelike" MgO described by Porter and Heuer[20] and the "MgO-rich pipes" described by Hannink et al.[35,36] are simply orthogonal views of the same cellular reaction product. Porter and Heuer reported that the m-ZrO$_2$ in the cellular product was coarsely twinned, whereas Hannink[36] described the m-ZrO$_2$ as polygonized. The spacing of the MgO within the growing cell varies with the solute content of the c-ZrO$_2$ matrix being consumed, as expected.[20,35,36]

The Displacive $c \rightarrow t$ Transformation in Y$_2$O$_3$-ZrO$_2$ Alloys

Scott[42] was the first to suggest that a displacive, composition-invariant $c \rightarrow t$ transformation could occur in rapidly cooled Y$_2$O$_3$-ZrO$_2$ alloys. Inasmuch as this transformation arises from the instability of c-ZrO$_2$, and occurs in a nonmartensitic fashion, we discuss it here rather than in the succeeding paper.[1]

---

*The term cellular decomposition is used in metallic systems to describe discontinuous ($\alpha \rightarrow \alpha' + \beta$) or eutectoid ($\alpha \rightarrow \beta + \gamma$) reactions in which the product phases grow in a coupled manner. The structural and chemical changes are confined to the interface that is advancing into, and thereby consuming, the parent matrix. Figure 4, due to Farmer et al. (Ref. 38), is a good example of a coupled eutectoid decomposition product.

†Hannink (private communication) recently performed microchemical analysis that also shows that the "MgO-rich" pipes are pure MgO.

Fig. 4. Eutectoidally decomposed region in Mg-PSZ. Pipes of pure MgO are included in the m-ZrO_2 (Refs. 37 and 38).

Scott's work involved phase relationships in the system Y_2O_3-ZrO_2. He emphasized that, for samples containing ≈2 to ≈7 mol% Y_2O_3, rapid cooling from the c phase field induced a transformation to a "multiply twinned t phase." Scott argued that any such composition would decompose diffusionally to a mixture of c- and t-ZrO_2 if heated long enough in the two-phase field; the compositions of the coexisting phases are then determined by the t and c solvuses. If the c phase is sufficiently poor in Y_2O_3 (<7 mol%), it could retransform on rapid cooling to t symmetry, whereas the low-Y_2O_3 t phase would transform to m symmetry.

Miller et al.[43] found this high-Y_2O_3 t phase in plasma-sprayed, 4.5 mol% Y_2O_3-ZrO_2 thermal barrier coatings, and described it as "nontransformable t'-ZrO_2," the nontransformability relating to its resistance to undergo the martensitic $t \rightarrow m$ transformation. They aged their samples at 1200°–1600°C for 1–100 h and observed decomposition of the t'-ZrO_2 to high-Y_2O_3 c-ZrO_2 and low-Y_2O_3 t-ZrO_2, in accordance with Scott's predictions.

Further evidence of high-Y_2O_3 t-ZrO_2 has been obtained by a number of workers,[19,30,31,44,45] but the mechanism of the $c \rightarrow t$ transformation has not been agreed upon. In the remainder of this section, we discuss the transmission electron microscopy evidence which convinces us of the nonmartensitic nature of the transformation, and suggest such transformations be called "homogeneous and massive," as they show some similarities to *massive*[‡] transformations in metallic alloys.[46] Similar transformations occur as $BaTiO_3$ goes ferroelectric[47] and in pyroxenes and anorthites[48]; the $\beta \rightarrow \alpha$ transformation in cristobalite is also similar.[48]

[‡]Massive transformations are solid state reactions in which the parent and product phases have the same composition and where the product forms nonmartensitically (i.e., without shape change) (Refs. 46 and 49). The growth kinetics are controlled by the diffusive jumping of atoms across the incoherent boundary separating the two phases (Ref. 46).

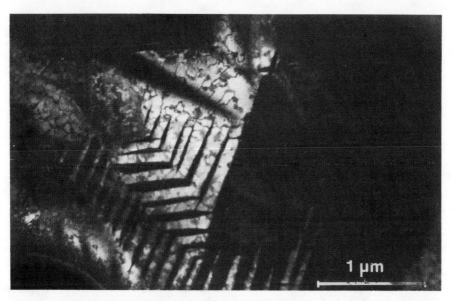

Fig. 5. Twins and anti-phase boundaries in t'-ZrO_2. See text for explanation (Ref. 31).

Figure 5 shows the microstructure of a 6.9 mol% Y_2O_3 sample that had been sintered at 1600°C for 2 h, slowly cooled through the two-phase ($c + t$) field (see Fig. 2) so that fine precipitates (≲5 nm in diameter) of t-ZrO_2 formed,[31] reheated to 1550°C for 1 h, and cooled rapidly to room temperature to induce the $c \rightarrow t$ transformation.[31] Diffraction contrast experiments revealed that the black bands in Fig. 5 are 90° twins, and arise because any of the **a** axes of c-ZrO_2 can become the **c** axis of t-ZrO_2.

Within each twin, a second type of planar defect is observed, which has the contrast and appearance of anti-phase domain boundaries (APBs) which develop in ordered alloys (Ca_3Au,[50] Fe_3Al[51]) when cooled through an order/disorder temperature.

Bender and Lewis[44] were the first to observe these APBs in a 3.4 mol% Y_2O_3 t-ZrO_2 single crystal which had been irradiated with a high-intensity laser. They showed that the APBs could be imaged with a fluorite-forbidden reflection and ascribed them to some form of "ordering." Lanteri et al.[19] observed similar APBs in a 4.5 mol% Y_2O_3, as-grown, t-ZrO_2 single crystal. We now suggest that these APBs form during the $c \rightarrow t$ transformation on cooling through a critical temperature T_0, and at the same time as the 90° twins.

As is shown in Fig. 1(a), c-ZrO_2 has the fluorite structure; its space group is $Fm3m$ with one formula unit (1 "molecule" of ZrO_2) associated with each lattice point; the unit cell is quadruply primitive, i.e., it contains four "molecules" of ZrO_2. We represent the lattice geometry schematically in Fig. 6(a). As shown in Fig. 6(b), t-ZrO_2 is conveniently represented with a C-centered cell, which has roughly the same volume, i.e., the same number of "molecules" per cell, but is only doubly primitive. It is thus clear that each lattice point in t-ZrO_2 has two "molecules" of ZrO_2 associated with it. In the conventional representation, the lattice points at the positions (1/2 0 1/2) and (0 1/2 1/2) have been "lost" during the $c \rightarrow t$ transformation.

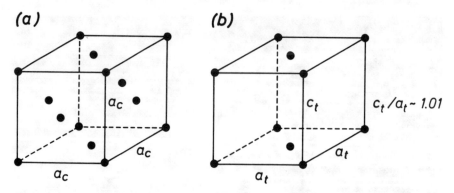

Fig. 6. Lattice geometry for c- and t-ZrO$_2$. Each dot is a lattice point and corresponds to one "molecule" of ZrO$_2$ in (a) and two "molecules" in (b).

Next, assume that the $c \rightarrow t$ transformation is displacive and can be nucleated at any of the lattice points of the parent c phase (which of course are all equivalent). If two adjacent "domains" of t-ZrO$_2$, one nucleated at 000 of the parent cubic phase and the other at (1/2 0 1/2) of the cubic phase, grow to impingement, they will be separated by an anti-phase domain boundary characterized by a displacement vector $\mathbf{R} = a/2\,[101]$ or $a/2\,[011]$ but not $a/2\,[110]$. Michel et al.[45] analyzed the number of possible planar defects that can arise from such a $c \rightarrow t$ transformation, and reached similar conclusions; they imaged ≈50 nm domains in a 3 mol% Y$_2$O$_3$-ZrO$_2$ single crystal but did not find such prominant twinning as shown in Fig. 5.

What determines the critical temperature T_0? Lanteri et al.[19] gave a simple argument based on models for the massive transformation in metallic alloys,[46] which we follow and elaborate on here. A schematic free energy-vs-composition curve for t- and c-ZrO$_2$ solid solutions is shown in Fig. 7 for a temperature T_0 such that *diffusion-controlled reactions cannot occur*. If diffusion were possible, however, the two-phase stable or metastable equilibrium would involve t-ZrO$_2$ of composition c_1 and c-ZrO$_2$ of composition c_2. Further, at this temperature, alloys of composition c_0 have the same free energy with either t or c symmetry. If c-ZrO$_2$ alloys are quenched from the single-phase field to T_0, alloys above and below c_0 in composition will behave differently: (i) Compositions richer in Y$_2$O$_3$ than c_0 must remain supersaturated with c symmetry indefinitely. (ii) Compositions leaner in Y$_2$O$_3$ than c_0 can undergo a composition-invariant displacive transformation to t symmetry. Within this latter group, compositions richer in Y$_2$O$_3$ than c_1 would undergo the displacive reaction in a stable or metastable two-phase field, while those leaner than c_1 would undergo this reaction in a single-phase field; in the former case, further decomposition to a two-phase mixture is possible in principle.

Whether the alloy composition is above or below c_1 is quite important in metallic systems undergoing this type of massive reaction, where it is now generally agreed that the transformation can proceed only for alloys leaner than c_1,[49,52,53] i.e., in the single-phase field. (The reader is referred to these references for the details of these arguments.) However, this restriction should apply only to those massive reactions which are nucleated heterogeneously (all known examples in metallurgy), and in which the nucleus *must* have a slightly different composition

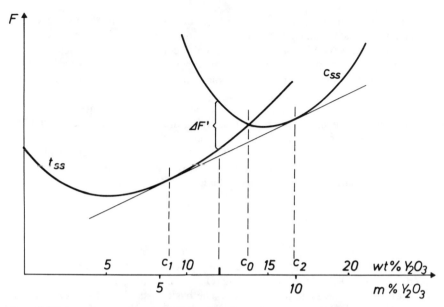

Fig. 7. Free energy vs composition. See text for explanation.

than the bulk parent (or product) phases.[49,52] We think the very slow cation diffusion kinetics in ZrO_2 lifts this restriction with regard to the $c \rightarrow t$ reaction, and believe that the small size of the APBs signifies a homogeneously nucleated transformation. We thus suggest the name "homogeneous massive" for such transformations, and note again that similar examples can be found in several mineral systems.[48] We do not know the supercooling required for such homogeneous nucleation. However, the structural similarities of c- and t-ZrO_2 — they differ only in a slight distortion of the oxygen sublattice — may lead to a modest nucleation barrier and relatively modest supercoolings for homogeneous nucleation. These supercoolings must have been achieved to produce the APBs.

To estimate T_0, we follow metallurgical practice and let c_0 lie halfway between c_1 and c_2. The locus of c_0 as a function of temperature, $T_0^{c \rightarrow t}$, is included on the phase diagram of Fig. 2. Extrapolating the t and c solvuses to lower temperatures, we estimate that the $c \rightarrow t$ reaction in the 3.4 mol% Y_2O_3 sample[44] occurred below 1350°C, and below 750°C in the 6.9 mol% Y_2O_3 sample.[31]

This type of construction for a $c \rightarrow t$ transformation was considered by Andersson and Gupta,[54] who drew a similar curve for a "diffusionless transformation phase boundary" between the c and t phases based on Scott's phase diagram. However, they imagined this transformation to be martensitic, a viewpoint they also express in this volume[55] (in fact, they label the $T_0^{c \rightarrow t}$ curve $M_S^{c \rightarrow t}$ in Ref. 55). In support of their claim for the martensitic nature of this transformation, they show electron microscopy evidence of t plates in a c matrix in a 4.5 mol% Y_2O_3 alloy quenched from 2300 K to room temperature in $\approx$2 min.

Further work is clearly needed to resolve this difference. Although it is possible for a single transformation to be martensitic or massive, depending on cooling rate (cf. the bcc $\rightarrow$ hcp massive transformation in metallic alloys[46]), it is unlikely that the homogeneous massive transformation described above could be avoided with a sufficiently rapid quench. However, the c-ZrO_2 matrix of the sintered 6.9 mol% Y_2O_3 sample of Chaim et al.[31] contained fine particles of t-ZrO_2 in a "tweed" microstructure, and did not appear to have undergone the $c \rightarrow t$ transformation; this sample had been cooled slowly through the two-phase field of Fig. 3.

We surmise that the strain energy associated with these precipitates, combined with the relief of supersaturation accompanying the precipitation, altered the free energy of the c matrix so that it could not undergo this displacive transformation. Similarly, the massive thermal strains attendant on a quench as severe as that used by Andersson et al.[55] could have suppressed the massive transformation, paving the way for a martensitic $c \rightarrow t$ transformation.

Summary and Conclusions

We have described the possible diffusion-controlled transformations that occur in various systems in c-ZrO_2—precipitation of t-ZrO_2 or one or more ordered defect fluorite intermediate compounds, eutectoid decomposition to m-ZrO_2 and MgO in the system MgO-ZrO_2, and a homogeneous massive $c \rightarrow t$ transformation in the system Y_2O_3-ZrO_2. The basic instability of c-ZrO_2, whether pure or as a solid solution, has been the underlying theme of our treatment. Although quantum calculations of pure c-ZrO_2 do show it to be less stable than the t polymorph, further work of this type is clearly needed, particularly for doped materials. From the diversity of phase transformations we have discussed here, it is also clear that c-ZrO_2 provides a fertile terrain for studying diffusion-controlled reactions.

Acknowledgments

We are grateful to a number of students and colleagues whose work has provided the stimulation and driving force to prepare this review. A. H. Heuer also acknowledges the Alexander von Humboldt Foundation for a Senior Scientist Award, which made possible his sabbatical leave at the Max-Planck-Institut für Metallforschung, during which time many of the ideas described in this paper were developed, and the NSF and AFOSR who have supported aspects of his research described herein.

References

[1] M. Rühle and A. H. Heuer; this volume, pp. 14–32.
[2] E. Ryshkewitch, Oxide Ceramics. Academic Press, New York, 1960; p. 350.
[3] R. C. Garvie, R. H. Hannink, and R. T. Pascoe, *Nature*, **258**, 703 (1975).
[4] D. L. Porter and A. H. Heuer, *J. Am. Ceram. Soc.*, **60** [3–4] 183–84 (1977).
[5] D. K. Smith and C. F. Cline, *J. Am. Ceram. Soc.*, **45** [5] 249–50 (1962).
[6] L. Pauling, Nature of the Chemical Bond, 3d ed. Cornell University Press, Ithaca, N. Y., 1960.
[7] R. D. Shannon and C. T. Prewitt, *Acta Crystallogr., Sect. B*, **25**, 925 (1969).
[8] M. Morinaga, H. Adachi, and M. Tsukuda, *J. Phys. Chem. Solids*, **44**, 301 (1983).
[9] E. C. Subbarao; pp. 1–24 in Advances in Ceramics, Vol. 3. Edited by A. H. Heuer and L. W. Hobbs. The American Ceramic Society, Columbus, OH, 1981.
[10] R. Ruh and H. J. Garrett, *J. Am Ceram. Soc.*, **50** [5] 257–61 (1967).
[11] C. F. Grain, *J. Am. Ceram. Soc.*, **50** [6] 288–90 (1967).
[12] C. Delamarre, *Rev. Int. Hautes Temp. Refract.*, **9**, 209 (1972).
[13] V. S. Stubican and S. P. Ray, *J. Am. Ceram. Soc.*, **60** [11–12] 534–37 (1977).
[14] V. S. Stubican, G. S. Corman, J. R. Hellmann, and Senft; this volume, pp. 96–106.
[15] J. Hangas, T. E. Mitchell, and A. H. Heuer; this volume, pp. 107–17.

[16] C. Pascual and P. Duran, *J. Am. Ceram. Soc.*, **66** [1] 23–27 (1983).
[17] V. S. Stubican, R. C. Hink, and S. P. Ray, *J. Am. Ceram. Soc.*, **61** [1–2] 17–21 (1978).
[18] M. Rühle, N. Claussen, and A. H. Heuer; this volume, pp. 352–70.
[19] V. Lanteri, T. E. Mitchell, and A. H. Heuer; this volume, pp. 118–30.
[20] D. L. Porter and A. H. Heuer, *J. Am. Ceram. Soc.*, **62** [5–6] 298–305 (1979).
[21] L. Schoenlein, M.S. thesis 1979; Ph.D. thesis 1983, Case Western Reserve University, Cleveland, OH.
[22] T. A. Witkowski, M.S. thesis 1981, Case Western Reserve University, Cleveland, OH.
[23] A. H. Heuer; pp. 98–115 in Advances in Ceramics, Vol. 3. Edited by A. H. Heuer and L. W. Hobbs. The American Ceramic Society, Columbus, OH, 1981.
[24] R. H. J. Hannink, *J. Mater. Sci.*, **13** [1] 2487–96 (1978).
[25] R. H. J. Hannink and M. V. Swain, *J. Austr. Ceram. Soc.*, **18**, 53 (1982).
[26] R. H. J. Hannink, K. A. Johnston, R. T. Pascoe, and R. C. Garvie; pp. 116–36 in Advances in Ceramics, Vol. 3. Edited by A. H. Heuer and L. W. Hobbs. The American Ceramic Society, Columbus, OH, 1981.
[27] P. Valentine, M.S. thesis 1982, Case Western Reserve University, Cleveland, OH.
[28] C. Minshall, M.S. thesis 1983, Case Western Reserve University, Cleveland, OH.
[29] A. H. Heuer, N. Claussen, W. M. Kriven, and M. Rühle, *J. Am. Ceram. Soc.*, **65** [12] 642–50 (1982).
[30] R. P. Ingel, D. Lewis, B. A. Bender, and R. W. Rice; this volume, pp. 408–14.
[31] R. Chaim, M. Rühle, and A. H. Heuer; unpublished work.
[32] J. M. Marder, T. E. Mitchell, and A. H. Heuer, *Acta Metall.*, **31**, 387 (1983).
[33] J. G. Allpress and H. J. Rossell, *J. Solid State Chem.*, **15**, 68 (1975).
[34] H. J. Rossell; pp. 47–63 in Advances in Ceramics, Vol. 3. Edited by A. H. Heuer and L. W. Hobbs. The American Ceramic Society, Columbus, OH, 1981.
[35] R. H. J. Hannink and R. C. Garvie, *J. Mater. Sci.*, **17**, 2637 (1982).
[36] R. H. J. Hannink, *J. Mater. Sci.*, **18**, 457 (1983).
[37] S. C. Farmer, L. H. Schoenlein, and A. H. Heuer, *J. Am. Ceram. Soc.*, **66** [2] 107–10 (1983).
[38] S. C. Farmer, T. E. Mitchell, and A. H. Heuer; this volume, pp. 152–63.
[39] H. J. Rossell and R. H. J. Hannink; this volume, pp. 139–51.
[40] D. Viechnicki and V. S. Stubican, *J. Am. Ceram. Soc.*, **48** [6] 292–97 (1965).
[41] M. V. Swain, R. C. Garvie, and R. H. J. Hannink, *J. Am. Ceram. Soc.*, **66** [5] 358–62 (1983).
[42] H. G. Scott, *J. Mater. Sci.*, **10**, 1527 (1975).
[43] R. A. Miller, J. L. Smialek, and R. G. Garlick; pp. 241–55 in Advances in Ceramics, Vol. 3. Edited by A. H. Heuer and L. W. Hobbs. The American Ceramic Society, Columbus, OH, 1981.
[44] B. A. Bender and D. Lewis; private communication.
[45] D. Michel, L. Mazeroles, M. Perez, and Y. Jorba, *J. Mater. Sci.*, **18**, 2618 (1983).
[46] T. B. Massalski, Phase Transformations. Edited by H. Aaronson. American Society for Metals, Metals Park, OH, 1970; p. 433.
[47] P. F. Gobin and G. Guenin, Solid State Phase Transformations in Metals and Alloys. Edited by D. de Fontaine. Les Editions de Physique, Paris, 1978; p. 573.
[48] A. H. Heuer and G. L. Nord, Jr., Electron Microscopy in Mineralogy. Edited by H.-D. Wenk. Springer-Verlag, New York, 1976; p. 274.
[49] D. A. Karlyn, J. W. Cahn, and M. Cohen, *Trans. Met. Soc. AIME*, **245**, 197 (1969).
[50] C. S. Barrett and T. B. Massalski, Structure of Metals. McGraw-Hill, New York, 1966.
[51] S. M. Allen and J. W. Cahn, *Acta Metall.*, **23**, 1017 (1975).
[52] E. S. K. Menon, M. R. Plichta, and H. I. Aaronson, *Scripta Metall.*, **17**, 1455 (1983).
[53] M. Hillert, *Met. Trans.*, **15A** [3] 411–19 (1984).
[54] C. A. Andersson and T. K. Gupta; pp. 184–201 in Advances in Ceramics, Vol. 3. Edited by A. H. Heuer and L. W. Hobbs. The American Ceramic Society, Columbus, OH, 1981.
[55] C. A. Andersson, J. Greggi, Jr., R. C. Kuznicki, and T. K. Gupta; this volume, pp. 78–85.

Phase Transformations in ZrO_2-Containing Ceramics: II, The Martensitic Reaction in t-ZrO_2

M. RÜHLE

Max-Planck-Institut für Metallforschung
Institut für Werkstoffwissenschaften
Stuttgart, Federal Republic of Germany

A. H. HEUER

Case Western Reserve University
Department of Metallurgy and Materials Science
Cleveland, OH 44106

Theoretical and experimental studies of the martensitic tetragonal (t) → monoclinic (m) transformation in ZrO_2 are reviewed. The reaction is nucleation-controlled, and the nucleation occurs homogeneously and nonclassically; it is invariably stress-assisted. This model explains the particle-size dependence of the M_s temperature, as well as the extent of the crack-tip stress field transformation, which governs the magnitude of transformation-toughening in ceramics containing t-ZrO_2.

The polymorphism of ZrO_2 is well known[1]; three crystallographic modifications exist, which possess, respectively, cubic (c), tetragonal (t), and monoclinic (m) symmetry and are stable at high, intermediate, and low temperatures.

$$\text{melt} \xrightarrow{2680°C} \text{cubic} \xrightarrow{2370°C} \text{tetragonal} \underset{1150°C}{\overset{950°C}{\rightleftarrows}} \text{monoclinic}$$

The high-temperature c phase has the fluorite structure, whereas the other polymorphs are distorted versions of this structure (see Fig. 1 of the preceding paper[2]). The space groups of these phases are well established (c: $Fm3m$, t: $P4_2/nmc$, m: $P2_1/c$), and it is also generally accepted[1] that the $t \to m$ transformation is martensitic in nature. The shape change and volume increase accompanying this transformation invariably causes cracking of bulk ZrO_2; the martensitic transformation under such conditions is deleterious.

The situation is different for fine t-ZrO_2 particles, particularly those included in a ceramic matrix. First, the start or M_s temperature of the martensitic transformation is lowered, in some cases below room temperature. Second, the martensitic transformation within confined t-ZrO_2 particles can be induced by stress, especially the stress field associated with propagating cracks, so that this transformation can dramatically increase the toughness of ZrO_2-containing ceramics.[3-7]

In this paper, we discuss the stability of t-ZrO_2 with regard to this martensitic transformation to the stable m form, i.e., the difficulty of nucleating the martensitic

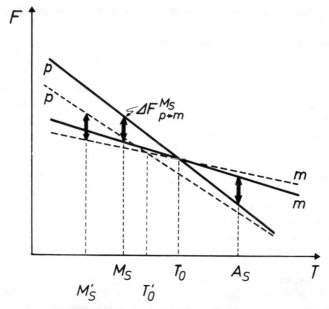

Fig. 1. Free energy (F) vs temperature (T) diagrams for martensitic reactions. p = parent phase, m = martensite phase. The solid and dashed curves are for pure ZrO_2 and ZrO_2 containing a stabilizer, respectively. T_0 and T_0' are the temperatures where parent and martensitic phases possess the same free energy. $\Delta F_{p \to m}^{M_S}$ is required for surmounting the nucleation barrier. M_S and M_S' are the martensite start temperatures, and A_S the temperature for the reverse transformation.

transformation. While the nucleation and growth aspects of martensitic transformations, indeed of phase transformations in general, are traditionally of great concern to metallurgists, they have been seriously underemphasized or ignored in some treatments of transformation-toughening in ceramics. Since the nucleation of martensitic transformations in particular is of paramount importance — once nucleation occurs, growth velocities of martensitic interfaces can approach the speed of sound — we emphasize this aspect of the $t \to m$ transformation of ZrO_2 in this review. This focus will be particularly important as we try to rationalize the resistance of t-ZrO_2, either in bulk form or as fine particles, to the martensitic transformation. The effect of various parametric variables (temperature, particle size, solute content, etc.) are all known phenomenologically, but a basic understanding of the nucleation of this transformation in ZrO_2 is still a topic of active current research.

This paper is organized as follows: the following section reviews current understanding of martensitic transformations in crystalline solids, and the section following, and the appendix, summarize the evidence concerning the martensitic character of the $t \to m$ transformation in ZrO_2, and how the transformation is used to toughen ceramics. The last section addresses the issues raised by the influence of the several variables on the stability of t-ZrO_2 particles, and the directions for future progress.

Martensitic Transformations in Crystalline Solids: The Significance of Nucleation

Characteristics of Martensitic Transformations

Many metallic and nonmetallic solids undergo martensitic phase transformations. The necessary and sufficient conditions for classifying a transformation as martensitic are that it is "a first-order, solid state structural change which is displacive, diffusionless, and dominated in its kinetics and morphology by the strain energy arising from shear or shear-like displacements."[8] The first-order character of the transformation implies the coexistence of the parent and product phases at an intermediate stage of transformation, and therefore the existence of an interface. Accommodation problems at this interface, due to the finite lattice strains accompanying transformation, must be modest, since the activation energy for moving the interface is small—martensitic transformations can occur at temperatures approaching 0 K. The interface must be planar and unconstrained, i.e., it must satisfy the conditions of an invariant plane strain, a set of conditions that are the foundation of the phenomenological crystallographic theory of martensitic transformations.[9]

Since martensitic reactions are diffusionless, the composition does not vary during the transformation. The system must therefore be considered as a single-component system exhibiting two solid phases which possess different energies.[8,10] As seen in Fig. 1, the two phases have the same free energy at temperature T_0. For temperatures $T > T_0$, the parent phase possesses the lower free energy, whereas the product (martensite) phase is the stable phase for $T < T_0$. In the absence of surface and strain energy contributions, the free-energy change may be expressed as

$$\Delta F_{p \to m} = F_m - F_p \tag{1}$$

where p and m refer to parent and martensitic product, respectively, and F is the Helmholtz free energy. The martensitic transformation occurs only if this chemical free-energy driving force, $\Delta F_{p \to m}$, is negative, and usually requires some supercooling. Actually, a strain-energy term must also be included, as elastic and/or plastic deformation of the parent and martensite phases invariably accompany the reaction, as well as an interfacial-energy term which may be important when considering the energetics of nucleation.

The back-transformation starts at $A_s > T_0$; phenomenologically, T_0 is usually given as

$$T_0 = (A_s + M_s)/2 \tag{2}$$

for transformation of bulk material. Use of Eq. (2) for T_0 is based on the assumption that the driving force necessary for nucleation and the nonchemical interfacial and strain energies are the same for the forward and reverse transformations. This approximation is satisfactory for transformation in bulk materials, but may not hold for confined particles.

The free-energy curves of the parent and martensite phases are of course a function of solute content, and solid solutions will have different values of T_0, M_s, and A_s. (For ZrO_2, alloying with HfO_2 results in higher T_0, M_s, etc. values, while alloying with MgO, Y_2O_3, or CaO decreases these characteristic temperatures.) As already mentioned, the transformation kinetics are controlled by nucleation and growth, and growth velocities are usually so fast that the reaction is nucleation-controlled. Unfortunately, acceptable theories of the nucleation of martensitic transformations,[11] in satisfactory agreement with experiment, have remained an

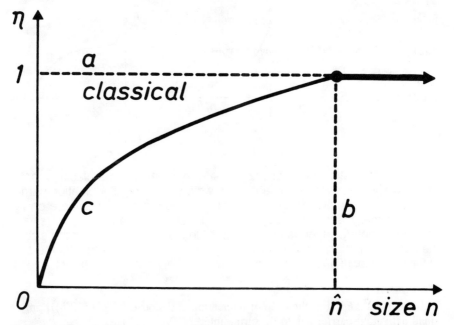

Fig. 2. Classical and nonclassical nucleation paths. A nucleus of size $\hat{n}$ of the martensite phase is characterized by $\eta = 1$. (a) shows the classical path and (b) and (c) show two examples of nonclassical paths.

elusive target for more than 30 years. To quote Olson and Cohen,[12] martensitic nucleation theories involve "nucleation on common defects by improbable physics or on improbable defects by common physics."

Classical and Nonclassical Nucleation Theories

Different models for nucleation of the martensitic product have been developed,[10–15] involving different types of "paths" that can give rise to a stable nucleus (Fig. 2). The classical nucleation path requires a nucleus of exactly the same structure as the martensitic product, even at the smallest nucleus size (path a in Fig. 2); the structure of the nucleus can be characterized by a structural parameter η, such that $\eta = 0$ represents the parent phase and $\eta = 1$ is the martensitic product. Nucleation on nonclassical paths allows for a continuous change of the structure from the parent to the product phase in a small but finite region, until a nucleus of the martensite phase of critical size is reached (e.g., paths b and c in Fig. 2).

In addition, theories of nucleation for both classical and nonclassical paths have been developed for processes without defects — homogeneous nucleation — and for processes involving interaction with defects — heterogeneous nucleation.

Classical Nucleation: Typical shapes of classical free-energy vs size ($F(n)$) curves for different types of nuclei are represented in Fig. 3. Figure 3(a) shows the case of classical homogeneous nucleation; ΔF^*, the nucleation barrier, is the maximum of $F(n)$. ΔF^* for homogeneous nucleation is very large, and this type of nucleation is essentially unobtainable in practice.[11,12,15] For classical heterogeneous nucleation, the surface energy and/or strain energy of the critical nucleus are reduced,

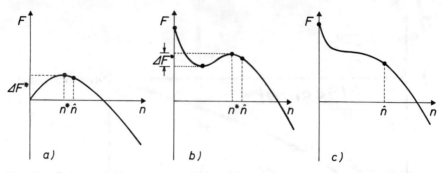

Fig. 3. Schematic free-energy curves for nucleation via a classical path. (a) shows homogeneous nucleation for which a very high nucleation barrier must be surmounted. (b) shows heterogeneous nucleation, where only a small thermally activated barrier, ΔF^*, must be surmounted. (c) shows barrierless heterogeneous nucleation. n^* in each sketch shows the size of the nucleus with equal probability of growth and decay, and $\hat{n}$ is the size of the operational nucleus that has zero probability of decay (after Ref. 13).

and thus so is ΔF^*. For the case of nucleation at dislocations (Fig. 3(b)), only a small activation barrier has to be surmounted.

Figure 3(c) represents the limiting case of classical heterogeneous nucleation. Here, the interaction between the nucleus and defect is so large that $\Delta F^* = 0$, i.e., barrierless nucleation is possible. The rate of arrival of additional atoms at $\hat{n}$ is controlled only by the motion of the interface. Actually, nucleation occurs in the stress field of the defect, and the misfit strain field associated with the formation of the new phase is complementary to the strain field of the defect.[16,17]

Nonclassical Nucleation: As mentioned already, nonclassical nucleation involves a continuous sequence of states along a reaction path — there is a locally continuous distortion of the parent phase into the martensite product as η goes from 0 to 1. Figure 4 shows an $F(\eta)$ curve (at constant n) for this case, in the absence of an applied stress and for $T = T_0$. Both parent and martensite structures are represented by local minima, and are separated by an energy barrier ϕ_0 (this barrier is not identical to the barrier height ΔF^* of homogeneous nucleation). A quantitative description of transformation involving nonclassical nucleation is possible by a Landau-Ginzburg theory.[18]

Increasing the chemical driving force by lowering the temperature shifts the free-energy curves, as is also shown in Fig. 4. For temperatures $T < M_s$ (corresponding to a finite chemical driving force ΔF_{chem}), regions of the parent phase with intermediate structures (strain embryos) may find themselves beyond a critical n, and undergo spontaneous nucleation. (Strain embryos may be present at defects or at regions of high stress concentrations.)

$F(\eta)$ can be modified by an applied stress, τ_{ij}, if the resulting strains scale with η. For a uniform applied stress, the free-energy curve can be modified by a work term

$$W = -\tau_{ij} e_{ij}^T \eta \tag{3}$$

where e_{ij}^T represents the unconstrained transformation strain of the martensitic transformation. The resulting change of the free energy curve is shown in Fig. 5,

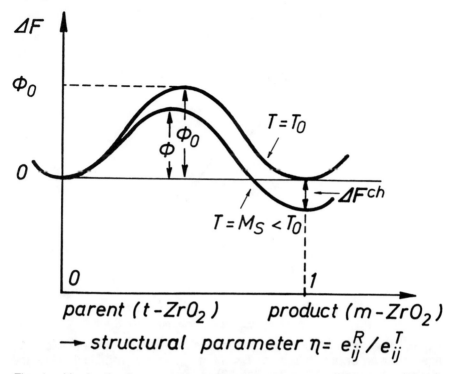

Fig. 4. Nucleation by a nonclassical path. The free energy ΔF associated with a continuous homogeneous distortion of the metastable parent phase to produce the martensite product is shown at temperature T_0 and at the M_S temperature. e_{ij}^T is the unconstrained transformation strain, whereas e_{ij}^R is the actual strain over the nucleus of size n (after Ref. 13).

and permits defining a "transformation stress," τ_r. The driving force given by this stress corresponds to the chemical driving force at M_s in an unstressed bulk material undergoing heterogeneous transformation.

Using Landau's theory,[18] a spinodal (or instability) point η_s, for which $d^2\Delta F/d\eta^2 = 0$, is given by

$$\eta_s = (1 - 1/\sqrt{3})/2 = 0.2113 \tag{4}$$

If the parent lattice is distorted by an applied or residual stress of this magnitude, then the parent lattice becomes mechanically unstable. For a pure shear transformation with transformation shear strains γ_T, the shear stress instability occurs at

$$\tau_{ij} = \frac{32\phi_0}{\gamma_T}\eta_s(1 - 3\eta_s - 2\eta_s^2) = 3.08\phi_0/\gamma_T \tag{5}$$

The barrier height ϕ_0 is actually not known *a priori* in this type of theory; however, estimates based on second- and third-order elasticity for different metallic systems[19–21] result in very small ϕ_0's and comparably small spinodal strains.

In this approach, it is assumed that a "back stress" from the surrounding parent phase does not influence the stress distribution. However, the free-energy change

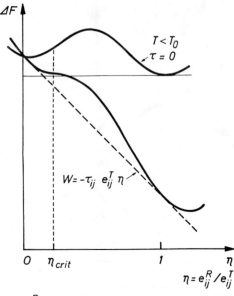

e_{ij}^R = residual strain over the nucleus for homogeneous nucleation (size $\hat{n}$)

Fig. 5. Free energy ΔF for homogeneous nucleation of a metastable lattice under an applied stress at $T < T_0$. If τ_{ij} exceeds a critical value, the nucleation is barrierless (after Ref. 13).

must be evaluated as a function of the size and shape of the nucleus, the interfacial energy, and the strain energy stored in the surrounding matrix.[13,14] The back stress actually opposes the transformation, and the onset of the mechanical instability may be delayed; a larger ΔF_{chem} is then required for nucleation. As in the case of classical nucleation theory, homogeneous nonclassical nucleation can be expected only if nucleation at defects is eliminated.[15]

We emphasize at this point that both homogeneous nonclassical and heterogeneous classical nucleation require significant strains. In the first case, the strains are associated with residual and/or applied stresses, whereas in the second case the strains are associated with defects. Under some conditions, for example in the case of confined ZrO_2 particles, it may be impossible to distinguish between the two types of nucleation.

The Martensitic $t \to m$ Transformation in ZrO_2

Crystallographic Aspects of the Transformation

The crystallographic aspects of the martensitic $t \to m$ transformation in ZrO_2 are well established and the basic results are summarized in the appendix. Experimental studies show that the lattice correspondence which occurs between t- and m-ZrO_2 is for the **c** axis of the t-ZrO_2 to be parallel to the **c** axis of m-ZrO_2. The nonvanishing components of the symmetrical unconstrained transformation strain tensor, e^T, defined with respect to the tetragonal lattice, are for pure ZrO_2[22]:

$$e_{11}^T = (\mathbf{a}_m/\mathbf{a}_t) \cos \beta - 1 = -0.00149$$
$$e_{22}^T = \mathbf{b}_m/\mathbf{b}_t - 1 = 0.02442$$
$$e_{33}^T = (\mathbf{c}_m/\mathbf{c}_t) - 1 = 0.02386$$
$$e_{13}^T = e_{31}^T = \tfrac{1}{2} \tan \beta = 0.08188$$

The data clearly show that the shear component is the dominant term in the strain tensor, and that the volume increase at room temperature ($e_{11}^T + e_{22}^T + e_{33}^T$) is about 4.7%.

Toughening Associated with the Transformation

Garvie et al.[3] first showed that $t \rightarrow m$ transformation of ZrO_2 precipitates in partially stabilized ZrO_2 (PSZ) increases the fracture toughness, and Porter and Heuer[4] showed that this was due to the $t \rightarrow m$ transformation occurring adjacent propagating cracks. These latter authors[23,24] followed this development by studying the precipitation of t-ZrO_2 in Mg-PSZ after annealing at elevated temperatures, and measured the strength σ and fracture toughness, K_{IC}. They found the highest values of σ and K_{IC} for those ceramics in which all the ZrO_2 precipitates had t symmetry prior to loading—the martensitic transformation had not occurred in the precipitates during cooling. A well-developed transformation zone ("wake"), about 0.5 μm in extent around arrested cracks,[23,24] exists in Mg-PSZ, and similar wakes have been found in *in situ* straining experiments in the HVEM.[25,26] Porter et al.[5] showed theoretically that an increase in fracture toughness by a factor of two or more could be realized from such crack-tip stress-induced martensitic transformations.

Claussen[27] likewise found an impressive increase in strength and toughness of $Al_2O_3/15$ vol% ZrO_2 dispersion-toughened ceramics if the ZrO_2 particles had retained their t symmetry. *In situ* straining experiments by Rühle and colleagues[28,29] showed that t-ZrO_2 particles in this matrix also transformed in front of the propagating crack, leaving a similar wake of m-ZrO_2 adjacent the propagating crack. It is now generally accepted that any ceramic matrix can be toughened by a $t \rightarrow m$ crack-tip stress-induced transformation if t-ZrO_2 of an appropriate size can be incorporated during processing.

Finally, Gupta et al.[30] were the first to describe nearly 100% t-ZrO_2 fine-grained polycrystals containing Y_2O_3 (Y-TZP), which also showed impressive strength and toughness. Recently, very strong, fully dense Y-TZP materials have become commercially available, and are described elsewhere in this volume.[31]

Energetics of the Martensitic Transformation in Confined Particles

The transformation of a confined particle is governed by the Helmholtz free-energy change of the entire system (particle plus matrix):

$$\Delta F = -V_p \Delta F_0 + \Delta U$$

where V_p is the particle volume, ΔF_0 the change in chemical free energy between parent and product, and ΔU the total elastic strain energy accompanying the transformation.[6] ΔU has two components, ΔU_T, the increase in strain energy in the particle and matrix due to transformation, and ΔU_I, the interaction term that couples to an applied stress. Using an Eshelby approach,[32] the minimum applied stress necessary to cause transformation was calculated by Porter et al.[5] and Evans and Heuer[6]; however, this simple approach leads to transformation stresses which are independent of particle size, and therefore inconsistent with experiment (see concluding section).

Size-dependent quantities must therefore be introduced. This can be accomplished, for example, through energy terms which have different functional dependencies on the volume and surface area of a particle.[33] The early approaches of this type were criticized by Porter et al.[5] A quite different type of "surface" term was introduced by Evans et al.,[34] who calculated the total free energy, ΔF_{tot}, before and after transformation, in the absence of an applied stress and ignoring nucleation, and found the right type of particle-size dependence. Their calculation assumed that transformation would give rise to twins or martensite variants such that stresses would arise that were localized at the particle-matrix interface, giving a short-range strain field. The energy stored in this strain field will scale as the particle surface area, thus yielding a size dependence in agreement with experiment. However, certain predictions of the theory, particularly the size dependence of the twin or variant spacing, were not confirmed by subsequent experimental work.

Lange[35] and Lange and Green[36] overcame this problem by introducing microcracking and twinning as essential processes governing the size dependence of the transformation. Twinning will certainly occur in the transformed particles, and microcracking may happen on occasion[29,31]; however, the use of such secondary *ex post facto* postulates is certainly not a satisfactory explanation for the fundamental particle-size dependence of the transformation. Of these theories, it should be said: *caveat emptor!*

It follows from the earlier section that the actual transformation behavior in small particles can be represented by reaction-rate diagrams, showing the Helmholtz free-energy changes as a function of transformed volume (Fig. 6) (see also Heuer et al.[37]). An initial energy barrier ΔF^* must be associated with the formation of a nucleus, as discussed above. The exact value of ΔF^* depends on any defects that may be present, and on the stress state of the particle. As transformation proceeds, the free-energy charge may fluctuate, as is suggested in Fig. 6, due to various accommodation processes such as twin or variant formation, until the particle is fully transformed.

Figure 6 clearly shows that the transformation is governed by two factors: (i) nucleation, whether classical or nonclassical, which is accomplished by whatever process allows the free energy barrier ΔF^* to be surmounted and (ii) the change of the total free energy of the confined ZrO_2 particle during transformation, $\Delta F = F_{mon} - F_{tet}$. It is in fact the magnitude of ΔF which determines the magnitude of the increase in toughness.[5]

We note that, to a first approximation, these two quantities are independent of each other, and that to satisfactorily understand the strength and toughness of transformation-toughened ceramics, an understanding and determination of both aspects of the transformation is essential.

Nucleation of the Martensitic Transformation in Confined *t*-ZrO_2 Particles

The important parameters controlling the nucleation of the martensitic $t \rightarrow m$ transformation in ZrO_2 are the size and shape of the particle, the structure of the interface between matrix and confined particle, and the chemistry of the system. However, it is difficult to study the influence of the different parameters independently. In the following, the observations and interpretations are discussed for different *t*-ZrO_2-containing systems; they are conveniently differentiated by the degree of coherency at the *t*-ZrO_2/matrix interface.

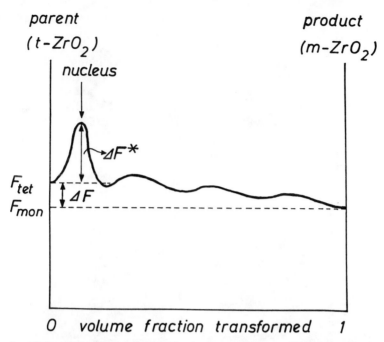

Fig. 6. Schematic illustration of the free energy plotted as a function of the extent of transformation. The two important parameters are the nucleation barrier ΔF^* and the chemical free-energy difference $\Delta F = F_{mon} - F_{tet}$.

Before starting the detailed description, it should be emphasized that the M_s temperature of small t-ZrO_2 particles is almost always lower than that of bulk ZrO_2, sometimes by as much as 1000°C. These marked supercoolings are found even for t-ZrO_2 formed by internal oxidation in metallic matrices.[38,39]

t-ZrO_2 With Incoherent Interfaces

t-ZrO_2 in Dispersion-Toughened Ceramics: In a dispersion-toughened ceramic, the ZrO_2 inclusions are distributed randomly; no orientation relationship exists between the ZrO_2 and the surrounding matrix.[37,40] In such ceramics, the interface is crystallographically incoherent, in that lattice sites are not conserved across the interface, but elastically coherent, in that the interface can transfer both shear stresses and hydrostatic stresses.[41]

The ZrO_2 can be either intra- or intergranular[37] (Fig. 7). Intragranular ZrO_2 particles are generally spherical or ellipsoidal, while intergranular ZrO_2 particles are of necessity faceted polyhedra, the faceting minimizing interfacial energetics.

The fraction of t-ZrO_2 particles with M_s above room temperature depends on the size and shape of the particles. Heuer et al.[37] showed that most intergranular ZrO_2 particles in Al_2O_3/15 vol% ZrO_2 with diameter $d(=d_c) > 0.6$ μm were monoclinic, while no well-defined critical particle size existed for intragranular particles. This *shape* factor can be explained by the differing stress concentrations within the t-ZrO_2. Residual stresses are formed in t-ZrO_2 particles dispersed in a ceramic matrix, as well as in the matrix itself, if a thermal expansion mismatch exists between the t-ZrO_2 and the surrounding matrix. The thermal stress in and

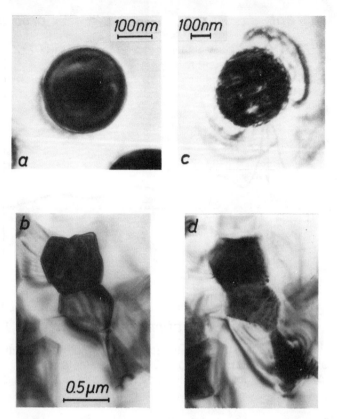

Fig. 7. Confined t- and m-ZrO$_2$ particles. (a) is an intragranular t-ZrO$_2$ particle, and (b) an intergranular t-ZrO$_2$ particle. (c) and (d) are comparably located twinned m-ZrO$_2$ particles.

around intragranular t-ZrO$_2$ particles in Al$_2$O$_3$ was studied by Rühle et al.[42–46] For spherical or ellipsoidal particles, the strain distribution within and adjacent the particle can be analyzed quantitatively. For such "regular" particles, the strain inside the particle is homogeneous, and independent of size(!), while the strains are anisotropic in the surrounding matrix. For ZrO$_2$ in Al$_2$O$_3$, the strain is dilatational, which actually should decrease the nucleation barrier. The constant dilatation strain for t-ZrO$_2$ inclusions is so low that it apparently does not influence the nucleation itself,[46] and we assume that *perfect* ellipsoidal particles, whatever their size, will never transform in the absence of stress.

Within a faceted particle, the strain is not constant and can vary markedly[47]; fortunately the stress distribution can be evaluated rigorously for cuboidal particles.[48–50] In such particles, logarithmic singularities exist at edges and corners, and the magnitudes of the strains scale with particle size. Faceted particles which are larger than a critical size will then have regions where the spinodal shear strain (Eqs. (4) and (5)) is reached over the volume of the *operational* nucleus. Thus, this model gives rise to a particle-size dependence of M_s. Furthermore, it can be shown that the strain levels in such faceted particles are insensitive to the actual shape of the inclusion, and that a mathematically "sharp" corner is not required.[48–50]

For intragranular particles with some shape inhomogeneity, and intergranular particles below the critical size, an applied stress can permit the nucleation barrier to be overcome. As mentioned earlier, the transformation shear strain component, e_{13}, is likely the most important in causing transformation.

Chemical effects can play an important role in changing M_s of t-ZrO_2 in dispersion-toughened ceramics. This has not been studied systematically but Y_2O_3 is known to markedly decrease M_s and HfO_2 to increase M_s. We discuss this chemical effect in more detail later.

t-ZrO_2 Formed by Internal Oxidation: Chen and Chiao[38,39] studied incoherent spherical ZrO_2 particles (diameter $d < 100$ nm) formed by internal oxidation of a Cu-Zr alloy; this system did not show a well-defined critical particle size, a result similar to that found for intragranular ZrO_2 in Al_2O_3. In fact, the t-ZrO_2 particles did not transform to m symmetry even after complete dissolution of the Cu matrix, but transformation could be induced by deformation of the Cu-ZrO_2 alloy. Their observations are striking confirmation that the transformation is nucleation-controlled, and that matrix constraints are not of prime importance when considering nucleation of the martensitic transformation within t-ZrO_2 particles.

Using essentially the same approach as Rühle and Kriven,[42] Chen and Chiao[38,39] assumed that the barrier to transformation was mainly due to the deviatoric transformation strain energy involved in forming an m-ZrO_2 nucleus within a t-ZrO_2 particle. They postulated that nucleation was assisted by interaction with a defect, which possesses a strain distribution equivalent to but of opposite sign than the unconstrained transformation strain.[51-53] For example, a sequence of $[001]_m$ screw dislocations, separated by a distance d, possess a strain field similar in magnitude to the strain of the martensite nucleus. Chen and Chiao showed that for ZrO_2 at room temperature, the presence of just one screw dislocation should permit the transformation to be nucleated.

t-ZrO_2 Polycrystals (TZP): TZP ceramics consist primarily of fine, faceted t-ZrO_2 grains (0.2–1.0 μm).[31] The t-ZrO_2 clearly cannot be considered as part of a dilute system in such ceramics, as in the examples of the dispersion-toughened ceramics and internally oxidized alloys. The chemical driving force in pure t-ZrO_2 is apparently so large that the stress concentration at the facets would result in an operational nucleus, and Y_2O_3 (or other rare earth oxides) is used as a stabilizer to reduce the chemical driving force. Defects such as small-angle grain boundaries and single dislocations are present, but such defects are not able to act as operational nuclei; additional applied stress is required for nucleation. *In situ* observations by Ma and Rühle[54] demonstrated that heterogeneous nucleation can occur at small-angle grain boundaries (Fig. 8(a)), or more frequently at the edges of the grains (Fig. 8(b)).

Autocatalytic nucleation is very important in these materials; the stresses caused by a transforming t-ZrO_2 region are usually transferred to the neighboring grains, thus causing further nucleation. The autocatalytic effect results in the transformation of discrete areas or "clusters," an effect which has been seen in *in situ* straining experiments as irregular "wakes."[28,29]

t-ZrO_2 with Coherent Interfaces

It is well established that small t-ZrO_2 coherent precipitates form if c-ZrO_2 solid solutions containing MgO, CaO, or Y_2O_3 are annealed in the two-phase $(c + t)$ field.[2] The solubility of these solutes in t-ZrO_2 (at an equivalent temperature of say 1600°C), is very different: it is <2 mol% for MgO,[55] ≈5 mol% for CaO,[56] and ≈5 mol% for $YO_{1.5}$.[2,31] The lattice parameters for t- and c-ZrO_2 con-

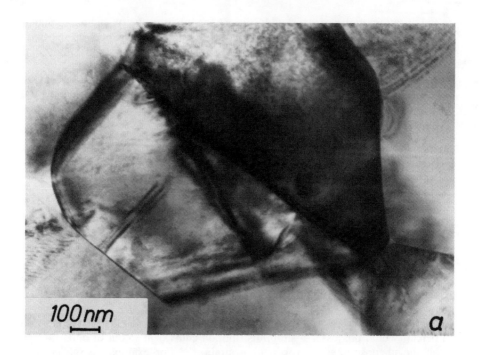

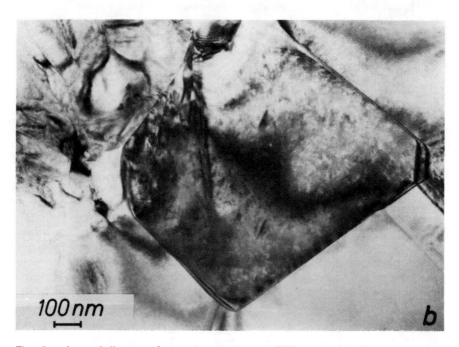

Fig. 8. A partially transformed grain in a Y-TZP ceramic. The nucleation starts in (a) at a small-angle grain boundary and at a grain-boundary facet in (b).

Table I. Lattice Parameters at Room Temperature for Zirconia Polymorphs

Stabilizer	$r_{stabilizer}/r_{Zr^{4+}}$*	Cubic a_c	Tetragonal a_t	c_t	c_t/a_t	a_t/a_c	c_t/a_c	V_t/V_c (%)
Pure ZrO_2		0.5127	0.5082	0.5189	1.021	0.991	1.012	+0.58
MgO	1.06	0.5080	0.5077	0.5183	1.021	0.999	1.020	+1.95
CaO	1.33	0.5130	0.5094	0.5180	1.017	0.993	1.009	−0.58
Y_2O_3	1.31	0.5132	0.5116	0.5157	1.008	0.997	1.005	+0.02

*Based on $r_{Zr^{4+}} = 0.084$ nm (Ref 59).

taining these solutes are likewise different; the observations of Hannink[57] are summarized in Table I. In addition, data are given in Table I for a hypothetical pure t-ZrO_2 containing no solute, obtained by extrapolating Scott's data for the system Y_2O_3-ZrO_2 to zero solute,[58] and a hypothetical c-ZrO_2 obtained by extrapolating Grain's data[55] for the system MgO-ZrO_2 to pure ZrO_2. The lattice contraction due to MgO, and the expansions due to CaO and Y_2O_3, can be understood on the basis of their ionic radii relative to ZrO_2,[59] which data are also included in Table I.

The data in Table I are for room temperature, whereas the important data controlling precipitate morphology are the lattice parameters of t- and c-ZrO_2 at the temperatures where precipitation is occurring. The thermal expansion data for these two ZrO_2 polymorphs[60,61] show that α for c-ZrO_2 and along $\langle 100 \rangle$ of t-ZrO_2 are similar, $\approx 10 \times 10^{-6}/°C$, while α along [001] of t-ZrO_2 is appreciably higher,[61] $\approx 15 \times 10^{-6}/°C$. Therefore, the axial ratio c_t/a_c shown in Table I must be considered as a lower bound when considering precipitate morphology.

The different lattice parameters of t-ZrO_2 relative to c-ZrO_2 in these three systems are the cause of the different precipitate morphologies observed. Before describing the observations, however, we note from Table I that t-ZrO_2 in Mg-PSZ will be under the largest compressive strain, that t-ZrO_2 in Ca-PSZ will be under a tensile strain, and that the strain in Y-PSZ will be small but compressive.

The precipitates in Mg-PSZ are lenticular ellipsoids; the tetragonal **c** axis is parallel to $\langle 100 \rangle$ of the matrix and is the rotation axis of the ellipsoid. The cross section of a typical precipitate is shown in Fig. 9, and compared with an oblate ellipsoid with the same major and minor axes.[43]

In Ca-PSZ, the precipitates are cuboids[57,62] with a {110} habit plane and the same orientation relationship as in Mg-PSZ. Distinctive "rafting" of the

Fig. 9. Comparison of t-ZrO_2 precipitate in Mg-PSZ with an oblate ellipsoid having the same principal axes.

particles along $\langle 100 \rangle_c$ is found,[62] presumably a result of elastic interactions between particles.[63]

In Y-PSZ, the t-ZrO$_2$ takes the form of twinned "colonies," the twin plane being {110},[64] and again shows the same orientation relationship. The twinned colonies are also believed to form to minimize coherency strains.

In principle, the precipitate morphology can be understood from the theory of Khachaturyan.[65] In the case of Mg-PSZ, where the tetragonality is large (Table I), the lenticular morphology is precisely what is predicted (see p. 249 of Ref. 65), although insufficient data are available to compare the observed aspect ratio of ≈4 (Fig. 9) with theory. The strain energy along c_t is minimized by the lenticular morphology, while the sharp tip of the precipitate (Fig. 9) occurs because the strain along a_t is so small (see Table I). The cuboids in Ca-PSZ and the twinned colonies in Y-PSZ form because the lattice strain on precipitation is smaller and more isotropic. Thus, in the case of precipitate morphology, agreement between theory and experiment is satisfactory.

A well-defined critical particle size exists for t-ZrO$_2$ in Mg-PSZ at a lens thickness of 50–100 nm (Fig. 9), and in Ca-PSZ at a cuboid size of ≈90 nm; it has not yet been possible to define a critical size for t-ZrO$_2$ in Y-PSZ, but a critical size, if it exists, must be considerably larger than these values. Hannink[57] explained the critical particle size by assuming that the coherency at the interface is lost if the particle size exceeds a certain value, and suggested that an incoherent particle transformed to m symmetry spontaneously on cooling. The nucleation problem was therefore transferred to understanding the processes leading to loss of coherency, although we show below that other explanations may be more likely.

Hannink[57] estimated the maximum extent of the coherency using the one-dimensional geometrical criterion of Brooks,[66] and suggested that coherency is lost when the total mismatch, Δd, along the interface of length δ, is equal to the Burgers vector, $\mathbf{b}$, of the interfacial misfit dislocation, i.e., $\Delta d \delta < \mathbf{b}$. An estimate of the maximum observed t-ZrO$_2$ precipitate size agreed within a factor of 2 or 3 with the predictions; the lack of better agreement may have resulted from uncertainties in the exact values of the lattice parameters, a too-simple model for coherency loss,[37] or possibly because this model is wrong!

Loss of coherency must generate an array of misfit dislocations, whose Burgers vector will have a component along the crystallographic axes of largest misfit. This should result in [001] edge-type dislocations on (100) planes for Mg-PSZ. However, this type of dislocation is not expected to act as an operational nucleus.

How then is the transformation nucleated? Significant shear stress concentrations will exist in t-ZrO$_2$ precipitates in Mg-PSZ, which scale with particle size, and the nucleation process should be similar to that described for t-ZrO$_2$ polyhedra in dispersion-toughened ceramics. For particles smaller than the critical size, only crack-tip stress-field transformation should be possible at a critical stress level, as is observed.

The cuboidal precipitates in Ca-PSZ behave in a similar fashion, although the volume strain is smaller for the Ca-PSZ than for the Mg-PSZ case (Table I). The similarity in the critical size in these two systems must arise because the t-ZrO$_2$ is under a biaxial tension in Ca-PSZ and under a uniaxial compression in Mg-PSZ; the former stress state should be more effective in nucleating the transformation.

The minimal volume strain for t-ZrO$_2$ in Y-PSZ is responsible for the stability of this polymorph in this system, but the ability to fabricate dislocation-containing Y-TZP ceramics is evidence that more subtle chemical effects are also important.

In part, this may be due to the valence of Y^{3+} compared to Mg^{2+} and Ca^{2+}, but further progress will require quantum mechanical calculations of the type recently reported by Morinaga et al.[67] for solute-free t- and c-ZrO_2.

Conclusions

The $t \to m$ martensitic transformation in bulk ZrO_2, confined t-ZrO_2 particles, and precipitates (and presumably in isolated t-ZrO_2 powder particles as well) is nucleation-controlled. Nucleation is difficult and requires large supercooling of the parent t phase. Nucleation probably occurs homogeneously and nonclassically, although we cannot exclude the possibility of heterogeneous classical nucleation in some (and possibly many) situations. Nucleation is invariably stress-assisted, the stresses arising from thermal expansion mismatch or morphology effects during precipitation. After nucleation has occurred, the transformation continues without any appreciable growth barrier until particles are completely transformed. For all particles other than perfect ellipsoids, the stress concentrations associated with facets and other shape inhomogeneities lead to a critical particle size, or to a critical stress level when transformation occurs in crack-tip stress fields.

In TZP ceramics, where t-ZrO_2 grains are in contact and not isolated in a matrix, autocatalytic effects play an important role in nucleation. Such autocatalytic effects do not appear to be important for dispersion-toughened and PSZ ceramics.

Appendix

The Crystallography of the $t \to m$ Transformation in ZrO_2

The crystal structures of t- and m-ZrO_2 are shown in Fig. 1 of Ref. 2. Three lattice correspondences (LCs) may arise between the t and m polymorphs, called for convenience $LC\ A$, B, or C, depending on which m axis, **a**, **b**, or **c**, is parallel to the **c** axis of t-ZrO_2.

The martensitic character of the $t \to m$ transformation in ZrO_2 was first suggested by Wolten.[68] Shortly afterward, Bailey[69] studied this transformation *in situ* in the electron microscope, using oxidized foils of Zr, and found twinning on $(110)_m$, $(1\bar{1}0)_m$, and $(\bar{1}00)_m$ planes, as well as direct evidence for all three lattice correspondences; however, $LC\ C$ was strongly pronounced.

Bansal and Heuer[70,71] examined the transformation in ZrO_2 single crystals by transmission electron microscopy and X-ray precession experiments. These authors performed a detailed analysis of the orientation relationship between parent and product, and also determined the orientations of the habit planes (the common plane between the t and m phases). The observations were compared to martensitic calculations using the phenomenological theory of Wechsler et al.[72] After this work, the martensitic nature of the $t \to m$ transformation was no longer in doubt. Kriven et al.[22] extended these calculations to different LCs and slip systems, and Choudhry and Crocker[73] most recently further refined the calculations and showed discrepancies with the experimental observations of Bansal and Heuer, due to different values of the lattice parameters used by the two groups of investigators. Resolution of the discrepancy between these calculations and experimental determination of the habit planes requires more detailed habit-plane determinations, and more accurate knowledge of lattice parameters, orientation relationships, and lattice-invariant deformation systems.

Little work has been performed to date on the crystallography of ZrO_2 particles transformed while confined within ceramic matrices. Porter and Heuer[23] showed that large t-ZrO_2 precipitates transformed above room temperature in overaged Mg-PSZ, and were invariably twinned. The twinning occurred on at least two systems: "midrib" twins parallel to $(100)_m$ (i.e., parallel to the long axes of the lenticular t-ZrO_2 precipitates) and "cross" twins parallel to $(110)_m$. The factors which govern the type of twins formed on cooling such particles in PSZ have not yet been determined.

Kriven[74] studied the crystallography of twins formed in transformed ZrO_2 particles in an Al_2O_3 matrix. Small ZrO_2 particles (<0.5 µm) usually contained one set of parallel twins, frequently on $(100)_m$. The situation in larger (>1 µm) particles was more complex, as different areas of the particle contained different martensite variants. More recently,[46] she showed that a particle transformed *in situ* in the HVEM transformed via *LC C* and was internally twinned; however, the twinning was inconsistent with the prediction of the phenomenological martensitic theory.[73] It is not known if this is due to a failure of the theory or to twinning occurring after the transformation.

Detailed TEM studies of transformed ZrO_2 inclusions in a mullite (3 $Al_2O_3 \cdot 2$ SiO_2) matrix by Bischoff and Rühle[75] showed that the twins present in large ZrO_2 particles reduced the transformation-induced stresses at the particle/matrix interface and in the surrounding matrix by forming "domains of closure." However, microcracks were frequently observed within such transformed particles, and it is still not established if post-transformation deformation twinning occurs to further reduce the strain energy of a transformed m-ZrO_2 inclusion.

Acknowledgments

We are grateful to Drs. I.-W. Chen and W. Mader and to a number of students for stimulating discussions during the preparation of this review. A. H. Heuer acknowledges the Alexander von Humboldt Foundation for a Senior Scientist Award, which made possible his sabbatical leave at the Max-Planck-Institut für Metallforschung, during which time many of the ideas discussed here started to develop. A. H. Heuer's research in this field is supported by the NSF (PSZ) and DOE (TZP).

References

[1] E. C. Subbarao; p. 1 in Advances in Ceramics, Vol. 3. Edited by A. H. Heuer and L. W. Hobbs. The American Ceramic Society, Columbus, OH, 1981.
[2] A. H. Heuer and M. Rühle; this volume, pp. 1–13.
[3] R. C. Garvie, R. H. J. Hannink, and R. T. Pascoe, *Nature (London)*, **258**, 703 (1975).
[4] D. L. Porter and A. H. Heuer, *J. Am. Ceram. Soc.*, **60**, 183 (1977).
[5] D. L. Porter, A. G. Evans, and A. H. Heuer, *Acta Metall.*, **27**, 1649 (1979).
[6] A. G. Evans and A. H. Heuer, *J. Am. Ceram. Soc.*, **63**, 241 (1980).
[7] A. G. Evans; this volume, pp. 193–212.
[8] M. Cohen and C. M. Wayman; p. 445 in Metallurgical Treatises. Metallurgical Society AIME, Warrendale, PA, 1981.
[9] C. M. Wayman, Introduction to the Crystallography of Martensitic Transformations. Macmillan, New York, 1964.
[10] P. F. Gobin and G. Guénin; p. 432 in Solid → Solid Phase Transformations in Metals and Alloys. Edited by D. deFontaine. Les Editions de Physique, Paris, 1981.
[11] U. Dehlinger, Theoretische Metallkunde, Vol. 2. Springer-Verlag, New York, 1968.
[12] G. B. Olson and M. Cohen, *Ann. Rev. Mater. Sci.*, 1 (1981).
[13] G. B. Olson and M. Cohen; p. 1145 in Solid → Solid Phase Transformations. Edited by H. I. Aaronson, D. E. Laughlin, R. F. Sekerka, and C. M. Wayman. The Metallurgical Society of AIME, 1982.
[14] G. Guénin and P. F. Gobin, *J. Phys.*, **43**, C4-57 (1982).
[15] G. B. Olson and M. Cohen, *J. Phys.*, **43**, C4-75 (1982).
[16] A. Kelly and R. B. Nicholson, *Prog. Mater. Sci.*, **10**, 151 (1963).

[17] R. B. Nicholson; p. 269 in Phase Transformations. Edited by H. I. Aaronson. American Society for Metals, Metals Park, OH, 1970.
[18] F. Falk, *J. Phys.*, **43**, C4-3 (1982).
[19] F. Milstein and B. Farber, *Phys. Rev. Lett.*, **44**, 277 (1980).
[20] M. J. Kelly, *J. Phys.*, **F9**, 1921 (1979).
[21] E. Esposito, A. E. Carlsson, D. D. Ling, H. Ehrenreich, and C. D. Gelatt, Jr., *Philos. Mag.*, **A41**, 251 (1980).
[22] W. M. Kriven, W. L. Fraser, and S. W. Kennedy; p. 82 in Advances in Ceramics, Vol. 3. Edited by A. H. Heuer and L. W. Hobbs. The American Ceramic Society, Columbus, OH, 1981.
[23] D. L. Porter and A. H. Heuer, *J. Am. Ceram. Soc.*, **62**, 298 (1979).
[24] A. H. Heuer; p. 98 in Advances in Ceramics Vol. 3. Edited by A. H. Heuer and L. W. Hobbs. The American Ceramic Society, Columbus, OH, 1981.
[25] L. H. Schoenlein and A. H. Heuer; p. 309 in Fracture Mechanics of Ceramics, Vol. 6. Edited by R. C. Bradt, A. G. Evans, D. P. H. Hasselman, F. F. Lange. Plenum, New York, 1983.
[26] L. H. Schoenlein, M Ruhle, and A. H. Heuer, this volume, pp. 275-82.
[27] N. Claussen, *J. Am. Ceram. Soc.*, **59**, 49 (1976).
[28] M. Ruhle, B. Kraus, A. Strecker, and D. Waidelich; this volume, pp. 256-74.
[29] M. Ruhle and A. H. Heuer; p. 359 in Proceedings of the 7th International Conference on High Voltage Electron Microscopy. Edited by R. M. Fisher, R. Gronsky, and K. H. Westmacott. National Technical Information Center, U.S. Department of Commerce, 1983.
[30] T. K. Gupta, F. F. Lange, and J. H. Bechtold, *J. Mater. Sci.*, **13**, 1464 (1978).
[31] M. Ruhle, N. Claussen, and A. H. Heuer; this volume, pp. 352-70.
[32] J. D. Eshelby, *Prog. Solid. Mech.*, **3**, 89 (1961).
[33] R. C. Garvie, *J. Phys. Chem.*, **82**, 218 (1978).
[34] A. G. Evans, N. Burlingame, M. Drory, and W. M. Kriven, *Acta Metall.*, **29**, 477 (1981).
[35] F. F. Lange, *J. Mater. Sci.*, **17**, 225, 235, 240, 247, 255 (1982).
[36] F. F. Lange and D. J. Green; p. 217 in Advances in Ceramics, Vol. 3. Edited by A. H. Heuer and L. W. Hobbs. The American Ceramic Society, Columbus, OH, 1981.
[37] A. H. Heuer, N. Claussen, W. M. Kriven, and M. Ruhle, *J. Am. Ceram. Soc.*, **65**, 641 (1982).
[38] I.-W. Chen and Y.-H. Chiao, *Acta Metall.* **31**, 1627 (1983).
[39] I.-W. Chen and Y.-H. Chiao; this volume, pp. 33-45.
[40] M. Ruhle, E. Bischoff, and N. Claussen; p. 1563 in Solid → Solid Phase Transformations. Edited by H. I. Aaronson, D. E. Laughlin, R. F. Sekerka, and C. M. Wayman. AIME, 1982.
[41] T. Mura, Micromechanics of Defects in Solids, Martinus Nijhoff, Amsterdam, 1982.
[42] M. Ruhle and W. M. Kriven, *Ber. Bunsenges. Phys. Chem.*, **87**, 222 (1983).
[43] M. Ruhle and W. M. Kriven; p. 1569 in Solid → Solid Phase Transformations. Edited by H. I. Aaronson, D. E. Laughlin, R. F. Sekerka, and C. M. Wayman. AIME, 1982.
[44] W. Mader and M. Ruhle; p. 358 in Electron Microscopy and Analysis, 1983, Conference Ser. No. 68. Edited by P. Doig. The Institute of Physics, London, 1983.
[45] W. Mader; Ph.D. Thesis, University of Stuttgart, 1984.
[46] W. M. Kriven; this volume, pp. 64-77.
[47] S. Schmauder, W. Mader, and M. Ruhle; this volume, pp. 251-55.
[48] G. Faivre, *Phys. Status Solidi*, **35**, 249 (1964).
[49] R. Sankaran and C. Laird, *J. Mech. Phys. Solids*, **24**, 251 (1976).
[50] K. P. Chiu, *J. Appl. Mech.*, **44**, 587 (1977).
[51] J. W. Christian; p. 129 in The Mechanisms of Phase Transformations in Crystalline Solids, Vol. 33. The Institute of Metals, 1969.
[52] L. Kaufmann and M. Cohen, *Prog. Met. Phys.*, **1**, 165 (1958).
[53] C. L. Magee; p. 115 in Phase Transformations. Edited by H. I. Aaronson. American Society for Metals, Metals Park, OH, 1970.
[54] L. Ma and M. Ruhle; unpublished work.
[55] C. F. Grain, *J. Am. Ceram. Soc.*, **50**, 288 (1967).
[56] J. R. Hellmann and V. S. Stubican, *J. Am. Ceram. Soc.*, **66**, 260 (1983).
[57] R. H. J. Hannink, *J. Mater. Sci.*, **13**, 2487 (1978).
[58] H. J. Scott, *J. Mater. Sci.*, **10**, 1527 (1975).
[59] R. D. Shannon and C. T. Prewitt, *Acta Crystallogr., Sect. B*, **25**, 925 (1969).
[60] O. J. Whittemore, Jr., and N. N. Ault, *J. Am. Ceram. Soc.*, **39**, 443 (1956).
[61] R. N. Patil and E. C. Subbarao, *J. Appl. Crystallogr.*, **2**, 281 (1969).
[62] J. M. Marder, T. E. Mitchell, and A. H. Heuer, *Acta Metall.*, **31**, 387 (1983).
[63] A. J. Ardell, R. B. Nicholson, and J. D. Eshelby, *Acta Metall.*, **14**, 1295 (1966).
[64] V. Lanteri, T. E. Mitchell, and A. H. Heuer; this volume, pp. 118-30.
[65] A. G. Khachaturyan, Theory of Structural Transformations in Solids. Wiley & Sons, New York, 1983.
[66] H. Brooks; p. 20 in Metal Interfaces. American Society for Metals, Metals Park, OH, 1957.
[67] M. Morinaga, H. Adachi, and M. Tsukuda, *J. Phys. Chem. Solids*, **44**, 301 (1983).

[68] G. M. Wolten, *J. Am. Ceram. Soc.*, **46**, 418 (1964).
[69] J. E. Bailey, *Proc. R. Soc. (London), Sect. A,* **279**, 359 (1964).
[70] G. K. Bansal and A. H. Heuer, *Acta Metall.*, **20**, 1281 (1972).
[71] G. K. Bansal and A. H. Heuer, *Acta Metall.*, **22**, 409 (1974).
[72] M. S. Weschsler, D. S. Lieberman, and T. A. Read, *Trans. AIME,* **197**, 1503 (1953).
[73] M. A. Choudhry and A. G. Crocker; this volume, pp. 46–53.
[74] W. M. Kriven; p. 168 in Advances in Ceramics, Vol. 3. Edited by A. H. Heuer and L. W. Hobbs. The American Ceramic Society, Columbus, OH, 1981.
[75] E. Bischoff and M. Rühle, *J. Am. Ceram. Soc., ***66**, 123 (1983).

Martensitic Transformations in ZrO$_2$ and HfO$_2$ — An Assessment of Small-Particle Experiments with Metal and Ceramic Matrices

I-Wei Chen AND Y-H. Chiao

Massachusetts Institute of Technology
Department of Nuclear Engineering
and Materials Science and Engineering
Cambridge, MA 02138

Experimental evidence from small-particle experiments, especially those of ZrO$_2$ and HfO$_2$ metal matrix composites, is considered in order to establish the mechanisms of martensitic transformation in ZrO$_2$ and HfO$_2$. The transformation is controlled by nucleation, which is often suppressed by the lack of potent defects. When such defects are available, nucleation of monoclinic particles occurs spontaneously in the internal stress field of the nucleating defects. The matrix constraint on the entire tetragonal particle is shown to be essentially irrelevant. Rather, it is the matrix constraint exerted by the tetragonal parent matrix on the monoclinic nucleus that is most significant. Without an appropriate array of interfacial misfit dislocations, which may form during the loss-of-coherency process in overaged, partially stabilized zirconia (PSZ), no fundamental particle-size effect is operational. These and other observations in bulk and dispersed ZrO$_2$ and HfO$_2$ are in quantitative agreement with the Olson–Cohen model of martensitic nucleation, with a modification to account for the coherent interface between the monoclinic nucleus and the parent matrix.

At the first Zirconia Conference in 1980,[1] much of the discussion on the martensitic transformations between tetragonal (t-ZrO$_2$) and monoclinic (m-ZrO$_2$) zirconia centered around the transformation crystallography, the transformation free energy, and the particle-size effect. Very little attention was given to the nucleation and growth mechanisms of the transformation from either an experimental or a theoretical viewpoint. In contrast, the Phase Transformation Conference of 1968[2] was devoted to the mechanisms of martensitic transformations (among other transformations). During the latter conference, the review by Christian on martensitic transformations concluded with a suggestion that the nucleation process must take place heterogeneously in the high internal stress field of a suitable dislocation-type defect; meanwhile, the small-particle experiment of Easterling et al.[2] furnished direct evidence that the transformation is indeed nucleation-controlled and that defects are necessary to nucleate martensite. The main reason for the apparently different emphasis in the two closely related fields, in our view, resides in the relative experimental complexity in the sample preparation and testing of ceramic materials, which often makes interpretation of experiments intended for fundamental inquiry concerning transformation mechanisms difficult. The study of ZrO$_2$-metal composites was therefore proposed[3] to overcome such difficulties and hopefully to bridge the information gap between ceramic martensitic transformations and metallic martensitic transformations.

Composites of ZrO_2 and metals can be obtained by the internal oxidation of zirconium-containing metallic alloys. With appropriate oxidation temperatures, equiaxed incoherent tetragonal ZrO_2 particles between a few nanometers to approximately 0.25 μm can be precipitated within the metallic matrix (Fig. 1(A)). The volume fractions of the oxide particles can be varied within the limits of Zr solubility. The zirconia transformation, under various thermal and mechanical conditions, can be conveniently studied with such materials. The metallic matrix simply serves as a load-bearing ductile carrier material for small, dispersed, transformable oxide particles.[3] Similar results are obtained for HfO_2, for which a micrograph of small HfO_2 particles is given in Fig. 1(B) for a comparison.

A detailed account of the first results on ZrO_2-metal composites using Cu-ZrO_2 specimens has been reported,[4] along with a proposed defect model, which is of the more general Olsen–Cohen type[5] but with a coherent nucleation path for the $t \rightarrow m$ ZrO_2 transformation. These results and those from our more recent experiments on Ta-HfO_2 will be discussed in the context of the nucleation mechanism. Model calculations along the line established in our previous paper[4] will be provided to compare the zirconia and hafnia systems in which the driving force varies substantially. On the basis of these results, we will attempt to identify the critical factors affecting the stability of t-ZrO_2 particles in ceramic matrices.

Martensitic Nucleation

Until recently, the stability of tetragonal ZrO_2 particles was considered by using a variety of energy arguments in which the total free energy of the transformable particle as a whole, with the attendant near-field response in the matrix, is compared before and after the transformation.[6] In this approach, the chemical free energy difference between the two phases, the elastic energy of the particle and the surroundings, the interfacial energies of the particle–matrix interface, the internal twin boundaries, and possibly the microcracks caused by transformation itself were taken into account in part or in toto. The above approach was criticized primarily from an experimental viewpoint, by Heuer, Claussen, Kriven, and Rühle[6] and need not be repeated here.

Mechanistically, a proper consideration of the stability of t-ZrO_2 should include the nucleation and the growth processes. The experimental evidence available at this juncture overwhelmingly suggests that the transformation is nucleation-controlled. Such evidence includes the stability of the retained tetragonal particles in Cu-ZrO_2 at cryogenic temperatures (ca. 4.2 K)[3,4] and the stability of the transformed monoclinic particles (induced by the plastic deformation of the metallic matrix) to a reheat temperature above the bulk M_s temperature (taken to be 950°C).[4] Together, it suggests that the stability of small particles of either phase is not determined by the type of equilibrium considerations scaled with the particle size, such as the one based on the small difference of surface energy of the two phases.[23] Also notable is the general absence of partially transformed tetragonal-monoclinic particles in our and others' experiments. Therefore, in most cases, growth of monoclinic nuclei must proceed to completion immediately after each nucleation event. Furthermore, the nucleation process is generally taken to be essentially athermal, to account for its occurrence at rather low temperatures. Thus, the nucleation barrier for this process must be very low.

The energetic argument outlined earlier is actually applicable to nucleation. However, the proper entities to be considered in energetics are the monoclinic

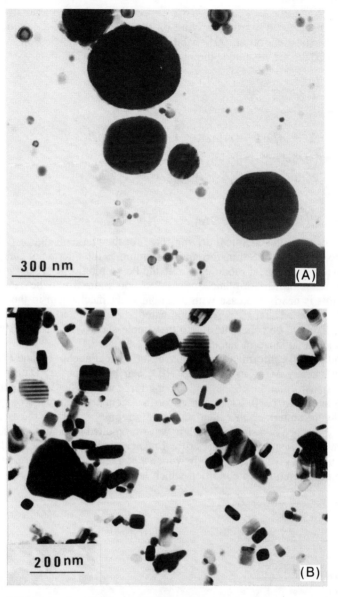

Fig. 1. (A) Spherical ZrO$_2$ particles obtained by internal oxidation of Cu-Zr (replica). (B) Cuboidal HfO$_2$ particles obtained by internal oxidation of Ta-Hf (replica).

nucleus and the parent tetragonal matrix in which the nucleus is embedded. For a coherent nucleus without microcracking or internal twinning, the nucleation energy in homogeneous nucleation was treated by Christian.[7] Since, for $t \rightarrow m$ ZrO$_2$ transformation, with the axes 1,2,3 assigned for the a,b,c crystallographic axes, respectively, the major distortion is that of e^T_{13} and only a comparatively small dilatation is left for e_{11}^T, e_{22}^T, and e_{33}^T, we expect the nucleus to assume an oblate spheroidal shape, with a radius r lying on the 1–2 plane and a semithickness t along the 3 axis, i.e., on an 001$_t$ habit. When identical elastic constants and a coherent interface are assumed, the nucleation energy ΔG is thus[7]

$$\Delta G = \frac{4\pi}{3} r^2 t (g_{ch} + g_{str}) + \frac{4\pi}{3} r t^2 K + 2\pi r^2 \gamma \tag{1}$$

where

$$K = [\pi(2 - \nu)/2(1 - \nu)]\mu e_{13}^{T2} + [\pi/4(1 - \nu)]\mu e_{33}^{T2}$$
$$- [\pi/32(1 - \nu)]\mu\{13(e_{11}^{T2} + e_{22}^{T2})$$
$$+ 2(16\nu - 1)e_{11}^T e_{22}^T - 8(1 + 2\nu)(e_{11}^T + e_{22}^T)e_{33}^T\} \tag{2a}$$

and

$$g_{str} \equiv [1/(1 - \nu)]\mu(e_{11}^{T2} + 2\nu e_{11}^T e_{22}^T + e_{22}^{T2}) \tag{2b}$$

In the above, g_{ch} is the chemical driving force for the phase change, g_{str} the strain energy due to the non-IPS transformation straining, γ the interfacial energy of the m/t interface, μ the shear modulus, and ν the Poisson ratio. If e_{11}^T and e_{22}^T were to vanish, an exact invariant-plane-strain (IPS) condition would be achieved. In zirconia, this is nearly the case with a simple shear mechanism in the a–b crystallographic plane. Thus, g_{str} is small compared to K. The interface between t- and m-ZrO$_2$ is likely to be coherent, due to the simple-shear-type lattice distortion, at least for the initial stage of nucleation.

With the above information, the size of the critical nuclei, r^* and t^*, and the nucleation barrier, ΔG^*, in ZrO$_2$ and HfO$_2$ can be computed. Data for elastic constants, transformation strains, and driving force were taken from the literature, with γ assumed to be 0.2 J/m^2 based on a recent measurement of γ for the tetragonal-cubic interface in calcia-stabilized zirconia.[22] In fact, the free energy contour in r–t space of a similar problem was first plotted by Kaufman and Cohen[8] for martensitic transformation; the critical nucleus is located at the saddle point. These results are given in Fig. 2, in which r^*, t^*, and ΔG^* are given by the saddle-point differentiation condition which leads to

$$r^* = \frac{4\gamma K}{(g_{ch} + g_{str})^2} \tag{3a}$$

$$t^* = -\frac{2\gamma}{(g_{ch} + g_{str})} \tag{3b}$$

$$\Delta G^* = \frac{32\pi\gamma^3 K^2}{3(g_{ch} + g_{str})^4} \tag{3c}$$

The preceding results, which indicate a large critical nucleus size and a very high nucleation barrier under all conditions, are obviously at variance with experimental observations in bulk ZrO$_2$ of a spontaneous martensitic transformation at M_s temperature. Preexisting embryos, a suggestion initially made by Kaufman and Cohen[8] for martensitic transformations in metals and lately invoked by Andersson and Gupta[1] conceptually for the application of ZrO$_2$ transformation, also appear unlikely, considering the rather large embryo size necessary for its effective operation. Indeed, for precisely the same reasons, the notion of homogeneous nucleation and preexisting embryos has been abandoned by metallurgists concerning martensitic transformations in metallic systems.[2,5] The above exercise, never-

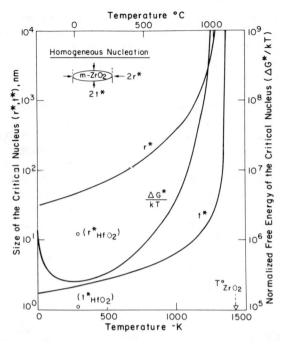

Fig. 2. Radius r^*, semithickness t^*, and free energy ΔG^* in units of kT for the $t \to m$ ZrO_2 transformation by the classical homogeneous nucleation mechanism. Corresponding dimensions for a critical nucleus in HfO_2 are given for 298 K by the two circles.

theless, demonstrates that the matrix constraint effect within the transformable parent tetragonal phase itself is still overwhelming, even if the surrounding inert matrix, ceramic or metallic, is removed. The constraint is exerted on the monoclinic nucleus by the immediate surrounding tetragonal matrix and is essentially independent of the material type at a distance. The most solid support for the existence of a constraint effect on the nucleus, and not a constraint effect on the entire tetragonal particle as used in the previous energy arguments, came from the small-particle experiments on Cu-ZrO_2,[3,4] and more recently on Ta-HfO_2. Many large (0.1 μm), equiaxed tetragonal particles remained untransformed (Fig. 1(A, B)) in these materials even after the metal matrices were removed electrolytically or chemically! Since the deviatoric strain energy is substantially higher than that of the hydrostatic part, it must be concluded that the large shear distortion suppresses the homogeneous nucleation of m-ZrO_2 from t-ZrO_2.

Similar observations of this nature led Christian,[2] and shortly afterward Magee,[9] to the assertion that martensitic nucleation must take place heterogeneously in the high internal stress field of some dislocation-type defect, to take advantage of the interaction energy between the martensitic nucleus and the existing defect to compensate the enormous elastic energy due to the deviatoric distortion accompanying the transformation. Qualitatively analogous mechanisms, in fact, were previously investigated for heterogeneous nucleation in twinning and precipitation. Olson and Cohen[5] advanced a quantitative application of this idea with a defect model containing a dislocation array. The nature of the dislocation array is such that its stress field, primarily its shear component, is sufficiently intense to nearly accomplish the desired lattice distortive transformation in the same shear mode. The required intensity of the stress field thus determines the ideal dislocation spacing within the array. It also proves necessary to have a dislocation array of an adequate spatial extent, to counteract the interfacial energy during nucleation. The latter condition thus determines the required potency of the suitable dislocation-array type of defects. In real materials, the onset of martensitic transformation is thus dictated by the availability of such potent defects.[5]

For $t \rightarrow m$ transformation in ZrO_2, we proposed earlier that an array of [001] screw dislocations stacked at a spacing of approximately 3 nm along essentially the a axis can provide precisely the IPS required for the transformation.[4] At the bulk M_s temperature, taken to be 950° or 200°C below the equilibrium temperature, we determined further that an array of 4 to 6 lattice dislocations should be sufficiently potent, with the chemical driving force available, to nucleate a monoclinic nucleus spontaneously and without a barrier. The above spacing, D, and the number of dislocations in the array, m, are given by[4]

$$D = \frac{\mathbf{b}}{2}\left(\frac{1}{\left(e_{13}^{T2} + \frac{1}{4}e_{33}^{T2}\right)^{1/2}}\right) \qquad (4)$$

and

$$m = \frac{-2\gamma}{D(g_{ch} + g_{str})} \qquad (5)$$

where $\mathbf{b}$ is the Burgers vector of [001] dislocations. Equations of this type were originally derived by Olson and Cohen[5] using an approximate description of the dislocation interaction energy; a rigorous analytic derivation based on continuously distributed dislocations is now available.[10] When these conditions are met, the prohibitively high deviatoric strain energy will be compensated adequately by the interaction energy of the nucleus with the defect stress field to facilitate nucleation.

Obviously, in a bulk tetragonal crystal, the M_s temperature is determined by the statistical availability of at least one defect containing m dislocations of the above description or equivalent defects with a qualitatively similar internal stress field and spatial extent. As M_s is approached, presumably the probability of finding at least one such defect of 4 to 6 dislocations in an array approaches unity in a macroscopic specimen; hence, the martensitic transformation begins spontaneously. Since the required m is dependent on the driving force, useful information can be obtained by comparing M_s under different thermodynamic conditions, g_{ch}. Two such examples exist in the literature. Whitney[11,12] reported a pressure dependence of the M_s temperature in ZrO_2. His analysis of the depression of M_s by a hydrostatic pressure indicated that the effective chemical driving force, including the $P\Delta V$ correction, remains a constant:

$$g_{ch}(M_s) + P\Delta V = \text{constant} \qquad (6)$$

at the M_s temperature, which itself varies according to the applied pressure. The preceding results can be interpreted in our defect model as a reflection of the constancy of the availability of defects of a given potency. Since the probability of finding m dislocations should remain the same in Whitney's samples, regardless of the external pressure, at M_s temperatures the effective driving force must be the same to attain the critical spontaneous nucleation condition (Eq. (5)). This is

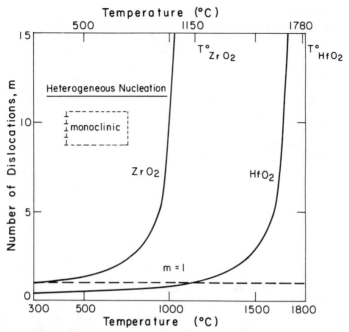

Fig. 3. Size of the critical nucleating defect by a heterogeneous nucleation mechanism in ZrO_2 and HfO_2. The size is given by m, which is the number of dislocations schematically shown in the inset.

certainly reasonable. The second example of varying the driving force came from HfO_2, in which the transformation enthalpy is 75% higher than that of ZrO_2, and the g_{str} is slightly smaller, whereas the shear is about the same. Taking the same value of m as that for ZrO_2, we compute the M_s temperature for bulk HfO_2 to be 1620°C, which is 160°C below the equilibrium temperature. Thus, we expect a smaller difference between M_s and the equilibrium temperature and, if the hysteresis loop is used as a qualitative measure, a smaller hysteresis for the HfO_2 transformation. These predictions are in accord with the available, though scanty, data on HfO_2 systems.[13]

The preceding discussion concerns the general nucleation mechanism and its verification for martensitic transformations in bulk tetragonal crystals. For better appreciation of the nature of the heterogeneous nucleation mechanism, the computed values of the critical m in ZrO_2 and HfO_2 at various temperatures are plotted in Fig. 3. Clearly, the nucleation condition is less stringent in hafnia below approximately 1200°C. Another notable feature in the figure is the steep increase of the critical m with temperature when the effective driving force, $g_{ch} + g_{str}$, is reduced to nearly zero. It was already noted that the very high nucleation barrier in homogeneous nucleation, and hence the need for an alternative heterogeneous nucleation mechanism, is essentially due to the large deviatoric strain energy rather than to the dilatational strain energy ($K \gg g_{str}$ in Eq. (1)). Yet, once heterogeneous nucleation occurs, the deviatoric strain energy is always compensated adequately by the decrease of the interaction energy of the defect-nucleus system and is rendered inconsequential. What remains as an important factor that affects the stability of the phases, provided there is already an abundant supply of heterogeneities for nucleation, is actually the dilatational strain energy given by g_{str}. Indeed, in bulk materials where defects of sufficient potency are presumably available, the M_s temperature is not much below the effective equilibrium temperature where $g_{ch} + g_{str} = 0$, due to the steep rise of the critical m values with the temperature in this regime. This observation suggests that, perhaps in many ZrO_2-containing materials in which nucleating defects are not severely limited, the volume constraint effect could nevertheless play an important role in controlling the stability of t-ZrO_2, despite the fact that heterogeneous nucleation is the dominant mechanism in transformation.

The effect of autocatalysis is also an important consideration. In a bulk specimen, once a sufficiently potent defect becomes available during cooling, a large portion of the entire specimen undergoes transformation due to the autocatalytic effect by which more nucleation defects are generated as the first nucleus becomes larger. This is avoided in small-particle experiments with isolated particles embedded in an inert and plastic matrix. As can be seen in Fig. 3, under no circumstances does m approach 0 even at 0 K. Thus, in small-particle experiments in which nucleation in individual particles is rendered independent and uncorrelated, we must find a nucleating defect (which need not be big at high undercooling) for each particle, or else almost all of the particles will remain untransformed. This type of experiment therefore provides a means to test the heterogeneous nucleation model proposed above and even allows systematic variations applied to the material to progressively sample the population of nucleating defects of decreasing potency. Such an experiment was performed using ZrO_2-Cu with the external stimulus provided by cold work.[4] The results of the experiment and our more recent work in search of the particle-size effect are discussed in the next section.

Particle Size Effect

Dispersed zirconia particles in ceramic matrices, such as PSZ and zirconia-alumina, as well as our ZrO_2-metal composites, are between 1 nm and 1 μm in size. This size range is smaller, by 2 to 4 orders of magnitude, than the small Fe-Ni particles investigated by Cech and Turnbull[14] in search of evidence of heterogeneous nucleation of martensite. Noting that the estimates of m at respective bulk M_s temperatures are essentially the same for ZrO_2[4] and Fe-Ni[5] and that a significant size effect was observed in Fe-Ni alloys below approximately the 50-μm size range, we therefore conclude that, in each dispersed zirconia particle, the probability of finding the nucleating defect is extremely small. Indeed, since dispersed zirconia particles reported in most experiments are actually single crystals themselves, the probability of retaining stable dislocations within them is virtually nil due to the difficulty of maintaining an equilibrium configuration for dislocations in small single crystals. From these considerations, it is not surprising that small tetragonal zirconia particles can be retained even at 4.2 K, as demonstrated by our ZrO_2-metal composite experiments.[3,4]

It is interesting to estimate the size of the nucleating defect, m, at such drastic undercooling. At room temperature, this is found from Fig. 4 to be $m = 0.9$ for ZrO_2[4] and $m = 0.4$ for HfO_2. Thus the mere generation of one dislocation, e.g., by cold work as shown in our Cu-ZrO_2 experiment, suffices to trigger the nucleation. If, however, insufficient external stimulus is applied for the generation of such a dislocation, the small tetragonal ZrO_2 may be indefinitely retained untransformed. Since particles are usually too small to have preexisting dislocation-

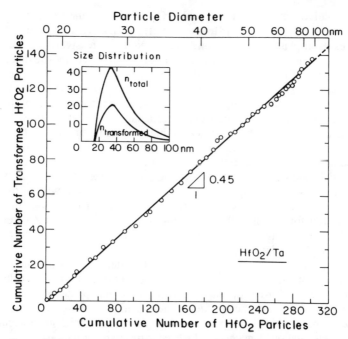

Fig. 4. Cumulative number of transformed HfO$_2$ particles, shown in the order of increasing sizes, vs the cumulative number of total HfO$_2$ particles shown in the same order. The actual size distributions for the two groups are given in the inset.

type defects in our own experiments, the particle-size effect, if any, is not expected to be strong or distinct, although perhaps a mild surface-area dependence for the susceptibility of external stimuli might be manifested if the stimulus due to handling is actually surface-sensitive, for example, in electrolytic or chemical dissolution of the carrier metal matrix. In the system Cu-ZrO$_2$, the stability of tetragonal ZrO$_2$ particles is so high that virtually all of the particles extracted remained tetragonal after dissolution of the matrix, and no distinct size effect for particles smaller than 0.1 μm was observed.[4] Indeed, transformation of ZrO$_2$ particles occurred only after substantial plastic deformation of the matrix, as will be discussed later.

Further direct support for our prediction of the lack of size dependence was furnished by the Ta-HfO$_2$ experiment. Since HfO$_2$ has a much higher driving force and therefore a substantially less stringent nucleation condition, it is hoped that even a relatively gentle stimulus, such as a slight straining or a mild quenching, may generate the nucleating defect, at least in some tetragonal HfO$_2$ particles. This prediction was indeed confirmed. As shown in Fig. 4, in which the cumulative number of hafnia particles examined in order of increasing size is plotted against the cumulative number of transformed particles found in the same order, a constant percentage of transformation, ca. 45%, was found for all the size ranges that we sampled. Thus, no size effect is evident in this HfO$_2$ experiment in which the nucleation, although via the same heterogeneous nucleation mechanisms as in ZrO$_2$, is made somewhat easier and more susceptible to external stimuli for the generation of nucleating defects.

The relative stability of tetragonal ZrO_2 particles in the Cu matrix affords a unique opporunity to study a different size effect, that we attributed to the interfacial stresses in plastic deformation.[4] We found a progressively pronounced size effect after substantial plastic deformation in the Cu-ZrO_2 composite that gradually lowered the size of the remaining stable tetragonal ZrO_2 particles. However, even after 90% cold work by rolling, a still significant fraction of very small tetragonal particles remained untransformed in the matrix, even after removal of the matrix. We attributed this to the dependence of interfacial stresses on plastic strain and on particle size, which will be expected at higher plastic strains (or high volume fraction of particles) when the mean free path of matrix dislocatons, which are generated at the particle matrix interface to accommodate the plastic deformation, is limited.[15,16] The detailed statistics of the particle-size effect and a theoretical analysis of ZrO_2-metal composites under plastic deformation will be found elsewhere.[17] At this time, we merely state that the general condition for inducing martensitic transformation by the nucleating defect mechanism in t-ZrO_2 embedded in a metal matrix is to reach a sufficiently high interfacial stress, which is the result of the matrix plastic deformation.

We note that, in the Cu-Fe experiment of Easterling et al.,[2,24] the nucleation condition is substantially less stringent. At least in principle, a matrix dislocation can cut through the Cu-Fe interface to enter the Fe precipitate, affording itself as the nucleating defect. For metal-oxide systems, this mechanism appears unlikely. Thus, for the more stable ZrO_2 particles, only a high interfacial stress accumulated by plastic deformation is necessary to trigger the transformation.

ZrO_2-Containing Ceramics

It is now well established that, in PSZ, tetragonal ZrO_2 particles transform during cooling from the aging temperature when the coherency is lost during growth.[6,18] Similar observations of the stability of t-ZrO_2 particles under a certain critical size in other ceramic matrices, notably Al_2O_3, were reported.[6] A recent assessment of these observations lead Heuer et al.[6] to conclude that the transformation is probably nucleation-controlled. It was suggested that perhaps the thermal stress during cooling causes the transformation, and that the coherency loss condition in PSZ may be necessary for the thermal stress to be effective. However, the maximal size of stable tetragonal particles appears to be generally smaller in PSZ than in ZrO_2-Al_2O_3, despite a very large thermal expansion mismatch between Al_2O_3 and t-ZrO_2.[6] Also significant is the fact that stable, tetragonal ZrO_2 particles are incoherent with the Al_2O_3 matrix. Clearly, it would be desirable if these features, as well as their connection with results found in ZrO_2-metal composites, could be reconciled on the same basis of a fundamental nucleation mechanism proposed above.

It is well known that, for complete loss of coherency, a plentiful supply of misfit dislocations must be available. Otherwise, a coherent precipitate may continue to grow to become metastable relative to an incoherent precipitate. This is often found to be the case experimentally. This point is well illustrated by the experiments of Brown and Woolhouse[19] and Weatherly and Nicholson,[20] in which they restored the equilibrium for oversized coherent precipitates and caused rapid loss of coherency by furnishing matrix dislocations with neutron irradiation[19] or with creep deformation.[20] Therefore, even at high temperatures, the nucleation of dislocations at the coherent particle-matrix interface is generally difficult. This difficulty, in our view, explains the stability of coherent t-ZrO_2 particles in PSZ during cooling. The nucleating defect simply cannot be generated by the thermal stress alone!

Loss of coherency leaves behind an array of misfit dislocations at the interface. The Burgers vector of these dislocations, under optimal conditions, tends to lie on the interface and along the crystallographic axis of the largest misfit. In PSZ-containing t-ZrO_2 in a cubic ZrO_2 matrix, the largest misfit is typically in the c direction.[6,18] Therefore, an array of misfit dislocations of the type [001] stacked along the c axis is expected from the loss-of-coherency operation. Although this dislocation array is not the kind expected for an ideal nucleating defect, only one [001] dislocation suffices at a sufficient undercooling. Thus, coherent particles in PSZ are prone to transformation during cooling. Other edge effects in the case of oblate precipitates further aggravate the situation due to thermal stress concentrations. In principle, even an array of misfit dislocations of the type [100] along the c axis could be sufficiently potent to compensate the deviatoric strain energy caused by e_{13}^T. This after-effect of even a partial loss of coherency, in our view, explains the instability of incoherent and semicoherent t-ZrO_2 particles in overaged PSZ during cooling.

Incoherent t-ZrO_2 particles in Al_2O_3 do not necessarily have the difficulty of the generation of dislocations at the interface. On the other hand, they do not necessarily have the heritage of [001] or [100] dislocation arrays at the interface. Thus, nucleating defects must be generated by external means, in very much the same way as in the case of ZrO_2-metal composites via raising the interfacial stresses, which in the present case are due to the thermal stress. Other secondary considerations such as faceting may also be effective. In general, however, we expect ZrO_2 particles in Al_2O_3 to be more stable than their incoherent counterpart in PSZ, despite the large thermal expansion mismatch.

Although it is tempting to offer the above picture as an extension of the defect nucleation mechanism we observed in ZrO_2-metal for the all-ceramic system, caution is clearly warranted in such a generalization. There are several important aspects that are similar and different between the two systems. Clarification of these points before closing should be helpful for achieving a better perspective. Concerning the mechanical state inducing the transformation in small particles, we reiterate our point that, in metal-ceramic systems, the metal matrix serves as an inert carrier material that transmits stresses to the particles but does not directly pump dislocations across the particle interface to trigger the transformation. In this respect, the experiment conducted with a Cu matrix is best regarded as a miniature mechanical test of small ZrO_2 single crystals using Cu as a gripping material, and the mechanical state involved is not fundamentally dissimilar to that caused by either an external stress or an internal (thermal) stress exerted onto isolated t-ZrO_2 crystals in an all-ceramic matrix. Concerning the transformation substructures, although microcracking is probably absent in ZrO_2-Cu transformation and routinely takes place in all-ceramic transformation, we found other structures, twinning in particular, in all of the transformed particles in our experiment. Therefore twinning is clearly not sensitive to the particular type of matrix used and is inherent in all the transformations down to very small particles. Concerning the matrix elastic-plastic constraint, we recall that, during the nucleation stage, it is the immediate surrounding of the m-ZrO_2 nucleus that controls the transformation strain energy; thus, the exact nature of the distant matrix plays no significant role in the nucleation consideration, if the nucleus is embedded entirely in a t-ZrO_2 matrix. (Even if the nucleus protrudes onto the particle-matrix interface, as long as the lenticular nucleus penetrates into the tetragonal matrix, the deviatoric strain energy in homogeneous nucleation still far exceeds the dilatational strain energy, as we showed in Martensitic Nucleation, and the elastic strain energy is still controlled by the tetragonal matrix regardless of the type of the surrounding matrix.) However, if transformation is predominantly surface-nucleated, one should expect significant differences in the potency of nucleating heterogeneities that are material-dependent and even history-dependent. Such differences are inherent in any surface nucleation phenomena and not necessarily unique for the case of a metal matrix as opposed to a ceramic matrix. Our proposed picture concerning the coherency state of PSZ in relation to the stability of t-ZrO_2, if confirmed, could become a good example of the surface nucleation phenomenon. Concerning the nature of the nucleating defect, it is in principle possible for any small defect, of a sufficient size and with a self-equilibrated internal stress field similar to that of a small twist boundary (or a shear crack), to serve equally well as the nucleating heterogeneity. It is not necessary for the defect to be mobile; thus, the high friction stress, which renders the dislocation glide difficult in ZrO_2, may not necessarily preclude the possibility of the operation of the defect-nucleation mechanism envisioned here and certainly may not pose a fundamental objection to comparing the ZrO_2 transformation in two matrices. In our view, further investigation to examine the validity of the defect nucleation model, including a stress-assisted one, could be profitably undertaken with both systems, but with additional variations designed to affect the state of nucleating defects. More pragmatic research to explore the alloying effects in mixed oxides on the nucleation process could possibly suggest other mechanisms of broad utility. We should then be able to assert whether or not a fundamental mechanism for martensitic transformation could be established.

In closing, we compare the critical martensitic nucleus in ZrO_2 and in ferrous alloys. Our model of martensitic nucleation suggests that the nucleus-matrix interface in ZrO_2 is coherent, due to the near simple-shear nature of the lattice distortion. This feature has the very significant advantage of avoiding a high interfacial energy γ, which directly controls the size of the nucleating defect by Eq. (5). If a higher interfacial energy (>0.2 J/m^2) is operative, the size of the required nucleating defect will become much too large to be available statistically or even to be generated by the interfacial stresses. In contrast, the Bain distortion in ferrous alloys during martensitic transformation can be obtained only by two superposed large shear components; hence, the nucleus interface has no choice but to become semicoherent in the latter case. Were a more complex lattic distortion such as the Bain distortion encountered in ZrO_2, we would expect that the martensitic transformation might be entirely suppressed and not available for ceramic-toughening applications. Whether this speculation of the absence of complex martensitic transformations in *oxide* ceramics may prove well founded or not remains to be seen in future investigations. Available evidence of ceramic martensitic transformations, as summarized by Kriven,[21] however, does lend some initial support to our contention.

Acknowledgment

Support for this research was provided by a Seed Grant from the National Science Foundation through the Center of Materials Science and Engineering at MIT under NSF Grant No. 81-19295. This support and the continuous interest of G. B. Olson and A. S. Argon during the research are gratefully acknowledged.

References

[1] A. H. Heuer and L. W. Hobbs (Editors); Advances in Ceramics, Vol. 3. Edited by A. H. Heuer and L. W. Hobbs. The American Ceramic Society, Columbus, OH, 1980.
[2] Monograph and Report Series, No. 33. Institute of Metals, London, 1968.
[3] I.-W. Chen and Y.-H. Chiao, for abstract see Am. Ceram. Soc. Bull., **61** [3] 812 (1982).
[4] I.-W. Chen and Y.-H. Chiao, Acta Metall., **31**, 1627–38 (1983).
[5] G. B. Olson and M. Cohen, Ann. Rev. Mater. Sci., **11**, 1–30 (1981).
[6] A. H. Heuer, N. Claussen, W. M. Kriven, and M. Rühle, J. Am. Ceram. Soc., **65** [12] 642–50 (1982).
[7] J. W. Christian; The Theory of Transformations in Metals and Alloys, Vol. 1, 2d ed. Pergamon, New York, 1975.
[8] L. Kaufman and M. Cohen; pp. 165–224 in Progress in Metal Physics, Vol. 7. Edited by B. Chalmers and R. King. Pergamon, London, 1958.
[9] C. L. Magee; pp. 115–56 in Phase Transformations. Edited by H. I. Aaronson. American Society Metals, Metals Park, OH, 1970.
[10] I.-W. Chen and G. B. Olson; unpublished work.
[11] E. D. Whitney, J. Am. Ceram. Soc., **45** [12] 612–13 (1962).
[12] E. D. Whitney, J. Electrochem. Soc., **112**, 91–94 (1965).
[13] E. C. Subbarao, H. S. Maiti, and K. K. Srivastava, Phys. Status Solidi, Sect. A, **21**, 9–40 (1974).
[14] R. E. Cech and D. Turnbull, Trans. AIME, **206**, 124–32 (1956).
[15] M. F. Ashby, Philos. Mag., **14**, 1157–78 (1966).
[16] A. S. Argon, J. Im, and R. Safoglu, Metall. Trans. A, **6A**, 825–37 (1975).
[17] I.-W. Chen and Y.-H. Chiao; unpublished work.
[18] R. H. J. Hannink, J. Mater. Sci., **13**, 2487–96 (1978).
[19] L. M. Brown and G. R. Woolhouse, Philos. Mag., **17**, 781–89 (1968).
[20] G. C. Weatherly and R. B. Nicholson, Philos. Mag., **17**, 801–31 (1968).
[21] W. M. Kriven; pp. 1507–32 in Proceedings of the International Conference of Solid-State Phase Transformations. Carnegie-Mellon University, Pittsburgh, PA, 1981.
[22] J. M. Marder, T. E. Mitchell, and A. H. Heuer, Acta Metall., **31**, 387–95 (1983).
[23] R. G. Garvie, J. Phys. Chem., **82**, 218–24 (1978).
[24] K. E. Easterling and H. M. Miekk-Oje, Acta Metall., **15**, 1133 (1967).

Theory of Twinning and Transformation Modes in Zirconia

M. Arshad Choudhry* and A. G. Crocker
University of Surrey
Department of Physics
Guildford, Surrey, England, GU2 5XH

Monoclinic zirconia particles are frequently observed to be twinned on (100), (001), {110}, and {011} planes. The operative twinning modes given in the literature were deduced from experimental observations, but unfortunately some of the quoted crystallographic elements are inconsistent. Therefore, possible deformation twinning modes of this structure have been analyzed using the theory due to Bilby and Crocker. The results are presented in terms of general monoclinic lattice parameters and also with use of the appropriate zirconia constants. The martensitic transformation between the tetragonal high-temperature and monoclinic low-temperature phases of zirconia is accompanied by a lattice invariant shear of the product, which may be slip or twinning. Because of the nature of the product structure and of the low symmetry of the twinning modes, no comprehensive investigation of the crystallography of the transformation has previously been carried out using the established theories. The CRAB theory is well-suited to such a study and has been used to obtain a complete set of predictions of the possible crystallographic features. In particular, in this study the twinning modes referred to above and their conjugates were used as lattice invariant shears with all three reported lattice correspondences. The results are discussed and compared with experimental observations.

Zirconia may exist in three crystallographic forms: cubic, tetragonal, and monoclinic. The tetragonal-to-monoclinic transformation occurs at approximately 950°C and exhibits martensitic characteristics. It is accompanied by a volume increase of about 3.5%, which is physically deleterious. Therefore it is important to elucidate the mechanisms involved.[1]

As part of a study of this transformation, induced in thin foils of zirconia, Bailey[2] discovered deformation twins on (100) and (110) planes of the monoclinic phase. He also showed that the observed shear strains on these planes were consistent with the following twinning modes:

$$K_1 = (100), \quad \eta_1 = [00\bar{1}], \quad K_2 = (00\bar{1}), \quad \eta_2 = [100]; \quad s = s_1 \quad (1)$$

$$K_1 = (110), \quad \eta_1 = [1\bar{1}\bar{P}], \quad K_2 = (1\bar{1}\bar{Q}), \quad \eta_2 = [110]; \quad s = s_2 \quad (2)$$

Here K_1 and η_1 represent the twinning plane and direction, K_2 and η_2 their conjugates, and s is the twinning shear strain. Using stereographic projections, Bailey[2] deduced that P, Q, s_1, and s_2, which are functions of the lattice parameters, are ≈ 8, ≈ 8, 0.35, and 0.25, respectively. Recently, Bischoff and Rühle,[3] adopting similar techniques, confirmed that modes 1 and 2 are operative but, using different lattice parameters, evaluated P, Q, s_1, and s_2 to be ≈ 8, ≈ 25, 0.328, and 0.228, re-

---

*On leave from the Department of Physics, Islamia University, Bahawalpur, Pakistan.

spectively. In addition, they observed twins on (001) planes, arising from the conjugate of mode 1, and on (011) planes from the mode

$$K_1 = (011), \quad \eta_1 = [\bar{R}1\bar{1}], \quad K_2 = (\bar{S}1\bar{1}), \quad \eta_2 = [011]; \quad s = s_3 \quad (3)$$

R, S, and s_3 being given as ≈ 25, ≈ 8, and 0.228. There are several surprising features of the reported irrational quantities associated with these modes. For example, there are no crystallographic reasons for P and Q (2), Q and R (3), and s_2 and s_3 (3) to be equal. Therefore, before using these modes as possible lattice invariant shear systems in the theories of martensite crystallography, it was necessary to obtain algebraic expressions for these irrational components. This has been done using the theory of the crystallography of deformation twinning due to Bilby and Crocker[4]; the results are presented later.

Bansal and Heuer[5] used the martensite crystallography theory of Bowles and Mackenzie[6] to investigate the transformation in zirconia. The theory requires knowledge of a lattice invariant shear of the product monoclinic phase which, of necessity, accompanies the transformations. Bansal and Heuer assumed this to be slip and used several possible low-index slip planes and directions. They also considered three possible correspondences relating the two phases. For one of these correspondences, (110) [001] and (1$\bar{1}$0) [110] slip were found to predict habit planes near those observed, which are close to {100} and {671} planes of the tetragonal phase. Recently Kriven et al.[7] repeated the Bansal and Heuer calculations with different lattice parameters and obtained somewhat different predictions. They also used twinning mode 1 as a lattice invariant shear and obtained real solutions which, however, were not consistent with the observations. It was concluded that similar calculations should be performed for the other observed twinning modes but that the available computer programs were not suitable. In the present study, programs based on the CRAB theory[8,9] of martensite crystallography have been used to perform these calculations. Before these results on twinning and transformations in zirconia are presented, the details of the crystal structures of the two phases and of the possible correspondences relating them are given.

Crystal Structures and Lattice Correspondences

The high- and low-temperature phases of zirconia are face-centered tetragonal (t) and monoclinic (m), respectively. The lattice parameters at 950°C, the transformation temperature, as deduced from the data of Patil and Subbarao,[10] are $a_t = b_t = 0.51398$ nm, $c_t = 0.52552$ nm, and $a_m = 0.51878$ nm, $b_m = 0.52141$ nm, $c_m = 0.53831$ nm and $\beta = 81.22°$. Note that both structures are approximately cubic and that the transformation from tetragonal to monoclinic involves a volume increase of 3.659%.

Bailey[2] showed that three lattice correspondences A, B, and C, may arise between the two phases. These have $\underline{c}_t$ transforming to $\underline{a}_m$, $\underline{b}_m$, and $\underline{c}_m$, respectively. When consistent right-hand bases are adopted for the unit cells, the matrix forms of these correspondences can be written:

$$A = \begin{bmatrix} 0 & 0 & 1 \\ 1 & 0 & 0 \\ 0 & 1 & 0 \end{bmatrix} \quad B = \begin{bmatrix} 1 & 0 & 0 \\ 0 & 0 & -1 \\ 0 & 1 & 0 \end{bmatrix} \quad C = \begin{bmatrix} 0 & 1 & 0 \\ -1 & 0 & 0 \\ 0 & 0 & 1 \end{bmatrix}$$

With the above lattice parameters, the magnitudes e_i and directions $\underline{d}_i$ of the principal distortions for the three correspondences were calculated and are listed in Table I. The quantity $\Sigma e_i^2 - 3$, which is a measure of the strain energy, was found to be 0.099, 0.099, and 0.098, respectively.

Table I. Magnitudes e_i and Directions $\underline{d}_i$ of the Principal Distortions for Lattice Correspondences A, B, and C*

	A	B	C
e_1	1.0144	1.1062	1.0144
$\underline{d}_1$	[100]	[0.676, 0.733, 0]	[100]
e_2	1.0958	0.9443	1.0939
$\underline{d}_2$	[0, 0.780, 0.625]	[−0.733, 0.676, 0]	[0, 0.728, 0.685]
e_3	0.9323	0.9921	0.9340
$\underline{d}_3$	[0, −0.625, 0.780]	[001]	[0, −0.685, 0.728]

*The directions are given relative to an orthonormal basis parallel to the axes of the parent tetragonal lattice.

Twinning Modes

To determine the full deformation twinning modes, the general analysis of Bilby and Crocker[4] was adopted. A direct primitive lattice basis $\underline{c}_i$ is chosen to define the crystal structure, whereas the reciprocal lattice basis is given by $\underline{c}^j$ ($i, j = 1$–3). The direct and reciprocal lattice metrics are denoted by $c_{ij} = \underline{c}_i \cdot \underline{c}_j$ and $c^{ij} = \underline{c}^i \cdot \underline{c}^j$, respectively. The directions η_1 and η_2 are represented by $u^i \underline{c}_i$ and $v^i \underline{c}_i$ and planes K_1 and K_2 by $h_i \underline{c}^i$ and $k_i \underline{c}^i$. The elements K_1 and η_2 are sufficient to define a twinning mode completely. The magnitude of the shear strain is then

$$g = 2\{(v^i h_i)^{-2} d^{-2} c_{pq} v^p v^q - 1\}^{1/2} \quad (4)$$

where d is the interplanar spacing given by $d = (c^{ij} h_i h_j)^{-1/2}$. The elements K_2 and η_1 can be derived using the following equations:

$$K_2: k_i = (c_{pq} v^p v^q) h_i - (v^j h_j) c_{ik} v^k \quad (5)$$

$$\eta_1: u^i = (v^j h_j) c^{ik} h_k - (c^{pq} h_p h_q) v^i \quad (6)$$

These equations indicate that, for type I twinning in which K_1 and η_2 are rational, so that h_i and v^i are integers, K_2 and η_1 are, in general, irrational. Degenerate cases, however, arise in which either or both of K_2 and η_1 are rational.

The metrics c_{ij} and c^{ij} for the monoclinic structure are given by

$$c_{ij} = \begin{bmatrix} a_m^2 & 0 & a_m c_m \cos \beta \\ 0 & b_m^2 & 0 \\ a_m c_m \cos \beta & 0 & c_m^2 \end{bmatrix}$$

$$c^{ij} = \operatorname{cosec}^2 \beta \begin{bmatrix} a_m^{-2} & 0 & -a_m^{-1} c_m^{-1} \cos \beta \\ 0 & b_m^{-2} \sin^2 \beta & 0 \\ -a_m^{-1} c_m^{-1} \cos \beta & 0 & c_m^{-2} \end{bmatrix}$$

These metrics have been used to determine the elements K_2, η_1, and s of the three observed twinning modes of zirconia, and the results are given as functions of the lattice parameters in Table II. In addition, the magnitudes of the irrational components of these modes are presented using values of the lattice parameters at 950°C, the transformation temperature. In particular, P, Q, R, and S of modes 2 and 3 were found to be 8.881, 31.160, 7.718, and 4.760, respectively, and the shear strains s_1, s_2, and s_3 were 0.309, 0.224, and 0.219. These accurate values, which are significantly different from those reported earlier,[2,3] were used in the martensite crystallography theories.

Table II. Elements of the Twinning Modes of Monoclinic Zirconia*

K_1	η_1	K_2	η_2	s
(100)	[00$\bar{1}$]	(00$\bar{1}$)	[100]	s_1
(110)	[1$\bar{1}$P]	(1$\bar{1}$Q)	[110]	s_2
(011)	[$\bar{R}$1$\bar{1}$]	($\bar{S}$1$\bar{1}$)	[011]	s_3

*Conjugate modes can be constructed by interchanging K_1 and K_2 and η_1 and η_2. The indices P, Q, R, and S and the shears s_i are given below using the elements c_{ij} and c^{ij} of the metric tensors and also the lattice parameters at 950°C: $P = -2c^{13}(c^{11} - c^{22})^{-1} = 8.881$; $s_1 = \{4(c^{11}c_{11} - 1)\}^{1/2} = 0.309$; $Q = -2c_{13}(c_{11} - c_{22})^{-1} = 31.160$; $s_2 = \{(c^{11} + c^{22})(c_{11} + c_{22}) - 4\}^{1/2} = 0.219$; $R = 2c^{13}(c^{33} - c^{22})^{-1} = 7.718$; $s_3 = \{(c^{22} + c^{33})(c_{22} + c_{33}) - 4\}^{1/2} = 0.224$; $S = -2c_{13}(c_{22} - c_{33})^{-1} = 4.760$.

Analysis of the Martensitic Transformation

For this analysis we shall use the parent metric p_{ij} given by

$$p_{ij} = \begin{bmatrix} a_t^2 & 0 & 0 \\ 0 & b_t^2 & 0 \\ 0 & 0 & c_t^2 \end{bmatrix}$$

the product metric c_{ij}, the correspondences between the two structures given earlier, and lattice invariant shear modes defined by the twinning modes of Table II and their conjugates. The calculated twinning modes for $(100)_m$, $\{110\}_m$, and $\{011\}_m$ twins were used as shear plane (m) and shear direction $[l]$ by transforming them to the parent basis by using

$$(m)_t = (m)_m (_mC_t) \tag{7}$$

$$[l]_t = (_tC_m)[l]_m \tag{8}$$

Here $(_mC_t)$ is the correspondence and $(_tC_m)$ is its inverse. In the monoclinic structure, $(1\bar{1}0)$ and $(0\bar{1}1)$ are equivalent to (110) and (011), respectively, and they also behave equivalently with respect to all three lattice correspondences. This is also true for $(1\bar{1}\bar{Q})$ and $(\bar{S}\bar{1}1)$; hence they do not represent distinct lattice invariant shears. The twinning modes referred to the monoclinic structure, and to the tetragonal structure using the three lattice correspondences, are given in Table III.

Table III. Six Twinning Modes in Monoclinic Zirconia Used as Lattice Invariant Shears (m) $[l]$ in the CRAB Theory of Martensite Crystallography[8]*

	M		A		B		C	
	(m)	$[l]$	(m)	$[l]$	(m)	$[l]$	(m)	$[l]$
1	(100)	[00$\bar{1}$]	(001)	[0$\bar{1}$0]	(100)	[0$\bar{1}$0]	(010)	[00$\bar{1}$]
2	(110)	[1$\bar{1}$P]	(101)	[$\bar{1}$P1]	(10$\bar{1}$)	[1$\bar{P}$1]	($\bar{1}$10)	[11$\bar{P}$]
3	(011)	[$\bar{R}$1$\bar{1}$]	(110)	[1$\bar{1}$R]	(01$\bar{1}$)	[$\bar{R}$11]	($\bar{1}$01)	[1$\bar{R}$1]
4	(00$\bar{1}$)	[100]	(0$\bar{1}$0)	[001]	(0$\bar{1}$0)	[100]	(00$\bar{1}$)	[010]
5	(1$\bar{1}$Q)	[110]	($\bar{1}$Q1)	[101]	(1Q1)	[10$\bar{1}$]	(11Q)	[$\bar{1}$10]
6	($\bar{S}$1$\bar{1}$)	[011]	(1$\bar{1}$S)	[110]	($\bar{S}$11)	[01$\bar{1}$]	($\bar{1}S\bar{1}$)	[$\bar{1}$01]

*The modes are referred to the monoclinic basis M and to the tetragonal basis using lattice correspondences A, B, and C. Note that modes 4–6 are the conjugates of modes 1–3.

Table IV. Predictions of the CRAB Theory of Martensite Crystallography[8]* when Modes 1–6 of Table II are [...] with Lattice Correspondences A, B, and C[†]

Mode	g	f	h					
			Correspondence A					
2	0.091	0.123	0.153	−0.978	0.140	0.		
			0.491	−0.394	−0.776	0.		
5	0.124	0.123	0.133	−0.983	0.124	0.		
			0.693	−0.343	−0.633	0.		
			Correspondence B					
2	0.120	0.126	0.140	0.981	−0.131	0.		
			0.510	0.381	0.770	0.		
3	0.053	0.130	0.073	0.947	−0.310	0.		
			0.951	0.301	0.056	0.		
4	0.122	0.052	0.418	0.824	0.381	0.		
5	0.101	0.118	0.142	0.981	−0.130	0.		
			0.683	0.353	0.638	0.		
6	0.040	0.135	0.043	0.978	−0.203	0.		
			0.947	0.310	0.070	0.		
			Correspondence C					
3	0.170	0.128	0.302	0.951	−0.064	0.		
			0.958	0.075	0.275	0.		
6	0.182	0.138	0.310	0.948	−0.062	0.		
			0.968	0.044	0.244	0.		

*Ref. 8. [†]The table gives the lower of the two values of both [...] values of the habit plane normal h and displacement direction u [...] u are unit vectors in the tetragonal basis.

All the twinning modes calculated from Eqs. (7) [...] put data along with metrics p_{ij}, c_{ij}, and all three reported [...] The computer program based on the CRAB theory enabl[es cal]culations [...] lattice invariant shear, the magnitude f, the habit plane [...] invariant plane strain, and the associated orientation rela[tion. For] a given set of data there are two values of g and f and fou[r...] that are likely to be operative in practice are expected to h[ave...] f, and when the lattice invariant shear is twinning, value[s of f] greater than the twinning shear are not allowed.

The results of these calculations are presented i[n ...for] modes 1–6 of Table III and lattice correspondences A, B[, and C. The] lower of the two values of g and the corresponding value [...] The habit planes are plotted in Fig. 1 on a unit triangle [referred] to the tetragonal basis. With use of correspondence A, o[nly modes... and] 6) gave real habit plane solutions and mode 6 is unacce[ptable because f is larger] than the twinning shear. One of the habit planes predic[ted...] is approximately $\{171\}_t$, and the second falls between [...] correspondence B, only mode 1 gave imaginary solut[ions...] Modes 2 and 5 predict habits very similar to those giv[en by...] predict one habit of type $\{130\}_t$, whereas the o[ther for] mode 6 is near $\{061\}_t$. The $\{121\}_t$ prediction of [...] calculations performed by Kriven et al.[7] With use o[f correspondence C,] [mode]s 3 and 6 gave acceptable predictions. These h[abits lie] [se]cond between $\{031\}_t$ and $\{061\}_t$. In all cases the [...]

[sym]metry of the monoclinic phase of zirconia res[ults...] [pr]ovides a challenging test for theories of defor[mation...] [t]he shear strains and some of the twinning elemen[ts are...] [inter]esting functions of the lattice parameters. When the [...] [exper]imentally,[2,3] their elements were deduced from g[...] [ster]eographic projections. The reported modes were [not en]tirely self-consistent. In the present paper, these m[odes...] [alg]ebraically with the well-established theory of the [...] [defor]m[atio]n twinning due to Bilby and Crocker[4] and present [...] [param]eters of the monoclinic unit cell. In addition, the v[...]

mental work on these plates would be welcome, and it would be particularly interesting if it is found that they involve a multiple lattice invariant shear that could be analyzed using the generalized CRAB theory.[8]

Acknowledgments

The authors are indebted to W. M. Kriven for introducing them to this problem and to the Government of Pakistan for financial assistance.

References

[1] E. C. Subbarao, H. S. Maiti, and K. K. Srivastava, *Phys. Status Solidi A*, **21**, 9–40 (1974).
[2] J. E. Bailey, *Proc. R. Soc. London*, [Ser.] A, **279**, 395–412 (1964).
[3] E. Bischoff and M. Rühle; *J. Am. Ceram. Soc.*, **66** [2] 123–27 (1983).
[4] B. A. Bilby and A. G. Crocker, *Proc. R. Soc. London*, [Ser.] A, **288**, 240–55 (1965).
[5] G. K. Bansal and A. H. Heuer, *Acta Metall.*, **22**, 409–17 (1974).
[6] J. S. Bowles and J. K. Mackenzie, *Acta Metall.*, **2**, 129–47 (1954); **2**, 224–34 (1954).
[7] W. M. Kriven, W. L. Fraser, and S. W. Kennedy; pp. 82–97 in Advances in Ceramics, Vol. 3. Edited by A. H. Heuer and L. W. Hobbs. The American Ceramic Society, Columbus, OH, 1981.
[8] A. F. Acton, M. Bevis, A. G. Crocker, and N. H. Ross, *Proc. R. Soc. London*, [Ser.] A, **320**, 101–33 (1970).
[9] A. G. Crocker, *J. Phys.*, **43**, C4-209–14 (1982).
[10] R. N. Patil and E. C. Subbarao, *J. Appl. Crystallogr.*, **2**, 281–88 (1969).
[11] A. G. Crocker, *J. Nucl. Mater.*, **16**, 306–26 (1965).
[12] D. M. M. Guyoncourt and A. G. Crocker, *Acta Metall.*, **16**, 523–34 (1968).
[13] A. G. Crocker, *J. Nucl. Mater.*, **41**, 167–77 (1971).
[14] P. E. J. Flewitt, P. J. Ash, and A. G. Crocker, *Acta Metall.*, **24**, 669–76 (1976).
[15] G. K. Bansal; Ph. D. Thesis, Case Western Reserve University, Cleveland, OH, 1973.

Acoustic Emission Characterization of the Tetragonal–Monoclinic Phase Transformation in Zirconia

D. R. CLARKE

Massachusetts Institute of Technology
Cambridge, MA   02139

A. ARORA

Rockwell International Science Center
Thousand Oaks, CA   91360

The processes accompanying the tetragonal–monoclinic ($t \rightarrow m$) phase transformation in zirconia (ZrO_2) have been studied using acoustic emission and electron microscopy in an attempt to characterize the different mechanisms by which the transformation can be accommodated in bulk materials. Experiments in which the acoustic emission is detected as specimens are cooled through the transformation, following densification by sintering, are described. For comparison, the acoustic emission from free, nominally unconstrained powders similarly cooled through the transformation is reported. The existence of distinct processes accompanying the phase transformation is established on the basis of postexperiment multiparametric correlation analysis of the acoustic emission. The most informative correlation determined to date is that of the number of events vs energy. From these and the electron microscopy observations, the occurrence of the transformation itself, twinning, and microcracking are distinguished.

The tetragonal-monoclinic ($t \rightarrow m$) phase transformation in zirconia (ZrO_2) is now well known to exhibit all the characteristics of a martensitic transformation, as was detailed previously and reviewed in this and the preceding international conference on ZrO_2. Less thoroughly investigated, but of particular importance in the design of transformation-toughened ceramics, is how the transformation is accommodated under constraint conditions, such as those that prevail in these materials. Microstructural examination of ZrO_2 particles that have been transformed near a propagating crack reveal that several accommodation processes may be involved; the particles are highly twinned (sometimes exhibiting a number of different variants) and, in some materials such as a number of the Al_2O_3-ZrO_2 composites, are accompanied by microcracks. In addition, there have been suggestions that an intermediate crystallographic phase forms as the t polymorph transforms under stress to m. Thus, three distinct accommodation mechanisms have been proposed to accompany a constrained $t \rightarrow m$ transformation, each of which may be expected to affect the magnitude of the toughening attainable. Determining the relative contribution of each and the extent to which they dissipate the energy of transformation becomes a major issue and one not easily addressed by postmortem studies, e.g., transmission electron microscopy, performed under ambient conditions.

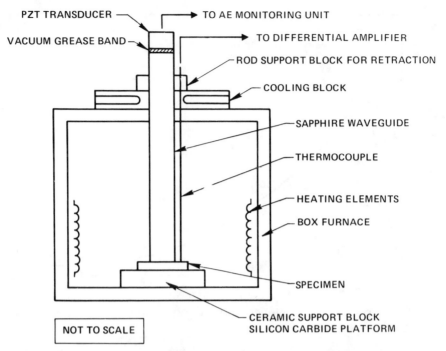

Fig. 1. Schematic diagram of the experimental arrangement used in obtaining acoustic emission signals at temperature (not to scale).

In principle, the techniques of acoustic emission overcome these problems as the transformation itself and the processes accompanying or succeeding it can be monitored directly and continuously during the transformation. The problems of characterizing the accommodation mechanisms then reduce to deciphering the recorded acoustic signals and identifying the individual processes involved.

The present paper demonstrates that significant acoustic emission is generated on cooling ZrO_2 through the $t \rightarrow m$ transformation and that a number of distinct processes accompany the transformation. Preliminary assignments to the nature of these processes are made by comparing the acoustic signals recorded from the constrained transformation in sintered pellets with those recorded from free powder and with the support of transmission electron microscopy (TEM) observations.

Experimental Procedure

Pure, unstabilized ZrO_2^* was used either in the form of loose powder (mean particle size <100 nm) or in the form of disk-shaped pellets (3-mm long by 10-mm in diameter) made by cold isostatic pressing, without a binder, to a pressure of approximately 10 000 psi (68.9 MPa).

In the arrangement adopted (Fig. 1), single pellets were placed on a dense SiC platform of a box furnace and sintered for 30 min at 1500°C. After the pellets were sintered, a retractable, single-crystal sapphire waveguide was loaded into contact with the sample, the furnace was shut off, and the acoustic emission was monitored during cooling to room temperature. A representative sintering and cooling cycle is reproduced in Fig. 2. The same arrangement was used to monitor the emission

---

*Zircar Products, Inc., Florida, NY.

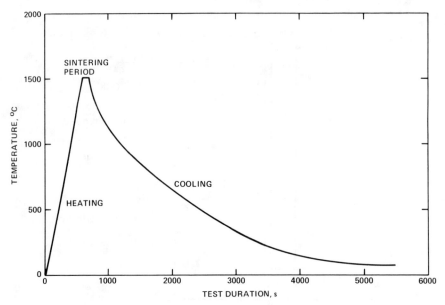

Fig. 2. Representative sintering and cooling cycle used during an in situ acoustic emission test.

from free powder except that (a) the sapphire waveguide was placed in direct contact with the SiC platform and the powder was loosely scattered on the platform and (b) the powder was heated to only 1250°C, sufficient to ensure transformation but preclude significant sintering. The waveguide was passed through a water-cooled collar (lubricated by vacuum grease) to keep the transducer at room temperature. Acoustic emissions were amplified with a gain of 40 dB (in part to compensate for the signal attenuation in the waveguide), filtered over a bandpass of 0.1–2 MHz, counted by a ring-down count totalizer, and recorded. The acoustic emission signal was also fed to a dedicated multiparametric analyzer for subsequent signal-correlation analysis.

Acoustic Emission Results

As the pellets were cooled from the sintering temperature, no acoustic emission was detectable until 1160°C, when "burst" type of emissions, consisting of small-duration (~1 ms) events, were observed. As cooling continued, the rate of emission increased, as did the event duration, until at rather low temperatures events with more than 3-ms duration were recorded. An example of the type of event recorded during cooling is reproduced in Fig. 3, with the parameters used to characterize an event defined on the figure.

A cumulative plot of the acoustic emission events recorded on cooling to close to ambient temperatures is shown in Fig. 4, indicating that acoustic emission is not only generated at the nominal transformation temperature but throughout the cooling cycle. A clearer indication of distinct phenomena occurring during cooling is gained from the same data but plotted in the form of the cumulative ring-down counts (Fig. 5). There is a sharp increase in emission at 1030°C (810 s in test time) for this particular sample, with fully half of the counts being recorded at this temperature. For all the pellets, the temperature at which the abrupt

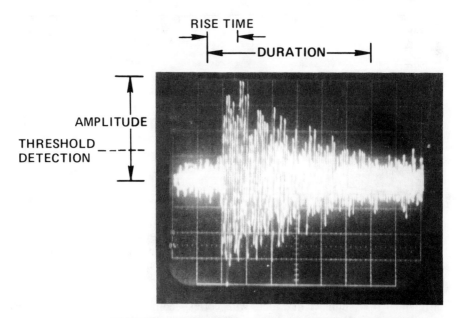

Fig. 3. A single acoustic emission event as displayed on an oscilloscope with characteristic features of the event indicated.

increase in emission occurred corresponded with the temperature of the $t \rightarrow m$ transformation indicated by differential thermal analysis (DTA) (Fig. 6).

In analyzing the raw acoustic emission data, a number of multiparametric correlations were attempted. The acoustic energy distribution-based correlations in

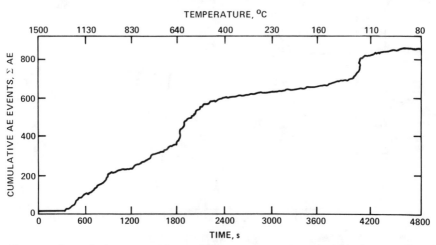

Fig. 4. Cumulative acoustic emission events recorded during cooling. The temperature is indicated on the top abscissa and the test duration on the bottom abscissa, pellet sample.

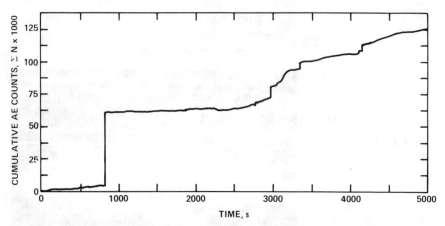

Fig. 5. Data of Fig. 4 expressed as the cumulative acoustic emission counts recorded during the test. The abrupt increase in the counts occurs at a test time corresponding to the phase transformation.

differential (i.e., number of events with corresponding energy) and decreasing cumulative (i.e., the number of events with energy more than the corresponding energy) form proved to be the most informative, providing evidence for a number of distinct processes occurring during the transformation and subsequent cooling.

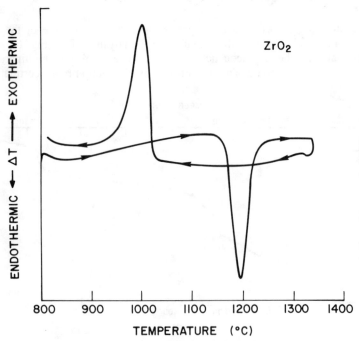

Fig. 6. Differential thermal analysis of a ZrO_2 pellet on heating through and then cooling down through the $t \rightarrow m$ phase transformation.

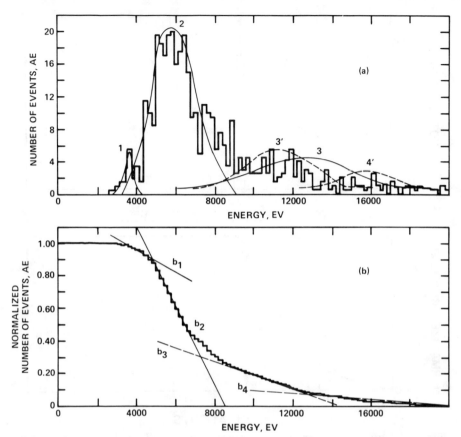

Fig. 7. (a) Distribution of the acoustic events according to their energy; (b) normalized number of events as a function of the event energy with the distinct b value slopes indicated.

For instance, in Fig. 7(a), at least three distinct (possibly four as shown by the dashed curve) distributions can be discerned in this histogram, indicating the existence of three discrete processes. The slopes in Fig. 7(b) are the values of the distribution exponent "b" in a power-law expansion of the event energies:

$$F(V) = F_0(V/V_0)^b \tag{1}$$

where $F(V)$ is the number of acoustic emission events having an energy greater than V, and F_0 and V_0 are constants related to the sample size and energy sensitivity of the monitoring system, respectively. (This power-law representation is the same as that used in classifying different types of seismic activity.) Three, and perhaps four, slope values in this figure are measurable, indicating that at least three mechanisms, each with a different b value and characteristic energies, contributed to the acoustic signal during the test.

In comparison, the signals recorded from the free powder, although considerably weaker (due to a smaller volume of material and poorer acoustic coupling), were simpler, and the emission was complete by the time the powder had cooled

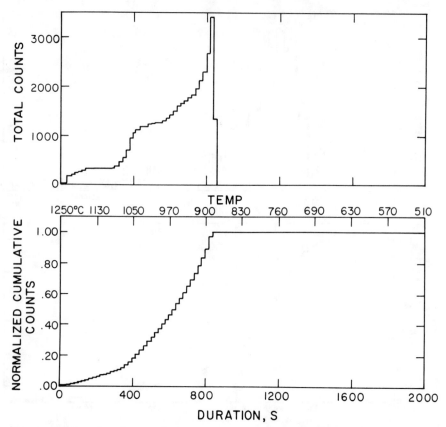

Fig. 8. Normalized cumulative counts recorded from the nominally unconstrained powder (compare with Fig. 5 for the constrained ZrO_2).

to 900°C. An example is shown in Fig. 8, which is a plot of the normalized cumulative counts. Unlike the distribution obtained in the constrained (pellet) case (Fig. 5), there is no sudden generation of the acoustic signal but rather a steady increase with decreasing temperature, consistent with an athermal transformation. The energy distribution of the recorded signals, shown in Fig. 9 for comparison with that of Fig. 7 for the pellet, also presents a simpler situation; the energy spectrum is rather narrow, and the cumulative event distribution exhibits only two b values.

Although, as in other acoustic emission experiments, there was considerable variation in the signal from one pellet to another and from sample to sample of powder, the correlation spectra all exhibited the characteristic features presented in Figs. 7 and 9.

Discussion

The correlation spectra of the acoustic emission signals reveal considerable differences between the free powder and constrained particle cases; two distinct, relatively low-energy, mechanisms operate during the transformation of nominally

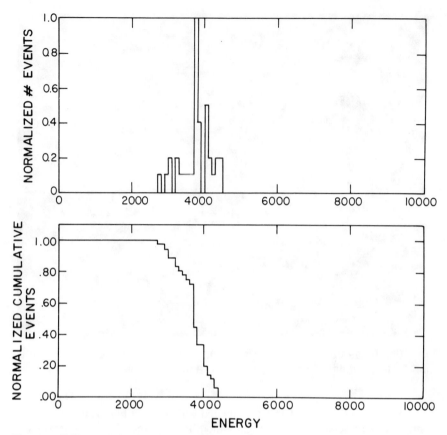

Fig. 9. Differential energy distributions obtained from the acoustic emission recorded from the free powder.

unconstrained powders, whereas at least three, two of which have higher energy, occur in the constrained material. The identity of the individual mechanisms cannot be established from these distributions, but some conclusions may be drawn on the basis of the microstructures of the samples after the tests. The powder was similar to that before the tests, i.e., monoclinic grains, a proportion of which were twinned. Examination of the pellets after the in situ tests showed that they were macroscopically cracked (being visible to the eye — Fig. 10), and also contained a high density of microcracks (the pellets crumbled easily and absorbed dye penetrant). Furthermore, the pellets were monoclinic, and each grain contained a high density of twins.

The post-test appearance of the pellets suggests that a variety of cracking occurred during cooling, ranging from microcracks to subsequent crack extension and branching. Using the assumption that the energy of these more macroscopic cracking phenomena would be expected to be higher than that of microcracking and also would occur at lower temperatures, the data of Fig. 4 were reanalyzed after truncating them to remove the data recorded below 900°C. The result, expressed in terms of the number of events vs the event energy, is reproduced in

Fig. 10. Appearance of the pellet samples after the in situ acoustic emission tests exhibiting large macroscopic and interconnected cracks.

Fig. 11. If these truncated data are compared with those of Fig. 7, the low-energy distributions are seen to be only slightly modified, whereas the high-energy event distributions have been eliminated, with a consequent decrease in the number of b values. This suggests that the lower-energy mechanisms are indeed intimately related to the transformation, whereas the higher-energy events were associated with the lower-temperature strain-energy relief mechanisms as proposed above. Furthermore, comparison of the truncated correlation spectrum with that obtained from the free powder indicates a number of similarities; both exhibit a relatively small, low-energy peak centered at 3200 eV and a larger peak at a slightly larger energy having a comparable b value. Further analysis is evidently required before the conclusion that they correspond to the same mechanisms is substantiated, but it is interesting to note that the samples exhibiting microcracking (the pellets) also had spectra which had the additional, higher-energy distribution of events.

Taken together, the acoustic emission analyses and the microscopy observations suggest that, during cooling of the free powder, the acoustic signal originated from both the transformation itself and the twinning. The observations from the constrained material provide evidence that, in addition to the mechanisms operating in the free powder, microcracking also occurs and gives rise to a distinguishable signal.

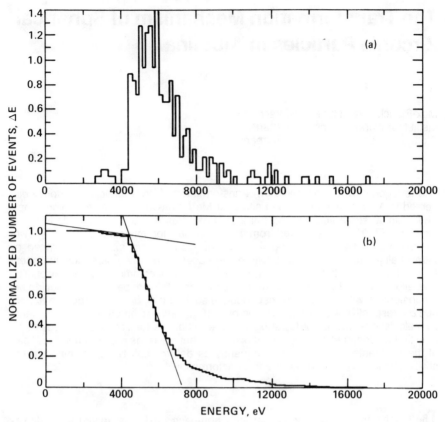

Fig. 11. Reanalysis of the differential energy distribution of Fig. 7 after truncating the data to remove the acoustic events recorded below 900°C.

Acknowledgments

The authors are grateful to the National Science Foundation 8007445 for support of this work, grant number DMR-8007445.

The Transformation Mechanism of Spherical Zirconia Particles in Alumina

W. M. KRIVEN*

Max-Planck-Institut für Metallforschung
Institut für Werkstoffwissenschaften
Stuttgart, Federal Republic of Germany

The tetragonal (t)-to-monoclinic (m) transformation of ZrO_2 particles randomly dispersed in Al_2O_3 was examined in situ by 1 MeV high-voltage electron microscopy. The residual strain field around tetragonal particles has a principal axis in the c_t direction of t-ZrO_2, as expected from thermal expansion mismatch with Al_2O_3. Two-dimensional analyses indicated that the tetragonal strain axis was replaced by parallel alignment of monoclinic twin plane traces, and hence perpendicular orientation of the monoclinic strain axis, suggesting self-accommodating behavior. The crystallography of the in situ transformation for an ~0.3 μm particle was elucidated. It transformed with a lattice correspondence such that a_t, b_t, and c_t became a_m, b_m, and c_m, respectively, and was twinned on $(110)_m$ planes. Comparison with martensitic calculations for bulk ZrO_2 suggests that either martensite theory needs to be modified for confined particle mechanisms, or that the twins are not an integral part of the mechanism and form subsequently as deformation twins, to minimize the shape change of the particle due to transformation.

The toughening effect of ZrO_2 particles homogeneously distributed in oxide ceramics is widely known[1-3] and is due to metastable tetragonal (t) ZrO_2 particles within crack-tip stress fields undergoing a martensitic transformation to a monoclinic (m) structure, leaving transformed regions or "wakes" at the sides of the crack.[4,5] In situ high-voltage electron microscopy was used to study the stress-induced transformation.[6-8] The monoclinic particles so formed were invariably twinned, and the transformation was irreversible on release of stress.

In ceramics, where at ambient temperature dislocations and other defects form with great difficulty, nucleation of the martensitic transformation is believed to occur at inhomogeneities or singularities in the internal strain field, e.g., at corners and edges of particles.[9-11] Such strains scale with particle size, leading to the "particle-size effect," in which only particles within a critical size range are transformable by crack-tip stresses. The condition for martensitic nucleation in ZrO_2 particles may be expressed quantitatively by[9,11]:

$$\frac{\varepsilon_{ij}^{R'}}{\varepsilon_{ij}^{T}} \geq \eta = 0.22 \tag{1}$$

where $\varepsilon_{ij}^{R'}$ represents the resolved shear strain within a particle, and ε_{ij}^{T} is the unconstrained transformation shear strain. The condition must be fulfilled over the volume of the operational nucleus. In ZrO_2, ε_{13}^{T} is probably the most important component of the transformation strain tensor. If a stress ε_{ij}^{A} is applied, then

---

*Now at the Materials Research Laboratory and Department of Ceramic Engineering, University of Illinois at Urbana-Champaign, Urbana, Illinois 61801.

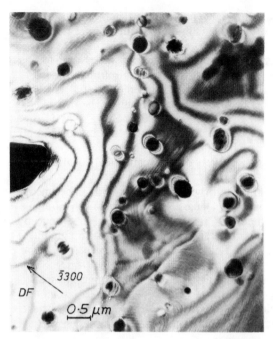

Fig. 1. HVEM electron micrograph of large-grained Al_2O_3 material containing randomly dispersed, intragranular spherical ZrO_2 particles. Tapering top and bottom foil surfaces gave rise to thickness contours which needed to be sufficiently wide for symmetrical strain analysis.

$$\varepsilon_{ij}^{R'} = \varepsilon_{ij}^{R} + \varepsilon_{ij}^{A} \qquad (2)$$

where ε_{ij}^{R} is the residual strain, for example due to thermal expansion mismatch ($\Delta\alpha\Delta T$), and ε_{ij}^{A} is the strain within the particle resulting from the applied stress.

The twins in m-ZrO_2 were identified as (100), (001), (110), and (011) types.[12–14] Using these as lattice-invariant shear systems, theoretical martensite calculations for the bulk structure changes of ZrO_2 gave plausible geometrical solutions.[15,16] A simple crystallographic model[13] postulated that particles of a given size may form one martensite variant with both short-range and long-range strain fields. Larger particles may achieve lower overall energy by a configuration of self-accommodating variants, but this requires very specific crystallography. As yet, however, experimental crystallographic observations of the stress-induced transformation of a particle embedded in a ceramic matrix are lacking, and are the overall aim of this work.

The material chosen for study[3,17,18] contained large Al_2O_3 grains with spherical or ellipsoidal t-ZrO_2 particles intragranularly dispersed in random orientation (Fig. 1). The ZrO_2 inclusions exhibited characteristic contrast oscillations or fringes (Fig. 1).[17,19,20] Such oscillations have not been previously observed. The theory for the diffraction contrast has been developed, and will be published elsewhere.[21–24] Essentially, the contrast can be explained as phase contrast due to intraband scattering of modified Bloch waves.

The contrast oscillations are due to elastic strain fields surrounding ZrO_2 particles and can be imaged in the HVEM under precisely set, dynamical diffraction conditions. It is possible to determine the three-dimensional shape of the strain field by systematic tilting with reference to the Al_2O_3 matrix and its Kikuchi patterns.

Even before transformation, an anisotropic strain field was found surrounding t-ZrO_2 particles.[19,20] It was attributed to anisotropic thermal expansion mismatch between Al_2O_3 and ZrO_2, arising when the material was sintered at 1500°C (in equilibrium) and cooled to room temperature. The thermal expansion coefficient of alumina[25] was $8.1 \times 10^{-6}/°C$ and for t-ZrO_2[26] was $11.6 \times 10^{-6}/°C$ in a_t and $16.8 \times 10^{-6}/°C$ in c_t. On transformation, the monoclinic strain field was also anisotropic and had both short-range and long-range components. The short-range strains were relatively much higher and localized at the particle-matrix interface.[19,27]

The aim of this work was, first, to provide experimental evidence that the residual strain around t-ZrO_2 particles originated from thermal expansion mismatch with the Al_2O_3 matrix. This could be most clearly done by showing that the anisotropy of the elastic strain field arose from maximum misfit in the c_t direction. To this end, the strain field would be three-dimensionally imaged and correlated to the internal crystallographic orientation of the t-ZrO_2. Second, it was proposed to study the transformation strains by determining the spatial relationship of the long-range monoclinic strain field to the previous tetragonal strain field around the same particle. Furthermore, it was hoped to elucidate the crystallography of the in situ transformation for comparison with martensitic calculations.[15,16]

Experimental Procedure

The theoretical basis for the experimental procedure, together with the computer simulation and evaluation of the data, will be published elsewhere[24,28]; here the experimental observations are emphasized. Stringent requirements needed to be fulfilled to perform this experiment, and they are briefly described as follows:

(1) Thin TEM specimens were prepared using standard techniques for ceramic materials. It was found necessary to anneal specimens at 1400°C for approximately 30 min to ensure adherence of particles to the matrix. This treatment removed residual bend contours and other artifacts due to foil preparation.

(2) The experiments were performed in an AEI 1.3 MeV high-voltage electron microscope operated at 1 MeV using a ±60° double-tilt stage. The microscope was fitted with an image intensifier to enable work at lowest beam intensity. Since specimen contamination limited the duration of the experiment, it was necessary to have a very good vacuum of 10^{-8} torr in the specimen chamber. This was achieved by liquid N_2 anti-contaminators and by an additional turbopump in the camera chamber of the microscope.

(3) Since $\{30\bar{3}0\}$ planes of α-Al_2O_3 were optimal for imaging and comparison of fringe contrasts, a grain lying close to [0001] in the untilted specimen needed to be found. Furthermore, to ensure spherical or ellipsoidal shape, the ZrO_2 particle was examined under tilts of ±45°.

(4) Only particles completely enclosed and preferably in the middle of the foil were chosen. This was indicated by symmetrical fringes in two-beam dark-field images. The depth position in the foil was measured by stereomicroscopy.[29,30]

(5) The projected thickness of the specimen at each tilt setting needed to be known, and was estimated (in units of extinction length, ξ_g) by counting thickness fringes from the edge of the grain imaged dynamically, using known **g** vectors. Top

and bottom foil surfaces were assumed parallel in the numerical evaluation of the data, but in a TEM specimen surfaces gradually taper toward the center. Only regions of Al_2O_3 having sufficiently wide thickness contours around a particle could be used (Fig. 1), since distortion or reversal of fringe contrast resulted from interference by narrow thickness contours.

(6) A systematic search for the strain field dipole was made in reciprocal space with reference to an α-Al_2O_3 Kikuchi map.[31] The planes intersecting at a major zone axis (e.g., [0001], [2$\bar{1}\bar{1}$2], [2$\bar{1}\bar{1}$1]) were systematically imaged near the zone axis. The contrast oscillations (fringes) could be seen only under precisely set bright- and dark-field, two-beam, dynamically diffracting conditions.

(7) The crystallographic orientation of the t-ZrO_2 zone axes within the particle was determined by selected area and microdiffraction techniques. The tetragonality was only 2% (a_t = 0.50815 ± 0001 nm, c_t = 0.51890 ± 0.0002 nm),[24] so that only low-index tetragonal zones could be used (see later). The t-ZrO_2 orientation was correlated to the (maximum) external strain field in the α-Al_2O_3 by stereographic analysis using computer-calculated stereograms.

(8) Ideally, it was hoped to transform the t-ZrO_2 particle to m symmetry by buckling the foil on beam heating and then repeat the analysis. However, it has not yet been possible to do a complete analysis on the same particle before and after transformation.

Each ZrO_2 particle lay in random orientation in the Al_2O_3 matrix, so that a detailed search was required to find a particle which fulfilled all of the above requirements. The particles had an unpredictable tendency to transform under the beam, which meant it was necessary to work at lowest beam intensity and use an image intensifier. When the transformation occurred during the tetragonal analysis, all the data collected on the strain field were useless without information on the internal orientation, and vice versa. Depending on particle orientation, the tilting experiment lasted up to 30 h, during which time contamination became a problem. In some cases particles would not transform, and attempts to force them by increasing the voltage to 1200 kV resulted in rapid radiation damage in the Al_2O_3 and disappearance of the strain field. All these problems were doubly compounded in attempting to record the whole tetragonal-to-monoclinic transformation. Numerous grains and particles needed to be examined. From more than 70 experimental runs, only 7 yielded both strain fields and diffraction patterns.

Additional difficulties were experienced in data evaluation. The 2% tetragonality in ZrO_2 meant that the lattice parameter difference lay just outside of measuremental errors in selected area diffraction patterns. Therefore, even theoretically, only high-index orientations such as [100], [110], or [011] projections could differentiate between a_t and c_t, and very precise measurements were involved. This was an important requirement in view of the aim of the experiment. The basic hypothesis was that the anisotropic strain field was due to thermal expansion differences between Al_2O_3 and t-ZrO_2, with the anisotropy due to maximum misfit in c_t.

Indexing of t-ZrO_2 diffraction patterns was complicated by double diffraction within each and between phases. This effect was particularly evident at major zone axes of Al_2O_3, and produced an apparent but deceptive low-index t-ZrO_2 pattern nearby. Five of the remaining experimental runs were eliminated by double diffraction, so that only in one run was a $[110]_{t\text{-}ZrO_2}$ projection found away from a major zone axis of Al_2O_3. ZrO_2-ZrO_2 double diffraction had occurred but could be recognized.

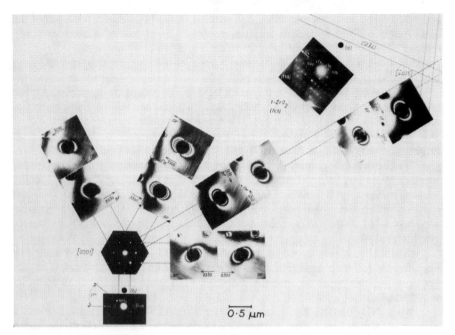

Fig. 2. Tetragonal strain analysis at [0001] of α-Al$_2$O$_3$ and correlation with internal orientation of t-ZrO$_2$ crystal axes.

Results

Tetragonal Strain Field

The results of the successful run in which the tetragonal strain field in the Al$_2$O$_3$ matrix was correlated to the internal crystallographic orientation of the t-ZrO$_2$ particle are presented in Fig. 2. A t-ZrO$_2$ particle of diameter $d = 300$ nm lay close to the middle of the foil. The thickness of the foil was 640 nm, which corresponds to 5.5 ξ_g where $\xi_g = 117$ nm for $\mathbf{g} = \{\bar{3}300\}$[22] (Fig. 1). The strain field was imaged at [0001] and at 48° away, near the $\langle 2\bar{1}\bar{1}1 \rangle$ type zone axis, as illustrated in the schematic α-Al$_2$O$_3$ Kikuchi map of Fig. 3. The arrows in Fig. 3 indicate the direction of the imaging bright-field $\mathbf{g}$ vectors, while the sites labeled (a) and (b) correspond to those in Fig. 2. It is clear that the strain field at [0001] in Fig. 2 is anisotropic because the contrast figure is not symmetrical for different $\mathbf{g}$ vectors. For equivalent $\{30\bar{3}0\}$ type vectors, the strain field causes a maximum displacement in the ($\bar{3}300$) planes.

A [110]$_{t\text{-ZrO}_2}$ projection was found at site (a), which was 0.94° below the ($12\bar{3}4$) plane and 9.4° along it from the [$\bar{4}405$] zone axis. With reference to the α-Al$_2$O$_3$ Kikuchi lines, in the sense schematically indicated in Fig. 2, the experimental orientation relation was determined:

[110]$_{t\text{-ZrO}_2}$ ^ [$\bar{4}405$]$_{\alpha\text{-Al}_2\text{O}_3}$ = 9.5°

(001)$_{t\text{-ZrO}_2}$ ^ ($12\bar{3}4$)$_{\alpha\text{-Al}_2\text{O}_3}$ = 74°

The symbol ^ denotes "angle between." This relation is shown stereographically in Fig. 4.

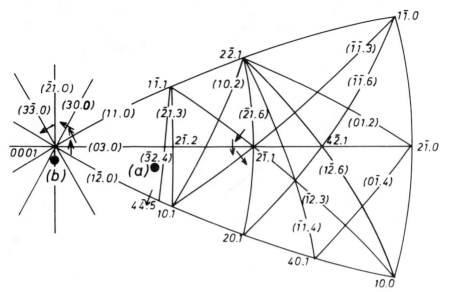

Fig. 3. Al_2O_3 Kikuchi map with imaging **g** vector types.

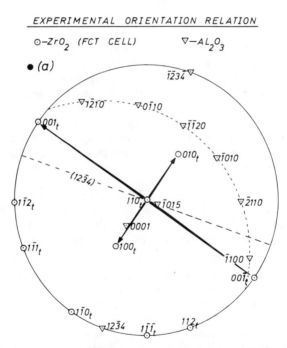

Fig. 4. Experimental orientation relation between t-ZrO_2 (fct cell) and α-Al_2O_3 matrix.

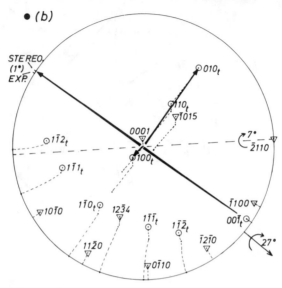

Fig. 5. Comparison of stereographically deduced and experimental orientation relations near [0001] of Al_2O_3.

At orientation (b) of Fig. 2, 6.9° from [0001], a row of reflections indexed as (001) of t-ZrO_2 were inclined at 37° to ($\bar{2}110$) of α-Al_2O_3. On a stereogram the experimental orientation (a) was tilted to (b), corresponding to the physical manipulations in the microscope. It is seen in Fig. 5 that the stereographically predicted and experimentally observed location of $c_{t\text{-}ZrO_2}$ at site (b) agreed to within 1°. This confirmed that the crystallographic orientation of t-ZrO_2 was correct.

Finally, the results of the tilting experiment may be summarized in Fig. 6. It was taken at 8° from [0001] by tilting along the {$3\bar{3}00$} plane which, from Fig. 5, brought the t-ZrO_2 almost into [100] projection. [010]$_{t\text{-}ZrO_2}$ was observed in the diffraction pattern in the expected orientation. The principal axis of the strain field is shown in Fig. 6, where the $c_{t\text{-}ZrO_2}$ axis is lying in the plane of the micrograph and inclined at 7° to the imaging ($3\bar{3}00$) vector. Thus the analysis shows that the strain field is anisotropic and the principal axis is parallel to the c_t axis. Unfortunately, contamination prevented transforming this particle and repeating the analyses.

Monoclinic Strain Field and Relationship to Tetragonal

Earlier work[19] showed that the monoclinic particle had both a short-range and a long-range strain field. The short-range field extended into the matrix by distances comparable to the spacing between twins. The long-range field was anisotropic with a principal axis apparently perpendicular to the internal monoclinic twin planes. This observation is confirmed in Fig. 7, where the strain field is imaged with mutually perpendicular **g** vectors of comparable extinction lengths. Although twins are not seen exactly edge-on, their traces are evident, particularly in the

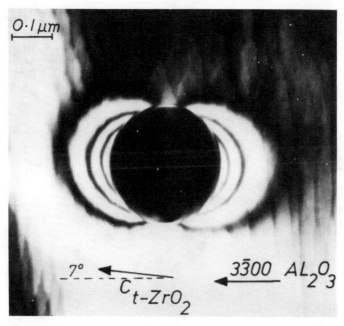

Fig. 6. Anisotropic elastic strain field around t-ZrO_2 in α-Al_2O_3 imaged in 2-beam dark-field dynamical conditions. The particle lies in a_t projection and shows maximum displacement in the c_t direction of ZrO_2.

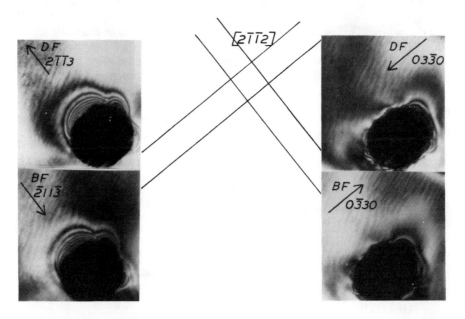

Fig. 7. Anisotropic monoclinic strain field with maximum matrix distortion approximately perpendicular to monoclinic twin planes.

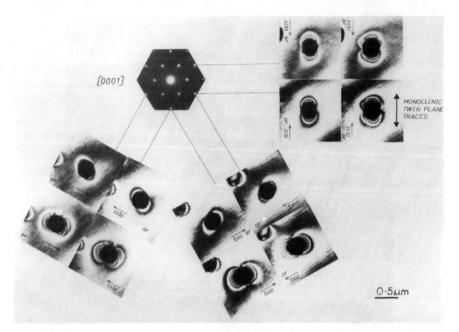

Fig. 8. Two-dimensional analysis of transformation strains around an m-ZrO$_2$ particle. The projection of the principal tetragonal strain axis is parallel to monoclinic twin plane traces.

($2\bar{1}\bar{1}3$) dark-field image. The monoclinic strain field appears to be markedly anisotropic in Fig. 7 and, in general, it is larger than the tetragonal field of a particle of comparable size. (Note that the intensity of the strain field is given by the fringe spacing, and not only by the spatial extent of the outermost fringe.)

A two-dimensional strain analysis around an [0001] zone axis using equivalent {30$\bar{3}$0} reflections is summarized in Fig. 8. The inner pairs of photographs showing untransformed particles reveal that the principal strain axis is projected approximately along the [30$\bar{3}$0] direction of α-Al$_2$O$_3$. The outer photographs show that the monoclinic twin plane traces also lie approximately parallel to [30$\bar{3}$0]. If the principal strain axis is perpendicular to the monoclinic twin planes in this case, as in the previous work,[19] then the principal strain axis is perpendicular to the principal axis of the residual strain field before transformation.

Transformation Crystallography

A fully enclosed spherical tetragonal particle of 0.3 μm diameter was oriented into a [1$\bar{1}$0]$_{fct}$ projection (Fig. 9(a)). Beam-induced buckling of the Al$_2$O$_3$ foil caused it to transform suddenly into a monoclinic [1$\bar{1}$0] projection twinned on the (110) plane (Fig. 9(b)). The experimental orientation relation in terms of the fct, $z = 4$ cell may be stated as:

$$[1\bar{1}0]_t \| [1\bar{1}0]_m, \quad (110)_t \| (110)_m$$

with twinning on (110)$_m$. Thus a topotaxial[32] relationship was preserved.

Several larger ZrO$_2$ particles in the sample transformed during cooling to room temperature. In one such particle, twinning on (100) was unambiguously

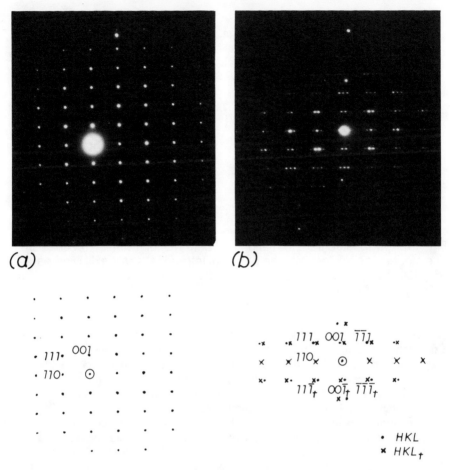

Fig. 9. Diffraction pattern of transforming ZrO_2 particle of ≈ 0.3 μm diameter. (a) [1$\bar{1}$0] projection of t-ZrO_2 (fct). (b) [1$\bar{1}$0] projection of m-ZrO_2 twinned on (110).

identified, and in another (001) twinning had occurred. Twinning on (100), (001), (110), and (011) has been previously reported in transformed ZrO_2.[12–14]

Discussion

A complete tetragonal strain field analysis was carried out in only one run. However, this example provides unambiguous crystallographic evidence that the thermal expansion mismatch ($\Delta\alpha\Delta T$) between t-ZrO_2 and α-Al_2O_3 causes the particle to be under residual tension at room temperature. An anisotropic strain field exists in the Al_2O_3 matrix, which has a principal symmetry axis parallel to [001] of t-ZrO_2. Based on this model the data will be quantitatively evaluated by calculating computer-simulated images for particles of corresponding size, depth position, and orientation.[24,28] Agreement between theoretically calculated and experimental images will then give values of the actual misfit parameters.

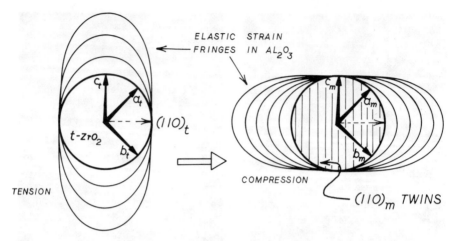

Fig. 10. Suggested transformation and deformation twinning mechanism. The model is consistent with the principal tetragonal strain axis lying parallel to $[001]_{fct}$, the occurrence of *LC C*, and a principal monoclinic strain axis perpendicular to $(110)_m$ twins.

It should be recalled that it was necessary to anneal thin foils prior to TEM examination because strain fields in unannealed foils were perturbed by foil preparation artifacts. This prevented actual determination of bulk strain fields; the present experiments actually determined the strain field in a micrometer-thick foil. It is believed that the strain field so determined is similar to that existing in bulk material, but is slightly smaller in magnitude.

Preliminary estimates of the residual strain, ε_{ij}^R, indicate a value of 0.008 in the $c_{t\text{-}ZrO_2}$ direction.[9,19] The maximum value of the unconstrained transformation strain, ε_{ij}^T, is 0.05.[33] In the absence of applied stress $\varepsilon_{ij}^{R'}/\varepsilon_{ij}^T = 0.16 < \eta$ (see Eq. (1)). Thus, the residual strain around t-ZrO_2 particles is insufficient to overcome the nucleation barrier to transformation. In this light, it is understandable that no transformation zone could be identified when this material was subjected to in situ straining experiments.[9] As the particles lay in random orientation, transformation occurred only when the shear component of the residual strain field was increased by the shear component of the applied strain field (i.e., that due to the crack).

The experimental observations on the transformational strains are preliminary, and a more detailed analysis of the monoclinic strain field needs to be made. However, the transformation mechanism suggested by this work is schematically shown in Fig. 10. It is pieced together from the following separate observations assumed to occur generally in small (≈ 0.3 μm) particles in Al_2O_3:

(1) The principal residual strain field axis lies in the $c_{t\text{-}ZrO_2}$ direction.
(2) During transformation *LC C* is maintained, and twinning occurs.
(3) The principal monoclinic strain field is perpendicular to the $(110)_m$ twin plane, in small particles.

Self-accommodation by twinning of the monoclinic particle is suggested by the analysis of Fig. 8. It is noteworthy that the principal strain axis of the Al_2O_3 matrix, in addition to being changed by the sign of the transformation, is also approximately perpendicular to the preexisting residual strain. Whether this is a causal effect, as would be suggested by the model in Fig. 10, or whether the direction of the principal strain is determined by the particle attempting to minimize its shape change during transformation, is not known at this time.

Transformation and nucleation studies have recently been made on small, spherical ZrO_2 and HfO_2 particles (20–100 nm wide) dispersed in a ductile copper matrix.[33,34] The intragranular particle size was an order of magnitude less than that of particles examined in this work. Deformation (cold-rolling) caused the transformation, forming $\{110\}_m$ twins. In comparison, the stress-induced transformation of ZrO_2 particles in Al_2O_3 described here also proceeded by $(110)_m$ twinning.

Martensitic calculations using the generalized CRAB (Crocker-Ross-Acton-Bevis)[35,36] theory were recently performed using lattice parameters for the bulk ZrO_2 transformation temperature at 950°C.[16] All the observed monoclinic twin systems and their conjugates, in conjunction with the three lattice correspondences (LC) A, B, and C^{15} were analyzed. In LC A, c_t becomes a_m, in LC B, c_t goes to $-b_m$, while in LC C, the c_t axis remains c_m.[15] Real solutions were obtained for LC A using $(110)_m$ twinning and its conjugates. For LC B (110), (011), their conjugates and $(00\overline{1})$ monoclinic twins yielded real solutions. In LC C only twinning on $(011)_m$ and conjugates gave solutions.

In the stress-induced transformation of ZrO_2 in Al_2O_3 (Fig. 9), LC C operated since c_t became c_m. However, the $(110)_m$ twinning mode does not agree with any of the martensitic calculations based on lattice parameters at 950°C. The martensitic solutions may be modified or changed by using room-temperature lattice parameters for the tetragonal phase, and correcting them by about 1% under the tensile stress applied by the Al_2O_3 matrix.[24] It is also possible that the phenomenological theory for a bulk transformation does not apply, or needs to be significantly modified, for transformations in a confined particle.

With these reservations in mind, however, the inconsistency between the current phenomenological theory and experimental analyses suggests that the observed twinning modes are not lattice-invariant shear twins of a martensitic mechanism (see Appendix). Rather, they may be post-transformational deformation twins, formed to minimize the particle shape change on transformation. A similar phenomenon was observed in ferrous lath martensites[37] where laths formed by lattice-invariant shear slip contained closely spaced deformation twins accommodating the shape deformation, between layers of retained austenite.

Conclusion

The transformation mechanism of spherical ZrO_2 particles randomly dispersed in Al_2O_3 was studied by 1 MeV high-voltage electron microscopy to simulate bulk conditions. The model of residual elastic strains around t-ZrO_2 particles arising from thermal expansion mismatch ($\Delta\alpha\Delta T$) with the Al_2O_3 matrix was confirmed. The strain field was anisotropic and caused a maximum displacement in the matrix parallel to [001] of t-ZrO_2. Preliminary estimates of its magnitude indicated that there was insufficient residual strain to overcome the nucleation barrier to transformation.

Two-dimensional strain analyses showed that the projection of the tetragonal strain axis was parallel to monoclinic twin plane traces. The monoclinic strain field was also anisotropic and apparently perpendicular to the internal twins. Hence, the principal axis of the monoclinic strain field was perpendicular to the principal axis of the preexisting (residual) strain field.

In a small (≈ 0.3 μm) spherical particle, the stress-induced transformation was followed in situ. A topotaxial lattice correspondence was preserved in which a_t, b_t, and c_t axes became a_m, b_m, and c_m axes, respectively. Twinning occurred on

$(110)_m$ planes. The crystallography of the stress-induced mechanism was compared with martensitic calculations[16] using the generalized CRAB theory, and found to be inconsistent. This implies that either the phenomenological theory for bulk transformations needs to be significantly modified for confined particle mechanisms, or that deformation twinning occurs after transformation to minimize the shape change due to transformation.

Appendix

An earlier interpretation of twinning in ZrO_2 particles as being martensitic may have been incorrect.[13,15] Twinning on $(001)_m$ was formulated into the (010) $\langle \bar{1}00 \rangle_{fct}$ lattice-invariant shear system by LCB and yielded a martensitic solution for $(001)_m$, in agreement with the CRAB calculations.[16] Due to the crystallographic equivalence of a_t and b_t, it was assumed that a $(100)_m$ solution also existed.[13,15] However, this deduction may have been in error, since a_t and b_t do become physically distinguishable in the context of a lattice correspondence, and the monoclinic β angle needs to be considered in the calculations. This demonstrates the limitations of performing martensitic calculations on noncubic systems using orthonormal formulations as was done earlier,[14] and illustrates the value of the generalized CRAB theory[36,37] for performing martensitic calculations on noncubic crystal structures.

Acknowledgments

The author gratefully acknowledges valuable discussions and collaboration with W. Mader and M. Rühle. The specimen was graciously provided by N. Claussen. H.-J. Schedler is thanked for care and maintenance of the AEI 1 MeV microscope "above and beyond the call of duty."

References

[1] R. C. Garvie, R. H. Hannink, and R. T. Pascoe, *Nature*, **258**, 703–704 (1975).
[2] Advances in Ceramics, Vol. 3, Edited by A. H. Heuer and L. W. Hobbs. The American Ceramic Society, Columbus, OH, 1981.
[3] A. H. Heuer, N. Claussen, W. M. Kriven, and M. Rühle, *J. Am. Ceram. Soc.*, **65** [12] 642–50 (1982).
[4] A. G. Evans and A. H. Heuer, *J. Am. Ceram. Soc.*, **63** [5–6] 241–48 (1980).
[5] A. G. Evans; this volume, pp. 193–212.
[6] D. L. Porter and A. H. Heuer, *J. Am. Ceram. Soc.*, **60** [3–4] 183–84 (1977).
[7] M. Rühle and B. Kraus, *Proc. 10th Int. Congress on Electron Microscopy*, **2**, 533–34 (1982).
[8] M. Rühle, B. Kraus, A. Strecker, and D. Waidelich; this volume, pp. 256–74.
[9] M. Rühle and W. M. Kriven, *Ber. Bunsenges. Phys. Chem.*, **87**, 222–28 (1983).
[10] S. Schmauder, W. Mader, and M. Rühle; this volume, pp. 251–55.
[11] M. Rühle and A. H. Heuer; this volume, pp. 14–32.
[12] J. E. Bailey, *Proc. R. Soc. London, Ser. A*, **279**, 359–412 (1964).
[13] W. M. Kriven; pp. 168–83 in Advances in Ceramics, Vol. 3, Edited by A. H. Heuer and L. W. Hobbs. The American Ceramic Society, Columbus, OH, 1981.
[14] E. Bischoff and M. Rühle, *J. Am. Ceram. Soc.*, **66** [2] 123–27 (1983).
[15] W. M. Kriven, W. L. Fraser, and S. W. Kennedy; pp. 82–97 in Advances in Ceramics, Vol. 3, Edited by A. H. Heuer and L. W. Hobbs. The American Ceramic Society, Columbus, OH, 1981.
[16] A. A. Chaudhry and A. G. Crocker; this volume, pp. 46–53.
[17] N. Claussen and M. Rühle; pp. 137–63 in Advances in Ceramics, Vol. 3, Edited by A. H. Heuer and L. W. Hobbs. The American Ceramic Society, Columbus, OH, 1981.
[18] N. Claussen, *Z. Werkstoff Tech.*, **13**, 138–47 and 185–96 (1982).
[19] M. Rühle and W. M. Kriven, "Analysis of Strain Around Tetragonal and Monoclinic Zirconia Inclusions," Proceedings of an International Conference on Solid → Solid Phase Transformations, Edited by H. I. Aaronson, D. E. Laughlin, R. F. Sekerka, and C. M. Wayman. AIME, Pittsburgh, 1982.

[20] W. M. Kriven and M. Rühle, *Proc. 10th Int. Cong. on Electron Microscopy,* **2**, 103–104 (1982).
[21] M. Wilkens, *Phys. Status Solidi,* **13**, 529–42 (1966).
[22] W. Mader and M. Rühle, *Proc. 10th Int. Congr. on Electron Microscopy,* **2**, 101–102 (1982).
[23] W. Mader and M. Rühle, Inst. Phys. Conf. Ser. No. 68, Chapter 10, paper presented at EMAG, Guildford, 30 August–2 September 1983; 385–88.
[24] W. Mader, Ph.D. Thesis, University of Stuttgart, 1984.
[25] J. B. Wachtman, T. G. Scuderi, and G. W. Cleek, *J. Am. Ceram. Soc.,* **45**, 319–23 (1962).
[26] R. N. Patil and E. C. Subbarao, *J. Appl. Crystallogr.,* **2**, 281–88 (1969).
[27] A. G. Evans, N. Burlingame, M. Drory, and W. M. Kriven, *Acta Metall.,* **29**, 447–56 (1981).
[28] W. Mader, W. M. Kriven, and M. Rühle; unpublished work.
[29] J. F. Nankivell, *Optik,* **20**, 171–97 (1963).
[30] B. Hudson and M. J. Makin, *J. Phys. E: Scientific Instruments,* **3**, 311 (1970).
[31] W. Mader, Diplom Thesis, University of Stuttgart, 1980.
[32] J. M. Thomas, *Philos. Trans., R. Soc. London,* **277** [1268] 251–86 (1974).
[33] I. W. Chen and Y. H. Chiao, *Acta Metall.,* **31**, 1627–38 (1983).
[34] I. W. Chen and Y. H. Chiao; this volume, pp. 33–45.
[35] A. F. Acton, M. Bevis, A. G. Crocker, and N. H. Ross, *Proc. R. Soc. London, Ser. A,* **320**, 101–33 (1970).
[36] A. G. Crocker, *J. Phys.,* **43**, C4209–14 (1982).
[37] B. P. J. Sandvik and C. M. Wayman, *Metallography,* **16**, 199–227 (1983).

Diffusionless Transformations in Zirconia Alloys

C. A. ANDERSSON, J. GREGGI, JR., AND T. K. GUPTA

Westinghouse Research and Development Center
Pittsburgh, PA 15235

Microscopic evidence of the diffusionless martensitic transformations between cubic and tetragonal, as well as between tetragonal and monoclinic structures, are presented and discussed.

Two categories of phase transformations occur in zirconia materials. Above ≈1500 K, the diffusion of both ionic species allows diffusion-controlled transformations to occur. This results in the usual phase equilibria being functions of temperature and composition, as predicted by conventional phase diagrams. Below ≈1500 K, cation diffusion becomes sluggish and transformations, when they occur, are of the diffusionless (martensitic) type. In this case, the parent and the transformed phase have the same composition. In practice, both transformation types can be observed. It is the diffusionless transformation in zirconia systems[1-5] which has attracted most attention in recent years. The improved strength and toughness in most of these material types have been attributed to the diffusionless transformation of the tetragonal phase to monoclinic symmetry ($t \rightarrow m$) in the vicinity of the propagating crack.[6-16]

Most of the studies to date[8-22] have been performed on two-phased materials: those in which the tetragonal (t) phase has been precipitated in a cubic (c) zirconia matrix[8,10,11,17,18] and those in which fine t particles of zirconia have been dispersed in another ceramic matrix.[5,9,19-22] In these cases, the tetragonal-(t) to-monoclinic (m) transformation occurs via a twinning mechanism. Another type of zirconia ceramic has also been developed, which is essentially single-phased polycrystalline tetragonal.[4,5,23-26] In contrast to transformations in two-phased materials, the diffusionless transformation in the single-phased material will be shown in this paper to occur by classical martensitic plate growth similar to that observed in steels.

Recently, we also observed the $c \rightarrow t$ diffusionless transformation in zirconia, which also occurs by a plate growth mechanism. Therefore, the object of this paper is to present microstructural observations of the diffusionless $c \rightarrow t$ and $t \rightarrow m$ transformations. The compositional and structural requirements for diffusionless phase transformation of both c and t phases are discussed elsewhere.[6]

Experimental Procedure

Composition

Four compositions of $ZrO_2 \cdot YO_{1.5}$ were selected for investigation: 2.6, 6.0, 8.7, and 14.4 mol% $YO_{1.5}$. The rationale for selecting these compositions is presented in Fig. 1, which is the combined diffusion-controlled and diffusionless phase diagram[6] of the system ZrO_2-$YO_{1.5}$. Superimposed on the solid line equi-

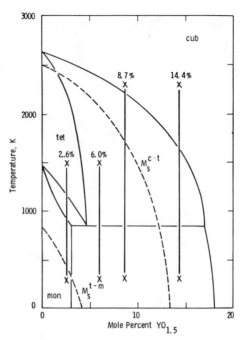

Fig. 1. Equilibrium phase diagram and estimated martensite start loci, M_s, for the ZrO_2-$YO_{1.5}$ materials in this investigation.

librium boundaries are dashed curves representing the boundaries for the start of diffusionless $c \rightarrow t$, $M_s^{c \rightarrow t}$, and $t \rightarrow m$, $M_s^{t \rightarrow m}$, transformations. The theory for the positioning of these boundaries was given previously.[6] The locations of dashed lines are estimated for the materials produced in this investigation. At compositions less than the $M_s^{t \rightarrow m}$, the monoclinic phase would dominate; at compositions greater than $M_s^{c \rightarrow t}$, the material would be cubic; between the two curves, the material would be predominantly tetragonal. On the basis of this phase diagram, the $ZrO_2 \cdot 2.6$ $YO_{1.5}$ and $ZrO_2 \cdot 8.7$ $YO_{1.5}$ materials will show the $t \rightarrow m$ and $c \rightarrow t$ martensitic transformations, respectively, and the $ZrO_2 \cdot 6.0$ $YO_{1.5}$ and the $ZrO_2 \cdot 14.4$ $YO_{1.5}$ materials will remain single-phased t and c, respectively.

Sample Fabrication

Ultrafine powders were hot-pressed in graphite dies at 1725 K for 1 h. Bars measuring 2.1 by 2.1 by 15 mm were diamond-machined from the disks. The 2.6 and 6.0 mol% $YO_{1.5}$ materials were heated in air to 1500 K for 10 min and rapidly air-cooled. The 8.7 and 14.4 mol% $YO_{1.5}$ materials were heated to 2300 K for 10 min and rapidly air-cooled. The bars were then diamond-sliced, ground, and hand-polished to form 2.1 by 2.1 by 0.08 mm plates. These were further thinned using standard ion-milling procedures and examined in a Phillips transmission electron microscope.

Diffusionless Phase Transformations

Cubic-to-Tetragonal Transformation

As shown in Fig. 1, the $ZrO_2 \cdot 14.4$ $YO_{1.5}$ material lies to the right of the $M_s^{t \rightarrow c}$ diffusionless phase boundary over the total temperature range and, therefore, no

Fig. 2. Transmission electron micrograph of $ZrO_2 \cdot 14.4YO_{1.5}$ heated into the c phase field and quenched to room temperature.

transformations are expected. The micrograph of this material in Fig. 2 confirms this. The grains are featureless except for the strain contrast lines. Examination at higher magnifications (not shown) in both bright and dark field indicate that incipient precipitation of a diffusion-controlled t phase might have initiated, indicating that the nose of the time–temperature–transformation curve might not have been totally avoided on quenching.

A photomicrograph of the quenched $ZrO_2 \cdot 8.7\ YO_{1.5}$ and its corresponding electron diffraction pattern are shown in Fig. 3. Since this material passed through the $M_s^{c \rightarrow t}$ boundary, extensive diffusionless transformations have occurred. The diffraction pattern is a superposition of the c and t diffraction spots. This is clarified in Fig. 4, which is a higher-magnification view of the central region of Fig. 3. Convergent-beam microdiffraction patterns were taken of the c matrix and the t plates. The t-ZrO_2 is identified by the weak (211) superlattice-type reflections. Reflections of the type (100) and (110) are unallowed for the t-ZrO_2 but occur by multiple diffraction. This was confirmed by tilting to orientations such as [001], where the (100) and (110) reflections disappear.[27]

The plates shown in Figs. 3 and 4 are formed by a diffusionless process and are not diffusion-controlled precipitates. The bar specimens were cooled from 2300 K to near room temperature in less than 1 min, which would allow insufficient time for diffusional growth. In addition, the composition of both the plates and the matrix, as determined by energy dispersive spectral analysis, are near the nominal 8.7 mol% $YO_{1.5}$. If diffusion had taken place, the t phase would be lower and the c phase higher in $YO_{1.5}$ content.

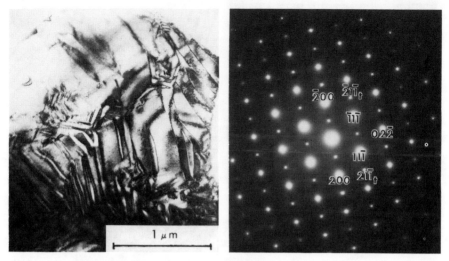

Fig. 3. Transmission electron micrograph and electron diffraction patterns of $ZrO_2 \cdot 8.7YO_{1.5}$ heated into the c phase field and quenched to room temperature.

Several variations of orientation of plates exist in each grain. Many of the plates have grown until they intersected with a grain boundary or another t plate. Others seem to end in the middle of a c grain.

Tetragonal-to-Monoclinic Transformation

The electron diffraction patterns of the polycrystalline $ZrO_2 \cdot 6.0\ YO_{1.5}$ and $ZrO_2 \cdot 2.6\ YO_{1.5}$ materials showed them to be almost entirely t phase, with only a small fraction of monoclinic phase. The photomicrograph and the diffraction pattern of the $ZrO_2 \cdot 2.6\ YO_{1.5}$ are shown in Fig. 5. As we previously[6] discussed in great detail, the t phase of this composition can be stabilized by suppressing the

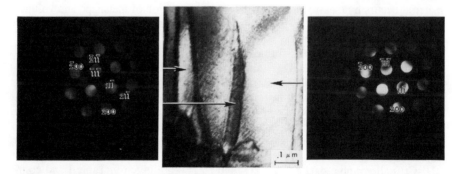

Fig. 4. Transmission electron micrograph and electron microdiffraction patterns of $ZrO_2 \cdot 8.7YO_{1.5}$ heated into the c phase field and quenched to room temperature.

 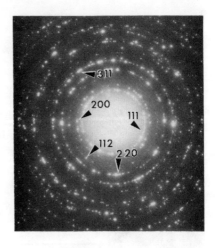

Fig. 5. Transmission electron micrographs and electron diffraction patterns of $ZrO_2 \cdot 2.6YO_{1.5}$ heated into the t phase field and quenched to room temperature.

$M_s^{t \to m}$ through proper control of the composition and grain size. For the 0.3-μm grain size in the figure, the $M_s^{t \to m}$ was estimated to be approximately that shown in Fig. 1. The 2.6 mol% material is just starting to transform, whereas the 6.0 mol% material (not shown) was relatively stable. Thus, cooling of the two materials in liquid nitrogen causes the lower yttria composition to transform more extensively, whereas the higher composition shows little change.[28]

The initial stage of martensitic transformation is shown in Fig. 6. Three of the numbered grains on the left have been induced to transform by electron-beam heating and bending of the foil, as shown on the right. Two martensite plates at two orientations have grown in grain 1. Grains 2 and 3 are more fully transformed, but again have transformed on different habit planes. Clearly, there was more than one nucleation event in each of the grains. The martensitic transformation in this material occurs by classic plate growth. Accommodation of the shape and volume change is not by a twinning mechanism but is probably by combined boundary dislocations and matrix strain.[28]

A more extensively transformed material of the same 2.6 mol% $YO_{1.5}$ material is presented in Fig. 7. This sample was fabricated with a larger grain size (0.5 μm), and its $M_s^{t \to m}$ was raised significantly above room temperature. Again, many of the grains show transformations at different orientations, implying different nucleation events. As in the previous figure, boundaries such as grain boundaries and other martensite plates impede the martensite plate growth, although the plates intersecting at a grain boundary may initiate plate growth in the adjacent grain.

Crystallographic data for m-ZrO_2 and t-ZrO_2 are given in Table I and compared with measured lattice spacings from the polycrystalline diffraction patterns in Figs. 5 and 7. It is very difficult to separate all of the rings in the diffraction pattern in Fig. 7 for the m-ZrO_2 due to the large number of closely spaced lattice planes. However, as shown in Table I, close matching for calculated, and a large

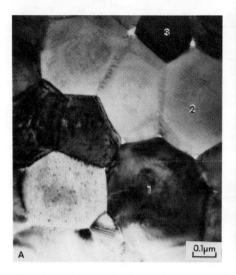

Fig. 6. Transmission electron micrographs of the same region of $ZrO_2 \cdot 2.6YO_{1.5}$ before (A) and after (B) electron-beam heating-induced transformation.

number of observed, lattice spacings occurs. The m-ZrO_2 is easily distinguished from the c and t forms by the presence of the large lattice spacings of the (100), (110), and ($\bar{1}$11) planes.

For the t-ZrO_2 the allowed reflections are indexed according to the larger face-centered tetragonal (fct) CaF_2-type cell instead of the body-centered

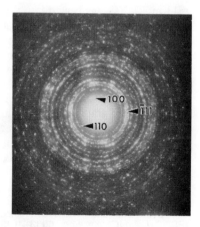

Fig. 7. Transmission electron micrographs of larger-grained $ZrO_2 \cdot 2.6YO_{1.5}$.

Table I. Calculated and Observed d Spacings for m- and t-ZrO$_2$

m-ZrO$_2$*			t-ZrO$_2$†		
hkl	d (Å)	$d_{obsd.}$ (Å)	hkl‡	d (Å)‡	$d_{obsd.}$ (Å)
100	5.079	5.08	111	2.996	2.98
011	3.694		002	2.635	
110	3.636	3.62	200	2.574	2.56
$\bar{1}$11	3.162	3.12	112¶	2.135¶	2.12¶
		2.93§	202	1.842	
111	2.839		220	1.821	1.82
002	2.621		113	1.582	
020	2.604	2.58	311	1.556	1.55
200	2.539	2.52	222	1.498	1.49

*JCPDS CARD 24–1165.
†G. Teufer, *Acta Crystallogr.* **15**, 1187 (1962).
‡Based on double fct CaF$_2$-type cell.
¶Not allowed for face-centered cubic symmetry.
§Possible evidence of t phase.

tetragonal (bct) cell. Again good correlation is obtained between actual and observed lattice spacings. Since the a and c axes of the fct ZrO$_2$ are very similar to each other, as well as to the lattice parameter of c-ZrO$_2$, the t-ZrO$_2$ would not be easily distinguishable from c-ZrO$_2$ if it were not for the presence of the (112) reflection, which is not allowed for c-ZrO$_2$.

The m pattern shows the possible presence of a small amount of t-ZrO$_2$ and, likewise, the t pattern shows the presence of weak rings or spots corresponding to small amounts of m-ZrO$_2$ (not included in Table I).

Summary

This paper presents microstructural observations related to the $c \rightarrow t$ as well as the $t \rightarrow m$ diffusionless transformations in the system ZrO$_2$–YO$_{1.5}$. The $c \rightarrow t$ martensitic transformation can be observed if a material containing less than ≈ 13 mol% YO$_{1.5}$ is heated into the c phase field and quenched through the $M_s^{c \rightarrow t}$. This transformation is similar to that which occurs when a lower YO$_{1.5}$ composition material is heated into the t phase field and quenched through the $M_s^{t \rightarrow m}$ to martensitically form m phase. In both cases in the materials under investigation, transformation occurs by plate growth which stops when the plate contacts a grain boundary, another martensite plate, or a free surface. Shape- and volume-change accommodation is probably accomplished by boundary dislocations and elastic straining of the grain, since no twinning is observed. Several orientations of plates in a single grain imply multiple nucleation events. Under conditions of extensive transformation, martensite plates impinging on a grain boundary can induce plate growth in the adjacent grain.

References

[1] R. C. Garvie, R. H. J. Hannink, and R. T. Pascoe, "Ceramic Steel?," *Nature (London)*, **258** [5337] 703–704 (1975).
[2] D. L. Porter and A. H. Heuer, " Mechanisms of Toughening Partially Stabilized Zirconia (PSZ)," *J. Am. Ceram. Soc.*, **60** [3–4] 183–84 (1977).
[3] N. Claussen, "Stress-Induced Transformation of Tetragonal ZrO$_2$ Particles in Ceramics Matrices," *J. Am. Ceram. Soc.*, **61** [1–2] 85–86 (1978).
[4] (a) T. K. Gupta, F. F. Lange, and J. H. Bechtold, "Effect of Stress-Induced Phase Transformation on the Properties of Polycrystalline Zirconia Containing Metastable Tetragonal Phase," *J. Mater. Sci.*, **13** [7] 1464–70 (1978). (b) T. K. Gupta, "Role of Stress-Induced Phase

Transformation in Enhancing Strength and Toughness of Zirconia Ceramics"; pp. 877–89 in Fracture Mechanics of Ceramics, Vol. 4. Edited by R. C. Bradt, D. P. H. Hasselman, and F. F. Lange. Plenum, New York, 1978.

[5]F. F. Lange, "Transformation Toughening, Parts I, II, III, IV, and V," *J. Mater. Sci.*, **17** [1] 225–63 (1982).

[6]C. A. Andersson and T. K. Gupta, "Phase Stability and Transformation Toughening in Zirconia"; pp. 184–201 in Advances in Ceramics, Vol. 3. Edited by A. H. Heuer and L. W. Hobbs. The American Ceramic Society, Columbus, OH, 1981.

[7]F. F. Lange and D. J. Green, "Effect of Inclusion Size on the Retention of Tetragonal ZrO_2: Theory and Experiments"; pp. 217–25 in Advances in Ceramics, Vol. 3. Edited by A. H. Heuer and L. W. Hobbs. The American Ceramic Society, Columbus, OH, 1981.

[8]A. H. Heuer, "Alloy Design in Partially Stabilized Zirconia"; pp. 98–115 in Advances in Ceramics, Vol. 3. Edited by A. H. Heuer and L. W. Hobbs. The American Ceramic Society, Columbus, OH, 1981.

[9]N. Claussen and M. Ruhle, "Design of Transformation-Toughened Ceramics"; pp. 137–63 in Advances in Ceramics, Vol. 3. Edited by A. H. Heuer and L. W. Hobbs. The American Ceramic Society, Columbus, OH, 1981.

[10]R. H. J. Hannink, K. A. Johnston, R. T. Pascoe, and R. C. Garvie, "Microstructural Changes During Isothermal Aging of a Calcia Partially Stabilized Zirconia Alloy"; pp. 116–36 in Advances in Ceramics, Vol. 3. Edited by A. H. Heuer and L. W. Hobbs. The American Ceramic Society, Columbus, OH, 1981.

[11]A. G. Evans, D. B. Marshall, and N. H. Burlingame, "Transformation Toughening in Ceramics"; pp. 202–16 in Advances in Ceramics, Vol. 3. Edited by A. H. Heuer and L. W. Hobbs. The American Ceramic Society, Columbus, OH, 1981.

[12]D. L. Porter, A. G. Evans, and A. H. Heuer, "Transformation Toughening in Partially Stabilized Zirconia (PSZ)," *Acta Metall.*, **27** [10] 1649–54 (1979).

[13]A. G. Evans and A. H. Heuer, "Review—Transformation Toughening in Ceramics: Martensitic Transformations in Crack-Tip Stress Fields," *J. Am. Ceram. Soc.*, **63** [5–6] 241–48 (1980).

[14]A. G. Evans, N. Burlingame, M. Drory, and W. M. Kriven, "Martensitic Transformation in Zirconia—Particle Size Effects and Toughening," *Acta Metall.*, **29** [2] 447–56 (1981).

[15]R. M. McMeeking and A. G. Evans, "Mechanics of Transformation-Toughening in Brittle Materials," *J. Am. Ceram. Soc.*, **65** [5] 242 (1982).

[16]D. J. Green, "Critical Microstructure for Microcracking in Al_2O_3–ZrO_2 Composites," *J. Am. Ceram. Soc.*, **65** [12] 610 (1982).

[17]D. L. Porter and A. H. Heuer, "Microstructural Development in MgO-Partially Stabilized Zirconia (Mg-PSZ)," *J. Am. Ceram. Soc.*, **62** [5–6] 298–305 (1979).

[18]R. H. J. Hannink, "Growth Morphology of the Tetragonal Phase in Partially Stabilized Zirconia," *J. Mater. Sci.*, **13** [11] 2487–96 (1978).

[19]N. Claussen, F. Sigulinski and M. Ruhle, "Phase Transformation of Solid Solutions of ZrO_2 and HfO_2 in an Al_2O_3 Matrix"; pp. 164–67 in Advances in Ceramics, Vol. 3. Edited by A. H. Heuer and L. W. Hobbs. The American Ceramic Soc., Columbus, OH, 1981.

[20]N. Claussen and D. P. H. Hasselman, "Improvement of Thermal Shock Resistance of Brittle Structural Ceramics by a Dispersed Phase of Zirconia"; pp. 381–95 in Thermal Stress in Severe Environments. Edited by D. P. H. Hasselman and R. A. Heller. Plenum, New York, 1980.

[21]N. Claussen, "Fracture Toughness of Al_2O_3 With an Unstabilized ZrO_2 Dispersed Phase," *J. Am. Ceram. Soc.*, **59** [1–2] 49–51 (1976).

[22]A. H. Heuer, N. Claussen, W. M. Kriven, and M. Rühle, "Stability of Tetragonal ZrO_2 Particles in Ceramic Matrices," *J. Am. Ceram. Soc.*, **65** [12] 642–50 (1982).

[23]T. K. Gupta, "Sintering of Tetragonal Zirconia and Its Characteristics," *Sci. Sintering*, **10** [3] 205–16 (1978).

[24]T. K. Gupta, J. H. Bechtold, R. C. Kuznicki, L. H. Cadoff, and B. R. Rossing, "Stabilization of Tetragonal Phase in Polycrystalline Zirconia," *J. Mater. Sci.*, **12**, 2421–26 (1977).

[25]T. K. Gupta, R. B. Grekila, and E. C. Subbarao, "Electrical Conductivity of Tetragonal Zirconia Below the Transformation Temperature," *J. Electrochem. Soc.*, **128** [4] 929–31 (1981).

[26]F. F. Lange, "Fracture Mechanics and Microstructural Design"; pp. 799–918 in Fracture Mechanics of Ceramics, Vol. 4. Edited by R. C. Bradt, D. P. H. Hasselman, and F. F. Lange. Plenum, New York, 1978.

[27]P. G. Valentine; M. S. Thesis, Case Western Reserve University, Cleveland, OH, 1981.

[28]C. A. Andersson, J. Greggi, Jr., and T. K. Gupta; unpublished work.

Short-Range Order Phenomena in ZrO$_2$ Solid Solutions

R. Chaim and D. G. Brandon

Technion, Israel Institute of Technology
Department of Materials Engineering
Haifa, Israel

The diffuse scattering intensity (DSI) of cubic (c)-ZrO$_2$ observed in electron diffraction patterns was used to construct the 3-dimensional reciprocal lattice. Qualitative comparison of DSI shapes with the calculated profiles expected for different defect symmetries revealed the existence of anisotropic defects along the ⟨111⟩ direct lattice.

The polymorphic nature of ZrO$_2$ is now fairly well understood. Cubic (c), tetragonal (t), and monoclinic (m) polymorphs represent the high-, medium-, and low-temperature equilibrium phases, respectively.

The improvement of mechanical properties in ZrO$_2$ has been related to the presence of metastable t precipitates in a c matrix at ambient temperatures. Partial stabilization of the c matrix is achieved by addition of oxides such as CaO, MgO, or Y$_2$O$_3$.

The ionic bonding in this system and the presence of cations of differing valency result in a large concentration of anion vacancies, which ensures local electrical charge compensation in the lattice.[1]

These isolated anion vacancies could exist as complexes,[2,3] presumably dispersed with some degree of order in the matrix solid solution. The local ordering of anion vacancies in the lattice is expected to cause lattice relaxation, which can give rise to diffuse scattering in electron diffraction patterns. Alpress and Rossell[4] related the maxima in diffuse contours to the contribution of 12 variants of a defect fluorite structure, CaZr$_4$O$_9$ (ϕ_1), which were crystallographically oriented in equivalent directions in the CaF$_2$-type lattice.

This microdomain model was supported by Hudson and Moseley,[5] who presented a reciprocal lattice model[6] containing diffuse rings centered at tetrahedral sites and perpendicular to the ⟨111⟩ directions. A defect structure of the CaZr$_4$O$_9$ type is characterized by anion vacancy strings along the ⟨111⟩ directions. However, the results of Faber et al.[7] indicated oxygen ion displacements along the ⟨100⟩ directions, a result that was confirmed by Morinaga et al.[8] These authors also found that in Ca-PSZ the oxygen ion vacancies and Ca ions are nearest neighbors rather than second nearest neighbors as in the CaZr$_4$O$_9$ structure. These discrepancies in previous work suggest that further investigation of the diffuse electron scattering intensity is necessary.

Experimental Procedure

Several commercial specimens in different heat-treatment conditions were examined by transmission electron microscopy (TEM), as indicated in Table I. The heat treatment was performed on disks 3-mm in diameter by 0.25-mm thick.

Table I. Composition and Heat Treatment of Stabilized ZrO_2 Samples

Composition (wt%)	Heat treatment	Designation
ZrO_2–3% MgO*	Annealed at 1250°C/2 h and air- or furnace-cooled	Mg–PSZ
ZrO_2–2% MgO–1% CaO[†]	As Mg–PSZ	Mg, Ca-PSZ
ZrO_2–12% Y_2O_3[‡]	Annealed 1600°C/2 h and furnace-cooled	Y–CSZ

*Chemical analysis.
[†]Thin-film EDS microanalysis.
[‡]Manufacturer's data.

After heat treatment, the specimens were prepared for TEM observation by standard ion-milling procedures followed by carbon coating. The TEM observations were made with a JEOL-100 CX microscope operating at 100 kV.

Results

Examination of various areas in the different specimens revealed the following characteristic features: (*a*) ellipsoidal *t* precipitates in a *c* matrix of ≈ 40 μm in Mg-PSZ and Mg, Ca-PSZ (Fig. 1). (*b*) *c* grains, ≈ 1 μm in diameter, and free of either *t* or *m* phase, in Y-CSZ.

Specimen tilting experiments in the microscope with dark-field images formed by diffusely scattered electrons (Fig. 2) showed that the *c* matrix was the origin of the diffuse scattering intensity. The existence or absence of the *t* phase had no influence on the DSI, and the *t* reflections were ignored relative to the DSI.

Careful examination from the hole edge to the specimen interior revealed considerable changes in the intensity of the DSI. In particular, there exists a critical thickness that corresponds to a maximum in DSI, due to the compromise between the dynamic diffraction conditions and absorption effects. However, the qualitative

Fig. 1. Ellipsoidal tetragonal precipitates in the cubic matrix of Mg, Ca-PSZ.

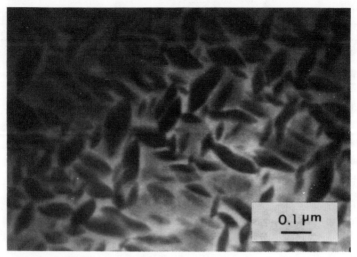

Fig. 2. Dark-field micrograph of the sample in Fig. 1 using the DSI, demonstrating that the cubic matrix is the origin of the DSI.

comparison of measured and calculated DSI produced by either electron or X-ray diffraction must consider the different characteristics of the two radiations, corresponding to dynamic and kinematic conditions, respectively. The influence of the dynamic conditions and the possible contribution of the strain field around the tetragonal precipitates on the DSI were checked using Y-CSZ (20 wt% of Y_2O_3), which were free of tetragonal precipitates.

TEM observations were carried out on thin areas, where the thickness was estimated to be about 0.5 on an extinction distance (ζ_g) for the appropriate reflection excited in the diffraction pattern (Fig. 3(a)). Although this thin area is highly curved, the comparison to the diffraction pattern from thicker areas (Fig. 3(b)) reveals no drastic change in the symmetry or the shape of the DSI. The only change was the decrease in the intensity of diffuse scattering, which necessitated long exposure times in the thin areas. The DSIs for all specimens are qualitatively similar and independent of both the heat treatment and the stabilizer ion. The intensity and distribution of DSI depend on zone axis (Figs. 4–6), which indicates the complex nature of the 3-dimensional intensity surface in reciprocal space. By recording the DSI on different low-index zone axes ([100], [110], [111]), this intensity surface was first constructed and then verified using the higher-index zone axes, as shown in Fig. 6. Care was taken to locate this intensity surface in accordance with the 3-dimensional information provided by a sequence of Bragg reflections.

The relative locations of the Ewald sphere and the reciprocal lattice are represented schematically in Fig. 7. For the cubic structure and 100-kV electrons, the Ewald sphere passes through different sections of the reciprocal lattice, until it reaches, at about the 4th-order reflection, the midpoint of the zero and first-order Laue zone. The detailed picture of the DSI could be deduced by observation of the continuous changes in scattering through different orders of the reflections, as shown schematically for a [100] zone axis. In addition, slight tilting from the exact Bragg conditions in different directions gives similar details. This process was used

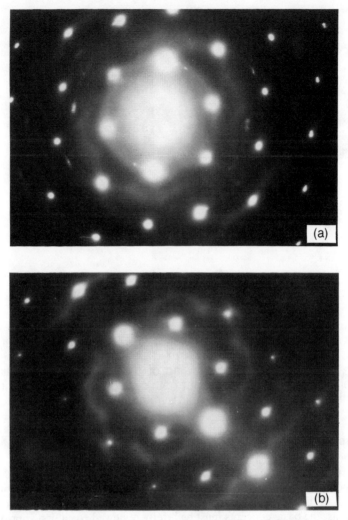

Fig. 3. (a) Diffraction pattern from a very thin area of Y-CSZ, showing the DSI in kinematical conditions. (b) Same zone axis as that in (a) but from a thicker area under dynamic conditions.

for sections containing high values of $\bar{g}$ vectors, and the resultant DSI surface, which has been determined relative to the cubic and the tetragonal reflections, is shown in Fig. 8.

The main feature of the DSI surface is the oval of revolution connecting the cubic reflections along the $\langle 111 \rangle$ direction of the reciprocal lattice. Tilting experiments showed the connection between the ring-shaped DSI, located between the conventional fluorite reflections, and the oval DSI, whose origin is at these reflections and which change intensity with distance from the Bragg reflections. Although the first was found to be a section of the latter, we deal only with the DSI near the Bragg reflections, which are believed to be influenced by the distortion field of the defects.

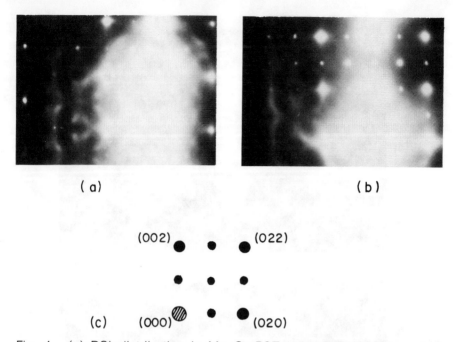

Fig. 4. (a) DSI distribution in Mg, Ca-PSZ, annealed and air-cooled. Z.A. = [100]. (b) DSI distribution in Y-CSZ, annealed and furnace cooled. Z.A. = [100]. (c) Schematic drawing of [100] zone axis diffraction. Note that some reflections forbidden for both c- and t-ZrO$_2$ are present and are thought to arise from orthorhombic symmetry (Ref. 9).

Discussion

The existence of a large concentration of oxygen vacancies is a familiar feature of the high ionic conductivity of partially stabilized zirconia.[10,11] X-ray DSI phenomena from structures containing point defects have been recognized and analyzed.[12,13] Before discussion of this phenomenon with regard to electron diffraction results, the basic concepts must be briefly reviewed. As a first approximation, the DSI can be written as[14]:

$$I_{DSI} = [f_D(\overline{K}) + if(\overline{K})(\overline{K}(\overline{U}(\overline{K})))]^2 \qquad (1)$$

where $\overline{U}(\overline{K})$ is the Fourier transform of the displacement field introduced by a point defect, $f_D(\overline{K})$ and $f(\overline{K})$ are the scattering amplitudes from the defect and the lattice atoms, respectively, and $\overline{K}$ is the wave vector of the scattered X-ray wave. In Eq. (1), the first term is designated as "Laue scattering" and originates from the scattering at the defects themselves. This DSI appears far from the Bragg reflections. The second term is the "Huang scattering" and gives rise to the DSI in the immediate vicinity of the Bragg reflections, which is associated with the distortion field*. This distortion, caused by the defect on its surroundings, results in displacement of the atoms. The Fourier amplitudes will be determined by the elastic

---

*Interference between the two terms is neglected here.

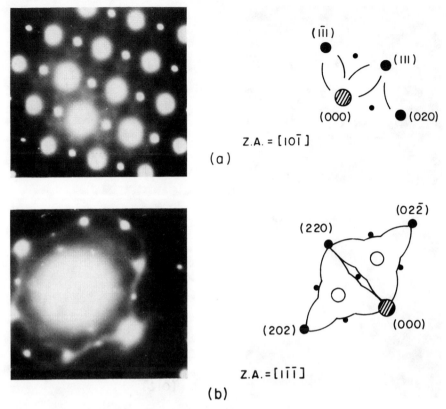

Fig. 5. Low-index diffraction patterns of Mg, Ca-PSZ annealed and air-cooled.

constants and the symmetry and strength of the defects.[15] The latter is determined by the nature of the interatomic forces in the lattice. By application of lattice statics, Kanzaki[16] concluded that isotropic cubic defects produce isointensity profiles with lemniscate shapes, whereas single double-force defects (such as a string of vacancies) should produce ellipsoidal profiles. This fact is significant in that the shape of the DSI near a Bragg reflection should provide information about the symmetry of the point defects present in a crystal. For an isotropic defect in an isotropic crystal, equal DSI is distributed on spherical surfaces touching the reciprocal lattice point (named "Huang spheres"). As the Huang scattering depends on the product $\bar{K} \cdot \bar{U}$, zero intensity is expected whenever $\bar{K} \cdot \bar{U} = 0$, which is the case of a plane through the reciprocal lattice point and perpendicular to the reciprocal lattice vector. For an anisotropic defect, this zero-intensity plane is no longer perpendicular to the reciprocal lattice vector. For different equivalent orientations of these defects in the crystal ("variants"), the DSI must be averaged. The averaging process might or might not give a zero-intensity plane or line due to the intersections of the individual orientations. The result of such averaging processes gives the complex features of apple shape and single-bubble shape[13] around certain reflections for a given defect symmetry. Although the detailed profile of the DSI is influenced by both the elastic constants and the strength of the defects, as

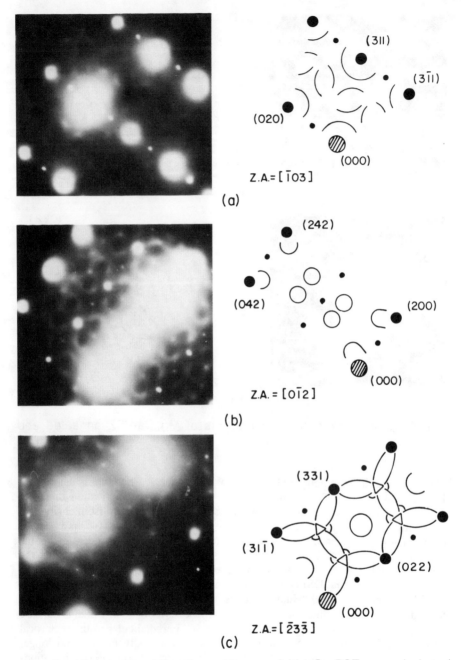

Fig. 6. High-index diffraction patterns of Mg, Ca-PSZ annealed and air-cooled.

mentioned earlier, the basic symmetry of the defects may be determined by comparing the measured DSIs with calculated isointensity profiles. The different isointensity profiles resulting from the defect symmetries of interest in the cubic

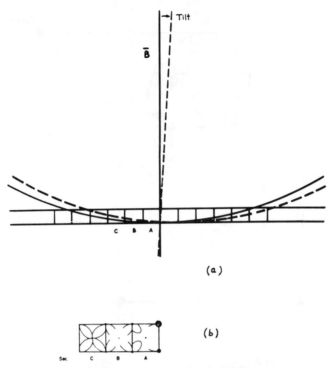

Fig. 7. (a) Schematic representation of selected sections of the reciprocal lattice congruent with the Ewald sphere. (b) Appropriate reciprocal lattice sections for Z.A. = [100].

lattice are compared with appropriate experimental DSIs in Table II. The subscripts dd, as, and sb relate to double-drop, apple shape, and single-bubble forms, as expected for decreasing defect symmetries.[14] The corresponding lattice vector normal to the zero-intensity plane or line is also indicated.

Through observation of $\{hhh\}$-type reflections on different zone axes (see Figs. 8 and 5(a)), the existence of a single-bubble form for the DSI was determined. However, this single-bubble shape is composed of several double-drop shapes in different $\langle 111 \rangle$ reciprocal directions; therefore, the sb contour around the $\{111\}$ reciprocal lattice points is on a section passing close to the center of this reciprocal lattice point.

For the $\{hh0\}$-type reflections, according to Fig. 8, the overlapping double-drop shapes along the $\langle 111 \rangle$ vectors result in an apple-shaped isointensity contour with $\langle hh0 \rangle$-type revolution axis, but in which the intensity along $\langle h00 \rangle$ reciprocal directions is missing (Fig. 4). This missing intensity presumably relates to the preferred orientation of the defects along certain equivalent directions and the resultant loss of continuity in the predicted apple-shaped DSI contour. Although the determination of the zero-intensity plane is in this case ambiguous (Fig. 4), if one superposes the contribution of the equivalent orientations to the DSI at the given reciprocal point (i.e., $(hh0)$ and $(h\bar{h}0)$), then there is no longer any contradiction to the existence of a zero-intensity plane perpendicular to the $\langle hh0 \rangle$-type reciprocal lattice vectors. However, referring to Fig. 8, zero-intensity planes per-

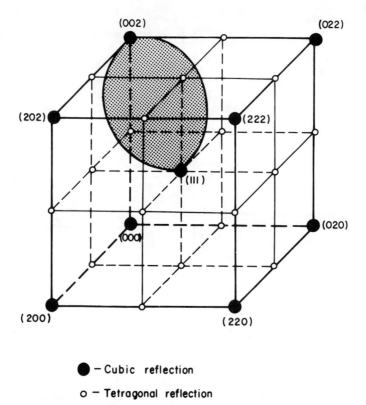

● – Cubic reflection

○ – Tetragonal reflection

Fig. 8. Three-dimensional DSI surface as derived relative to the cubic bcc reciprocal lattice.

pendicular to $\langle h00 \rangle$-type reciprocal vectors are also possible. The more important evidence for defect symmetry determination (Table II) is the DSI about the $\{h00\}$-type reflections. Referring to the 3-dimensional reciprocal lattice DSI surface (Fig. 8), there are no DSI features along the $\langle h00 \rangle$-type lattice vectors at $\{h00\}$ reflections. This negates the possibility of cubic edge symmetry for the defects (Table II). Moreover, by the choice of the reciprocal lattice section passing close to the center of the $(h00)$ Bragg reflections, the single-bubble contour appears (and is readily visible in Fig. 6(a)). Again this *sb* shape is composed of different

Table II. Different Isointensity Profiles at Various Reflections for Several Defect Symmetries (from Ref. 14) Compared with Observed Shape

Defect symmetry	Reflection		
	$(h00)$	$(hh0)$	(hhh)
Cubic edge	$dd[100]$	$as[001]$	sb
Cubic body diagonal	sb	$as[1\bar{1}0]$	sb
Experimental isointensity shape	sb	$as[1\bar{1}0]$ or $[001]$	sb

double-drop forms in different ⟨111⟩ reciprocal directions. Thus, this qualitative analysis of DSI by comparison with calculated intensity profiles confirms the preferred body-diagonal orientation, rather than cubic edge symmetry, for the existing defects. Finally, although this phenomenon relates to the short-range order (SRO) in the cubic solid solution, no microdomain contrast was detected, at least to the limit of the conventional electron microscope resolution.

Conclusions

The diffuse scattering intensity (DSI) of the cubic matrix observed in electron diffraction patterns was used to construct the 3-dimensional reciprocal lattice. Qualitative comparison of DSI shapes in distinct reciprocal lattice sections with the calculated profiles revealed the existence of anisotropic defect displacement fields along the ⟨111⟩ directions in the real lattice. The diffuse intensity in the cubic matrix observed with DSI for the dark-field imaging mode is evidence for fine-scale short-range ordering. This diffuse intensity is not due, at least in the examples studied here, to resolvable microdomains of an ordered defect-fluorite compound.

Acknowledgment

The authors are grateful to A. H. Heuer and M. Rühle, both for their critical comments and for suggesting major improvements in the text of this paper.

References

[1]T. H. Etsell and S. N. Flengas, "The Electrical Properties of Solid Oxide Electrolytes," *Chem. Rev.*, **70** [3] 339–76 (1970).
[2]D. K. Hohnke, "Ionic Conduction in Doped Oxides with the Fluorite Structure," *Solid State Ionics*, **5**, 531–34 (1981).
[3]N. G. Eror and V. Balachandran, "Point Defect Complexes and Their Association Energies in a Acceptor-Doped Oxide," *Solid State Commun.*, **44** [7] 1117–19 (1982).
[4]J. G. Alpress and H. J. Rossell, "A Microdomain Description of Defective Fluorite-Type Phases $Ga_xM_{1-x}O_{2-x}$ (M = Zr, Hf; x = 0.1–0.2)", *J. Solid State Chem.*, **15**, 68–78 (1975).
[5]B. Hudson and P. T. Moseley, "On the Extent of Ordering in Stabilized Zirconia," *J. Solid State Chem.*, **19**, 383–89 (1976).
[6]B. Hudson and P. T. Moseley; pp. 385–88 in Proc. EMAG 75. Edited by J. A. Venables, Academic Press, New York, 1976.
[7]J. Faber, Jr., M. H. Mueller, and B. R. Cooper, "Neutron-Diffraction Study of $Zr(Ca, Y)O_{2-x}$: Evidence of Differing Mechanisms for Internal and External Distortions," *Phys. Rev. B*, **17** [12] 4884–88 (1978).
[8]M. Morinaga, J. B. Cohen, and J. Faber, Jr., "X-Ray Diffraction Study of $Zr(Ca, Y)O_{2-x}$; II. Local Ionic Arrangements," *Acta Crystallogr., Sect. A*, **36**, 520–30 (1980).
[9]L. H. Schoenlein, "Microstructural Studies of an Mg-PSZ"; Ph. D. Thesis, Case Western Reserve University, Cleveland, OH, 1982; Chapter 4.
[10]W. L. Worrell, Solid Electrolytes; p. 143. Edited by S. Geller. Springer-Verlag, Berlin, 1977.
[11]E. Van der Voort, "Anion Vacancy Migration in CaF_2 Lattices and in Monoclinic ZrO_2," *J. Phys. C: Solid State Phys.*, **7**, L395–99 (1974).
[12]A. N. Goland and D. T. Keating, "Diffuse X-Ray Scattering from Doublet Defects in Martensite and Ferrite," *J. Phys. Chem. Solids*, **29**, 785–97 (1968).
[13]P. H. Dederiches, "The Theory of Diffuse X-Ray Scattering and its Application to the Study of Point Defects and Their Clusters," *J. Phys. F: Metal Phys.*, **3**, 471–96 (1973).
[14]H. Trinkaus, "On Determination of the Double-Force Tensor of Point Defects in Cubic Crystals by Diffuse X-ray Scattering," *Phys. Status Solidi B*, **51**, 307–19 (1972).
[15]J. W. Flocken and J. Hardy, "Calculations of the Intensity of X-Ray Diffuse Scattering Produced by Point Defects in Cubic Metals," *Phys. Rev. B*, **1** [6] 2472–83 (1970).
[16]H. Kanzaki, "Point Defects in fcc Lattice, II. X-Ray Scattering Effects," *J. Phys. Chem. Solids*, **2**, 107–14 (1957).

Phase Relationships in Some ZrO$_2$ Systems

V. S. STUBICAN, G. S. CORMAN, J. R. HELLMANN, AND G. SENFT

The Pennsylvania State University
Department of Materials Science and Engineering
University Park, PA 16802

The phase relationships in the system ZrO$_2$-CaO were compared with those in the system HfO$_2$-CaO. Two ordered phases, CaZr$_4$O$_9$ and Ca$_6$Zr$_{19}$O$_{44}$, were found in the system ZrO$_2$-CaO. CaZr$_4$O$_9$ seems to be metastable, and the Ca$_6$Zr$_{19}$O$_{44}$ phase, which is an equilibrium phase, disorders at 1355° ± 15°C. In the system HfO$_2$-CaO, three ordered phases were found. The CaHf$_4$O$_9$ phase is metastable and decomposes into Ca$_2$Hf$_7$O$_{16}$ and a hafnia solid solution. Ca$_2$Hf$_7$O$_{16}$ and Ca$_6$Hf$_{19}$O$_{44}$ are equilibrium phases in the system HfO$_2$-CaO. It was confirmed that Zr$_3$Y$_4$O$_{12}$, which appears in the system ZrO$_2$-Y$_2$O$_3$, disorders at 1382° ± 5°C into a fluorite-type solid solution. This result was used to refine the existing ZrO$_2$-Y$_2$O$_3$ phase diagram. Similar results were obtained in the system ZrO$_2$-Yb$_2$O$_3$, where the Zr$_3$Yb$_4$O$_{12}$ phase was found. This phase is stable up to 1637° ± 12°C. Phase relationships in the ternary system ZrO$_2$-Y$_2$O$_3$-Yb$_2$O$_3$ were determined for three temperatures, 1200°, 1400°, and 1650°C, for compositions containing >50 mol% ZrO$_2$.

This paper briefly summarizes recent developments in phase relationships in a few important ZrO$_2$ systems. Establishing the correct phase diagrams in binary ZrO$_2$ systems is hampered by several major difficulties.[1] At very high temperatures, the presence of impurities and selective vaporization of one component may influence the equilibria. At relatively low temperatures, the reactions in zirconia systems are slow and equilibrium is difficult to achieve. However, considerable progress has been made in the last 10 years in our understanding of precipitation, eutectoid decomposition, phase transition, and ordering processes in zirconia systems. This paper summarizes progress since ZrO$_2$-80.[2]

Zirconia Binary Systems

It is well known that the most commonly used oxides to form cubic solid solutions with zirconia are CaO, MgO, Y$_2$O$_3$, and rare earth oxides. All of them form fluorite-type phases that are stable over wide ranges of composition and temperature. In a recent study, Hellmann and Stubican[3] redetermined the chemical equilibrium for the system ZrO$_2$-CaO for temperatures above ≈1000°C by heating reactive powders for long times. The eutectoid decomposition of cubic solid solution was found to occur at 1140° ± 40°C and 17 ± 0.5 mol% CaO, as shown in Fig. 1. Decomposition reactions in sintered cubic solid solutions were studied using electron diffraction and transmission techniques. It was found that decomposition follows metastable extensions of the boundary lines that separate two-phase regions from the adjacent cubic solid solution region. This type of behavior is usually connected with a kinetic barrier for nucleation, which prevents nucleation of the missing phase.[4] Similar results could be expected if single crystals of ZrO$_2$-CaO solid solutions were annealed below the eutectoid temperature. Recently

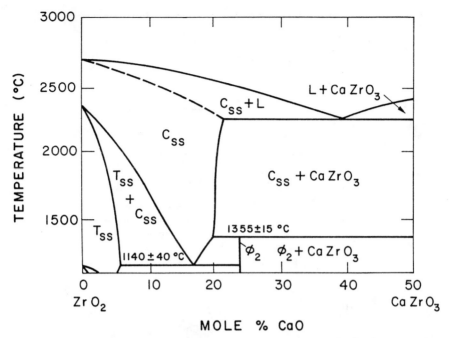

Fig. 1. Part of the phase diagram for the system ZrO_2-CaO: ϕ_2 is the $Ca_6Zr_{19}O_{44}$ phase, C_{ss} the fluorite-type solid solution, T_{ss} the tetragonal solid solution, and M_{ss} the monoclinic solid solution (Ref. 3).

Marder et al.[5] studied precipitation phenomena in the system ZrO_2-CaO using two samples heat-treated at high temperature and decomposing them at lower temperatures. From the results obtained, these authors concluded that the eutectoid in the system ZrO_2-CaO is at 1000°C and 15 mol% CaO. However, chemical equilibrium in this system is very difficult to achieve if decomposition of solution-heat-treated samples is attempted, and the equilibrium at relatively low temperatures (1000°–1400°C) is more closely approximated by using reactive powders and heating them for a long time at a desired temperature.

The development of ordered phases in the system ZrO_2-CaO was studied during prolonged heating of reactive powders derived from gels using a Guinier–Hagg parafocusing X-ray diffraction camera.[6] The pure phase ϕ_2 was obtained at 1250°C with powders containing 24 mol% CaO, which indicated that the correct composition of this phase was $Ca_6Zr_{19}O_{44}$. This phase has rhombohedral symmetry, space group $R\bar{3}c$, and is analogous to the ϕ_2 phase in the system HfO_2-CaO.[7] At 1355° ± 15°C, the ϕ_2 phase decomposes to a cubic ZrO_2 solid solution and $CaZrO_3$. The pure phase ϕ_1 of the composition $CaZr_4O_9$ was obtained by heating reactive powders containing 20 mol% CaO at 1180°C. This phase often appears as a precursor to the formation of the ϕ_2 phase. It is interesting to note that the ordered compound $CaZr_4O_9$, after prolonged heating (116 days) at 1100°C, rejects zirconia, indicating that this phase is metastable. The final products of decomposition of the ϕ_1 phase could not be established due to the sluggishness of the decomposition reaction. Further work is necessary to establish the equilibrium products of the decomposition of the $CaZr_4O_9$ compound below 1150°C.

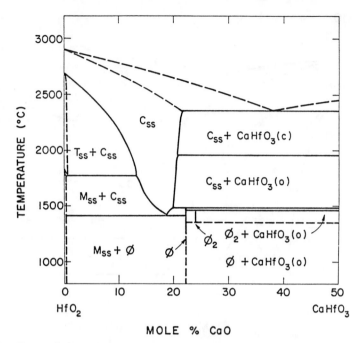

Fig. 2. Part of the phase diagram for the system HfO_2-CaO: ϕ is the $Ca_2Hf_7O_{16}$ phase, ϕ_2 the $Ca_6Hf_{19}O_{44}$ phase, C_{ss} the fluorite-type solid solution, T_{ss} the tetragonal solid solution, and M_{ss} the monoclinic solid solution (Ref. 8).

It is interesting to note that we never observed the $Ca_2Zr_7O_{16}$ (ϕ) phase, analogous to the phase $Ca_2Hf_7O_{16}$, which forms in the system HfO_2-CaO.[7] This may possibly be explained by the fact that a high degree of cation ordering is necessary to form this phase. Due to the low cation diffusivities in the system ZrO_2-CaO below 1150°C, the formation of the phase $Ca_2Zr_7O_{16}$ (ϕ) may be difficult to achieve in a reasonable time.

It could be quite useful in further research to compare results obtained from the investigation of the system ZrO_2-CaO with the results recently obtained for the system HfO_2-CaO.[8] The recently established phase diagram for the system HfO_2-$CaHfO_3$ is shown in Fig. 2. The eutectoid decomposition of the cubic solid solution in this system was found to occur at 19 ± 0.5 mol% CaO and 1415° ± 7°C. Three ordered compounds exist in this system. The ordered compound $CaHf_4O_9$ (ϕ_1) was found to be metastable at all temperatures investigated. It decomposes to a hafnia solid solution and the $Ca_2Hf_7O_{16}$ (ϕ) phase. The compound $Ca_2Hf_7O_{16}$ (ϕ) is stable with an upper limit of 1473° ± 5°C. The compound $Ca_6Hf_{19}O_{44}$ (ϕ_2) is stable from ≈1350° ± 50°C to its upper limit of 1463° ± 5°C.

As was mentioned previously, it is possible that the ordered compound $Ca_2Zr_7O_{16}$ (ϕ) exists in the system ZrO_2-CaO, and the $CaZr_4O_9$ (ϕ_1) compound is metastable. However, thus far all attempts to synthesize the $Ca_2Zr_7O_{16}$ (ϕ) phase have been unsuccessful.

The system ZrO_2-Y_2O_3 has a relatively large cubic solid solution field. The recently established phase diagram by Stubican et al.[9] shows the presence of the

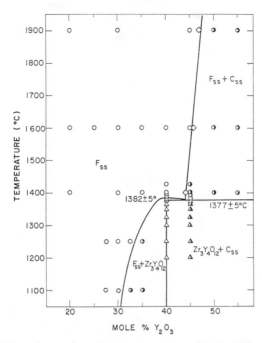

Fig. 3. Part of the phase diagram for the system ZrO_2-Y_2O_3 showing disordering of the compound $Zr_3Y_4O_{12}$ into a fluorite-type solid solution at 1382° ± 5°C (Ref. 12).

stable ordered compound $Zr_3Y_4O_{12}$. Long-range ordering in the system ZrO_2-Y_2O_3 occurs at 40 mol% Y_2O_3, as described by Ray and Stubican.[10] Scott also observed long-range ordering at 40 mol% Y_2O_3 in a concurrent study.[11]

Recent work to further refine the phase diagram ZrO_2-Y_2O_3 was done by Stubican and Corman.[12] Special attention was paid to determination of the stability limits of the $Zr_3Y_4O_{12}$ compound and to the possibility of formation of other ordered compounds at high concentrations of yttria. Figure 3 shows that the compound $Zr_3Y_4O_{12}$ is stable up to 1382° ± 5°C, where it disorders to a fluorite-type solid solution. This result is in fair agreement with the previous results obtained by Scott[13] and Pascual and Duran.[14] In a recent paper, Pascual and Duran proposed a tentative phase diagram for the system ZrO_2-Y_2O_3 that shows a very high solubility of ZrO_2 in Y_2O_3 (≈31 mol% ZrO_2 at 2000°C) and a new hexagonal phase Y_6ZrO_{11} at 75 mol% Y_2O_3. According to these authors, the homogeneity range of the Y_6ZrO_{11} phase is between ≈63 and 90 mol% Y_2O_3 at low temperature and from 75 to 80 mol% Y_2O_3 at 1750°C. This hexagonal phase disorders into an yttria cubic-type rare-earth solid solution above 1750°C. However, recent results[12] show that the solubility of zirconia in yttria is 21 mol% at 1950°C and 21.9 mol% ZrO_2 at 2100°C, which agrees with previous results by Stubican et al.[9] and disagrees with the results by Pascual and Duran. Furthermore, we could not confirm the existence of the ordered compound Y_6ZrO_{11},[12] at least as an equilibrium phase. Several mixtures of reactive powders containing 55, 65, 75, and 85 mol% Y_2O_3 were heated at 1300°C for 6 months in sealed Pt tubing to avoid contamination from the furnace. After these mixtures were annealed, X-ray Guinier photographs

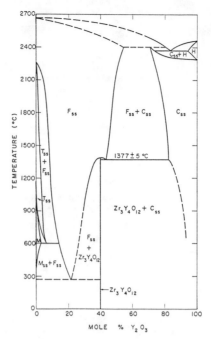

Fig. 4. Phase diagram for the system ZrO_2-Y_2O_3: F_{ss} is the fluorite-type solid solution, T_{ss} the tetragonal solid solution, M_{ss} the monoclinic solid solution, and C_{ss} the yttria cubic-type solid solution (Ref. 12).

showed only mixtures of the $Zr_3Y_4O_{12}$ compound and a cubic-type yttria solid solution, which is in agreement with the phase diagram by Stubican et al.[9] It is possible that the weak X-ray diffraction lines attributed by Pascual and Duran to the Y_6ZrO_{11} compound were due to the presence of impurities.

Figure 4 shows our slightly corrected phase diagram for the system ZrO_2-Y_2O_3, which shows that the $Zr_3Y_4O_{12}$ compound disorders at 1382° ± 5°C to a fluorite-type solid solution.

Similar results were obtained with the system ZrO_2-Yb_2O_3. Part of the phase diagram for this system is shown in Fig. 5. Only one ordered compound, $Zr_3Yb_4O_{12}$, was found, and its upper limit of stability is at 1637° ± 12°C.

Because the $Zr_3Yb_4O_{12}$ compound orders relatively rapidly, a sharp concentration boundary between short-range and long-range order is readily observed in the system ZrO_2-Yb_2O_3. Cubic solid solutions containing 29 and 31 mol% Yb_2O_3 were formed at 2000° and then annealed for 30 days at 1300°C. Electron diffraction and transmission photographs shown in Fig. 6 indicate the presence of a fluorite-type phase with diffuse maxima in the diffraction pattern and the absence of any second phase in transmission photographs. A similarly treated solid solution containing 31 mol% Yb_2O_3 shows the presence of the second phase, e.g., $Zr_3Yb_4O_{12}$ domains (Fig. 7). These results indicate that there is a sharp concentration boundary between a fluorite phase with short-range order and a mixture of the fluorite phase and $Zr_3Yb_4O_{12}$ precipitates, as indicated in the phase diagram in Fig. 5.

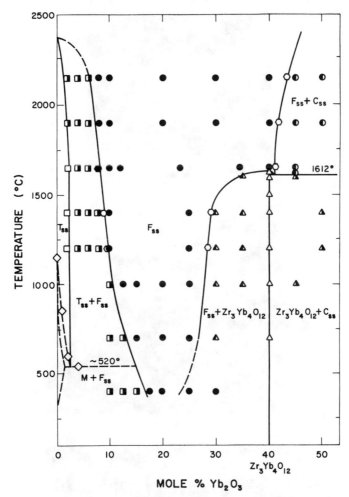

Fig. 5. Part of the phase diagram for the system ZrO_2-Yb_2O_3: F_{ss} is the fluorite-type solid solution, T_{ss} the tetragonal solid solution, M_{ss} the monoclinic solid solution, and C_{ss} the ytterbia cubic-type solid solution (Ref. 12).

No ordered compound other than $Zr_3Yb_4O_{12}$ could be detected in the system ZrO_2-Yb_2O_3. Several mixtures of reactive powders containing 55, 65, 75, and 85 mol% Yb_2O_3 were heated at 1300°C for 6 months in sealed Pt tubing. After the mixtures were annealed, X-ray Guinier photographs showed only mixtures of the $Zr_3Yb_4O_{12}$ compound and ytterbia cubic-type solid solution.

Ternary Zirconia Systems

The extent to which the stabilizing effects of yttria and ytterbia are additive can be evaluated by studying ternary phase diagrams. Knowledge of phase boundaries in the ternary system ZrO_2-Y_2O_3-Yb_2O_3 is important if, for example, the controlled preparation of stabilized zirconia with high electrical conductivity is sought.

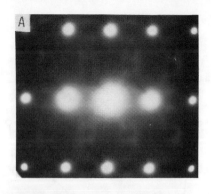

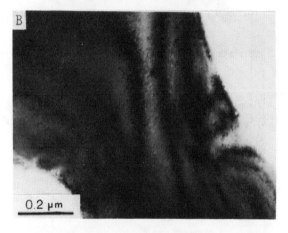

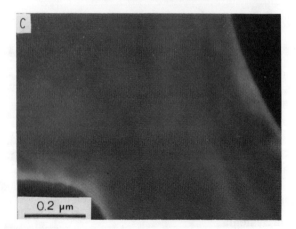

Fig. 6. (A) Selected area diffraction pattern of a 29 mol% Yb_2O_3 specimen quenched from 2000°C and annealed for 30 days at 1300°C. $[211]_F$ is the zone axis. (B) Bright-field image with beam direction parallel to $[211]_F$. (C) Dark-field image obtained using diffuse maxima as the principal beam (Ref. 12).

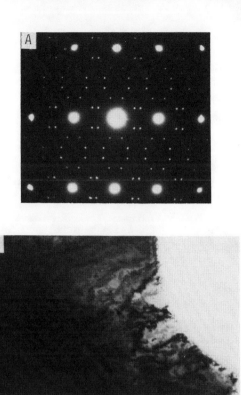
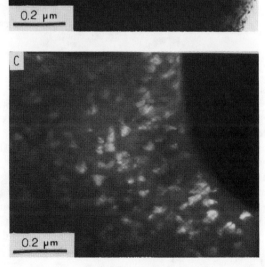

Fig. 7. (A) Selected area diffraction pattern of a 31 mol% Yb_2O_3 specimen quenched from 2000°C and annealed for 30 days at 1300°C. $[211]_F$ is the zone axis. (B) Bright-field image with beam direction parallel to $[211]_F$. (C) Dark-field image obtained using one reflection of the ordered compound $Zr_3Yb_4O_{12}$ (Ref. 12).

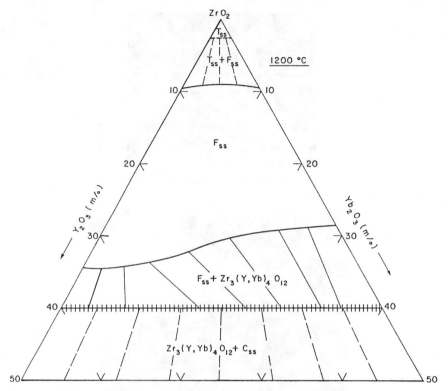

Fig. 8. Part of the ternary phase diagram for the system ZrO_2-Y_2O_3-Yb_2O_2 at 1200°C: F_{ss} is fluorite-type solid solution and T_{ss} is tetragonal solid solution (Ref. 12).

Recently, phase relationships in the ternary system ZrO_2-Y_2O_3-Yb_2O_3 were studied in our laboratory. Three isothermal sections for compositions containing >50 mol% ZrO_2 were determined using reactive powders heated at given temperatures for extended periods (1–2 months).

Figure 8 shows the phase diagram ZrO_2-Y_2O_3-Yb_2O_3 at 1200°C. Figures 9 and 10 show the phase diagrams for the same system at 1400° and 1650°C, respectively. The 1200°C isotherm shows two single-phase regions and three two-phase regions. There is complete solubility between the $Zr_3Yb_4O_{12}$ and $Zr_3Y_4O_{12}$ compounds. The binary compound $Zr_3Y_4O_{12}$ does not appear at 1400°C in the ternary diagram because it is not stable above 1382° ± 5°C. At 1650°C, a very large field of fluorite-type solid solution is present, and no ordered compound appears in the phase diagram.

Acknowledgment

This work was sponsored by National Science Foundation Grant No. DMR8107143.

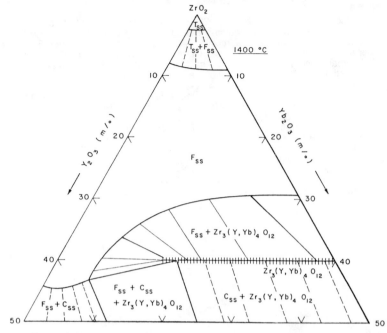

Fig. 9. Part of the ternary phase diagram for the system ZrO_2-Y_2O_3-Yb_2O_3 at 1400°C. C_{ss} is ytterbia cubic-type solid solution (Ref. 12).

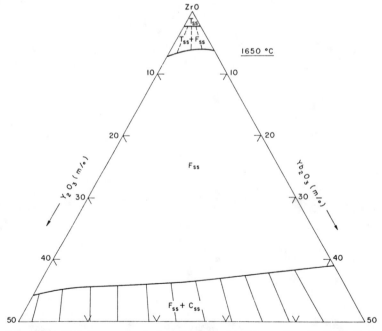

Fig. 10. Part of the ternary phase diagram for the system ZrO_2-Y_2O_3-Yb_2O_3 at 1650°C (Ref. 12).

References

[1]V. S. Stubican and J. R. Hellmann; pp. 25–36 in Advances in Ceramics, Vol. 3. Edited by A. H. Heuer and L. W. Hobbs. The American Ceramic Society, Columbus, OH, 1981.
[2]"Science and Technology of Zirconia"; Advances in Ceramics, Vol. 3. Edited by A. H. Heuer and L. W. Hobbs. The American Ceramic Society, Columbus, OH, 1981.
[3]J. R. Hellmann and V. S. Stubican, *J. Am. Ceram. Soc.*, **66** [4] 260–64 (1983).
[4]J. W. Christian; p. 672 in The Theory of Transformation in Metals and Alloys. Pergamon, London, 1965.
[5]J. M. Marder, T. E. Mitchell, and A. H. Heuer, *Acta Metall.*, **31** [3] 387–95 (1983).
[6]J. R. Hellmann and V. S. Stubican, *Mater. Res. Bull.*, **17**, 459–65 (1982).
[7]J. G. Allpress, H. J. Rossell, and H. G. Scott, *J. Solid State Chem.*, **14** [3] 264–73 (1975).
[8]G. Senft and V. S. Stubican, *Mater. Res. Bull.*, **18**, 1163–70 (1983).
[9]V. S. Stubican, R. C. Hink, and S. P. Ray, *J. Am. Ceram. Soc.*, **61** [1–2] 17–21 (1978).
[10]S. P. Ray and V. S. Stubican, *Mater. Res. Bull.*, **12**, 549–56 (1977).
[11]H. G. Scott, *Acta Crystallogr., Sect. B*, **33**, 281–82 (1977).
[12]V. S. Stubican and G. S. Corman; unpublished results.
[13]H. G. Scott, *J. Mater. Sci.*, **12** [2] 311–16 (1977).
[14]C. Pascual and P. Duran, *J. Am. Ceram. Soc.*, **66** [1] 23–27 (1983).

Ordered Compounds in the System CaO-ZrO$_2$

J. Hangas, T. E. Mitchell, and A. H. Heuer

Case Western Reserve University
Department of Metallurgy and Materials Science
Cleveland, OH 44106

Single crystals of cubic (c)-ZrO$_2$ containing between 16 and 22 mol% CaO were heat-treated at 1200°C to study the precipitation of ordered intermediate fluorite-type phases. ϕ_1 (CaZr$_4$O$_9$) was observed in 18–20 mol% CaO crystals, in the form of precipitates which coarsened with increasing aging time. ϕ_2 (Ca$_6$Zr$_{19}$O$_{44}$) was the predominant precipitate phase in crystals containing more than 20 mol% CaO, but only a few crystals contained both ϕ_1 and ϕ_2. The absence of significant quantities of ϕ_1 in samples containing more than 20 mol% CaO could have either a kinetic or a thermodynamic origin, but attainment of equilibrium must require more than 340 h annealing at 1200°C.

Phase equilibrium in the system CaO-ZrO$_2$ has proved to be difficult to determine with certainty. Since the phase diagram by Stubican and Ray was published in 1977,[1] four revisions have been suggested.[2-5] Most of the problems have been concerned with the temperature and composition of the invariant eutectoid reaction involving the decomposition of the cubic ZrO$_2$ solid solution (c-ZrO$_2$) in the neighborhood of 15 mol% CaO (the latest determination[5] gives 1140 ± 40°C and 17 ± 0.5 mol% CaO) and the existence of ordered compounds in the composition region between 20 and 24 mol% CaO. We have investigated the precipitation and stability of these ordered compounds, using single crystals containing between 16 and 22 mol% CaO; our results confirm many details of the latest phase diagram of Hellman and Stubican[5] but do raise questions about the stability of one of the ordered phases.

Literature Review

The ordered compounds in the system CaO-ZrO$_2$ are all defect fluorite types and are related to analogous compounds in the system CaO-HfO$_2$.[6] In this latter system the structure of the following three compounds has been determined[7,8]: ϕ_1-CaHf$_4$O$_9$, ϕ_2-Ca$_6$Hf$_{19}$O$_{44}$, and ϕ-Ca$_2$Hf$_7$O$_{16}$, with cation "order" apparently being least in CaHf$_4$O$_9$ and greatest in Ca$_2$Hf$_7$O$_{16}$. The latest phase diagram for this system[9] does not show CaHf$_4$O$_9$ as a stable phase.

The corresponding ϕ_1 and ϕ_2 phases in the system CaO-ZrO$_2$ were discovered by Michel[10] and are included in the latest CaO-ZrO$_2$ diagram of Hellmann and Stubican[5] (Fig. 1). No analog of Ca$_2$Hf$_7$O$_{16}$ is known in the system CaO-ZrO$_2$.

As seen in Fig. 1, both ϕ_1 and ϕ_2 are formed by peritectoid reactions, ϕ_1 being stable below 1235 ± 15°C and ϕ_2 below 1355 ± 15°C. The stability of these phases was determined by Hellman and Stubican[5] by heating very fine powders derived from gels for long times. ϕ_1 was also observed by these workers as a

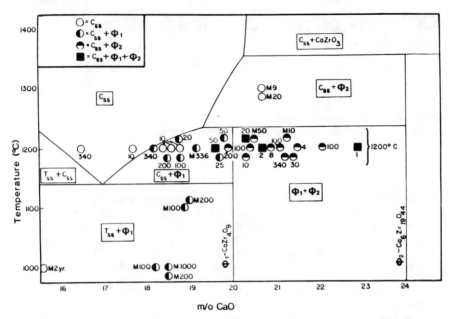

Fig. 1. Hypereutectoid part of the CaO-ZrO$_2$ phase diagram of Hellmann and Stubican. A single temperature of 1200°C was used for heat treatment in the present study. The compositions of foils from this study are plotted with the length of heat treatment in hours marked beside them. Those foils used in the study of Marder et al. have the preface 'M' before the duration of the heat treatment at various temperatures.

precipitate phase in two hypereutectoid sintered ceramics containing 18.5 and 19.8 mol% CaO and aged, respectively, at 1000°C for 64 days and at 1100°C for 64 days.

Marder et al.[4] also studied precipitation of ϕ_1 from CaO-ZrO$_2$ cubic solid solutions in the temperature range 1000°–1300°C, in this case using a single-crystal boule that had been grown by skull melting. The nucleation of ϕ_1 could be described in terms of a conventional C-shaped T–T–T (time–temperature–extent of transformation) curve. The coarsening of ϕ_1 obeyed Ostwald ripening kinetics, with an interfacial energy $\gamma = 0.1$ J/m^2 and an activation energy of the coarsening rate constant of 376 ± 50 kJ/mol, in reasonable agreement with the interdiffusion kinetics in this system.

ϕ_1 is monoclinic (space group $C2/c$): its lattice parameters (in the system CaO-ZrO$_2$) are $a = 1.7819$ nm, $b = 1.4612$ nm, and $c = 1.2065$ nm with $\beta = 119.5°$.[10] Because of the orientation relationship[7] and the symmetry of the phases, 12 precipitate variants can form in the cubic matrix.[11]

ϕ_2 is rhombohedral (space group $R3/c^7$): on a hexagonal basis, its lattice parameters are $c = 1.7711$ nm and $a = 1.8331$ nm. From the structural similarity to the fluorite structure of c-ZrO$_2$, we expect that the threefold $\langle 0001 \rangle$ axis will be parallel to any of the cubic $\langle 111 \rangle$ directions, with $\langle 11\bar{2}0 \rangle \| \langle 110 \rangle$; therefore, only four variants of ϕ_2 should form in the c-ZrO$_2$ matrix.

Results

Identification of Phases

Figure 2 shows three zone axis diffraction patterns, corresponding to [111], [112], and [110] of c-ZrO_2. The extra spots are due to ϕ_1, as first reported by Allpress and Rossell[11]; detailed analysis[12] confirms that there are 12 variants of ϕ_1 in the c-ZrO_2 matrix. (Some minor quantity of ϕ_2 is also present, as is shown by the arrowed reflections.)

Comparable diffraction patterns of ϕ_2 are shown in Fig. 3 (which also shows evidence for a small amount of ϕ_1, as will be discussed below). The identification of these patterns as a mixture of c-ZrO_2 and ϕ_2 is illustrated in Fig. 4. The [110] c-ZrO_2 zone axis diffraction pattern is shown by the filled circles in Fig. 4. As noted above, we expect the c axis of two ϕ_2 variants to be parallel to two $\langle 111 \rangle$ directions in the (110) plane of c-ZrO_2 and, thus, these two variants should produce the patterns shown in Figs. 4(a) and (b). Figure 4(c) is actually produced by both of the other two precipitate variants, whose c axes are symmetrically inclined to the [110] zone axis. Superposition of these three patterns results in the diffraction pattern of Fig. 4(d), which is identical to the [110] pattern of Fig. 3(c) and confirms the identification of ϕ_2.

Composition Dependence of the Phase Assemblage

Skull-melted single crystals containing 16–22 mol% CaO, obtained from Hrand Djevahirajian, Monthey, Switzerland, were available for this study. The foils used by Marder et al.[4] were also available for reexamination.

Crystals were heated in air for up to 336 h at a single temperature of 1200°C. Thin foils were prepared for transmission electron microscopy by standard means, and the composition of each foil was analyzed by X-ray energy dispersive spectroscopy. In this technique, absolute accuracies better than ±5% of the amount of the element present are possible.[13] The amount of CaO in a CaO-ZrO_2 thin-foil sample can be obtained from the ratio of the intensities, I, of the CaK and ZrL X-ray spectra:

$$C_{Ca}/C_{Zr} = k_{CaZr}(I_{CaK}/I_{ZrL}) \qquad (1)$$

provided the constant k_{CaZr} is known sufficiently accurately. In the present experiments, k_{CaZr} was determined (in wt%) as 0.49 ± 0.01, using a eutectic sample of $CaZrO_3$-ZrO_2 solid solution, assuming the $CaZrO_3$ was stoichiometric; we believe that quoted compositions shown in Fig. 1 are accurate to ±0.5 mol% (the annealing time in hours is included for each datum). We also reanalyzed the foils studied by Marder et al.[4] and found their 16.1 mol% hyperstoichiometric crystal was actually variable, containing from 18 to 21 mol% CaO; these data are plotted on Fig. 1 (the symbol M followed by annealing time in hours).

Of the samples studied, two crystals containing 16.5 and 17.7 mol%, annealed at 1200°C for 340 and 10 h, respectively, showed no precipitates, in agreement with Hellmann and Stubican's phase diagram (Fig. 1). Diffuse intensity was observed in the c-ZrO_2 diffraction patterns, as discussed by Marder et al.[4] and Hellmann and Stubican.[5]

With only two exceptions to be discussed below, ϕ_1 was the only precipitate phase found in the crystals containing 18–19.8 mol% CaO, as is also expected from the phase diagram of Fig. 1. The precipitate morphology did vary with aging time. At short times, the ϕ_1 precipitates were striated (Fig. 5), presumably due to

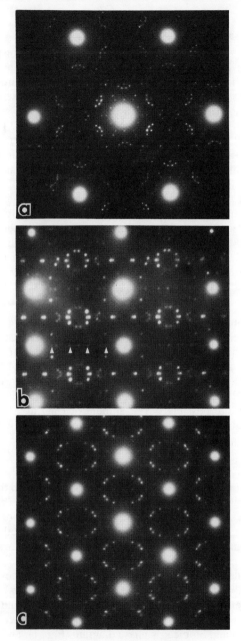

Fig. 2. (a) [111], (b) [112], and (c) [110] zones of 20.4 mol% CaO specimen aged 20 h at 1200°C; the intense reflections are from c-ZrO$_2$, whereas the less intense reflections are primarily from ϕ_1 precipitates. The arrowed reflections in (b) are from ϕ_2.

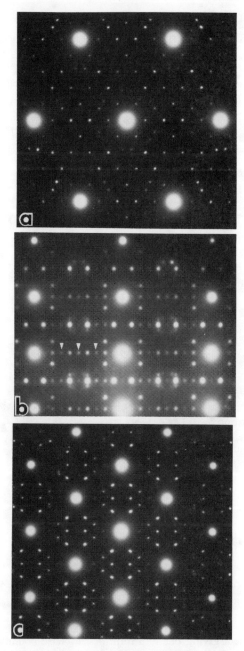

Fig. 3. (a) [111], (b) [112], and (c) [110] zones of 22.1 mol% CaO aged 100 h at 1200°C, showing ϕ_2 reflections. At this composition there should be 30% ϕ_1, but only some faint reflections (arrowed) are present.

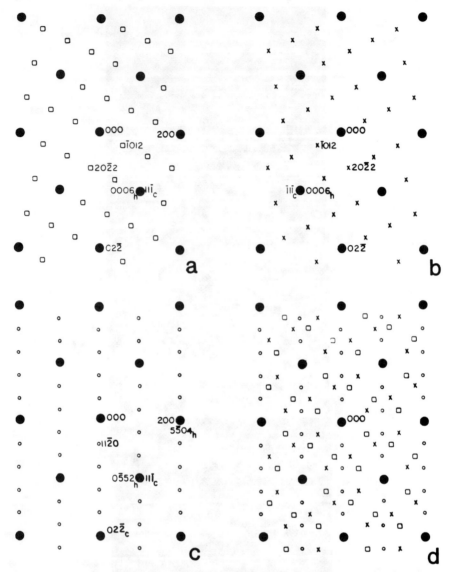

Fig. 4. [110] zone with ϕ_2 precipitate reflections identified. The composite pattern in (d) is identical to Fig. 3(c). See text for further discussion.

particle impingement giving rise to a form of antiphase boundary (APB).[4] In the [112] c zone of Fig. 5, the particles are elongated. The fully coarsened ϕ_1 precipitates lack such striations and appear equiaxed, as shown in Fig. 6 (this example is from a foil taken from the study of Marder et al.). The mechanism by which the microstructure of Fig. 5 evolves into that of Fig. 6 is not well understood at this time.

ϕ_1 and ϕ_2 have been observed together in four 1200°C foils, a 22.9 mol% sample aged for 1 h, a 20.7 mol% sample aged for 2 h, a 20.4 mol% sample aged

Fig. 5. Dark-field micrograph of ϕ_1 precipitates in 18.6 mol% CaO specimen aged 100 h at 1200°C.

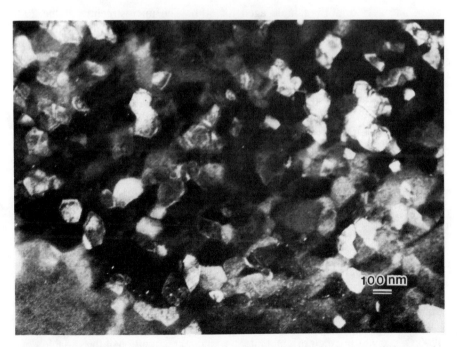

Fig. 6. Dark-field micrograph of fully coarsened ϕ_1 precipitates in 19.2 mol% CaO crystal aged 336 h at 1200°C.

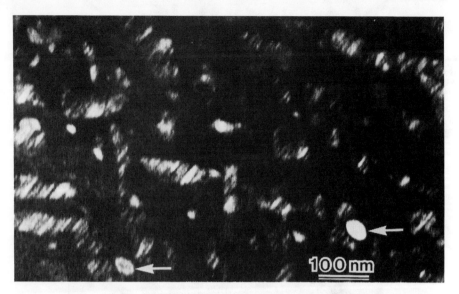

Fig. 7. Dark-field micrograph showing both ϕ_1 and ϕ_2 (arrowed) in a 20.4 mol% CaO sample aged 20 h at 1200°C. The ϕ_1 precipitates retain the underaged striated morphology. The ϕ_2 precipitates are smaller and oval in shape.

20 h, and a 19.6 mol% sample aged for 50 h. The microstructure of the 20-h sample is shown in Fig. 7. The ϕ_1 is striated and is present in much greater quantities than ϕ_2, the oval particles which are arrowed (the diffraction pattern of Fig. 3 is from this sample). Although the diffraction pattern of the 50-h annealed 19.6 mol% crystal was similar to Fig. 3, the ϕ_1 and ϕ_2 were more similar in size, ≈40 nm (Fig. 8). On the basis of Fig. 1, it is thought that either the ϕ_2 particles in this sample are metastable (see below) or this crystal actually contains more than 20 mol% CaO. The ϕ_1 and ϕ_2 particles in the 1- and 2-h annealed samples were much smaller, <10 nm in diameter, and were present in roughly comparable volume fractions.

The remaining samples contained from 20.3 to 22.1 mol% CaO. Aside from one fully cubic specimen, the rest appeared to contain only ϕ_2 precipitates (Fig. 9), although weak ϕ_1 reflections were visible in electron diffraction patterns (Fig. 3). Figure 1 predicts that, at equilibrium, such crystals should contain 55–90% ϕ_1. The ϕ_2 precipitates are equiaxed, exhibit high coherency strains in bright-field micrographs, and contain planar defects which are most probably APBs arising from particle impingement during growth. Finally, they appear to be "rafted" or aligned along $\langle 100 \rangle$ of the c-ZrO_2 matrix, presumably a result of significant coherency stresses.

Discussion

Although our studies shed some light on the subsolidus phase equilibrium at 1200°C, that was not their original intent — in fact, recognition of the variable CaO content in the large single crystals grown by skull melting came as an unpleasant surprise. In any event, we confirm the findings of Hellmann and Stubican[3,5] that

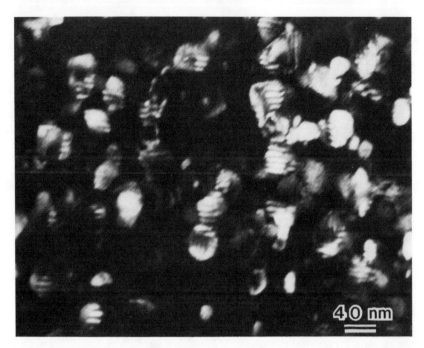

Fig. 8. Dark-field micrograph of ϕ_1 and ϕ_2 precipitates in 19.6 mol% CaO sample aged 50 h at 1200°C.

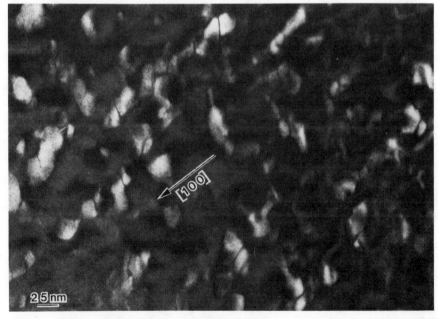

Fig. 9. Dark-field micrograph of ϕ_2 precipitates in 22.1 mol% CaO sample aged 100 h at 1200°C.

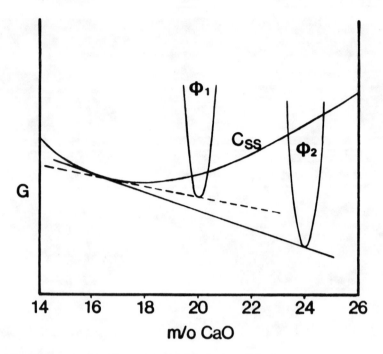

Fig. 10. Free energy diagram showing plausible free energy curves under conditions where ϕ_1 is metastable.

ϕ_2 is a stable phase in this system and that the upper limit of CaO content in c-ZrO_2 at 1200°C is closer to their limit than to the solvus shown by Marder et al.[4] Furthermore, in agreement with both these groups, ϕ_1 was found to nucleate and coarsen readily, at least in compositions between 18 and 20 mol% CaO. The major question raised by our study is whether ϕ_1 is an equilibrium phase or not. We note again that the comparable phase in the system CaO-HfO$_2$ appears not to be an equilibrium phase.[9] The presence of ϕ_2 in crystals containing less than 20 mol% CaO could be a result of uncertainties in the chemical analysis, such crystals actually being richer in CaO (>20 mol%). However, the sparcity of ϕ_1 in crystals containing more than 20 mol% CaO cannot be so easily explained away. One possibility is that ϕ_1 is metastable but nucleates readily, and then disappears as the ϕ_2 nucleates and grows. This would be in keeping with Hellmann and Stubican's[5] reactive gel experiments, where ϕ_1 was observed to nucleate more easily than ϕ_2. Another possibility is that ϕ_2 nucleates more easily in the higher-CaO content crystals because the free energy of the cubic solid solution increases with increasing CaO content, and the driving force for nucleation of ϕ_2 precipitates is thereby increased. For such crystals, precipitation of ϕ_1 would increase the CaO content of the c-ZrO_2 matrix and thereby increase the free energy of the cubic solid solution. A problem with this explanation is that, once the ϕ_2 has precipitated, the lower-CaO content of the cubic solid solution should then provide suitable conditions for the formation of ϕ_1. We therefore prefer the explanation that ϕ_1 is a metastable phase at 1200°C; a schematic free energy vs composition plot consistent with this view is shown in Fig. 10. However, it is difficult to fault Hellmann and Stubican's conclusion that ϕ_1 is a stable phase, given its presence in their 20 and

22.2 mol% samples heated 116 days at 1180°C. The final resolution of this dilemma will likely come only after extended annealings of crystals containing 20–24 mol% CaO. Such experiments are in progress.

Acknowledgments

This research was supported by the NSF under Grant No. DMR 77-19163. We are grateful to D. B. Williams of Lehigh University for the donation of the $CaZrO_3$-containing eutectic sample, and for helpful discussions concerning determination of k_{CaZr} in the system CaO-ZrO_2. A. H. Heuer also acknowledges the Alexander Von Humboldt Society for a Senior Scientist Award, which made possible his sabbatical leave at the Max-Planck-Institut für Metallforschung, Stuttgart, W. Germany, where the first draft of this paper was written.

References

[1] S. P. Ray and V. S. Stubican, *Mater. Res. Bull.*, **12**, 549–56 (1977).
[2] V. S. Stubican and J. R. Hellmann; pp. 25–31 in The Science and Technology of Zirconia, Advances in Ceramics, Vol. 3. Edited by A. H. Heuer and L. W. Hobbs. The American Ceramic Society. Columbus, OH, 1981.
[3] J. R. Hellmann and V. S. Stubican, *Mater. Res. Bull.*, **17** [4] 459–65 (1982).
[4] J. M. Marder, T. E. Mitchell, and A. H. Heuer, *Acta Metall.*, **31**, 387 (1983).
[5] J. R. Hellmann and V. S. Stubican, *J. Am. Ceram. Soc.*, **66** [4] 260–64 (1983).
[6] C. Delamarre, *Rev. Int. Hautes Temp. Refract.*, **9**, 209 (1975).
[7] J. G. Allpress, H. J. Rossell, and H. G. Scott, *J. Solid State Chem.*, **14** [3] 264–73 (1975).
[8] H. J. Rossell and H. G. Scott, *J. Solid State Chem.*, **13**, 345–52 (1975).
[9] G. B. Senft and V. S. Stubican, *Mater. Res. Bull.*, **18**, 1163–70 (1983).
[10] D. Michel, *Mater. Res. Bull.*, **8**, 943 (1973).
[11] J. G. Allpress and H. J. Rossell, *J. Solid State Chem.*, **15**, 68 (1975).
[12] J. Hangas, M. S. Thesis, Case Western Reserve University, Cleveland, OH, 1984.
[13] J. Goldstein; pp. 83–120 in Introduction to Analytical Electron Microscopy. Plenum, New York, 1979.

Tetragonal Phase in the System ZrO_2-Y_2O_3

V. Lanteri, A. H. Heuer, and T. E. Mitchell

Case Western Reserve University
Department of Metallurgy and Materials Science
Case Institute of Technology
Cleveland, OH 44106

Analytical electron microscopy has been used to obtain information on tetragonal (t) ZrO_2 in the system ZrO_2-Y_2O_3. Evidence of two tetragonal solid solutions (t and t') has been found in single crystals obtained from skull melting. Of the two, t'-ZrO_2 has the higher Y_2O_3 content and has been described by previous workers as "nontransformable"; it is present in as-grown crystals but is metastable. After annealing at high temperatures (1600°C), t'-ZrO_2 decomposes into a mixture of the two equilibrium phases: a low-Y_2O_3-content t-ZrO_2 and a high-Y_2O_3-content cubic (c) phase, c-ZrO_2. The t-ZrO_2 appears as precipitate colonies consisting of two twin-related variants in contact along the coherent {101} twin plane.

Precipitates of the high-ZrO_2 tetragonal distorted-fluorite phase in the system ZrO_2-Y_2O_3 appear to be quite different from those observed in other well-studied partially stabilized zirconias (Mg-PSZ, Ca-PSZ).[1-3] In this paper, we discuss the nature of the tetragonal phase in Y_2O_3 partially stabilized zirconia (Y-PSZ) single crystals. We note that three versions[4-6] of the phase diagram exist, which differ in detail mainly at the high-ZrO_2 region; our work thus also sheds light on the subsolidus phase equilibria in this system.

Scott[4] described a metastable "nontransformable" high-Y_2O_3-content tetragonal ZrO_2 solid solution (t-ZrO_2), which he and other workers[7,8] suggested formed from the cubic ZrO_2 solid solution (c-ZrO_2) via a displacive phase transformation; the nontransformability relates to its reluctance to undergo the stress-assisted martensitic transformation to monoclinic symmetry found in lower-Y_2O_3-content t-ZrO_2 (Fig. 1). This form of nontransformable t'-ZrO_2 is widely encountered in plasma-sprayed Y-PSZ[7-9] and may form only in nonequilibrium situations, as suggested by Scott. For example, t'-ZrO_2 was not found by Stubican et al.,[5] who used reactive gels and powders in the experiments that led to their construction of the phase diagram for this system.

Virtually all prior studies of this nontransformable t-ZrO_2 have utilized X-ray studies of powders. One of the goals of this study has been the electron microscopy examination of this form of ZrO_2 in single crystals. According to Miller et al.,[7] the nontransformable t'-ZrO_2 formed in plasma-sprayed Y-PSZ is unstable; if these materials are annealed in the two-phase (t-ZrO_2 + c-ZrO_2) phase field, t-ZrO_2 forms via a diffusion-controlled precipitation reaction and may subsequently transform martensitically to monoclinic symmetry on cooling.

Attention has been given in this work to a single composition (8 wt% Y_2O_3) in the two-phase field, which should be single phase with cubic symmetry at high temperatures but which, according to the experiments of Miller et al.,[7] should transform to t'-ZrO_2 on cooling from high temperature. The stability of t'-ZrO_2

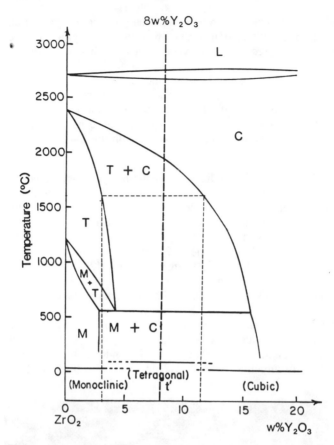

Fig. 1. Phase diagram of the ZrO_2-rich region of the system ZrO_2-Y_2O_3 (Ref. 4).

has been investigated with respect to annealing at high temperatures (1600°C), and the phases present after this heat treatment have been characterized.

Experimental Procedure

As mentioned above, the single-crystal sample used contained 8 wt% Y_2O_3 and was grown by skull melting.* The crystals were opaque and white, a physical appearance characteristic of nonreduced zirconia. For X-ray diffraction, the sample was crushed into fine powder in an agate mortar. Plates (1 by 10 by 10 mm) were annealed in air at 1600°C for 24, 50, and 100 h and quenched by removal from the furnace. The microstructure of the as-received and annealed specimens was characterized by analytical electron microscopy using a Philips EM400T microscope with standard dark-field techniques, supplemented by energy dispersive X-ray analysis (EDX).

Before proceeding to the results, it is appropriate to describe the techniques used to form dark-field images of t-ZrO_2. To distinguish this phase from c-ZrO_2,

---

*This crystal was furnished by NRL and manufactured by CERES Corp., Waltham, MA. For further information, see Ref. 10.

only certain foil orientations can be used, where reflections that are unique to the t phase are present. When transformation from c to t symmetry occurs in this system, whether by precipitation or via a displacive phase transformation, three t variants result, the t **c** axis being parallel to any of the three original $\langle 100 \rangle_c$ directions. To image all the t variants, the most convenient foil orientation is $\langle 111 \rangle$. Also, t-ZrO_2 can be described in terms of either a nonprimitive base-centered unit cell, which in fact is a slightly distorted version of the c fluorite unit cell, or a conventional primitive t unit cell.[11,12] For convenience in this paper, the former C-centered t cell will be used for comparison with the fluorite cell of c-ZrO_2.

Results

Figure 2 is a typical $[\bar{1}11]$ diffraction pattern of the as-received material. The most intense ("fundamental") reflections are those allowed by both c- and t-ZrO_2, whereas the weaker reflections are due to the t phase and have the subscript t. For all $\{1\bar{1}2\}_t$-type reflections to be present in a $[\bar{1}11]$ zone axis selected-area diffraction (SAD) pattern, three t variants must be present in the area of foil giving rise to the SAD, each variant giving rise to one $\{1\bar{1}2\}_t$ reflection.[13] It is thus possible to image all the t variants in the dark field using the three $\{1\bar{1}2\}_t$ reflections (see below). X-ray diffraction performed on the as-received specimen showed that only a very small amount of c-ZrO_2 was present, indicating that the material contains almost 100% t-ZrO_2.

Figure 3 shows dark-field micrographs of the as-received material using the three $\{1\bar{1}2\}_t$ reflections just described. The t-ZrO_2 particles imaged in Fig. 3(a) appear to be precipitates ≈ 10 nm thick by ≈ 50 nm long, lying on two $\{101\}$ habit planes. The t-ZrO_2 precipitates in Y-PSZ thus appear to be different from t-ZrO_2

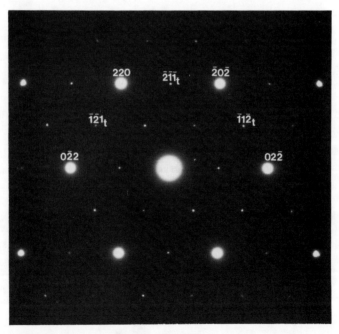

Fig. 2. Diffraction pattern, $[\bar{1}11]$ zone axis.

precipitates in Mg-PSZ, where the precipitate/matrix habit plane is invariably $\{100\}$, and Ca-PSZ, where the precipitates tend to be equiaxed. Figure 3(b) is obtained by using a second $\{1\bar{1}2\}_t$ reflection in the same area of the foil. A second tetragonal variant is imaged in this case and is also present as well-defined precipitates with a $\{101\}$ habit plane. The third $\{1\bar{1}2\}_t$ tetragonal reflection (Fig. 3(c)) gives rise to a different microstructure in the same area of the foil — the "matrix" in which the precipitates of Fig. 3(a) and (b), are contained. If the three micrographs are superimposed, all the foil area has been illuminated; it appears that the entire foil area has t symmetry, confirming the X-ray analysis.

It thus appears that the single crystal consists of a t matrix containing t precipitates! We believe that, during the slow cooling that is part of the skull-melting crystal-growth process, a high-temperature displacive transformation from c to t symmetry occurred, giving rise to the t' matrix structure of Fig. 3(c). The precipitates of Fig. 3(a) and (b) are believed to be low-Y_2O_3 t-ZrO_2, which formed via a conventional diffusion-controlled reaction, either from the c phase or from the t'-ZrO_2 itself. The small amount of c-ZrO_2 indicated by the X-ray analysis was not detectable in the area of foil imaged in Fig. 3. It has been suggested[14,15] that the $c \rightarrow t'$ displacive reaction should give rise to antiphase domain boundaries (APBs), which have been imaged by TEM by Bender and Lewis[16] in 3 and 6 wt% Y-PSZ single crystals after laser irradiation, by Chaim et al.[15] in a 12 wt% Y-PSZ polycrystal after suitable heat treatment, and by Michel et al.[17] in 5.4 wt% single crystals. The arrowed features in Fig. 3(c) are also thought to be APBs arising from this transformation.

EDX analysis of the matrix of Fig. 3(c) showed a Y_2O_3 content of ≈ 8 wt%. If, as has been suggested, t-ZrO_2 is a precipitate that formed in a c or t' matrix, then it should have a lower Y_2O_3 content than the parent t' matrix (see Fig. 1). Unfortunately, the small size of the t-ZrO_2 precipitates and the difficulty of distinguishing the precipitates during conventional bright-field imaging prevent confirming this prediction by EDAX analysis. Therefore, we studied further decomposition of the t' matrix during high-temperature (1600°C) annealing, with the expectation that the precipitates would coarsen.

From the phase diagram (Fig. 1), t'-ZrO_2 should decompose into a low-Y_2O_3 t-ZrO_2 solid solution containing ≈ 3 wt% Y_2O_3 plus a c-ZrO_2 solid solution containing ~ 12 wt% Y_2O_3. Dark-field micrographs of the same area of a foil, again obtained with the three $\{1\bar{1}2\}_t$ reflections, are shown in Fig. 4 from a specimen which had been aged 24 h at 1600°C. Figure 4(a) and (b) indicates that the t-ZrO_2 precipitates have grown at the expense of the t' matrix, the precipitates occurring in the form of large "colonies"; each colony actually consists of plates of two twin-related variants. The two variants have different **c** axes but share a habit plane, which is the coherent twin plane. Small colonies can actually be discerned at the arrowed region of Fig. 3(a); it is clear the colonies coarsen during annealing. The remaining t'-ZrO_2 is imaged in Fig. 4(c); notice that it is located at the periphery of the colonies and has a **c** axis different from that of the two variants constituting the colony. The t'-ZrO_2 is clearly a minor component, confirming that t'-ZrO_2 is unstable with respect to diffusion at high temperatures. We thus expect the decomposition products of t'-ZrO_2 to be a mixture of equilibrium phases, namely, c-ZrO_2 and the colonies of t-ZrO_2. Such a microstructure can be seen in Fig. 5 from a specimen aged for 50 h at 1600°C. Figure 5(a) is a bright-field micrograph and shows coarse colonies set among c regions. Figure 5(b) is the corresponding dark-field micrograph, taken with a t reflection.

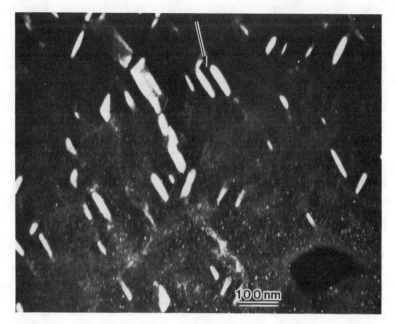

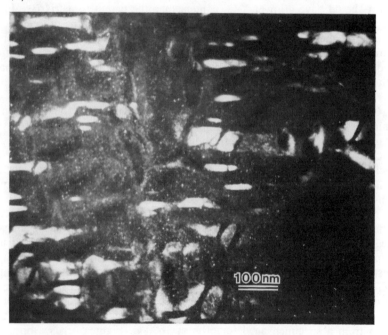

(c)

Fig. 3. Dark-field micrographs of as-received single crystal (8 wt% Y_2O_3): (a) **g** = $\bar{1}1\bar{2}$, (b) **g** = 211, and (c) **g** = $12\bar{1}$.

Microchemical EDX analysis was performed to compare the Y_2O_3 content of the colonies and the c-ZrO_2. The two spectra are compared in Fig. 6, where it is clear that the c phase is significantly richer in Y_2O_3 than are the t colonies. Quantification of these data revealed a Y_2O_3 content of 4.1 ± 0.1 and $10.6 \pm 0.1\%$ for the colonies and the c-ZrO_2, respectively, close to but not identical with the predictions of Scott's phase diagram (Fig. 1).

The twin relation of variants within the same colony is best illustrated in Fig. 7 from a specimen aged at 1600°C for 100 h; the same features as in Fig. 5 are present, and the coarsening of the tetragonal colonies has progressed further. The selected-area diffraction pattern from one colony shows that only two variants are present. The third variant, which in fact was the original t'-ZrO_2 in the starting material, is not contained in this area of foil. By imaging the colony with the two remaining $\{1\bar{1}2\}_t$ reflections, reverse contrast is observed, showing that the colony indeed contains two twin-related variants.

Discussion

This study has shown evidence of two tetragonal solid solutions, t-ZrO_2 and t'-ZrO_2. These two phases have a quite different microstructure and composition. In the as-received crystal, metastable bulky t'-ZrO_2 is present, while after annealing, the equilibrium t-ZrO_2 precipitates are in the form of colonies. The colony orientation appears to be closely related to the original orientation of the t'-ZrO_2. Within each colony, there are two twin-related variants, whose respective **c** axes

(a)

(b)

Fig. 4. Dark-field micrographs of specimen aged 24 h at 1600°C: (a) **g** = $\overline{11}2$, (b) **g** = 211, and (c) **g** = $12\overline{1}$.

are along the **a** and **b** axes of the original t'-ZrO_2. These colonies have also been observed in polycrystalline sintered ceramics with the same Y_2O_3 content[18] and in a 12 wt% Y_2O_3 polycrystal.[15] (In the former material the tetragonal colonies nucleated and grew in a cubic matrix.) The formation of colonies at the expense of the t' matrix, and in c-ZrO_2 grains in a sintered ceramic, and their stability after lengthy annealing suggest that the twin interface within the colonies has an unusually low interfacial energy and that formation of colonies in both t' and c-ZrO_2 minimize lattice strain very effectively. In fact, as the habit plane of the colony is perpendicular to the **a** axis of t'-ZrO_2 and of c-ZrO_2, the "fit" between the t colony and its matrix is very similar in both situations.

Earlier workers also pointed out that t'-ZrO_2 is nontransformable, in the sense that it does not undergo the martensitic transformation to monoclinic symmetry even under stress (e.g., during grinding). However, the t colonies formed from t'-ZrO_2 are also quite stable in that no monoclinic phase was found by X-ray diffraction in crushed specimens that had been annealed for 50 h. This is quite different from what has been observed in other partially stabilized zirconias and is not well understood at this time.

Finally, it is interesting to provide a thermodynamic/kinetic rationale for the existence of two t-ZrO_2 solid solutions in this system. We assume that (i) c-ZrO_2 does form in 8 wt% materials from the melt, (ii) during cooling following skull melting, the cooling history is such that diffusional decomposition to c- and t-ZrO_2, which should occur at $T < 2000°C$ (Fig. 1), does not occur, and (iii) the c-ZrO_2 undergoes a displacive transformation to tetragonal symmetry at some critical

(a)

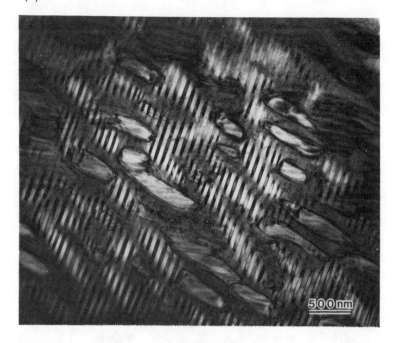

(b)

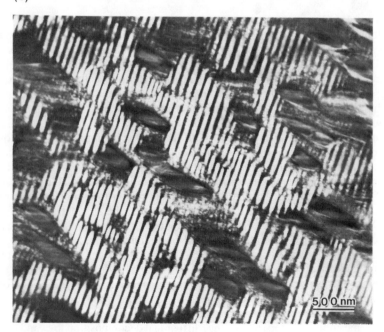

Fig. 5. Micrographs of specimen aged 50 h at 1600°C: (a) bright field and (b) dark field, **g** = $\bar{1}1\bar{2}$.

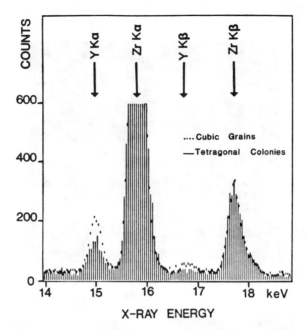

Fig. 6. Energy dispersive X-ray spectra of c matrix and t colonies.

temperature T_0. Plausible free energy vs composition curves above and below T_0 are depicted in Fig. 8. Although a two-phase assemblage has the minimum free energy for an 8 wt% material for the situation depicted in Fig. 8(a), supersaturated c-ZrO_2 solid solution will persist until t-ZrO_2 of the equilibrium composition can be nucleated. If cooling below T_0 without nucleation of t-ZrO_2 occurs (Fig. 8(b)), a displacive transformation in an 8 wt% material from supersaturated c-ZrO_2 to supersaturated t'-ZrO_2 can occur and will lower the free energy from G_g to G_t ($\Delta G'$). The solid solution t'-ZrO_2 can further lower its energy by forming t-ZrO_2 of free energy G_t and c-ZrO_2 of free energy G_c for a further energy reduction of $\Delta G''$, but this will be a sluggish transformation, as considerable diffusion is required.

Further study is under way on the mechanism of the displacive $c \rightarrow t'$ transformation and on the variation of T_0 with composition. These experiments should also assist in determining the actual boundaries of the two-phase (c-ZrO_2 + t-ZrO_2) field by using EDX microchemical analysis of the equilibrium microstructures (c-ZrO_2 and colonies of t-ZrO_2).

Acknowledgments

The authors thank the Naval Research Lab in Washington, DC, for providing the single crystals and R. F. Hehemann for informative discussions. This research was supported by AFOSR under Contract 82-0227. A. H. Heuer also acknowledges the Alexander von Humboldt Foundation for a Senior Scientist Award, which made possible his sabbatical leave at the Max-Planck-Institut Für Metallforschung in Stuttgart, Federal Republic of Germany, where this paper was written.

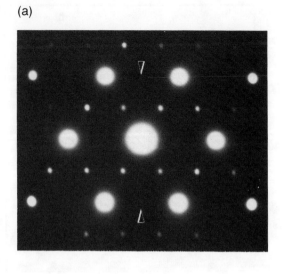

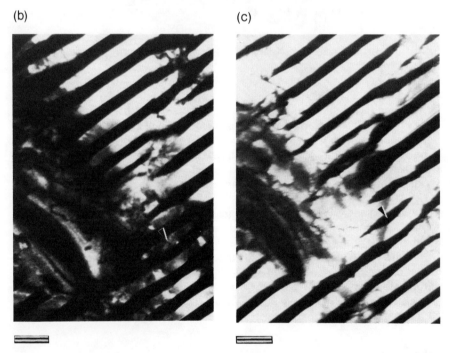

Fig. 7. Specimen aged 100 h at 1600°C: (a) SAD pattern, [$\bar{1}11$] zone axis from one colony, (b) dark-field micrograph, **g** = $\bar{1}1\bar{2}$, and (c) dark-field micrograph, **g** = 211. The arrows in (b) and (c) point to the same feature. Bar = 500 nm.

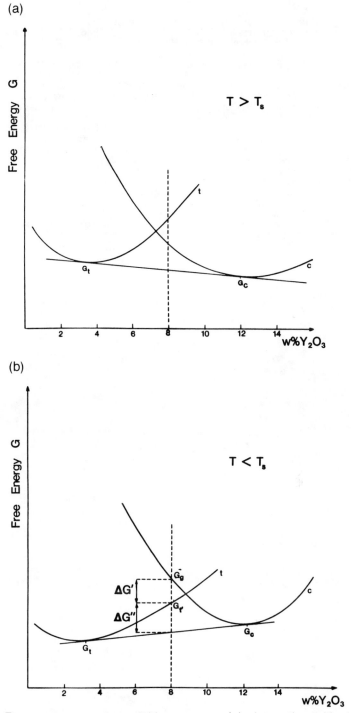

Fig. 8. Free energy vs composition curves: (a) above the critical temperature T_0 and (b) below the critical temperature T_0.

References
[1] R. H. J. Hannink, *J. Mater. Sci.*, **13**, 2487–97 (1978).
[2] D. L. Porter and A. H. Heuer, *J. Am. Ceram. Soc.*, **62** [5–6] 298–305 (1979).
[3] J. M. Marder, T. E. Mitchell, and A. H. Heuer, *Acta Metall.*, **31** [3] 387–95 (1983).
[4] H. J. Scott, *J. Mater. Sci.*, **10** [9] 1527–35 (1975).
[5] V. S. Stubican, R. C. Hink, and S. P. Ray, *J. Am. Ceram. Soc.*, **61** [1–2] 17–21 (1978).
[6] C. Pascual and P. Duran, *J. Am. Ceram. Soc.*, **66** [1] 23–27 (1983).
[7] R. B. Miller, J. L. Smialek, and R. G. Garlick; pp. 241–53 in Advances in Ceramics, Vol. 3. Edited by L. W. Hobbs and A. H. Heuer. The American Ceramic Society, Columbus, OH, 1981.
[8] C. A. Anderson and T. K. Gupta; pp. 184–201 in Advances in Ceramics, Vol. 3. Edited by L. W. Hobbs and A. H. Heuer. The American Ceramic Society, Columbus, OH, 1981.
[9] R. J. Bratton and S. L. Lau; pp. 226–40 in Advances in Ceramics, Vol. 3. Edited by L. W. Hobbs and A. H. Heuer. The American Ceramic Society, Columbus, OH, 1981.
[10] R. P. Ingel; Ph. D. Thesis, The Catholic University of America, Washington, DC (1982).
[11] G. Teufer, *Acta Crystallogr.*, **15** [11] 1187 (1962).
[12] J. Lefevre, *Ann. Chim.*, **8**, 117–49 (1963).
[13] L. H. Schoenlein; Ph. D. Thesis, Case Western Reserve University, Cleveland, OH, 1982.
[14] A. H. Heuer and M. Rühle; this volume, pp. 1–13 .
[15] R. Chaim, A. H. Heuer, and M. Rühle; unpublished results.
[16] B. A. Bender and D. Lewis; private communication.
[17] D. Michel, L. Mazerolles, M. Perex, and Y. Jorba; this volume, pp. 131–38.
[18] P. G. Valentine; M.S. Thesis, Case Western Reserve University, Cleveland, OH, 1982.

Polydomain Crystals of Single-Phase Tetragonal ZrO$_2$: Structure, Microstructure, and Fracture Toughness

D. MICHEL, L. MAZEROLLES, AND M. PEREZ Y JORBA

C.N.R.S.-L.A. 302
Vitry, France

Crystals (5 by 5 by 15 mm) of single-phase tetragonal ZrO$_2$ were prepared by skull-melting at the composition ZrO$_2$-3 mol% M$_2$O$_3$ (M = Y, Yb, Gd). Crystals displayed a domain microstructure induced by the cubic $(c) \rightarrow (t)$ phase transition at about 2000°C (domain size $\approx$ 50 nm). Fracture of tetragonal ZrO$_2$ crystals occurred differently than for c-ZrO$_2$ (ZrO$_2$-9 mol% Y$_2$O$_3$). Fracture toughness was much higher for specimens and cleavage planes were not the same as for fluorite crystals. Studies on fracture surfaces indicated that stress-induced transformation was not involved in the toughening mechanism, but that crack deflections along {001} planes could account for the observed toughening.

Zirconia displays three polymorphic varieties: monoclinic (m) form ($P2_1/c$, $Z = 4$) at low temperature, a tetragonal (t) phase ($P4_2/nmc, Z = 2$) stable between 1100° and 2400°C, and a cubic (c) phase ($Fm3m, Z = 4$) at higher temperatures.

For pure ZrO$_2$, the transition from t to m symmetry is diffusionless (martensitic type) and the t phase cannot be retained at room temperature even by quenching. The addition of a stabilizing oxide lowers the transition temperatures $c \rightleftharpoons t \rightleftharpoons m$ and favors the retention of t- and c-ZrO$_2$ at room temperature.

Single-phase t samples can be prepared only in well-defined preparative conditions.[1-8] Lefevre[2] determined first that the best stabilization of this phase is obtained with M$_2$O$_3$ oxides (M = Sc, Y, or lanthanides) in a narrow range of composition (2–6 mol%) by quenching from high temperature ($T > 2000$°C).

This paper deals with structural features and fracture behavior of t-ZrO$_2$ crystals (3 mol% M$_2$O$_3$) grown by skull-melting.

According to Rouanet[9] (Fig. 1), the temperature for the lower limit of the cubic stability range at the considered composition is about 2000°C.

Samples were grown from the melt as single crystals of the cubic high-temperature phase. During the cooling process, the crystals underwent the $c \rightarrow t$ phase transition. This transformation induced a domain microstructure which was studied by transmission electron microscopy (TEM). The microtwinned nature of these samples was related to the particular fracture aspect observed by scanning electron microscopy (SEM).

The t polydomain crystals were much tougher than crystals of fully stabilized ZrO$_2$ (9 mol% M$_2$O$_3$) with a c structure at all temperatures.

A model is proposed and discussed to explain the important enhancement in toughness observed for polydomain samples.

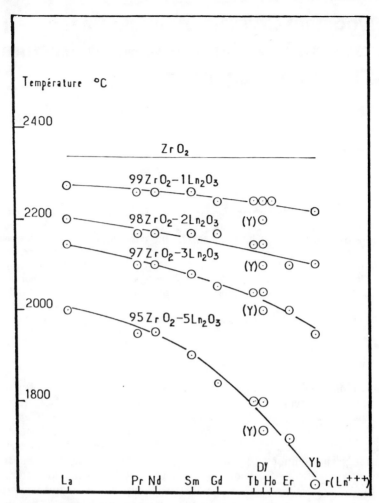

Fig. 1. Temperatures for the lower limit of the cubic stability range in the systems ZrO_2-M_2O_3 (after Rouanet, Ref. 9).

Experimental Details

Preparation of t-ZrO_2 Crystals

Starting materials were high-purity ZrO_2 (99.9 P.C.U.K.) and Y_2O_3, Gd_2O_3, or Yb_2O_3 (Rhone Poulenc 3N).

In a previous paper on the crystal growth in the systems ZrO_2-Ln_2O_3,[10] we reported the synthesis of t-ZrO_2 crystals using the skull-melting technique developed in our laboratory.[11,12]

The mean size of crystals grown from a 500 g melt was about 15 by 5 by 5 mm. At the composition 3 mol% M_2O_3 (M = Y, Gd, Yb) samples consisted only of the t phase when the directional solidification of ingots was performed with sufficiently rapid cooling rates (>400°C/h).

As-grown crystals were free of m-ZrO_2 as checked by X-ray powder diagrams on ground crystals.

Electron Microscopy

Transmission electron microscopy experiments were performed on a JEOL 200 CX equipped for high-resolution observation. Platelets oriented along (001) and (110) planes were prepared from crystals. Thinning was achieved by a mechanical polishing followed by argon-ion bombardment.

Fracture Study

Cracks were introduced into polished oriented specimens, around Vickers indents, using a Leitz "Microdurimet" with loads ranging from 50 to 2000 g.

The lengths of indents and cracks were measured for different orientations. Stress intensity factors (K_c) were calculated from these values using the calibration curve proposed by Evans and Charles.[13]

On fractured crystals, cleavage orientations were determined from X-ray diffraction (Laue and Debye-Scherrer diagrams) and the fracture surfaces were examined by SEM.

Results

Structure and Microstructure of t-ZrO₂ Crystals

Unit-cell parameters of t-ZrO_2 samples containing 3 mol% M_2O_3 are given in Table I. To allow comparison with the fluorite prototype, it is convenient to refer to a C-centered (pseudocubic) t cell with $a' = a \cdot 2^{1/2}$.

The structure of ZrO_2 (3 mol% M_2O_3) crystals can be derived from the c fluorite type with: (1) a lattice distortion of about 1% (see values of c/a' in Table I) and (2) shifts of oxygen atoms from ideal fluorite positions (0.026 nm for the Y_2O_3 stabilized t phase).[14]

The $c \rightarrow t$ transition defines three possible orientations for the 4-fold axis. In addition, equivalent lattice points in the fcc structure are nonequivalent in the t structure. This leads to six equiprobable orientation-translation states from one c single crystal.[14]

Consequently, t crystals which have undergone this transformation have to exhibit a domain structure ruled by the previous symmetry considerations.

Observations by TEM actually revealed a microdomain structure in the samples. The orientation of variants was determined by dark-field studies with suitable diffraction spots. For instance, Fig. 2 shows the micrograph, for a {001} platelet, obtained with a fluorite-forbidden diffraction spot. In these conditions, only domains with their 4-fold axis normal to the observation plane are brightened. Antiphase boundaries appear as lines between domains having the same contrast for the same orientation. These lines have opposite contrast to that of the domains. A high-resolution study[15] indicated that antiphase boundaries follow strictly {001} fluorite planes. The domains have an elongated morphology with {001} habit planes. Their mean size is about 500 nm in crystals obtained with our experimental conditions.

Table I. Unit-Cell Parameters (in nm) for the Studied Tetragonal Zirconia Crystals ZrO₂-3 mol% M₂O₃

Stabilizing element	a	a' = a·1/2	c	c/a
Yb	0.3605	0.5098	0.5166	1.013
Y	0.3610	0.5105	0.5168	1.012
Gd	0.3614	0.5111	0.5174	1.012

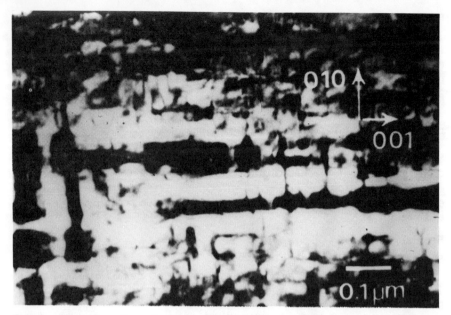

Fig. 2. Domain microstructure of a tetragonal zirconia crystal (3 mol% Y_2O_3). Observation by TEM on oriented {001} platelet (indexing relative to the fluorite lattice). Dark field **g** = 110.

Fracture Study of c and t Crystals

The fracture behavior of t-ZrO_2 crystals (3 mol% M_2O_3) was compared to that of fully stabilized c crystals (ZrO_2-9 mol% Y_2O_3) prepared by the same technique.

Fracture Toughness: For the same crystallographic orientation, the initiation of cracks by Vickers indentation on t crystals required loads at least ten times higher than for c crystals. The t crystals were found to be much tougher than c ones. From the ratio of the crack length to the indent size, the following values of stress intensity factors were estimated: $K_c = 1.8$ MN·m$^{-3/2}$ for the c phase with 9 mol% Y_2O_3, and $K_c = 6$ MN·m$^{-3/2}$ for the t phase with 3 mol% Y_2O_3.

Figure 3 illustrates the significant difference for the initiation and propagation of cracks on (001) planes in the two materials.

Fractographic Study: For both their crystallographic orientations and their aspects, fracture surfaces were basically different for the two kinds of crystals.

The fracture of c fully stabilized ZrO_2 crystals occurred along {111} planes by the usual cleavage of crystals with a fluorite structure. On the contrary, for t polydomain crystals, the surface orientation corresponded to {001} fluorite planes. X-ray diffraction diagrams on fracture surfaces consisted only of lines derived from 002 and 004 fluorite reflections (Fig. 4).

Very rough aspects were displayed on the fracture surfaces, as shown in Fig. 5. Microkinks of about 50–100 nm were formed by fracture along directions of the three cubic {001} planes.

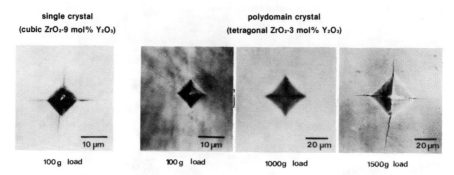

Fig. 3. Vickers indents on cubic and tetragonal zirconia crystals. Orientation of surfaces {001}.

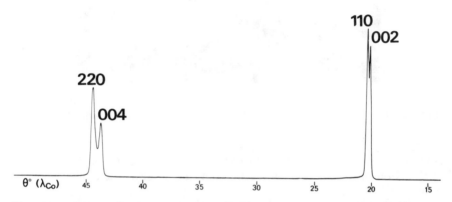

Fig. 4. X-ray diffraction diagram obtained on the fracture surface of a tetragonal zirconia crystal (3 mol% Y_2O_3). Cobalt target (indexing with the primitive tetragonal cell).

Discussion

Partially stabilized ZrO_2 (PSZ) is well known to display high strength and high fracture toughness.[16-18] Ceramics prepared from sintering processes generally consist of two (or three) phases with tetragonal zirconia as a major constituent.

Toughening in these materials has been assigned to stress-induced transformation.[19-24] Different workers have demonstrated that the metastable tetragonal phase can be transformed into the monoclinic stable form near cracks. Zones of monoclinic zirconia were actually revealed by different techniques (TEM, X-ray diffraction, and Raman spectroscopy) around cracks or on fracture surfaces.[25-28]

In this study, tetragonal ZrO_2 samples were prepared from the melt as single crystals without monoclinic or cubic phases. Similar to PSZ, these materials exhibit a high fracture toughness. The enhancement with respect to the properties of fully stabilized zirconia cannot be explained in this case by stress-induced transformation. Contrary to PSZ, the monoclinic phase is not easily obtained by

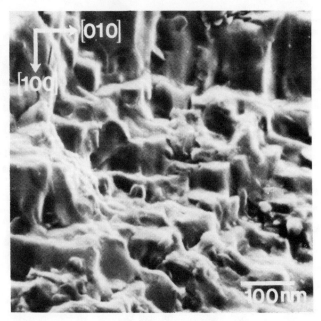

Fig. 5. Scanning electron micrograph of a fracture surface of a tetragonal zirconia crystal (3 mol% Y_2O_3) (indexing relative to the fluorite lattice).

polishing or milling treatments on these materials. Analyses on surface fractures (TEM, X-ray diffraction, Raman spectroscopy) did not reveal detectable amounts of monoclinic ZrO_2.[14]

The fractographic studies show that toughening is induced by the domain microstructure of samples, which gives rise to crack deflections at a submicrometric scale.

This observed fracture mode along well-defined crystallographic planes can be explained in two ways: (1) internal stress fields associated with the t lattice distortion which can induce crack deflection at domain boundaries, or (2) preferential cleavage along {001} planes. This cleavage is the only one which does not affect the stronger metal–oxygen bonds of the tetragonal structure. As shown on Fig. 6, only weaker Zr (or Y)–O bonds are broken by a cleavage perpendicular to the 4-fold axis.

Changes in direction of crack propagation would correspond in this model to changes in the preferred fracture plane from one domain to another with a different orientation of the 4-fold axis.

Acknowledgments

The authors wish to thank A. Kahn (ENSCP, Paris) for structure refinement, C. Haut (Laboratoire de Métallurgie, Orsay) for SEM observations, J. Rzepski and R. Portier (CECM, Vitry) for TEM observations, and J. P. Dallas for crystal elaboration.

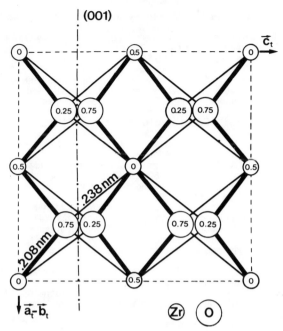

Fig. 6. Proposed cleavage model for tetragonal zirconia crystals. $(110)_t$ projection of the structure; the dashed line represents cleavage along $(001)_t$.

References

[1]J. Lefevre, R. Collongues, and M. Perez y Jorba, "On the Cubic → Tetragonal Transition in Systems Zirconia-Rare Earth Oxides," *C. R. Acad. Sci.*, **249**, 2329–31 (1959).
[2]J. Lefevre, "Structural Modifications of Fluorite-Related Phases Involving ZrO_2 or HfO_2," (in Fr.), *Ann. Chim.*, **8**, 117–49 (1963).
[3]H. G. Scott, "Phase Relationships in the Zirconia-Yttria System," *J. Mater. Sci.*, **10**, 1527–35 (1975).
[4]T. K. Gupta, J. H. Bechtold, R. C. Kuznicki, L. H. Cadoff, and B. R. Rossing, "Stabilization of Tetragonal Phase in Polycrystalline Zirconia," *J. Mater. Sci.*, **12**, 2421–26 (1977).
[5]T. K. Gupta, F. F. Lange and J. H. Bechtold, "Effect of Stress-Induced Phase Transformation on the Properties of Polycrystalline Zirconia Containing Metastable Tetragonal Phase," *J. Mater. Sci.*, **13**, 1464–70 (1978).
[6]C. A. Andersson and T. K. Gupta, "Phase Stability and Transformation Toughening in Zirconia"; pp. 184–201 in Advances in Ceramics 3. Edited by A. H. Heuer and L. W. Hobbs. The American Ceramic Society, Columbus, OH, 1981.
[7]F. F. Lange and D. J. Green, "Effect of Inclusion Size on the Retention of Tetragonal ZrO_2: Theory and Experiments"; pp. 217–25 in Advances in Ceramics 3. Edited by A. H. Heuer and L. W. Hobbs. The American Ceramic Society, Columbus, OH, 1981.
[8]F. F. Lange, "Transformation Toughening," *J. Mater. Sci.*, **17**, 225–45 (1982).
[9]A. Rouanet, "High Temperature Phase Diagrams in Systems ZrO_2-Ln_2O_3 (Ln = lanthanide)" (in Fr.) *Rev. Int. Hautes Temp. Refract.*, **8**, 161–80 (1971).
[10]D. Michel, M. Perez y Jorba, and R. Collongues, "Growth from Skull-Melting of Zirconia-Rare Earth Oxide Crystals," *J. Cryst. Growth*, **43**, 546–48 (1978).
[11]D. Michel, M. Perez y Jorba, and R. Collongues, "Growth of Stabilized Zirconia Crystals and Properties of Cubic Phases ZrO_2-CaO" (in Fr.), *C. R. Acad. Sci.*, **C266**, 1602–1604 (1968).
[12]A. M. Anthony and R. Collongues, "Modern Methods of Growing Single Crystals of High Melting-Point Oxides"; pp. 147–249 in Preparative Methods in Solid State Chemistry. Edited by P. Hagenmuller. Academic Press, New York, 1972.
[13]A. G. Evans and E. A. Charles, "Fracture Toughness Determination by Indentation," *J. Am. Ceram. Soc.*, **59**, 371–72 (1976).

[14] D. Michel, L. Mazerolles, and M. Perez y Jorba, "Fracture of Metastable Tetragonal Zirconia Crystals," *J. Mater. Sci.*, **18**, 2618–28 (1983).

[15] D. Michel, L. Mazerolles, and R. Portier, "Electron Microscopy Observation of the Domain Boundaries Generated by the Cubic → Tetragonal Transition of Stabilized Zirconia"; pp. 809–12 in Solid State Chemistry 1982, Studies in Inorganic Chemistry, Vol. 3. Edited by R. Metselaar, H. J. M. Heijligers, and J. Schoonman, Elsevier, New York, 1983.

[16] R. C. Garvie and P. S. Nicholson, "Structure and Thermomechanical Properties of Partially Stabilized Zirconia in the $CaO\text{-}ZrO_2$ System," *J. Am. Ceram. Soc.*, **55**, 152–57 (1972).

[17] R. C. Garvie, R. H. Hannink, and R. T. Pascoe, "Ceramic Steel," *Nature (London)*, **258**, 703–704 (1975).

[18] Advances in Ceramics 3. Edited by A. H. Heuer and L. W. Hobbs. The American Ceramic Society, Columbus, OH, 1981.

[19] D. L. Porter and A. H. Heuer, "Mechanisms of Toughening Partially Stabilized Zirconia (PSZ)," *J. Am. Ceram. Soc.*, **60**, 183–84 (1977).

[20] D. L. Porter and A. H. Heuer, "Microstructural Development in MgO-Partially Stabilized Zirconia (Mg-PSZ)," *J. Am. Ceram. Soc.*, **62**, 298–305 (1979).

[21] D. L. Porter, A. G. Evans, and A. H. Heuer, "Transformation-Toughening in Partially Stabilized Zirconia (PSZ)," *Acta Metall.*, **27**, 1649–54 (1979).

[22] A. G. Evans, N. Burlingame, M. Drory, and W. M. Kriven, "Martensitic Transformations in Zirconia-Particle Size Effects and Toughening," *Acta Metall.*, **29**, 447–56 (1981).

[23] A. G. Evans and A. H. Heuer, "Review — Transformation Toughening in Ceramics: Martensitic Transformations in Crack-Tip Stress Fields," *J. Am. Ceram. Soc.*, **63**, 241–48 (1980).

[24] N. Claussen, "Stress-Induced Transformation of Tetragonal ZrO_2 Particles in Ceramic Matrices," *J. Am. Ceram. Soc.*, **61**, 85–86 (1978).

[25] T. Kosmac, R. Wagner, and N. Claussen, "X-Ray Determination of Transformation Depths in Ceramics Containing Tetragonal ZrO_2," *J. Am. Ceram. Soc.*, **64**, C-72–C-73 (1981).

[26] N. Claussen and M. Rühle, "Design of Transformation-Toughened Materials"; pp. 137–63 in Advances in Ceramics 3. Edited by A. H. Heuer and L. W. Hobbs. The American Ceramic Society, Columbus, OH, 1981.

[27] R. C. Garvie, R. H. Hannink, and M. V. Swain, "X-Ray Analysis of the Transformed Zone in Partially Stabilized Zirconia (PSZ)," *J. Mater. Sci. Lett.*, **1**, 437–40 (1982).

[28] D. R. Clarke and F. Adar, "Measurement of the Crystallographically Transformed Zone Produced by Fracture in Ceramics Containing Tetragonal Zirconia," *J. Am. Ceram. Soc.*, **65**, 284–88 (1982).

The Phase $Mg_2Zr_5O_{12}$ in MgO Partially Stabilized Zirconia

H. J. ROSSELL AND R. H. J. HANNINK

CSIRO, Division of Materials Science
Advanced Materials Laboratory
Melbourne, Victoria, Australia 3001

Evidence is presented to show that the cubic phase in Mg-PSZ which has been aged at 1100°C contains microdomains of the fluorite-related superstructure $Mg_2Zr_5O_{12}$ which are 3–7 nm in diameter or greater and oriented in all the ways that preserve a common substructure with the matrix. Some aspects of the formation of these microdomains and their influence on the microstructural development of this material are discussed.

It is now well recognized that ZrO_2 partially stabilized (PSZ) with oxides such as CaO, MgO, and Y_2O_3 can be subjected to thermal treatment so that the thermomechanical properties of the final product are greatly enhanced.[1] Such materials generally consist of a cubic (*c*) stabilized ZrO_2 (CSZ) matrix containing a dispersion of precipitates which are predominantly tetragonal (*t*) ZrO_2. The PSZ materials are sintered and solution-treated at temperatures high enough for them to be in the cubic single-phase field and then cooled at a sufficiently rapid rate so that the precipitated zirconia is in the tetragonal form at room temperature. They then are aged under suitable conditions to allow the tetragonal precipitates to grow to such a size that they will transform, at room temperature, to the monoclinic (*m*) form in the presence of an applied tensile stress field. The volume expansion and shear strains associated with this transformation impart the enhanced thermomechanical properties to these PSZ materials, which are termed transformation-toughened ZrO_2 (TTZ).

Until recently, it was thought that Mg-PSZ materials could not be aged below 1400°C, because at that temperature the cubic solid solution phase begins to decompose eutectoidally to the component oxides,[2] which results in thermal stress cracking when the body is cooled from the aging temperature. However, it has been shown that suitably prefired and cooled Mg-PSZ materials can acquire enhanced thermomechanical properties after aging treatments at 1100°C.[3,4] Such materials not only possess mechanical properties superior to those of Mg-PSZ aged above 1400°C but also exhibit *R*-curve behavior,[5] i.e., a rising resistance to cracking as cracks propagate.

It has been proposed[6] that the microstructural change indirectly responsible for these enhanced mechanical properties is the formation of small regions or microdomains within the cubic phase in which the anion vacancies are ordered. Detailed consideration of the geometry and intensities of electron diffraction patterns revealed that these microdomains were composed of $Mg_2Zr_5O_{12}$, a compound possessing the fluorite-related superstructure known as δ. However the published report of this work was only in abstract form,[7] while in a subsequent verification[8] based on geometrical diffraction data only, a differing interpretation for the origin of the diffuse scattering shown by these materials was presented. This paper therefore presents the original confirmatory work on the presence of δ phase in Mg-PSZ and Mg-CSZ more fully.

Experimental Procedure

Mg-PSZ containing 9 mol% MgO was prepared from mixed powders of the component oxides. These were sintered and solution-treated at 1700°C, cooled at 275°C/h to 1000°C, and then at 70°C/h to room temperature. The samples were reheated to 1100°C in air, aged for various times, usually 0.5–2 h, and then furnace-cooled. ZrO_2 containing 14 mol% MgO, here designated Mg-CSZ, was prepared in a similar manner.

Samples for electron microscopy were mechanically polished,[9] thinned by Ar-ion milling, and examined at 200 kV in a JEM 200B electron microscope fitted with a tilting stage.

A specimen of ZrO_2–28.6 mol% MgO was melted in an argon-arc furnace, quenched on the water-cooled hearth, and reoxidized by heating in air at 900°C for 0.5 h. Crystal fragments from the crushed specimen were dispersed on carbon-coated grids for electron microscopy.

Electron Optical Observations

The electron diffraction patterns from apparent single crystals of Mg-PSZ all contain very strong reflections due primarily to the face-centered c-ZrO_2 solid solution phase (Fig. 1). They also may contain less strong "half-order" reflections due to the presence of t-ZrO_2 (Fig. 1(a)–(d), crystallites of which have grown coherently in the cubic matrix crystal, or closely spaced groups of spots in these positions due to similarly aligned semicoherent monoclinic zirconia precipitates (Fig. 1(e)). The very strong reflections also contain contributions from the tetragonal or monoclinic precipitates. Confirmation of this interpretation is provided from dark-field images formed from those diffracted beams arising specifically from these ZrO_2 precipitates[6] (Fig. 2(a)).

In addition to the above, the diffraction patterns exhibit either faint spots which do not lie on a simple lattice (Fig. 1(c)–(e)) or diffuse features (Fig. 1(a)) which occupy much the same regions as the faint spots. The two effects are related, since patterns suggestive of an intermediate stage may be observed (Fig. 1(b)).

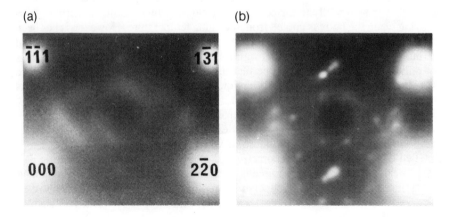

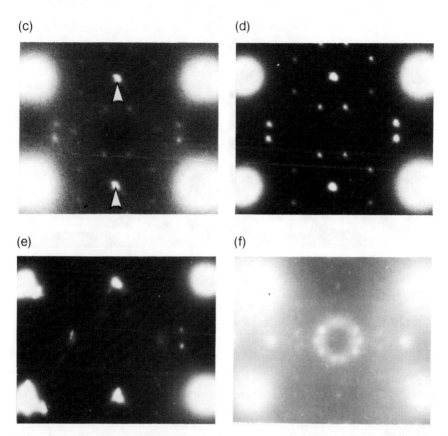

Fig. 1. Typical electron diffraction patterns from apparent single crystals of the cubic phase in the system MgO-ZrO_2. The patterns illustrated are from the [112] orientation. (a) Pattern from 14 mol% Mg-CSZ. The strong indexed reflections are those expected from an ideal fluorite structure and correspond to those in the patterns (b)–(e). The diffuse features are due to extra ordering effects. (b) Mg-PSZ aged at 1100°C—diffuse features as in (a) appear, but some sharpening into discrete spots has occurred. (c) Mg-PSZ aged 0.5 h at 1100°C—arrowed reflections are those expected, in addition to the strong ones, for the t-ZrO_2 structure. The diffuse features of (a) have been replaced by faint but sharp spots. (d) As in (c), but aging time is 2 h. Growth of oriented crystallites of the t-ZrO_2 phase and the extra ordered regions is indicated by the increase in intensity and sharpness of the arrowed spots and of the extra faint spots. (e) As in (c), but aging time is 16 h. The spots from the tetragonal phase have become split into closely spaced groups, indicating that semicoherent crystallites of the m-ZrO_2 phase have formed. (f) c-ZrO_2-20 mol% CaO aged so that the diffuse intensity characteristic of this phase is becoming sharpened into spots. This pattern may be compared to (a) and (b): The spot placements and the proportions of the diffuse features (e.g., the diameter of the diffuse central ring) are different. Hence, the structure giving rise to these diffraction features in CaO-ZrO_2 must be different from that in MgO-ZrO_2.

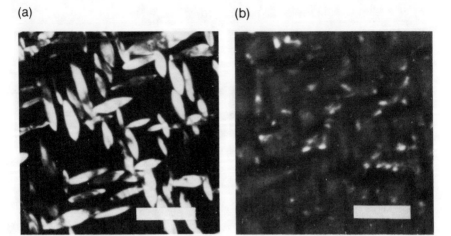

Fig. 2. (a) Dark-field image formed with a t-phase reflection from as-fired (not aged) 9 mol% Mg-PSZ, depicting the form of the coherent tetragonal precipitates. (b) Dark-field image formed with a group of faint sharp spots from a material as in Fig. 1(c), showing that the extra ordering is confined to small regions of the matrix adjacent the tetragonal precipitates. These regions are ≈10 times larger in linear dimension in materials such as in Fig. 1(d). The length bars represent 0.2 μm.

Dark-field images formed with beams corresponding to the faint spots show that they arise from regions ≈10 nm in diameter in the cubic matrix crystal immediately adjacent the boundary of a tetragonal or monoclinic precipitate (Fig. 2(b)). The widths of the diffuse features suggest that they arise from regions of ≈3 nm in diameter within the bulk of the cubic phase. No localization could be seen in dark-field images because of the overlap of the features of interest in the projection of the crystal, which may have been 50–100 nm thick. Neither of these last two electron diffraction effects could be seen in Guinier powder X-ray diffraction patterns, which, even after massive exposures, showed only the reflections expected for the cubic phase with admixed tetragonal or monoclinic phase. When sequences of diffraction patterns such as Figs. 1(a)–(d) are considered together with the direct observation that the discrete-ordered regions will grow on aging, it is reasonable to conclude that the extra diffraction effects arise from ordered regions that have the same structure, and that the diffraction spots are broadened into diffuse bands when the regions are very small.

Structure of the Cubic Phase in MgO-ZrO$_2$

Preamble

The electron diffraction patterns from this phase resemble those obtained from the cubic phase in the systems ZrO$_2$-CaO and HfO$_2$-CaO, and therefore a similar description of the structure or constitution of the cubic material[10] is implied, viz., that the solute cations and formal anion vacancies are not distributed randomly on the cation and anion sites of the fluorite structure, but instead there exist small discrete regions of the crystal where their distribution is ordered. In the case of the Zr(Hf)O$_2$-CaO cubic phase, the ordering scheme within the small regions is the

Table I. Unit Cell Data for $Mg_2Zr_5O_{12}$*

Rhombohedral, $R\bar{3}$	$r=0.6183(1)$ nm	$\alpha=99.592(5)°$
Hexagonal representation:	$a=0.9445(1)$ nm	$c=0.8745(2)$ nm
Axial relation to fluorite subcell:		
$r_1=(1/2)[12\bar{1}]$	$r_2=(1/2)[\bar{1}12]$	$r_3=(1/2)[2\bar{1}1]$
$a_{hex}=(1/2)[21\bar{3}]$	$b_{hex}=(1/2)[\bar{3}21]$	$c_{hex}=[111]$

*The rhombohedral supercell contains 1.75 fluorite M_4O_{8-x} subcells.

same as that of the fluorite-related superstructure phase ϕ_1, $CaZr(Hf)_4O_9$.[11] However, despite the resemblances in electron diffraction patterns, that particular ordering scheme cannot apply in the present case, since the geometric proportions of the relevant sharp or diffuse diffraction features are different from those of the CaO-ZrO_2 specimens (Fig. 1(f)). Moreover, a macroscopic phase $MgZr_4O_9$ of ϕ_1 structure has never been reported.

The only fluorite-related ordering scheme or superstructure reported in the system ZrO_2-MgO is that of the phase $Mg_2Zr_5O_{12}$.[12] This phase appears in material of ≈ 28 mol% MgO which has been rapidly cooled from the melt or from temperatures above $\approx 1900°C$ and is sufficiently well formed as to yield powder X-ray diffraction or single-crystal electron diffraction patterns. Such patterns show strong reflections due to the fluorite-type substructure and fainter reflections due to the superstructure ordering. Unit cell data are collected in Table I. The lattice parameters were determined from a Guinier photograph of $Mg_2Zr_5O_{12}$ that had been quenched from the melt. This unit cell is identical to that exhibited by a number of oxides M_7O_{12}[13] which possess the fluorite-related superstructure known as δ.[14,15] The superstructure $Mg_2Zr_5O_{12}$ does not survive annealing at temperatures below $\approx 1500°C$ and, since it forms rapidly, it is speculated that the formal anion vacancies only are ordered in the structure.[13] The corresponding phase $Mg_2Hf_5O_{12}$ appears to be more stable and has been well characterized.[16]

The electron-diffraction pattern from ordered $Mg_2Zr_5O_{12}$ in the [100] orientation is shown in Fig. 3(a). The fluorite-based indexes of the subcell pattern are [$12\bar{1}$]. All the supercell spots in this pattern occur as a subset of the faint spots in the [112] pattern from the nominally cubic phase in Mg-PSZ (Fig. 3(b)), an observation which supports the idea that δ-phase ordering is feasible in the latter material.

Structure Model

It is proposed that within a single crystal of the cubic MgO-ZrO_2 solid solution phase, microdomains of δ structure ($Mg_2Zr_5O_{12}$) are formed in such a way that the basic fluorite substructure remains continuous throughout. Within the microdomains, there may be some relaxation of atomic positions as a consequence of the ordering (Fig. 4). Under these conditions, the unit cell of the δ phase can exist in eight orientations for a given subcell orientation (Table II). It is proposed that, in a single crystal of cubic material, microdomains of δ phase have formed in all these orientations in approximately equal numbers. Thus any "single-crystal" electron diffraction pattern will be an equally weighted superposition of eight different patterns of δ phase, each of which has a subcell pattern essentially coincident with the diffraction pattern of the matrix. This model for the structure is tested by calculation of the electron diffraction patterns it would give for various microdomain sizes and comparison of these calculations with the corresponding experi-

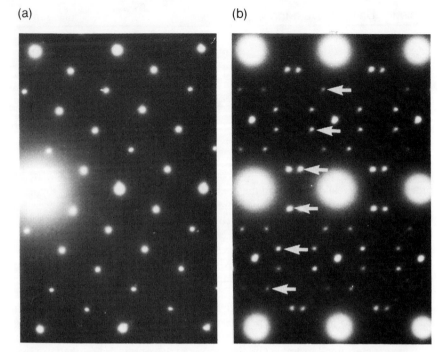

Fig. 3. (a) Single-crystal electron diffraction pattern from specially prepared ordered $Mg_2Zr_5O_{12}$ in the $[100]_{supercell}$ ($=[12\bar{1}]_{subcell}$) orientation. (b) $[112]_c$ electron diffraction pattern from Mg-PSZ aged at 1100°C for 2 h. The reflections of (a) are part of this pattern: one row of such spots is arrowed.

mental patterns. In these calculations it is assumed that the δ-phase substructure has the same cubic basis as the matrix material between the microdomains so that the strong reflections from both structures will coincide exactly.

Calculation of Electron Diffraction Patterns

Since the microdomains are small, the intensity distribution associated with their reciprocal lattice points will be extended; such a distribution could be intersected by the Ewald sphere if the reciprocal lattice point lies close to it and some intensity could be registered in the resultant diffraction pattern. In the calculations, the Ewald sphere was assumed to be planar and coincident with the required reciprocal lattice plane (rel plane) of the cubic basis. For all eight δ-phase orientations, reciprocal lattice points (rlp's) at distances ≤ 1.0 nm^{-1} (≤ 0.1 Å^{-1}) from the required rel plane and within the bounds of the observed zero-order Laue zone were collected. Their kinematic intensities were calculated by assuming that the cations are disordered in $Mg_2Zr_5O_{12}$ and using atomic coordinates determined for the isostructural $Zr_3Sc_4O_{12}$.[13] The kinematic approximation was considered valid in view of the small volumes of the δ-phase microdomains and the expectation that dynamic effects would predominate in the subcell reflections only. The resultant diffraction intensities at the Ewald surface were calculated by assuming spherical microdomains associated with a Gaussian distribution of intensity in reciprocal space $I = I_0 \exp(-\pi d^2 r^2)$, where r is the distance from the rlp and d the micro-

(a)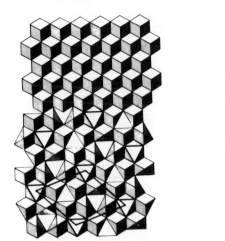

(b)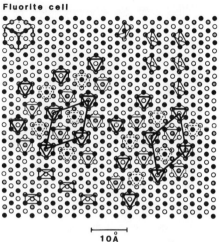

Fluorite cell

10 Å

Fig. 4. (a) Schematic derivation of the δ-phase superstructure ordering from the basis fluorite structure, which is depicted here as edge-sharing MO_8 cubes in the [111] orientation. In the δ phase, formal anion vacancies occur in pairs across the body diagonal of one-seventh of the parent MO_8 cubes, producing an equivalent number of MO_6 octahedra, which are ordered as shown. As a result, all remaining cubes have one vertex missing. The basal plane of the hexagonal supercell is outlined. (b) A [111] projection of the fluorite basis structure, showing schematically the formation of microdomains of δ-phase structure in several possible orientations. The fluorite unit cell is outlined. Cations appear at three levels; anions are above and below each cation.

Table II. Possible Orientations of the M_7O_{12} Supercell in a Given Subcell*

No.	Rhombohedral			Hexagonal			No.	Rhombohedral			Hexagonal		
1	1/2	−1	−1/2	3/2	−1/2	−1	5	−1/2	1	−1/2	−3/2	1/2	−1
	−1	−1/2	1/2	−1/2	−1	3/2		1	1/2	1/2	1/2	1	3/2
	−1/2	1/2	−1	−1	−1	−1		1/2	−1/2	1	1	1	−1
2	1/2	1	−1/2	1	1/2	−3/2	6	−1/2	−1	−1/2	−1	−1/2	−3/2
	−1/2	1/2	1	−3/2	1	1/2		1/2	−1/2	1	3/2	−1	1/2
	1	−1/2	1/2	1	1	1		−1	1/2	1/2	−1	−1	1
3	−1	−1/2	−1/2	−1/2	−3/2	−1	7	1	1/2	−1/2	1/2	3/2	−1
	−1/2	1	1/2	−1	1/2	3/2		1/2	−1	1/2	1	−1/2	3/2
	1/2	1/2	−1	−1	1	−1		−1/2	−1/2	−1	1	−1	−1
4	1	−1/2	−1/2	1/2	−1	−3/2	8	−1	1/2	−1/2	−1/2	1	−3/2
	1/2	1/2	1	1	3/2	1/2		−1/2	−1/2	1	−1	−3/2	1/2
	−1/2	−1	1/2	1	−1	1		1/2	1/2	1/2	−1	1	1

*Supercell axes are given, in terms of the subcell axes, for the obverse settings that correspond to the atomic coordinates (Ref. 13).

domain diameter (Laue's approximation). Multiple-scattering processes, whereby a diffracted beam acts as a source for further diffraction, were simulated by accumulating the contributions of the primary pattern when translated and centered at each of the subcell spots in turn and weighted according to the estimated relative intensity of that spot in the observed pattern. It was sufficient to consider only the subcell reflections as sources for multiple scattering, although in one case ([210], Fig. 5(b)), unusually prominent spots due to the coexisting tetragonal zirconia phase occurred in the experimental pattern, and it was necessary to include the effect of these in the calculated pattern also.

Electron diffraction patterns calculated for the model structure with δ-phase microdomains >6 nm in diameter consisted of spots, and the reflections were plotted using circles with areas proportional to the intensity. The match of observed and calculated patterns was excellent for microdomains of ≈ 10 nm diameter (Fig. 5(a), (c)). In general, the calculated patterns were relatively insensitive to variations of up to 50% in the size of the microdomains. For the [210] orientation, however, rlp's lay closer to the rel plane than usual, and consequently the intensities were influenced noticeably by a 10% variation in the microdomain size. A best-fit value of 6.5 nm was determined for the particular [210] pattern studied (Fig. 5(b)).

When the microdomains were reduced to <3 nm in diameter, the patterns became more diffuse because of the particle-size broadening effect and were output with half-tone symbols. The best fit of the calculated patterns with those observed occurred for ≈ 2 nm microdomains, but the match was not good; the essential features were present and their relative intensities were substantially correct, but the detailed shapes of the diffuse features were not well reproduced. A considerable improvement was effected if the microdomains were assumed to be cylinders 1.5–2.0 nm in diameter and 2.0–2.5 nm long, with axes parallel to their respective unique crystallographic axes (vectors of the form $\langle 111 \rangle_{cub}$: see Table II), i.e., the intensity distribution in reciprocal space assumed to be disklike with a Gaussian profile. The match with the observed patterns was considered sufficiently good for the purposes of this study. No further refinement of the domain shape was sought, although a more appropriate shape function undoubtedly would improve the match further (Fig. 6). The use of a shape function in the calculation of the spot patterns (Fig. 5) would be valueless, because, as noted above, these patterns were relatively insensitive to microdomain dimensions.

The calculation of various sections was repeated assuming that, in the structure of δ phase, the cations were ordered so that either Mg or Zr occupied the special site of octahedral coordination by O^{13}: The match of calculated and observed intensities was inferior to that obtained when disordered cations were used, especially in the case of the spot patterns. Thus this study has provided some knowledge about the structure of $Mg_2Zr_5O_{12}$.

The rhombohedral fluorite-related superstructure phase $Zr_{10}Sc_4O_{26}$ has a structure, known as γ, that is related to that of the δ phase.[15] In particular, the hexagonal representation of the unit cell in terms of the subcell is $a = (1/2)[12\bar{3}]$, $b = (1/2)[\bar{3}12]$, and $c = [222]$ so that the a and b axes of the γ phase have the same lengths as those of the δ phase while the c axis of γ is twice that of δ. If the γ phase appeared in the system MgO-ZrO_2, it would have the composition $Mg_2Zr_{12}O_{26}$, which is very close to the eutectoid composition in this system and much closer to the composition of the materials studied here than is $Mg_2Zr_5O_{12}$. Some evidence for the existence of the γ superstructure in the system MgO-ZrO_2

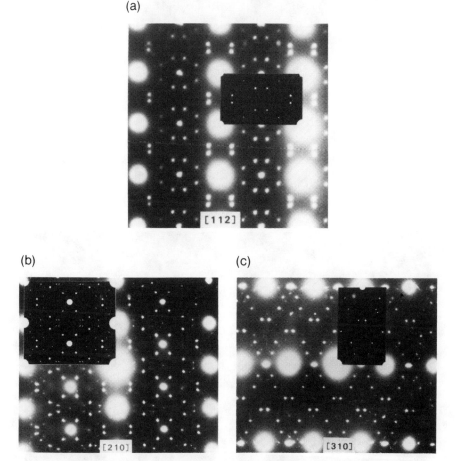

Fig. 5. Observed electron diffraction patterns from specimens of Mg-PSZ aged at 1100°C for 2 h with patterns calculated for 10 nm (a), (c), and for 6.5 nm diameter (b) microdomains of δ phase. Reflections due to the t-ZrO_2 phase have been ignored in the calculated patterns in (a) and (c).

was obtained. Some crystals in the specially prepared specimen of 28.6 mol% MgO content that had been quenched from the melt gave electron diffraction patterns similar to those of δ phase but with a doubled c_{hex} axis (Fig. 7(a)). The presence of γ phase in this specimen may have been a consequence of evaporation of MgO during melting. No attempts were made to establish the conditions for existence of the γ phase in this system.

Therefore, the above calculations were repeated assuming that the microdomains were isostructural with the γ-phase $Sc_4Zr_{10}O_{26}$,[15] and were oriented in all the (eight) possible ways that preserved a continuous substructure-matrix array. All possible variations in cation order were investigated. The diffraction features of interest in the resultant calculated patterns had the same geometric proportions as those in the patterns calculated for δ-phase domains, as expected, but the matches of intensity were noticeably inferior to those obtained above. By chance, most of

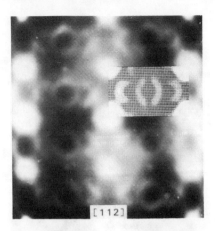
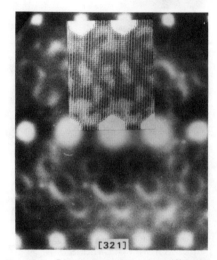
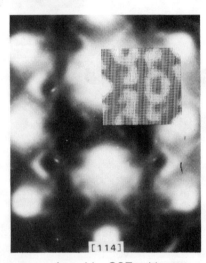

Fig. 6. Observed electron diffraction patterns from Mg-CSZ, with patterns calculated for δ-phase microdomain 2.2 nm long and 1.7-nm in diameter. Reflections due to t-ZrO_2 have been ignored.

these patterns contained only those γ-phase reflections of the type $hk2l$ (hexagonal axes), which are geometrically indistinguishable from δ-phase reflections. However, patterns calculated for the substructure orientation [310] (Fig. 7(b)) contained reflections peculiar only to the γ phase, and since these were not present in the corresponding experimental pattern (Fig. 7(c)), the hypothesis of γ-phase microdomains could be rejected with confidence.

Discussion

The electron diffraction effects observed in the Mg-PSZ and Mg-CSZ materials that have been subjected to subeutectoid aging can be explained in terms of microdomains of the metastable ordered superstructure $Mg_2Zr_5O_{12}$ existing within the cubic phase. The MgO content of the microdomains (28.6 mol%) is much higher than that of the material as a whole. This effective collection of Mg cations

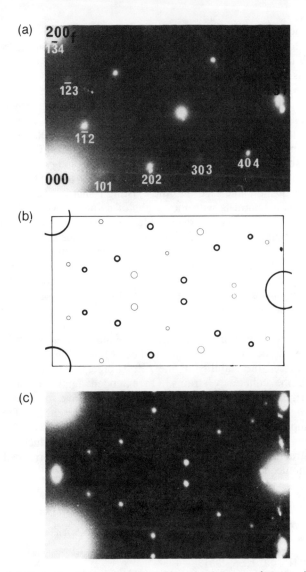

Fig. 7. (a) Experimental electron diffraction pattern from a single crystal of γ phase ($Mg_2Zr_{12}O_{26}$) that appeared in the special preparation of ordered $Mg_2Zr_5O_{12}$. The orientation is $[11\bar{1}]_\gamma$ (=$[013]_{substructure}$), and the indexing is based on the hexagonal representation of the γ-phase supercell. The fainter set of supercell reflections hkl, l odd will not be present in the equivalent pattern from δ phase. (b) [310] pattern calculated for 10 nm microdomains of γ phase in all eight orientations. The heavier spots are reflections of the type hk2l, and are geometrically indistinguishable from the spots expected for δ phase: the lighter reflections (hkl, l odd) are characteristic of the γ phase. Areas of spots are proportional to the expected intensity. (c) Experimental [310] electron diffraction pattern from Mg-PSZ, showing that the faint spots must be from microdomains of δ phase in 8 orientations, and not γ phase.

into discrete regions of higher concentration occurs after aging at 1100°C for times as short as 0.5 h and is clearly a precursor state in the eutectoid decomposition reaction, which ultimately must produce MgO + m-ZrO_2.[2]

This state is stabilized to some extent because of the energy gain from anion ordering in the δ phase.[13] The true eutectoid decomposition eventually does occur after longer aging times, but it is initiated at grain boundaries and other surfaces.[6]

Mg-PSZ that has been fired and cooled but not aged at 1100°C contains lenticular precipitates of t-ZrO_2, ≈150 nm long, which have nucleated and grown within the CSZ matrix crystals (Fig. 2(a)). During aging at 1100°C, nucleation and growth of δ phase is most likely at the interface between t- and c-ZrO_2, as observed (Figs. 2(b) and 1(c), (d)). The MgO for the formation of the δ phase is provided by the extension of the t interface into the existing c matrix and the consequent MgO rejection. This contrasts with the results of aging the Mg-CSZ material at 1100°C. Here the δ phase must nucleate at sites of local Mg enrichment due either to random fluctuations or to the precipitation of the very fine dispersion of t-ZrO_2 that generally occurs in this material[6] (Fig. 1(a)). Therefore the microdomains are distributed more uniformly; their smaller size, compared to the PSZ case, is due to a larger density of nuclei and the absence of relatively extended "nutrient" zones. This state persists unchanged during aging until the material is consumed by the front of the eutectoid decomposition reaction, which advances from grain boundaries and pores.

Mg-PSZ materials containing t-ZrO_2 precipitates surrounded by the larger (10 nm) δ-phase microdomains have an enhanced MOR.[4] The δ phase, therefore, must have influenced the stability of the t-ZrO_2 precipitates, probably through modification of the coherency strains. On extended aging of such materials (1100°C for 2 h), the t precipitates acquire m intergrowths and some will transform completely to m symmetry on cooling to room temperature; further, the expected eutectoid decomposition reaction begins to occur as described above. The presence of microdomains of δ phase, therefore, contributes to the production of a very diverse microstructure in Mg-PSZ[6]. This in turn has been correlated with the enhanced thermomechanical properties of this material.[17]

References

[1] Advances in Ceramics, Vol. 3. Edited by A. H. Heuer and L. W. Hobbs. The American Ceramic Society, Columbus, OH, 1981.

[2] C. F. Grain. "Phase Relations in the ZrO_2-MgO System," *J. Am. Ceram. Soc.*, **50** [6] 288–90 (1967).

[3] R. C. Garvie, R. H. J. Hannink, and N. A. McKinnon, "Partially Stabilized Zirconia Ceramics," U.S. Pat. No. 4 279 655, 1981.

[4] R. H. J. Hannink and R. C. Garvie, "Sub-Eutectoid Aged Mg-PSZ Alloy with Enhanced Thermal Up-shock Resistance," *J. Mater. Sci.*, **17**, 2637–43 (1982).

[5] M. V. Swain, R. C. Garvie, R. H. J. Hannink, R. R. Hughan, and M. Marmach, "Material Development and Evaluation of Partially Stabilized Zirconia for Extrusion Die Applications," *Proc. Br. Ceram. Soc.*, **32**, 343–53 (1982).

[6] R. H. J. Hannink, "Microstructural Development of Sub-Eutectoid Aged MgO-ZrO_2 Alloys," *J. Mater. Sci.*, **18**, 457–70 (1983).

[7] R. H. J. Hannink and M. J. Rossell, "An Electron-Optical Study of the Cubic MgO-Stabilized Zirconia Decomposition," *Micron*, **11**, 36 (1980), Suppl. 1.

[8] S. C. Farmer, L. H. Schoenlein, and A. H. Heuer, "Precipitation of $Mg_2Zr_5O_{12}$ in MgO-Partially-Stabilized ZrO_2," *J. Am. Ceram. Soc.*, **66** [7] C-107–C-109 (1983).

[9] A. P. Smith, "Polishing of Hard Materials," *Am. Ceram. Soc. Bull.*, **62** [8] 886–88 (1983).

[10] J. G. Allpress and H. J. Rossell. "A Microdomain Description of Defective Fluorite-Type Phases $Ca_xM_{1-x}O_{2-x}$ (M = Zr, Hf; x = 0.1–0.2)," *Solid State Chem.*, **15**, 68–78 (1975).

[11] J. G. Allpress, H. J. Rossell, and H. G. Scott. "Crystal Structures of the Fluorite-Related Phases $CaHf_4O_9$ and $Ca_6Hf_{19}O_{44}$," *J. Solid State Chem.*, **14**, 264–73 (1975).

[12] O. Yovanovitch and C. Delamarre, "Contribution to the Study of Compounds of Fluorite-derived Structure in the Systems ZrO_2-MgO, HfO_2-MgO and HfO_2-Sc_2O_3," *Mater. Res. Bull.*, **11**, 1005–10 (1976).

[13]H. J. Rossell, "Crystal Structures of Some Fluorite-Related M_7O_{12} Compounds," *J. Solid State Chem.*, **19**, 103–11 (1976).

[14]J. Lefevre, "Contribution to the Study of Different Structural Modifications of Fluorite-Type Phases in the Systems based on Zirconia or Hafnia," *Ann. Chim. (Paris)*, **8**, 117–49 (1963).

[15]M. R. Thornber, D. J. M. Bevan, and J. Graham, "Mixed Oxides of the Type MO_2 (Fluorite)-M_2O_3. III. Crystal Structures of the Intermediate Phases $Zr_5Sc_2O_{13}$ and $Zr_3Sc_4O_{12}$," *Acta Crystallogr., Sect. B*, **24**, 1183–90 (1968).

[16]A. M. Gavrish and E. I. Zoz, "Intermediate Compounds in the Systems ZrO_2 (HfO_2)-MgO," *Izv. Akad. Nauk SSSR, Neorg. Mater.*, **14**, 181–83 (1978).

[17]R. H. J. Hannink and M. V. Swain, "Magnesia-Partially Stabilized Zirconia: The Influence of Heat Treatment on Thermomechanical Properties," *J. Aust. Ceram. Soc.*, **18**, 53–62 (1982).

Diffusional Decomposition of c-ZrO$_2$ in Mg-PSZ

S. C. FARMER, T. E. MITCHELL, AND A. H. HEUER

Case Western Reserve University
Department of Metallurgy and Materials Science
Cleveland, OH 44106

Decomposition of Mg-PSZ containing 8.1 and 11.3 mol% MgO was studied by transmission electron microscopy; specimens were annealed between 800° and 1100°C, either as-sintered, after solution annealing above 1800°C, or after intermediate heat treatments at 1400°–1600°C. The equilibrium decomposition product is a mixture of monoclinic (m)-ZrO$_2$ and MgO; however, the latter is difficult to form and a number of metastable intermediate decomposition products are observed. For example, Mg$_2$Zr$_5$O$_{12}$ forms if solution annealing above 1800°C is followed by heat treatment at 1100°C or below. However, the dominant decomposition reaction involves formation of a grain-boundary "cellular" reaction product of m-ZrO$_2$ and MgO. Furthermore, the monoclinic phase so produced shows unusual microstructures.

Partially stabilized ZrO$_2$ (PSZ) has emerged during the last decade as a structural ceramic with unusually attractive mechanical properties.[1] Those PSZs that give optimal combinations of strength and toughness are two-phase ceramics, in which matrix grains consisting of solute-rich cubic (c)-ZrO$_2$ contain coherent particles of low solute content tetragonal (t)-ZrO$_2$.[2-4] Recent studies[4-6] showed that improvement in mechanical behavior, particularly improved thermal shock resistance against up-quenching, can be achieved by annealing PSZ at relatively low temperatures ("subeutectoid" aging[5,6]). This result has focused attention on the microstructural changes which occur as the cubic matrix decomposes by various diffusional reactions. Knowledge of the structure and morphology of the phases developed at these lower aging temperatures is requisite to exploit fully the potential for further property improvement,[6] as well as to limit property degradation through "destabilization" via these same reactions, an old but still worrisome problem.

We studied the decomposition of Mg-PSZ in the temperature range 800°–1100°C. The equilibrium phase diagram[7] (Fig. 1) indicates that, at these temperatures, monoclinic (m)-ZrO$_2$ and MgO are the stable phases. It is well known, however, that it is easy to retain t-ZrO$_2$ metastably, and that it is difficult to form MgO by eutectoid decomposition.[8]

Given these circumstances, it is not surprising to find metastable intermediate compounds which are not found in the equilibrium phase diagram.[4-6,9] Delamarre[10] was the first to report the occurrence of Mg$_2$Zr$_5$O$_{12}$ in sintered materials with compositions containing between 25 and 35 mol% MgO. Mg$_2$Zr$_5$O$_{12}$ is a defect fluorite phase with rhombohedral symmetry (lattice parameters are $a = 0.618$ nm and $\alpha = 99°\ 35'$ for single crystals of stoichiometric composition), and is isostructural with δ-Sc$_3$Zr$_4$O$_{12}$.[11] Mg$_2$Zr$_5$O$_{12}$ has been shown to develop metastably in

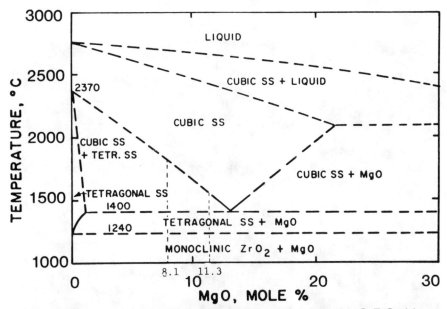

Fig. 1. Phase diagram of the ZrO$_2$-rich region in the MgO-ZrO$_2$ binary system. Dashed lines indicate the two compositions studied.

Mg-PSZ during low-temperature annealing[6,9]; its development is discussed in detail in these references. Decomposition of c-ZrO$_2$ in this temperature range was studied previously by Porter and Heuer[8] and Hannink and coworkers,[4–6] and our results are in general accord with theirs. However, certain differences do exist between our studies and Hannink et al.'s; they are discussed below.

Experimental Procedures

The materials used for this study were (i) an 8.1 mol% commercial polycrystalline Mg-PSZ ceramic,* which was heat-treated at 1850°C in argon for 2 h to form a single-phase c-ZrO$_2$ solid solution, then air-quenched; and (ii) an 11.3 mol% MgO body (in the as-sintered form containing both t- and m-ZrO$_2$ particles[8]), which was sintered for 2 h at 2000°C in an oxidizing ambient atmosphere in an oxyacetylene furnace. The furnace floor was then dropped and the material left on the hot refractory to cool to room temperature. Some of the 8.1 and 11.3 mol% samples were used in the solution-annealed state; others were given intermediate heat treatments for 1–6 h at 1400°–1600°C in air. For the low-temperature decomposition heat treatments, samples were packed in powder of the same composition and heat-treated in air between 800° and 1100°C for various times. They were then ground, mechanically polished, and thinned by argon ion bombardment to produce foils for transmission electron microscopy at 120 kV.

Results

The decomposition of c-ZrO$_2$ in 8.1 and 11.3 mol% Mg-PSZ at temperatures from 800° to 1100°C gives rise to numerous decomposition products, the particular diffusional phase transformation observed depending sensitively on the previous

---

Supported by the National Science Foundation under Grant No. DMR 77–19163.
*No. 1027, Zirconia Div., Corning Glass Works, Solon, OH.

(A)

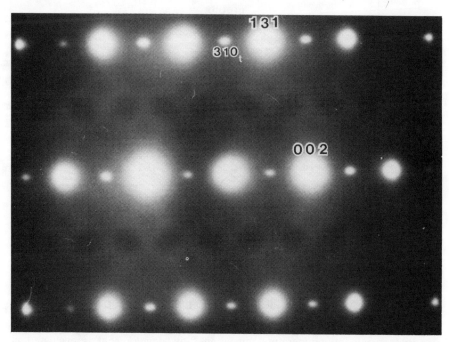

131
310$_t$
002

(B)

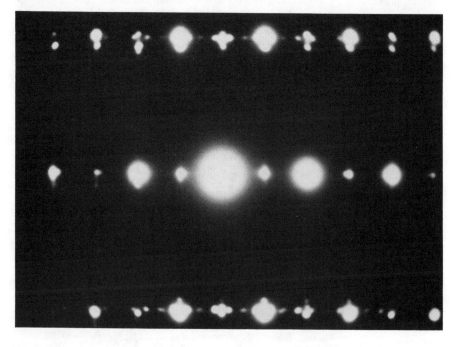

(C)

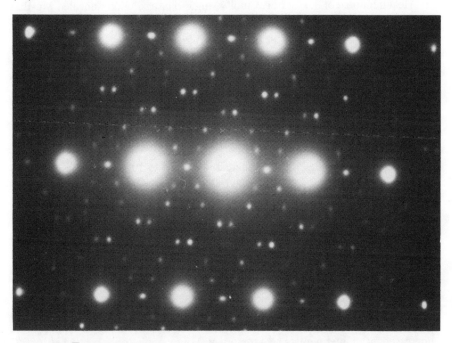

Fig. 2. [1$\bar{3}$0] electron diffraction patterns from cubic grains in three samples having different thermal histories: (A) c-ZrO$_2$ grain containing t precipitates (11.3 mol% heat-treated 20 h at 1100°C), (B) c-ZrO$_2$ grain with t and m precipitates (as-received 8.1 mol% sample heat-treated 1 h at 800°, 5 h at 1100°C; other grains in this sample contained a minor amount of Mg$_2$Zr$_5$O$_{12}$); (C) c-ZrO$_2$ grain with t precipitates showing additional reflections from two variants of the ordered phase Mg$_2$Zr$_5$O$_{12}$ (typical of either 8.1 or 11.3 mol% materials, solution-annealed and heat-treated 1 h at 800° and 5 h at 1100°C).

thermal history of the materials. Figure 2 (A)–(C) presents diffraction patterns of the cubic [1$\bar{3}$0] zone from three samples (no sample contained all features of Fig. 2). The [1$\bar{3}$0] zone was found to be the most convenient for differentiating between the different phases formed in decomposed PSZ.

The diffraction pattern in Fig. 2(A) shows the strong "fundamental" reflections of c-ZrO$_2$, as well as the less-intense extra reflections of the t phase, which have odd-odd-even indices and which are forbidden for the cubic fluorite structure.[12] Appreciable diffuse intensity associated with the c-ZrO$_2$ solid solution is visible, particularly on the original negative. The diffraction pattern in Fig. 2(B) contains c and t reflections plus additional reflections due to a considerable amount of m-ZrO$_2$ (presumably formed from t-ZrO$_2$ by the martensitic $t \rightarrow m$ transformation). The diffraction pattern in Fig. 2(C) shows c- and t-ZrO$_2$ and additional reflections from two variants of Mg$_2$Zr$_5$O$_{12}$.

Formation of $Mg_2Zr_5O_{12}$

Heat treatment of 8.1 and 11.3 mol% single-phase solution-annealed Mg-PSZ at 800°C for 1 h, followed by annealing at 1100°C for 5 h, resulted in growth of coherent t-ZrO_2 precipitates *and* precipitates of the ordered phase, $Mg_2Zr_5O_{12}$, within the c-ZrO_2 matrix. The area shown in the dark-field micrograph of Fig. 3 is typical of the matrix grains observed after this heat treatment; a single variant of $Mg_2Zr_5O_{12}$ is strongly diffracting. The tetragonal phase present in the sample of Fig. 3 (but not visible there) nucleated and grew to ~5 nm diameter during cooling after solution annealing, and coarsened to 50–80 nm diameter during the 800°–1100°C heat treatment. The equilibrium phase at these temperatures, m-ZrO_2, is not present within the cubic grains after heat treatment at 1100°C for 5 h.

If the aging time at 1100°C is increased from 5 to 10 h, $Mg_2Zr_5O_{12}$ does not coarsen appreciably, but a much greater amount of m-ZrO_2 is found at the grain boundaries (see the following section).

$Mg_2Zr_5O_{12}$ is a metastable decomposition product (although it may have a region of stability at some other temperature on the equilibrium phase diagram). The conditions under which $Mg_2Zr_5O_{12}$ formed in our experiments were very specific — solution annealing at $T \geq 1800°C$ followed by the 800°–1100°C heat treatment. The lower-temperature heat treatment was necessary to nucleate this phase; 800°C was usually used, although 900°C was also found to be suitable. Without this heat treatment, the competitive reactions occurring at 1100°C (to be described next) did not permit the formation of $Mg_2Zr_5O_{12}$. Even with such a specific heat treatment, however, nucleation and growth of $Mg_2Zr_5O_{12}$ could be suppressed by an intermediate heat treatment. For example, prior heating at 1600°C for 1 h (followed by furnace cooling) prevented the formation of $Mg_2Zr_5O_{12}$ during the 800°–1100°C aging; at 1500°C, a 2-h heat treatment (with cooling to room temperature in $\approx$20 min) did not prevent its formation, whereas a 4-h heat treatment lessened the amount formed during the 800°–1100°C cycle. Such behavior was noted previously in the formation of ordered defect-fluorite phases in some rare earth oxide systems,[13] in which a suitable quenched-in high-temperature cation arrangement is necessary to permit nucleation of particular ordered phases.

To further investigate the effects of the high-temperature thermal history, the batch of 11.3 mol% MgO solution-annealed material, which had been heat-treated for 1 h at 1600°C and then furnace-cooled, was given a second solution-annealing treatment (2 h at 2100°C in an oxyacetylene furnace and air-cooled to 20°C in $\approx$5 min). When subsequently heat-treated at 1100°C, $Mg_2Zr_5O_{12}$ formed readily, with no need for a lower-temperature nucleation heat treatment.

Formation of m-ZrO_2 as a Grain-Boundary Decomposition Product

The most abundant decomposition product developed at 1100°C is m-ZrO_2, which forms as a grain-boundary reaction product rather than within the cubic grains. The m-ZrO_2-containing decomposition product advances as a transformation front into the surrounding grains and consumes the c-ZrO_2, along with its t-ZrO_2 and $Mg_2Zr_5O_{12}$ precipitates. Formation of this m-ZrO_2 proceeded more rapidly in the 11.3 mol% material than in the 8.1 mol% sample, confirming Hannink's[6] result. More significantly, electron diffraction patterns invariably disclosed the presence of other phases associated with m-ZrO_2. First, the advancing planar transformation front actually produces a coupled decomposition product (Fig. 4(A)). The lamellar spacing of the minor phase is $\approx$150 nm. The monoclinic

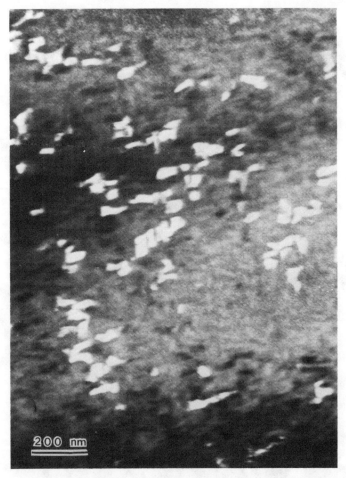

Fig. 3. Dark-field electron micrograph of $Mg_2Zr_5O_{12}$ precipitates in c-ZrO_2 matrix containing t-ZrO_2 precipitates (8.1 mol%, solution-annealed and heat-treated 1 h at 800° and 10 h at 1100°C).

regions between the minor phases consist of multiple variants, as becomes apparent when this region is tilted off the zone axis (Fig. 4(B)). The diffraction pattern for this region is shown in Fig. 4(C). It consists of a superposition of a monoclinic zone of the type $[10\bar{3}]_m$ and two zones of the type $[0\bar{2}1]_m$. This pattern also contains extra spots, corresponding to the presence of MgO in a $[11\bar{2}]_{MgO}$ zone axis orientation. A dark-field micrograph taken using the (222) MgO reflection is shown in Fig. 4(D). Correlation with the diffraction pattern shows that the growth direction of the MgO is ⟨110⟩, and all rods have the same orientation. In other orientations, the rodlike morphology of the MgO is unmistakable (Fig. 5); the rods lie in well-defined planes. Figure 5 shows virtually identical features to that obtained by Porter and Heuer[8] in their 8.1 mol% Mg-PSZ decomposed for 135 h at 1000°C (their Fig. 9), except that our rod spacing is smaller because of the

(A)

(B)

158

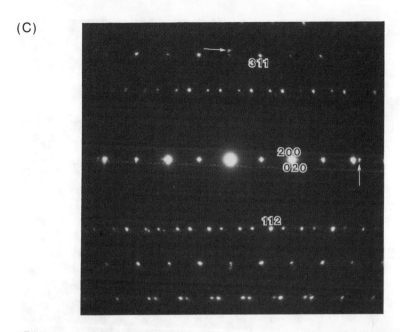

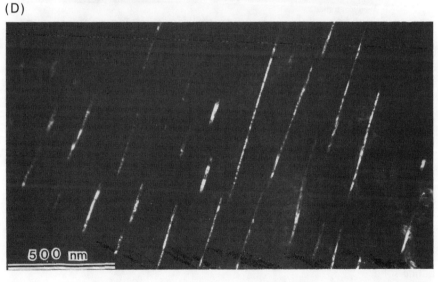

Fig. 4. (A) Bright-field electron micrograph showing the "monoclinic" phase advancing into a c-ZrO_2 grain containing t-ZrO_2 precipitates; rods of the MgO phase form perpendicular to the transformation front (11.3 mol%, heat-treated 10 h at 1100°C). (B) Area of Fig. 4(A) tilted off the zone axis. The multiple orientations within this region are apparent. (C) Diffraction pattern for this region consisting of a superposition of two $[0\bar{2}1]_m$ zones showing (100) twinning, a $[10\bar{3}]_m$ zone, and a $[\bar{1}12]_{MgO}$ zone (the arrowed reflections).

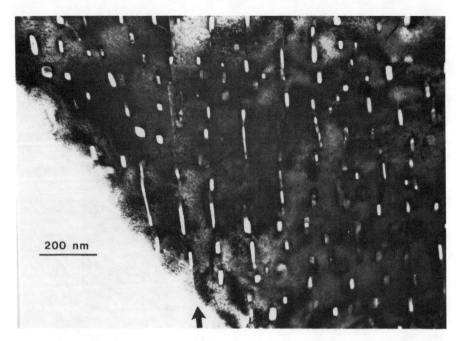

Fig. 5. Bright-field electron micrograph from a region oriented roughly perpendicular to that shown in Fig. 4. The rodlike morphology is apparent; the rods lie in well-defined planes. This region contains a single twin boundary (arrowed).

greater MgO content. We note that this region of m-ZrO_2 has a single twin boundary (arrowed), which we think arose from the impingement of two separately nucleated regions.

Figure 6(A) shows another orientation of the grain-boundary decomposition product growing into a c grain containing t precipitates. The grain-boundary "phase" has been tilted to an approximately [1̄01] monoclinic orientation and shows a periodic modulation with a periodicity of ≈ 3.5 nm. The corresponding electron diffraction pattern, Fig. 6(B), shows "side bands" around each fundamental reflection, the side-band spacing being the same for all orders of reflections, in agreement with the periodicity in the image. The diffraction pattern of Fig. 6(B) is, to all intents and purposes, that of a periodically decomposed crystalline solid.

Detailed tilting experiments demonstrated the existence of a second set of fringes with half the spacing (Fig. 7); they may be Moiré fringes, approximately parallel to $[101]_m$, arising from an extra reflection with a smaller d spacing. The extra reflection presumably comes from a different crystalline periodicity within the modulations, but this has not yet been identified. Both modulations are independent of the cooling rate from the 1100°C decomposition temperature. We have observed similar modulations, confirmed by Bischoff and Rühle,[14] in a pure skull-melted m-ZrO_2 single crystal. From a preliminary examination of such material, the modulations appear to be identical to those observed in the grain-boundary decomposition reaction of 8.1 and 11.3 mol% MgO material.

(A)
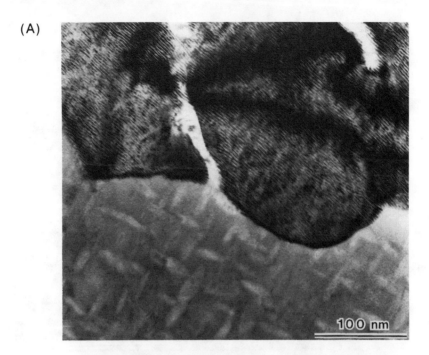

(B)
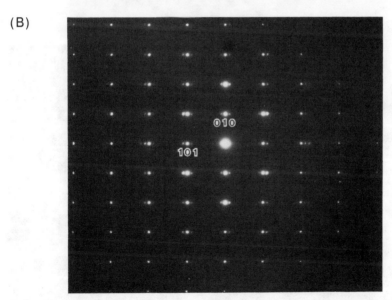

Fig. 6. (A) Bright-field electron micrograph of the grain-boundary decomposition product growing into a c grain containing t precipitates. The orientation is approximately [101] and shows a modulation with a periodicity of ≈3.5 nm. (B) Diffraction pattern of $[\bar{1}01]_m$ zone. Side bands occur about each fundamental reflection, the side-band spacing being the same for all orders of reflections, in agreement with the periodicity in the image.

(A)

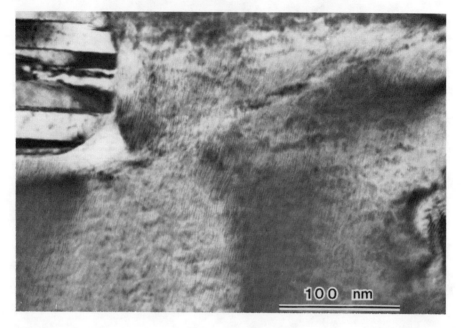

(B)

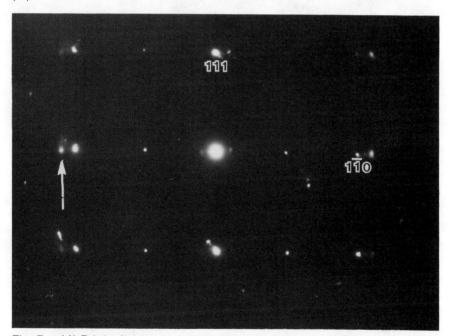

Fig. 7. (A) Bright-field electron micrograph of grain-boundary monoclinic region showing what appear to be Moiré fringes; (B) Diffraction pattern of $[11\bar{2}]_m$ for the above micrograph. The unknown reflection giving rise to the fringes is arrowed.

Interpretation

Various heat treatments of Mg-PSZ have been shown to produce low-temperature precipitation of several phases — t-ZrO_2, m-ZrO_2, and $Mg_2Zr_5O_{12}$ — as well as eutectoid decomposition of c-ZrO_2 into m-ZrO_2 plus MgO rods. Both t-ZrO_2 and $Mg_2Zr_5O_{12}$ are metastable phases at the temperatures studied. The eutectoid decomposition product nucleates at grain boundaries and grows readily at the expense of c- and t-ZrO_2; Fig. 4(A) is an almost "classical" example of a cellular transformation front. We are, however, still puzzled over the modulated structure found in this decomposition product and in single-crystal m-ZrO_2. (Another puzzle is that the ability to nucleate $Mg_2Zr_5O_{12}$ apparently depends on quenching in a high-temperature ($>1600°C$) "state.")

Some remarks should also be made about the apparent disagreements that exist between this work and that of Hannink and coworkers.[4-6] First, the thermal histories of the materials used differed significantly, which is expected to lead to differences in the progress of phase transformations. All of our differences related to the formation of $Mg_2Zr_5O_{12}$ are attributed to this factor. Second, Hannink et al. did not report that the m-ZrO_2 is modulated. However, this feature of the microstructure can be overlooked without detailed tilting experiments, and Fig. 7 is actually of a 9 mol% sample donated to us by Hannink. Third, Hannink reported that the rods in the decomposition product were "MgO-rich"; however, he now (private communication) has obtained microchemical data that they are pure MgO. Finally, we have not observed intergrowths of m-ZrO_2 within the t-ZrO_2 precipitates, as Hannink reports. This aspect has not really been explored in depth by either Hannink or ourselves and is the subject of continuing work by both groups.

Acknowledgments

The authors thank R. H. J. Hannink for critical discussion of portions of this work and A. H. Heuer thanks the Alexander Van Humboldt Society for a Senior Scientist Award which made possible his sabbatical leave at the Max-Planck-Institüt in Stuttgart, West Germany, where the first draft of this paper was prepared.

References

[1] A. H. Heuer and L. W. Hobbs; Advances in Ceramics 3, Science and Technology of Zirconia. Edited by A. H. Heuer and L. W. Hobbs. The American Ceramic Society, Columbus, Ohio, 1981.
[2] R. H. J. Hannink, *J. Mater. Sci.*, **13**, 2487–96 (1978).
[3] A. H. Heuer; pp. 98–115 in Advances in Ceramics 3, Science and Technology of Zirconia. Edited by A. H. Heuer and L. W. Hobbs. The American Ceramic Society, Columbus, Ohio, 1981.
[4] R. H. J. Hannink and M. V. Swain, *J. Austl. Ceram. Soc.*, **18** [2] 53–62 (1982).
[5] R. H. J. Hannink and R. C. Garvie, *J. Mater. Sci.*, **17**, 2637–43 (1982).
[6] R. H. J. Hannink, *J. Mater. Sci.*, **18**, 457–70 (1983).
[7] C. F. Grain, *J. Am. Ceram. Soc.*, **50** [6] 288–90 (1967).
[8] D. L. Porter and A. H. Heuer, *J. Am. Ceram. Soc.*, **62** [5–6] 298–305 (1979).
[9] S. C. Farmer, L. H. Schoenlein, and A. H. Heuer, *J. Am. Ceram. Soc.*, **66** [7] 107–109 (1983).
[10] C. Delamarre, *Rev. Hautes Temp. Refract.*, **9** [2] 209–24 (1972).
[11] M. R. Thornber, D. J. M. Bevan, and J. Graham, *Acta Crystallogr.*, Sect. B, **24**, [Pt. 9] 1183–90 (1968).
[12] L. H. Schoenlein, L. W. Hobbs, and A. H. Heuer, *J. Appl. Crystallogr.*, **13**, 375–79 (1980).
[13] D. J. M. Bevan and E. Summerville; pp. 401–523 in Handbook of the Physics and Chemistry of Rare Earths, Vol. 3. Edited by C. A. Gschneider and L. Eyring. North-Holland, Amsterdam, 1979.
[14] E. Bischoff and M. Rühle; private communication.

High-Resolution Microscopy Investigation of the System ZrO_2–ZrN

G. VAN TENDELOO

RUCA, University of Antwerp
B-2020, Antwerpen, Belgium

G. THOMAS

University of California
Department of Materials Science and Mineral Engineering
and the National Center for Electron Microscopy
Lawrence Berkeley Laboratory
Berkeley, CA 94720

The system ZrO_2-ZrN has been investigated using different electron microscopy techniques. In the range between 2.5 and 75 mol% ZrN, an incommensurate modulated structure is formed, having rhombohedral $R\bar{3}$ symmetry based on the $Zr_7O_{11}N_2$ structure which is isostructural to $Zr_5Sc_2O_{13}$. The modulation is most probably due to pseudoperiodic composition fluctuations of nitrogen with respect to oxygen. The final structure can be described as a completely coherent mixture of pure Zr_7O_{14} layers alternating with $Zr_7O_{11}N_2$ layers perpendicular to the rhombohedral threefold axis. The incommensurability arises from local deviations in spacing and/or orientation of these layers. Monoclinic precipitates embedded in this rhombohedral matrix are characterized with the help of high-resolution electron microscopy combined with optical microdiffraction.

The three polymorphs of ZrO_2 have been well established.[1] At temperatures near the melting point, ZrO_2 has the cubic fluorite structure; on cooling, this structure first undergoes a slight tetragonal distortion and, around 1000°C, it transforms martensitically toward the monoclinic room-temperature phase (space group $P2_1/c$; lattice parameters $a = 0.517$, $b = 0.523$, $c = 0.534$ nm, $\beta = 99°15'$.[2,3]

The latter transformation may be completely or partially suppressed by alloying pure zirconia with additions of other oxides such as CaO, MgO, or Y_2O_3.[4,5] The present study is an electron microscopy investigation of a range of ZrO_2-ZrN alloys in order to investigate a possible stabilization by ZrN[6] and to elucidate the microstructures.

The formation of different oxynitrides in the system Zr-O-N has been studied by Gilles.[7,8] He established a pseudobinary phase diagram between ZrO_2 and $ZrN_{4/3}$ (not ZrN) and found three structurally related oxynitrides denoted β, β', and γ with nominal compositions: $\beta Zr_7O_8N_4\square_2$; β', $Zr_7O_{11}N_2\square_1$; γ, $Zr_2ON_2\square_1$, where $\square$ denotes a vacancy on the oxynitride sublattice. γ can be described on a cubic lattice with $a = 1.01$ nm, and β has rhombohedral symmetry and is isostructural with $Zr_3Sc_4O_{12}$ or $Zr_3Yb_4O_{12}$.[7] β' is very closely related to β; it also has rhombohedral symmetry and the same a parameter but twice the c parameter.

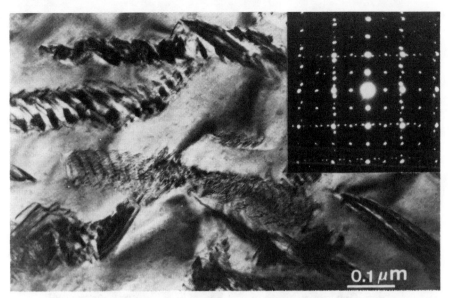

Fig. 1. Typical morphology of most ZrO_2-ZrN samples. The electron diffraction pattern over a precipitate is shown as an inset.

Outline of the Investigation

The present investigation refers to materials with the following nominal compositions: ZrO_2 + 0, 5, 10, 15, 20, 30, 40, 50, 75, and 100% ZrN which were investigated "as-received" or heat-treated at 1000°C. For more experimental details, see Ref. 9.

Most of the investigated samples showed the typical morphology illustrated in Fig. 1, i.e., large amounts of precipitates embedded in the matrix. From the existing literature, it could be assumed that they were monoclinic ZrO_2 particles in a cubic matrix. However, a closer investigation showed that the matrix was not cubic but rhombohedrally distorted and, moreover, it showed a complicated incommensurate superstructure.[9] The precipitates were not always monoclinic but in some cases also tetragonal. The monoclinic precipitates were always heavily twinned. A selected area diffraction pattern (along one of the fluorite cube axes) over such a precipitate is shown as the inset in Fig. 1. Apart from the matrix spots, all other reflections can be attributed to different variants of the monoclinic ZrO_2. From the "conventional" image of Fig. 1, it is impossible to determine twin planes or habit planes. The main aim of the present study was, therefore, to (a) determine the structure of the matrix and (b) gain information about the internal structure of the different precipitates.

Structure of the ZrO_2-ZrN Matrix

Electron Diffraction Evidence

A series of diffraction patterns taken from a matrix area is shown in Fig. 2. From these patterns it can be seen that the structure is not cubic. However, the most intense reflections are close to the cubic fluorite positions, e.g., in Fig. 2(f) the

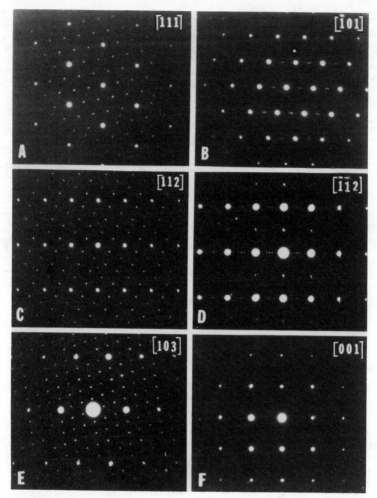

Fig. 2. Diffraction series taken from Zr-O-N matrix.

angle is not exactly 90° but approximately 89°, suggesting a rhombohedral distortion. The other diffraction patterns can be divided into two groups: (a) reciprocal lattice sections exhibiting a clear superstructure such as [111], [112], or [103] when the fluorite cubic indices are used (Fig. 2(a), (c), and (e)) and (b) sections exhibiting extra reflections at incommensurate positions such as [$\bar{1}$01] and [$\bar{1}\bar{1}$2] (Fig. 2, (b) and (d)).

In the [$\bar{3}$21] cubic section of Fig. 3(a) originally cubic reflections are encircled while reflections at commensurate superstructure positions are indicated by arrows. The remaining reflections are at incommensurate positions but concentrated around commensurate positions such as (½ ½ ½), which suggests that they might be due to a splitting of a commensurate reflection. The direction of the incommensurate modulation is always along [111]*, i.e., along one of the threefold axes of the originally cubic matrix. The average distance between successive

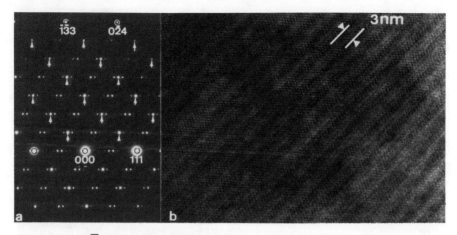

Fig. 3. (a) $[\bar{3}21]_{cub}$ diffraction pattern; cubic reflections are indicated by dots and commensurate superstructure reflections are indicated by arrows; (b) High-resolution image corresponding to (a); note the orientation deviations at the lower right corner.

satellites is somewhat variable but always corresponds to a real space periodicity around 3.2 ± 0.3 nm.

The commensurate reflections suggest the formation of a superstructure independent of that of the modulation. This becomes evident if one tilts 90° away from the modulation toward a [111] section (Fig. 2(a)). The modulation direction is now along the zone axis.

High-Resolution Microscopy Evidence

Two essentially different orientations have been examined: (a) the [111] zone (perpendicular to the modulation) and (b) a $[\bar{3}21]$ zone perpendicular to [111], containing the [111]* row. In the [111] zone, a hexagonal network of bright dots is observed separated by 0.55 nm, indicating that the structure has a truly threefold axis and that the hexagonal pattern of Fig. 2(a) is not the result of superposition. In the $[\bar{3}21]$ orientation, a modulated structure with a periodicity of about 3 nm can be easily recognized. The modulation is not exactly periodic, and in some areas spacing as well as orientation deviations can be observed. At a higher magnification (Fig. 4), it becomes evident that the modulation is not of the interface-modulated kind. There is no well-defined defect plane; the image appears more like a sinusoidal intensity modulation. However, when observed along the $[1\bar{1}5]_{cub}$ direction (e.g., in Fig. 4), a shift in intensity maxima is observed between subsequent modulations, separated by 3 nm.

In very restricted areas, mostly close to a monoclinic precipitate, the unmodulated structure is observed (Fig. 4(a), left side) in contact with the modulated structure (Fig. 4(a), right side). Since the areas are too small for conventional electron diffraction and since convergent beam diffraction is too difficult due to beam damage which causes the superstructure to disappear, selected area laser optical diffraction from the matrix as well as from the modulated structure were used (Fig. 4(b) and (c)). They prove that both structures are closely related and that

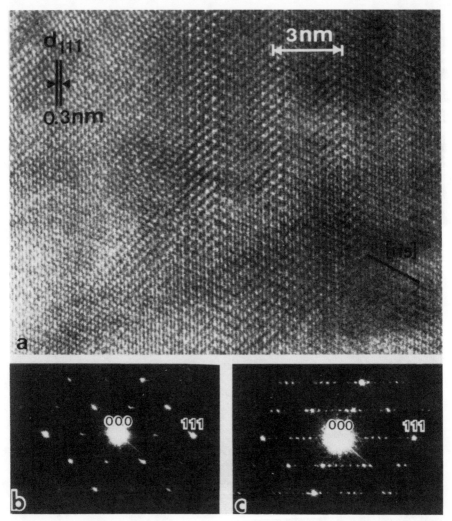

Fig. 4. (a) High-resolution image of the basic superstructure in contact with the modulated structure. The orientation is $[\bar{3}21]_{cub}$ (see Figs. 3(a) and 4(a) and (b)). Laser optical microdiffraction pattern from the (b) unmodulated as well as from the (c) modulated structure.

some reflections which are unsplit in the commensurate superstructure such as (½ ½ ½) are split up into different satellites due to the modulation.

Structure of the Unmodulated Phase

With the assumption, as is evident from Fig. 2(f), that the structure is a slight rhombohedral modification of the fluorite-type structure and taking into account the doubling of the periodicity along the rhombohedral axis (Fig. 4(b)), the unit cell parameter along the $\bar{c}$ axis should be around 1.8 nm (assuming a hexagonal description of the rhombohedral structure). This would fit excellently with the β' phase described by Gilles[7] which seems to be isostructural to $Zr_5Sc_2O_{13}$.[10]

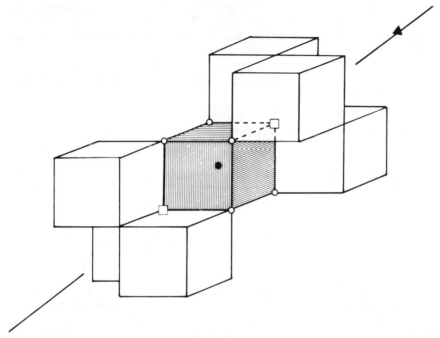

Fig. 5. The Bevan cluster, consisting of seven cubes; the center of each cube contains a Zr atom. The central Zr has a sixfold coordination $ZrO_6\square_2$ while all other Zr atoms have a sevenfold coordination $ZrO_7\square_1$.

More details can be found in Ref. 9. The measured unit-cell parameters are $a = 0.96$ nm and $c = 1.76$ nm, and the complete relation between fluorite structure $(\bar{a}_1, \bar{a}_2, \bar{a}_3)$ and the superstructure $(\bar{A}_1, \bar{A}_2, \bar{A}_3)$ (omitting the rhombohedral distortion) can be derived from the diffraction patterns to be

$$\begin{bmatrix} \bar{A}_1 \\ \bar{A}_2 \\ \bar{A}_3 \end{bmatrix} = \begin{bmatrix} 1 & -3/2 & 1/2 \\ 1/2 & 1 & -3/2 \\ 2 & 2 & 2 \end{bmatrix} \begin{bmatrix} \bar{a}_1 \\ \bar{a}_2 \\ \bar{a}_3 \end{bmatrix} \quad (1)$$

In the cubic fluorite structure, every Zr atom is surrounded by eight oxygens, and the structure can be described as built up from ZrO_8 cubes sharing edges with each other.[9,10] When the anion-to-cation ratio deviates from the ideal value of 2 (e.g., in $Zr_5Sc_2O_{13}$ or $Zr_3Sc_4O_{12}$), vacancies are introduced and some of the cations have only a sixfold or a sevenfold coordination. The basic building block turns out to be the "Bevan cluster"[11] represented in Fig. 5. It consists of a sixfold coordinated metal atom (hatched cube) surrounded by six sevenfold coordinated cubes sharing edges with the central cube. Both vacancies in this cluster necessarily are at diagonally opposite sites of the central cube in order to retain the rhombohedral threefold symmetry along [111]. The β-phase $Zr_7O_8N_4\square_2$ is exclusively built up by stacking layers of such clusters perpendicular to the threefold axis while the β' phase $Zr_7O_{11}N_2\square_1$ (isostructural to $Zr_5Sc_2O_{13}$) would be built up by alternate layers of such Bevan clusters and layers of "ideal" clusters where all metal atoms have an eightfold coordination. In this sense the β' phase can be considered as midway between the cubic fluorite phase and the β phase. The cubic fluorite as well as the β phase has a three-layer periodicity along $[111]_{cub}$ or $[0001]_{hex}$; the first one is completely built up from perfect I (ideal) clusters while the second is completely built up from B (Bevan) clusters, all oriented along the unique threefold axis. The β' phase having a six-layer periodicity is a mixture of equal amounts of B and I cluster layers such that the structure can be written as $BIBIBI$, while the β phase is BBB, and the cubic ZrO_2 phase is III.

Structure of the Modulated Phase

Analysis of the modulated structure must take into consideration the following experimental data: It is a modulation of the rhombohedral β'-$Zr_7O_{11}N_2$ structure; the modulations are roughly parallel to $(00.1)_\beta$, with an average distance of 3.2 nm but not strictly bound to this plane; the high-resolution microscopy shows a modulation which is not of the interface-modulated type (e.g., different from long-period superstructures), but after one wavelength an out-of-phase shift is observed (Fig. 4).

With knowledge of the basic β' structure and from the shift of the satellites with respect to their original position, one can obtain more information about this last point. Indeed, although there is no discrete interface visible in the high-resolution micrographs, the boundary can be described as in Fig. 6(a), which is in fact the same as the purely interface-modulated structure (e.g., for a long-period antiphase boundary structure) (Fig. 6(b)), except for the fact that the boundary is smeared out. The final displacement, however, between A and B is the same as that between A' and B'. Neglecting the reason for the origin for such a smeared-out interface, one can determine the displacement vector $\overline{\mathbf{R}}$ for this boundary in the same way as one can do for periodic antiphase boundaries or shear planes.[12]

Careful inspection of the diffraction patterns shows that all satellites undergo a shift which is 0 (mod 1) or ½ with respect to the β' positions. From the shift of three independent reflections the displacement vector can be determined as

$$\overline{\mathbf{R}} = \pm[\tfrac{2}{3}\,\tfrac{1}{3}\,\tfrac{1}{6}] \tag{1}$$

Since the $\overline{\mathbf{R}}$ vector is not in the defect plane, this displacement is nonconservative and implies a deviation from ideal stoichiometry.

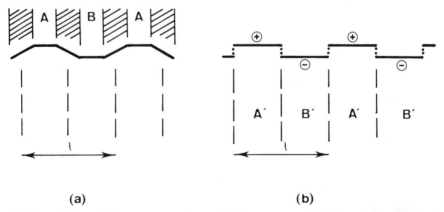

Fig. 6. Schematic representation of (a) wavy modulation compared to (b) a strict long-period interface modulated structure.

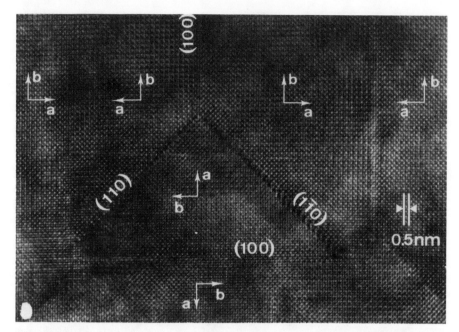

Fig. 7. High-resolution observation of heavily twinned monoclinic ZrO_2. All variants have a common c axis parallel to the electron beam. The orientation of the axes as determined from HREM and optical diffraction is indicated.

The high-resolution observations (Fig. 4) indicate that the modulation has more of a wave character than an antiphase character. Similar considerations have been made for $(Au, Ag)Te_2$,[13] $Mo_{2+x}S_3$,[14] and the "chimney ladder" structure in $MnSi_{2-x}$.[15]

Precipitation in the Rhombohedral Zr-O-N Matrix

Precipitation in the rhombohedral Zr-O-N matrix is a complex phenomenon. Tetragonal as well as monoclinic ZrO_2 particles can be present in the matrix.[9]

In this summary only the monoclinic precipitation is considered; internal twinning occurs frequently in these precipitates, and on the basis of the type of twinning, two essentially different twin structures can be considered: (a) precipitates with all twin variants having a common $\bar{c}$ axis and (b) precipitates with twin variants having no common axis.

A high-resolution example of the first type is shown in Fig. 7 when viewed along the c_m axis. The precipitate is heavily twinned, and all twin boundaries are edge-on and strictly along crystallographic planes. Such configurations are very similar to the ones observed by Kriven for 50% Al_3O_2-50% ZrO_2[16] and for ZrO_2 particles in a mullite matrix.[17] The frequent and almost regular (100) twinning of such precipitates occurs to minimize strain and shape deformation effects built up as a result of the differences in lattice parameters due to the monoclinic shear transformation.

The second type of precipitate is more complex and produces diffraction patterns (inset, Fig. 1) that are very complex. High-resolution microscopy com-

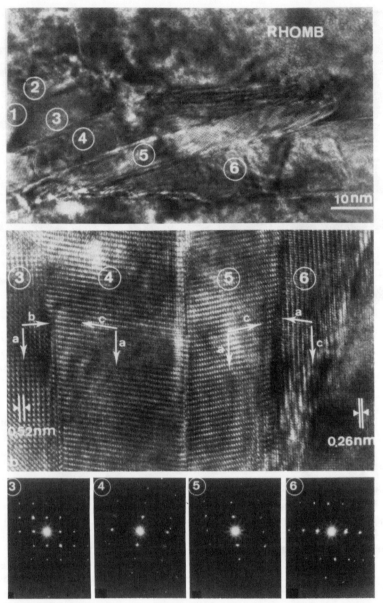

Fig. 8. A complex twinned area analyzed with the aid of high-resolution microscopy and laser optical diffraction of the areas indicated 3–6.

bined with optical diffraction seems to be the only way to correlate the different diffraction spots with areas in direct space. The usual dark-field technique, i.e., selecting successively all diffracted reflections to form an image, fails in this case because the reflections are too close to each other to be individually imaged. The lattice image (Fig. 8(*a*) and (*b*)) establishes the presence of six orientation variants and, with the aid of optical diffraction over the different domains (bottom), one can deduce the orientation relationship as indicated in Fig. 8(*b*).

Conclusions

High-resolution elecron microscopy combined with electron diffraction and laser optical microdiffraction has been shown to be a powerful technique to interpret the complex microstructures obtained in alloys of ZrO_2-ZrN. These complexities include the formation of an incommensurate modulated rhombohedral matrix containing twinned ZrO_2 precipitates. The easy beam damage to these materials means that high-voltage microscopy is essential. Microchemical analyses at 100 kV using EELS for mapping oxygen and nitrogen is almost impossible.[18]

Acknowledgments

This work was supported by the Director, Office of Energy Research, Office of Basic Energy Sciences, Materials Sciences Division of the U.S. Department of Energy under Contract No. DE-AC03-76SF00098. The authors are grateful to L. Anders for preparing electron microscopy samples. Financial support has been provided by the NSF. G. V. T. is also indebted to NATO for partial financial support. We also thank S. Amelinckx, U. Dahmen, R. Gronsky, and A. H. Heuer for valuable discussions.

References

[1] E. C. Subbarao; pp. 1–24 in Advances in Ceramics, Vol. 3. Edited by A. H. Heuer and L. W. Hobbs. The American Ceramic Society, Columbus, OH, 1981.
[2] J. D. McCullough and K. N. Trueblood, Acta Crystallogr., **12**, 507 (1959).
[3] D. K. Smith and H. W. Newkirk, Acta Crystallogr., **18**, 983 (1965).
[4] V. S. Stubican and J. R. Hellman; pp. 25–36 in Advances in Ceramics, Vol. 3. Edited by A. H. Heuer and L. W. Hobbs. The American Ceramic Society, Columbus, OH, 1981.
[5] D. L. Porter and A. H. Heuer, J. Am. Ceram. Soc., **62** [5–6] 298–305 (1979).
[6] A. Claussen, R. Wagner, L. J. Gauckler, and G. Petzow, J. Am. Ceram. Soc., **61** [7–8] 369–70 (1978).
[7] J.-C. Gilles, Rev. Hautes Temp. Refract., **2**, 237 (1965).
[8] J.-C. Gilles and R. Collongues, C. R. Hebd. Seances Acad. Sci., **254**, 1084 (1962).
[9] G. van Tendeloo and G. Thomas, Acta Metall., in press.
[10] M. R. Thornber, D. J. M. Bevan, and J. Graham, Acta Crystallogr., Sect. B, **24**, 1183 (1968).
[11] B. Hudson and P. T. Moseley, J. Solid State Chem., **19**, 383 (1976).
[12] J. Van Landuyt, R. De Ridder, R. Gevers, and S. Amelinckx, Mater. Res. Bull., **5**, 353 (1970). Also see R. De Ridder, J. Van Landuyt, and S. Amelinckx, Phys. Status Solidi A, **9**, 551 (1972).
[13] G. van Tendeloo, P. Gregoriades and S. Amelinckx, J. Solid State Chem., in press.
[14] R. Deblieck, G. A. Wiegers, K. D. Bronsema, D. Van Dyck, G. van Tendeloo, J. Van Landuyt, and S. Amelinckx, Phys. Status Solidi A, **77**, 249 (1983).
[15] R. De Ridder, G. van Tendeloo, and S. Amelinckx, Phys. Status Solidi A, **33**, 383 (1976).
[16] W. M. Kriven; pp. 168–83 in Advances in Ceramics, Vol. 3. Edited by A. H. Heuer and L. W. Hobbs. The American Ceramic Society, Columbus, OH, 1981.
[17] E. Bischoff and M. Rühle, J. Am. Ceram. Soc., **66** [2] 123 (1983).
[18] M. Sarikaya, P. Rez, and G. Thomas; unpublished work.

Compatibility Relationships of Al_2O_3 and ZrO_2 in the System ZrO_2-Al_2O_3-SiO_2-CaO

P. Pena and S. de Aza

Instituto de Cerámica y Vidrio
C.S.I.C. Departamento de Cerámica
Arganda del Rey
Madrid, Spain

Melting relationships in that part of the diagram in which alumina or zirconia occur as primary phases have been studied. Solid-state compatibility of ZrO_2 and Al_2O_3 inside the system has been established as well as the character, composition, and temperature of the different invariant points where ZrO_2 and Al_2O_3 coexist with other phases. The results show that, in the subsystems ZrO_2-Al_2O_3-A_3S_2-CAS_2 and ZrO_2-Al_2O_3-CA_6-CAS_2, the solid solution of CaO in ZrO_2 at the subsolidus temperature is negligible.

When a material is designed, the most important source of information is its phase diagram. Recently, there has been increased interest in ZrO_2-based ceramics, which made particularly interesting all those multicomponent systems where ZrO_2 is one of the phases. Consequently, in our institute, quaternary ZrO_2-oxide systems are being studied that are particularly pertinent to problems of ZrO_2-containing alkali-resistant glasses, refractories, ceramics, and zirconia-toughened ceramics. They are all part of the very complicated nine-component system ZrO_2-Al_2O_3-SiO_2-TiO_2-CaO-MgO-Na_2O-Fe_2O_3-FeO. The work reported in this paper is a part of this research program, and it is related to the system ZrO_2-Al_2O_3-SiO_2-CaO.

Literature

The solid-state phase equilibria in the system ZrO_2-Al_2O_3-CaO were established by Bannister,[1] who found that at 1700°C the cubic solid solution of CaO in ZrO_2 is not stable in the presence of Al_2O_3 but decomposes to form CA_6 and unstabilized ZrO_2. Espinosa and White[2] outlined the phase regions and reported the temperature and composition of five invariant points.

The system Al_2O_3-CaO-SiO_2 has been the subject of many investigations, and the diagram of Criado and de Aza[3] incorporates the data of several workers.

The most recent phase diagram proposed for ZrO_2-CaO-SiO_2 is that due to Quereshi and Brett,[4,5] who reported the existence of the compound $Ca_3ZrSi_2O_9$ (C_3ZS_2),* a primary phase field of $ZrSiO_4$ (ZS) and a two-liquid region. In their investigation of the system ZrO_2-MgO-CaO-SiO_2, de Aza et al.[6] have since shown that Ca_2SiO_4 (C_2S) and ZrO_2 are not compatible in the solid state but that a connection exists between $CaZrO_3$ (CZ) and the ternary compound C_3ZS_2.

---

*Solid phases are described by abbreviated formulas, e.g., calcium silicozirconate $Ca_3ZrSi_2O_9$, zircon $ZrSiO_4$ by ZS, etc.

The system ZrO_2-Al_2O_3-SiO_2 has been subjected to many investigations, but the most recent phase diagram is that due to Quereshi and Brett,[7] who reported the existence of a primary phase field of $ZrSiO_4$ (ZS), which incorporates the data of previous workers.

Solacolu et al.[8] published a phase diagram of the system ZrO_2-Al_2O_3-SiO_2-CaO showing solid-state compatibility relations.

Experimental Procedure

The starting materials were washed Belgian sand (99.9% purity), calcium carbonate[†] (99.5% purity) as the source of CaO, low-hafnium-grade ZrO_2[‡] (99.8% purity), and low-Na Al_2O_3[§] (99.998% purity). All these materials, except the $CaCO_3$, were ground to pass a 35-μm sieve before use.

Batch compositions weighing ≈ 1 g (after loss of CO_2) were mixed in an isopropyl alcohol medium, and pellets weighing 0.2–0.3 g were pressed, loaded into Pt-foil capsules, and then prefired at 950°C to remove CO_2 prior to firing at the selected temperature, in a high-temperature furnace[¶] with an electronic temperature controller** (± 1°C). Firing times ranged from 2–24 h in order to reach equilibrium; specimens were then air-quenched, removed from the Pt foil, mounted in an epoxy resin, and then progressively polished down to 0.1-μm diamond paste and examined by reflected light microscopy.[††] X-ray diffraction[‡‡] and energy-dispersive X-ray microanalysis were used to assist identification of crystalline phases and to establish their compositions. EDX microanalyses have been performed on the sample polished sections by SEM–EDX equipment.[§§] The calibration was carried out with ZrO_2, Al_2O_3, $ZrSiO_4$, and $CaO \cdot Al_2O_3$ as reference materials. The appropriate corrections for atomic number, absorption, and fluorescence were made by a computer program.[¶¶]

Results and Discussion

Solid-State Phase Diagram

The solid-state compatibility relationships of ZrO_2 and Al_2O_3 in the quaternary system ZrO_2-Al_2O_3-SiO_2-CaO are shown in Fig. 1. Solid solubilities have been ignored. This was constructed using previously published diagrams[1-8] and differs slightly from that of Solacolu et al.[8] According to de Aza et al.,[6] the solid-state compatibility tetrahedrons ZrO_2-C_2S-CZ-C_2AS and ZrO_2-C_3ZS_2-C_2S-C_2AS do not exist, but the compatibility tetrahedron ZrO_2-C_3ZS_2-CZ-C_2AS is found.

Boundary Surfaces of Primary Phase Volumes of Al_2O_3 and ZrO_2 in the Quaternary System

To obtain a knowledge of the melting relationship inside the quaternary system, the effect of composition on melting behavior of mixtures lying on a plane of constant Al_2O_3 or ZrO_2 content within the primary phase volume of either alumina or zirconia, respectively, were investigated. For this purpose, selected

---

[†]E. Merck, Darmstadt, Federal Republic of Germany.
[‡]Koch-Light Lab. Ltd., Colnbrook, Bucks, England.
[§]Fluka AG, Buchs SG., Switzerland.
[¶]Bulten-Kanthal, S-73481 Hallstahammar, Sweden.
**Eurotherm Controller, Eurotherm Corp., Reston, VA.
[††]Photomicroscope II, C. Zeiss 7082 Ober-kochen, W. Germany.
[‡‡]Diffractometer Model PW 1140/00, Philips Gloeilampenfabrieken NV, Eindhoven, The Netherlands.
[§§]Kevex 7077 (Kevex Corp., Burlingame, CA) attached to a SEM super 3A ISI, Berkeley, CA.
[¶¶]Quantex ZAF corrections via Magic V.

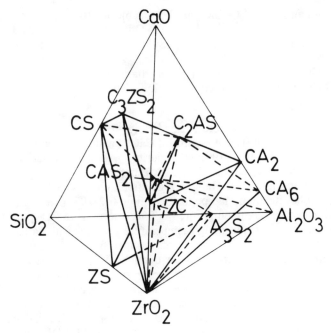

Fig. 1. Solid-state compatibility relationships of ZrO_2 in the quaternary system ZrO_2-Al_2O_3-SiO_2-CaO.

mixtures containing 80 wt% Al_2O_3 or 70 wt% ZrO_2 were examined after quenching from various temperatures.

Figures 2 and 3 show the crystallization fields and initial crystallization temperatures of the second phases to freeze on cooling, Al_2O_3 or ZrO_2 being the primary phase, plotted on a ternary diagram on which the composition of the mixtures are expressed in terms of their ZrO_2, SiO_2, and CaO or Al_2O_3, SiO_2, and CaO content respectively, recalculated to 100 wt%. This procedure is the same as that employed by the authors[9] in studying the system ZrO_2-Al_2O_3-SiO_2-TiO_2 and is equivalent to projecting the composition of the mixture through the Al_2O_3 or ZrO_2 corner onto the opposite face of the composition tetrahedron. Figure 2 gives a true (undistorted) projection of the boundary surface of the primary phase of Al_2O_3 since solid solution of the other components in Al_2O_3 is negligible (EDX microanalysis shows <0.1 wt%); that is, the freezing paths from the compositions of the mixtures investigated to the boundary surface will then lie along straight lines radiating from the Al_2O_3 corner. However, in Fig. 3, because of the appreciable solubility of CaO in ZrO_2, the latter condition is not *generally* fulfilled since solid solution of varying composition will separate during freezing and the diagram obtained will be a distorted version of the boundary surface of the primary crystallization volume of ZrO_2. It will, however, show the actual phases separating during freezing of mixtures in the composition plane investigated and the temperature at which they separate, while the true projection of the boundary surface will not do so.

It is worth noticing however, that in Fig. 3 those compositions lying in the primary phase volume of ZrO_2 within the compatibility tetrahedrons

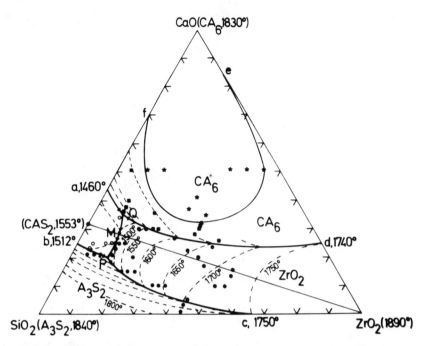

Fig. 2. Projection through the Al_2O_3 corner, onto the opposite face of the quaternary tetrahedron ZrO_2-Al_2O_3-SiO_2-CaO, of the boundary surface of the primary phase volume of alumina, showing phase boundaries, isotherms, and invariant points. The various symbols ■, *, *, ●, and ○ represent experimental compositions.

ZrO_2-Al_2O_3-A_3S_2-CAS_2 and ZrO_2-Al_2O_3-CA_6-CAS_2 did not show any appreciable solid solubility of CaO, Al_2O_3, or SiO_2 in ZrO_2 (EDX microanalysis shows <0.1 wt%); then the projected ZrO_2 boundary surface corresponding to these two compatibility tetrahedrons will be a true (undistorted) projection. Consequently, it is possible to calculate, with both projections (Figs. 2 and 3), the real composition of the invariant points of these compatibility tetrahedrons (points P and Q in Figs. 2 and 3).

Table I shows the character, temperature, and composition of these invariant points as well as those characteristic of point M (Figs. 2 and 3), the invariant point of the subsystem ZrO_2-Al_2O_3-CAS_2.

In Fig. 2, it should be noted that the field CaO-e-f is a field of primary, not secondary, crystallization of CA_6. It occurs because in this range of composition the 80 wt% Al_2O_3 section lies in the primary crystallization volume of CA_6. The existence of a field of primary CA_6 within the 80 wt% Al_2O_3 projection is in accordance with the published data[2,3] on ZrO_2-Al_2O_3-CaO and Al_2O_3-CaO-SiO_2.

Although Fig. 3 relates specifically to mixtures containing 70 wt% ZrO_2, except for those compositions located in the quaternary subsystems ZrO_2-Al_2O_3-A_3S_2-CAS_2 and ZrO_2-Al_2O_3-CA_6-CAS_2, Fig. 2 would apply to all mixtures lying within the primary phase volume of Al_2O_3 because the solid solubility of the other components in Al_2O_3 is negligible within the melting range. Thus, in bodies to which Fig. 2 refers, and probably to a reasonable approximation in

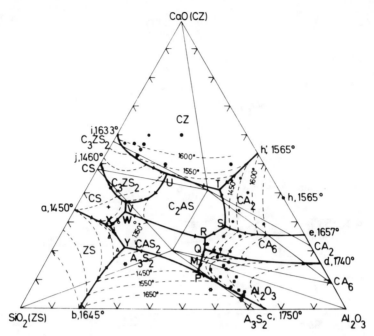

Fig. 3. Projection through the ZrO$_2$ corner, onto the opposite face of the quaternary tetrahedron ZrO$_2$-Al$_2$O$_3$-SiO$_2$-CaO, of the boundary surface of the primary phase volume of zirconia, showing phase boundaries, isotherms, invariant points, and secondary phases crystallizing during freezing from ZrO$_2$-Al$_2$O$_3$-SiO$_2$-CaO mixtures containing 70 wt% ZrO$_2$. The various symbols *, *, and ■ represent experimental compositions.

Fig. 3, the second solid phase to crystallize from the liquid phase during cooling (Al$_2$O$_3$ or a ZrO$_2$ solid solution being the first) can be obtained by plotting the composition of the mixtures in terms of ZrO$_2$, SiO$_2$, and CaO or Al$_2$O$_3$, SiO$_2$, and CaO content. The second phase will then be that in whose crystallization field the point lies, while its temperature of initial crystallization on cooling (or complete solution on heating) will be indicated by the isotherms.

As the temperature falls below that of initial crystallization of the second phase, the freezing path will now lie in the boundary surface of the Al$_2$O$_3$ or ZrO$_2$ primary crystallization volume. If the solid solution in the primary and secondary phases is assumed to be negligible, the direction of the freezing path in the projection at this stage will be a straight line passing through the point representing the composition of the mixture and the secondary phase.

Table I. Character, Temperature, and Composition of the Invariant Points

Invariant points	Character of the invariant point	Temperature (°C)	Composition (wt%)			
			Al$_2$O$_3$	ZrO$_2$	CaO	SiO$_2$
P	peritectic	1440 ± 5	44.0	7.0	12.0	36.5
Q	peritectic	1380 ± 5	44.5	5.0	19.5	31.0
M	maximum	1520 ± 10	44.0	7.0	15.0	34.0

When a boundary line is intersected, a third phase will start to separate at the temperature indicated by the corresponding isotherms, and the freezing path will follow the boundary in the direction of falling temperature to the nearest invariant point where (a) freezing will be completed if the three phases coexisting with either Al_2O_3 or ZrO_2 and liquid at the point are the same as those coexisting in the solid mixture or (b) (if this is not the case) one of the solid phases will disappear and a new phase will be formed by peritectic reaction, leaving excess liquid. The freezing path will then follow that boundary whose temperature falls away from the invariant point until the appropriate invariant point is reached, where freezing will be completed. Alternatively, the freezing path may leave the boundary before it reaches an invariant point. This will occur when the boundary line is a reaction line, if the secondary phase that had separated previously is completely redissolved before the invariant point is reached. The freezing path will then leave the boundary in the direction of the line passing through the point representing the composition of the solid phase now crystallizing along with either ZrO_2 or Al_2O_3, and that representing the composition of the mixture. Freezing will continue along this line until it intersects another boundary line where a third solid phase will appear, and then the freezing path will proceed as in (a) or (b).

In addition, in Fig. 3 appreciable solubility of CaO occurs in the ZrO_2 phase outside the quaternary subsystems ZrO_2-Al_2O_3-A_3S_2-CAS_2 and ZrO_2-Al_2O_3-CA_6-CAS_2; one consequence is that, although the sequence of events will in general be similar, the freezing path in the various phase fields will tend to be curved and it is not possible to predict their course exactly without further information. Second, compositions projecting into certain areas of the diagram may freeze completely before the freezing path reaches an invariant point. Outside these regions, even though the precise freezing paths are not known, the invariant points at which freezing will be completed, and hence the temperature of initial melting, can be directly deduced from either Fig. 2 or Fig. 3.

Table II summarizes the temperature and character of all those quaternary invariant points corresponding to those subsystems where ZrO_2 is one of the phases.

Conclusions and Practical Implications

The compatibility relationships of alumina in the system ZrO_2-Al_2O_3-SiO_2-CaO have been investigated. Since solid solution of the other components in Al_2O_3 are negligible, a true projection of the boundary surface of the primary phase

Table II. Temperature and Character of Quaternary Invariant Points

Invariant point	Phases at the invariant point	Character of the invariant point	Temperature (°C)
P	ZrO_2 + Al_2O_3 + A_3S_2 + CAS_2	Peritectic	1440 ± 5
Q	ZrO_2 + Al_2O_3 + CAS_2 + CA_6	Peritectic	1380 ± 5
R	ZrO_2 + CA_6 + CAS_2 + C_2AS	?	1350 ± 10
S	ZrO_2 + CA_6 + CA_2 + C_2AS	Peritectic	1400 ± 10
T	ZrO_2 + CA_2 + CZ + C_2AS	?	1350 ± 10
U	ZrO_2 + CZ + C_2AS + C_3ZS_2	Peritectic	1425 ± 10
V	ZrO_2 + C_2AS + C_3ZS_2 + CS	Peritectic	1325 ± 10
W	ZrO_2 + C_2AS + CAS_2 + CS	Peritectic	1300 ± 10
X	ZrO_2 + CS + ZS + CAS_2	?	1220 ± 10
Y	ZrO_2 + ZS + A_3S_2 + CAS_2	Peritectic	1260 ± 10

volume of Al_2O_3 through the Al_2O_3 corner has been made. Four secondary crystallization fields of phases coexisting with Al_2O_3 and liquid were established, viz., those of ZrO_2, $3Al_2O_3 \cdot 2SiO_2$, $CaAl_2Si_2O_8$, and $CaO \cdot 6Al_2O_3$. The temperature, composition, and character of the quaternary invariant points of the subsystems where Al_2O_3 is one of the coexisting phases have been determined.

Because of the number of the phase fields involved, the work on the compatibility relationships of ZrO_2 in the system was necessarily of an explanatory nature.

An important feature of the system is the existence of the extensive volumes of primary crystallization of Al_2O_3, ZrO_2, and CA_6. These meet along the boundary line dQ of Figs. 2 and 3, and each shares two of its boundary surfaces with the other two. Thus, the primary phase volume of Al_2O_3 shares faces with the phase volumes of zirconia and CA_6, which gives rise to the possibility of producing ZrO_2-bonded and CA_6-bonded Al_2O_3-based refractories.

Similarly, the primary phase volume of ZrO_2 shares faces with the phase volumes of Al_2O_3 and CA_6, which give rise to the possibility of obtaining Al_2O_3 bonding and CA_6 bonding in ZrO_2 of suitable composition.

Analogous considerations can also be applied to the Al_2O_3, ZrO_2, and $3Al_2O_3 \cdot 2SiO_2$ phases, which also meet along the boundary line cP in Figs. 2 and 3.

Another practical implication of the work follows from the fact that solid solutions of CaO in ZrO_2 are not stable in the presence of Al_2O_3.

The phase relationships established are also of interest from the standpoint of using ZrO_2 additions to confer alkali resistance on glasses. Thus, it can be seen from Figs. 2 and 3 that glass-forming compositions containing 35–50 wt% SiO_2 will be saturated with ZrO_2 at 1450°–1550°C when they contain on the order of 10–15 wt% ZrO_2.

Acknowledgment

The authors are indebted to F. Guitian Rivera for EDX microanalyses. The work was supported by CAICYT under Contracts Nos. 3172/79 and 0079/81.

References

[1] M. J. Bannister, "Development of the $SiRO_2$ Oxygen Sensor: Ternary Phase Equilibria in the System ZrO_2-Al_2O_3-CaO," *J. Austral. Ceram. Soc.*, **17** [1] 21–24 (1981).
[2] J. Espinosa and J. White, "Compatibility Relationships in the System ZrO_2-Al_2O_3-CaO," *Bol. Soc. Esp. Cerám. Vidr.*, **12** [4] 237 (1973).
[3] E. Criado and S. de Aza, "Phase Relationships in the Subsystem $CaO \cdot Al_2O_3 \cdot 2SiO_2$-$2CaO \cdot Al_2O_3 \cdot SiO_2$-$CaO \cdot 6Al_2O_3$," *Refracttari e Laterizi*, **2** [6] 285–89 (1977).
[4] M. H. Quereshi and N. H. Brett, "Phase Equilibria in Ternary Systems Containing Zirconia and Silica. I. The System CaO-ZrO_2-SiO_2," *Trans. Br. Ceram. Soc.*, **67** [6] 205–19 (1968).
[5] M. H. Quereshi and N. H. Brett, "Phase Equilibria in the System $CaO \cdot SiO_2$-$2CaO \cdot SiO_2$-ZrO_2"; pp. 275–91 in Science of Ceramics, 4th ed. Edited by G. H. Stewart. The British Ceram. Society, Great Britain, 1968.
[6] S. de Aza, C. Richmond, and J. White, "Compatibility Relationships of Periclase in the System CaO-MgO-ZrO_2-SiO_2," *Trans. Br. Ceram. Soc.*, **73** [4] 109–16 (1974).
[7] M. H. Quereshi and N. H. Brett, "Phase Equilibria in Ternary Systems Containing Zirconia and Silica. II. The System Al_2O_3-ZrO_2-SiO_2," *Trans. Br. Ceram. Soc.*, **67** [11] 569–78 (1968).
[8] S. Solacolu, R. Dinescu, A. Barbulescu, and M. Keul, "Phase Thermal Equilibrium in the System CaO-Al_2O_3-SiO_2-ZrO_2 and Its Application to the Stabilized Zirconia-Based Refractories," *Revue Roumaine de Chimie*, **16** [4] 519–26 (1971).
[9] P. Pena and S. de Aza, "The System ZrO_2-Al_2O_3-SiO_2-TiO_2 and Its Practical Implication"; pp. 247–55 in Science of Ceramics, 9th ed. Edited by K. J. de Vries. The Nederlandse Keramische Vereniging, Amsterdam, 1977.

Phase Relations in the Ternary System ZrO_2-Al_2O_3-SiO_2 by the Slow-Cooling Float-Zone Method

I. SHINDO, S. TAKEKAWA, AND K. KOSUDA

National Institute for Research in Inorganic Materials
Ibaraki, 305, Japan

T. SUZUKI AND Y. KAWATA

Daiichi Kigenso Kagaku Kogyo Co., 5–17, Kohraibashi
Higashi-ku, Osaka, Japan

Phase relations in the ternary system ZrO_2-Al_2O_3-SiO_2 were investigated by the slow-cooling float-zone method (hereafter called SCFZ method) using a lamp image furnace. The eutectic composition between ZrO_2 and Al_2O_3 was determined to be $ZrO_2 : Al_2O_3 = 37.3 : 62.7$ (mole ratio). The melting nature of mullite was observed to be incongruent. The ternary eutectic composition determined by previous workers should be interpreted as the ternary peritectic. A revised phase diagram involving the results for the binary systems ZrO_2-Al_2O_3 and Al_2O_3-SiO_2 was also established.

The refractories involving zirconia-alumina-silica have successfully helped the glass industry improve corrosion resistance and their fused refractories have been widely used.

Phase relations in this ternary system, therefore, have been investigated by many workers.[1-3] Budnikov and Litvakovskii[1] reported the presence of the ternary eutectic composition, and the same result was also reported by Cevales.[2] It is of much interest that this ternary eutectic composition lies close to that of the Corhart ZAC* refractories; however, their microtexture revealed the colony structure that was the characteristic phenomenon of the peritectic solidification.

To further improve fused and casted refractories, it is of great importance to control their microtexture. The hardness or toughness of a refractory should be strongly affected by its own texture. The starting composition and cooling rate are the main factors that influence the formation of texture, and the melting nature is also important.

When the eutectic composition is used as the starting composition, the eutectic lamellar structure should appear, and its width and interface can be varied by the cooling rate. On the contrary, a colony texture should appear when the peritectic composition is used. Therefore, the phase diagram involving liquid phases performs an important role in producing refractories with well-controlled textures.

It is well known that there are many difficulties in determining the precise phase diagram involving liquid phases using ordinary methods such as the quench-

---

*Corhart Refractories Div., Corning Glass Works, Solon, OH.

ing or optical technique in high-temperature regions because of the lack of suitable containers.

One of the present authors (I.S.) has devised the SCFZ method, which is a modified floating-zone method and has been successfully applied to the system MgO-TiO_2.[4] The SCFZ method needs no containers, such as crucibles, and is expected to be a favorable technique. Thus, application of the SCFZ method was attempted to this ternary system, and both the melting nature of mullite, which remains a problem, and the presence of the ternary eutectic composition have been carefully investigated.

The SCFZ Method

The theoretical background of the SCFZ method is fundamentally based on that of the maximum fractional crystallization principle, as suggested by Bowen,[5] and is similar to that of the normal freezing method.[6,7] The detailed procedures of this method have been described elsewhere,[4] and a brief description is given here. In the ideal maximum fractional crystallization process, the following conditions are maintained: (1) The melt is completely homogeneous, (2) diffusion in the solid is negligible, and (3) the solid and the liquid are in equilibrium at their interface during solidification.

To realize these conditions, it is convenient to use a floating-zone apparatus. The continuous counterrotation of the upper and lower shafts allows the melt to be completely mixed and kept homogeneous. Equilibrium between the solid and liquid phases can be achieved by controlling the advance rate of the solidification front. Figure 1 is a schematic illustration of a floating melt system before the cooling process in a SCFZ experiment (A) and the resulting zonal structure in solidified bodies (B). The notations C_1, S_1, S_2, and S_3 all correspond to those in Fig. 2, which is a phase diagram of a hypothetical binary system containing a peritectic compound. The solidification sequence that is determined by the EPMA technique may be plotted on a composition-temperature diagram where the temperature scale is arbitrary. In order to complete a phase diagram, the temperatures necessary for delineating the liquidus and the solidus curves must be determined for the specimen fragments with selected compositions, which are separated from the solidified bodies.

Experimental Procedure

Powders of ZrO_2 (99.9% pure),[†] Al_2O_3 (99.9% pure),[‡] and SiO_2 (99.9% pure)[§] were mechanically mixed in various ratios. Each mixed batch was placed in a sealed rubber tube and hydrostatically pressed at a pressure of approximately 700 kg/cm² to form a rod of desired diameter (5–7 mm) and length (40–50 mm). The rods were sintered in an oxygen atmosphere at 1600°C.

The floating-zone apparatus used was of an infrared radiation convergence type[¶] with two 3.5-kW quartz halogen lamps or a 10-kW xenon lamp as the radiation source. The details of this apparatus were published elsewhere.[8]

Two sintered rods were used: One rod was suspended from the lower end of the upper shaft and the other was fixed at the top of the lower shaft. The facing ends of the upper and the lower sintered rods were melted and connected. The shafts were counterrotated at 30 rpm. Approximately 30 min later, the electric

[†]Daiichi Kigenso Co., Osaka, Japan.
[‡]Iwatani Sangyo Co., Tokyo, Japan.
[§]Rare Metallic Co., Ltd., Tokyo, Japan.
[¶]Nichiden Kikai Co., Tokyo, Japan.

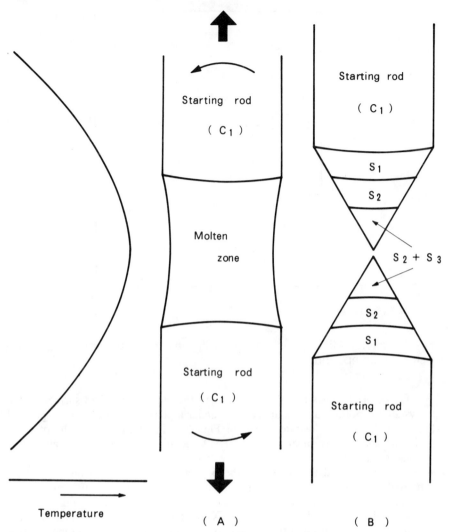

Fig. 1. Floating melt system of the SCFZ method (A) and the resulting zonal structure (B). The notation corresponds to that in Fig. 2.

power supply was gradually reduced at a constant rate, while the two shafts were gradually separated simultaneously. When the melt was completely condensed, two boules, each with a cone shape, were obtained.

The boules were bisected parallel to the longitudinal direction and polished mechanically. The polished sections were examined by means of optical microscopy and EPMA technique.

Results and Discussion

The Binary System ZrO_2-Al_2O_3 (Starting Composition; ZrO_2:Al_2O_3 = 20:80 (Mole Ratio))

Each SCFZ run gave a pair of boules, each having a cone-shaped recrystallized portion at one end.

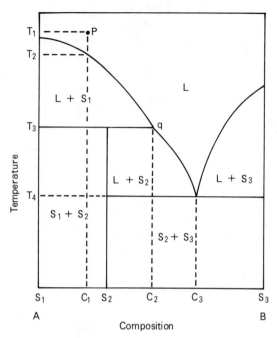

Fig. 2. Hypothetical phase diagram in the binary system containing a peritectic compound.

Figure 3(A) shows a magnified view of the cut and polished surface of the recrystallized portion and its appearance under a phase contrast microscope, and Fig. 3(B) shows an illustrated view of Fig. 3(A) involving the results of compositional analysis by means of the EPMA technique. Figure 3 indicates three regions that form a zonal structure. The first region corresponds to the original sintered portion, and the second is the portion that was melted once and later solidified as a single phase of alumina. In the third region, a typical eutectic lamellar texture is seen; a magnified view of this region is seen in Fig. 4.

According to the EPMA and the X-ray powder diffraction analyses, the lamellar texture consisted of two phases, Al_2O_3 and ZrO_2, with their compositional ratio being $Al_2O_3:ZrO_2 = 62.7:37.3$ (mole ratio).

The results revealed that, in the binary system ZrO_2-Al_2O_3, none of the compounds existed with only one eutectic point. The composition of this eutectic point determined by this experiment agrees very well with previously reported ones by Fisher et al.[9] or Schmid and Viechnicki.[10]

The System Al_2O_3 and SiO_2 (Starting Composition; $Al_2O_3:SiO_2$ = 3:1 (Mole Ratio))

Figure 5(A) shows part of magnified view of the initially solidifed region of the polished surface, and Fig. 5(B) shows a typically observed lamellar texture in the finally solidifed region. As shown in Fig. 5, four regions formed the zonal structure. The first region was the originally sintered rod, the second was once melted and solidified with coprecipitation of a foreign phase, and the third was a single phase. In the fourth region, the typical lamellar texture was observed. Using EPMA, the composition of the precipitated phase was identified as alumina, and that of the matrix was also found to correspond to $2Al_2O_3 \cdot SiO_2$. In the third region, the composition was varied with its position. The content of Al_2O_3 in this region

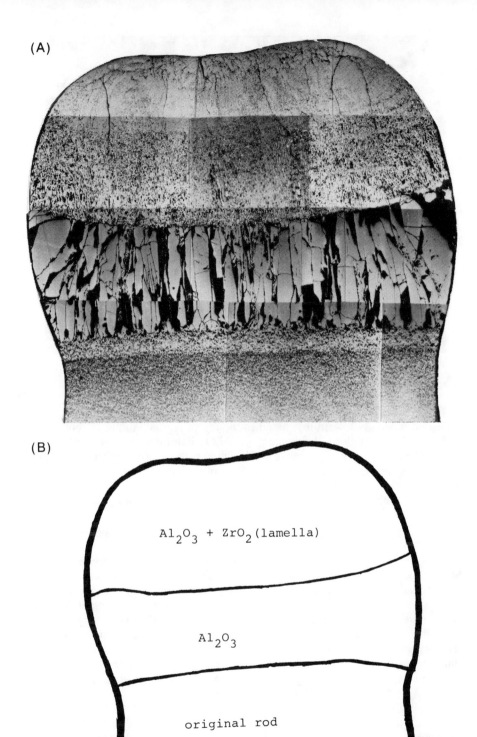

Fig. 3. Zonal structure of the solidified portion (A) and its illustrated view involving the results of compositional analysis (B).

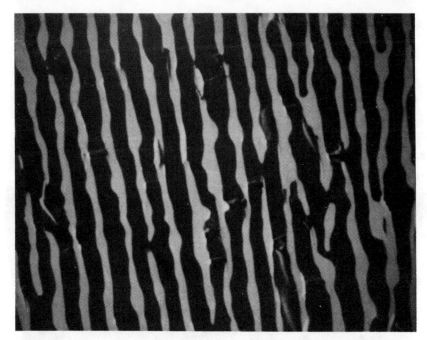

Fig. 4. Magnified view of the lamellar structure observed on the finally solidified portion.

decreased gradually toward that found in the fourth region. And at the boundary with the fourth region, the content of Al_2O_3 exhibited the smallest value ($\approx 3Al_2O_3 \cdot 2SiO_2$). The X-ray powder diffraction pattern revealed that the solid in the third region was mullite. The composition of each lamella in the fourth region was analyzed, and the results revealed that the lamellar texture consists of two phases, mullite and silica-rich glass, of which the composition was measured to be approximately $Al_2O_3 \cdot 9SiO_2$.

The Ternary System ZrO_2-Al_2O_3-SiO_2 (Starting Composition; ZrO_2:Al_2O_3:SiO_2 = 31:42:27 (Mole Ratio))

Figure 6 shows the cut and polished surface of the remaining boule after the SCFZ experiment. By means of the EPMA technique, the order of solidified bodies was measured as ZrO_2, ZrO_2 + Al_2O_3, ZrO_2 + mullite, and ZrO_2 + mullite + glass (silica-rich glass). It was a significant finding that coprecipitation of alumina and mullite was not detected. This phenomenon revealed that the ternary eutectic point that was reported by previous investigators should be interpreted as the ternary peritectic point. Other experiments in which the starting compositions were illustrated in Fig. 7 as numbers 1, 2, and 3 revealed the same results.

Thus, a revised phase diagram in this ternary system was obtained as shown in Fig. 7 by the SCFZ experiment.

In this SCFZ experiment, the melting temperatures have not been yet measured and, therefore, the temperature scale is not given in Fig. 7. As has already been pointed out, the temperature measurement is necessary to complete the phase diagram; however, fundamental phase relations can be obtained from the solidi-

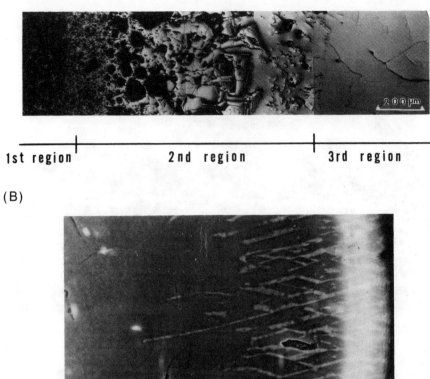

Fig. 5. Part of the magnified view of the initially solidified region (A) and the lamellar structure in the finally solidified region (B) (starting composition: $Al_2O_3:SiO_2 = 3:1$ (mole ratio)).

fication scheme regardless of the lack of the temperature measurement. The presence of the compounds and/or their melting nature are clearly detected. Furthermore, the existence and range of solid solubility, if any, can also be identified easily as shown in the case of the system $MgO-TiO_2$.[4]

The most characteristic results of this SCFZ experiment are both the identification of the ternary peritectic point which had been believed to be the ternary eutectic point and the confirmation of the incongruent melting nature of mullite.

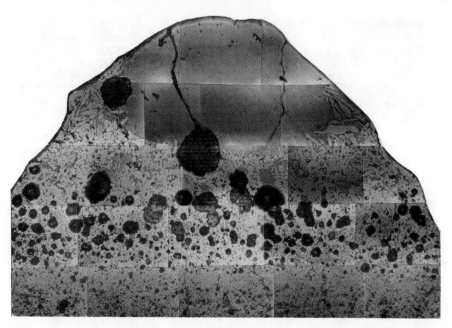

Fig. 6. Surface structure of the remaining boule after the SCFZ experiment (starting composition; $ZrO_2 : Al_2O_3 : SiO_2 = 31 : 42 : 27$ (mole ratio)).

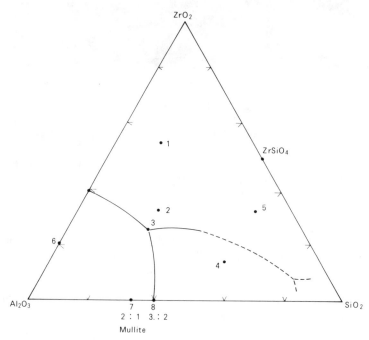

Fig. 7. Revised phase diagram in the system ZrO_2-Al_2O_3-SiO_2 by the SCFZ experiment. The number corresponds to the starting composition.

Acknowledgment

The authors thank Drs. Kimura and Okamura for their critical reading of the manuscript and also Mr. Bannai for providing samples.

References

[1] P. P. Budnikov and A. A. Litvakovskii, *Dokl. Akad. Nauk SSSR*, **106**, 268 (1956).
[2] Glacomo Cevales, *Ber. Dtsch. Keram. Ges.*, **52**, 319 (1975).
[3] M. H. Qureshi and N. H. Brett, *Trans. Br. Ceram. Soc.*, **67**, 569 (1968).
[4] I. Shindo, *J. Crys. Growth*, **50**, 839 (1980).
[5] N. L. Bowen, *Proc. Natl. Acad. Sci. U.S.A.*, **27**, 301 (1941).
[6] W. G. Pfann, Zone Melting, 2d ed. Wiley & Sons, New York, 1966.
[7] W. A. Tiller, *AIME*, **215**, 555 (1959).
[8] T. Akashi, K. Matsumi, T. Okada, and T. Mizutani, *IEEE Trans. Magn.*, **5**, 285 (1969).
[9] G. R. Fisher, L. J. Manfredo, R. N. McNally, and R. C. Doman, *J. Mater. Sci.*, **16**, 3447 (1981).
[10] F. Schmid and D. Viechnicki, *J. Mater. Sci.*, **5**, 470 (1970).

This study was supported by special coordination funds for promoting science and the technology of science and the Technology Agency of the Japanese Government.

Section II
Transformation Toughening—Mechanical Aspects

Toughening Mechanisms in Zirconia Alloys.................. 193
  A. G. Evans

A Thermodynamic Approach to Fracture Toughness in PSZ.... 213
  R. J. Seyler, S. Lee, and S. J. Burns

R-Curve Behavior in Zirconia Ceramics..................... 225
  M. V. Swain and R. H. J. Hannink

Residual Surface Stresses in Al_2O_3-ZrO_2 Composites.......... 240
  D. J. Green, F. F. Lange, and M. R. James

Calculations of Strain Distributions in and around ZrO_2 Inclusions... 251
  S. Schmauder, W. Mader, and M. Rühle

In-Situ Observations of Stress-Induced Phase Transformations in ZrO_2-Containing Ceramics............................... 256
  M. Rühle, B. Kraus, A. Strecker, and D. Waidelich

In-Situ Straining Experiments of Mg-PSZ Single Crystals...... 275
  L. H. Schoenlein, M. Rühle, and A. H. Heuer

Theoretical Approach to Energy-Dissipative Mechanisms in Zirconia and Other Ceramics............................ 283
  W. Pompe and W. Kreher

Microcracking Contributions to the Toughness of ZrO_2-Based Ceramics.................................... 293
  K. T. Faber

Microcrack Extension in Microcracked Dispersion-Toughened Ceramics.......................... 306
  F. E. Buresch

Toughening Mechanisms in Zirconia Alloys

A. G. Evans

University of California
Department of Materials Science and Mineral Engineering
Berkeley, CA   94720

Three predominant toughening mechanisms are operative in ZrO_2 alloys: transformation-toughening, microcrack-toughening, and deflection-toughening. Transformation- and microcrack-toughening are shown to involve crack shielding. Such toughening mechanisms are temperature- and particle-size-sensitive. Transformation-toughening is the more potent of the crack-shielding mechanisms, primarily because of the nondamaged character of the transformed material ahead of the crack tip. Deflection-toughening exhibits less potential than transformation-toughening but has the advantage that the toughness is retained to elevated temperatures and is particle-size-insensitive.

Toughening in ZrO_2 alloys can proceed by several mechanisms, acting independently or in concert. The predominant mechanisms include transformation toughening,[1,2] microcrack toughening,[3,4] and crack-deflection toughening.[5] However, the microstructural characteristics that determine toughening differ for each mechanism. It is imperative, therefore, that the mechanisms be well quantified in order to distinguish their realms of predominance and, hence, to design optimal microstructures. The intent of the present article is to provide a critical appraisal of toughening mechanisms, with the primary intent of establishing the quantitative basis needed to optimize the microstructures of ZrO_2 alloys.

Two fundamentally different approaches have been used to model toughening: a mechanics approach[2,6] and a thermodynamic approach.[1,6-8] Recently, it was demonstrated[6] that the two approaches yield equivalent predictions of transformation-toughening, as required, of course, from any self-consistent treatment. Naturally, therefore, the self-consistent models are afforded primary attention and are used to compare and contrast the various toughening mechanisms.

Transformation Toughening

Mechanics

Background: From the mechanics perspective, transformation-toughening can be regarded as a crack-shielding process[2,6] (Fig. 1), or, equivalently, a residual stress-related toughening. The characteristics of crack-shielding are predicated on the nonlinearity of the stress-strain curve at stresses exceeding a critical transformation stress, σ_{ij}^c (Fig. 1(b)). Specifically, on transformation, the strain exhibits a discontinuity related to the transformation strain, e_{ij}^T. This strain discontinuity is reflected in a corresponding change in the stress ahead of the crack tip (Fig. 1(c)), within the transformation zone (Fig. 1(a)). The crack-tip stress is lowered by the transformation, resulting in crack shielding, provided that the transformation strain has the same sign as the crack-tip strain. Thus, the dilatational transformation strain must be *positive* (i.e., a volume increase) in order to induce crack-shielding.

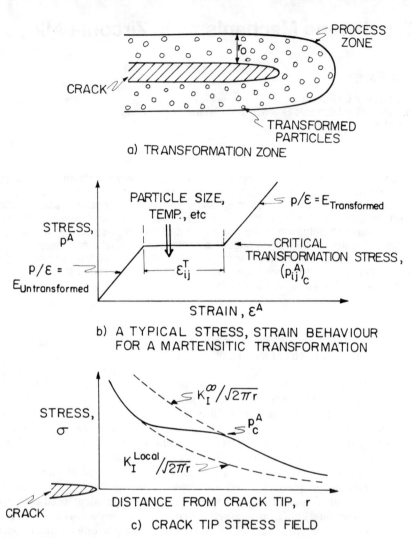

Fig. 1. Schematic illustrating the basis for crack shielding by a transformation zone: (a) the fully developed zone; (b) the stress-strain curve of the transforming material; (c) the modified stress field ahead of the crack tip.

The association between crack shielding and the change in toughness is determined by the crack-advance mechanism. For brittle materials, crack growth generally occurs by direct advance of the crack tip[9] and, hence, the crack-tip stress is of primary significance. In more ductile materials, the stress ahead of the crack tip may be more significant.[10] In the near-tip region, the linearity of the stress-strain curve at large strains (Fig. 1(b)) dictates that the near-tip stress σ_t be characterized by a stress intensity factor, K_I. This stress intensity is lower than the stress intensity,

K_∞, associated with the applied field.* The reduced stress intensity

$$\Delta K \equiv K_\infty - K_I \qquad (1)$$

corresponds directly with an increase in toughness. Specifically, since the crack propagates when K_I attains the toughness of the fully transformed material ahead of the crack tip, K_T^c, the applied stress intensity at crack advance becomes

$$K_\infty^c = K_T^c + \Delta K_c \qquad (2)$$

The measured toughness increase due to transformation is primarily determined by ΔK_c. However, it is important to recognize that ΔK_c superposes on K_T^c and *not* on the toughness of the untransformed, tetragonal ZrO_2. This issue will be more comprehensively addressed in subsequent sections.

Determination of ΔK: The crack-shielding process is most conveniently examined for a *dilatational* transformation strain, e^T. A brief discussion of the additional effects of shear is presented later. The problem of a crack embedded in an untransformed solid is examined first. Load application results in a frontal transformation zone, of dilatational profile, intersecting the crack tip (Fig. 2). In this

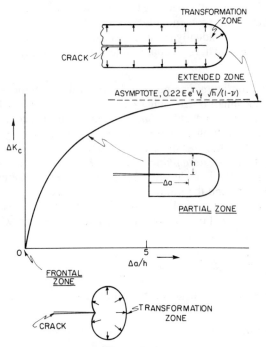

Fig. 2. Schematic showing the R curve and the associated transformation zone shapes.

---

*A similar reduction in the near-tip crack opening δ_t is another necessary consequence of crack shielding, as governed by the scaling relations,[11] $\sigma_t \approx \delta_t \approx K_I$.

instance, ΔK, determined by integrating the tractions along the transformation-zone boundary, induced by the transformation, reveals that ΔK is zero.[2] Hence there is *no toughening*.

The second problem concerns a transformation zone that extends fully over the crack surfaces (Fig. 2). In this case, closure tractions are exerted over the crack surface, in addition to the zone boundary tractions. Integration of these tractions, over the transformation zone, yields evidence of crack shielding, such that[2,6]

$$\Delta K_c = 0.21 E e^T V_f \sqrt{h}/(1 - \nu) \tag{3}$$

where E is Young's modulus of the material,[†] ν Poisson's ratio, h the transformation zone width, V_f the volume fraction of material susceptible to transformation, and 0.21 the constant determined by the dilatational shape of the transformation zone.

The widely differing influences of these two zone configurations can be rationalized from residual stress considerations. The transformation zone is residually *compressed* after transformation, while the untransformed matrix is subject to tangential tension. Furthermore, since the crack-tip stress, $K/\sqrt{r}$, is substantially larger than the residual stress, as $r \rightarrow 0$, the near-tip stress is influenced by only that portion of the residual field perturbed by the crack (which concentrates at the crack tip). Consequently, because the crack does not enter the transformation zone in the frontal zone problem (Fig. 2), significant crack shielding does not occur. Conversely, the residually compressed extended zone (Fig. 2) is *perturbed* by the crack and thus transfers an appreciable compressive stress concentration to the crack tip.

R-Curve Effects: Configurations midway between the two zone-shape extremes result in intervening values of ΔK_c (Fig. 2), resulting in R curve behavior, wherein ΔK_c increases with crack advance, Δa.[2] Normally, the introduction of a stable crack, e.g., during machining, generates a transformation zone *over* the crack surface, at least when the transformation is *irreversible*. The R curve is then restricted in extent, and the *asymptotic* value of ΔK_c given by Eq. (4) should apply.

If an annealing step is included, after incorporation of the crack, the transformation zone reverts and the full R curve should be experienced. Then, the ΔK at instability depends on the applied loading, according to the usual requirement that $dK/da > d\Delta K_c/da$. For example, an analytic expression with the approximate form of Fig. 2 is given by

$$\Delta K_c \approx \left[\frac{0.44}{\pi(1-\nu)}\right] E e^T V_f \sqrt{h} \tan^{-1}\left(\frac{\Delta a}{h}\right) \tag{4}$$

Furthermore, for a small crack subject to uniform tension, σ_∞,[11] the stress intensity is given by

$$K = (2/\pi)\sigma_\infty(a_o + \Delta a)^{1/2} \tag{5}$$

where a_o is the initial crack radius. For this case, *instability* occurs when

$$\Delta K \approx \left[\frac{0.44}{\pi(1-\nu)}\right] E e^T V_f \sqrt{h} \tan^{-1}\left[\frac{2}{\sqrt{\pi}}\sqrt{\frac{a_o}{h}}\right] \tag{6}$$

[†]The modulus E is the *average* modulus of the composite, which is assumed to be about the same before and after transformation. The average is applicable provided that h is larger than the spacing between ZrO_2 particles. Equation (4) can thus be used for *any* ZrO_2-containing toughened material (PSZ, TZP, ZTA).

Thermodynamics

The thermodynamics of crack advance can be examined by applying conventional Griffith concepts[1,7,11] or adopting energy-balance integrals.[6] It will be demonstrated that the latter approach permits a more rigorous analysis and also offers the physical insights needed to address more complex toughening situations.

The Griffith Approach: In the Griffith approach, the total energy change associated with a crack increment da is given by[7]

$$dU = -\mathcal{G}\, da + 2\Gamma\, da + d\phi + dU_D \tag{7}$$

where $\mathcal{G}\, da$ is the mechanical energy released, $2\Gamma\, da$ the energy required to create new fracture surfaces, $d\phi$ the change in potential associated with the transformed particles, and dU_D the energy dissipated by the transforming particles. At Griffith equilibrium, $dU/da = 0$ and $\mathcal{G} = \mathcal{G}_c$ and, hence

$$\mathcal{G}_c = 2\Gamma + d\phi/da + dU_D/da \tag{8}$$

Furthermore, the quantity 2Γ is the toughness in the absence of transformation, and thus the increase in strain energy release due to the transformation becomes

$$\Delta \mathcal{G}_c = d\phi/da + dU_D/da \tag{9}$$

Further progress is achieved by noting that the potential change associated with the transformation of individual tetragonal particles is[1,7]

$$\Delta\phi = -\Delta F_{chem} + \Delta U_T - \Delta U_I \tag{10}$$

where ΔF_{chem} is the chemical free energy change, ΔU_T includes both the strain energy change and the changes in interface energy, and ΔU_I is the interaction energy with the local stress. A plot of the changes in energy that accompany crack advance subject to *irreversible* transformation (Fig. 3) indicates that the net energy increase occurs solely behind the crack tip,[7] manifest as a change in the interaction energy[‡] (region I in Fig. 3). In front of the crack, a balance *always* exists between the decrease in interaction energy (region II) and the energy dissipation (region III). The interaction energy per unit volume is given, at *constant* stress, by[1,7]

$$\Delta U_I = \sigma e^T \tag{11}$$

Hence, if it is assumed (see later) that the transformation at the zone boundary occurs at constant stress, the strain energy release rate can be related to the interaction energy U_I^c at the critical transformation stress, σ_c (Fig. 3), by[7]:

$$\Delta \mathcal{G}_c = V_f \int_{-h}^{h} \Delta U_I^c\, dy$$

$$\equiv 2V_f e^T \sigma_c h \tag{12}$$

By making the association that the critical stress is related to the zone height by[2,6]

$$h = \left(\frac{\sqrt{3}(1+\nu)^2}{12\pi}\right)\left(\frac{K}{\sigma_c}\right)^2 \tag{13}$$

[‡]Note that thermodynamic equilibrium requires energy conservation at the transformation condition and, hence, from Eq. (10), $\Delta U_I^c \approx -\Delta F_{chem} + \Delta U_T + \Delta U_D$. Consequently, the interaction energy can, if desired, be specifically related to the chemical-energy and strain-energy changes and the energy dissipation. Furthermore, if the transformation zone is small (about one particle diameter in width), $\Delta U_T \approx 0$, because the crack surface relaxes the transformation strain. Then, in the absence of dissipation, $\Delta U_I \approx -\Delta F_{chem}$, and the change in crack-propagation resistance relates primarily to the change in chemical free energy.[8]

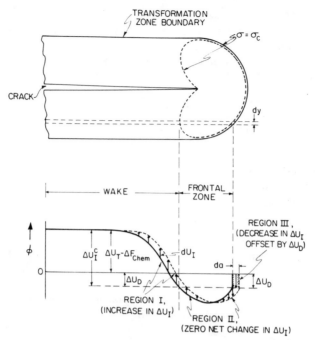

Fig. 3. Change in energy in a strip, of width dy, that occurs during a crack advance, da. Region I is a zone of interaction energy change behind the crack tip. Region II is a zone ahead of the crack tip in which the increase and decrease in interaction energy exactly counteract. Region III is a zone in which there is a decrease in interaction energy exactly equal to the energy dissipation ΔU_D. The interaction energy changes are indicated by the arrows. Also shown are the various energy-change magnitudes.

and recalling that[11] $K^2(1 - \nu^2)/E = \mathcal{G}$, it has been demonstrated by Budiansky et al.[6] (Appendix) that $\Delta \mathcal{G}_c$ given by Eq. (12) is *identical* to ΔK_c predicted by Eq. (3).

The existence of an R curve is also predicted,[7] because of the predominant influence of the interaction energy *behind* the crack tip on $\Delta \mathcal{G}$. It is also evident that, if the transformation is fully *reversible,* the transformation zone must be confined ahead of the crack tip and, hence, toughening does not occur.

Energy-Balance Integrals: Analogous results can be derived by examining energy-balance integrals.[6] Specifically, for the frontal zone (Fig. 2), since the volume elements within the transformation zone do not experience unloading, the path-independent J integral applies. Consequently, the relation[6,11]

$$J = (1 - \nu^2)K^2/E \qquad (14)$$

pertains for all line contours around the crack tip. Hence, since the elastic modulus of the transformed and untransformed materials are essentially the same, contours around the tip (which give K_l) and remote from the tip (which give K_∞) yield identical values of K, giving $K_l = K_\infty$.

Conversely, when a fully developed zone exists (Fig. 2), the material within the zone, behind the crack tip, has experienced unloading and a path-independent

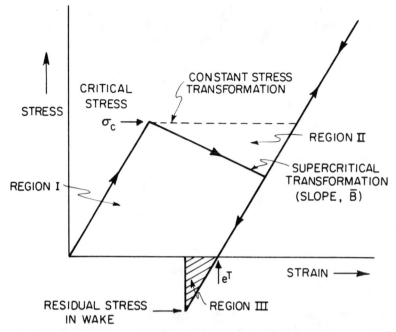

Fig. 4. Stress-strain curve for an element of material subject to martensitic transformation. Region I is an area attributed to transformation at constant stress. Region II is an area that allows for the stress decrease due to supercritical transformation. Region III is the strain energy stored in the wake.

J does not apply. In this case the appropriate energy-balance integral, I,[6] has the same form as J at the tip

$$I = (1 - \nu^2)K_I^2/E \tag{15a}$$

but remote from the tip becomes

$$I = (1 - \nu^2)K_\infty^2/E - 2\int_0^h U(y)\,dy \tag{15b}$$

where $U(y)$ is the residual energy density in the wake. Equating the magnitudes of the conservation integral for the near tip and remote paths thus gives

$$K_\infty^2 = K_I^2 + [2E/(1 - \nu^2)]\int_0^h U(y)\,dy \tag{16}$$

or

$$\Delta \mathcal{G} = 2\int_0^h U(y)\,dy \tag{17}$$

Comparison of Eq. (17) with Eq. (12) indicates that $U(y)$ is related to the interaction energy. In fact, the determination of U involves procedures similar to the interaction energy calculations described above. However, the evaluation of $U(y)$ can be conducted more rigorously than ΔU_I by relating $U(y)$ to features of the stress-strain curve (Fig. 4). In addition, a superior physical appreciation of the toughening can be gained. Specifically, it is recognized that the material in the wake undergoes a complete loading and unloading cycle as the element translates from the front to the rear of the crack tip during crack advance. Hence, each element in the wake is subject to the residual stress work contained by the hysteresis loop (Fig. 4). Consequently, by appreciating that the wake is subject to a residual compression, due to transformation (Fig. 4), given by[6]

$$\sigma_{11} = \sigma_{33} = Ee^T V_f / [3(1 - \nu)] \qquad (18)$$

the residual energy density can be readily evaluated as

$$U(y) = \sigma_c e^T V_f + \frac{\overline{B}(e^T V_f)^2}{2(1 - \overline{B}/B)} + \frac{E(e^T V_f)^2}{9(1 - \nu)} \qquad (19)$$

where B is the bulk modulus and $\overline{B}$ is the slope of the stress-strain curve of the transforming material (Fig. 4). The three terms in Eq. (18) derive from the areas I, II, and III under the stress-strain curve depicted in Fig. 4. The latter two terms in Eq. (19) cancel when

$$\overline{B} = -2E/[3(1 + \nu)] \qquad (20)$$

and Eq. (17) for the toughening is then *identical* to Eq. (12), viz,

$$\Delta \mathcal{G}_c = 2V_f e^T \sigma_c h$$

The above condition for $\overline{B}$ has been determined to obtain at the supercritical condition,[6] wherein *all* susceptible regions within the transformation zone are *fully transformed*. Partial transformations within the zone yield a reduced toughening,[6] as summarized in Fig. 5.

The prediction of toughness based on the hysteresis loop represents a useful device for analyzing more complex toughening situations, as noted later. In addition, it can now be appreciated that the solution for $\Delta \mathcal{G}_c$ determined from the interaction energy change, ΔU_I, was perhaps fortuitous, because the supercritical transformation does not occur at constant stress. However, the energy deficit accompanying the stress drop during transformation (region II in Fig. 4) is exactly offset by the stored strain energy in the wake (region III in Fig. 4) and the residual stress work is, in fact, identical to the interaction energy change.

Finally, it is reemphasized that any stress-strain curve that involves a transition slope less than $\overline{B}$ in Eq. (20) still gives a residual stress work characterized by Eq. (12).[6]

Trends in Toughness

It is immediately evident from Eqs. (3) and (12) that the transformation-induced toughening increases as the transformation zone size increases, or, equivalently, as the critical stress decreases. Specific comparisons with experiments, based on the measured zone size, indicate that the dilational component of transformation-toughening consistently accounts for ½–⅔ of the measured toughness.[2,6] The remaining source of toughness derives from crack deflection and perhaps, in some cases, from the deviatoric component of the transformation strain, as discussed in subsequent sections. It is pertinent here to discuss the trends in transformation-toughening that derive for corresponding trends in the critical stress, as a basis for interpreting experimental observations.

The critical stress is dictated by the nucleation of the martensitic transformation,[12] a phenomenon that is not well understood at the quantitative level. Nevertheless, it seems reasonable to suppose that the critical stress diminishes with

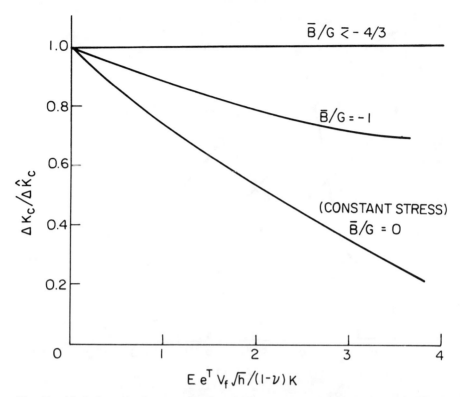

Fig. 5. Variations in the toughening ΔK_c, compared with the supercritical toughening $\Delta \hat{K}_c$, for subcritical transformations, with $\overline{B}/G > -4/3$ (Ref. 6).

supercooling and tends toward its minimum value at temperatures just above M_s. Consequently, ΔK_c must decrease as the temperature increases, as commonly observed[8,13] (Fig. 6(a)). However, at temperatures below M_s, the thermally transformed material does not contribute to crack shielding (i.e., does not result in a long-range residual compression over the crack surface). Hence, ΔK_c must diminish, as depicted in Fig. 6(a). The toughness trend is modified by the presence of solute, consistent with the effect of the solute on M_s (Fig. 6(a)). More quantitative predictions of trends in ΔK_c with temperature await further understanding of the nucleation process.

The M_s temperature is also influenced by particle size and shape, such that larger or more angular particles exhibit larger M_s temperatures.[14] Again, therefore, trends in particle size can be predicated on the analogous trends in M_s, as summarized in Fig. 6(b). Optimum toughening is thus contingent on the development of a system containing particles with a narrow range of M_s, just below the use temperature.

Microcrack-Toughening

General Analysis

The toughening induced by a microcrack zone or a profuse zone of bifurcation involves crack shielding.[4] The shielding magnitude can be most conveniently

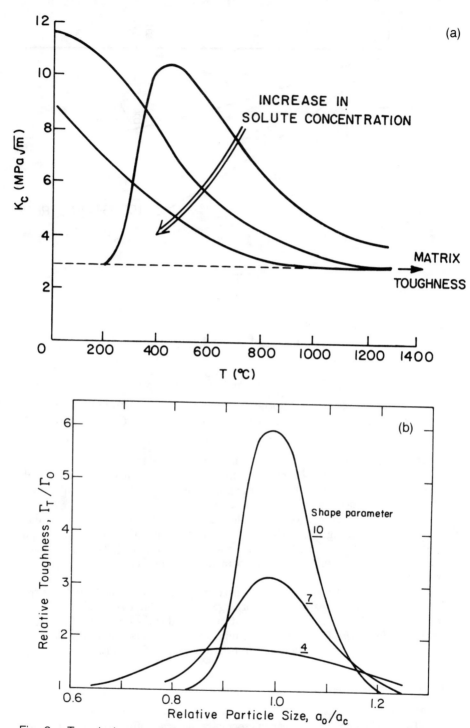

Fig. 6. Trends in transformation-toughening: (a) effects of temperature and solute content; (b) effect of particle size.

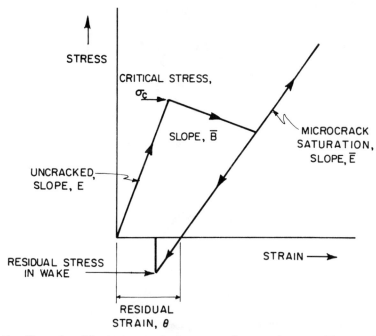

Fig. 7. The simplified stress-strain curve for a microcracking material.

analyzed using the energy-balance integrals described in the preceding section. For this purpose, the stress strain-curve depicted in Fig. 7 is used. Several principal features of the curve dictate the magnitude of the crack shielding, ΔK. First, microcracking invariably occurs in response to a localized residual tension, due to thermal expansion mismatch or a phase transformation. Hence, a residual microcrack opening (Fig. 8) and an accompanying dilatation necessarily obtain. This effect appears as the permanent strain in Fig. 7. Second, microcracked solids exhibit a lower elastic modulus $\bar{E}$ than that associated with uncracked solids, E. This phenomenon results in one of the important differences between microcrack-toughening and transformation-toughening, elucidated below. The slope of the unloading curve in Fig. 7 reflects this modulus reduction. Third, for simplicity of analysis and for ready comparison with transformation-toughening, the microcracking is assumed to be supercritical, as characterized by the transformation slope, $\bar{B}$, specified in Eq. (20). Implicit in this premise is that microcracking occurs at all potential microcrack sites within the microcrack process zone. An *upper bound* crack shielding is thus implied. Finally, a restriction is imposed on the microcrack site density to allow the material to assume linearity,[§] at large strains, when microcracking saturates. The large strain linearity allows the near-tip crack field to be characterized by K_I and thus permits the crack shielding to be expressed in terms of ΔK. When microcrack saturation does not occur, an alternative description, based on J, will be required, and the problem becomes substantially more complex.

[§]The specific allocation of a microcrack site density probably restricts the analysis to multiphase materials, such as ZrO_2 alloys, that microcrack at second-phase sites.

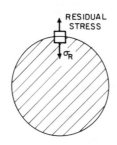

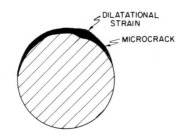

Fig. 8. A schematic showing the dilatation that accompanies microcracking, following relief of a residual tension.

When the process zone is contained ahead of the crack, unloading does not occur, and J is essentially path-independent. Hence, from Eq. (14)

$$K_I^2 = K_\infty^2(\overline{E}/E)[(1 - \overline{\nu}^2)/(1 - \nu^2)] \tag{21}$$

Crack shielding thus occurs, by virtue of the reduced modulus $\overline{E}$ of the process zone. However, this reduction is largely offset by the reduced crack propagation resistance of the microcracked material within the process zone. The effective toughness can be estimated from the following relations. The degradation in crack-growth resistance is given approximately by[17]

$$\overline{K}_c = K_c(1 - f_s) \tag{22}$$

where $\overline{K}_c$ is the critical stress intensity of the microcrack-damaged material, K_c the toughness of the solid without microcracks, and f_s the saturation density of microcracks.¶ The modulus decrease is given by[16]

$$\overline{E}/E \approx 1 - 16(1 - \overline{\nu}^2)(10 - 3\overline{\nu})f_s/45(2 - \overline{\nu}) \tag{23}$$

where $\overline{\nu} = \nu[1 - 16f_s/9]$. Equating K_I to $\overline{K}_c$, Eqs. (21), (22), and (23) yield a toughness

$$K_\infty^c = K_c(1 - f_s)[1 - 16(1 - \overline{\nu}^2)(10 - 3\overline{\nu})f_s/45(2 - \overline{\nu})]^{-1/2} \tag{24}$$

The toughness is thus exclusively dependent on the site density of microcracks near

---

¶More specifically, f_s is a dimensionless quantity given by the product,[16] $N_s\langle a\rangle^3$, where N_s is the maximum number of microcracks that can form, at saturation, per unit volume, and $\langle a\rangle$ is the average radius of the microcracks.

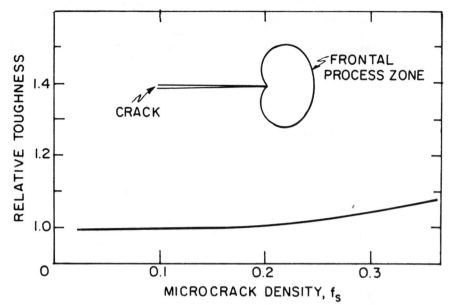

Fig. 9. Effects of a frontal microcrack process zone on toughening.

the crack tip. A plot of K_∞^c/K_c as a function of f_s (Fig. 9) reveals[4] that the measured toughness K_∞^c is essentially unchanged by microcracking, at least for the range of $f_s (\gtrsim 0.3)$ wherein Eqs. (22) and (23) are regarded as reasonable approximations.

For the fully developed process zone, the energy-balance integral gives

$$\frac{K_\infty^2(1-\nu^2)}{E} = \frac{K_I^2(1-\bar{\nu}^2)}{\bar{E}} + 2\int_0^h U(y)\,dy \tag{25}$$

For consistency with the transformation-toughness calculation, it is again assumed that a hydrostatic stress, σ_c, activates the microcracks.** Then, the hysteresis loop in Fig. 7 dictates that[4]

$$U(y) = \sigma_c\theta + E\theta^2/9(1-\nu) + \left(\frac{3\sigma_c(1-2\nu)}{2}\right)\left(\frac{1}{\bar{E}} - \frac{1}{E}\right) \tag{26}$$
$$+ \bar{B}\frac{[\theta + 3\sigma_c(1-2\nu)(1/\bar{E} - 1/E)]^2}{2[1 - 3\bar{B}(1-2\nu)/\bar{E}]}$$

where θ is the dilatational strain that results from partial relief of the residual stress by microcracking. The additional complexity of this function, vis-a-vis the transformation problem (Eq. (19)), derives from the change in modulus induced by the microcracks. Combining Eqs. (25) and (26) and inserting the appropriate relations for $\bar{E}$ (Eq. (23)) and $\bar{B}$ (Eq. (20)) allows the toughness to be expressed in the form

$$\Delta K_\infty^c/K_c = F(\theta f_s\sqrt{hE}/K_c, f_s) \tag{27}$$
$$\approx 0.4f_s + 0.25f_s\sqrt{hE}\theta/K_c$$

**A critical value of the principal tensile stress is probably more pertinent for randomly oriented residual stress sites. However, the results for this stress condition are similar to those deduced for the hydrostatic stress.

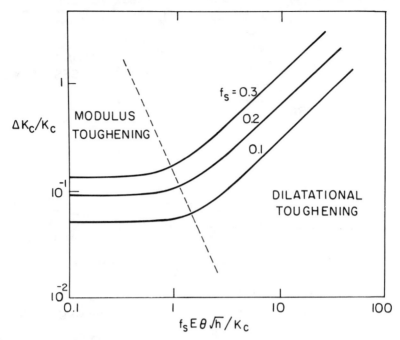

Fig. 10. Trends in toughness with zone size, h, or dilatation, θ, for several saturation microcrack densities, f_s.

where F is the function plotted in Fig. 10.[4] Comparison with Fig. 9 reveals that the presence of the process-zone wake again enhances the toughness. Hence, microcrack-toughening must also be characterized by an R curve. In fact, recognition of this R-curve behavior rationalizes some measured effects of specimen geometry on K_∞^c.[4,18,19]

Trends in Toughness

Reasonable choices of the microcrack site density ($f_s \approx 0.3$) indicate that the crack shielding which emanates from the modulus change is relatively small, $K_\infty^c \sim 1.1 K_c$, and independent of the process zone size. More substantial, zone-size-dependent toughening emerges from the dilatation. Specifically, for a zone size of 10 μm and a dilatational strain $\theta \approx 0.02$ (pertinent to ZrO$_2$), $\theta\sqrt{hE}/K_c \approx 6$, giving $K_\infty^c \approx 1.5 K_c$. Toughening levels of this magnitude have been measured in Al$_2$O$_3$/ZrO$_2$[3] and attributed to a microcrack process zone.

The modulus-toughening, being zone-size-independent, should be insensitive to particle size and temperature. However, the dilatational-toughening depends on particle size and, to some extent, on temperature, via the dilatation, θ, and the zone size, h, reflecting the influence of temperature on the thermal expansion mismatch between tetragonal ZrO$_2$ and the matrix material (presumably, e^T is temperature-insensitive).

Crack-Deflection-Toughening

Cracks can be deflected either by localized residual fields or by fracture-resistant second phases. The deflection results in a toughening dictated by the reduced driving force on the deflected portion of the crack.[5] In particular, the

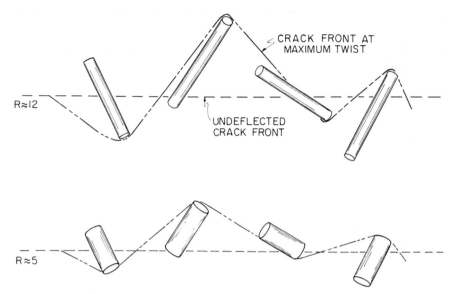

Fig. 11. Schematic illustrating the twisting of a crack between rod-shaped particles.

twisting of the crack between deflecting particles (Fig. 11) provides an appreciable reduction in driving force. The resultant toughening, for a randomly oriented second phase, depends only on the volume concentration and shape of the deflecting particles. Predictions for rod-shaped deflecting particles[5] are presented in Fig. 12. Note that high-aspect-ratio rods induce the maximum toughening, by virtue of their influence on the twist angles. These toughnesses are essentially independent of temperature and particle size.

Crack-deflection toughening in ZrO_2-containing materials has been proposed in two primary instances. The toughening observed in ZnO containing *monoclinic* ZrO_2[20] exhibits a variation with volume concentration of ZrO_2 (Fig. 13), characteristic of deflection-toughening. The deflection in this case has been induced by the localized residual stress associated with the monoclinic ZrO_2. Note that factor-of-two increases in K_c have been measured for volume fractions exceeding ≈ 0.2. Furthermore, such effects could influence the base toughness, K_c^T, on which transformation-toughening is superposed (Eq. (2)). This issue is amplified in a subsequent section.

Another probable example of deflection-toughening obtains at elevated temperatures in PSZ single crystals.[13] The tetragonal precipitates in PSZ, although resistant to transformation at elevated temperatures, evidently exhibit an increased toughness vis-a-vis fully stabilized, cubic ZrO_2. The toughening appears to depend on the shape of the tetragonal precipitates and has a magnitude reasonably attributed to crack deflection.

Discussion

Comparison of Toughening Mechanisms

Transformation- and microcrack-toughening can be explicitly compared, from Eqs. (4) and (27), for equivalent values of $Ee^T V_f \sqrt{h}$ (or $E\theta f_s \sqrt{h}$). It is immediately

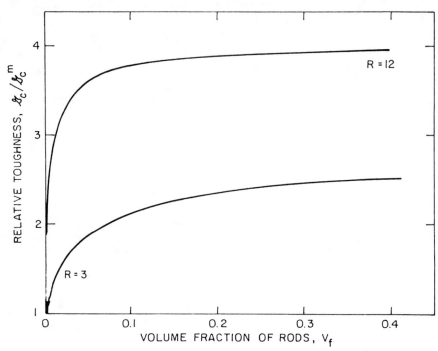

Fig. 12. Effects of rod-shaped particles on the deflection-toughening magnitude. R is the aspect ratio of the rods.

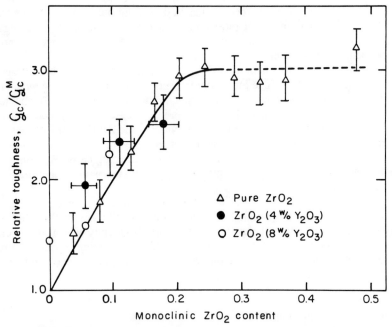

Fig. 13. Effect of monoclinic ZrO_2 on the toughness of ZnO.

evident that microcrack-toughening prevails for small $\theta\sqrt{h}$, because of the existence of zone-size-independent modulus toughening. However, the toughening levels involved are relatively small ($K^c/K_c \lesssim 1.1$). At larger $\theta\sqrt{h}$, where the toughening levels become appreciable, transformation-toughening dominates. In this range, the increased toughening in the transformation case can be considered to originate from the essentially invariant modulus (before and after transformation) as well as the nondegrading character of the material ahead of the crack tip.

The predominance of transformation-toughening is further amplified by recognition of the relatively large values of e^T/θ. Specifically, for ZrO_2, $e^T \approx 0.06$, leading to K_c values (Eq. (4)) for transformation zones several micrometers in size, approaching 10–14 MPa·$\sqrt{m}$. Such toughness levels are typical of PSZ materials with transformation-zone sizes in this range.[21] However, the effective dilatation due to release of the residual strain by microcracking is not expected to exceed 0.02. Consequently, ΔK_c values, for several micrometer-sized zones (Eq. (27)) are in the range 2–4 MPa·$\sqrt{m}$, characteristic of toughness changes in some Al_2O_3/ZrO_2 materials.[3]

Deflection-toughening has a potency similar to microcrack-toughening. However, this toughening mechanism is temperature-insensitive and offers major advantages at elevated temperatures.

Multiple Mechanisms

Several toughening mechanisms can operate simultaneously in ZrO_2 alloys. Mechanism concurrence can result in additional or counteracting influences on the net toughness. In the former category, transformation-toughening and deflection-toughening are strictly additive. The additivity is specified by Eq. (2), wherein K_c^T is the toughness attributed to deflection by the monoclinic ZrO_2. With $K_c^T \approx 4$ MPa·$\sqrt{m}$, as measured for ZrO_2 in ZnO,[20] and ΔK_c given by Eq. (2), most of the available toughness data for PSZ can be explained by a combination of deflection- and dilatational transformation-toughening.

Conversely, transformation- and microcrack-toughening are nonadditive, as schematically illustrated in Fig. 14. Thus, although the dilatation from the transformation and microcracking are essentially additive at the transformation-zone boundary (Fig. 14), the lower modulus of the microcracked material results in a permanent strain of smaller magnitude and a reduced hysteresis. The onset of microcracking, after stress-induced transformation, thus results in a toughening level *smaller* than transformation-toughening in the absence of microcracking.

The Effects of Transformation Shear Strains

Transformed monoclinic ZrO_2 particles are generally twinned.[14,15] The alternating shear strains between twins tends to minimize long-range shear stresses for transformed particles. Furthermore, the limited observations of twins in particles transformed in crack-tip stress fields[14] indicates no obvious shear strain bias induced by the crack. This combination of shear stress minimization by twinning and orientation insensitivity suggests that the transformation shear stress does not contribute importantly to the crack shielding, at least in PSZ. This implication appears to be consistent with the previously noted conclusion that the toughening is accountable in terms of the additivity of crack-deflection toughening and dilatational crack shielding.

However, it is also recognized that the shear may exert a dominant influence on crack shielding in other systems. Then, a shear strain analysis comparable to that described earlier would be required. Very preliminary calculations[22] indicate

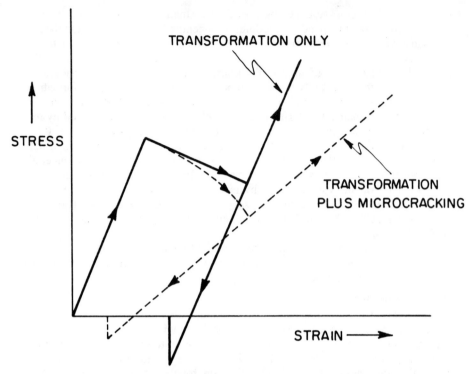

Fig. 14. The relative hysteresis loop areas for transformed material and transformed plus microcracked material.

that, when the transformation shear strain always orients with the crack-tip shear, the asymptotic crack shielding becomes

$$\Delta K \approx 0.2 \gamma_T G V_f \sqrt{h} \qquad (28)$$

where γ_T is the *net* transformation shear strain in a transformed particle and G is the shear modulus. This shielding level is, of course, an upper bound.

Concluding Remarks

It has been demonstrated that models of the toughening mechanisms pertinent to ZrO_2 alloys have attained a relatively advanced stage of development. Quantitative comparison of test data with the models is now feasible and permits the relative contributions from the various mechanisms to be distinguished.

Transformation-toughening emerges as the mechanism with the major potential because of the substantial crack shielding induced by the transformation zone. Toughness levels approaching ≈ 20 MPa·$\sqrt{m}$ seem attainable when the zone width has been optimized. However, the toughness is strongly dependent on temperature, solute concentration, particle size, and particle shape. The use temperature of the materials is thus restricted, and the processing constraints are significant.

Microcrack-toughening can also occur by means of crack shielding. However, microcracking degrades the material ahead of the crack tip and, hence, does not allow large toughness increases. Residual strain energy effects on toughening are also reduced, compared with transformation-toughening, because of the diminished elastic moduli of the microcrack process zone. Toughnesses in the range 6–10 MPa·$\sqrt{m}$ are typical of microcrack process zone development.

Deflection-toughening is particle size- and temperature-insensitive and thus has advantages at high temperatures. However, the toughening expectation is only ≈6–8 MPa·$\sqrt{m}$, about twice the toughness of the base material. Toughening is optimized in the presence of rod-shaped deflecting particles that cause appreciable crack twist. Deflection-toughening superposes on the crack-shielding mechanisms and can thus be used in conjunction with transformation-toughening.

Appendix

Equivalence of ΔK_c and $\Delta \mathcal{G}_c$

Commencing with the relation for the composite toughness

$$\mathcal{G}_c = K_c^2(1 - \nu^2)/E \tag{A-1}$$

we obtain an expression for the increase in toughness, $\Delta \mathcal{G}_c$, given by

$$\Delta \mathcal{G}_c + \mathcal{G}_c^m = \frac{(\Delta K_c + K_c^m)^2(1 - \nu^2)}{E} \tag{A-2}$$

Furthermore, since the matrix toughnesses $\mathcal{G}_c^m$ and K_c^m are similarly related ($\mathcal{G}_c^m = (K_c^m)^2(1 - \nu^2)/E$), Eq. (A–2) becomes

$$E \Delta \mathcal{G}_c = (1 - \nu^2)[2 \Delta K_c K_c^m + \Delta K_c^2] \tag{A-3}$$

Inserting $\Delta \mathcal{G}_c$ for Eq. (12) gives

$$2EV_f e^T h \sigma_c = (1 - \nu^2)[2 \Delta K_c K_c^m + \Delta K_c^2] \tag{A-4}$$

Then, substituting σ_c from Eq. (13), we obtain,

$$2\left[\frac{\sqrt{3}}{12\pi}\right]^{1/2}\left(\frac{EV_f e^T \sqrt{h}}{(1 - \nu)}\right)K = 2 \Delta K_c K_c^m + \Delta K_c^2 \tag{A-5}$$

Budiansky et al.[6] demonstrated by numerical analysis[††] that the K which defines the zone boundary via Eq. (13) is the average between K_c and K_c^m, viz,

$$K = K_c^m + \Delta K_c/2 \tag{A-6}$$

Hence, substituting Eq. (A–6) into Eq. (A–5), ΔK_c becomes

$$\Delta K_c = \left(\frac{\sqrt{3}}{12\pi}\right)^{1/2}\left[\frac{EV_f e^T \sqrt{h}}{(1 - \nu)}\right]$$

$$\equiv \frac{0.21 EV_f e^T \sqrt{h}}{(1 - \nu)}$$

which is identical to Eq. (3).

Acknowledgments

The author wishes to thank D. B. Marshall, R. M. Cannon, J. W. Hutchinson, B. Budiansky, A. H. Heuer, M. Rühle, and K. T. Faber for many useful discussions on this topic. Financial support for this work was provided by the Office of Naval Research under Contract No. N00014–81–K–0362.

[††]Although requiring numerical analysis (appendix in Ref. 6) of the influence of the inner transformations on the crack-tip field to ascertain this result, the emergence of the average is intuitively satisfying.

References

[1] A. G. Evans and A. H. Heuer, *J. Am. Ceram. Soc.*, **63** [5–6] 241–48 (1980).
[2] R. McMeeking and A. G. Evans, *J. Am. Ceram. Soc.*, **65** [5] 242–46 (1982).
[3] N. Claussen, *J. Am. Ceram. Soc.*, **59** [1–2] 49–51 (1976).
[4] A. G. Evans and K. T. Faber, *J. Am. Ceram. Soc.*, **67** [4] 255–60 (1984).
[5] K. T. Faber and A. G. Evans, *Acta Metall.*, **3**, 565 (1983).
[6] B. Budiansky, J. Hutchinson, and J. Lambroupolos, *Int. J. Solids Struct.*, **19**, 337 (1983).
[7] D. B. Marshall, M. D. Drory, and A. G. Evans; pp. 289–307 in Fracture Mechanics of Ceramics, Vol. 5. Edited by R. C. Bradt, A. G. Evans, F. F. Lange, and D. P. H. Hasselman. Plenum, New York, 1983.
[8] F. F. Lange, *J. Mater. Sci.*, **17**, 225 (1982).
[9] B. R. Lawn, B. J. Hockey, and S. M. Wiederhorn, *J. Mater. Sci.*, **15**, 225 (1980).
[10] R. O. Ritchie, J. Knott, and J. R. Rice, *J. Mech. Phys. Solids*, **21**, 393 (1973).
[11] B. R. Lawn and T. R. Wilshaw; Fracture of Brittle Solids. Cambridge University Press, New York, 1974.
[12] A. H. Heuer, N. Claussen, W. Kriven, and M. Ruhle, *J. Am. Ceram. Soc.*, **65** [12] 642–50 (1982).
[13] R. P. Ingel, D. Lewis, B. A. Bender, and R. W. Rice, *J. Am. Ceram. Soc.*, **65** [9] C-150–C-152 (1982).
[14] D. L. Porter and A. H. Heuer, *J. Am. Ceram. Soc.*, **60** [3–4] 183–84 (1977).
[15] A. G. Evans, N. H. Burlingame, W. M. Kriven, and M. D. Drory, *Acta Metall.*, **29**, 447 (1981).
[16] B. Budiansky and J. O'Connell, *Int. J. Solids Struct.*, **12**, 18 (1976). 81.
[17] J. Zwissler and M. Adams; pp. 211–41 in Fracture Mechanics of Ceramics, Vol. 5. Edited by R. C. Bradt, A. G. Evans, F. F. Lange, and D. P. H. Hasselman. Plenum, New York, 1983.
[18] Y. Fu and A. G. Evans; unpublished work.
[19] N. Claussen, B. Mussler, and M. V. Swain, *J. Am. Ceram. Soc.*, **65** [1] C-14–C-16 (1982).
[20] H. Ruf and A. G. Evans, *J. Am. Ceram. Soc.*, **66** [5] 328–32 (1983).
[21] M. V. Swain, R. H. J. Hanninck, and R. C. Garvie; pp. 339–54 in Fracture Mechanics of Ceramics, Vol. 5. Edited by R. C. Bradt, A. G. Evans, F. F. Lange, and D. P. H. Hasselman. Plenum, New York, 1983.
[22] R. M. McMeeking; private communication.

A Thermodynamic Approach to Fracture Toughness in PSZ

R. J. SEYLER, S. LEE, AND S. J. BURNS

University of Rochester
Materials Science Program
Department of Mechanical Engineering
Rochester, NY 14627

A thermodynamic analysis relating changes in fracture toughness to thermal length changes and the enthalpy of the tetragonal-to-monoclinic phase transformation in partially stabilized zirconia (PSZ) was proposed previously. The Clausius-Clapeyron equation is used to relate volume and shear phase changes to the crack-tip deformation zone. Shear transformations, which have a unique role in fracture toughening of PSZ, shield the crack tip from the applied stress intensity factor for a crack that is held stationary. Crack-tip shields from volume and normalized strain transformations are computed to be zero for stationary cracks and this deformation zone is much smaller than the shear transformation zone. Shear transformations in bulk specimens are deduced by measuring length changes vs temperature on samples that were previously subjected to a uniaxial compressive stress, i.e., with a shear component, during a thermal cycle. Thus, the constraint on the shear strain transformation in the bulk sample is directly included in the shear stress-biased thermally cycled samples undergoing tetragonal-to-monoclinic transformations. Subsequent monoclinic-to-tetragonal transformations during heating, with zero applied stress, record the shear stress-biased length changes.

The rare earth oxide ZrO_2, zirconia, is a ceramic known for its high-temperature chemical inertness. Zirconia and partially stabilized zirconia (PSZ) may be microstructurally modified by thermal treatment to improve their fracture toughness by either microcracking and/or phase transformation toughening. At present, specific guidelines for optimum phase transformation toughening of PSZ (and other materials) are not available. However, phase transformation toughening of PSZ, which is the subject of this paper, has been previously considered.[1-14]

Crack-Tip Transformation Zone

The Clausius-Clapeyron equation is used to relate a phase transformation temperature change to a hydrostatic pressure change for a specified volume change between thermodynamic phases in equilibrium.[15] Conceptually, the Clausius-Clapeyron equation represents the equality of the Gibbs free energy of both phases during transformation where pressure and volume are conjugate-state variables. A stress-induced phase transformation near a crack tip will satisfy this Gibbs equality of phases at the transformation zone boundary.

A crack-tip stress field contains significant shear stresses, and the zirconia phase change of interest, tetragonal (t) to monoclinic (m), is a shear transformation that includes a volume change. Although the shear strains have no associated volume change, a shear stress work term exists. Consequently, for the Clausius-

Clapeyron equation to adequately represent the stress-induced martensitic transformation zone in zirconia, it must include the shear components as well as the volume change.

The solid-state, crack-tip, phase transformation is derived below by equating the incremental Gibbs free energy for the tetragonal and monoclinic phases using strain volumes[16] and stresses as conjugate pairs of state variables. Define the incremental tensorial strain volume, $d\Omega_{ij}$, as the product of the volume times the incremental tensorial strains. Thus, Ω_{ij} has components for deviatoric and dilatational strains, the conjugate variables being the stresses, σ_{ij}.

At a crack tip, the stresses are not uniform so the tetragonal-to-monoclinic transformation first occurs near the crack tip in the high-stress region. At the boundary between the transformed to untransformed material, the incremental Gibbs free energy (per unit mass) will be assumed to be equal. Therefore with tetragonal as phase (1) and monoclinic as phase (2) it follows that

$$-S_{(1)}dT + \Omega_{(1)ij}d\sigma_{ij} = -S_{(2)}dT + \Omega_{(2)ij}d\sigma_{ij} \tag{1}$$

where S is the entropy and T is the temperature. The intrinsic variables T and σ_{ij} are the same in both phases. The nine strain volume stress terms are best separated into hydrostatic and deviatoric components. Recall that the sum of the dilatational strains is $\Delta v/v$ and, for plane strain, that the sum of the two nonzero deviatoric strain volume terms is the volume times the change in angle, $\Delta\gamma$, between phases. Equation (1) after rearrangement then becomes

$$(S_{(2)} - S_{(1)})dT = \frac{\Delta H}{T}dT = (v_{(2)} - v_{(1)})d\sigma + v\Delta\gamma d\tau \tag{2}$$

The entropy difference between the phases has been related to the enthalpy, ΔH, which is the heat of the transformation. $d\sigma$ is $1/3 d\sigma_{ii}$, and $d\tau$ is the incremental shear stress. Equation (2) relates the stress changes to temperature changes in the crack-tip region. The stress change is also related to the position of the phase transformation at the crack tip. A linear elastic crack-tip stress field[17] for mode I is now assumed:

$$\frac{d\sigma}{dT} = -\frac{K_1}{(2\pi)^{1/2}}\left(\frac{1}{3}\right)\frac{(1+\nu)\cos(\theta/2)}{r^{3/2}}\frac{dr}{dT} \tag{3}$$

$$\frac{d\tau}{dT} = -\frac{K_1}{(2\pi)^{1/2}}\left(\frac{\cos(\theta/2)(1 - 2\nu + \sin(\theta/2))}{4r^{3/2}}\right)\frac{dr}{dT} \quad 0° < \theta < \theta_i \tag{4a}$$

or

$$\frac{d\tau}{dT} = -\frac{K_1}{(2\pi)^{1/2}}\left(\frac{\sin\theta}{4r^{3/2}}\right)\frac{dr}{dT} \quad \theta_i < \theta < 180° \tag{4b}$$

where K_1 is the applied stress intensity factor, r the radial distance from the crack tip, ν Poisson's ratio, and θ the angle from the crack plane to the field point. The shear stress given by Eq. (4) represents the maximum shear stress in the crack-tip region under the conditions of plane strain. Equations (4a) and (4b) are equivalent at a critical angle θ_i, where $\theta_i = 2\sin^{-1}(1 - 2\nu)$. Direct substitution of Eqs. (3) and (4a) or (4b) into Eq. (2) gives a pair of differential equations for $r(T)$, which can be integrated. Choosing the integration limits as $r \to \infty$ for $T \to T_r$, the transformation temperature, gives

$$r = \left(\frac{K_I^2}{2\pi}\right)\left(\frac{v_{avg}}{\Delta H}\right)^2$$
$$\times \left\{\frac{2/3(1+\nu)\cos(\theta/2)\Delta v/v_{avg} + (1-2\nu+\sin(\theta/2))\cos(\theta/2)/2\Delta\gamma}{\ln(T/T_r)}\right\}^2$$
$$0° \leq \theta < \theta_i \quad (5a)$$

$$r = \left(\frac{K_I^2}{2\pi}\right)\left(\frac{v_{avg}}{\Delta H}\right)^2$$
$$\times \left\{\frac{2/3(1+\nu)\cos(\theta/2)\Delta v/v_{avg} + 1/2\sin(\theta)\Delta\gamma}{\ln(T/T_r)}\right\}^2 \quad \theta_i < \theta < 180° \quad (5b)$$

with $\Delta v = v_{(2)} - v_{(1)}$ and $v_{avg} = 1/2(v_{(2)} + v_{(1)})$. Figure 1 is a polar plot of r vs θ from Eq. (5), assuming a maximum shear stress and $\nu = 0.30$ ($\theta_i \approx 47°$). Figure 1(a) shows three plots: the dilatational component, assuming $\Delta v/v = 4.6\%$ with no shear strains; the shear strain component (the largest plot) with $\Delta\gamma = 16\%$

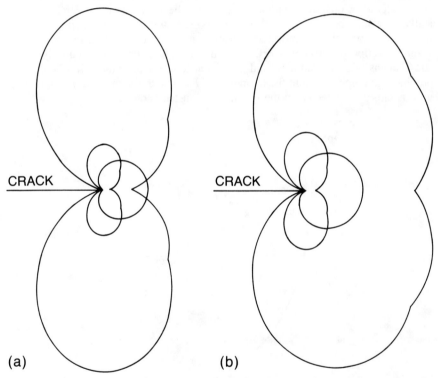

Fig. 1. Crack-tip deformation contours. (a) The large contour is for 16% shear strain only, the smallest contour is for 4.6% volume change only, and the middle contour is for 8% shear only. (b) The 4.6% and 8% contours are replotted. The large contour is both 4.6% volume and 8% shear given by Eq. (5).

and no volume term; finally, a reduced shear strain component for $\Delta\gamma = 8\%$. Selection of an 8% shear strain value is based on the assumptions that some shear strains are relieved by twinning and that precipitates are randomly oriented. Figure 1(b) again shows the 4.6% volume term and the 8% shear term, while the largest contour is the sum of the two terms as given by Eq. (5). The size and shape of each contour have been maintained for the conditions outlined above.

The size of the deformation zone, w, is obtained by considering the crack to move to the right, leaving a wake of deformed material next to the fracture surface. Equation (5) can be used to find w if a fracture criterion is used to determine the applied K_1 value that would make the crack propagate. The next section describes such a fracture criterion using crack-tip shielding concepts. However, it is worthwhile to consider the size of the deformation zone for a specified value of applied K_I. Choosing $w = r$ for $\theta = 84°$ in Eq. (5b) with $K_1 = 6$ MPa/m, $\Delta H/v_{avg} = 2.6 \times 10^8$ J/m³,[18] $\nu = 0.302$,[19] $\Delta v/v_{avg} = 0.046$,[3] and $\Delta\gamma = 0.08$ permits w vs T/T_r to be plotted as Fig. 2. T_r is the $t \rightleftharpoons m$ phase transformation temperature. Note in Eq. (5) that ΔH, $\Delta v/v_{avg}$, and $\Delta\gamma$ are approximately proportional to the volume fraction of the tetragonal phase. Since $\Delta v/v_{avg}$ and $\Delta\gamma$ are divided by ΔH, w is approximately independent of the volume fraction of transformable material. Constraint will alter ΔH and the transformation strains. Physically, it would be argued that, for an applied stress intensity factor (with no fracture criteria), the transformation occurs at some stress level, independent of volume fraction of tetragonal ZrO_2, and that the stress level is given only by the stress intensity factor applied and the distance from the crack tip. Figure 2 is presented to emphasize that the temperature T at which the material is going to be used is closely connected to the transformation temperature, T_r.

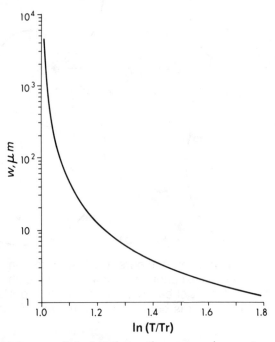

Fig. 2. Plot of the predicted deformation zone size vs temperature; see text for details.

Crack-Tip Shielding for a Phase Transformation

A more complete understanding of phase transformation toughening must incorporate a fracture criterion. The crack-tip shielding concept[20] where the crack tip is protected from the applied stress intensity factor by crack-tip deformation will be used. The true crack tip is assumed to propagate when the local crack-tip force, the strain energy release rate, $\mathcal{G}$, is equal to twice the true surface energy. The crack-tip shield is the reduction of the applied stress intensity factor by crack-tip deformation. This concept was used by McMeeking and Evans to determine the crack-tip shields for a cylindrical volume expansion using Eshelby transformations and equations from fracture mechanics.[6]

In this section, the crack-tip shields will be found using dislocation dipole models. Dislocation models of cylindrical volume expansions, normalized strain centers, and shear centers are given below. A dislocation near a crack tip interacts with the crack tip and gives a negative stress intensity factor on the sharp crack tip[21] when the dislocation relieves stresses. The dislocation therefore reduces the applied stress intensity factors by shielding the crack tip. Let the total shield due to deformation be ΔK_I^T, ΔK_{II}^T, and ΔK_{III}^T. Typically, these are negative values, and in general all three fracture modes are generated by crack-tip deformation.

Stress intensity factors at a crack tip add since stresses add. A sharp crack with an applied stress intensity factor, K_I, with crack-tip deformation has a true crack-tip driving force, $\mathcal{G}$, of

$$\mathcal{G} = \frac{(1-\nu^2)}{E}\left\{(K_I + \Delta K_I^T)^2 + (\Delta K_{II}^T)^2 + \frac{1}{(1-\nu)}(\Delta K_{III}^T)^2\right\} \tag{6}$$

where E is Young's modulus. Note that a negative ΔK_I^T will reduce the K_I applied at the crack tip, and for a given $\mathcal{G}$, the material will have a larger fracture toughness, K_{Ic}. ΔK_{II}^T and ΔK_{III}^T, if nonzero, will aid the crack in propagating for a given $\mathcal{G}$. If consideration is restricted to mode I cracks, then for symmetric deformation ΔK_{II}^T and ΔK_{III}^T must be zero. The sign and magnitude of ΔK_I^T are calculated below.

The mode I crack-tip stress intensity value from dislocation (1) with Burgers vector $\mathbf{b}_e$ in Fig. 3 in isotropic elasticity is known:[21]

$$\Delta K_I = -\frac{Eb_e}{4(1-\nu^2)(2\pi r)^{1/2}}\left\{3\sin\phi\cos\left(\frac{\theta}{2}\right) - \sin(\phi-\theta)\cos\left(\frac{3\theta}{2}\right)\right\} \tag{7}$$

Modes II and III are also known[21] but will not be considered in this text. The angle ϕ that the slip plane makes with the crack plane, see Fig. 3, can be changed for a dislocation at a fixed r, θ position.

Dislocation (2) has the opposite Burgers' vector from dislocation (1). This dipole pair has a net stress intensity value given by

$$\sum_i^2 \Delta K_I^i = \Delta K_I(\mathbf{b}, r, \theta) + \Delta K_I(-\mathbf{b}, r + \Delta r, \theta + \Delta\theta)$$

$$= \Delta K_I(\mathbf{b}, r, \theta) - \Delta K_I(\mathbf{b}, r + \Delta r, \theta + \Delta\theta) \tag{8}$$

In the incremental limit

$$d(\Delta K_I) = \left.\frac{\partial(\Delta K_I)}{\partial r}\right|_\theta dr + \left.\frac{\partial(\Delta K_I)}{\partial \theta}\right|_r d\theta$$

$$= \left.\frac{\partial(\Delta K_I)}{\partial r}\right|_\theta dr + \left(\frac{1}{r}\right)\left.\frac{\partial(\Delta K_I)}{\partial \theta}\right|_r rd\theta \tag{9}$$

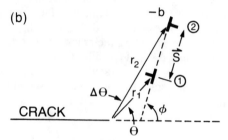

Fig. 3. (a) A dislocation dipole for expansion transformations. (b) A dislocation dipole for shear transformations.

with

$$\mathbf{S} = \Delta r \hat{\mathbf{e}}_r + r\Delta\theta \hat{\mathbf{e}}_\theta \tag{10}$$

where $\hat{\mathbf{e}}_r$ is the unit vector in the radial direction and $\hat{\mathbf{e}}_\theta$ is the unit vector in the plane of the paper perpendicular to $\hat{\mathbf{e}}_r$. The vector $\mathbf{S}$ gives the spacing between the dislocations used to model the defect. It follows from Eqs.(7), (9), and (10) that

$$\sum_i^2 \Delta K_I^i = \left.\frac{\partial(\Delta K_I)}{\partial r}\right|_\theta r\Delta\theta + \frac{1}{r}\left.\frac{\partial(\Delta K_I)}{\partial \theta}\right|_r r\Delta\theta \tag{11}$$

$$\sum_i^2 \Delta K_I^i = \frac{h(\theta, \phi)}{r^{3/2}} \tag{12}$$

where $h(\theta, \phi)$ is only a function of the defect geometry, i.e., how the defect is oriented and the space between the dislocations. The total deformation shield is found by integrating $\Sigma_i^\infty (\Delta K_I^i)$ over the crack-tip deformation zone:

$$\Delta K_I^T = \iint_{\text{area}} \sum_i^2 (\Delta K_I^i) r \, dr \, d\theta \tag{13}$$

In Eq. (13) the integration over r is done first. The stress contours, for a particular stress σ^*, are always in the form

$$\sigma^* = \frac{K_I}{r^{1/2}} \mathbf{g}(\theta) \tag{14}$$

where σ^* is typically a hydrostatic stress, shear stress, or a composite stress such as that shown in Fig. 1. Equation (13) is integrated to the edge of the stress boundary given by σ^*:

$$\Delta K_I^T = 2\frac{K_I}{\sigma^*}\int_{-\pi}^{\pi} h(\theta)g(\theta)d\theta = \frac{K_I}{\sigma^*}I \qquad (15)$$

with

$$I = 2\int_{-\pi}^{\pi} h(\theta)g(\theta)d\theta$$

From Eq. (14) it also follows that $w^{1/2} = (K_I/\sigma^*)g(90°)$. Typically, $\theta \approx 90°$ for w, although there are examples where this is not correct. Finally the total crack-tip shield due to deformation is

$$\Delta K_I^T = \frac{w^{1/2}}{g(90°)}I \qquad (16)$$

Examples

(i) Cylindrical volume expansion, Fig. 4(a): The dislocation dipole is integrated around all ϕ values to form a cylindrical expansion. This solution is already known[6]

$$\sum_{i=1}^{\infty}\Delta K_I = -\frac{Eba}{(1-\nu^2)}\left(\frac{\pi}{8}\right)^{1/2}\frac{\cos(3\theta/2)}{r^{3/2}} \qquad (17)$$

with

$$\frac{\Delta v}{v} = 2\pi a\mathbf{b} = e^T$$

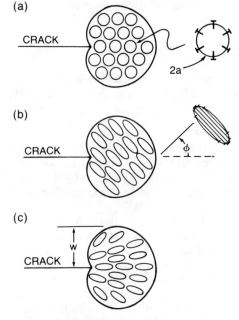

Fig. 4. Schematic representation of (a) cylindrical transformation, (b) normalized strain transformation, and (c) oriented normalized strain.

so

$$\Delta K_I^T = -\frac{Ee^T w^{1/2}}{(1-\nu^2)} I \quad I = 0 \tag{18}$$

There is no transformation-toughening due to a cylindrical volume expansion when the crack is stationary. The functional analysis in Eq. (18) agrees with Ref. 6 except for a factor of $(1 + \nu)$, which is the difference between plane stress and plane strain volume expansion.

(ii) Normalized Strain: The dislocation dipole pair in Fig. 4(b) has a normalized strain, as might be found from a lenticular precipitate that has only a normal strain due to expansion upon transformation. The value of ΔK_I for contributions inside a hydrostatic stress line is

$$\Delta K_I^T = -\frac{Ee^T w^{1/2}}{(1-\nu^2)} I \tag{19}$$

where $I = 0$ for all $\phi = $ constant.

(iii) Orientated normalized strain: Figure 4(c) shows precipitates that are oriented around the crack tip. In this example $\phi = \theta + \alpha$ and a hydrostatic stress is used:

$$\Delta K_I^T = -\frac{Ee^T}{(1-\nu^2)} w^{1/2} 0.66 \cos(2\alpha) \tag{20}$$

Thus, volume transformations do not produce crack-tip shields except for the special case of an orientated transformation.

(iv) Shear centers: Figure 3(b) shows a dislocation dipole that represents a shear center. The shear centers are placed inside a shear stress contour such as that shown in Fig. 5. It is found that

$$\Delta K_I^T = -\frac{E\gamma^T}{(1-\nu^2)} w^{1/2} 0.21 \sin(2\phi) \tag{21}$$

where γ^T is the shear strain. Thus, the shear center has a real contribution to the crack shield. The coefficient 0.21 is smaller than the value in Eq. 20, but it should be realized that the zone size, w, as given in Fig. (1), is much larger.

Measurements on Shear-Stressed PSZ Samples

It has been shown in the previous two sections that shear transformations have a unique role in fracture-toughening stationary cracks. The shear transformation strains in PSZ materials will be constrained by the matrix, the cubic phase. The measurements of constrained shear phase transformations are discussed in this section.

Consider a piece of material that is thermally cycled through the monoclinic → tetragonal → monoclinic transformation. The length of the specimen will change and, for ZrO_2, this is known to be hysteretic. When the material is subjected to a uniaxial stress, which always has a shear stress component, then the shear stresses will bias the transformation, provided the direction of the applied stress interacts with the shear component of the transformation. The applied stress should produce two effects: It should cause a shift in the transformation temperature, and the magnitude of the thermal length change should be larger since the shear strain is an order of magnitude larger than the dilatational length change.

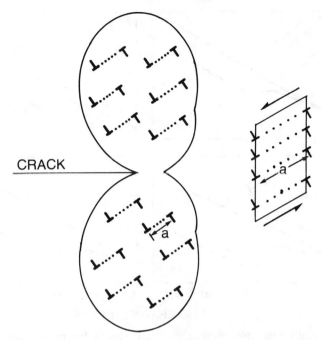

Fig. 5. Schematic representation of shear transformation.

Recall that conventional thermal expansion measurements are made with zero applied stress; so for polycrystalline samples there are no shear terms.

In Fig. 6, PSZ samples were stressed to about 200 MPa during a specified portion of an initial thermal cycle. The stress was applied quickly and held approximately constant until removed; then, all samples were quenched in liquid nitrogen ($-150°C$) to give a reference state. Specimen length change measurements were determined using a thermal mechanical analyzer (TMA) during heating in subsequent thermal cycles under approximately zero stress. Note that the absolute length changes at 30°C are arbitrary for each specimen. Figure 6(a) is for a sample with no prestressed thermal cycle. Note that no apparent $m \rightarrow t$ transformation is observed during the initial zero stress thermal cycle (upper trace) but is observed during the second.

Figure 6(b) is for a sample that was stress thermal cycled from 23°C $\rightarrow$ 400°C $\rightarrow$ 23°C. The first subsequent zero stress thermal cycle (upper trace) indicates a slight negative length change. The next zero stress thermal cycle results in a modest m to t transformation near 200°C, which is of greater magnitude than the nonprestressed sample.

The specimen in Fig. 6(c) was stressed only during the 23°C $\rightarrow$ 400°C portion of the thermal cycle. As in the case of Fig. 6(b), the first subsequent zero stress thermal cycle exhibits only a limited $m \rightarrow t$ transformation above 210°C. The second zero stress thermal cycle (lower trace) reveals a large negative length change above 180°C. The magnitude of this transformation and the thermal expansion coefficient of the lower-temperature phase seem to have been substantially stress-biased.

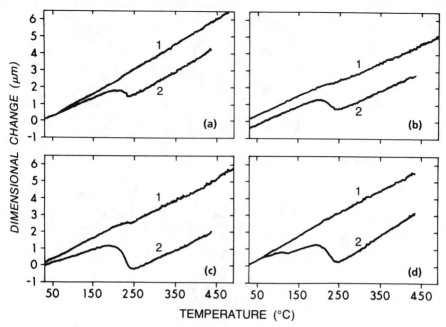

Fig. 6. Length changes during thermal cycles for previously stressed specimens.

Stress was applied during the 400°C → −55°C portion of the initial thermal cycle for the specimen in Fig. 6(d). The first subsequent TMA curve exhibits no transformation, suggesting that the tetragonal phase may have been further stabilized by the stress. A second zero stress TMA curve from −150°C for this sample shows a more complicated $m \rightarrow t$ transformation involving two negative length changes between 100° and 230°C. The total negative length change for the two transitions is again measurably greater than for the unstressed sample in Fig. 6(a).

The specific interaction of the stress with precipitates during the $m \rightarrow t$ transformation that gives rise to the two transformation steps in Fig. 6(d) has not been determined. The effect is, however, repeatable, as illustrated in Fig. 7. In this example both specimens were stressed during cooling from 400°C but differ in the temperature where the stress was removed (25°C vs −55°C). Although both specimens reveal two-step transformations of comparable magnitudes during the second TMA heating curve, the lesser initial negative length changes are observed at temperatures that differ by approximately 35°C. This shift in the lesser transformation temperature could be a stress-bias effect associated with the removal temperature. A second explanation for the temperature shift is the rate effect in the $t \rightarrow m$ transformation on cooling, observed by Matsui et al.[22] A measurably faster cooling rate may have been imposed on the sample cooled to −55°C because cold nitrogen vapors rather than room air were used to cool the sample and press.

Ideally, length changes should be measured during both the heating and cooling cycles on samples with significant shear stresses present. In our less than ideal experiments, we see evidence that the samples seem to be stress-biased in recorded TMA traces on subsequent heating. The stress bias is, we believe, a shear

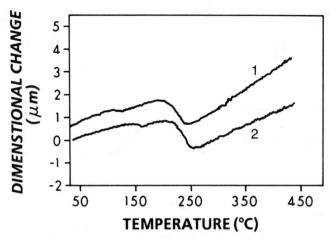

Fig. 7. Stress effects observed in length changes during thermal cycles for specimens stressed while being cooled to different temperatures.

stress bias and during these constrained transformations seems to be about as large as the volume transformation. The initial no-stress TMA traces that show no length changes are not understood but suggest an initial stabilization of the tetragonal phase imparted by the prestress.

Conclusions

An attempt has been made to provide a thermodynamic basis for evaluation of the crack-tip shielding effects imparted by the stress-induced $t \rightarrow m$ phase transformation in partially stabilized zirconia. This thermodynamic basis involves the Clausius-Clapeyron equation, which has been rederived to include both hydrostatic and deviatoric stress contributions to transformation toughening in plane strain. The hydrostatic, shear, and composite stress contours ahead of a stationary crack have been predicted, as well as the transformation-zone dimensions relative to the transformation temperature.

Four cases of transformation-induced crack-tip shielding ahead of a stationary crack were examined. The case of a hydrostatic volume change of a cylindrical particle gives no toughening (shielding), in direct agreement with the fracture mechanics arguments of McMeeking and Evans.[6] Similarly, the normalized strain case gives no transformation toughening. However, for oriented normalized strain, hydrostatic shielding was found to exist as a special case. Finally, centers of shear that can represent a martensitic transformation and therefore the large shear component of the $t \rightarrow m$ transformation in ZrO_2 shield and therefore transformation-toughen stationary cracks.

Preliminary experiments involving prestressing of commercial PSZ material in unique thermal-mechanical cycles result in measurable differences in negative length changes for $m \rightarrow t$ transformations as well as thermal expansion coefficients. These differences are believed to result from shear stress bias. Therefore, shear transformations contribute a unique role in fracture-toughening cracks in PSZ materials.

Acknowledgments

This work was supported by U.S. Department of Energy under Contract No. DE-AC02-81ER10856 at the University of Rochester. D. L. Evans kindly supplied the PSZ. Discussions with J. C. M. Li were helpful.

References

[1] R. C. Garvie, R. H. Hanninck, and R. T. Pascoe, *Nature (London)*, **258** [5537] 703–704 (1975).
[2] D. L. Porter, A. G. Evans, and A. H. Heuer, *Acta Metall.*, **27**, 1649 (1979).
[3] R. Stevens, *Trans. Br. Ceram. Soc.*, **80**, 81 (1981).
[4] A. G. Evans and A. H. Heuer, *J. Am. Ceram. Soc.*, **63** [5–6] 241–48 (1980).
[5] (a) F. Lange, *J. Mater. Sci.*, **17**, 247 (1982).
(b) F. Lange, *J. Mater. Sci.*, **17**, 255 (1982).
[6] R. M. McMeeking and A. G. Evans, *J. Am. Ceram. Soc.*, **65** [5] 242–46 (1982).
[7] R. T. Pascoe, R. H. Hannink, and R. C. Garvie, *Sci. Ceram.*, **9**, 447 (1977).
[8] R. W. Rice, R. C. Pohanka, and W. J. McDonough, *J. Am. Ceram. Soc.*, **63** [11–12] 703–10 (1980).
[9] L. K. Lenz and A. H. Heuer, *Comm. J. Am. Ceram. Soc.*, **65** [11] C-192–C-104 (1982).
[10] N. Claussen, R. L. Cox, and J. S. Wallace, *J. Am. Ceram. Soc.*, **65** [11] C-190–C-191 (1982).
[11] H. P. Kirchner, R. M. Gruver, M. V. Swain, and R. C. Garvie, *J. Am. Ceram. Soc.*, **64** [9] 529–33 (1981).
[12] N. Claussen and M. Ruhle; p. 137 in Science and Technology of Zirconia. Edited by A. Heuer and L. Hobbs. The American Ceramic Society, Columbus, OH, 1981.
[13] C. A. Andersson and T. Gupta; pp. 184–201 in Science and Technology of Zirconia. Edited by A. Heuer and L. Hobbs. The American Ceramic Society, Columbus, OH, 1981.
[14] A. G. Evans, D. B. Marshall, and N. H. Burlingame; pp. 202–16 in Science and Technology of Zirconia. Edited by A. Heuer and L. Hobbs. The American Ceramic Society, Columbus, OH, 1981.
[15] H. B. Callen; p. 161 in Thermodynamics. Wiley, New York, 1960.
[16] J. C. M. Li, *Metall. Trans. A*, **9A**, 1353–80 (1978).
[17] B. R. Lawn and T. R. Wilshaw; p. 53 in Fracture of Brittle Solids. Cambridge University Press, New York, 1975.
[18] R. C. Garvie, *J. Phys. Chem.*, **69** [4] 1238 (1965).
[19] D. L. Porter and A. H. Heuer, *J. Am. Ceram. Soc.*, **62** [5–6] 298–305 (1979).
[20] B. S. Majumdar and S. J. Burns, *Acta Metall.*, **29**, 579–88 (1981).
[21] J. R. Rice and R. Thomson, *Philos. Mag.*, **29**, 73–97 (1974).
[22] M. Matsui, T. Soma, and I. Oda; this volume, pp. 371–81.

R-Curve Behavior in Zirconia Ceramics

M. V. SWAIN AND R. H. J. HANNINK

CSIRO Division of Materials Science
Melbourne, Victoria, Australia

Rising crack resistance with crack extension, or R-curve behavior as it is termed, has been observed in a variety of magnesia-partially stabilized zirconia (Mg-PSZ) ceramics. Three distinct types of materials displaying this behavior are identified. They are (i) material containing highly metastable tetragonal (t-ZrO_2) precipitates in a cubic (c-ZrO_2) matrix, (ii) material containing considerable monoclinic zirconia (m-ZrO_2) formed by partial or complete decomposition of the c-ZrO_2 matrix phase, and (iii) highly overaged "refractory-grade" Mg-PSZ materials containing large precipitates and coarse particles of m-ZrO_2. The form of the R curve for (i) agrees with recent theoretical predictions. It is also found that substantial "plastic"-like behavior occurs during the initial loading phase. In type (ii) material, microcracking and crack branching appear to be responsible for the R-curve behavior, whereas for type (iii) material crack deflection by precipitates, together with crack branching, appears to be the mechanism responsible.

A number of recent theoretical treatments of transformation-toughening of partially stabilized zirconia (PSZ) ceramics have predicted R-curve behavior.[1-3] This R-curve behavior or rising fracture toughness with crack extension is a consequence of crack-tip shielding by the compressive transformed zone developed about the crack tip prior to extension into this zone. The extent of crack growth over which this behavior is predicted is only a few times the width of the transformed zone, typically less than 20 μm.

Observations of R-curve behavior in strong ceramics have been limited. Recent studies investigating the grain-size dependance of the fracture toughness of alumina[4,5] revealed increasing R-curve behavior with increasing grain size. Studies of PSZ materials have shown that this material exhibits considerable R-curve behavior. Green et al.[6] found that in Ca-PSZ R-curve behavior was associated with cracking at grain boundaries adjacent the main crack tip. This material had very low strength and relatively low fracture toughness. More recently, Swain[7] presented evidence for profound R-curve behavior in Mg-PSZ, particularly after "subeutectoid" heat treatment. This occurred in materials with high strength (>600 MPa) and high fracture toughness (>500 J/m^2). However, these latter observations suggested that the rise in toughness with crack extension occurred over distances as large as 500 μm. It was found that the resistance to strength degradation resulting from thermal shock was also much greater in these materials.

The aim of the present study was to identify some of the mechanisms responsible for the R-curve behavior in PSZ materials. An additional goal was to measure, if possible, the rate of change of the critical stress intensity factor or fracture toughness with crack extension.

Experimental Details

While R-curve behavior has been observed in both Mg-PSZ and Ca-PSZ, the present work will concentrate on Mg-PSZ exclusively because of the more profound effects observed in this material. The results presented here will be on three Mg-PSZ materials, containing nominally 9.5 mol% MgO fired at $\approx 1700°C$ and then cooled in a controlled manner.

Measurements of the fracture toughness were obtained in two ways from notched-bend (NB) tests: (*a*) using load, crack length, and standard fracture mechanics equations, and (*b*) from the total energy consumed during crack extension, or work-of-fracture (WOF), that is, the area under the force displacement curve divided by the projected area of surface created (2γ WOF). Additional measurements on a number of specimens were carried out using the double cantilever beam (DCB) technique and in a few cases the strength of indented specimens in bending (ISB) test. Previous work by Mussler et al.[4] indicated the usefulness of the latter test in resolving the conflict between NB and DCB results on alumina of varying grain sizes.

In one particular series of experiments, designed to examine more closely the rate of change of toughness with crack extension, relatively large plates (30 by 80 by 3.5 mm) were fractured using the DCB technique. A shallow guiding groove ≈ 0.5-mm deep was machined on each side of the plate. This was subsequently annealed at $\approx 1000°C$ for 15 min prior to testing to remove the surface compressive stresses introduced by machining. A starting notch was also introduced prior to annealing, approximately 10–15 mm beyond the loading points. In general, crack extension was very stable, and crack growth was monitored either directly — the material was sufficiently translucent for fiber optic light sources to unambiguously reveal the crack tip — or using a dye penetrant and following changes in load line compliance. It was found that the presence of dye penetrant in the crack tip increased the toughness by $\approx 10\%$.

After DCB testing, the arms of the fractured specimen were notched with a 100-μm diamond saw blade to at least 0.5 of the specimen section for NB testing after a similar annealing treatment. NB tests were performed in a 3-point bend arrangement with a moment arm of 20 mm at a crosshead displacement of 0.05 mm/min. Compliance of the specimen was monitored with a very sensitive linear variable displacement transducer (LVDT) between loading platen and load cell. It has been found that the NB test is more sensitive than the DCB technique for monitoring R-curve behavior as small changes in crack growth are more readily measurable by changes in NB compliance than by DCB compliance.

Observations and Discussion

R-curve behavior has been observed in a range of zirconia materials with critical stress intensity factors (K_{Ic}) varying from 3 to 12 MPa·m$^{1/2}$. This included materials with highly metastable t-ZrO$_2$ precipitates in the c-ZrO$_2$ matrix to materials containing only m-ZrO$_2$. For the present paper, attention will be focused on three particular types of materials that highlight the extremes of behavior. They are (1) PSZ materials containing highly metastable t-ZrO$_2$ precipitates in a stabilized matrix, in this case MgO-stabilized, (2) "subeutectoid" aged materials containing substantial amounts of decomposed m-ZrO$_2$, and (3) refractory-type PSZ materials containing large m-ZrO$_2$ precipitates in a cubic matrix together with large undissolved m-ZrO$_2$ particles. These three types of PSZ materials will now be discussed in more detail in the order outlined above. However, it is emphasized that the microstructures of many PSZ materials may be a mixture of more than one of these types.

PSZ Containing Metastable t-ZrO₂ Precipitates

These materials were prepared by firing in the solid solution c-ZrO_2 phase field at $\approx 1700°C$, cooled rapidly enough that all the precipitates were of t-ZrO_2 symmetry, and then heat-treated in either the c-MgO, t-ZrO_2 or subeutectoid region of the Mg-PSZ phase diagram. Previous work in this laboratory showed that the subeutectoid heat treatment is the more desirable as it imparts the greatest improvement in the thermomechanical properties.[8-10] These materials contain precipitates straddling the critical size range, and up to $\approx 25\%$ of the precipitate population may have been converted to m-ZrO_2 during firing and subsequent heat treatment.

Typical observations of the load displacement curves of materials with highly metastable t-ZrO_2 precipitates when NB-tested are shown in Fig. 1(a). In all cases there is a distinct break from the normal loading curve well before peak load is achieved. Most engineering ceramics fracture catastrophically under these conditions, with little deviation from linearity unless they are very deeply notched. The deviation from linearity of the compliance is more readily appreciated on subtraction of the machine and loading system compliance Fig. 1(b); details of this technique are described elsewhere.[7] The point of deviation occurs at the same displacement corresponding to almost identical stress intensity factor values at the notch tip.

Analysis of the results in Fig. 1 would usually proceed by determining secant compliance at various points on the load-displacement curve. Then, using either a computed or experimentally determined compliance crack-length relationship, the stress intensity factor with crack extension may be determined. For crack depth-to-specimen width ratios greater than 0.6, K_{Ic} values were determined with Eq. (1).[11]

$$K_{Ic} = 1.99Pl/b(w - c)^{3/2} \tag{1}$$

where P is the applied load, l the moment arm, b the breadth of the specimen, and $(w - c)$ the remaining web of material for a beam of width w containing a crack of length c. The consequence of such an approach is shown in Fig. 2. These results suggest that rising crack resistance occurs over some 300–400 μm of crack extension and that the maximum K_{Ic} is higher in an inert environment than air. The maximum values correspond closely to independent K_{Ic} evaluations on the same material with DCB and WOF (NB) techniques. However, the above approach is wrong!

The error in such an approach is readily appreciated by comparing the results in Fig. 1 with those generated in Fig. 3. In these experiments, the specimen was completely unloaded at various stages of the load deflection curve. This enabled the compliance at various stages of the loading curve to be determined and the crack extension calculated. It is immediately obvious from Figs. 3(a) and 3(b) that substantial residual crack-opening displacement occurs even prior to achievement of the peak load and that there has been little change in the compliance of the specimen despite this plastic-like behavior. In the present tests, after every loading-unloading cycle the specimen, which had been polished on both sides prior to testing, was removed from the loading system and the notch tip was examined for crack emanation and extension. When crack extension occurred it was measured to confirm the compliance calibration. Figure 4 is a micrograph of the notch tip and

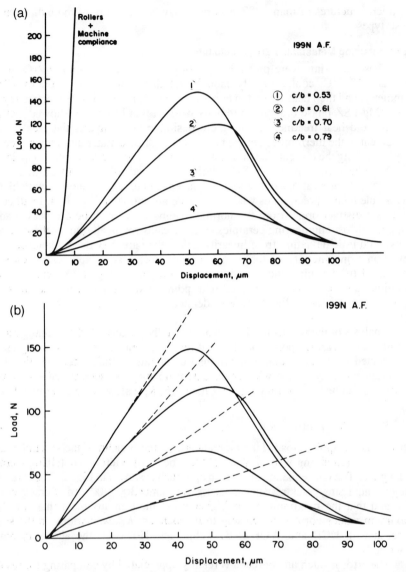

Fig. 1. Observations of the NB load deflection curve of material containing highly metastable t-ZrO_2 precipitates: (a) raw data, and (b) after subtraction of machine and testing rig compliance.

emanating crack after the third loading cycle. This figure reveals that the crack has traversed what appears to be a twinned grain at the notch tip. Observations of other crack tips in the same material supported this conclusion. Another feature of the observations in Fig. 3(b) is that the compliance curves, particularly for the longer cracks and more compliant specimen, show deflection from linearity and a residual crack opening. These results suggest that there is an expanded zone about the crack

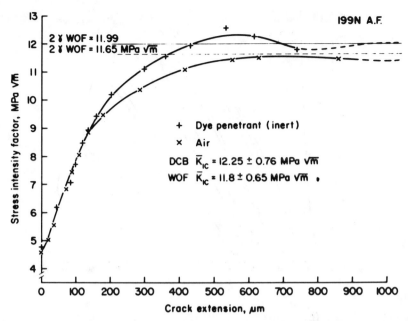

Fig. 2. Calculated R curve for the material tested in Fig. 1, using the secant compliance technique. Note that this curve is wrong, for the reasons outlined in the text.

tip preventing crack closure. Such a zone is indeed expected for transformation-toughened materials. Micrographs taken with an interferometer attached to a microscope are shown in Fig. 5 and confirm that there is a substantially uplifted zone adjacent the crack. The amount of uplift is greater nearer the crack tip, as it is this region that is preventing crack closure, whereas areas further away from the crack tip would not be in intimate contact.

Analysis of the observations in Fig. 3 was carried out as follows. The crack length, or remaining web section, was determined from the compliance which had been previously calibrated for such a specimen. The stress intensity factor was estimated from Eq. (1) with the load, P, equal to the peak load before unloading. This was repeated for all load-unloading cycles; the results are shown in Fig. 6. Two sets of data obtained on similarly sized specimens with notches of different depths ($c/w = 0.55$ and 0.58) are presented. The paucity of results on the steeply rising section of the R curve reflects the difficulty of manually unloading the specimen during the initial loading curve. The rise in toughness occurs over a crack extension of ≈ 50 μm; thereafter, the slope of the toughness crack-extension curve decreases dramatically, becoming almost horizontal. The initial phase of the R curve is significantly different from the results presented in Fig. 2, indicating the importance of permanent offset during the loading. The physical significance of this permanent offset is that, during the initial loading of the system, work is done by the loading system in generating the transformation zone about the crack tip. This work may not involve crack extension, as is implied in the interpretation of the results in Figs. 1 and 2.

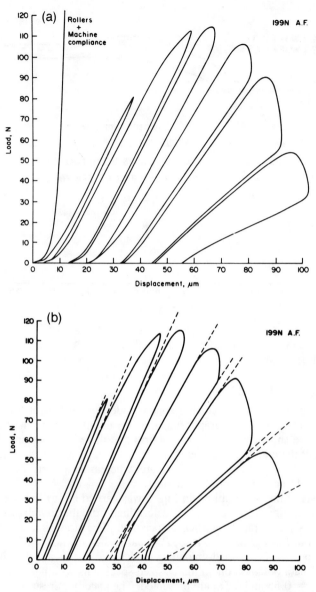

Fig. 3. Observations of the NB load deflection curve for the same material as in Fig. 1. The specimen was unloaded and reloaded at various stages during the test and a small offset introduced before reloading: (a) raw data, and (b) on removal of machine and testing rig compliance. Note the permanent offset on the initial loading curves and the nonlinearity of the compliance curves after some crack extension.

Observations of the crack-initiation region of the NB fracture surface by scanning electron microscopy (SEM) at high magnification revealed no major difference from the remainder of the fracture surface. The crack path was highly

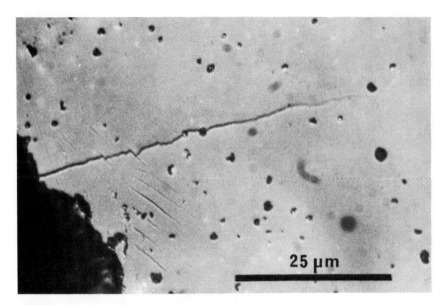

Fig. 4. Micrograph of a crack emanating from the notch root after the third load-unloading cycle. Notice the deformed grain through which the crack has propagated.

tortuous and what appeared to be precipitate relief appeared on the fracture surface. Minor crack branching was very common and it seemed to originate where a crack deviated about a precipitate. Examples of such behavior as observed with both scanning and transmission electron microscopy (TEM) are shown in Fig. 7. The TEM observation is not from the same specimen as the SEM micrograph but does contain metastable precipitates.

Fig. 5. Interferometric micrograph (green filter) of fringes about a crack in the material considered in Figs. 1–4. The extent of uplift increases approaching the crack tip from the starting notch.

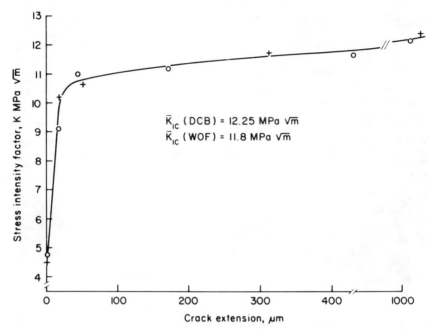

Fig. 6. Calculated R curve from the results in Fig. 3. They are considerably different from those presented in Fig. 2.

As mentioned earlier, the t-ZrO_2 precipitates in the material tested were highly metastable; a 1-h subeutectoid heat treatment resulted in all the t-ZrO_2 precipitates losing coherency and transforming to m-ZrO_2 on cooling. The fracture toughness of such material was greatly reduced; NB and DCB tests gave comparable values of $K_{Ic} = 4.4 \pm 0.2$ MPa·m$^{1/2}$ with virtually no evidence of R-curve behavior.

The R-curve predictions of Marshall et al.[1] and McMeeking and Evans[2] indicate that the steep rise in toughness should occur over a crack extension of between 1 and 2 times the transformed zone size and then asymptotically approach a limiting value. Unfortunately, it has not been possible to measure the transformed zone size directly to compare the observed behavior with that predicted. The results in Fig. 6 would agree with predictions if the zone size were ≈20–30 μm. This is a very large zone size, almost an order of magnitude greater than previously observed in PSZ systems.[8] However, a number of indirect means are available for estimating the zone size: theoretically, from expressions relating the increase in K_{Ic} to the zone size,[2] the two-wavelength X-ray technique as recently outlined by Garvie et al.,[12] and from observations of the discontinuity in the fringe patterns about the crack tip. An example of the latter technique is shown in Fig. 5; notice the well-defined discontinuity of the fringe about the crack tip. The width of this discontinuity is between 25 and 35 μm, suggesting that the width of the transformation zone is 12–18 μm. At the surface of the bar, plane stress conditions prevail which may lead to slight overestimation of the zone size. X-ray analysis of the fracture surface was made with Cr$K\alpha$ and Cu$K\alpha$ radiation. The volume concentrations of the m-ZrO_2 phase were virtually identical, the ratio of the Cr$K\alpha$ to the Cu$K\alpha$ varying from 1.09 to 1.01, implying that the depth of the transformed

(a)

(b)

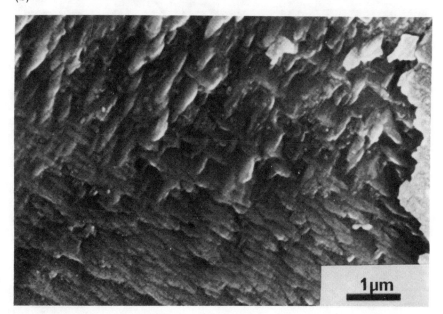

Fig. 7. Transmission (a) and scanning (b) electron micrographs of transformation-toughened Mg-PSZ showing crack deflection about precipitates in the matrix.

zone was greater than 6 μm. The theoretical expression relating the increase in K_{Ic} to the transformed zone size is given by

$$\Delta K = 0.22 V_f e^T E h^{1/2}/(1 - \nu^2) \qquad (2)$$

where ΔK is the transformation-toughening increment, e^T the unconstrained dilational transformation strain, E Young's modulus, ν Poisson's ratio, V_f the volume fraction of transformable t-ZrO_2 precipitates, and h the width of the transformation zone. Substituting the following values, $V_f = 0.4$, $e^T = 0.06$, $E = 205$ GPa, $\nu = 0.25$, and $\Delta K = 6$ MPa·m$^{1/2}$ into Eq. (2) provides a value, $h = 27$ μm. This prediction is somewhat higher than the experimental observations, which may imply that the shear component of the transformation needs to be included in a complete theoretical description of transformation-toughening. The other feature of the results in Fig. 5 is the permanent offset; this aspect will be considered in more detail elsewhere.

Before other materials that exhibit R-curve behavior are considered, mention should be made of indentation tests made in materials containing highly metastable t-ZrO_2 precipitates. Hannink and Swain[13] pointed out that it was possible to induce transformations in these materials by Vicker's pyramid indentations or by scratching with sharp objects. The transformation was shown to be a cooperative phenomenon and led to bands of precipitates in particular orientations all transforming to m-ZrO_2. Subsequent work by Lankford[14] suggests that strain-induced transformation, as distinct from stress-induced transformation, may account for this behavior. A typical observation of this deformation about a 30-kg force indentation is shown in Fig. 8. Notice the virtual absence of radial cracks emanating from the

Fig. 8. Vicker's pyramid indentation in Mg-PSZ containing readily transformable t-ZrO_2 precipitates. The indentation was formed at a load of 30 kg force and shows little evidence of radial cracking emanating from the corners of the impression.

corners of the impression. ISB tests on the material shown in Figs. 3–6 gave values of K_{Ic} between 12 and 18 MPa·m$^{1/2}$, and often fracture did not originate from the indentation with loads of 50 kg force. Obviously, the indentation-induced transformation prevents or hinders crack initiation and growth.

Subeutectoid Aged Materials

The materials referred to here contain relatively large amounts of continuous m-ZrO$_2$ phase formed by subeutectoid heat treatment. As discussed elsewhere,[15] MgO-fully stabilized ZrO$_2$ (Mg-CSZ) displays considerable R-curve behavior when partially or fully decomposed to m-ZrO$_2$. Similar behavior occurs in Mg-PSZ, although the rate of decomposition is much slower.[8]

Observations of a crack in decomposed Mg-CSZ are shown in Fig. 9. This material had been heat-treated at 1100°C for 8 h and was almost completely m-ZrO$_2$. The main features evident in Fig. 9 are that both crack branching and microcracking have occurred. It is thought that both these features are responsible for the R-curve behavior. m-ZrO$_2$ possesses considerable thermal expansion anisotropy, which undoubtedly assists in microcracking development.

The decomposition reaction in Mg-PSZ is more subtle than in Mg-CSZ and leads to the destabilization of t-ZrO$_2$ precipitates.[9] The decomposition process initiates at grain boundaries and pores. After many hours of aging, a well-developed decomposed zone may be seen "decorating" the grain boundaries. Crack propagation in this material occurs in a stable manner and the crack has a tendency to follow the grain-boundary phase, although occasional branching and deflection of the crack front through a grain may be seen, Fig. 10. TEM observations show that the grain-boundary phase consists of m-ZrO$_2$ with MgO pipes that developed during decomposition. A crack propagating through this phase displays substantial branching as it interacts with these pipes, Fig. 11.

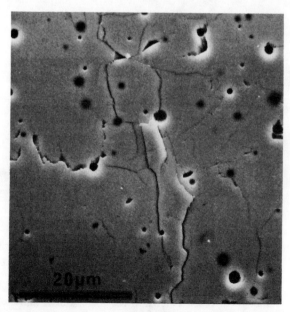

Fig. 9. Scanning electron micrograph of an extended crack in m-ZrO$_2$ formed by subeutectoid decomposition reaction.

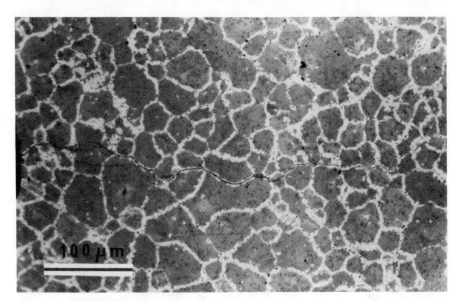

Fig. 10. Micrograph of a crack propagating in a partially decomposed Mg-PSZ material containing a thick m-ZrO_2 grain-boundary phase.

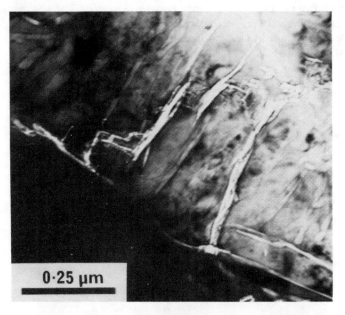

Fig. 11. Transmission electron micrograph of a crack interacting with microstructural detail within the decomposed m-ZrO_2 phase.

Crack resistance with crack extension for the material shown in Fig. 10 is depicted in Fig. 12. The rate of change of K_{Ic} with crack extension is substantially different from that shown in Fig. 5. The other difference is the absence of the large

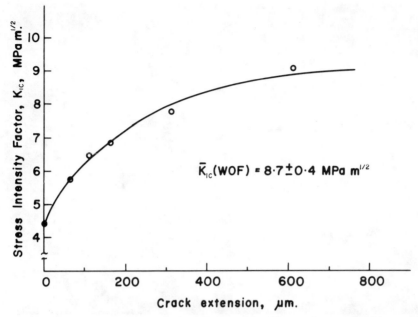

Fig. 12. *R* curve for the material depicted in Fig. 9. Compare this curve with that for a material containing transformable *t*-ZrO$_2$ precipitates.

initial offset observed with material containing highly transformable *t*-ZrO$_2$ precipitates. However, the extent of permanent offset does increase with crack extension, although there is little evidence of uplift about the crack, as shown in Fig. 5. The permanent offset is probably a consequence of fracture debris preventing crack closure.

Overaged Refractory PSZ

A third distinct type of PSZ material in which *R*-curve behavior is observed is what might be termed overaged, refractory-grade PSZ. Such material usually consists of very large *m*-ZrO$_2$ precipitates in a cubic matrix, together with coarse undissolved lumps of *m*-ZrO$_2$ and considerable porosity. Although the precise *R* curve for such a material has not been determined, there is at least a factor of 2 between initiation and steady-state propagation values of K_{Ic}. These materials display excellent tolerance to severe thermal shock damage.

Observations of crack interaction with the microstructure of such materials are shown in Fig. 13. These micrographs indicate that the coarse precipitates can severely impede the progress of the crack. The large precipitates act as crack deflectors, causing crack branching. They almost behave as fibers bridging the crack behind the crack tip. The large *m*-ZrO$_2$ grains also act as crack-trapping sites due to the long-range tensile hoop stress surrounding them.

Conclusions

The present study has identified three distinct PSZ material types that display significant *R*-curve behavior. The most pronounced increase in toughness with crack extension occurred in material containing highly metastable *t*-ZrO$_2$ precipitates. These materials displayed an increase in K_{Ic} from 4.5 to 12 MPa·m$^{1/2}$ with

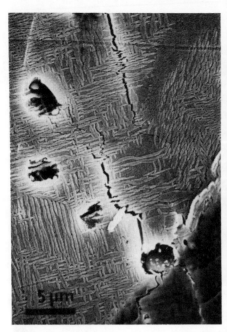 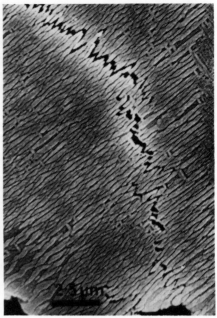

Fig. 13. Scanning electron micrographs of cracks interacting with coarse precipitates in overaged Mg-PSZ.

the major portion of this rise occurring in the first 50 μm of crack extension. This observation agrees very well with recent theoretical predictions of R-curve behavior in PSZ materials. It was also found that care needs to be exercised in interpreting load deflection curves to allow for permanent offset or plastic-like behavior, particularly during the initial stages of crack growth.

To summarize, R-curve behavior may occur in three distinct ways: (1) crack extension into the transformed zone surrounding a crack tip in material containing highly metastable t-ZrO_2 precipitates, (2) crack branching and microcracking in subeutectoid aged material, and (3) crack deflection and crack branching caused by very large precipitates in overaged material.

References

[1] D. B. Marshall, A. G. Evans, and M. D. Drory, "Transformation Toughening in Ceramics"; pp. 289–307 in Fracture Mechanics of Ceramics, Vol. 6. Edited by R. C. Bradt, D. P. H. Hasselman, F. F. Lange, and A. G. Evans. Plenum, New York, 1983.

[2] R. McMeeking and A. G. Evans, "Mechanics of Transformation Toughening in Brittle Materials," *J. Am. Ceram. Soc.*, **65** [5] 242–45 (1982).

[3] L. R. F. Rose; unpublished work.

[4] B. Mussler, M. V. Swain, and N. Claussen, "Dependence of Fracture Toughness of Alumina on Grain Size and Test Technique," *J. Am. Ceram. Soc.*, **65** [11] 566–72 (1982).

[5] R. Steinbrech, R. Khehans, and W. Schaarwachter, "Increase of Crack Resistance During Slow Crack Growth in Alumina Bend Specimens," *J. Mater. Sci.*, **18**, 265–70 (1983).

[6] D. J. Green, P. S. Nicholson, and J. D. Embury, "Fracture Toughness of Partially Stabilized ZrO_2 in the System CaO-ZrO_2," *J. Am. Ceram. Soc.*, **56** [12] 619–23 (1973).

[7] M. V. Swain, "R-Curve Behavior of Magnesia Partially Stabilized Zirconia and its Significance to Thermal Shock"; pp. 355–69 in Fracture Mechanics of Ceramics, Vols. 5 and 6. Edited by R. C. Bradt, D. P. H. Hasselman, F. F. Lange, and A. G. Evans. Plenum, New York, 1983.

[8]M. V. Swain, R. H. J. Hannink, and R. C. Garvie, "The Influence of Precipitate Size and Temperature on the Fracture Toughness of Calcia and Magnesia-Partially Stabilized Zirconia"; pp. 339–53 in Ref. 7.

[9]R. H. J. Hannink, "Microstructural Development of Sub-eutectoid Aged MgO-ZrO_2 Alloys," *J. Mater. Sci.*, **18**, 457–70 (1983).

[10]R. H. J. Hannink and M. V. Swain, "Magnesia-Partially Stabilized Zirconia: the Influence of Heat Treatment on Thermomechanical Properties," *J. Aust. Ceram. Soc.*, **18** [2] 53–62 (1982).

[11]W. K. Wilson, "Stress Intensity Factors for Deep Cracks in Bending and Compact Tension Specimens," *Eng. Fract. Mech.*, **2**, 169–71 (1970).

[12]R. C. Garvie, R. H. J. Hannink, and M. V. Swain, "X-Ray Analysis of the Transformed Zone in Partially Stabilized Zirconia (PSZ)," *J. Mater. Sci. Lett.*, **1**, 437–40 (1982).

[13]R. H. J. Hannink and M. V. Swain, "A Mode of Deformation in Partially Stabilized Zirconia," *J. Mater. Sci.*, **16**, 1428–30 (1981).

[14]J. Lankford, "Plastic Deformation of Partially Stabilized Zirconia," *J. Am. Ceram. Soc.*, **66** [11] C-212–C-213 (1983).

[15]M. V. Swain, R. C. Garvie, and R. H. J. Hannink, "The Influence of Thermal Decomposition on the Mechanical Properties of Magnesia-Stabilized Cubic Zirconia," *J. Am. Ceram. Soc.*, **66** [5] 358–62 (1983).

Residual Surface Stresses in Al_2O_3-ZrO_2 Composites

DAVID J. GREEN, FREDERICK F. LANGE, AND MICHAEL R. JAMES
Rockwell International Science Center
Thousand Oaks, CA 91360

Compressive surface stresses can be introduced into transformation-toughened ZrO_2 ceramics by a variety of techniques. These techniques are discussed and compared for Al_2O_3-ZrO_2 composites, particularly with respect to the transformation and residual stress profiles. Previous work on the direct measurement of the surface stresses and their influence on mechanical properties, such as strength, is reviewed.

It is recognized that the transformation of t-ZrO_2 to m-ZrO_2 monoclinic structure can be used to toughen certain ceramics.[1-8] The toughening mechanism has been analyzed theoretically by several authors.[9-13] Hence, this transformation toughening has the potential to strengthen these materials. In addition, however, it has been shown that the ZrO_2 phase transformation can be induced at the surface of these materials. Such surfaces would be expected to be placed in compression as the unconstrained transformation occurs martensitically and involves a volume increase.[14,15] In such situations, if the material fails from flaws within the compression zone, an additional strengthening mechanism exists, i.e., compressive surface strengthening. The surface phase change was first observed by Garvie and his coworkers,[1,16] when they noted that surface grinding increased the amount of m-ZrO_2 at the surface. Moreover, they found that removal of the transformed layer by polishing led to a slight strength decrease ($\approx 20\%$). The influence of surface grinding on the strength of transformation-toughened materials has been studied by other groups.[17-19] For example, in Al_2O_3-ZrO_2 composites, it has been shown that annealing of ground surfaces can lead to a dramatic strength decrease.[18] This was interpreted as the removal of the surface compressive stresses. To optimize the influence of surface grinding, Gupta[19] studied the effect of changing the grit size of the grinding wheel on the strength of transformation-toughened ZrO_2. In this work it was found that a maximum occurred in the strengthening as a function of the grit diameter. The increase in strength was interpreted in terms of an increasing depth of the compressive zone, whereas for very coarse grit, strength-degrading microcracks were observed in the surface.

In addition to strengthening, compressive surface stresses are expected to lead to improvements in other properties of transformation-toughened materials, such as improved resistance to contact damage.

The aim of this paper is to review the techniques used by the authors to (1) introduce surface compression, (2) measure the depth and magnitude of these stresses, and (3) predict the influence of surface compression on strength.

Techniques for Introducing Surface Compression

As indicated earlier, surface grinding can be used to introduce surface stresses in transformation-toughened materials. In addition, other techniques have been suggested, such as impact,[20] surface chemical reactions to form m-ZrO_2,[21] and low-temperature quenching.[22] These techniques are best understood in terms of the thermodynamics of the constrained ZrO_2 phase transformation. It has been shown that the minimum work (W) required to cause the transformation of a constrained ZrO_2 inclusion is given by[23]

$$W = -|\Delta G_c| + \Delta U_{SE} f + \Delta U_s/D \tag{1}$$

where ΔG_c is the chemical free energy change of the unconstrained ZrO_2 phase transformation, ΔU_{SE} the strain energy change, $(1 - f)$ the reduction of strain energy caused by accompanying surface phenomena, ΔU_s the change in surface energy associated with the surface phenomena, and D the inclusion diameter.

To induce the surface phase transformation, therefore, W could be supplied by external sources such as grinding or impact stresses. Alternatively, techniques for increasing $|\Delta G_c|$ (e.g., low-temperature quenching) or grain size (D) could provide the means of inducing the phase change. Indeed Eq. (1) suggests a myriad of ways that the surface transformation could be introduced. One technique that appears to have great potential was recently investigated by the authors.[24] In this technique, the alloying oxide (stabilizer) such as Y_2O_3 or CeO_2, which has been added to the ZrO_2 in order to retain the tetragonal phase, is removed by a heat treatment in unstabilized ZrO_2 powder. This technique eliminates sources of damage that can accompany techniques such as surface grinding or impact and, as shown in Table I, leads to a larger fraction (f_s) of the ZrO_2 that transforms to m-ZrO_2. The data in Table I were obtained by measuring the areas under the {111} m-ZrO_2 and the (111) t-ZrO_2 X-ray diffraction peaks ($CuK\alpha$ radiation). The apparent fraction of m-ZrO_2, not correcting for penetration depth, was obtained from the fractional areas under the peaks. For specimens heat-treated at higher temperatures, dye penetration indicated the presence of microcracks, but it is expected that the Y_2O_3 removal technique should allow control over the depth of the compressive zone. It is worth noting that Eq. (1) should also describe the parameters that control the depth from the surface for which the transformation can occur.[23]

Determination of Transformation Profile

It is important in work on inducing phase transformations at surfaces to determine the transformation profile, which, for transformations that involve volume changes, will be related to the residual stress profile. In the early work of Pascoe and Garvie,[16] an iterative process of X-ray diffraction and polishing was used. In this type of process, the m-ZrO_2 content is the mean of the zone penetrated by the X rays, weighted exponentially by the proximity to the surface. Figure 1, for example, compares the apparent m-ZrO_2 profiles for a ground surface and a surface in which the Y_2O_3 has been removed. These types of polishing experiments do give information on the transformation depth, i.e., the depth at which the apparent surface m-ZrO_2 content is the same as the bulk m-ZrO_2 content, but the difficulties in obtaining the true profile, their accuracy, and their laborious nature make the approach rather unattractive. More rapid techniques for estimating the transformation depth[25,26] and the true surface m-ZrO_2 content[26] have been suggested, but they depend on assumptions about the transformation profile.

Table I. Fraction of ZrO$_2$ Transformed to Monoclinic by Various Techniques for a Sintered Al$_2$O$_3$-30 vol% ZrO$_2$ Composite*

Technique	f_s
Sintered surface	<0.02
Ground surface (320-grit diamond)	0.11
Impacted surface (220-grit SiC)	0.10
Y$_2$O$_3$ removal	
1400°C/1 h	0.31
1400°C/4 h	0.60
1400°C/16 h	0.67

*ZrO$_2$ contained 2.2 mol% Y$_2$O$_3$.

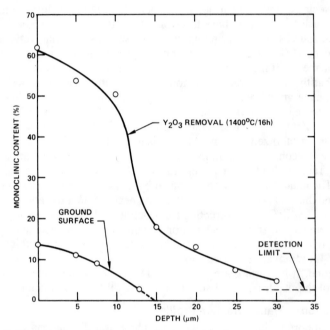

Fig. 1. Comparison of apparent monoclinic content as a function of distance from surface for a ground surface and a surface from which Y$_2$O$_3$ was removed (Al$_2$O$_3$-30 vol% ZrO$_2$). X-ray radiation, CuKα.

An alternative approach for obtaining the transformation profile is to cross-section the specimen and use a technique such as Raman microprobe analysis to measure the m-ZrO$_2$ content. This approach was used previously to determine the ZrO$_2$ phase transformation in the vicinity of fracture surfaces.[27] Table II summarizes the transformation depth for the Y$_2$O$_3$ removal specimen, using various techniques. It was found that the two calculation techniques[25,26] underestimate the transformation depth, presumably because they both assume a uniform transformation zone.

Table II. Transformation Depths for Al_2O_3-30 vol% ZrO_2 (2.2 mol% Y_2O_3)*

Technique	Transformation depth (μm)
Polishing	~32
Kosmac et al.[25]	6
Garvie et al.[26]	8

*Heat-treated in ZrO_2 powder to remove Y_2O_3 (1400°C/16 h).

Measurement of Magnitude and Depth of Surface Stresses

The mechanical response of materials containing residual surface stresses is expected to depend on the profile of these stresses. This profile could be calculated theoretically. For example, the transformation profiles discussed in the last section could be used to calculate the stress profile, using the volume change associated with the transformation. The stresses can be calculated from thermal stress theory because of the complete correspondence between stresses developed by concentration differences and stresses developed by temperature differences. For example, for an infinite slab whose surfaces have been transformed, the residual stresses parallel to the surface are biaxial and are given by[28]

$$\sigma_R(x) = \sigma_R(y) = \frac{V_v E \varepsilon}{(1-\nu)}[f_a - f_p] \qquad (2)$$

where E is Young's modulus of the slab, ν its Poisson ratio, ε the strain associated with volume change, V_v the volume fraction of ZrO_2 in the composite, f_p the fraction of ZrO_2 transformed at a point, and f_a the average fraction of ZrO_2 transformed. For the ZrO_2 transformation, one expects the surface to be in compression, whereas the interior of the body will be under a compensating tension. An alternate approach is to measure the residual stresses directly. One technique, based on X-ray diffraction, has been used to measure the residual stresses due to grinding on both Al_2O_3 and transformation-toughened Al_2O_3-ZrO_2 composites.[23,29] In this technique, a given set of diffracting planes in the surface of a specimen is examined as a function of their angular rotation with respect to the surface. For this work, it was found necessary to use $CrK\alpha$ radiation so that the penetration depth of the X rays was small.[23,29] To determine the stress profile, the compressive layer can be removed by polishing the specimen and redetermining the residual stress in successive steps. An example of this is shown in Fig. 2 for the ground surface of Al_2O_3-30 vol% ZrO_2,[23] along with a correction that was applied to the raw data to compensate for stress relaxation when a portion of the surface is removed.[30] It should be noted that the depth of the stresses is similar to the measured transformation depth, as discussed earlier (Fig. 1). It was also found that the magnitude of the residual stresses at the surface measured by the X-ray diffraction technique did correlate reasonably well with those predicted by Eq. (2).[23] For this calculation, the value of f_p (at the surface) was calculated using the technique of Garvie et al.[26] described earlier, which was then substituted in Eq. (2), assuming $f_a = 0$ (maximum value of residual stress). The calculated and measured residual stress for a variety of Al_2O_3-ZrO_2 composites is compared in Table III.[23,24] As can be seen, compressive surface stresses as high as 1 GPa have been measured in these materials.

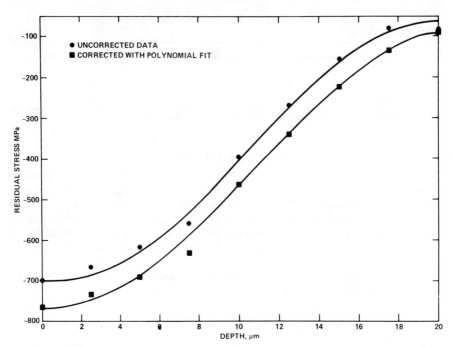

Fig. 2. Measured residual surface stress for a ground Al_2O_3-30 vol% ZrO_2 composite as a function of layer thickness removed by polishing.

Table III. Comparison of Measured and Calculated Residual Surface Stress in Al_2O_3-ZrO_2 Composites

Vol% ZrO_2 (mol% Y_2O_3)	f_p	Type of Surface	Surface stress (MPa)	
			Measd.	Calcd.
7.5 (1.4)	0.53	Ground	−440	−300
15 (1.4)	0.59	Ground	−570	−440
30 (2.4)	0.16–0.25	Ground	−450 to −680	−310 to −480
50 (2.4)	0.17	Ground	−780	−480
60 (3.6)	0.41	Ground	−1010	−1300
30 (2.2)	0.85	Y_2O_3 removal	−550	−1600

Influence of Surface Compression on Strength

For materials that fail from surface flaws, it is expected that compressive surface stresses would give rise to strengthening. Using the simple residual stress distribution in Fig. 3, Lawn and Marshall[31] showed that the residual stress component of the stress intensity factor (K_I^R) is given by

$$K_I^R = m\sigma_0(\pi a_0)^{1/2} B \qquad (3)$$

where

$$B = \left(\frac{2}{\pi} \sin^{-1} \delta_1\right)\left(1 + \frac{\delta_1}{2d_1}\right) - \frac{2}{\pi \delta_1}$$

$$(1 - (1 - \delta_1^2)^{1/2}) - \frac{\delta_1}{2d_1} \quad \text{(when } \delta_1 < 1\text{)}$$

$$B = \left(1 - \frac{2}{\pi\delta_1}\right) \quad \text{(when } \delta_1 \geq 1\text{)} \tag{4}$$

where m is a free surface correction, $\delta_1 = \delta/a_o$, and $d_1 = d/a_0$. During failure there will also be an applied component of the stress intensity factor K_I^A that is given by

$$K_I^A = mY\sigma_A(\pi a_0)^{1/2} \tag{5}$$

where σ_A is the applied stress and Y is a constant that depends on the crack loading and geometry. For conditions where the crack is completely open at failure, i.e., when there is no contact between the opposing crack faces, the total stress intensity factor (K_I^T) is given by simply adding Eqs. (3) and (5). For example, if in a tensile test $Y = 1$ and if the free surface correction is ignored ($m = 1$), one obtains

$$K_I^T = K_I^A + K_I^R = \sigma_0(\pi a_0)^{1/2}B + \sigma_A(\pi a_0)^{1/2} \tag{6}$$

At failure, when $K_I^A + K_I^R = K_c$, $\sigma_A = \sigma_f$, and with substitution $K_c = \sigma_f^0(\pi a_0)^{1/2}$, where σ_f^0 is the strength in the absence of surface compression, one obtains

$$\frac{\sigma_f}{\sigma_f^0} = \left[1 - B\left(\frac{\sigma_0}{\sigma_f^0}\right)\right] \tag{7}$$

This equation is illustrated in Fig. 4 as a function of (σ_0/σ_f^0) and δ_1, for the case

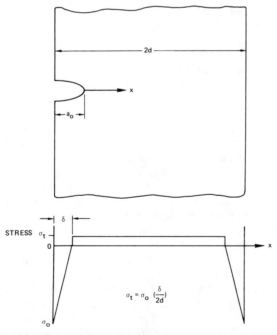

Fig. 3. Simple residual stress distribution for a plate under surface compression.

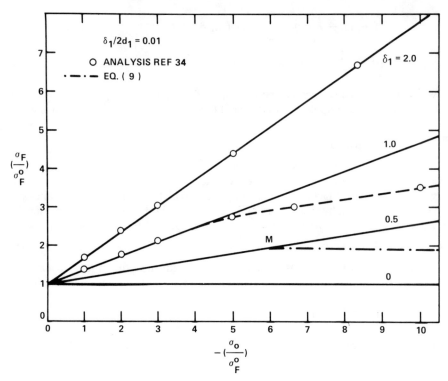

Fig. 4. Potential strengthening for a material subjected to a simple residual stress distribution (Fig. 3).

where $d_1 = 50$. This example is similar to that analyzed by Swain[32] and, as can be seen, the strengthening is linearly dependent on (σ_0/σ_f^0). It is worth noting, however, that, when $d_1 < 4$, the residual tensile stresses in the body can give rise to weakening (i.e., $\sigma_f/\sigma_f^0 < 1$).

As pointed out earlier, this type of approach does not account for partial crack closure when simple superposition of K_I^R and K_I^A cannot be used. This effect is expected to be important when σ_0 is large, i.e., when the compressive surface stress keeps the crack closed at the surface. Such effects have been analyzed by one of the present authors; it was found in general that K_I^T is greater than that predicted by superposition when the crack is partially closed. The superposition and crack-closure solutions merge once the surface crack becomes completely open.[33,34] In terms of strengthening, this implies that the solution for a partially closed crack predicts less strengthening than that predicted in Eq. (7). Using a previous analysis,[34] the strengthening incorporating the influence of crack closure is included in Fig. 4 for the cases where $\delta_1 \geq 1$. For the case where $\delta_1 = 2.0$, the crack was found to be completely open at the failure condition for $\sigma_0/\sigma_f^0 \leq 10$. The small differences between Eq. (7) and the closure analysis are presumably a result of ignoring the free surface correction. For the case where $\delta_1 = 1.0$, however, the surface crack is partially closed at failure for $\sigma_0/\sigma_f^0 > 5$ and the strengthening is less than that predicted by Eq. (7). To approximately determine the influence of crack closure when $\delta_1 < 1$, the following approach was taken. When σ_0 is very

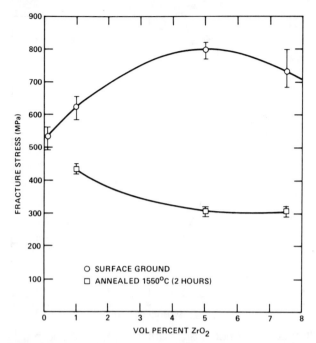

Fig. 5. Comparison of strengths of hot-pressed Al_2O_3-ZrO_2 composites after surface grinding and subsequent annealing.

large compared to σ_f^0, the crack will be closed to a depth δ. In the limit, therefore, the surface crack will be so strongly pinned at the surface that it will act more like an internal crack of length $(a_0 - \delta)$. For this case, the total stress intensity factor will be

$$K_I^T = \left(\frac{\sigma_0 \delta_1}{2d_1} + \sigma_A\right)\left(\pi\left(\frac{a_0 - \delta}{2}\right)\right)^{1/2} \tag{8}$$

and the strengthening will be

$$\frac{\sigma_f}{\sigma_f^0} = \left(\frac{2}{1 - \delta_1}\right)^{1/2} - \frac{\sigma_0 \delta_1}{2\sigma_f^0 d_1} \tag{9}$$

This equation is plotted in Fig. 4 for the case where $\delta_1 = 0.5$ and is found to intersect that predicted by Eq. (7) at point M. The strengthening, taking account of crack closure, is therefore expected to be given by Eq. (7) at low values of σ_0/σ_f^0 and by Eq. (9) at high values. From this approximate approach it appears that σ_f/σ_f^0 must go through a maximum for cases when $\delta_1 < 1$.

As discussed previously, the depth of the compressive zone in transformation-toughened ZrO_2 ceramics is relatively shallow (≈ 20 μm). Therefore, to obtain substantial strengthening, failure must occur from rather small surface cracks, perhaps ≈ 50 μm or less. For sintered Al_2O_3-ZrO_2 composites, this has not been found to be too common as several alternative flaw populations exist, e.g., agglomerates or voids.[35] Strengthening effects due to surface compression can, however, be more striking on hot-pressed materials. Figure 5 shows the difference in

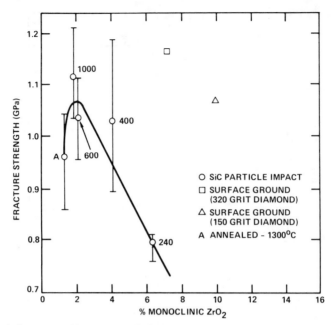

Fig. 6. Influence of impact of SiC particles on the strength of hot-pressed Al_2O_3-30 vol% ZrO_2 (2 mol% Y_2O_3). The numbers indicate grit size of SiC used. Included are average strength values for ground specimens.

strength obtained for hot-pressed Al_2O_3-ZrO_2 composites that are surface-ground and those that are surface-ground and subsequently annealed. The use of SiC particle impact to induce surface compression has also been examined, and the influence on strength is shown in Fig. 6. For small SiC particles, some strengthening was observed but, for larger sizes, the strength was found to decrease. In these latter materials, more surface ZrO_2 was transformed to monoclinic, but presumably the cracks introduced by the impact must be larger than those previously present and hence led to strength degradation.

The presence of surface compressive stresses is expected to be beneficial for improved resistance to contact damage, e.g., wear resistance. Table IV compares the radius of the radial cracks produced at Vickers hardness indentations for annealed specimens and specimens in which the Y_2O_3 was removed from the surface. It is clear that the surface compression has led to a substantial decrease in crack size, particularly at low loads. At the higher loads, when the crack sizes are substantially larger than the compressive zone size, the difference between the two surfaces becomes less pronounced.

Finally, it is worth remembering that the surface compression will be relieved if the material is subjected to a high temperature. Figure 7 shows the apparent fraction of m-ZrO_2 on a ground surface after it was annealed for 2 h at various temperatures. It is generally expected that, once the material is raised above the transformation temperature ($\approx 1100°C$), the m-ZrO_2 would transform to t-ZrO_2 and, provided no grain growth occurs during annealing, the t-ZrO_2 would remain as low as room temperature when the specimen is cooled, as it did in the original

Table IV. Comparison of Indentation Crack Sizes for As-Fabricated and Heat-Treated Surfaces

Surface treatment	P (N)	c (μm)*
As-fabricated	98	134
	147	191
	196	229
	294	290
Y_2O_3 removal		
−1400°C/1 h	98	91.6
	147	129
	196	176
	294	223
−1400°C/4 h	98	82.3
−1400°C/16 h	98	80.0

*Radial crack radius.

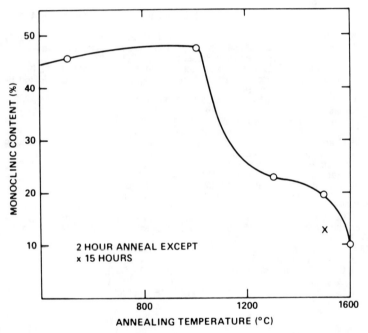

Fig. 7. Influence of annealing on apparent surface monoclinic content for a hot-pressed Al_2O_3-7.5 vol% ZrO_2.

fabrication. Although this is true for a substantial amount of the m-ZrO_2, annealing temperatures as high as 1600°C were required before the m-ZrO_2 content returned to its bulk value. This process was also found to be time-dependent, and previous work[36] indicated that this was probably a result of some microcrack healing that must occur before all the m-ZrO_2 produced by grinding is retransformed back to t-ZrO_2 by annealing.

Acknowledgments

The work was funded by the Office of Naval Research, Contract N00014-17-C-0441, and the authors wish to acknowledge the technical assistance of Milan Metcalf and Lynn Watanabe.

References

[1] R. C. Garvie, R. H. Hannink, and R. T. Pascoe, *Nature (London)*, **458** [12] 703–704 (1975).
[2] D. L. Porter and A. H. Heuer, *J. Am. Ceram. Soc.*, **60** [3–4] 183–84 (1977).
[3] N. Claussen, *J. Am. Ceram. Soc.*, **59** [1–2] 49–51 (1976).
[4] T. K. Gupta, F. F. Lange, and J. H. Bechtold, *J. Mater. Sci.*, **13** [7] 1464–70 (1978).
[5] F. F. Lange, *J. Mater. Sci.*, **17** [1] 247–54 (1982).
[6] N. Claussen and J. Jahn, *J. Am. Ceram. Soc.*, **63** [3–4] 228–29 (1980).
[7] F. F. Lange, *J. Mater. Sci.*, **17** [1] 255–62 (1982).
[8] F. F. Lange, B. I. Davis, and D. O. Raleigh, *J. Am. Ceram. Soc.*, **66** [3] C-50–C-52 (1983).
[9] A. G. Evans and A. H. Heuer, *J. Am. Ceram. Soc.*, **63** [5–6] 241–48 (1980).
[10] F. F. Lange; pp. 255–74 Fracture Mechanics of Ceramics, Vol. 6. Edited by R. C. Bradt, A. G. Evans, D. P. H. Hasselman, and F. F. Lange. Plenum, New York, 1983.
[11] A. G. Evans, D. B. Marshall, and N. H. Burlingame; pp. 202–16 in Advances in Ceramics, Science and Technology of Zirconia, Vol. 3. Edited by A. H. Heuer and L. W. Hobbs. The American Ceramic Society, Columbus, OH, 1981.
[12] R. M. McKeeking and A. G. Evans, *J. Am. Ceram. Soc.*, **65** [5] 242–46 (1982).
[13] B. Budiansky, J. W. Hutchinson, and J. C. Lambropoulos; Rept. Mech-25. Harvard University, Feb. 1982.
[14] E. C. Subbarao, H. S. Mati, and K. K. Srivastava, *Phys. Status Solidi A*, **21** [1] 9–40 (1974).
[15] G. K. Bansal and A. H. Heuer, *Acta Metall.*, **20** [11] 1281–89 (1972); **22** [4] 409–17 (1974).
[16] R. T. Pascoe and R. C. Garvie; pp. 774–84 in Ceramic Microstructures-'76. Edited by R. M. Fulrath and J. A. Pask. Westview Press, Boulder, CO, 1977.
[17] D. J. Green and F. F. Lange; for abstract see *Am. Ceram. Soc. Bull.*, **58** [9] 883 (1979).
[18] N. Claussen and G. Petzow; pp. 680–91 in Energy and Ceramics. Edited by P. Vincenzini. Elsevier, Amsterdam, 1980.
[19] T. K. Gupta, *J. Am. Ceram. Soc.*, **63** [1–2] 117 (1980).
[20] F. F. Lange and A. G. Evans, *J. Am. Ceram. Soc.*, **62** [1–2] 62–65 (1979).
[21] F. F. Lange, *J. Am. Ceram. Soc.*, **63** [1–2] 38–40 (1980).
[22] N. Claussen, *Z. Werkstofftechnik*, **13** [4] 113–48 (1982).
[23] D. J. Green, F. F. Lange, and M. R. James, *J. Am. Ceram. Soc.*, **66** [9] 623–29 (1983).
[24] D. J. Green, *J. Am. Ceram. Soc.*, **66** [10] C-178–C-179 (1983).
[25] T. Kosmac, R. Wagner, and N. Claussen, *J. Am. Ceram. Soc.*, **64** [4] C-72–C-73 (1981).
[26] R. C. Garvie, R. H. J. Hannink, and M. V. Swain, *J. Mater. Sci. Lett.*, **1** [10] 437–40 (1982).
[27] D. R. Clarke and F. Adar, *J. Am. Ceram. Soc.*, **65** [6] 284–88 (1982).
[28] W. D. Kingery; Introduction to Ceramics. Wiley & Sons, New York, 1960.
[29] F. F. Lange, M. R. James, and D. J. Green, *J. Am. Ceram. Soc.*, **66** [2] C-16–C-17 (1983).
[30] J. B. Cohen, H. Dolle, and M. R. James; pp. 453–78 in Proc. Symposium on Accuracy in Powder Diffraction, June 11–15, 1980. *Natl. Bur. Stand. (U.S.), Spec. Publ.*, No. **567**, 1980.
[31] B. R. Lawn and D. B. Marshall, *Phys. Chem. Glasses*, **18** [1] 7–18 (1977).
[32] M. V. Swain, *J. Mater. Sci. Lett.*, **15** [6] 1577–79 (1980).
[33] D. J. Green; to be published in *Journal of Materials Science*.
[34] D. J. Green, *J. Am. Ceram. Soc.*, **66** [11] 807–10 (1983).
[35] F. F. Lange, *J. Am. Ceram. Soc.*, **66** [6] 396–407 (1983).
[36] D. J. Green, *J. Am. Ceram. Soc.*, **65** [12] 610–14 (1982).

Calculations of Strain Distributions in and around ZrO₂ Inclusions

S. SCHMAUDER, W. MADER, AND M. RÜHLE

Max-Planck-Institut für Metallforschung
Institut für Werkstoffwissenschaften
Stuttgart, Federal Republic of Germany

It is assumed that the nucleation of the martensitic tetragonal → monoclinic transformation within a ZrO₂ inclusion is governed by the strain distribution inside the inclusion. If the strain reaches a critical level (over a certain volume), then a stable nucleus can form. Strains within and in the surroundings of ZrO₂ particles generally result from the thermal expansion mismatch between the particle and the surrounding matrix and from the stress field in front of a crack tip. Strain distributions are calculated for ZrO₂ particles of different shapes. Although the (internal) strains are homogeneous for particles with spherical or ellipsoidal shape, they are very inhomogeneous for faceted particles. For cuboidal inclusions, the strain distributions can be evaluated by using the theory of Y. P. Chiu [*J. Appl. Mech.*, **44**, 587 (1977)]. The stress distribution for tetrahedral inclusions is calculated by utilizing an isotropic finite element calculation. Stress concentrations are observed for faceted particles at specific edges and corners. The effect of particle size and shape on the stress distribution will also be discussed.

It is well established that the martensitic transformation of zirconia, from tetragonal (t) to monoclinic (m) symmetry, can be utilized to increase the toughness of a ceramic matrix in which t-ZrO₂ inclusions are embedded. A necessary prerequisite for the utilization of this effect is that ZrO₂ remains tetragonal at room temperature. This condition is fulfilled if ZrO₂ is included in a ceramic matrix and if the inclusions have a diameter which is smaller than a critical value, d_c, which is defined here as that size above which particles transform spontaneously to the m symmetry on cooling to room temperature without the application of external stresses.[1] The size of d_c depends on the ceramic matrix in which ZrO₂ is embedded, the shape of the particle, and the chemical composition (solute content) of the t-ZrO₂. Microstructural studies reveal that the $t \to m$ transformation of confined ZrO₂ particles is essentially nucleation-controlled.[2] An energy barrier has to be overcome before the particles transform to m symmetry, at a rather high velocity.[3] For metallic systems,[4] it is well established that nucleation occurs heterogeneously. This means that the transformation to the new phase nucleates at defects present in the material (e.g., dislocations). In t-ZrO₂ inclusions no defects are observed by transmission electron microscopy (TEM). Therefore, nucleation can be caused either by the formation of a defect which then acts as a nucleus for the transformation or by a stress-assisted nucleation process in the regions of high stress concentration.[4,5]

We will present data to show that under certain circumstances a stress-assisted nucleation at the matrix-inclusion interface is possible. For this kind of nucleation the stresses must exceed a certain value over the extent of a possible nucleation site, of size n.

Stress and strain distributions were calculated for ZrO_2 inclusions of different shapes, embedded in a ceramic matrix. In this paper, alumina is considered as the matrix. The results of the elastomechanical calculations are summarized here; an extended version will be published elsewhere.[5]

Transformation Strains at Room Temperature

Crystallographic studies indicate that there probably exists a certain lattice correspondence, C, in which the c_t and c_m axes coincide and which operates during a $t \rightarrow m$ transformation.[6] For such a lattice correspondence, the nonvanishing components of the symmetrical unconstrained transformation strain tensor, $\mathbf{e}^T$, have been obtained with respect to the tetragonal lattice (cf. Ref. 7):

$$e^T_{11} = (a_m/a_t)\cos\beta - 1 = -0.00149 \quad (1)$$

$$e^T_{22} = (b_m/b_t) - 1 = 0.02442 \quad (2)$$

$$e^T_{33} = (c_m/c_t) - 1 = 0.02386 \quad (3)$$

$$e^T_{13} = e^T_{31} = (1/2)\tan\beta = 0.08188 \quad (4)$$

The room-temperature crystallographic data for m-ZrO_2 were obtained from Patil and Subbarao,[8] who give $\beta = 9.3°$. The lattice constants of the face-centered tetragonal (fct) unit cell at room temperature can be obtained from X-ray studies of Y_2O_3-stabilized t-ZrO_2 powders.[9] The values obtained by extrapolation to zero Y_2O_3 content are $a_t = b_t = 5.0815 \pm 0.0010$ Å and $c_t = 5.1890 \pm 0.0020$ Å. These lattice parameters are in good agreement with those obtained by Scott.[10]

Thermal Mismatch in the System Al_2O_3/t-ZrO_2

Internal stresses and strains in ZrO_2, as well as in the Al_2O_3 matrix, can result from the different thermal expansion coefficients of ZrO_2 and Al_2O_3. The ceramic composite is fabricated at $\approx 1550°C$. It is assumed that the t-ZrO_2 inclusion is in equilibrium with the Al_2O_3 matrix. On cooling to room temperature the misfit values e^{p*}_{ij} between inclusion and matrix are calculated as

$$e^{p*}_{11} = e^{p*}_{22} = -0.00456 \quad (5)$$

$$e^{p*}_{33} = -0.01122 \quad (6)$$

$$e^{p*}_{ij} = 0 \; (i \neq j) \quad (7)$$

Elastic Constants

For alumina, the isotropic elastic moduli $E = 396$ GPa and $\nu = 0.24$ were calculated by applying the Voigt–Reuss–Hill approximation. The six elastic constants for t-ZrO_2 are not known. Therefore, the values $E = 192$ GPa and $\nu = 0.3$ of fully stabilized polycrystalline c-ZrO_2[11] are used for the t-ZrO_2 inclusions.

Shapes of Particles

Three inclusion shapes were assumed: (1) regular-shaped inclusion (sphere, ellipsoid)—the solution for the regular-shaped inhomogeneous inclusion was derived by Eshelby;[12] (2) cuboidal inclusion—Chiu[13] derived, for homogeneous cuboidal inclusions, a mathematically closed solution assuming isotropic media;[14] (3) tetrahedral inclusion—numerical solutions for tetrahedral inhomogeneous inclusions were obtained by the finite element method.

Results

Equations (1)–(4) show that the main component of the transformation strain tensor is the shear component, $e^T_{13} \approx 0.08$. From the theory of stress-assisted nucleation, it is known that the elastic strain, **e**, acting over the size of the nucleus must reach ≈ 0.21 of $\mathbf{e}^T$ to form a stable nucleus.[4] This means that the elastic shear strain must attain a value of about 1.8% over the nucleus volume.

Stress and strain distributions were calculated for the three inclusion shapes. The results are summarized below.

Regular-Shaped Inclusions

Spherical t-ZrO_2 inclusions embedded in an approximately isotropic alumina matrix possess zero shear strains (with respect to the t lattice) when the initial strains are uniform and acting parallel to the t cell axes (see Eqs. (5)–(7)). The same result is obtained for ellipsoidal inclusions when the t axes coincide with the ellipsoidal axes.

Cuboidal Inclusions

Chiu[13] derived the elastic fields in and around cuboidal homogeneous isotropic inclusions. Our detailed results (Table I) reveal that logarithmic singularities of the different strain components are expected along edges and at corners of the cuboidal inclusion (provided edges E_i and coordinate axis x_i are parallel). Figure 1 illustrates the strain distribution of the shear strain component e_{13} around the center of an edge, E_2; the shear strain approaches infinity at the site of the edge. The theory[13] allows a calculation to be made of a potential nucleus volume, v, within which certain values of the strains are exceeded. The volume v scales with the size of the cuboidal inclusion.

Tetrahedral Inclusions

Displacements and stress distributions for tetrahedral t-ZrO_2 inclusions, embedded in an Al_2O_3 matrix, are calculated by utilizing an isotropic finite element calculation (for details see Schmauder et al.[5]). Both inclusion and matrix are assumed to have isotropic, but different, elastic stiffnesses. The calculations were performed with the displacement method of the ASKA (*A System for Kinematical Analysis*) program package,[15] with the possibility of treating discontinuous stresses at locations where the material constants change.

The calculations reveal that the stresses are extremely inhomogeneous in the inclusion and the matrix. Due to the misfit values (see Eqs. (5)–(7)) normal stresses inside the inclusion are positive. Inside the inclusions, magnitudes of the normal and shear stress components are less than 10 and 1 GN/m^2, respectively.

Table I. Singularities and Discontinuities at the Cuboidal Edges Due to Initial Strains

Initial strain component	Singularity for elastic strain component	Discontinuity for elastic strain component	At edges
e^p_{11}	e_{23}	e_{11}, e_{22}, e_{33}	E_1
	e_{13}	e_{11}, e_{33}	E_2
	e_{12}	e_{11}, e_{22}	E_3
e^p_{12}	e_{13}	e_{12}	E_1
	e_{23}	e_{12}	E_2
	e_{11}, e_{22}	e_{12}	E_3

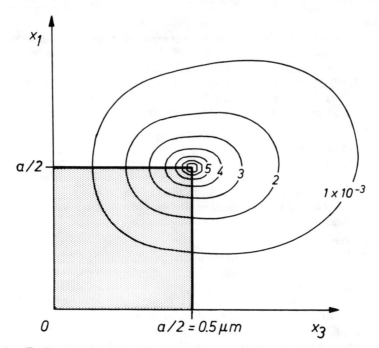

Fig. 1. Typical distribution of strain component e_{13} in the plane $x_2 = 0$ if initial strains are acting parallel to the cuboidal axes. There is a logarithmic singularity at an edge E_2.

Normal components showed discontinuities at the interface. Shear stress components appear to be continuous across the inclusion-matrix interface, except at the tetrahedron edges. This indicates that shear strain components can have singularities at the tetrahedral edges.

Discussion

The present work has shown the potential of applying continuum mechanics to geometrically relevant models of real microstructures, such as zirconia-toughened ceramics.

The well-defined critical particle size for intercrystalline faceted particles and the undefined critical particle size of intracrystalline, regular-shaped particles[2] can be explained by using the results of the calculations. These will now be discussed.

It is assumed that a continuous size distribution of equally shaped particles exists within the Al_2O_3 matrix. For particles of arbitrary size and shape, internal stresses and strains are created due to thermal mismatch between matrix and inclusion. *Intracrystalline* spherical and ellipsoidal inclusions possess a constant strain.[12] Within the tetragonal lattice no shear strains exist while the *t*-particle cell axes coincide with the ellipsoidal axes, irrespective of the size of the inclusion.[5,12] This lack of strain means that stress-assisted nucleation is not possible. Theoretical calculations predict that regular-shaped particles should not transform, independent of size. This result is a contradiction to experimental observations.[1] However, shear strains could be produced by (1) random orientation of the t-ZrO_2 particle with respect to the matrix, (2) noncoinciding tetragonal and ellipsoidal axes, and

(3) large inclusions which show deviation from the spherical or ellipsoidal shape, which in turn results in stress and strain inhomogeneity inside the inclusion that can cause nucleation.

For *intercrystalline* cuboidal or tetrahedral inclusions, stresses show singularities at edges and corners. The region over which a certain critical value of the strain can be overcome scales with the size of the inclusion. Therefore, inclusions which are above a certain size develop a stable nucleus which allows the nucleation of the martensitic transformation. In smaller inclusions, the critical shear strain (acting over the size of a potential nucleus) cannot be reached. If a stress is applied at the specimen, then the applied strain superposes the strain existing within the inclusion. For certain edges and/or corners, the critical component of strains increases, resulting in a higher level of strain. One facet of the cuboidal or tetrahedral inclusion is always in a favorable orientation such that the shear strain component of the crack tip adds to the existing residual strain inside the particle. The stress level required for the formation of a stable nucleus can then be derived by the superposition of the residual stress (due to thermal mismatch) and the applied stress.

References

[1] A. H. Heuer, N. Claussen, W. M. Kriven, and M. Rühle, *J. Am. Ceram. Soc.*, **65** [12] 642 (1982).

[2] M. Rühle and W. M. Kriven, *Ber. Bunsenges. Phys. Chem.*, **87**, 222 (1983).

[3] J. W. Christian; The Theory of Transformations in Metals and Alloys, 2d ed. Pergamon Press, Elmsford, NY, 1975.

[4] (a) G. B. Olson and M. Cohen; p. 1145 in Proceedings of an International Conference on Solid-Solid Phase Transformations. Edited by H. I. Aaronson, D. E. Laughlin, R. F. Sekerka, and C. M. Wayman. Carnegie Mellon University, Pittsburgh, PA, 1983.

(b) G. B. Olson and M. Cohen; Proceedings of the International Conference on Martensitic Transformations (ICOMAT). Edited by L. Delay and M. Chandrasekaran. les Editions de Physique, Les Ulis, 1982.

[5] S. Schmauder, W. Mader, and M. Rühle; unpublished work.

[6] W. M. Kriven, W. L. Fraser, and S. W. Kennedy; p. 82 in Advances in Ceramics, Vol. 3. Edited by A. H. Heuer and L. W. Hobbs. The American Ceramic Society, Columbus, OH, 1981.

[7] U.-W. Chen and Y.-H. Chiao, *Acta Metall.*, in press.

[8] R. N. Patil and E. C. Subbarao, *J. Appl. Crystallogr.*, **2**, 281 (1969).

[9] H. Schubert; unpublished work.

[10] H. G. Scott, *J. Mater. Sci.*, **10**, 1527 (1975).

[11] D. L. Porter, A. G. Evans, and A. H. Heuer, *Acta Metall.*, **27**, 1649 (1979).

[12] J. D. Eshelby; p. 88 in Progress in Solid Mechanics, Vol. 2. Edited by I. N. Sneddon and R. Hill. Wiley-Interscience, New York, 1961.

[13] Y. P. Chiu, *J. Appl. Mech.*, **44**, 587 (1977).

[14] T. Mura; Micromechanics of Defects in Solids. Martinus Nijhoff Publishers, Amsterdam, The Netherlands, 1982.

[15] J. H. Argyris, *Aircr. Engi.*, **26**, 347 (1954).

In Situ Observations of Stress-Induced Phase Transformations in ZrO$_2$-Containing Ceramics

M. Rühle, A. Strecker, and D. Waidelich

Max-Planck-Institut für Metallforschung
Institut für Werkstoffwissenschaften
Stuttgart, Federal Republic of Germany

B. Kraus

Gatan Europa
München, Federal Republic of Germany

In situ straining experiments were performed in a high-voltage electron microscope fitted with a double-tilting straining stage. The stress-induced tetragonal (t)-to monoclinic-(m) transformation was studied for t-ZrO$_2$ inclusions confined in different ceramic matrices (Al$_2$O$_3$, spinel, cubic (c)-ZrO$_2$) and for fine-grained t-ZrO$_2$ polycrystals (TZPs). Crack propagation could easily be observed, during which t-ZrO$_2$ inclusions close to the crack tip transformed to m symmetry. The sizes and shapes of the transformation zones were evaluated for several transformation-toughened ceramics and are discussed with regard to their mechanical properties.

The $t \rightarrow m$ martensitic transformation in ZrO$_2$ can be utilized for the toughening of ceramics.[1] To understand the mechanisms leading to toughening and to determine important parameters (e.g., the size of the transformation zone), it was of interest to perform *in situ* straining experiments of ZrO$_2$-containing ceramics in a high-voltage electron microscope. The experimental observations of the *in situ* straining experiments for different ZrO$_2$-containing ceramics are reported in this paper.

Experimental Details

The following specimens were investigated:

Al$_2$O$_3$-ZrO$_2$ Dispersion Ceramics (ZrO$_2$-Toughened Al$_2$O$_3$: ZTA)

All ZTA specimens studied contained 15 vol% ZrO$_2$. The Al$_2$O$_3$/ZrO$_2$ powders were milled for different times (16 h, 12 h, 10 h, and 10 min) and sintered for 1 h at 1550°C.[2] The size distributions of the Al$_2$O$_3$ and ZrO$_2$ grains, as well as the fraction of ZrO$_2$ particles with m symmetry after fabrication, were determined by TEM. A typical micrograph of ZTA is shown in Fig. 1. In these ZTA specimens, only intercrystalline ZrO$_2$ inclusions are observed.[3] For the quantitative evaluation of the size distributions, the diameter d of the grains was taken as $A^{1/2}$, where A is the projected area of the grain. The size distributions were determined with a particle-size analyzer; Fig. 2 and Table I contain the results of the evaluations. *In situ* straining experiments were performed for specimen I (98% t-ZrO$_2$), specimen II (82% t-ZrO$_2$), and specimen III (78% t-ZrO$_2$).

 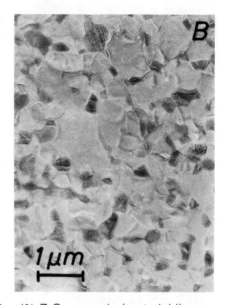

Fig. 1. TEM micrographs of Al_2O_3-15 vol% ZrO_2 ceramic (material I) containing faceted ZrO_2 inclusions. (A) Bright-field electron micrograph. (B) Image taken with objective aperture removed. ZrO_2 particles appear dark because of absorption contrast.

Table I. Results for Zirconia-Toughened Alumina (ZTA) 15 vol% ZrO_2

Material	Milling time of powders	Fraction of t-ZrO_2 (X rays) (%)	Mechanical data		Grain size and width of size distribution				Transformation zone
			Flexural strength (MPa)	K_{Ic} MPa·m$^{-1/2}$	Al_2O_3 $\bar{d}$ (μm)	$\Delta\bar{d}$ (μm)	ZrO_2 $\bar{d}$ (μm)	$\Delta\bar{d}$ (μm)	w_{112} (μm)
I	16 h	98	610	4.85	0.44	±0.19	0.21	±0.09	1.8
II	12 h	82	900	6.55	0.48	±0.18	0.25	±0.10	4.5
III	10 h	78	1150	6.89	0.46	±0.20	0.28	±0.10	4.9
IV	10 min	37	620	5.25	0.64	±0.31	0.37	±0.17	
Al_2O_3			350						

All grain boundaries were wetted with an amorphous film (Fig. 3) which could be imaged easily by dark-field imaging.[4] Qualitative analysis using analytical electron microscopy revealed that Si, Mg, and Ca are present within the amorphous film.

Mg-PSZ

t-ZrO_2 inclusions in Mg-PSZ are formed by precipitation in a cubic matrix[5] and are homogeneously distributed. The precipitates are lens-shaped with an aspect ratio of about $a:b:c = 4:4:1$ and a maximum extension along the tetragonal c axis of about 70 nm (optimum aging). Three variants exist with their c axis parallel to the three {100} directions of the cubic matrix. The grain size of the c matrix is ≈50 μm; a crystalline grain-boundary phase also exists (Fig. 4), which contains a high density of impurities such as Si, Fe, Ca, and Al. The Mg-PSZ specimen possesses ≈5% porosity in the form of numerous small pores.

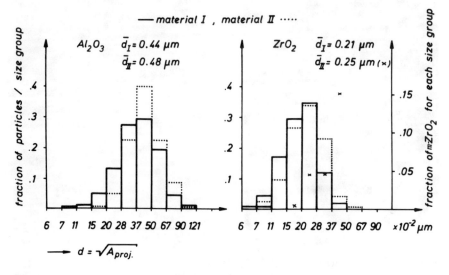

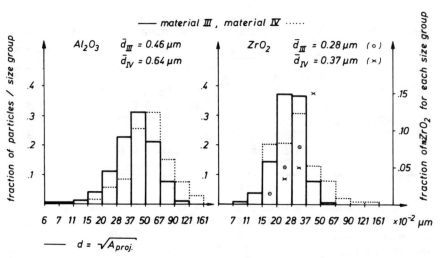

Fig. 2. Size distribution of Al_2O_3 and ZrO_2 grains in four Al_2O_3-15 vol% ZrO_2 ceramics. The fraction of m-ZrO_2 is noted for each size group of the different materials. (A) Material I (16-h milling time); material II (12-h milling time). (B) Material III (10-h milling time); material IV (10-min milling time).

Spinel-ZrO_2 Dispersion Ceramics (ZrO_2-Toughened Spinel : ZTS)

These compositions were prepared from Al_2O_3 and MgO powders and Zr acetate which, during heat treatment at ≈900°C, decomposed before the spinel formed.[6] After sintering, the t-ZrO_2 fraction was about 0.7 (determined by X-ray

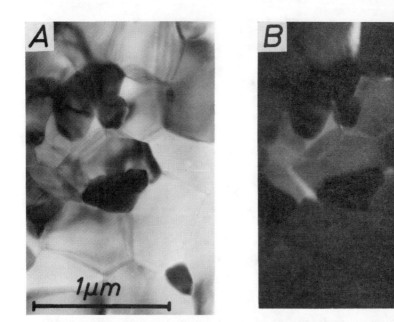

Fig. 3. Amorphous grain-boundary film in Al_2O_3-15 vol% ZrO_2. (A) Regular bright-field image; (B) dark-field image taken with part of the diffuse ring caused by the amorphous film; the grain-boundary phase appears bright.

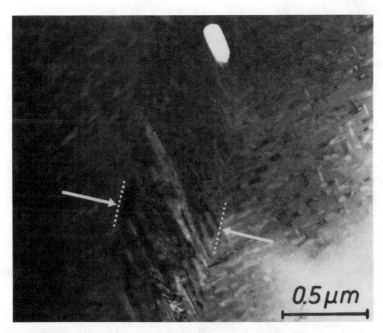

Fig. 4. Crystalline grain-boundary phase in Mg-PSZ. The phase is an Mg silicate but contains Fe, Ca, and Al impurities.

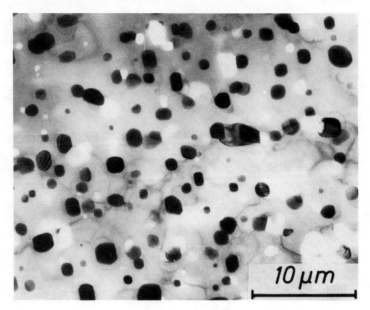

Fig. 5. Bright-field micrograph of ZrO_2-toughened spinel. The dark areas are the t-ZrO_2 inclusions.

methods), the remainder having m symmetry. The t-ZrO_2 inclusions in the large-grained spinel were mainly intracrystalline (Fig. 5). The t-ZrO_2 inclusions are slightly faceted and possess a mean diameter of ≈ 1 μm, whereas the mean diameter of the spinel matrix grains is of the order of ≈ 200 μm. Large intercrystalline m-ZrO_2 inclusions ($d \gtrsim 1$ μm) are present at grain boundaries.

Tetragonal Zirconia Polycrystals (TZP)

Two commercial materials containing 4.1 wt% Y_2O_3 were studied. One material was hot-pressed (TZP-HP) at 1400°C and contained only t-ZrO_2 grains with a mean diameter of ≈ 0.5 μm (Fig. 6(A)), whereas the material sintered at 1600°C contained both c- and t-ZrO_2 solid solution grains (Fig. 6(B)), their compositions being 3.5 and 9.3 wt% Y_2O_3, respectively.[7] All grain boundaries were wetted with an amorphous film.

Specimen Preparation

Specimens (3 by 5 mm) were carefully polished to a thickness of 40–80 μm. The central area of the specimen was prethinned with a commercial ion thinner for about 2 h. During the prethinning, the specimen was rotated. Two parallel elongated dimples (distance ≈ 0.5 μm) were next introduced by ion-thinning. The specimen was not rotated during this part of the thinning. The distance of the elongated dimples was adjusted by displacing the specimen parallel to the impinging ion beam. The final ion-thinning was performed while the specimen was rotated until small holes were introduced in the areas of the two dimples. A schematic cross section of the specimen suitable for straining experiments is shown in Fig. 7(A). The specimen was glued into "grips" (Fig. 7(B)) which could be inserted in the deformation stage. The deformation was performed in the double-

Fig. 6. TEM micrographs of t-ZrO_2 polycrystals (TZP). (A) Hot-pressed TZP, containing 100% t-ZrO_2 ($\approx$4.5 wt% Y_2O_3). (B) Sintered TZP. Cubic grains (9.3 wt% Y_2O_3) and t-ZrO_2 grains are present. The t-ZrO_2 are already transformed to m symmetry (under the electron beam).

tilting straining stage described by Loretto and Brooks[8] (Fig. 7(C)). Figure 8 shows HVEM micrographs taken of the electron transparent area before and after straining. The relative deformation velocity could be varied between 10^{-8} and 10^{-6} m/s, resulting in strain rates 4×10^{-5} to 4×10^{-3}. Forces applied to the specimen could not be measured.

Experimental Results

Micrographs of the unstrained specimens were taken of the entire transparent part of the ridge of the specimen. In the Mg-PSZ and TZP specimens, no m-ZrO_2 inclusions or grains were present prior to deformation. In ZTS, all intracrystalline ZrO_2 inclusions possessed t symmetry, whereas in ZTA those t-ZrO_2 grains above the critical particle size[3] had m symmetry.

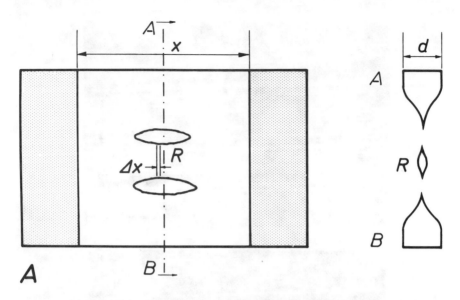

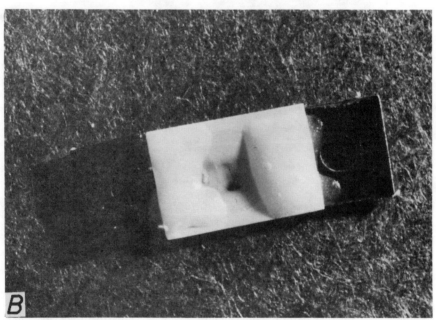

Usually, a crack started to propagate at the thinnest part of the specimen. However, the specimens were elastically strained before the crack formation started. The elastic deformation could be observed in the HVEM by moving bend contours.

Zirconia-Toughened Alumina (ZTA)

Some t-ZrO_2 particles transformed to m-ZrO_2 during elastic deformation prior to any crack formation. Those t-ZrO_2 grains were usually close to one of the foil

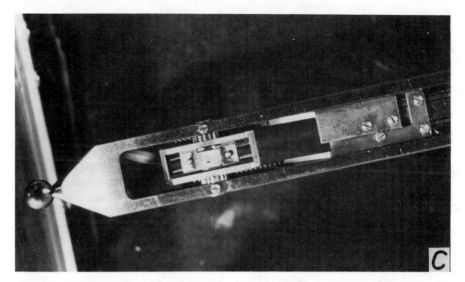

Fig. 7. Specimen and straining stage for *in situ* straining experiments. (A) Cross section through specimen (original thickness $d \approx 40$–70 μm). (B) Specimen glued on specimen holder. (C) Double-tilting straining stage (Ref. 8).

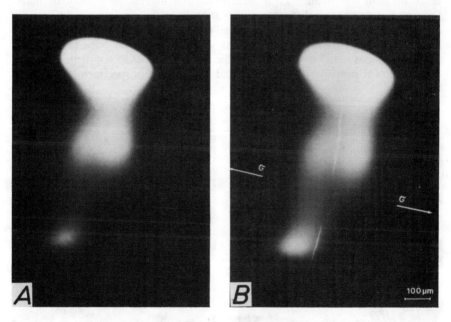

Fig. 8. Specimen for *in situ* straining experiments. The ridge between the two holes is (nearly) transparent for 1 MeV electrons (low magnification TEM photos). (A) Unstrained specimen; (B) deformed specimen with crack.

surfaces. The crack started to propagate from the regions of highest stress concentrations in the thinnest area of the ridge. In the fine-grained ZTA materials I to III only intercrystalline cracking was observed, the t-ZrO_2 grains transforming in front of the crack tip. The distance of the transforming particles from the crack tip depended on the size of the particle, the depth of the particle inside the foil, and the foil thickness.

Straining of the specimen was stopped after the crack propagated for several micrometers and micrographs were taken of the entire electron transparent area. Then the straining was continued, the crack arrested, etc. for 4 to 7 times. Micrographs were taken after each discrete crack propagation. On the micrographs the transformed m-ZrO_2 particles could easily be identified by the presence of mechanical twins (Fig. 9). Figure 10 shows one sequence of such a deformation experiment. The end of the crack is clearly marked on the micrographs.

It is possible[9] that microcracks also form in the transformation zone, and detailed studies were performed attempting to identify such microcracks. Until now, microcracks at grain boundaries in ZTA were observed in only one sequence. In this particular case, the distance of two transforming ZrO_2 grains was very small (Fig. 11). Tangential microcracks were frequently observed at the interface between a ZrO_2 grain and an Al_2O_3 grain when the ZrO_2 grain diameters were ≥ 0.5 μm, but not for those with $d < 0.3$ μm.

Micrographs of the *in situ* straining experiments (Fig. 10) were evaluated, and the occurrence of transformations of individual t-ZrO_2 grains noted for each stage of crack propagation. Frequently, large t-ZrO_2 inclusions (with diameters closer to the critical diameter d_c (Ref. 3) transform at larger distances from the crack tip than the smaller ZrO_2 inclusions. The highest probability of transformation at these large distances occurs at $\approx 40°$ to the crack plane. In Fig. 12, all ZrO_2 inclusions present in the specimen of Fig. 10 are drawn in schematically; hatching indicates those ZrO_2 inclusions which transformed prior to the crack advancing to position 3 to 4. In front of the crack tip, most t-ZrO_2 particles within a distance of 5 μm from the crack tip had transformed in specimen I.

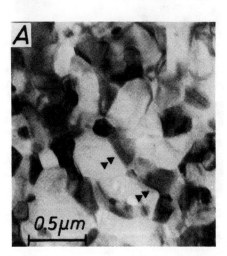 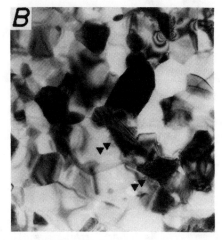

Fig. 9. Transformation of t-ZrO_2 grain in front of a crack tip. (A) Micrograph taken prior to straining; (B) after straining. Marked t-ZrO_2 inclusions transformed to m symmetry. Twins can be observed in the m-ZrO_2.

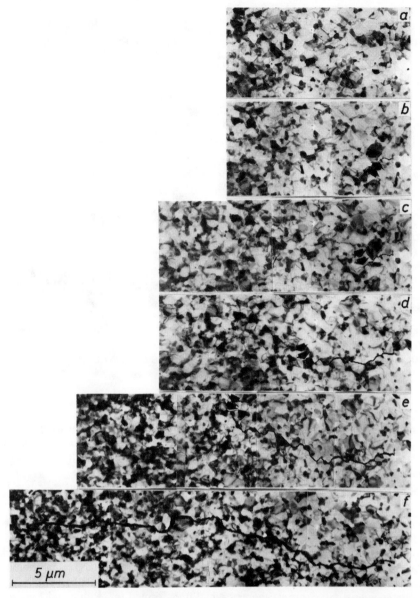

Fig. 10. Sequence of straining experiments (specimen I). (a) TEM micrograph taken prior to straining. Micrographs of the complete transparent ridge of the specimen is inclined. No m-ZrO_2 inclusions are present. (b) Crack starts to propagate. The crack is marked on all micrographs. (c) to (f) Straining experiment. Crack propagates by increased loading. t-ZrO_2 inclusion transformed to m symmetry.

The experiments clearly demonstrate the presence of a transformation zone ahead of the crack tip. At the side of the crack, a transformation zone can be also evaluated. In Fig. 13, the fraction of transformed particles in materials I and II are

Fig. 11. Microcracking in the surroundings of m-ZrO_2 inclusion. Tangential microcracks (T) can frequently be observed, radial microcracks are usually not observed. Radial microcracking (R) can be observed only if two adjacent m-ZrO_2 grains are close together.

plotted as a function of their distance from the crack, as measured over a crack length of ≈ 30 μm (see also Table I). All t-ZrO_2 particles close to the crack transformed, the fraction of the transformed particles decreasing with increasing distance perpendicular to the crack (Fig. 13); t-ZrO_2 inclusions with sizes close to the critical diameter transform far ahead of the crack tip.

Mg-PSZ

During *in situ* straining experiments of the porous Mg-PSZ specimen (5% porosity), the crack jumped from pore to pore within the thinned transparent area of the straining specimen. It was thus not possible to map out the transformation zone in front of an arrested crack in the cubic matrix, although this was possible for single-crystal specimens of Mg-PSZ.[10] However, a well-defined transformation zone perpendicular to the crack does exist, and Fig. 14 shows that all t-ZrO_2 precipitates within ≈ 0.5 μm of the crack had transformed (cf. Rühle and Heuer[11]).

Zirconia-Toughened Spinel (ZTS)

In ZTS, the crack propagated in the thin area of the foil approximately parallel to a (100) or (110) plane of the spinel matrix; in thicker foils ($\gtrsim 0.9$ μm), the crack propagated perpendicular to the applied stress. Only those intracrystalline t-ZrO_2 inclusions had transformed through which the crack had propagated, although sometimes the crack propagated around the inclusion. One example of a particle before and after transformation is shown in Fig. 15.

Tetragonal Zirconia Polycrystals (TZP)

The two commercial TZP ceramics show transformation behavior which is distinctively different from the observations in the t-ZrO_2-containing ceramics. The

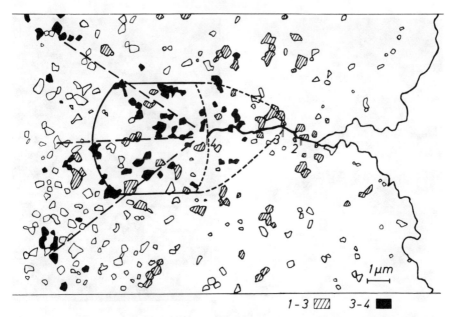

Fig. 12. Schematic representation of (partial) evaluation of straining experiment of Fig. 10 (specimen I). The specimen was strained so that the crack propagated to different positions (1, 2, 3, 4). All ZrO_2 inclusions are marked on the schematic drawing. Hatched areas: ZrO_2 grains transformed during propagation of crack from hole to position 3. Dark areas: ZrO_2 grains transformed during crack propagation from position 3 to 4. The transformation zone is outlined.

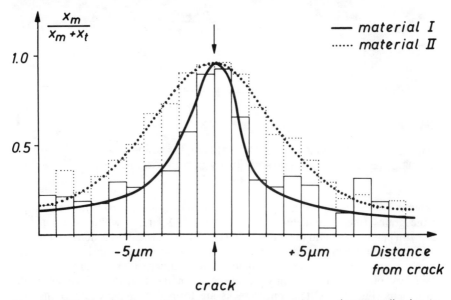

Fig. 13. Determination of transformation zone size w (perpendicular to propagated crack). The fraction of transformed ZrO_2 grains is plotted as a function from distance of the crack for materials I and II.

267

Fig. 14. Transformation zone in Mg-PSZ. (A) Bright-field image. The transformed precipitates can be analyzed readily by the presence of twins. (B) Dark-field image with reflection of m-ZrO_2: the transformed m-ZrO_2 particles are bright.

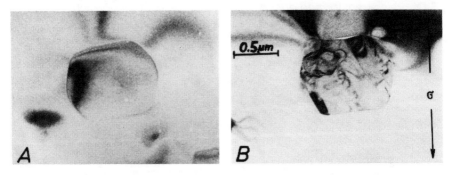

Fig. 15. Transformation of ZrO_2 inclusion in spinel. (A) t-ZrO_2 prior to straining; (B) after crack passed. Twins can be observed in m-ZrO_2.

sintered TZP-SI contained tetragonal grains (mean diameter ≈ 0.9 μm) and larger cubic grains (mean diameter ≈ 2 μm). All t-ZrO_2 grains within a distance of 2–3 μm perpendicular to the propagating crack transformed, as is noted in Fig. 16. Additional observations indicate that the transformation zone *ahead* of the crack is ≤ 1.5 μm wide. In addition, intercrystalline microcracks (along the grain-boundary plane) can frequently be observed in the transformed zones of the specimens (Fig. 17). Often, microcracks are formed near the cubic grains, and sometimes the crack runs transgranularly through these large c-ZrO_2 grains.

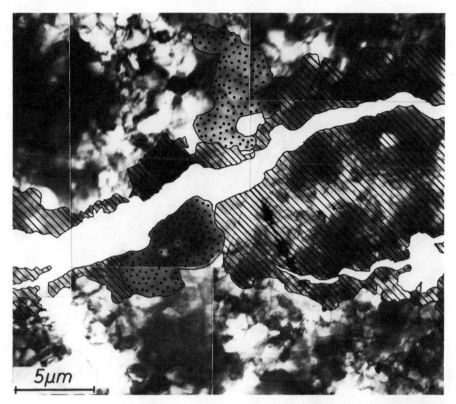

Fig. 16. *In situ* straining of sintered TZP. The transformed m-ZrO_2 grains are hatched, the c-ZrO_2 grains are dotted.

The situation in the TZP-HP ceramic is very different. This material contains no c-ZrO_2. Transformed areas (≈ 3 μm in diameter) of m-ZrO_2 alternate with untransformed areas, as can be seen in Fig. 18. A continuous transformation zone thus cannot be defined, an observation which was obtained for three different TZP-HP specimens. However, microcracks are also observed in the transformed regions of the foil and are similar to those shown in Fig. 17.

Discussion

The *in situ* straining experiments demonstrate that metastable confined t-ZrO_2 particles or precipitates transform in front of a crack tip. The experiments allow determination of the width w of the transformation zone perpendicular to the crack and also the extension, a, in front of a crack. However, the results are obtained for a thin foil which approximates a plane stress situation. In a bulk ceramic, plane strain usually obtains. It is well known from fracture mechanics and plasticity theory that the (quasi) plastic zone sizes (in front and at the side of a crack) are different for plane strain and plane stress (Fig. 19) and the relative zone size in each case has been calculated.[12] A similar result for transformation zone sizes in t-ZrO_2 toughened materials might be expected, since the stress-induced martensitic transformation ahead of a crack tip also represents a form of plastic deformation.

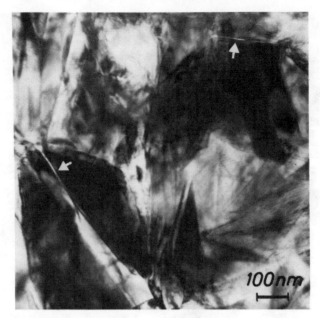

Fig. 17. Microcracking in transformed TZP. The microcracks are marked.

Fig. 18. Straining experiment of hot-pressed TZP. Only "clusters" of m-ZrO_2 grains can be observed. The transformed m-ZrO_2 grains are hatched.

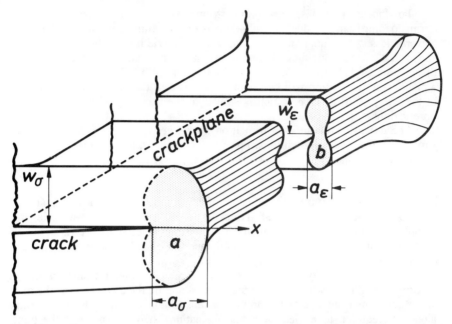

Fig. 19. Plastic zone around a crack for (a) plane strain and (b) plane stress configuration. a_σ, a_ε maximal zone width ahead and w_σ, w_ε in the wake of a crack.

However, the t-ZrO_2 particles or inclusions in the different materials present different geometries to the far-field applied stress, which needs to be taken into account. The different materials will be discussed separately.

ZrO₂-Toughened Alumina (ZTA)

We have studied only ceramics which contained intergranular ZrO_2 inclusions.[13] Those ZrO_2 inclusions are faceted and possess an inhomogeneous stress distribution.[14] The t-ZrO_2 particles are under a large tensile and inhomogeneous strain due to thermal expansion mismatch. Most particles which intersect at free surfaces will transform spontaneously or under very low applied stresses, which is actually observed during the straining experiments. We therefore consider only particles completely enclosed in the foil. If the foil thickness is larger than three times the diameter of the confined inclusion, then it can be accepted that a plane strain condition is a proper approximation, i.e., the measured transformation zone sizes correspond to the bulk situation. The inclusions close to a foil surface may be under plane stress.

For a completely quantitative evaluation, certain basic problems remain. Prior to loading, the stresses in particle and matrix arising from thermal expansion mismatch can be calculated *only* for spherical and ellipsoidal inclusions using an Eshelby-type solution.[15] The relaxation of the stresses in foils as a function of distance from the foil surfaces has been calculated by Mura[16] and Mader.[17] However, the stress (strain) distribution is not yet calculated for the case where particles and matrix with different elastic moduli are under residual stresses and are, in addition, subject to far-field loading.

For faceted particles, the strain distributions within inclusions (before and during loading) can be calculated numerically only by finite element calculations.[14] Nevertheless, a semiquantitative evaluation of the width w of the transformation zone is not critical. Our situation represents an intermediate state which is close to plane strain (w_ε). Since $w_\sigma/w_\varepsilon = 1.5$ (for Poisson ratio 0.25) the corrections for a plane (strain) situation will be minor. It is expected that the corrections will be of the order of 10 to 15%, which is similar to the accuracy of the experimental observations.

The extension of the transformation zone *ahead* of the crack, a, depends strongly on the loading conditions. For mode I cracking, it is the maximum width *ahead* of the crack tip

$$a_\sigma/a_\varepsilon = 5 \tag{1}$$

It is difficult to correct from the situation of a thin foil, a_σ, to a plane strain (bulk) situation, a_ε, especially since the foil thickness is not constant. The assumption that completely confined inclusions are under plane strain loading may be valid, but the correction can easily change the measured zone size ahead of the crack by a factor of 2.

The three-dimensional shape of the transformation zone in front of a mode I crack is determined by (i) the strain distribution around the crack and (ii) the main strain component necessary to form a martensite nucleus. If hydrostatic strains alone are required to nucleate the transformation, then the shape of the transformation zone will be considerably different from the case where shear components are required (see Fig. 7 of Ref. 1 and Fig. 1 of Ref. 18). The evaluation of Fig. 12 suggests that the transformation requires a large shear component but hydrostatic components of the stress field near the crack are also important.

The width of the transformation zone, w, must depend strongly on the size of the t-ZrO_2 particles. Those t-ZrO_2 particles, which are close in size to the "critical" particle size,[3] require smaller stresses for the formation of a stable nucleus than the smaller t-ZrO_2 inclusions. The mean value of the projected particle diameter of material I, $\bar{d}_I$, is 0.21 μm, whereas $\bar{d}_{II} = 0.25$ μm. This small change in the mean diameter results in a change in the half-width of the transformation zone w from 1.8 to 4 μm, which demonstrates clearly that w depends sensitively on the size distribution of the ZrO_2 inclusion. The increase in w results in an increase of K_{Ic}. Theoretical considerations predict

$$\Delta K_{Ic} \approx w^{1/2} \tag{2}$$

The experimental results fit quite well with the predictions (Table I) even though it is not evident that the half-width of the transformation zone size can be used for the quantitative evaluation of the experimental results.

Mg-PSZ

A well-defined transformation zone can be outlined in Mg-PSZ (Fig. 8). The distinction between the plane strain and the plane stress situation is much easier for this material. For example, in a $\langle 100 \rangle$ zone axis foil orientation, two of the three precipitate variants (those with c axes in the foil plane) extend throughout the foil, intersecting both foil surfaces. These particles are then subject to actual plane stress loading. For this case, it is possible to use the plane stress solutions and to show that the size of the transformation zone a_σ *ahead* of the crack is about 5 times larger than for the plane strain case a_ε. For the transformation zone normal to the crack, the differences are smaller, $w_\sigma/w_\varepsilon = 1.52$. Fortunately, and as noted above, the width w_ε of the transformation wake, w_ε, is the parameter needed for testing existing theories of transformation-toughening in two-phase ceramics. Therefore, the *in situ* result should give appropriate results.

The third variant, with *c* axis normal to the foil plane, is different. Coherency stresses are undoubtedly present due to the precipitation, but will be modified by stresses arising from thermal expansion mismatch between particle and matrix (*c*- and *t*-ZrO_2) during cooling from precipitation temperatures. The residual stresses in both particle and matrix will be relaxed during foil preparation, the effect being greater the closer the particle is to the foil surface. It is possible that particles far removed from the foil surface do approximate to plane strain loading during the *in situ* deformation; as far as matrix constraints are concerned, therefore, the situation is similar to bulk material. However, it is also possible that a loading situation between plane strain and plane stress exists for this case, which would not allow a closed-form solution. Thus, real caution is required when evaluating the zone dimensions for this orientation or particles.

If those considerations are correct, then an essential difference in the shape and size of the transformation zone ahead and at the wake of the crack should be observed for the three precipitate variants. In situ straining experiments of single-crystal specimens will resolve those problems.[10]

ZrO_2-Toughened Spinel (ZTS)

The straining experiments were performed in large-grained spinel material in which *t*-ZrO_2 inclusions were formed by chemical reactions. All intracrystalline ZrO_2 grains were tetragonal. During the straining experiments, only those ZrO_2 inclusions transformed to *m* symmetry which were directly in the path of the propagating crack, which usually propagated through the ZrO_2 particles. Therefore, the width of the transformation zone is very small. The reduction in "transformability" of the ZrO_2 inclusion in this case can be correlated with either (i) the regular shape of the inclusion, which results from the processing material; (ii) the presence of solute (Mg) ions in the ZrO_2 inclusions (those solutes usually increase the chemical driving force for the martensitic transformation and thereby stabilize the *t*-ZrO_2 by making nucleation more difficult); or (iii) with the small difference in the elastic moduli between ZrO_2 and the surrounding matrix.

Tetragonal Zirconia Polycrystals (TZP)

In TZP ceramics the nucleation of the transformation is governed not only by the size of the grains but also by the Y_2O_3 distribution in the ceramics, since all materials contain 2 to 5 wt% Y_2O_3. Analytical electron microscopy with high spatial resolution showed that the Y_2O_3 concentration is not constant throughout the specimen.[7] The *sintered* material that was studied was fired at 1600°C in the cubic-tetragonal two-phase field of the phase diagram, which results in larger *c*-ZrO_2 and smaller *t*-ZrO_2 grains. The ratio of grain size and Y_2O_3 concentrations allows the nucleation of the transformation and the transformation itself happens in a continuous layer close to the crack. The $t \rightarrow m$ transformation, as well as the formation of microcracks in the transformed region, must lead to increased toughness, whereas the fine grain size results in a high strength. The influence of the large cubic grains on strength and toughness is not yet understood.

The hot-pressed material of the same overall composition (4.5 wt% Y_2O_3) was fabricated at a lower temperature (1400°C) so that the specimen was processed in a single-phase field of the ZrO_2-Y_2O_3 phase diagram; only *t*-ZrO_2 was present. The Y_2O_3 concentration of the *t*-ZrO_2 is higher than that of the *t*-ZrO_2 grains in the sintered material. At the same time, the mean grain size in the hot-pressed material is smaller than the mean grain size of the stress-induced material. Both changes impede the nucleation of the martensitic transformation. A nucleus can be created only under special conditions (high stress concentrations in the material, presence of defects,[19] special orientation of the grain with respect to the propagating crack). The stresses and strains associated with the transformation frequently induce a transformation in neighboring grains (autocatalytic effects). Thus, clusters of grains transform close to the crack. Experimental observations suggest that the transformation of these "clusters" is completed after the crack has passed. The toughness of the hot-pressed t-ZrO_2 material is rather low, whereas the strength is rather high. Further investigations on the mechanical properties and the microstructural studies must be performed to fully correlate mechanical properties with microstructure.

Conclusion

In situ straining experiments of t-ZrO_2-containing ceramics allow determination of the transformation zone size ahead and at the side of the propagating crack. The transformation to m symmetry can be observed in the microscope. The quantitative evaluation of the zone size that may be appropriate for bulk material requires exact evaluation of the elastic state of the t-ZrO_2 particles before and after the transformation. Evaluation of the results suggests that, if the t-ZrO_2 inclusion is completely confined in the surrounding matrix, the influence of the relaxation at the thin foil can be neglected as a first approximation.

Acknowledgment

The authors acknowledge valuable discussions with A. H. Heuer. The work was partially supported by the Deutsche Forschungsgemeinschaft.

References

[1] A. G. Evans and A. H. Heuer, *J. Am. Ceram. Soc.*, **63** [5–6] 241–48 (1980).
[2] N. Claussen and J. Jahn, *Ber. Dtsch. Keram. Ges.*, **55** [11] 487 (1978).
[3] A. H. Heuer, N. Claussen, W. M. Kriven, and M. Rühle, *J. Am. Ceram. Soc.*, **65** [12] 642–50 (1982).
[4] M. Rühle, C. Springer, L. J. Gauckler, and M. Wilkens, p. 641 in Proceedings of the 5th International Conference on High Voltage Electron Microscopy, Kyoto, Japan. Edited by T. Imura and H. Hashimoto. Japanese Society of Electron Microscopy, Tokyo, Japan, 1977.
[5] D. L. Porter and A. H. Heuer, *J. Am. Ceram. Soc.*, **60** [3–4] 183–84 (1977).
[6] F. Sigulinski and N. Claussen; unpublished research.
[7] M. Rühle, N. Claussen, and A. H. Heuer; this volume, pp. 352–70.
[8] M. H. Loretto and J. W. Brooks; University of Birmingham, private communication and Ph. D. Thesis of J. W. Brooks.
[9] N. Claussen; Proceedings of a European Colloquium on Ceramics in Advanced Energy Technologies, Petton, Sept. 1982; to be published.
[10] L. H. Schoenlein, A. H. Heuer, and M. Rühle; this volume, pp. 275–82.
[11] M. Rühle and A. H. Heuer; p. 359 in Proceedings of the 7th International Conference on High Voltage Electron Microscopy. Edited by R. M. Fisher, R. Gronsky, and K. H. Westmacott. Berkeley, CA, 1983.
[12] F. A. McClintock and G. R. Irwin; p. 85 in Symposium on Fracture Toughness Testing and its Application, STP 381, ASTM, Philadelphia, 1965.
[13] M. Rühle and W. M. Kriven, *Ber. Bunsenges. Phys. Chem.*, **87**, 222 (1983).
[14] S. Schmauder, W. Mader, and M. Rühle, this volume, pp. 251–55.
[15] J. D. Eshelby; p. 89 in Progress in Solid Mechanics, Vol. 2. Edited by I. N. Sneddon and R. Hill. North-Holland, Amsterdam, 1961.
[16] T. Mura, Micromechanics of Defects in Solids. Martinus Nijhoff Publishing, The Hague, 1982.
[17] W. Mader, Ph. D. Dissertation, Universität Stuttgart, 1984.
[18] R. J. Seyler, S. Lee, and S. J. Burns; this volume, pp. 213–24.
[19] L. Ma and M. Rühle; unpublished work.

In Situ Straining Experiments of Mg-PSZ Single Crystals

L. H. Schoenlein and M. Rühle

Max-Planck-Institut für Metallforschung
Institut für Werkstoffwissenschaften
Stuttgart, Federal Republic of Germany

A. H. Heuer

Case Western Reserve University
Department of Metallurgy and Materials Science
Cleveland, OH 44106

In situ straining experiments were performed in an HVEM on Mg-PSZ single-crystal specimens containing tetragonal (t)-ZrO_2 precipitates with both the foil normal and applied stress in $\langle 100 \rangle$ directions. In this orientation, the stress state around t-ZrO_2 precipitates intersecting the top and bottom foil surfaces approximates a plane stress condition, while precipitates completely included in the foil are in a plane strain configuration. Transformed particles next to the crack were analyzed to reveal any difference in the transformation mechanism for t-ZrO_2 particles in these two stress states.

The stress-induced tetragonal (t)-to-monoclinic (m) martensitic transformation in ZrO_2 particles is the origin of the enhanced strength and toughness of transformation-toughened ceramics.[1-3] The most important parameter for current quantitative theories of transformation-toughening[4] is the width of the transformation "wake." Previous transmission electron microscopy (TEM) measurements of the transformation wake in Mg-PSZ have primarily utilized thin foils prepared by "back-thinning" bulk specimens containing indent-initiated cracks.[1,5] However, Rühle and coworkers[6] have demonstrated both the feasibility and the desirability of determining the extent of transformation adjacent propagating cracks *in situ* in the high-voltage electron microscope. To date, these *in situ* measurements have concentrated on dispersion ceramics such as ZrO_2-toughened Al_2O_3 and ZrO_2-toughened spinel (ZTA and ZTS, respectively), although one example of *in situ* transformation in polycrystalline Mg-PSZ has been previously published.[5] The purpose of this paper is to report preliminary results of an *in situ* determination of the transformation wake in single-crystal Mg-PSZ.

Experimental Procedure

Mg-PSZ single crystals with a bulk composition of 8.1 mol% MgO, grown by the skull-melting technique,[7] were provided by a commercial supplier.* Colum-

---

*Ceres Corp., Waltham, MA.

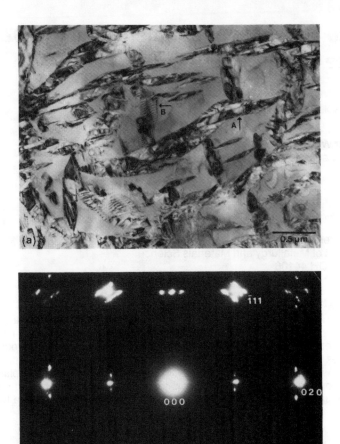

Fig. 1. (a) Bright-field micrograph of a foil of the as-received Mg-PSZ single crystal cut normal to the growth direction. Arrowed section of particle A is twinned on $(001)_m$. Particle B is twinned on $(100)_m$. (b) SAD pattern of foil normal; $B = [101]_c$.

nar single crystals, 1–2 cm in diameter and 5 cm long, were obtained from the skull and could be easily separated. All crystals were white, opaque, and severely cracked. By illuminating the crystals with a bright light source, internal cracks could be seen and crack-free regions identified, from which suitable specimens were obtained.

The microstructure of the as-grown single crystal is shown in Fig. 1(a). The material is two phase; the matrix is a MgO-rich cubic (c)-ZrO_2 solid solution and contains twinned particles of m-ZrO_2 up to ≈5 μm long, which are discussed further in the appendix.

The selected-area diffraction (SAD) pattern from the region of Fig. 1(a) is shown in Fig. 1(b); it contains c and twin-related m reflections. The diffraction pattern reveals that the foil normal, and hence the growth direction, is $\langle 101 \rangle_c$.

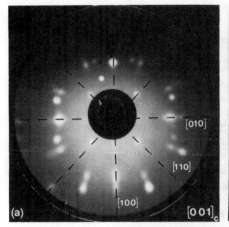
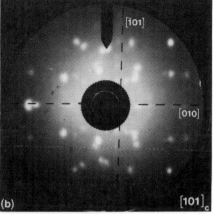
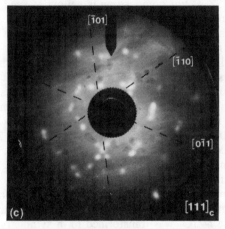

Fig. 2. Back-reflection Laue photographs of as-received Mg-PSZ single crystals. Fluorite diffraction symmetry, as indicated by mirror lines (dashed lines), is preserved near major zone axis orientations despite the presence of monoclinic reflections: (a) $[001]_c$ orientation (4-mm symmetry), (b) $[101]_c$ orientation (2-mm symmetry), and (c) $[111]_c$ orientation (3-mm symmetry).

Portions of the as-grown crystal required heat-treatment before they could be used for *in situ* experiments, as will be discussed below. Orientation of the single crystals was accomplished using the back-reflection Laue method. Unfortunately, reflections from the m-ZrO$_2$ precipitate complicated the Laue patterns to the extent that patterns off of a low-index zone axis could not be indexed. However, the fluorite diffraction symmetry could be recognized in major-zone axis orientations (4 mm for $|001|_c$, 2 mm for $[101]_c$, and 3 mm for $[111]_c$ (Fig. 2(a), (b), and (c)), despite the presence of monoclinic reflections, and the following procedure was used to orient the crystals.

Random searching near the crystal-growth direction provided a Laue photograph containing two mirror lines (Fig. 2(b)). Trial and error analysis proved them to be parallel to $[010]_c$ and $[\bar{1}01]_c$ and, by noting the orientation of these mirror lines with regard to fiducial marks on the crystal, the latter could be rotated into other low-index zones (Fig. 2(a), [001]$_c$; Fig. 2(c), [111]$_c$). Having thus identified two cubic directions in the crystal, specimens of any orientation could be produced.

Small rectangular parallelepipeds were sectioned from the oriented crystals using a diamond blade on a Buehler low-speed saw such that the faces were normal to $\langle 100 \rangle_c$ and were then polished with 6-μm diamond paste. The dimensions of the parallelepipeds were kept small ($\sim$3 by 5 by 5 mm^3) to minimize the chance of thermal fracture during the cooling cycle of the subsequent heat treatment.

We attempted to dissolve the precipitates shown in Fig. 1(a) by heating specimens at 1850°C for 4 h; this temperature should yield single-phase c-ZrO$_2$ (the solvus for this composition is at $\approx$1800°C, according to the equilibrium phase diagram of Grain[8]). This heat treatment had been successful in earlier work on polycrystalline Mg-PSZ of the same composition.[9] Both vacuum and Ar ambients were used in the present experiments for this solution heat treatment, but the specimens became black in both environments, indicating substantial reduction. (Commercial Mg-PSZ of this composition is yellow as-fired but becomes brown on solution-annealing in Ar.) Attempts to reoxidize the reduced single crystals by heating to 800°C in air caused severe cracking and loss of structural integrity. Such cracking was not observed with the polycrystalline Mg-PSZ and must arise from gradients in lattice parameters as the surface reoxidizes; the absence of cracking in polycrystalline PSZ may be due to the grain-boundary networks providing a short circuit diffusion path to the diffusing species responsible for reoxidation, thus keeping composition gradients to a minimum. This problem was obviated by heating the specimens at 1900°C for 4 h in air in a ZrO$_2$ furnace.[†] After this annealing treatment, the specimens were cooled from 1900° to 1000°C in $\approx$1 min and furnace-cooled to room temperature in $\approx$2 h. The microstructure obtained by this annealing treatment (Fig. 3) consists of a high number density of t-ZrO$_2$ precipitates in a c-ZrO$_2$ matrix; the particles range from 0.05 to 0.15 μm in length. The volume fraction of t-ZrO$_2$ was determined by crushing a small piece of crystal and determining the m/c ratio by conventional X-ray means[9,10] (it being assumed that all the t-ZrO$_2$ was transformed to m symmetry by crushing). This analysis indicated that the air-annealed material contains $\approx$65% t-ZrO$_2$. This very high volume fraction of t-ZrO$_2$ could arise from continued metastable precipitation at temperatures below the 1400°C eutectoid. For example, if the solvi defining the two-phase $c + t$ field are extrapolated to 1000°C, the diagram of Grain predicts an equilibrium ZrO$_2$ volume fraction of $\approx$60% for materials of this composition. In addition, the MgO content of the air-annealed material could be lower than that of the starting powder—MgO can readily evaporate during heat treatment of Mg-PSZ. We suspect both phenomena are occurring, as qualitative energy dispersive X-ray analysis on the foil of Fig. 3 indicated a slightly smaller Mg-Zr ratio than a foil known to be 8 mol% Mg-PSZ. The particle morphology is similar to t-ZrO$_2$ precipitates in polycrystalline Mg-PSZ in that the particles are lens-shaped in cross section, with the tetragonal c axis being the minor axis of revolution. The length-to-width (aspect) ratio varies from 2.5 to 3.5 in the single crystal; this is smaller than has been found by either Porter et al.[11] or Hannink[12] and undoubtedly is a sensitive function of the heat treatment used to form the precipitates. (Note that the transformed m-ZrO$_2$ precipitates in the as-received crystals (Fig. 1(a)) had a much larger aspect ratio.)

---

†Centres de Recherches sur la Physique des Hautes Temperature, Orleans Cedex, France.

Fig. 3. Bright-field micrograph of an air-annealed Mg-PSZ single crystal. Small, densely packed t-ZrO_2 precipitates are present, indicating that precipitation occurred on cooling from the annealing temperature.

Specimens for HVEM *in situ* straining experiments, 3 × 5 mm^2 wide and 150 μm thick, were cut from the heat-treated single crystals such that the foil normal and straining directions were along $\langle 100 \rangle_c$. The specimens were then carefully polished to ≈80 μm thickness with 6 μm diamond paste. Two elongated holes were ion-milled in the specimens, the area between the holes being electron-transparent (Fig. 4). (This was done by shielding the central ridge during ion-thinning using a small piece of wire.) The thinned specimens were then mounted on deformation holders with epoxy and inserted in the double-tilting straining stage described by Brooks and Loretto.[13]

Results and Discussion

Experiments conducted to date can best be described as partially successful. We have fractured specimens *in situ,* but we have not been able to propagate a crack a short distance, map out the transformation zone, move the crack a little further, etc., as has been done in ZTA, for example.[6] Nevertheless, useful information has been obtained.

In thin regions of the specimen, cracks initiated at points of stress concentration (i.e., at the corner of the foil edge in Fig. 5, where two $\langle 100 \rangle$ cleavage cracks intersect) and propagated in directions in which the specimen was thinnest; they occasionally arrested in thicker regions of the foil. In initially thicker regions of the foil, the cracks propagated in a direction normal to the applied stress, as expected, but the particle density was so high that it was difficult to map out the transformation zone. We thus will describe the transformation wake (Fig. 5) near a crack which arrested in a reasonbly thin area.

As seen in Fig. 5, all t-ZrO_2 precipitates within 0.5–1 μm of the crack have transformed to *m* symmetry; the edge of the transformation zone is quite sharp, as

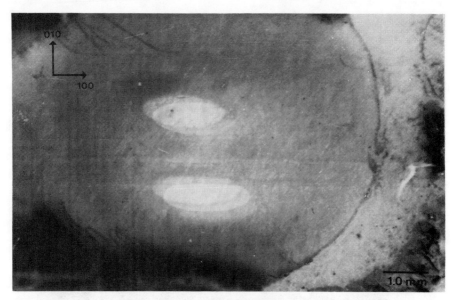

Fig. 4. Optical micrograph of oriented single-crystal Mg-PSZ specimen for *in situ* straining experiments. The area between the holes is electron-transparent. The foil normal and applied stress are in $\langle 100 \rangle_c$ directions.

Fig. 5. Bright-field micrograph of the transformation wake (outlined) near a crack arrested in a thin region of the foil. Particles within ≈0.5–1.0 μm of the crack have transformed to monoclinic symmetry and are internally twinned. Transformed particles which intersect the foil surface (e.g., at A) are twinned on $(001)_m$.

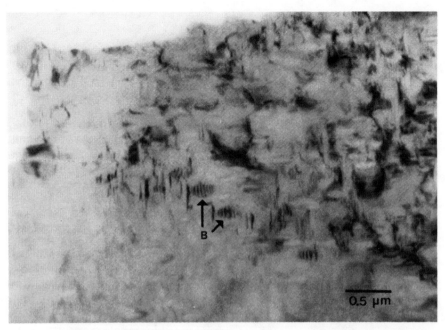

Fig. 6. Magnified image of transformation wake near foil edge in Fig. 5. Transformed particles which may be included in the foil (e.g., at B) are twinned differently, probably on $(100)_m$. Note the sharp transition between the transformed (top) and untransformed (bottom) regions.

shown in greater detail in Fig. 6. Precipitates intersecting both foil surfaces, e.g., A in Fig. 5, have transformed such that they are twinned on $(001)_m$, while a second group of particles, e.g., B in Fig. 6, are twinned on a different system, probably $(100)_m$. We suspect that the latter type of particles are completely included in this foil.

The two t variants that intersect the foil surface in the $\langle 001 \rangle_c$ orientation of Fig. 5 are those that have their c axes in the plane of the foil and are in a state of approximate plane stress during straining. The third variant, with its c axis normal to the foil plane, would appear to be disk-shaped in this orientation but generally cannot be easily distinguished from the matrix in the untransformed state.[14] This variant, if completely included in the foil, may possibly be in a state of plane strain during straining if far from the foil surface or in a poorly defined state of loading if near the foil surface. By analogy with ductile metals, where plane stress and plane strain configurations give rise to different plastic zone sizes adjacent cracks, Rühle and Heuer[15] suggested that a similar effect may occur near cracks in ZrO_2-containing ceramics — particle transformation wakes will be different for plane stress and plane strain, and it is the latter which are appropriate for bulk samples.

Figures 5 and 6 do suggest that the transformation mechanism for t-ZrO_2 precipitates undergoing crack-tip-induced transformation in thin foils may depend on whether the particles intersect the foil surfaces, in that the type of twinning found is different, but the factor of 5 difference in zone size possible for plane-stress and plane-strain conditions[16] was not observed — transformation wakes for both types of twinned particles had roughly the same spatial extent. The magnitude of the zone size was in acceptable agreement with previous data.[1,5]

Appendix
Twinning During the $t \to m$ Transformation

The particles of m-ZrO_2 shown in Fig. 1(a) twinned during the $t \to m$ transformation. Such twinning is ubiquitous for this martensitic transformation in ZrO_2 and can be used to readily distinguish t- from m-ZrO_2 during TEM examination. As in the case of the particles transformed *in situ*, two types of twins are observed. The particles of Fig. 1(a) are similar to the m-ZrO_2 particles that are present in overaged polycrystalline Mg-PSZ[9,17]; in the single crystal, they result from precipitation of t-ZrO_2 during the slow cooling that follows crystallization in the skull-melting process and subsequent transformation to m symmetry at an M_s temperature estimated to be between 500° and 900°C. The m-ZrO_2 invariably twins during this transformation. $(001)_m$ twins can be seen in particle A in Fig. 1(a); this type of twinning occurs in particles just larger than the "critical" size for which t-ZrO_2 is found at room temperature in polycrystalline Mg-PSZ[5] and also in particles transformed under plane strain loading (e.g., A in Fig. 5). The second type of twinned particles, e.g., B in Fig. 6, exhibits $(100)_m$ twinning, a type of twinning found to occur in larger, more severely overaged particles.[9] Available literature[18] indicates that the M_s temperature can vary between 250° and 900°C for 9 mol% Mg-PSZ. We believe that there is a correlation of $(001)_m$ twinning with low M_s temperatures and $(100)_m$ twinning with higher M_s temperatures, but this change in twinning mode with M_s temperature is not understood in detail.

Acknowledgment

A. H. Heuer thanks the Alexander-von-Humboldt Foundation for a Senior Scientist Award which made possible his stay at the Max-Planck-Institut für Metallforschung, where this paper was written, and the National Science Foundation (Grant No. DMR77-19163) for supporting his research on PSZ.

References

[1] D. L. Porter and A. H. Heuer, *J. Am. Ceram. Soc.*, **60** [3–4] 183–84 (1977).
[2] N. Claussen, *J. Am. Ceram. Soc.*, **61** [1–2] 85–86 (1978).
[3] T. K. Gupta, F. F. Lange, and J. H. Bechtold, *J. Mater. Sci.*, **13** [7] 1464 (1978).
[4] R. M. McMeeking and A. G. Evans, *J. Am. Ceram. Soc.*, **65** [5] 242–46 (1982).
[5] L. H. Schoenlein and A. H. Heuer; pp. 309–26 in Fracture Mechanics of Ceramics, Vol. 6. Plenum, New York, 1983.
[6] M. Rühle, B. Kraus, A. Strecker, and D. Waideich; this volume, pp. 256–74.
[7] K. Nassau, *Lapidary J.*, **35** [6] 1194 (1981).
[8] C. F. Grain, *J. Am. Ceram. Soc.*, **50** [6] 288–90 (1967).
[9] D. L. Porter and A. H. Heuer, *J. Am. Ceram. Soc.*, **62** [5–6] 298–305 (1979).
[10] R. C. Garvie and P. S. Nicholson, *J. Am. Ceram. Soc.*, **55** [6] 303–305 (1972).
[11] D. L. Porter, A. G. Evans, and A. H. Heuer, *Acta Metall.*, **27**, 1649 (1979).
[12] R. H. J. Hannink, *J. Mater. Sci.*, **13**, 2487 (1978).
[13] J. W. Brooks and M. H. Loretto, University of Birmingham, England; private communication.
[14] L. H. Schoenlein, M. Rühle, and A. H. Heuer; unpublished work.
[15] M. Rühle and A. H. Heuer; p. 359 in Proceedings of the 7th International Conference on High Voltage Electron Microscopy, Berkeley, CA, 1983.
[16] F. A. McClintock and G. R. Irwin; p. 85 in Symposium on Fracture Toughness Testing and its Application; STP 381. ASTM, Philadelphia, 1965.
[17] G. K. Bansal and A. H. Heuer, *J. Am. Ceram. Soc.*, **58** [5–6] 235–38 (1975).
[18] R. H. J. Hannink and M. V. Swain, *J. Aust. Ceram. Soc.*, **18** [2] 53 (1982).

Theoretical Approach to Energy-Dissipative Mechanisms in Zirconia and Other Ceramics

WOLFGANG POMPE AND WOLFGANG KREHER

Zentralinstitut für Festkörperphysik und Werkstofforschung
DDR-8027 Dresden
Akademie der Wissenschaften, German Democratic Republic

A theoretical model for the fracture toughness of ceramics is developed which takes into account such energy-dissipative mechanisms as stress-induced microcracking or phase transformations. To establish the general fracture criterion, a Griffith-type energy balance is employed. This energy balance comprises the elastic energy release rate, the fracture surface work consumed in the process zone at the crack tip, the energy dissipated in the dissipation zone, and the energy stored by residual stresses. Within this model, the effect of stress-induced phase transformations and microcracking on the increase of fracture toughness can be calculated on the basis of micromechanical data. The influence of residual stress states is especially studied, whereby other work is critically examined.

Microcracking or stress-induced phase transitions affect the behavior of a ceramic material much as plasticity does. These mechanisms have in common that they proceed at the expense of elastic energy, turning it into heat and other forms of energy irreversibly stored. They are therefore called energy-dissipative mechanisms. In front of a crack tip, they are capable of absorbing energy that otherwise would be available for crack propagation. A review of relevant work concerning ceramics with energy dissipation may be found in Refs. 1 and 2.

To theoretically investigate the toughness and strength of such ceramics, we have to consider the interaction of a macroscopic crack with a cloud of microcracks around its tip, or a cloud of particles that underwent a stress-induced phase transformation. As Fig. 1 shows, we must distinguish two interaction zones. First is the process zone where direct interaction processes between the macrocrack and the structural elements take place when the macrocrack propagates (e.g., coalescence with microcracks and intersection of particles). The second, larger zone is characterized by indirect interaction processes, i.e., elastic field interaction of the macrocrack with microcracks or transformed particles. This zone consumes energy and moves together with the macrocrack through the material; it is called the dissipation zone.

This separation has two advantages: It simplifies the theoretical analysis, and it allows for a unified description of different energy-dissipative mechanisms. On the other hand, failing to distinguish these zones may lead to wrong conclusions (cf. the comment by Kreher and Pompe[3] on a paper by Hoagland and Embury[4]).

It is the aim of the following theoretical considerations to derive relevant qualitative dependencies between toughness and structural properties and to determine under what conditions the energy-dissipative mechanisms should really increase toughness.

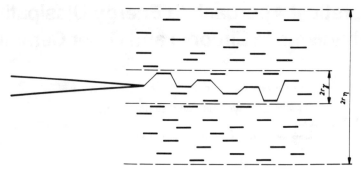

Fig. 1. Dissipation zone ($2r_\eta$) and process zone ($2r_\gamma$) at the tip of a macrocrack.

Theoretical Model

It has been now widely accepted that a Griffith-type energy balance is the most appropriate description of crack propagation in materials with energy dissipation.[1,5-7] According to this approach, a macroscopic crack becomes unstable if, for a small variation dl of its length l, the energy released from the elastic stress field balances the energy consumed in the surroundings of the crack tip. Usually, we may assume that the dissipation zone is small if compared with macroscopic lengths (such as macrocrack size or sample size). This may be called small-scale dissipation, analogous to small-scale yielding in elastic-plastic fracture mechanics. The release of energy (per unit sample thickness) is given by $\mathcal{G}\,dl$, where $\mathcal{G}$ is the ordinary elastic energy release rate. The energy consumed during crack propagation consists of different parts. The effective fracture surface work $2\gamma_p\,dl$ is consumed in the process zone, the energy dU_d is dissipated in the dissipation zone, and additionally there may be a change dU_s of elastically stored energy since the residual stress field may change. Thus, we obtain for the critical energy release rate or fracture toughness $\mathcal{G}c$:

$$\mathcal{G}_c = 2\gamma_p + \frac{dU_d}{dl} + \frac{dU_s}{dl} \qquad (1)$$

Let us see how far we can proceed without considering details of the dissipation process. The specific fracture surface work γ_p, effectively consumed in the process zone, may be less than or greater than the local fracture surface energy γ_0 of the plain ceramic material. This may be a result of coalescence of the macrocrack with microcracks or branching, respectively. A quantitative description of γ_p will be given later. Now let us look at the dissipation zone. We may consider it as an approximately homogeneous zone which can be described by constant specific values of dissipated and stored energy. Let η_d be the energy dissipated per unit volume, η_s the elastic energy stored by the residual stress field inside the dissipation zone, and η_{so} the elastically stored energy outside that zone. Then we can easily calculate the increments dU_d and dU_s by simply multiplying the specific energies by the newly created volume of the dissipation zone if the macrocrack advances. In the first order of dl, this volume increment is equal to $2r_\eta\,dl$ (per unit sample thickness). Hence, we have

$$dU_d = 2r_\eta\,\eta_d\,dl \qquad dU_s = 2r_\eta(\eta_s - \eta_{so})\,dl \qquad (2)$$

Finally, within the small-scale dissipation approach, the size r_η is given by[2,4]

$$r_\eta = 1/3 \frac{\mathcal{G}_c E_0}{\sigma_c^2} \tag{3}$$

Here σ_c denotes that local tensile stress necessary to initiate energy dissipation, and E_0 is Young's modulus of the ceramic material in the initial state. Substituting Eqs. (2) and (3) into Eq. (1) gives

$$\mathcal{G}_c = \frac{2\gamma_p}{1 - 1/3 \frac{\eta_d + \eta_s - \eta_{so}}{\sigma_c^2/2E_0}} \tag{4}$$

This is a general expression for the toughness of a material which is capable of dissipating energy by limited microstructural instabilities.

In the following, energy dissipation by stress-induced phase transformations and microcracking is considered in more detail. The corresponding changes of the microstructure are described by two scalar parameters. For the case of phase transformations, the volume fraction v_{tr} of particles that underwent the transformation is used, whereas microcracking is described by a so-called damage parameter ω, which measures the average weakening of the material. ω is defined by the drop of effective Young's modulus

$$E/E_0 = 1 - \omega \tag{5}$$

where E_0 is the modulus of the undamaged material and E is the modulus of the microcracked material. For instance, we may consider parallel oriented penny-shaped microcracks that are statistically arranged and have a statistically varying radius c. The effective modulus may be then calculated (Hoenig[8]), and we obtain for the damage parameter

$$\omega = \frac{16}{3} N_v \overline{(c^3)} \tag{6}$$

Here, N_v is the number of microcracks per unit volume and $\overline{(\ldots)}$ depicts averaging with respect to the crack radius. Other microcrack geometries or topological arrangements may lead to relations different from Eq. (6). Therefore, it is more convenient to discuss matters in terms of the general damage parameter ω. Now the quantities in Eq. (4) must be related to ω and v_{tr} by the help of a micromechanical model. We shall proceed on the basis of two general suppositions. First, we assume that the energy released by the microscopic instabilities is completely dissipated and cannot induce further changes of the microstructure. For instance, during microcracking, a certain amount of elastic energy is transformed into fracture surface energy of these microcracks, whereas another amount of energy is converted into elastic wave energy and is finally dissipated into heat. Hence, the total amount of dissipated energy is larger than the simple fracture surface energy. Second, we assume that a stable state is reached after the microscopic instabilities have taken place. This final state, characterized by ω or v_{tr}, is considered as adjusted by grain size, fraction of stabilizing additives, and other technological parameters. This final state is maintained until the macroscopic instability occurs.

It is now possible to calculate γ_p, η_d, η_s, η_{so}, and σ_c for different types of energy-dissipative mechanisms. In this paper, details of the calculation procedure are avoided; a more extensive treatment is presented elsewhere.[2,9] The results are summarized in Table I.

We consider three examples: (1) energy dissipation by phase transformations, (2) grain-boundary microcracking in a polycrystalline ceramic, and (3) microcracking in a ceramic composite under the influence of high residual stresses due to second-phase inclusions (v, volume fraction of the second phase). Actually, these three mechanisms may be acting simultaneously. It should be noted that in such a case they are mutually amplifying; i.e., the resulting toughness is greater than the simple sum of individual toughness values. This synergism occurs since the energy dissipated by microcracking, for instance, is proportional to $\mathcal{G}_c$ itself (cf., Eq. (2) and (3)); thus, any other toughness-increasing mechanism simultaneously amplifies energy dissipation by microcracking.

Now the special mechanisms and the resulting formulas will be briefly discussed. As mentioned above, the fracture surface energy γ_p, which is effectively consumed in the process zone, depends on the actual microstructure at the very crack tip. It may increase because of second-phase inclusions or local branching, or it decreases if the macrocrack absorbs microcracks which formed earlier (cf. Fig. 1). In this paper we consider only the latter effect, which is proportional to the degree of damage ω. The coefficient μ_γ (Table I) depends on microcrack geometry and arrangement. Usually μ_γ is of the order of unity, but it may be reduced in materials with laminar structure (e.g., graphite).

The energy η_d that is dissipated per unit volume may be calculated by using the stress–strain diagram of the dissipating material. We may assume that the phase transformation or the formation of microcracks proceeds at approximately constant stress, namely at the initiation stress σ_{pc} and σ_{mc}, respectively. So we obtain η_d from the transformation strain ε_{tr} (which is assumed to be isotropic) or from the change of effective Young's modulus due to the damage ω (cf. Eq. (5)). This method has the advantage that it leads immediately to the total dissipated energy; it includes also the energy of acoustic waves emitted during the nonequilibrium process of phase transformation or microcracking. That means that the dissipated energy is larger than the change of equilibrium free energy due to phase transformation or fracture surface energy of microcracks alone.

The change $\eta_s - \eta_{so}$ of energy stored by residual stresses can be calculated using a model for spherical particles embedded in a matrix. In the first example there may exist a thermal expansion mismatch $\Delta\varepsilon_T$ between the transforming particles and the matrix before the phase transformation occurs. Due to the transformation, the particles additionally expand by ε_{tr} and the stored energy changes as indicated in Table I. For the polycrystalline ceramic the residual stress state before microcracking is, in the present theory, characterized by the maximum tensile stress σ_t arising perpendicular to the grain boundary between neighboring grains (this stress component is mainly responsible for grain-boundary microcracking). The energy stored by the corresponding residual stress field is partly released when weakening due to microcracking occurs. For the third example the residual stress state before microcracking is caused by the thermal expansion mismatch $\Delta\varepsilon_T$ between particles and matrix; it arises when the material is cooled from the fabrication temperature down to room temperature (usually we may neglect the residual stresses due to thermal expansion anisotropy inside the matrix in comparison with the particle/matrix stresses caused by $\Delta\varepsilon_T$).

The last line of Table I shows how the initiation stress (σ_{pc} or σ_{mc}) is modified by the residual stress. In the first example, the actual critical stress σ_c is higher due

Table I. Theoretical Results for Different Energy-Dissipative Mechanisms (the Formulas are Discussed in the Text)

Mechanism:	Stress-induced phase transformation (1)	Microcracking (2)	Microcracking (3)
Origin of residual stresses:	thermal expansion mismatch	thermal expansion anisotropy	second-phase inclusions
Example:	$ZrO_2(t \to m)$	Al_2O_3 (polycryst.)	$Al_2O_3 + ZrO_2(m)$
Structural parameter:	v_{tr}	ω	ω, v
Initiation stress:	σ_{pc}	σ_{mc}	σ_{mc}
$\gamma_p =$	γ_0	$\gamma_0(1 - \mu_\gamma \omega)$	$\gamma_0(1 - \mu_\gamma \omega)$
$\eta_d =$	$v_{tr}\sigma_{pc}\varepsilon_{tr}$	$\dfrac{\sigma_{mc}^2}{2E_0}\dfrac{\omega}{1-\omega}$	$(1-v)\dfrac{\sigma_{mc}^2}{2E_0}\dfrac{\omega}{1-\omega}$
$\eta_s - \eta_{iso} =$	$\dfrac{5}{4}v_{tr}(1-v_{tr})E_0\varepsilon_{tr}(\varepsilon_{tr} + 2\Delta\varepsilon_T)$	$-\dfrac{\sigma_t^2}{2E_0}\omega$	$-\dfrac{5}{4}v(1-v)E_0\Delta\varepsilon_T^2\dfrac{(1+v)\omega}{2-\omega+v\omega}$
$\sigma_c =$	$\sigma_{pc} + \dfrac{5}{6}(1-v_{tr})E_0\Delta\varepsilon_T$	$\sigma_{mc} - \sigma_t$	$\sigma_{mc} - \dfrac{5}{12}(1+2v)E_0\Delta\varepsilon_T$

to the compressive stress inside the particles (for positive $\Delta\varepsilon_T$), whereas the second and third examples show a decrease of σ_c because of the tensile residual stresses between the grains or in the vicinity of the particles.

Discussion

In the following, the theoretical results are applied to various ceramics. The first example concerns the tetragonal(t)-monoclinic(m) transition in ZrO_2. This transition is connected with a volume increase corresponding to a linear expansion strain ε_{tr} of $\approx 1.4\%$. The transformation is assumed to be isotropic; i.e., the shear strain is neglected. To utilize the transformation as an energy-dissipative mechanism, it is necessary to retain the t phase at room temperature. This can be achieved by different means. Consideration is given here to experiments by Lange,[10] who used different mole fractions of Y_2O_3 (between 2.5 and 7.5 mol%) so that it was possible to retain the t and cubic(c) phases at room temperature. It is assumed that the t phase completely transforms into the m modification at the critical stress. Figure 2 shows the comparison between the present theory and experimental results (the critical stress intensity K_c and the fracture toughness are related by $K_c/K_o = (\mathcal{G}_c/2\gamma_0)^{1/2}$). The theoretical curve was calculated by using Eq. (4) and Table I(1) with parameters $E_0 = 200$ GPa, $\varepsilon_{tr} = 1.4\%$, $\sigma_{pc} = 2500$ MPa, and $\Delta\varepsilon_T = 0$ (no residual stresses before the transformation). These values are in accordance with data provided by Lange.[10] Figure 2 shows that there is a good

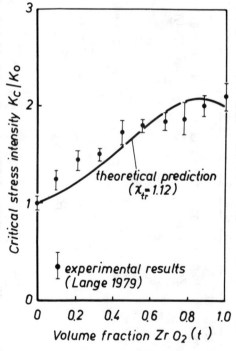

Fig. 2. Normalized critical stress intensity K_c/K_o as a function of volume fraction of transforming particles v_{tr}. Experimental results by Lange (Ref. 10) correspond to $K_o = 3$ MPa·m$^{1/2}$. The dimensionless parameter χ_{tr} is defined by $\chi_{tr} = E_0\varepsilon_{tr}/\sigma_{pc}$.

correspondence between theory and experiment. Lange[10] also presented theoretical results. Lange, however, used a constant radius r_η for the dissipation zone (fitted to the experimental results for the toughness) instead of the variable radius given by Eq. (3).

Polycrystalline Al_2O_3 is an example of energy dissipation by grain-boundary microcracking (see, e.g., Krell[11]). Residual stresses originate because of the thermal expansion anisotropy of the hexagonal grains. Different theoretical investigations have shown that $\sigma_t = 150$ MPa is a reasonable choice for the tensile residual stresses at room temperature. It is usually assumed that these stresses do not depend on grain size D (at least they are only weakly dependent). Contrary to this, the critical stress σ_{mc}, at which microcracking starts, shows an inverse square-root dependence on D:

$$\sigma_{mc} = AD^{-1/2} \tag{7}$$

For Al_2O_3, Krell and Kreher[12] assessed the constant $A = 2.34$ MPa·m$^{1/2}$. Thus, the toughness becomes a function depending on grain size D. Figure 3 shows the comparison between the present theory and experimental results.[13] The theoretical curve has been calculated using Eq. (4) and Table I(2) with $\mu_\gamma = 1.0$. We have assumed a constant damage parameter $\omega = 0.5$, i.e., the grain-boundary topology, which is responsible for the attainable amount of stable damage, remains constant for varying grain size. A better correspondence between theory and experiment would be obtained if a slightly decreasing ω with increasing grain size were

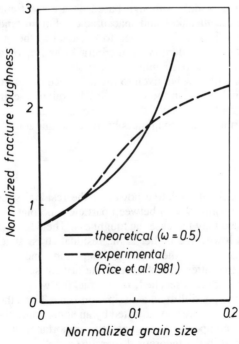

Fig. 3. Normalized fracture toughness $\mathcal{G}_c/2\gamma_0$ as a function of normalized grain size D/D_0. D_0 is given by $D_0 = (A/\sigma_t)^2 = 245$ μm. Experimental results by Rice et al. (Ref. 13) correspond to $\gamma_0 = 20$ J/m^2.

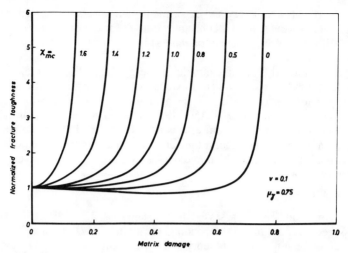

Fig. 4. Normalized fracture toughness $Y_c/2\gamma_0$ as a function of damage ω for different residual stress states.

assumed. At this point, more detailed investigations are necessary; nevertheless, $\omega = 0.45$–0.50 seems reasonable.

In connection with their experimental results, Rice and Freiman[14] also tried to develop a theoretical model. Although based on a correct energy balance, their theory involved some mistakes and inaccuracies. For instance, the size of the dissipation zone was implicitly assumed to be equal to the grain size. Also, the proportionality between the density of microcracks and σ_t/σ_A must be questioned (σ_t, tensile residual stress; σ_A, applied sample stress).

Consideration will now be given to the influence of residual stresses created by second-phase inclusions on the toughening by microcracking. To this end we make use of the formulas given in Table I(3). For constant-volume fraction v of the second phase, the normalized toughness becomes a function of damage ω and the dimensionless parameter

$$\chi_{mc} = E_0 \Delta \varepsilon_T / \sigma_{mc} \tag{8}$$

which is a measure for the relative power of the residual stress state due to the thermal expansion mismatch $\Delta \varepsilon_T$ between particles and matrix. As Fig. 4 shows, microcracking indeed may increase the toughness. The degree of damage required for this increase is lower, the higher is the residual stress state.

This effect arises due to the growing dissipation zone for increasing residual stress. Obviously, the latter cannot exceed the critical stress for microcrack formation. If this stress level is reached, microcracking would occur throughout the sample, excluding the possibility of a dissipation zone. Thus, the theory shows that residual stresses must be carefully adjusted by an appropriate choice of the components of the ceramic composite. Seemingly, this importance of residual stresses has scarcely been noticed in the theoretical literature.

As an example, let us discuss Al_2O_3 with unstabilized ZrO_2 particles. This system is especially promising since the volume increase due to the $t \rightarrow m$ transition of ZrO_2 creates high residual stresses. On the other hand, the Al_2O_3 matrix is

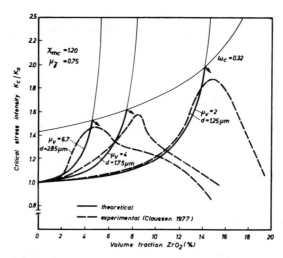

Fig. 5. Normalized critical stress intensity as a function of volume fracture v of m-ZrO_2 particles for various particle diameters d. Experimental results (Claussen, Ref. 16) correspond to $K_o = 5$ MPa·m$^{1/2}$.

strong enough to withstand these stresses so that energy dissipation by microcracking occurs during loading. Such composites were first investigated by Claussen.[15,16] To compare his experimental results and the theory, we use the following parameters: $\Delta\varepsilon_T = 1.6\%$ (thermal expansion mismatch plus transformation strain), $E_0 = 390$ GPa, and $\sigma_{mc} = 5000$ MPa (according to an assumed flaw size for microcrack formation of 1 μm). This yields $\chi_{mc} = 1.2$. Furthermore, we use Eq. (6) and assume proportionality between the number of microcracks and the number of zirconia particles. The resulting proportionality constant μ_v for the relation between damage ω and volume fraction v depends on the diameter d of the ZrO_2 particles and has been adjusted to the experimental curves (for details see Kreher and Pompe[2]).

Figure 5 shows that the increase in toughness is predicted fairly well by theory. The sudden decrease above a certain volume fraction may be due to coalescence of microcracks in the process zone, which has not been considered in the present model. Still, a remarkable conclusion can be drawn: The experimental results are compatible with the assumption of a critical degree of damage ω_c, independent of particle size d, which limits the increase of toughness. The resulting theoretical curve for constant $\omega = \omega_c$ is also plotted on Fig. 5. It may be seen that $\omega_c = 0.32$ fits well to the maxima of the experimental curves. In comparison with $\omega = 0.5$, which was used to explain the increasing toughness of polycrystalline Al_2O_3 (Fig. 3), the lower values of necessary damage indicate the positive influence of residual stresses.

The experimental evidence discussed above shows that the theoretical model, though it may seem somewhat sophisticated, can serve as a heuristic means to promote understanding of material behavior. It may give a clue how to combine parameters of the system, where to place adjustable constants, etc. in order to obtain a formula that, first, describes the experimental data sufficiently well, and second, allows extrapolation into regions where no experiments have yet been done.

References

[1] F. F. Lange, "Fracture Mechanics and Microstructural Design"; pp. 799–819 in Fracture Mechanics of Ceramics, Vol. 4. Edited by R. C. Bradt, D. P. H. Hasselman, and F. F. Lange. Plenum, New York, 1978.

[2] W. Kreher and W. Pompe, "Increased Fracture Toughness of Ceramics by Energy-Dissipative Mechanisms," *J. Mater. Sci.*, **16** [3] 694–706 (1981).

[3] W. Kreher and W. Pompe, "Comment on 'A Treatment of Inelastic Deformation Around a Crack Tip due to Microcracking'," *J. Am. Ceram. Soc.*, **65** [7] C-117 (1982).

[4] R. G. Hoagland and J. D. Embury, "A Treatment of Inelastic Deformation Around a Crack Tip due to Microcracking," *J. Am. Ceram. Soc.*, **63** [7–8] 404–10 (1980).

[5] S. D. Antolovich, "Fracture Toughness and Strain-Induced Phase Transformations," *Trans. Metall. Soc. AIME*, **242** [11] 2371–73 (1968).

[6] W. Pompe, H. A. Bahr, G. Gille, and W. Kreher, "Increased Fracture Toughness of Brittle Materials by Microcracking in an Energy-Dissipative Zone at the Crack Tip," *J. Mater. Sci.*, **13** [12] 2710–23 (1978).

[7] D. L. Porter, A. G. Evans, and A. H. Heuer, "Transformation Toughening in Partially Stabilized Zirconia (PSZ)," *Acta Metal.*, **27** [10] 1649–54 (1979).

[8] A. Hoenig, "Elastic Moduli of a Non-Randomly Cracked Body," *Int. J. Solid Struct.*, **15** [2] 137–54 (1979).

[9] W. Kreher and W. Pompe, "Strength of Ceramics"; in Current Topics in Materials Science. Edited by E. Kaldis. North Holland, Amsterdam; in press.

[10] F. F. Lange, "Phase Retention and Fracture Toughness of Materials Containing Tetragonal ZrO_2"; pp. 45–56 in Proceedings of the 3rd International Conference on Mechanical Behaviour of Materials (ICM3), Cambridge, UK, Aug. 1979, Vol. 3. Edited by K. J. Miller and R. F. Smith. Pergamon, Oxford, 1980.

[11] A. Krell, "Alumina Structure with Improved Fracture Properties," *Phys. Status Solidi (A)*, **63** [1] 183–92 (1981).

[12] A. Krell and W. Kreher, "On Subcritical Crack Growth in Ceramics as Influenced by Grain Size and Energy-Dissipative Mechanisms," *J. Mater. Sci.*, **18** [8] 2311–18 (1983).

[13] R. W. Rice, S. W. Freiman, and P. F. Becher, "Grain-Size Dependence of Fracture Energy in Ceramics: I, Experiment," *J. Am. Ceram. Soc.*, **64** [6] 345–50 (1981).

[14] R. W. Rice and S. W. Freiman, "Grain-Size Dependence of Fracture Energy in Ceramics: II, A Model For Noncubic Materials," *J. Am. Ceram. Soc.*, **64** [6] 350–54 (1981).

[15] N. Claussen, "Fracture Toughness of Al_2O_3 with an Unstabilized ZrO_2 Dispersed Phase," *J. Am. Ceram. Soc.*, **59** [1–2] 49–51 (1976).

[16] N. Claussen, "Erhöhung des Riβwiderstandes von Keramiken durch gezielt eingebrachte Mikrorisse," *Ber. Dtsch. Keram. Ges.*, **54** [12] 420–23 (1977).

Microcracking Contributions to the Toughness of ZrO_2-Based Ceramics

K. T. Faber

The Ohio State University
Department of Ceramic Engineering
Columbus, OH 43210

Analyses are presented to evaluate two microcrack toughening situations in ZrO_2-based ceramics. The first analysis considers toughening due to stress-induced microcracking of residually strained monoclinic ZrO_2 particles. The second assesses microcracking that occurs as a consequence of the stress-induced tetragonal-monoclinic transformation. Both toughening mechanisms are characterized by a permanent dilatational strain and a modulus reduction; however, only the latter may result in a toughness enhancement or reduction. Conditions under which toughness enhancement and reduction are observed are established. The magnitudes of toughening increases from transformation-toughening and microcrack-toughening are also compared.

In recent analyses of transformation-toughening in ZrO_2-based systems,[1,2] the predicted fracture toughnesses fall short of the impressive increases demonstrated in these materials, either in partially stabilized ZrO_2 or in the alumina zirconias. It has been speculated that, in addition to the dilatant crystallographic transformation, additional toughening processes occur concomitantly. These include microcracking[3,4] and crack deflection.[5,6] It is the intent of this paper to analyze the contribution to toughening that microcracking represents. Two possible scenerios will be presented. First, the simpler microcrack-toughening event that occurs in materials containing stable residually strained monoclinic ZrO_2 particles will be examined for background development. The second analysis will consider microcracking that occurs as a consequence of the stress-induced martensitic transformation, of prime interest here. Finally, a comparison will be made between the magnitudes of toughening afforded by transformation-toughening and microcrack-toughening.

The methodology used here is that proposed by Budiansky et al.[1] as a continuum approach treating the dilatant, irreversible transformation of a second-phase particle in a linear elastic matrix, a review of which is worthwhile at this point. The toughening is assumed to arise from crack-shielding processes, which are stress-induced in a process zone (either the transformation or microcracking process zone) ahead of the crack tip and in the wake over the crack surfaces. For a frontal process zone only, the crack-shielding effect may be characterized by the following relation:

$$\frac{K^{\infty 2}}{E_1}(1 - \nu_1^2) = \frac{K_{\text{tip}}^2}{E_2}(1 - \nu_2^2) \tag{1}$$

where K^∞ is the stress intensity associated with the applied field, K_{tip} the stress intensity at the crack tip, E the elastic modulus, ν Poisson's ratio, and the subscripts 1 and 2 refer to the far-field and the near-tip material, respectively. At fracture, when $K^\infty = K_c^\infty$, K_{tip} may be described by the fracture toughness of the near-tip field and is given by K_c (monoclinic) for transformation-toughened materials, since crack propagation occurs through a transformed and, hence, monoclinic region. For microcracked materials, K_{tip} may be evaluated as $K_c(1 - f_s)$, where K_c is the toughness of the crack-free material and f_s the microcrack density. Only under rare circumstances, such as large microcrack densities, ($f_s > 0.3$) will the frontal process zone provide any significant toughening, as noted in Ref. 7.

Consider now the toughening increment, which includes the shielding behind the crack tip in the region known as the wake. An energy balance may be established for conditions of steady-state crack growth, such that

$$\frac{K^{\infty 2}}{E_1}(1 - \nu_1^2) = \frac{K_{tip}^2}{E_2}(1 - \nu_2^2) + 2\int_0^h U(y)\,dy \qquad (2)$$

where $U(y)$ is the work or residual energy left behind in the wake. Physically, the energy balance implies that the energy from the remote field, K^∞, can be partitioned into that deposited in the wake and that released during crack advance at the crack tip through K_{tip}. The term $U(y)$ can be evaluated by considering the work performed during the transformation (or microcracking) over the loading in the region in front of the advancing crack and the unloading of the same region behind the crack tip. Hence, the hysteresis in the stress-strain response of the process zone, of height h, will provide adequate information for work-density analysis.

Both the transformation-toughened materials and materials that are susceptible to stress-induced microcracking demonstrate R-curve behavior, whereby the toughness depends on crack-growth history.[2,7-9] For the present purpose, consideration will be given only to conditions of maximum toughening, i.e., when the R curve reaches its asymptotic limit, for a fully developed wake.

Microcracking of Monoclinic Zirconia

The first case considers the microcracking that arises from residually strained monoclinic ZrO_2 particles that transform on cooling from the processing temperature. This situation was recently observed experimentally by Claussen et al.[10] who note significant toughness increases ($K_c^\infty/K_c = 2$) when stress-induced microcracking occurs in monoclinic ZrO_2-containing materials.* A similar situation was recently modeled by Evans and Faber[7] for stress-induced microcracking in materials having significant thermal expansion anisotropy or mismatch. The model is readily extended to the ZrO_2 (volume expansion) case.

To calculate the work density associated with the microcrack process zone, consider the stress-strain curve that derives on loading ahead of the crack tip and unloading behind the crack tip (Fig. 1) and demonstrates a characteristic hysteresis. The stress-strain behavior can be partitioned into two distinct components. The first derives from a permanent strain, θ, due entirely to the dilatant microcracking that occurs due to stress enhancement in the near-tip field in regions of residual tension.† A permanent microcrack opening is expected. The stress-strain

---

*An earlier observation by Claussen (Ref. 3) may also be considered as an example of stress-induced microcracking in the Al_2O_3–monoclinic ZrO_2 systems. The example noted above presents more clear-cut evidence for the specific toughening mechanism.
†The microcrack zone is actually subjected to an effective strain, θ_{eff}, of magnitude $\theta_{eff}f_s$.

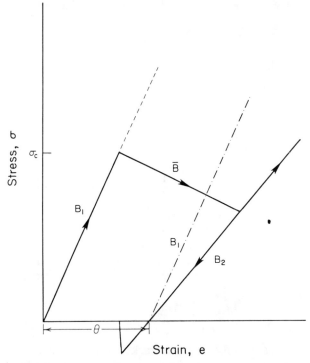

Fig. 1. Idealized stress-strain response for process zone undergoing stress-induced microcracking.

hysteresis is modified by the increase in compliance of the microcracking zone, which is shown in Fig. 1 as the bulk modulus reduction (B_1 to B_2).

Evaluation of $U(y)$ is performed by geometrically analyzing the stress-strain curve in Fig. 1 and may be written as[7]

$$U(y) = \sigma_c \theta_{\mathit{eff}} + \frac{\sigma_c^2}{2}\left(\frac{1}{B_2} - \frac{1}{B_1}\right) + \frac{\bar{B}\left[\theta_{\mathit{eff}} + \sigma_c\left(\frac{1}{B_2} - \frac{1}{B_1}\right)\right]^2}{2[1 - \bar{B}/B_2]}$$
$$+ \frac{B_2(1 - 2\nu_2)\theta_{\mathit{eff}}^2}{3(1 - \nu_2)} \qquad (3)$$

where σ_c is the mean hydrostatic stress for microcracking, B is the bulk modulus, and $\bar{B}$ refers to the stress-strain gradient during microcracking. The first three terms are directly obtained from the tensile portion of the stress-strain hysteresis, and the fourth term is associated with the compressive constraints of the surrounding matrix that arise on unloading.[‡] The above expression, similar in form to that derived by Budiansky et al.,[1] must take into account the bulk modulus reduction, as noted in the second and third terms. The critical stress for microcracking, σ_c,

---

[‡]A deviatoric shielding term has been omitted here from the original microcracking analysis (Ref. 7) for simplicity. Toughness variations of less than 5% occur when the deviatoric component is omitted.

can be related to the zone height, h, and the applied stress intensity, K^∞, by Eq. (4):

$$h = \frac{3^{1/2}(1 + \nu_1)^2}{12\pi}\left(\frac{K^\infty}{\sigma_c}\right)^2 \quad (4)$$

Integrating the work density over the zone height, h, yields

$$2\int_0^h U(y)\,dy = \frac{(1 + \nu_1)K^\infty \theta_{\text{eff}}(h)^{1/2}}{(3)^{1/4}\pi^{1/2}}$$

$$+ \frac{\left[\theta_{\text{eff}} + \frac{(1 + \nu_1)K^\infty}{(3)^{1/4}\pi^{1/2}}(1/B_2 - 1/B_1)\right]^2}{1 - (B/B_2)}\bar{B}h$$

$$+ \frac{[(1 - \nu_1)K^\infty]^2}{4(3)^{1/2}\pi}\left(\frac{1}{B_2} - \frac{1}{B_1}\right) + \frac{2B_2(1 - 2\nu_2)\theta_{\text{eff}}^2 h}{3(1 - \nu_2)} \quad (5)$$

By incorporating Eq. (5) into Eq. (2) and equating $\bar{B}$ to that determined for the transformation problem $[= -2B_1(1 - 2\nu_1)/1 + \nu_1]$, the relative toughness may be established through solution to the quadratic[7] to provide a toughening increment dependent on two dimensionless parameters: $\theta(h)^{1/2}E_1/K_c$ and the modulus ratio, E_2/E_1. The latter reduces to a dependence strictly on the microcrack density, f_s,[11] as given by the following:

$$\frac{E_2}{E_1} = 1 - \frac{16}{45}\frac{(1 - \nu_2^2)(10 - 3\nu_2)}{(2 - \nu_2)}f_s \quad (6)$$

where ν_2 is Poisson's ratio for the microcracked material. Hence:

$$K_c^\infty/K_c = Q[\theta(h)^{1/2}E_1/K_c, f_s] \quad (7)$$

where the function, Q, is plotted in Fig. 2 for prediction of the relative toughness increases. For high permanent strain values or large zone sizes, the predictions lie well within the experimental observations of Claussen et al.[10]

Microcracking as a Consequence of the Stress-Induced Martensitic Transformation

To assess the extent of toughening afforded by microcracking in the vicinity of the transformed monoclinic particles, the same approach is utilized. The martensitic transformation/microcracking phenomena can be conveniently viewed as two specific stress-strain responses: particle and matrix. Figure 3(a) shows the stress-strain behavior associated purely with the dilatant particle transformation. Unlike the microcracking response discussed in the previous section, only a permanent strain component derives, and the unloading stress-strain curve is parallel to the loading curve. The permanent strain associated with the transformation, shown in the hysteresis, corresponds to the constrained strain, e_c^T.[12,§] Due to the resultant volumetric increase on transformation, the particle, now compressively strained, may cause radial microcracking in the matrix. The stress-strain behavior associated with the matrix can be characterized by that shown in Fig. 3(b), similar to that described in the previous section. Rather than a straightforward addition of the two responses shown in Fig. 3, further considerations must be made to two features: alteration of the total (composite) strain with the additional microcracking and change in the compliance of the system, vis-à-vis the compliance of the matrix.

---

§The constrained strain e_c^T for a spherical particle has been described by Eshelby (Ref. 12) as $e^T/2$, for Poisson's ratio of 0.2.

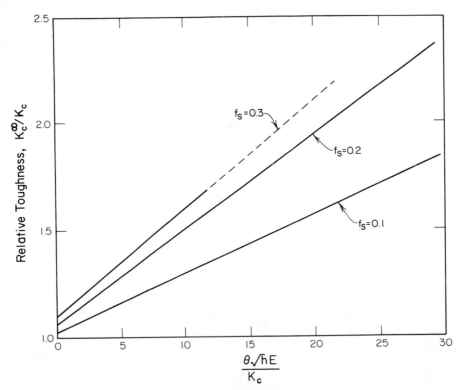

Fig. 2. Relative toughness increases vs the dimensionless stress intensity parameter for a variety of microcrack densities, f_s.

To examine the increase in the total strain of the system, a simple analysis is presented. Consider a cube of dimension l which contains a particle of radius r. On transformation of the particle, the strain associated with the particle is defined by the constrained strain, e_C^T (Fig. 4(a)). The effective transformation strain of the *system* is the product of the unconstrained strain and the volume fraction of the transforming particles $(e^T V_f)$.[2]

As the limiting case for microcracking and transformation, consider a series of microcracks in orthogonal directions formed as a consequence of the transformation (Fig. 4(b)). Imagine that the crack-opening displacements are large enough such that the particle strain may now be represented by the unconstrained transformation strain, e^T. The particle radius is now increased by an amount $e^T/3$. Consequently, the crack opening, δ, of each of the orthogonal cracks may be defined in terms of the new particle radius:

$$\delta = (2/3) r e^T \tag{8}$$

such that the linear strain, $e_l (= \delta/l)$, may be written as

$$e_l = (2/3)(r/l) e^T \tag{9}$$

The total strain of the system, e, follows:

$$e = 2(r/l) e^T \tag{10a}$$

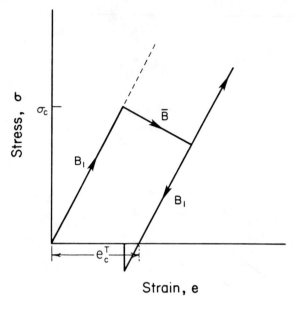

a) Transformation of Particle

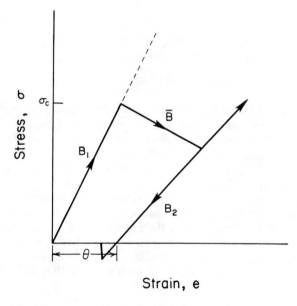

b) Microcracking of Matrix

Fig. 3. Idealized stress-strain diagrams for (a) a particle undergoing a dilatant transformation and (b) the matrix undergoing microcracking.

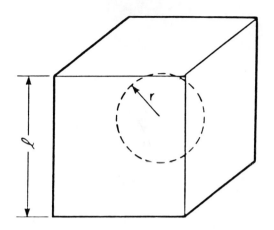

(a) Transformed Particle in Matrix

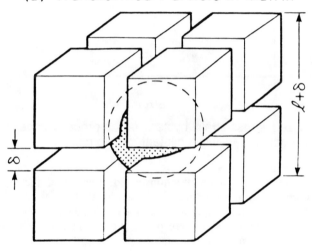

(b) Transformed Particle in Microcracked Matrix

Fig. 4. Schematics of (a) a transformed particle in a matrix and (b) transformed particle in microcracked matrix with three orthogonal microcracks.

The ratio r/l may be related to the volume fraction, V_f, such that:

$$e = 2\left(\frac{3V_f}{4\pi}\right)^{1/3} e^T \tag{10b}$$

This equation now affords a comparison between the total strain from the transformation alone and from the transformation plus microcracking. The ratio

$$\frac{e^{trans+micro}}{e^{trans}} = 2\left(\frac{3}{4\pi}\right)^{1/3} V_f^{-2/3} \tag{11}$$

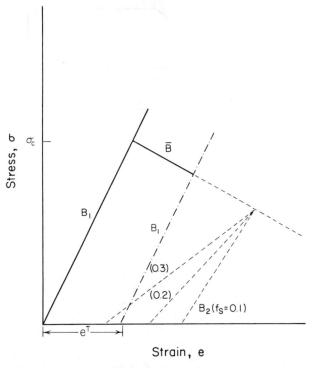

Fig. 5. Idealized stress-strain hysteresis for process zone undergoing microcracking accompanying a dilatant transformation.

is greater than unity for all volume fractions and decreases with increasing volume fractions. Incorporation of the increase in dilatational strain along with modulus reduction of the composite allows for prediction of the new toughening increment. The general trend for the stress-strain response of the composite is shown in Fig. 5, where it has been assumed that the transformation and microcracking occur at identical stresses σ_c and follow the identical stress-strain gradient $\bar{B}$. One can anticipate from the schematic of Fig. 5 that either a toughening or a detoughening may be anticipated on the basis of the amount of strain increase (from the accompanying microcracking process) as compensated by the modulus reduction. Reasonable predictions are afforded only when the magnitude of the strain increase is known.

It is possible, however, to define the limits of strain increase and microcrack density (= modulus reduction) over which transformation-toughening will be enhanced or reduced compared to transformation-toughening alone. Again, geometric arguments are utilized. Consider the stress-strain schematic shown in Fig. 6. The toughening increment may be enhanced *only* if the area fraction depicted in Fig. 6 as A is larger than B. Physically, this concept is indicative of the opposing effects of the increase in strain and the reduction in modulus. If area B is larger than area A, the increase in compliance (and reduction in modulus) has the overriding effect of lowering the overall toughness of the material. Alternatively, if area A is larger than area B, the increase in the total permanent strain supports the increase in toughness.

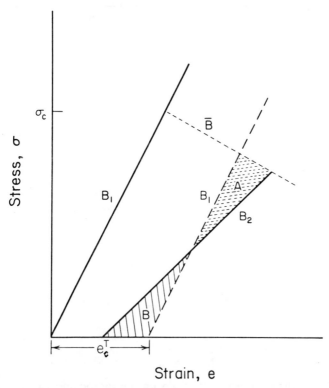

Fig. 6. Idealized stress-strain hysteresis for process zone undergoing microcracking accompanying the dilatant transformation depicting regions that define the minimum strain increase for enhanced toughening.

The toughness increases or decreases may then be computed numerically by considering the geometric areas A and B. The toughness enhancement is bounded by the limits:

$$\frac{\theta'}{\theta} > F[E_2/E_1, \sigma_c] \qquad (12)$$

where θ' is the increase in dilatational strain. Equation (12) is plotted in Fig. 7 to delineate regions of toughening enhancement and reduction. The permanent strain must be significantly increased for any substantial toughening increases due to microcracking occurring along with the stress-induced martensitic transformation.

Experimental Correlation

Complimentary observations of microcracking and pertinent measurements of microcracking strain data and microcrack densities that accompany transformation-toughening are few, making correlation with the model difficult. To allow some cursory predictions concerning the toughness changes, the recent data of Claussen et al.[10] can be analyzed in the light of the proposed model.

From Claussen's observations, two assessments may be made. First, microcrack densities are available from the experimental elastic modulus measurements,

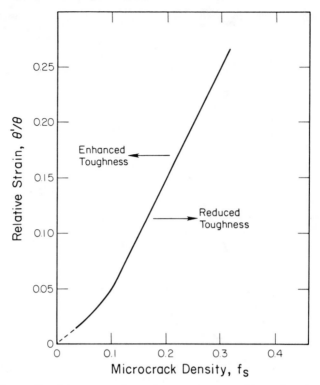

Fig. 7. Normalized increased strain vs microcracking density to define conditions for enhanced or reduced toughening when microcracking accompanies the martensitic transformation.

as discussed in the second section. Elastic modulus measurements on the Al_2O_3-ZrO_2 studied suggest that the microcrack density is approximately 0.185.

Second, crack-opening displacement measurements from Fig. 8 provide an estimate of the upper-bound increase in the dilatational strain. With the assumption that the microcrack density in this fully microcracked body is equivalent to that found in a typical process zone, the additional strain may be calculated by equating the crack-opening displacement (δ) to Δl. The new dilatational strain may now be written as

$$e = \delta/3l \tag{13}$$

where l is the length of the matrix element [$= (r^3/V_f)^{1/3}$] considered. The total dilatational strain must be modified by the microcrack density to account for regions where no microcracking occurs. This modified strain value may then be compared with the total transformation strain (the unconstrained strain modified by the volume fraction of zirconia particles) to determine a total strain increase of approximately 14%.

With Fig. 7, the microcrack densities measured by Claussen from elastic modulus measurements and the increase in permanent strain estimates may be compared to the toughening criteria. The data suggest that these predictions fall just into the regime of toughening enhancement, such that a slight increase in

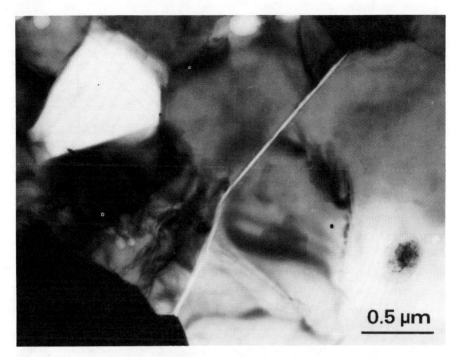

Fig. 8. Bright-field transmission electron micrograph of Al_2O_3 + 15 vol% monoclinic ZrO_2 showing grain-boundary microcrack between ZrO_2 particles (courtesy W. M. Kriven).

toughness over and above that of transformation-toughening might be expected. Recall, however, that this result is expected to be an upper bound since no elastic relaxation of the dilatational strain across the sample has been considered. A more realistic prediction would be that no substantial toughening would occur.

Microcracking vs Transformation-Toughening

Although microcracking has been suggested as a significant toughening mechanism, it is worthwhile to compare the relative toughnesses that can be expected from the two crack-shielding mechanisms. At the asymptotic limit, the fracture toughness for transformation-toughening can be written as[2]:

$$\Delta K^T = [0.22/(1 - \nu)]EV_f e^T (h)^{1/2} \qquad (14)$$

Likewise, the toughening from stress-induced microcracking, described in the second section, can be modified to achieve the same form as Eq. (14). By curve-fitting Eq. (7), shown in Fig. 2, the *change* in toughness can be written as

$$\Delta K^M = 0.25 E_1 f_s \theta (h)^{1/2} \qquad (15)$$

The two equations are now in a convenient form for comparison, which is provided in Fig. 9. The specific plot is for transformation volume fractions of 0.15 over a range of three microcrack densities. Only for permanent transformation strains less than the microcracking strain will microcrack-toughening be more effective than transformation-toughening. Even for identical values of strain, transformation-

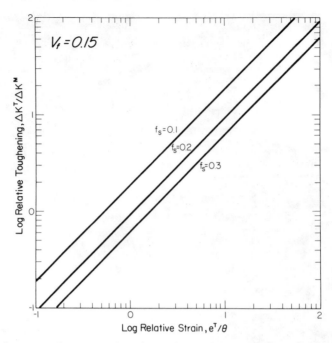

Fig. 9. Log toughness ratios (transformation to microcracking) vs log strain ratio (transformation to microcracking) plotted to compare transformation- and microcrack-toughening.

toughening may afford up to twice the toughening of stress-induced microcracking due to the enhanced compliance of the microcrack zone.

The comparison may not be a fair one in one respect, since microcracking materials are characterized by zones that increase in size as crack advance occurs. The comparison is made here only for identical zone heights that remain constant throughout crack propagation. Comparable toughening increases for microcracking, and transformation-toughening situations may result if the microcrack zone is significantly larger than the transformation zone. Zone estimates in the transformation case are approximately 1 μm. Observations of microcrack zones in materials such as Al_2O_3 range over several grain diameters.

Summary and Implications

The recent continuum stress intensity approach for dilatant transformations developed by Budiansky et al.[1] has provided a means of modeling two additional crack-shielding phenomena in brittle materials: stress-induced microcracking and microcracking during a stress-induced phase transformation. Both of these mechanisms rely on a wake effect for toughening enhancement, whereby the magnitude of the increase is determined in part by the increase in the permanent strain due to the dilatational crack opening. The enhancement is modified by the increase in compliance of the process zone, which alters the total work density associated with the wake region.

Stress-induced microcracking has been found to provide a significant toughening effect, based on the continuum model. However, to assess how much toughness is afforded when microcracking occurs in the vicinity of the tetragonal monoclinic transformation, qualifications for toughening have been developed by comparing the relative increase in strain with the decrease in modulus (or microcrack density increase). Significant increases in strain with low microcrack densities must be accomplished in order to observe any substantial increases in the fracture toughness. Because large increases in strain are not normally accompanied by only low microcrack densities, it is more than likely that microcracks provide no observable change in the fracture toughness.

Using the above ideas for material design of toughened materials by transformation and microcracking provides some insight. Ideally, one would prefer small microcrack densities which may be attained through the presence of a large volume fraction of ZrO_2 particles. At large volume fractions, however, the chance for particle-particle interactions is increased such that the transformation may occur on cooling, hence defeating the original purpose. In the unlikely event that such a microstructure could be developed whereby large fractions of tetragonal particles could be retained, a toughness enhancement over and above transformation-toughening is expected.

Acknowledgments

This research was supported in part by The Ohio State University Office of Research and Graduate Studies under the University Small Research Grant Program.

References

[1] B. Budiansky, J. W. Hutchinson, and J. C. Lambropoulos, "Continuum Theory of Dilatant Transformation Toughening in Ceramics," *Int. J. Solids & Struct.*, **19** [4] 337–55 (1983).
[2] R. M. McMeeking and A. G. Evans, "Mechanics of Transformation-Toughening in Brittle Materials," *J. Am. Ceram. Soc.*, **65** [5] 242–45 (1982).
[3] N. Claussen, "Fracture Toughness of Al_2O_3 with an Unstabilized ZrO_2 Dispersed Phase," *J. Am. Ceram. Soc.*, **59** [1–2] 49–51 (1976).
[4] D. J. Green, P. S. Nicholson, and J. D. Embury, "Fracture Toughness of a Partially Stabilized ZrO_2 in the System $CaO-ZrO_2$," *J. Am. Ceram. Soc.*, **56** [12] 619–23 (1973).
[5] K. T. Faber, Y. Fu, and A. G. Evans, "Multimechanistic Toughening in Ceramic Materials"; for abstract, see *Am. Ceram. Soc. Bull.*, **61** [8] 812 (1982).
[6] R. P. Ingel, D. Lewis, and R. W. Rice, "High Temperature Mechanical Properties of ZrO_2-Y_2O_3 Single Crystals"; for abstract, see *Am. Ceram. Soc. Bull.*, **61** [8] 816 (1982).
[7] A. G. Evans and K. T. Faber, "On the Crack-Growth Resistance of Microcracking Brittle Materials," *J. Am. Ceram. Soc.*, **67** [4] 255–61 (1984).
[8] H. Hübner and W. Jillek, "Sub-critical Crack Extension and Crack Resistance in Polycrystalline Alumina," *J. Mater. Sci.*, **12** [1] 117–25 (1977).
[9] R. Knehens and R. Steinbrech, "Memory Effects of Crack Resistance During Slow Crack Growth," *J. Mater. Sci. Lett.*, **1** [8] 327–29 (1982).
[10] N. Claussen, R. L. Cox, and J. S. Wallace, "Slow Growth of Microcracks: Evidence for One Type of ZrO_2 Toughening," *J. Am. Ceram. Soc.*, **65** [11] C-190–C-191 (1982).
[11] (a) J. R. Bristow, "Microcracks, and the Static and Dynamic Elastic Constants of Annealed and Heavy Cold-Worked Metals," *Br. J. Appl. Phys.*, **11**, 81 (1960).
(b) R. J. O'Connell and B. Budiansky, "Seismic Velocities in Dry and Saturated Cracked Solids," *J. Geophys. Res.*, **79** [35] 5412–26 (1974).
[12] J. D. Eshelby, "Determination of the Elastic Field of an Ellipsoidal Inclusion and Related Problems," *Proc. R. Soc., London, Ser. A*, **241**, 376–96 (1957).

Macrocrack Extension in Microcracked Dispersion-Toughened Ceramics

F. E. Buresch

Institute for Reactor Materials
Kernforschungsanlage Jülich GmbH, D-5170 Jülich 1
Federal Republic of Germany

Microcracking is a well-known process occurring in stressed ceramics. Microcracking is mostly affected by the microstructure, that is, the grain- and pore-size distribution, especially the size and distribution of dispersed particles which undergo phase transformations. Spontaneous and/or stress-induced microcracking are governed by a critical grain-particle size. The influence of the grain-/particle-size distribution on the kinetics of the stress-induced microcrack distribution will be shown quantitatively. The distribution, length, density orientation, and elastic interaction of microcracks inside the process zone determine the toughness of a specific material. A parallel array of cracks can enhance the toughness. Fracture takes place when the colinear microcrack density increases up to a critical value.

The front of a crack in ceramics and similar materials can be surrounded by an inelastic region called the process or damage zone.[1-7] This zone of inelastic deformation can be a consequence of microcracking. Microcracks are generated in these materials by residual strains. It is generally recognized that residual strains will occur as a result of constrained internal volume and shape changes. Microcracks accumulate into crack systems. Characteristic features of microcrack systems such as the density, orientation, length, and elastic interaction of microcracks are influenced by both the microstructure of the material and the loading conditions, including temperature and atmosphere. At a critical microcrack density, favorably oriented contiguous microcracks coalesce inside the process zone, forming a macrocrack. These processes are quantified by the concentration criterion for microcrack enlargement[8] and by the energy density criterion of the process zone.[9]

Microcracks change the constitutive law of a brittle material. The deformation becomes nonlinear nonelastic. It is recognized that the mechanical and fracture behavior of the microcrack system of the material inside the process zone determines the technological parameters of the material.

Microcracks proceed from crack nuclei. In normal fine- and coarse-grained ceramics, crack nuclei are stress concentrators like intergranular pores in facets of high internal strain energy. The density of these microcrack nuclei is very high. Normally each facet acts as a crack nucleus, with different amounts of internal strain energy. Therefore, facets will break uncontrolled at different external stress levels and stress states. Depending on the strength of the pinning points of microcracks, contiguous microcracks will coalesce at relatively low external stress intensity levels.

To circumvent these unfavorable properties of conventional ceramics, dispersion-strengthened ceramics have been developed in the past few decades.

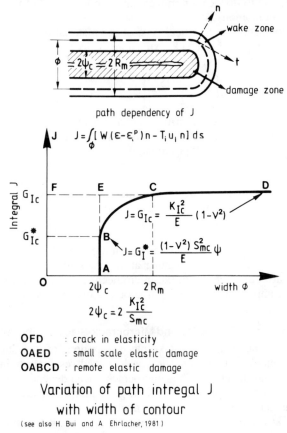

Fig. 1. Influence of microcrack zones on path dependence of inelastic energy release rate J of brittle materials.

Transformation-toughened ceramics are a special class with improved properties. These materials have a relatively low controlled density of active microcrack nuclei of distinct strength, and thus enable a high-strength, that is, a low-stress intensity factor of the crack system inside the process zone, hindering contiguous microcracks to coalesce to or with the macrocrack at low-stress levels.

Macrocrack Extension in Microcracked Ceramics

We first explain crack extension in ceramics in terms of nonlinear fracture mechanics terminology. The J integral introduced by Rice[10] and extended by Broberg[11] and Bui and Ehrlacher[12] is a measure of the elastic dissipative energy release rate during fracture. The value of J depends on the way ϕ confines the crack tip. As first pointed out by Neuber[13] and Irwin,[1] there is a region of nonlinear nonelastic deformation of size $2R_m$ in front of a critical crack of all technological materials (Fig. 1). In this region the linear elastic law of rigid brittle materials is not valid due to both the high stress gradient, which changes strongly over microstructural dimensions such as the grain size, and the energy dissipation, such as microcracking, due to dislocation movement. The inner part of the region of size 2ψ (Fig. 1) is autonomous.[11]

It was recognized that a critical state of the material inside the region 2ψ is responsible for macroscopic crack extension. Thus the J integral for an area $\phi \cong 2\psi$ enclosing this region is a specific property of the material and is given by

$$J(\phi \cong 2\psi) = G_{Ic}^x = \left[\frac{(1-v^2)S_{mc}^2}{E}\right]\psi^c \tag{1}$$

labeled B in Fig. 1. S_{mc} is the critical notch fracture strength or the cohesion strength of the damage/process zone. Equation (1) can be expressed in terms of the critical stress intensity factor approximately as

$$K_{Ic} = (S_{mc}/2)(\pi\psi_c)^{1/2} \tag{2}$$

if, as in high-strength fine-grained ceramics such as Al_2O_3 cutting tools, the inelastic region is very narrow ($2\psi \sim 2R_m$) and

$$G_I^x = G_{Ic} = (K_{Ic}^2/E)(1-v^2) \tag{3}$$

(point E in Fig. 1). Stresses and strains which are caused by the inelastic deformation in $2\psi_c$ behave in a finite manner at the crack tip.

We now turn our attention to the micromechanisms which take place in the region 2ψ and compute the cumulative microcrack distribution as a function of the grain-size distribution.

Relationship between Grain-Size and Microcrack-Size Distribution

The micromechanisms which take place in regions of inelastic deformation are caused by microcrack nuclei, such as localized residual strains inherent in ceramics and similar materials. Thus microcrack nuclei depend on microstructural features, such as the grain-size distribution, grain morphology, and impurity content, as well as elastic parameters of the material.

Microcracks extend from crack nuclei either by spontaneous microcracking or by stress-induced microcracking, depending on the fabrication and loading conditions. Spontaneous microcracking occurs at a critical combination of values of Young's modulus E, Poisson's ratio v, residual strain ε, anisotropy factor A, grain size d, and surface energy γ_s. The critical residual stress-induced energy release rate for microcracking is then given as

$$G_{Ir} = \frac{E\varepsilon^2 A^2 d}{24(1-v^2)} \geq 2\gamma_s \tag{4}$$

Here $\varepsilon = \Delta\alpha \Delta T$ and A are functions of the directions of the crystallographic axes of adjacent grains with respect to their common facet. $\Delta\alpha$ is the difference in thermal expansion coefficients of grains parallel to their common facet, and ΔT is the difference between the freezing temperature of grain-boundary creep during cooling from sintering temperature T_f and working temperature, e.g., room temperature.

With constant integral values for E, v, γ_s, and ε, the value of G_{Ir} depends only on the grain-size d. There is a critical grain-size d_{cr} above which spontaneous microcracking will occur during fabrication. However, special attention must be given to γ_s and E, which will change locally due to pore or impurity segregation, which are reasons for different fracture modes varying the Weibull distribution.

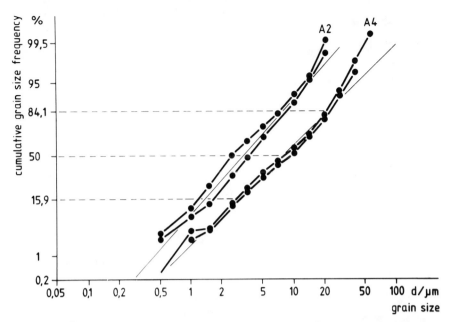

Fig. 2. Cumulative grain-size distribution of two Al_2O_3 ceramics with mean grain sizes of 3 and 9 μm.

Microcracking will occur if the condition in Eq. (4) is fulfilled, that means at a critical combination of values of E, $\varepsilon = \Delta\alpha\,\Delta T$, A, and d. A normalized criterion is given as[14,15]

$$A^2 d_n^+ \geq 1 \qquad (5)$$

where d_n^+ is the normalized grain size.

For the grain-size distribution shown in Fig. 2, the cumulative microcrack density distribution is as shown in Fig. 3 for Al_2O_3-ceramic A2, with a mean grain size of 3 μm and a standard deviation of grain-size distribution $S = 0.8$. In addition, Fig. 3 shows the cumulative microcrack-size distribution for a hypothetical grain-size distribution with $S = 0.5$. Figure 2 clearly shows that the mean grain size of $d = 3$ μm for this material has only a marginal influence on fracture.

The normalized fraction of cracked facets increases with increasing grain size. The curves in Fig. 4 indicate that, if the grain size is larger than 3.5, 2.5, 1.5, and 1 μm, respectively, the grain facets are cracked. Thus up to 90% of grain facets larger than 14 μm and about 60% of mean grain sizes of 3 μm are cracked if grain facets larger than 1 μm are involved in the fracture process (curve R_8 in Fig. 3). This holds for a standard deviation of grain size distribution of 0.8.

Spontaneous and Stress-Induced Microcracking

Uncontrolled Microcracking in an Al_2O_3 Matrix

Local spontaneous microcracking will occur if Eq. (4) is fulfilled. This is different for a one-phase and a multiphase material, where specific $\varepsilon = \Delta\alpha\,\Delta T$ values must be taken into account. During loading of a ceramic, microcracks

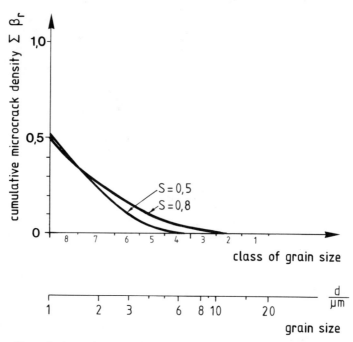

Fig. 3. Cumulative microcrack density vs grain size for Al_2O_3-A2 and two standard deviations of grain-size distributions S.

accumulate inside the process zone, building up a microcrack system. The quantitative computations of the evolution of a microcrack system for specific loading conditions on the basis of a quantitative microstructural analysis are beyond the scope of the present paper due to the unknown elastic crack interaction effects which take place in systems of cracked facets. Here, conditions are worked out for local microcracking of an Al_2O_3-ZrO_2 composite on the basis of quantitative microstructural analysis. This is done separately for a one-phase Al_2O_3 and a ZrO_2 dispersion in an Al_2O_3 matrix due to the unknown elastic interaction effect between the two crack systems.

Whether the evolution of a microcrack system is spontaneous or stress induced depends on the stress level at which the microcrack nuclei are activated. Crack nuclei in the one-phase Al_2O_3 are inhomogenities in strained facets. Thus the energy release rate of all nuclei for extension depends on their specific activation potential. For constant bulk values for E, v, and ε, this depends only on A and d. Therefore for $A = 1$ the distribution of the potential microcrack nuclei which will first extend is given by G_{Ir} as a function of the grain-size distribution (Eq. (4)). This is shown on the lower right side of Fig. 5 with the straight line up to the maximum measured grain size of 20 μm with $E = 360$ GPa and $\varepsilon = 0.8 \times 10^{-3}$ for Al_2O_3-A2. The microstructure of this material is equivalent to those which are used as a matrix for Al_2O_3-ZrO_2 composites. The material parameters are described elsewhere.[9]

This material will not microcrack spontaneously. However, for a hypothetical grain-size distribution extending beyond 20 μm, microcracking will occur at the

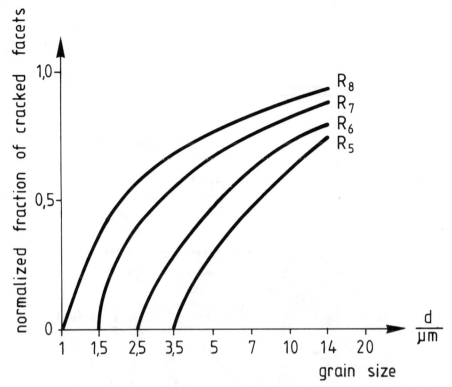

Fig. 4. Normalized fraction or cracked facets of Al_2O_3-A2 vs grain size if grains larger than 1, 1.5, and 3.5 μm are involved in the fracture process.

intersection of the dashed line with the value of the surface energy for intergranular cracks of about 1–2 J/m², that is at a grain size in the range of 80–100 μm, as observed in the literature.[16,17]

Stress-induced microcracking is activated by an external stress field which is defined by the stress-intensity factor K_I and given by the energy release rate

$$G_I = \frac{K_I^2(1 - v^2)}{E} + G_{Ir} = \frac{K_I^2(1 - v^2)}{E} + \frac{E\varepsilon^2 A^2 d}{24(1 - v^2)} \geq 2\gamma_s \qquad (6)$$

The evolution of a stress-induced microcrack system is also a function of the distribution of potential microcrack nuclei, that is, the grain-size distribution. The external stress field lowers the effective surface energy for intergranular microcracking to some extent. This is seen in the convex curve on the right side of Fig. 5, where G_I (Eq. (5)) is evaluated as a function of the measured grain-size distribution for Al_2O_3-A2 in the range of 0.5–20 μm (see Fig. 2 and Table I). The dashed curves are valid for a hypothetical grain-size distribution between 20 and 100 μm. Thus stress-induced microcracking will first occur at the largest facet corresponding to the largest grain, $d_{max} = 20$ μm, that is, on the convex curve at a stress intensity factor of about 0.5 MPa·m$^{1/2}$ which is shown on the right side of Fig. 5. With increasing stress intensity, the microcrack system inside the process

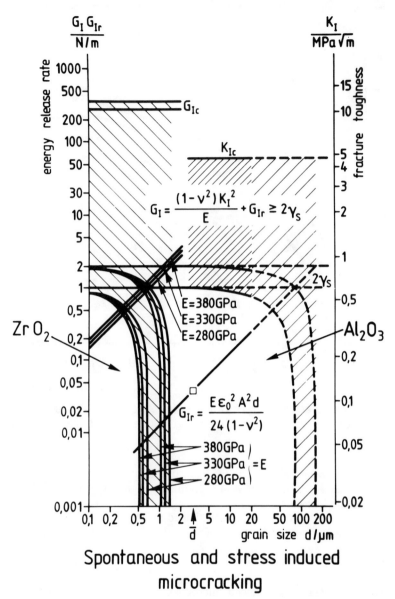

Fig. 5. Fracture energy release rate of spontaneous and stress-induced microcracking (straight lines and convex curves) vs grain size for Al_2O_3 and ZrO_2 dispersion ceramics.

zone is built up, and the microcrack density increases, following the distribution function which was evaluated for this material in Fig. 3 up to the critical value of $K_{Ic} = 4.7$ MPa·m$^{1/2}$.

With G_I as a function of the grain size (Eq. (6)), it is interesting to note that the dashed curve shows the transition from spontaneous to stress-induced microcracking for the hypothetical grain-size distribution beyond 20 μm.

Table I. Material Parameters of Al_2O_3-A2*

Composition	99.7% Al_2O_3	S_c	195 MPa
ρ	3.85 g/cm³	K_{Ic}	4.7 MPa·m$^{1/2}$
p	4%	G_{Ic}	64 N/m
E	358 GPa	S_{mc}	362 MPa
$\bar{d}$	3 μm	$2\psi_c$	600 μm

*Purity, density, porosity, Young's modulus, mean grain size, bending strength, stress intensity factor, energy release rate, notch fracture strength, process zone.

Controlled Microcracking of ZrO_2 Dispersions

The phase-transformation temperature of dispersed t-ZrO_2 can be reduced by the mechanical constraint of the matrix if the particle sizes are below a critical value.[18] If a ZrO_2 dispersion is distributed in an α-Al_2O_3 matrix, anisotropic contraction generates residual strains between the a and c axes of both the hexagonal α-Al_2O_3 and the t-ZrO_2 grains. Furthermore, the residual strains are fortified by the volume expansion during the $t \rightarrow m$ transformation of ZrO_2.

For an explicit computation of the residual tension stresses in a facet which can activate crack nuclei, the differences in linear thermal expansion coefficients between the various crystallographic axes of the different phases are introduced in Eq. (6). These values are included in Table II. Residual strains are then a result of the difference between the freezing temperature T_f for grain-boundary creep and the working temperatures, e.g., room temperature. Rühle and Kriven[6] and Heuer et al.[18] use a $T_f = 1550$ K for the freezing temperature.

This value may be changed to some extent by impurities or additives which act as a grain-boundary glassy phase.[19] Besides these inconsistencies, a value of $T_f = 1550$ K will be used to compute the residual strains which act in facets. These values are included in Table II. It follows from this that radial residual tension stresses act in a t-ZrO_2 grain. Conversely, tangential residual compressive stresses act in the surrounding Al_2O_3 matrix. However, the residual stresses are anisotropic. Maximum and minimum residual tension stresses exist specifically in cases where the respective a and c axes of α-Al_2O_3 and t-ZrO_2 grains are parallel to each other in a facet (see Table II).

A volume expansion of 3–5% of ZrO_2 grains occurs during the $t \rightarrow m$ phase transformation. Thus linear strains of more than 1% generate residual compressive stresses in an m-ZrO_2 grain, whereas in surrounding Al_2O_3-matrix grains residual tangential tension stresses arise. These tension stresses generate residual radial

Table II. Thermal Expansion Coefficients, Their Differences, and Residual Compressive Strains with Respect to the Matrix in an Al_2O_3-ZrO_2 Composite for a Temperature Difference $\Delta T = 1500$ K

		α-Al_2O_3	
t-ZrO_2		$\alpha_a = 8.2 \times 10^{-6}$ K^{-1}	$\alpha_c = 9.1 \times 10^{-6}$ K^{-1}
$\alpha_c = 11.6 \times 10^{-6}$ K^{-1}		$\Delta\alpha = -3.4 \times 10^{-6}$ K^{-1} $\varepsilon = -5.1 \times 10^{-3}$	$\Delta\alpha = -2.5 \times 10^{-6}$ K^{-1} $\varepsilon = -3.8 \times 10^{-3}$
$\alpha_a = 16.8 \times 10^{-6}$ K^{-1}		$\Delta\alpha = -8.6 \times 10^{-6}$ K^{-1} $\varepsilon = -12.9 \times 10^{-3}$	$\Delta\alpha = -7.7 \times 10^{-6}$ K^{-1} $\varepsilon = -11.6 \times 10^{-3}$

microcracks if the resulting energy release rate is equal to or greater than the intergranular specific surface energy $2\gamma_s$ (Eq. (6)).

To assess microcracking in a ZrO_2-Al_2O_3 composite, we will look for the highest residual stress-induced energy release rate G_{Ir} during the $t \rightarrow m$ transformation which can activate a crack nucleus. Irregular or regular-shaped ZrO_2 grains are randomly oriented at either intergranular or intragranular sites in a polycrystalline Al_2O_3 matrix. As can be seen from Table II, the highest tangential tensile stresses arise around ZrO_2 grains at intergranular sites if the boundary of a t-ZrO_2 grain adjacent an α-Al_2O_3 grain before the transformation is oriented in such a way that the two c axes of Al_2O_3 and ZrO_2 grains are parallel. Then the tangential compressive stresses around a ZrO_2 grain are lowest. Thus with a value of $\varepsilon = 4 \times 10^{-3}$ for the tangential compressive strain and a volumetric expansion of 4.5% during the $t \rightarrow m$ transformation, of a maximum tangential tensile strain of approximately $\varepsilon = 1.1 \times 10^{-2}$ results.

The following calculations based on experiments conducted by Lange,[21] Green,[22] Kosmac et al.,[5] Rühle and Kriven,[6] and Claussen et al.[23] The measured critical data at fracture of $K_{Ic} = 11$ MPa·m$^{1/2}$ and $G_{Ic} = 300$ N/m are the upper bounds of the microcrack region for the specific ZrO_2 composite as labeled on the ordinates in Fig. 5.

As shown in an earlier work, Young's modulus has a strong influence on residual stress-induced microcracking.[20] This is a consequence of porosity. In particular, intergranular pores or cracks relax residual stresses. To a first approximation, we use Young's moduli given by Lange,[21] Green,[22] and Claussen et al.,[23] with values of $E = 380$ and 280 GPa before and after complete transformation, respectively. A value of $E = 330$ GPa, the value given by Claussen et al., is also included in the computations for a 0.15 ZrO_2-Al_2O_3 composite with some spontaneous microcracking after fabrication.

With these values of E, ε, and $A = 1$, we can evaluate from Eq. (4) the local microcracking and the evolution of microcrack systems as a function of microstructure in terms of a quantitative microstructural analysis based on stereological methods. With the specific grain-size distribution in the range of 0.1–2 μm, Fig. 5 shows, in the lower left part with three straight lines increasing from left to right, G_{Ir} as a function of grain size, with three values for Young's modulus as parameter. This holds good for the maximum tangential residual tensile stress around m-ZrO_2 grains in the Al_2O_3 matrix, with parallel orientation of the respective c axes of t-ZrO_2 and α-Al_2O_3 grains.

The critical ZrO_2 grain size for spontaneous microcracking depends on the specific intercrystalline surface energy $2\gamma_s$; a value of 1–2 N/m is assumed, as in the case of the Al_2O_3 matrix.

Thus spontaneous microcracking occurs if the residual stress-induced energy release rate G_{Ir} as a function of grain-size d equals $2\gamma_s = 1$ N/m, that is, at the critical grain size $d_{cr} = 1$ μm, Young's modulus has its highest value of 380 GPa. The critical grain size for spontaneous microcracking increases by an amount equal to about 30%, with decreasing Young's modulus in the given range, that is, 280 GPa.

External tensile stresses lower the critical grain size for microcracking. For the present grain-size distribution of the dispersed ZrO_2 phase, the convex curves on the left side of Fig. 5 are evaluated for G_I (Eq. (6)) as a function of grain size d for the above-mentioned three values of Young's modulus. Thus, stress-induced microcracking will first be expected at very low K_I values for a grain size

$d_{cr} = 0.5$–0.8 μm, depending on the Young's modulus. With increasing stress intensity, K_I up to 0.6 MPa·m$^{1/2}$, that is, increasing energy release rate G_I up to 1 N/m, the evaluation of Eq. (6) shows that small facets of grains in the range of $d \approx 0.1$ μm with $A = 1$ are also involved in the fracture process.

With increasing stress intensity, between 0.6 MPa·m$^{1/2}$ and the critical value of 11 MPa·m$^{1/2}$, the residual stresses in facets will activate microcrack nuclei with $A < 1$. A microcrack system will be built up with a high degree of oriented cracks normal to the principal tensile stress up to the critical microcrack density. It is assumed that the above-mentioned computations are representative for a macrocrack extension in an Al$_2$O$_3$-ZrO$_2$ composite recently exemplified by Claussen et al.[23]

Assuming that the ZrO$_2$ grains are nearly equiaxial, the distribution function of the anisotropy factor A given by Davidge et al.[15] is valid. Then the kinetics of microcracking of the ZrO$_2$ dispersion are nearly the same as mentioned earlier for the α-Al$_2$O$_3$ matrix phase. This, in turn, means that the evaluated facet size distribution function shown in Figs. 3 and 4 is valid with respect to a factor of 10 for the smaller ZrO$_2$ grains and the volume fraction of the dispersed phase.

It may be further noticed that, with respect to Fig. 5, residual strain-induced microcracking in Al$_2$O$_3$-ZrO$_2$ composites is activated only by the ZrO$_2$ dispersion. This is due to the G_{Ir} values (left side) which are about a factor of 10 higher than the G_{Ir} values of the Al$_3$O$_3$-matrix grains (right side). Thus stress-induced microcracking is shifted to smaller grains.

Influence of Microcrack Systems on the Mechanical and Fracture Properties of Al$_2$O$_3$-ZrO$_2$ Composites

It must be assumed that the high fracture toughness of Al$_2$O$_3$-ZrO$_2$ composites is mainly a consequence of the high density of oriented metastable microcracks of the crack system inside the process zone of size $2\psi_c$. This will be explained in this section.

As mentioned (Eq. (2)), two features of the process zone determine the fracture toughness of a ceramic. These are the size of the process zone and the strength of the crack system inside the process zone. According to recent work by Kosmac et al.,[5] the size of the process zone of a dispersion-toughened composite was estimated to about 1–5 μm. With a value of 1 μm and a fracture toughness of 6 MPa·m$^{1/2}$, the cohesive strength is in the range of 7 GPa. This value seems very high when compared to the theoretical strength of this material, $E/10 \approx 30$ GPa. However, the value of 1–5 μm for the size of the process zone is evaluated, assuming that all t-ZrO$_2$ grains were transformed to the monoclinic state. This seems very optimistic for the above-mentioned distribution function of cracked facets (Fig. 3) which shows clearly that mostly large facets are cracked during macrocrack extension. Thus, with values measured by Claussen et al.,[23] that after fracture about 40% of the ZrO$_2$ grains remain tetragonal, the size $2\psi_c$ of the process zone can be estimated to be 30 μm. With the above-cited fracture toughness of 6 MPa·m$^{1/2}$, the cohesive strength of the process zone is about 1 GPa. This seems more realistic with respect to the measured values of Al$_2$O$_3$ ceramics which were characterized in an earlier work.[9] Exemplary typical values for this material are included in Table I. This material has a comparable microstructure to the Al$_2$O$_3$ ceramics which are used as a matrix for ZrO$_2$ composites.

A high cohesive strength of the process zone is responsible for the high fracture toughness of ZrO$_2$ composites. We will now show how the strength of a microcrack system is determined by the density, orientation, and elastic interaction of the microcracks. For computation, literature values of K_{Im} and Young's moduli for crack systems are used.

For an understanding of the strengthening and toughening mechanisms which arise in the process zone, mechanical parameters of crack systems as a function of the characteristic parameters are explored. Following the energy density criterion of the process zone,[9] it is assumed that this region is a homogeneous continuum with elastic parameters E_m, G_m, ν_m, and the strength S_{mc}, that is, the cohesion strength of the process zone. S_{mc} is given by two terms; one is equivalent to the Griffith equation for a linear elastic material in plain strain, and the other describes the inelastic behavior of microcrack systems:

$$S_{mc} = 2\left[\frac{2\gamma_s E}{(2.5 - 1.5\nu - 4\nu^2)a}\right]^{1/2}\left(\frac{\beta E_m}{E - E_m}\right)^{1/2} \quad (7)$$

where β is the microcrack density; E_m and E are Young's moduli of the material with and without microcracks, respectively. In Eq. (7) the orientation and elastic interaction of the cracks are included in E_m.

The stress-intensity factor of crack systems is evaluated from Eq. (7) as

$$\frac{K_{Im}}{S_m(\pi a)^{1/2}} = I_m^{-1} = \left(\frac{E - E_m}{\beta E_m}\right)^{1/2} \quad (8)$$

From this it follows that the Young's modulus of microcrack systems is given as

$$\frac{E_m}{E} = \frac{1}{1 + \beta K_{Im}^2/(S_m^2 \pi a)} \quad (9)$$

With these equations, stress-intensity factors from the literature for specific crack systems can be used to compute the corresponding Young's moduli and vice versa. With the Young's moduli for statistically oriented noninteracting cracks derived by Budiansky and O'Connell,[24] strength of this crack system is evaluated using Eq. (8) and shown in Fig. 6. Figure 7 shows strength values of the crack system with parallel oriented cracks in rows and stacks where K_{Im} values from Gross[25] are used. Figure 8 shows the corresponding Young's moduli which were evaluated by using Eq. (9). Figures 9 and 10 show strength and Young's moduli for the diamondlike crack arrangement where K_{Im} values computed by Gross[25] are used. Both parameters have a minimum at densities of colinear cracks of 0.5 if the crack tips are arranged one on another. Then stress concentration is highest. With increasing density of colinear cracks, unloading effects take place. As a consequence strength and Young's moduli increase.

Strengths of colinear crack systems are derived with the K_{Im} values computed by Koiter,[26] Paris and Sih,[27] and Feddersen[28] with $\beta_y = 0$ and are shown in Figs. 7 and 9. The corresponding Young's moduli are shown in Fig. 10.

From the above-discussed parameters of different crack systems it is evident that the mechanical and fracture behavior of ceramics which are damaged by microcracks are governed by structural parameters of the crack configurations. These parameters determine the strength, crack resistance, and thermal shock behavior of ceramics.[14]

Parallel-oriented cracks diminish the stress-intensity factor of the crack system of the process zone. Thus the possibility of coalescence of favorably oriented cracks decreases.

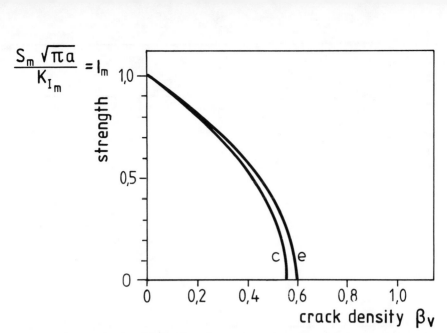

Fig. 6. Strength of statistically oriented crack systems vs volume fraction of crack density (circular and elliptical cracks).

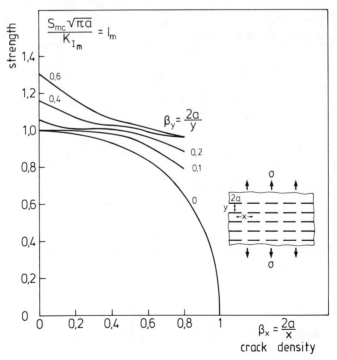

Fig. 7. Strength of orthogonal crack system vs density of parallel (β_y) and colinear (β_y) oriented cracks.

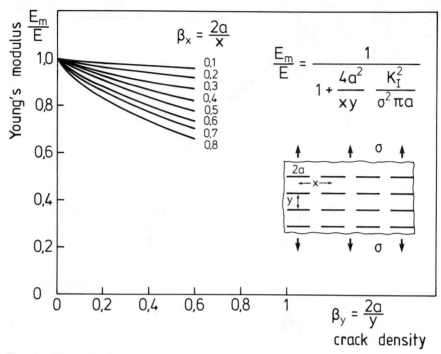

Fig. 8. Young's modulus of orthogonal crack system vs density of parallel (β_y) and colinear (β_y) oriented cracks.

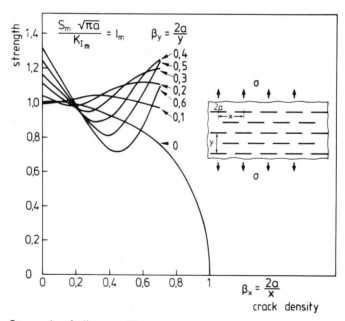

Fig. 9. Strength of diamondlike crack system vs density of parallel (β_y) and colinear (β_y) oriented cracks.

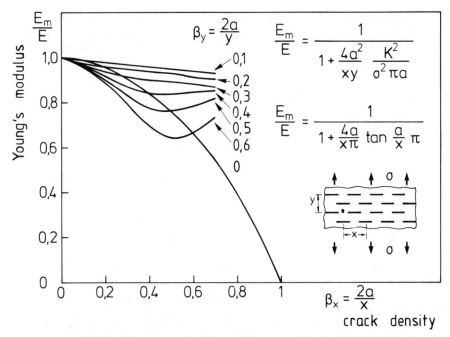

Fig. 10. Young's modulus of diamondlike crack system vs density of parallel (β_y) and colinear (β_y) oriented cracks.

Discussion

The toughness of ceramics and similar materials is governed by energy-dissipative processes which take place in front of the crack tip. Microcracking is the most important mode. These mechanisms can be explored with the model of the energy density criterion of the process zone.

Transformation-toughening within ZrO_2 composites is of great technological importance. Experimentally it was found that grain sizes smaller than 1 μm of t-ZrO_2 dispersions in an Al_2O_3 matrix are most effective. This agrees very well with the theoretical evaluations of Fig. 5. The convex curves on the left side define the faceted grain size for microcracking as a function of Young's modulus and K_1 value of the external stress field. Therefore these curves define all stress levels between spontaneous and stress-induced microcracking.

The toughness of a ceramic increases with increasing strength S_{mc} and the size of the damaged zone $2\psi_c$. According to the present calculations S_{mc} increases mainly with increasing density of parallel oriented displaced microcracks. It must be concluded that with increasing microcrack density this crack arrangement has a high self-stabilizing performance, limiting small process zone size. This is experimentally verified with domain microstructures.[29] On the contrary, colinearly and especially statistically oriented microcracks have a detrimental effect on toughness. This is the case when spontaneous nonoriented microcracking arises in a ZrO_2 composite, as observed experimentally.[23] These cracks extend along the Al_2O_3 grain boundaries. The extension may be enhanced with decreasing temperature due to the low thermal expansion coefficients of the monoclinic b axis (Table II).

Parameters of dispersion-toughened ceramics may be compared with characteristic features of damaged zones of conventional Al_2O_3 ceramics (Table II) which were measured in an earlier work,[9] keeping in mind the mechanisms which influence the toughness of ceramics and similar materials. As pointed out by Neuber[13] and Irwin,[1] the K_{Ic} value is defined in terms of the notch root radius ψ as

$$K_{Ic} = \lim_{\psi \to 0}(S_{mc}/2)(\pi\psi)^{1/2} \tag{10}$$

With this physical interpretation it was found experimentally by Irwin[1] and others that the K_I value of large notched specimens decreases with decreasing notch root radius down to a critical value. This is given in Eq. (2).

For a given notch length and notch root radius, $\psi < \psi_c$, no further decrease of K_I and alternatively of the energy release rate G_I can be measured. Then $G_I = G_{Ic}^+$, as shown in Eq. (1) and Fig. 1. This is the case for an ideal small-scale elastic damage (OAED in Fig. 1).

An experimental interpretation of this phenomenon was given by Weiss,[30] keeping in mind that the K_I value is inversely proportional to the stress-concentration factor K_T. He found that, for $\psi < \psi_c$ and constant notch length, no further increase of stress concentration is possible. This was also verified for Al_2O_3 ceramics by the author.

The limited stress concentration of sharp notches is due to microcracking. In the above-mentioned experiments, only a rigid stable shape of the notch root can transform stress concentration. For large notch root radii, the stress concentration in front of the notch is low, and microcracking spread over the ligament of the specimen. With decreasing notch root radius, stress concentration in front of the notch root increases, and microcracking is more confined.

With further decreasing notch root radius, that is, increasing stress concentration, the surface of the notch root deteriorates by microcracking and becomes less rigid. Thus the notch root cannot transfer stress concentrations. However, the shape of the process wake zone of size $2\psi_c$ is fundamentally equivalent to a rigid notch root radius of size 2ψ and determines the highest possible stress concentration of the material (Fig. 1). Thus the surface of the wake zone is free of normal stresses.

The relatively low K_{Ic} values of conventional Al_2O_3 ceramics are mainly due to uncontrolled microcracking of an extended grain-size distribution ranging from 0.1 to 100 μm. As the microcrack size of ZrO_2 composites is comparable to the Al_2O_3 facet size,[22] both grain-size distributions must be as narrow as possible, with the maximum grain size less than 2 μm to get very tough, strong composites. Because the critical density for macrocrack extension was found to be equivalent to a colinear crack density of 0.5, the volume concentration of ZrO_2 dispersion should be less than 0.2.[14]

References

[1]G. R. Irwin, "Structural Aspects of Brittle Fracture," *Appl. Mater. Res.*, **3**, 65 (1964).
[2]A. G. Evans and K. T. Faber, "Toughening of Ceramics by Circumferential Microcracking," *J. Am. Ceram. Soc.*, **64** [7] 394–98 (1981).
[3]R. G. Hoagland, G. T. Hahn, and A. R. Rosenfield, "Influence of Microstructure on the Fracture Propagation in Rock," *Rock Mech.* **5** [2] 77–106 (1973).
[4]R. G. Hoagland and J. D. Embury, "A Treatment of Inelastic Deformation around a Crack Tip due to Microcracking," *J. Am. Ceram. Soc.*, **63** [7–8] 404–10 (1980).
[5]T. Kosmac, R. Wagner, and N. Claussen, "X-ray Determination of Transformation Depth in Ceramics Containing Tetragonal ZrO_2," *J. Am. Ceram. Soc.*, **64** [4] C-72–C-73 (1981).
[6]M. Rühle and W. M. Kriven, "Stress-Induced Transformations in Composite Zirconia Ceramics"; unpublished work.

[7]N. Claussen, "Umwandlungsverstärkte keramische Werkstoffe," *Z. Werkstoff. Technol.,* **13**, 138–47 (1982).
[8]V. S. Kuksenko, "Concentration Criterion of Volume Fracture Solids"; pp. 383–404 in Fracture Mechanics and Technology, Vol. 1. Edited by G. C. Sih and C. C. Chow. Sythoff and Noordhoff, Rockville, MD. 1977.
[9]F. E. Buresch, "About the Process Zone Surrounding the Crack Tip of Ceramics"; p. 929 in Fracture '77, Vol. 3. Edited by D. M. R. Taplin. University of Waterloo Press, Waterloo, Ontario, Canada, 1977.
[10]J. R. Rice, "Mathematical Analysis in the Mechanics of Fracture"; p. 191 in Fracture II. Academic Press, New York, 1968.
[11]K. B. Broberg, "On the Treatment of the Fracture Problem at Large Scale of Yielding"; pp. 837–59 in Fracture Mechanics and Technology, Vol. 2. Edited by G. C. Sih and C. C. Chow. Sythoff and Noordhoff, Rockville, MD, 1977.
[12]H. D. Bui and A. Ehrlacher, "Propagation of Damage in Elastic and Plastic Solids"; pp. 533–51 in Advances in Fracture Research. Edited by D. Francois et al., ICF5, 1981.
[13]H. Neuber, "Ober die Berücksichtigung der Spannungskonzentration bei Festigkeitsberechnungen," *Konstruktion,* **20**, 245–51 (1968).
[14]F. E. Buresch, "Relations between Microstructure, Fracture Toughness and Thermal Shock Resistance," *Sci. Ceram.,* **12** [6] 513–22 (1983).
[15]R. W. Davidge, J. R. LcLaren, and I. Titchell, "Statistical Aspects of Grain Boundary Cracking in Ceramics and Rocks"; pp. 495–506 in Fracture Mechanics of Ceramics, Vol. 5. Edited by R. C. Brodt, A. G. Evans, D. P. H. Hasselman, and F. F. Lange. Plenum, New York, 1983.
[16]F. F. Lange, "Fracture Mechanics and Microstructural Design"; pp. 799–819 in Fracture Mechanics of Ceramics, Vol. 4. Edited by R. C. Bradt, D. P. H. Hasselman, and F. F. Lange. Plenum, New York, 1978.
[17]N. M. Parikh, "Factors Affecting Strength and Fracture of Nonfissionable Ceramic Oxides," Proceedings of the Conference on Nuclear Applications of Nonfissionable Ceramics. Washington, DC. *Am. Nucl. Soc.*
[18]A. H. Heuer, N. Claussen, W. K. Kriven, and M. Rühle, "Stability of Tetragonal ZrO_2 Particles in Ceramic Matrices," *J. Am. Ceram. Soc.,* **65** [12] 642–50 (1982).
[19]R. J. Charles; private communication.
[20]F. E. Buresch, R. Hecker, and W. Rixen, "The Effect of Cyclic Fast Neutron Irradiation Damage and Thermal Treatment on Modulus of Rupture of BeO," *Sci. Ceram.,* **5**, 195–218 (1970).
[21]F. F. Lange, "Transformation Toughening, Part 4: Fabrication, Fracture Toughness and Strength of Al_2O_3/ZrO_2-Composites," *J. Mater. Sci.,* **17**, 247 (1982).
[22]D. J. Green, "Critical Microstructures for Microcracking in Al_2O_3-ZrO_2-Composites," *J. Am. Ceram. Soc.,* **65** [12] 610–14 (1982).
[23]N. Claussen, R. L. Cox, and J. S. Wallace, "Slow Crack Growth of Microcracks: Evidence for One Type of ZrO_2-Toughening," *J. Am. Ceram. Soc.,* **65** [11] C-190–C-191 (1982).
[24]B. Budiansky and R. J. O'Connell, "Elastic Moduli of a Cracked Solid," *Int. J. Solids Struct.,* **12**, 81–97 (1976).
[25]D. Gross, "Spannungsintensitätsfaktoren von Rißsystemen," *Inge. Arch.,* **51**, 301–40 (1982).
[26]W. T. Koiter, "An Infinite Row of Colinear Cracks in an Infinite Elastic Sheet," *Ing. Arch.,* **28**, 163–72 (1959).
[27]P. C. Paris and G. C. Sih, "Stress Analysis of Cracks"; pp. 30–83 in Symposium on Fracture Toughness Testing and its Application, STP. 381. American Society for Metals, Metals Park, OH, 1965.
[28]C. E. Fedderson, "Discussion to Place Strain Crack Toughness Testing," STP 410, p. 77 in American Society for Metals, Metals, Park, OH, 1967.
[29]D. Michel, L. Mazerolles, M. Perez, and Y. Jorba, "Polydomain Crystals of Single-Phased Tetragonal Zirconia—Structure, Microstructure and Fracture Toughness"; this volume, pp. 131–38.
[30]V. Weiss, "Notch Analysis of Fracture"; p. 227 in Fracture III. Academic Press, New York, 1971.

Section III
Transformation Toughening—Microstructural Aspects

Microstructural Design of Zirconia-Toughened Ceramics (ZTC)... 325
    N. Claussen

Microstructural Studies of Y_2O_3-Containing Tetragonal ZrO_2 Polycrystals (Y-TZP)............................ 352
    M. Rühle, N. Claussen, and A. H. Heuer

Effect of Microstructure on the Strength of Y-TZP Components 371
    M. Matsui, T. Soma, and I. Oda

Thermal and Mechanical Properties of Y_2O_3-Stabilized Tetragonal Zirconia Polycrystals............. 382
    K. Tsukuma, Y. Kubota, and T. Tsukidate

Aging Behavior of Y-TZP................................... 391
    W. Watanabe, S. Iio, and I. Fukuura

Phase Stability of Y-PSZ in Aqueous Solutions................ 399
    K. Nakajima, K. Kobayashi, and Y. Murata

Physical, Mircostructural, and Thermomechanical Properties of ZrO_2 Single Crystals.................. 408
    R. P. Ingel, D. Lewis, B. A. Bender, and R. W. Rice

Ripening of Inter- and Intragranular ZrO_2 Particles in ZrO_2-Toughened Al_2O_3.............................. 415
    B. W. Kibbel and A. H. Heuer

Anomalous Thermal Expansion in Al_2O_3-15 Vol% $(Zr_{0.5}Hf_{0.5})O_2$ .. 425
    W. M. Kriven and E. Bischoff

Improvement in the Toughness of β''-Alumina by Incorporation of Unstabilized Zirconia Particles........ 428
    J. G. P. Binner, R. Stevens, and S. R. Tan

Microstructure and Property Development of in Situ-Reacted Mullite-ZrO$_2$ Composites.................................. 436
   J. S. Wallace, G. Petzow, and N. Claussen

Size Effect on Transformation Temperature of Zirconia Powders and Inclusions................................. 443
   I. Müller and W. Müller

Relationship Between Morphology and Structure for Stabilized Zirconia Crystals............................. 455
   D. Michel

Microstructural Design of Zirconia-Toughened Ceramics (ZTC)

NILS CLAUSSEN

Max-Planck-Institut für Metallforschung
Institut für Werkstoffwissenschaften
Stuttgart, Federal Republic of Germany

Zirconia toughening of ceramics is defined here in a rather broad sense: "Toughening" encompasses all types of mechanical property enhancement which can be achieved by utilizing either the tetragonal-to-monoclinic phase transformation of ZrO_2 particles dispersed in a ceramic matrix or other effects not directly associated with the transformation. Also special measures for high-temperature applications, using ZrO_2-containing ceramics, should in many cases be part of zirconia toughening. A schematic grouping of the large variety of ZTC is presented. These materials are based on characteristic microstructural features and divide naturally into three main groups: (a) partially stabilized ZrO_2 ceramics, (b) dispersed ZrO_2 ceramics, and (c) complex ZrO_2 systems. Microstructural design criteria are presented based on experimental fabrications and results, together with prospects for future developments.

Ceramic materials can be toughened and strengthened by utilizing the tetragonal (t) $\rightarrow$ monoclinic (m) phase transformation of ZrO_2 particles dispersed or precipitated in a ceramic matrix.[1-14] The toughening originates from the volume and shape change associated with the transformation. Even though this fact has been well recognized, the exact mechanisms, i.e., the micromechanics of the toughness increase, still remain the subject of active discussion. This is particularly due to the fact that at least two types of toughness enhancement have been found. In the first case, the martensitic transformation of ZrO_2 particles near the advancing crack tip is directly involved in the energy absorption (stress-induced transformation).[8,14] In the second case, nucleation and extension of matrix microcracks caused by the transformation of particles on cooling prior to specimen loading are responsible for increased energy absorption during crack propagation (microcrack nucleation and extension).[15-18] Also, a combined effect of both mechanisms seems possible, i.e., first a stress-induced transformation and then microcrack nucleation.[12] Furthermore, crack deflection and crack blunting at m and t particles may account for a further increase in toughening.[8] Zirconia (transformation) toughening, which was originally thought to be a phenomenon inherent only in partially stabilized ZrO_2 (PSZ) in which t particles were coherently precipitated within the cubic (c) stabilized matrix, has also been shown to be applicable to Al_2O_3 and other ceramic matrices in which ZrO_2 can be incorporated.[3,12,13,19] Furthermore, fine-grained, fully t-ZrO_2 has been developed with extremely high toughness.[5,20,21] Another important aspect of transformation-toughening is the generation of compressive surface stresses (surface-toughening) resulting in considerable strength increases.[12,19,22,23] Although all of the work on ZTC has been centered on ZrO_2- or

HfO$_2$-alloyed ZrO$_2$ as "toughening agent," it is possible that phase transformations in other materials, not necessarily martensitic in nature, may be suitable for toughening of ceramics, e.g., transformations in BN,[24] enstatite,[25] etc.

The objective of this paper is to review the present data on ZTC microstructures and their relationship to mechanical properties, to give microstructural design criteria, and to indicate new developments which may overcome the characteristic deficiencies of ZTC. In this presentation, the broader definition of "zirconia toughening of ceramics" will also include other effects of ZrO$_2$ not associated with the phase transformation; these miscellaneous effects form something like a base in the schematic diagram of fractional mechanical property improvement vs application temperature (area 1 in Fig. 1). As demonstrated in this diagram, improvements due to phase transformations (area 2) vanish at the equilibrium temperature (T_0), hence, the intrinsic properties of the matrix become increasingly more important with increasing application temperature (area 3). Therefore, microstructural design of ZTC must also encompass the development

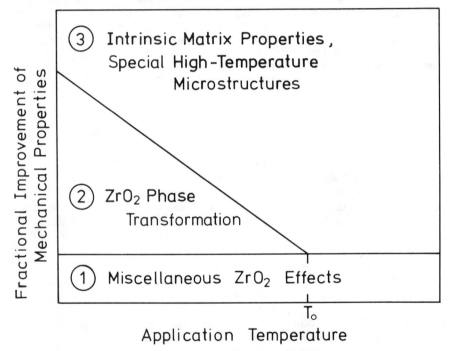

Fig. 1. Areas of specific microstructural design which contribute to mechanical property enhancement at different application temperatures. (1) Miscellaneous effects of ZrO$_2$ not associated with the phase transformation; the various specific effects will naturally vary with temperature; for simplicity, however, the combined contribution to the mechanical properties has been assumed to be linear. (2) The contribution of transformation-toughening will decrease to zero at the equilibrium temperature T_0. (3) Area where special microstructures, utilizing the intrinsic matrix properties, become increasingly important with increasing application temperature.

of special high-temperature microstructures if these materials are to be used at elevated temperatures.

Characterization of ZTC Microstructures

The different ZTC microstructures are schematically grouped into three classes (Fig. 2): (A) ceramics based on partially stabilized zirconia (PSZ); (B) dispersed zirconia-containing ceramics, and (C) complex zirconia systems encompassing all other composites not fitting into either group A or B. In all cases ZrO_2 can be partially or fully substituted with HfO_2. The microstructural classification in Fig. 2 is not thought to be complete, but from a materials scientist's point of view it is considered more useful than other classifications based on compositions, applications, characterization of the various properties, technological state, etc. Most of the materials have been prepared only on a small scale, e.g., in research laboratories some are at a more advanced stage and have found commercial applications or are being studied in pilot plants, while others merely indicate future possible trends. Most of the microstructural forms presented in Fig. 2 and their combinations can be prepared by different techniques, i.e., starting with different types of powders, different heat treatments, etc. A few, however, are typical of particularly specialized preparation methods.

Ceramics Based on Partially Stabilized ZrO_2 (PSZ)

A1: Conventional ZrO_2 ceramics, partially stabilized by the addition of MgO, CaO, Y_2O_3, or rare-earth oxides, are usually sintered in the *c* solid solution range, i.e., at relatively high temperatures ranging between 1600° and 1900°C.[2,26–28] With appropriate cooling, the microstructure contains large (50–100 μm) *c* grains within which are dispersed coherent precipitates of *t* symmetry. The precipitates are then coarsened or modified by aging at temperatures between 1300° and 1500°C in order

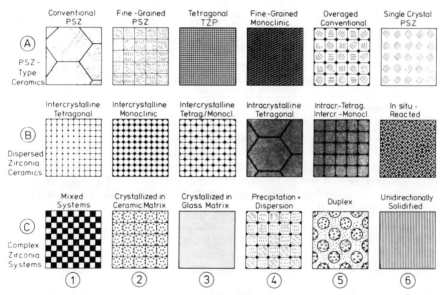

Fig. 2. Classification of ZrO_2-toughened ceramics (ZTC) based on typical microstructural features. The different ZTC types are presently being produced or studied.

to optimize their transformability under stress, i.e., the stress field of a propagating crack front must be able to transform large numbers of t particles in order to realize the optimum toughness of the two-phase material. Sintering and aging can also be combined in a single-step heat treatment in the c/t phase field, yielding similar microstructures and properties. Due to the large grain sizes of conventional PSZ, the strength is limited to about 700–800 MPa, a fact which requires the development of A2 types.

A1 PSZ ceramics (mostly Mg-PSZ) are technologically rather advanced ZTC, usually optimized for high toughness and wear resistance. They are being applied as metal forming tools, dies, bearings, etc. and being tested as automotive components, such as cam follower inserts, rocker faces, valve guides and seats, cylinder liners, piston caps, hot plates, etc.

A2: Fine-grained PSZ ceramics, also consisting of c grains with t precipitates, have so far found almost no attention, mainly because of the technological fabrication difficulties. The preparation of fine-grained (<10 μm) PSZ should, however, be feasible through a more refined powder technology. For instance, starting with highly reactive (c, t, or even amorphous) PSZ powders prepared by evaporation-decomposition of solutions,[29] water or air atomization of fused material,[30,31] solid-solution annealing and regrinding, etc., would allow rapid densification such that grain growth is rather limited, while allowing for optimization of the t precipitate size. Furthermore, grain-growth inhibition, e.g., by fine-particle additions of Al_2O_3 or Mg spinel to Mg- or Y-PSZ, or isostatic hot-pressing could be used to achieve the same goal. The great advantage of A2 PSZ lies in improved room- and high-temperature strength properties (>1000 MPa at room temperature), particularly after introducing compressive surface stresses by grinding. Applications are to those for A1 types and also competitive to A3 with the advantage of improved medium ($\approx$250°C) temperature stability.

A3: Tetragonal ZrO_2 polycrystals (TZP) are a fine-grained, predominantly single-phase material with Y_2O_3 or rare-earth oxide dopings in which the retention of the t form is due to the constraint imposed by grains on one another.[5,13,20,21,32] The critical grain size is usually between 0.1–1 μm depending on alloying, e.g., Y_2O_3 solute content and density. Similar preparation conditions prevail as in A2 materials, though to a less critical extent, and sintering can be carried out in the t single-phase field (1.5 to 3.5 mol% Y_2O_3) at reasonable temperatures (1300°–1450°C). This type of material can presently be considered the toughest and strongest of all ZrO_2 ceramics, probably of all polycrystalline ceramics made so far. The reasons are twofold: The grain size is extremely fine and the "active" transformable phase may reach $\approx$100%. The main deficiency of tough Y-TZP ceramics is a $t \rightarrow m$ transformation at medium (200°–400°C) temperatures in humid atmospheres[20,34] or hot aqueous solutions.[35] One suggested possibility of preventing this type of degradation is presented later.

As for A1 materials, A3 Y-TZP ceramics are presently being tested as components in engines and metal forming tools, furthermore many new applications are being developed, such as tape scissors, knives, textile cutters, golf club inserts, highly wear-resistant milling balls, etc.

A4: Fine-grained, m-ZrO_2 can be prepared by hydrothermal synthesis[36]; however, the technique is rather limited for the manufacture of useful components. In a more recent technique,[37] coarse-grained, c Mg-stabilized (<15 mol%) material was decomposed by aging at temperatures of $\approx$1100°C. The resulting microstructure consists of 1–5-μm m grains with highly dispersed MgO-rich pipes. The decomposition does not involve a transitional t phase; the m phase is not twinned. The extremely high thermal expansion anisotropy causes enhanced grain-boundary microcracking and crack branching. Even though the strength is relatively low, the thermal shock properties are good, which is, to some extent, due to the reduced Young's modulus and effective coefficient of thermal expansion compared to other PSZ types.[38]

A5: Subeutectoid aging of suitably presintered conventional Mg-PSZ produces essentially a three-phase material where at least part of the t precipitates have been transformed to m symmetry by aging at temperatures of $\approx 1100°C$.[39] At this low temperature, little precipitate-coarsening takes place; however, the tendency to loose coherency increases with annealing time. Furthermore, m lamellae are formed at the boundaries of the c grain. This complex microstructure, offering a variety of modifications between the A1 and A4 types, combines various toughening mechanisms. The materials exhibit very low A_s temperatures (300°–600°C) and excellent thermal shock properties, especially under up-shock conditions.[40]

A6: Single-crystal PSZ provides the opportunity to utilize the intrinsic properties of PSZ ceramics without the limitations encountered in polycrystals, i.e., without the influence of grain-boundary phases, impurities, pores, etc. Y-PSZ single crystals (up to ≈ 60-mm long) have been prepared by skull-melting with similar t precipitates as in Y-PSZ (A1).[41] The growth of Mg- and Ca-PSZ single crystals seems to be technologically more difficult due to rapid overaging of the less-stable precipitates. The great advantage of single-crystal PSZ lies in the high strength (>1400 MPa). Even at temperatures of 1500°C, values of up to ≈ 700 MPa have been measured.[42] Hot-forging of these materials offers the possibility of shaping parts without significant loss of the good properties inherent in the single-crystal material.[43] A6 materials, however, will probably be limited to specialty applications.

Dispersed-ZrO$_2$ Ceramics

This group contains all systems in which zirconia particles are dispersed in a ceramic matrix other than ZrO_2.

B1: The intercrystalline dispersion of t-ZrO_2 particles in a fine-grained matrix usually leads to very high strengths compared to the matrix alone. This is due to an increased fracture toughness (K_{Ic}), reduced matrix grain size, and to grinding-induced compressive surface stresses.[12] The latter case is especially effective for high Young's modulus materials (e.g., Al_2O_3). The intercrystalline dispersion can be obtained either by homogenizing techniques, i.e. fine grinding and mixing (attrition milling)[12,19,44,45] or by wet-chemical techniques.[14,46,47] The retention of all ZrO_2 particles in their t form is, however, difficult to achieve unless great care is taken to achieve a narrow particle-size distribution. For ZrO_2 volume fractions >0.15, small amounts of stabilizers, e.g., 0.5–1 mol% Y_2O_3 are usually required to have most particles retain their t symmetry at room temperature.[13]

B2: Fine-grained ceramics with an intercrystalline dispersion of m (at room temperature) particles are made in the same way as B1 but with a slightly larger ZrO_2 particle size (or without stabilizer at higher volume fraction).[12,18] Ceramics with this microstructure exhibit lower strengths than the B1 type; however, the toughness (K_{Ic}) may be higher. Their advantage lies in the ability of microcrack-induced stable crack growth (R-curve behavior), and in improved thermal shock resistance under severe conditions. B2 ceramics (zircon-ZrO_2) have found applications in nozzles for continuous casting.[48]

B3: The intercrystalline dispersion of both *t* and *m* particles is actually the most common microstructure of the dispersion-type ZTC. This type is usually obtained through conventional ceramic mixing techniques and, as with the microstructure, the properties are also between B1 and B2. Most Al_2O_3-based ZTC cutting tools exhibit B3 microstructures.[49]

B4: Intracrystalline t-ZrO_2 particles can, mainly for geometrical reasons, be best dispersed in coarse-grained ceramics. This type of microstructure has so far been prepared only for TEM analysis to study the transformation behavior of ZrO_2 in ceramic matrices,[14,50,51] so that the anisotropy effects associated with particles located at the grain boundaries are excluded. It is yet unclear whether the B4 type offers any advantages over B1 to B3, especially since intragranular particles are far less sensitive to stress-induced transformation,[52] i.e., they are more stable and may therefore contribute little to the toughening. Because of the relatively large matrix grains, this material is an analog of A1 PSZ and may be expected to exhibit useful properties only when most features of A1 PSZ are met (e.g., transcrystalline fracture).

B5: A medium-grain-sized ceramic with intracrystalline *t* particles and *m* particles at the grain boundaries, which should exhibit properties comparable to A5 PSZ. Grinding-induced surface compressive stresses would cause high initial strength and enhanced intercrystalline microcracking may create crack branching. B5-type microstructures (Al_2O_3-ZrO_2) have been obtained through sol-gel techniques.[53]

B6: In situ-reacted microstructures seem to be strongly dependent on the manufacturing process used.[54–58] They range somewhat between the B-type microstructures and the more complex zirconia systems (C-type); however, the technological simplicity of fabrication techniques based on the dispersion of zircon or other Zr compounds with oxide reactants may justify this grouping. The typical features of this microstructure are rounded ZrO_2 particles which are randomly located within the grains and at grain boundaries. The ZrO_2 phase (*m* or *t*) present at room temperature is determined by the size (e.g., the critical size in mullite is ≈ 1.2 μm),[58] which again can be controlled by the sintering and reaction conditions.

Complex ZrO_2 Systems

Most of the microstructures presented in this group are in a very early state of investigation, and some may never be successfully realized. It is obvious that the following microstructure types represent only a small cross section of the many possibilities to be developed in the future.

C1: A typical microstructure of mixed ZTC systems can clearly not be described in a schematic form. Evidently, the variations would be too large. However, for present purposes, the C1 type of mixed systems may be defined as representing any microstructure which is derived from a simple combination of A or B types. For instance, a combination of A3 and B1 has been prepared by mixing high volume fractions of ZrO_2 (≈ 2 mol% Y_2O_3) with alumina.[13] The resulting properties are correspondingly intermediate, i.e., the material exhibits a higher Young's modulus and hardness than A3 and a higher fracture toughness than B1. Many applications may require such combinations of properties. An important role will probably be played by certain surface property needs, i.e., low friction and wear, etc.

C2: The analog of PSZ type materials (A1) based on ceramic matrices other than zirconia is a highly attractive goal because of the nearly ideal distribution of the t- (or m)-ZrO_2 phase obtained by A1 manufacture. This microstructure, in which the ZrO_2 "precipitates" may be optimized by aging as in A1, has not yet been fully realized. However, efforts are being made, for instance, by the use of rapidly quenched near-eutectic Al_2O_3-ZrO_2 compositions which, in the form of amorphous powders, may be densified and crystallized to the desired microstructure.[30] Figure 3 shows a TEM micrograph of an Al_2O_3-33 vol% ZrO_2 sample hot-pressed from partly amorphous powders with both t and m microstructural components. At lower ZrO_2 contents, however, some of the B-type microstructures may also result.

C3: The crystallization of t- (or m)-ZrO_2 in a glass or glass ceramic may be a useful tool to improve the toughness of this class of materials. It may be technologically even more feasible than C2 microstructures. A first step toward this type has been indicated by plasma dissociation of zircon: small (≈ 10 nm) t and larger m ZrO_2 particles have crystallized in the SiO_2 glass matrix during cooling after the dissociation process.[59]

C4: This microstructure is obtained by adding ZrO_2 to a ceramic matrix in which a third phase can be precipitated. ZrO_2 in either m or t form would function as a low-temperature ($T < A_s$) toughening agent, and the (intragranular) precipitates would act in the conventional way of precipitation strengthening at high temperatures ($T > A_s$). An example would be Al_2O_3-rich spinel with dispersed t-ZrO_2 particles which, after annealing at $T > 1000°C$, may contain Al_2O_3 precipitates with no change in the morphology of the ZrO_2 particles.

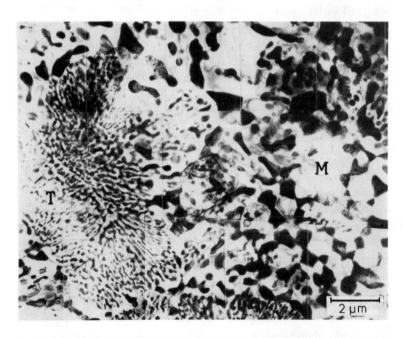

Fig. 3. Bright-field TEM photograph of Al_2O_3-33 vol% ZrO_2 (C2, cf. Fig. 2), crystallized during hot-pressing of rapidly quenched, amorphous particles; left tetragonal (*T*), right monoclinic (*M*) ZrO_2 phase.

C5: For the present purpose, duplex microstructures are defined such that they consist of a high-strength matrix (e.g., B1 type) with relatively large (<50 μm), spherical inclusions consisting of a material with a high volume fraction of m particles (high M_s). The intent of this microstructure is to combine high resistance for crack initiation (e.g., grinding-induced compressive stresses) with the ability for crack branching, i.e., increasing crack propagation resistance (R-curve behavior). Although very few experiments have been performed with this type of microstructure, it seems to offer good possibilities for the development of thermal and mechanical shock-resistant materials.[60]

C6: Directionally solidified eutectics involving ZrO_2 as one phase were prepared at a time when the effect of transformation-toughening had not yet been discovered. These aligned eutectic microstructures have been shown to exhibit good high-temperature strength properties.[61] A combination of fiber reinforcement with ZrO_2 toughening may be possible in such eutectic systems as CaO-ZrO_2, MgO-ZrO_2, Al_2O_3-ZrO_2, etc.

Experimental Data and Correlation to Microstructure

Considerable difference exists in the mechanical behavior of the various ZTC types. This can be partly attributed to the inherent matrix properties. However, closely controlled microstructural design within the framework of the various ZTC groups enables optimal application-oriented properties to be specified.

In the following section, some general experimental facts concerning the t-to-m transformation and toughening behavior are presented. Some of the interpretations provided must be treated as somewhat speculative at this stage. Additional difficulties in proposing specific mechanisms arise from the fact that practically all variables have interacting effects.

Size Dependency of the Transformation Temperature (M_s)

In all systems studied to date, it has been found that M_s decreases with decreasing ZrO_2 particle or grain (A3) size, a fact which can be explained by nucleation arguments.[52,62] Since M_s determines the type of toughening or whether toughening is possible at all, it is of great technological interest to precisely control the precipitate or particle size. This is usually done by alteration of the sintering and aging schedule (A1, A2, A5) and milling duration (B1–B3), etc. Furthermore, additions of stabilizers, especially the high-solvent types (Y_2O_3 and most rare-earth oxides), will shift M_s to lower temperatures for a given particle size. Higher volume fractions of ZrO_2 in B- and C-type materials tend to increase M_s.

It is convenient to define a critical size, d_c, above which, at room temperature, spontaneous transformation will occur and below which it will not. The d_c values vary, depending on the system between ≈0.1 (Si_3N_4-ZrO_2) and 1.2 μm (mullite-ZrO_2).[12,63] Optimum toughness by stress-induced transformation is usually achieved when the size of particles, precipitates, or grains is just below d_c.

Temperature Dependence of the Stress-Induced Transformation

The availability for toughening is associated with the chemical driving force for the transformation. Since the driving force decreases with increasing temperature, the toughening effect is also expected to decrease. This temperature dependence has been confirmed by experiment.[13,64]

Optimum Amount of Stabilizer

The chemical driving force for the t-to-m transformation decreases with amounts of solute present in t-ZrO_2; this is especially pronounced for the stabilizing elements, such as Y_2O_3; HfO_2 is the notable exception to this behavior. This fact implies that, for PSZ and high volume-fraction ZrO_2 ceramics, an optimum addition must exist for which the retained volume fraction and the transformability are maximum. The optimum amount depends very much on the type of microstructure, the solute used, and especially on the homogeneity of the particle, precipitate, or grain distribution, i.e., with increasing uniformity of size and spacing, the optimum K_{Ic} is shifted toward lower solute contents. Figure 4 shows K_{Ic} of PSZ ceramics vs mole fraction of stabilizer.[6,13,41,65] Naturally, the curves are not directly comparable since many parameters, such as microstructure, test technique, etc. are different for the materials shown. It is obvious that for some ZTC types the technological development must be directed toward lowering the amount of solute in t-ZrO_2 down to zero.[66]

Optimum Volume Fraction of ZrO_2

All theoretical derivations for the fracture toughness predict increasing K_{Ic} with increasing volume fraction of the dispersed ZrO_2 phase. However, since all microstructures except A3 impose a geometrical and homogeneity-dependent limit on the volume fraction, an optimum develops. For mechanically mixed B1 types, this occurs at ≈ 15 vol% and probably somewhat higher for those made by wet-chemical or fusion-quench techniques. The optimum volume fraction of t precipitates of A1 and A2 PSZ results from a combined optimization of stabilizer content and cooling rate from the solid solution temperature. Optimum toughness is then achieved through the precipitate size by aging. At peak age conditions the t precipitate phase will have attained the equilibrium volume fraction.[39]

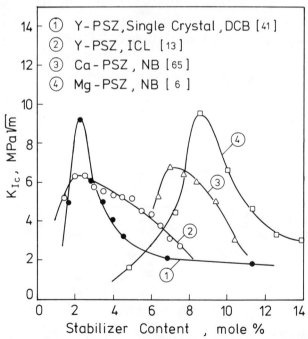

Fig. 4. Fracture toughness of various PSZ types with stabilizer content. The curves are not directly comparable since microstructure, test techniques, etc. are different for the materials shown.

Surface Strengthening

The transformation of t-ZrO_2 can be utilized to introduce residual surface compressive stresses which increase the bend strength of ZTC considerably. There are various ways of producing surface compressive stresses (see Table II in Ref. 68), e.g., by grinding, destabilization, low-temperature treatment, laser induction, etc. Even though techniques that do not change the surface morphology have the greater strengthening potential, grinding-induced transformation seems to be the technologically most useful technique. In this case, the extent of strengthening depends not only on the volume fraction of transformable t-ZrO_2, the transformational depth, and Young's modulus, but also on the matrix grain size.[12,67] Therefore, high Young's modulus B1 materials or the fine-grained A2 or A3 PSZ exhibit much greater strengthening than, for instance, A1 PSZ, especially when high Young's modulus materials (e.g., Al_2O_3, WC, etc.) are dispersed. The influence of increasing grinding severity, causing an increasing fraction of transformed ZrO_2 in the surface region, is particularly pronounced in B1 Al_2O_3.[12,68,69] Although rough grinding introduces deep surface flaws, the transformation depth (proportional to the compressive layer) obviously extends further. Some transformational profiles measured in different ZTC are shown in Fig. 5.[13,70–72] From the curves it becomes evident that reducing the transformability by stabilizing additives (e.g., Y_2O_3 in material B1) decreases the transformation depth which may be compensated by higher volume fractions of transformable t-ZrO_2.

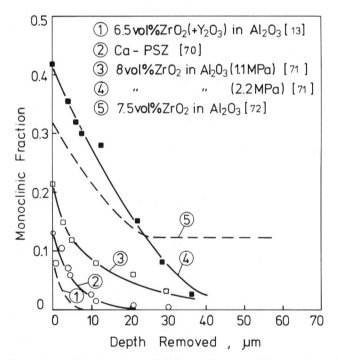

Fig. 5. Transformation depth profiles in different ZTC produced by surface grinding. The grinding pressure indicated with materials 3 and 4 was controlled on a disk grinding machine. Grinding conditions for material 5 were more severe than for 3 and 4.

Although grinding-induced transformation is comparable to the transformation in the crack-tip stress field, the exact mechanisms have not been fully investigated. There are a number of possible mechanisms which may be simultaneously active: (*a*) Transformation is triggered by the high shear strains caused by the grinding process; (*b*) a high density of parallel flaws, induced by the abrasive media, creates similar stress-field transformations to those prevailing at single sharp crack tips; (*c*) the dislocations produced in the process pile up at particles and consequently cause their transformation. From various experiments with B1 ceramics using different matrices it becomes obvious that surface grinding results in a much larger transformed depth than that due to crack propagation.

Improvement of Thermal Shock Resistance

Appropriate design of ceramics for the many different applications requiring thermal shock resistance is rather complex. However, the variety of ZTC microstructures provides a unique opportunity to select a suitable material/microstructure type. In a recent paper, the various possibilities of ZrO_2 toughening for improving the thermal shock resistance of ceramics were outlined.[73] In addition to the ZrO_2 toughening mechanisms the effective thermal coefficient of expansion of a body can be reduced by using ZrO_2 particles with a wide size distribution which essentially increases the M_s-temperature range, or by subeutectic aging of A1 microstructures[74] (leading to A5 type with reduced A_s temperatures). By changing the elastic behavior through the introduction of a high density of spontaneous microcracks or by the occurrence of transient compressive stresses which oppose the thermal tensile stresses, thermal shock resistance may also be improved.[12,73] Principally, there are two directions of microstructural developments: (*a*) prevention of the initiation of thermal stress failure and (*b*) arresting propagating cracks in thermal stress fields. Generally, the high-strength types (e.g., A1–A3, B1, C1) are suitable for case (*a*) and those where microcracking can be induced (e.g., A4, A5, B3, B5, and especially C5) are suitable for case (*b*).

Selected Examples of ZrO_2 Toughening

Some ZrO_2-containing systems of major interest will be discussed in the following sections to demonstrate the various aspects of ZrO_2 toughening (cf. Fig. 1). Miscellaneous effects of ZrO_2 not associated with the phase transformation will be treated first, then microstructures that utilize the transformation are presented and finally the possibilities for improving the high-temperature properties are outlined. In some cases, the results are generalized and, hence, applicable also to other systems.

Miscellaneous Effects of ZrO_2

Mullite (B3): Powder mixtures of fused mullite (3 $Al_2O_3 \cdot 2\ SiO_2$) and ZrO_2, attritor-milled for 8 h, were isopressed and sintered for 3 h at 1610°C.[75] The fracture toughness of a 10 vol% ZrO_2 composite was ≈ 4 $MPa \cdot m^{1/2}$ compared to 2.5 $MPa \cdot m^{1/2}$ for the ZrO_2-free mullite. An interesting observation in this system was that ZrO_2 had a significant effect on the amount and distribution of the amorphous phase. In pure mullite, thin films and small triangular pockets of an amorphous phase at grain-boundary junctions were frequent. In ZrO_2-containing specimens, on the other hand, minute, isolated islets of an amorphous phase were located at ZrO_2-mullite interfaces but not at mullite-mullite grain boundaries (Fig. 6). This distribution most probably results from interfacial energy differences. High-temperature mechanical properties have not yet been studied, but the glass scavenging by ZrO_2 particles may advantageously affect the creep resistance.

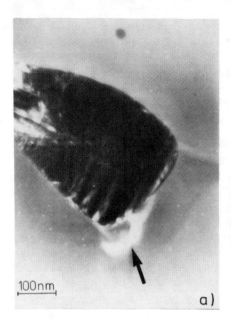 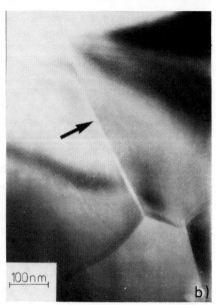

Fig. 6. Dark-field TEM micrographs of (a) sintered mullite-ZrO_2 and (b) ZrO_2-free mullite (Ref. 75). The glassy phase (arrows) in (a) is mostly associated with the ZrO_2 particles while the mullite grain boundaries are essentially glass-free. In (b) all grain boundaries are covered with an amorphous film.

A further positive effect of ZrO_2 was to prevent exaggerated grain growth and to promote the densification.[75]

Spinel (B1): $MgO \cdot Al_2O_3$, was attritor-milled for 4 h with ZrO_2 powder and then sintered for 2 h at temperatures between 1500° and 1600°C. Strength and toughness were increased to 565 MPa and 4.15 MPa·$m^{1/2}$ for a 15 vol% composite compared to 270 MPa and 2.0 MPa·$m^{1/2}$ for the pure spinel.[76] As shown in Fig. 7, the grain-size distribution was considerably narrowed by the ZrO_2 dispersion, which certainly contributed to the strength enhancement. As with fused mullite, exaggerated grain growth was strongly reduced and the sintering rate increased.

Y-PSZ Single Crystals (A6): It has been shown that single crystals of Y-PSZ exhibit high strength (>1200 MPa), which decreases with increasing temperature, like A1-type ceramics.[41,43] However, at $T > 800°C$, i.e., where the stress-induced transformation no longer occurs, the strength remains approximately constant up to ≈1500°C. This was attributed to direct crack-precipitate interaction comparable to the dislocation-precipitate interaction strengthening observed in metals.

Microstructures Utilizing the ZrO_2 Transformation

The phase transformation of ZrO_2 can be utilized in a number of ways by designing microstructures which are optimized to achieve specific mechanical properties (Fig. 8). Some typical examples are presented below.

Al_2O_3 (B1, high strength): It has yet to be made clear for dispersed-ZrO_2 systems whether the microstructural requirements for optimum strength and toughness are identical. When only stress-induced transformation is the active toughening

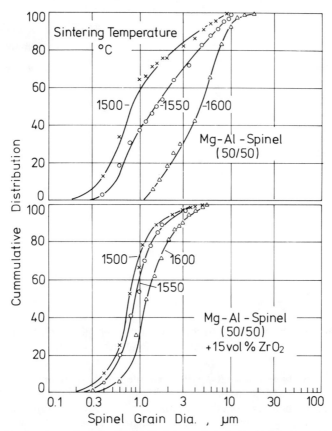

Fig. 7. Grain-size distribution of sintered spinel (1 MgO·1 Al_2O_3) with (lower diagram) and without (upper diagram) ZrO_2 particles.

mechanism, both parameters appear to exhibit the same trend. In such systems the characteristics of the ZrO_2 particles would be: t structure, equally sized with $d \leq d_c$, intercrystalline and uniformly dispersed, irregular shape, and high volume fraction. For the Al_2O_3 matrix: small equiaxed grain size (d_A), where d_A may be $\ll d_c$.

In real Al_2O_3-ZrO_2 systems, however, it has been found that these requirements are more applicable for strength optimization. High fracture toughness was usually found with intermediate (≈ 5 μm) Al_2O_3 grain sizes[77] and with both t and m particles present.[3,78] This may result from a "nonideal," i.e., broad size distribution, where, in composites containing m-ZrO_2, a larger fraction of the t particles have $d \approx d_c$ and are thus transformable in the crack-tip stress field or can contribute to microcracking or other toughening mechanisms which could be more effective with m particles. Present studies should clarify this question.

Al_2O_3-ZrO_2 ceramics (ZTA) with strengths of >1000 MPa have been produced by isostatically hot-pressing attritor-milled oxide powders.* For instance, a 15 vol% ZrO_2 composite, milled for 4 h with Al_2O_3-15 wt% SiO_2 milling media, sintered at 1450°C for 1 h to $\approx 96\%$ theoretical density isostatically hot-pressed

---

*The Al_2O_3 (Alcoa A16) contained $\approx 0.07\%$ MgO. For all ZTA ceramics, as for pure Al_2O_3, small amounts of MgO are added to improve densification (Ref. 72).

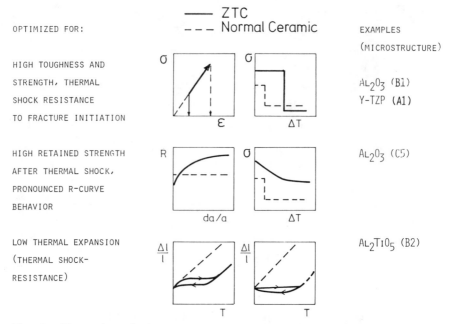

Fig. 8. Examples of microstructures which are optimized to achieve specific mechanical properties utilizing the ZrO_2 phase transformation.

at 1550°C for only 1 min at 200 MPa, exhibited peak strengths of ≈1300 MPa and K_{Ic} of only 5.5 MPa·m$^{1/2}$ in the as-ground surface state. It is assumed that a glassy phase, originating from the milling wear, (≈0.1 wt% SiO_2), contributes to the complete densification under these "fast-firing-type" conditions. The main strengthening factors in this 100% t-ZrO_2-Al_2O_3 composite are the grinding-induced surface transformation (compressive stresses) along with the fine Al_2O_3 (<1 μm) and ZrO_2 (<0.5 μm) grain sizes. Sintering the same material to full density requires 1550°C for 2 h, resulting in grain sizes at least twice as large and a strength of <600 MPa with K_{Ic} unchanged (≈5.5 MPa·m$^{1/2}$) (Fig. 9). In the case of the isostatically hot-pressed material the relatively low K_{Ic} implies a surface transformation-zone depth which is larger than the deepest machining flaw (critical size ≈40 μm). The essential effect of the isostatic hot-pressing process is in reducing the Al_2O_3 grain and ZrO_2 particle size rather than in the elimination of pores or flaws. The apparent discrepancy between strength and toughness may be due to the fact that the stresses and strains in rough grinding cause the transformation of a much wider t-ZrO_2 particle size range and to a greater depth than can crack-tip stresses. It seems obvious that, for applications where severe surface damage may occur, microstructures optimized for toughness must be preferred.

Bend strengths of >2000 MPa appear reasonable for optimally processed ZTA ceramics. In mechanically mixed systems, 15 to 20 vol% of retained t-ZrO_2 is feasible. Higher volume fractions of t-ZrO_2 and greater homogeneity of the dispersion may be obtained by wet-chemical,[77] CVD,[47] or fusion-quenched[30] powder mixtures. An aging treatment to acquire optimal properties, as in PSZ ceramics, may also be a necessary step for ZTA materials, especially when starting from amorphous powders. In all cases, "colloidal" powder processing should improve the mean strength values.[79]

Fig. 9. Scanning electron micrographs of Al_2O_3-15 vol% t-ZrO_2 composite; (a) sintered at 1550°C for 2 h, strength 570 MPa; (b) presintered at 1450°C for 2 h and isostatically hot-pressed at 1600°C for 10 min, strength 1050 MPa.

Y-TZP (A3, high strength and toughness): A somewhat similar phenomenon as in the B1 ZTA materials is observed with the new generation of superstrong TZP ceramics.[21,80] In TZP, strength increases with decreasing size of t grains, while toughness is maximized at the largest possible t grain size for a given Y_2O_3 content (e.g., ≈0.4–0.5 μm at 3 mol% Y_2O_3). A further increase in fracture toughness is achievable with decreasing Y_2O_3 solution content. A reduction in Y_2O_3 content must, however, be coupled to a reduction in the t grain size.

The schematic phase diagram in Fig. 10 depicts a few typical cases, namely three realistic microstructures with 3 and 5 mol% Y_2O_3 and one Y_2O_3-free TZP (4 in Fig. 10) which, with refined processing techniques (i.e., direct isostatic hot-pressing of microcrystalline amorphous powders with SiO_2 added to develop a glassy grain-boundary phase), might exhibit strengths of >3000 MPa and toughnesses of >20 MPa·m$^{1/2}$. The grain size must, however, be extremely small (<0.03 μm).[66] Materials 2 and 3 in Fig. 10 are based on identical powders with the same Y_2O_3 composition. Sample 2 was hot-pressed at the upper limit of the t single-phase field boundary and sample 3 was sintered at a higher temperature in the c/t region. Material 2, consisting of 100% t grains of ≈0.2 μm exhibits a strength of >1500 MPa, while the toughness is only slightly greater than 6 MPa·m$^{1/2}$. In material 2, the fine t grains were rather stable, i.e., almost no transformation was detected by TEM in in situ straining experiments.[33] Material 3, however, had a t grain size of ≈0.4 μm and contained less Y_2O_3 than the t phase in material 2 (reference the positions of 2 and 3 with respect to the phase boundary at the various temperatures). These grains are highly transformable in the stress

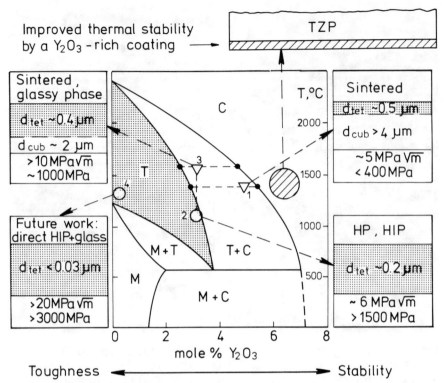

Fig. 10. ZrO_2-Y_2O_3 phase diagram with experimental results (positions 1 to 3) and a future perspective (position 4) given in the adjacent blocks. Low-temperature (200°–300°C) stability increases with decreasing grain size and increasing Y_2O_3 content. A PSZ-type protective layer is indicated above (see also Fig. 12).

field of a crack-tip, leading to $K_{Ic} > 10$ MPa·m$^{1/2}$. Some large c grains also developed and the strength was ≈1000 MPa. In composition 1, consisting of relatively fine t grains and a higher fraction of larger c grains, both strength and K_{Ic} are relatively low. However, this material is rather stable within the critical temperature range of 200° to 400°C.

The medium-temperature mechanical property degradation is due to a transformation to m symmetry starting at the surface and is particularly enhanced in humid atmospheres.[21,35] The degradation is a characteristic of all tough Y-TZP[†] (Fig. 11). Small grain sizes along with high Y_2O_3 concentrations yield degradation-resistant but weaker materials. One possible way of retaining both toughness and structural stability is by enriching a thin (10–40 μm) surface layer with stabilizing oxides, such as Y_2O_3, MgO, CeO_2, etc., as indicated by the shaded composition in Fig. 10. One of the different methods of obtaining a PSZ-type surface layer is by slip-coating green TZP bodies in a Y_2O_3 slurry before sintering or by sintering in a Y_2O_3-powder bed.

A 3Y-TZP sample sintered for 2 h at 1400°C in a Y_2O_3-powder bed is shown in Fig. 12; a 10-μm c layer has developed at the surface. In a severe autoclave test

---

[†]It is interesting to note here that the internal friction of 3Y-TZP exhibits a very pronounced peak at 200°C (Ref. 92), which correlates well with the degradation phenomenon.

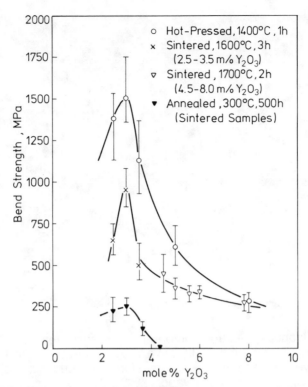

Fig. 11. Bend strength of A3-type TZP vs Y_2O_3 content. Note the drastic degradation after low-temperature annealing for long times (see Ref. 34).

(250°C, 6 bar water vapor, 1 h) no surface decomposition was detected.[81] Other potential ways of preventing the degradation appear to be the stabilization of the t grains by other rare-earth oxides, e.g., Ce-TZP,[82] or by adding other stabilizers to Y-TZP, e.g., Mg/Y-TZP.[83]

Al_2O_3 (C5, R-curve behavior): Duplex microstructures in which compressive zones B are dispersed in a matrix A (Fig. 13) have been found to show excellent resistance to severe thermal shock,[60,73] i.e., where initiation cannot be avoided, but where the retained strength is high due to an enhanced ability for crack arrest. A hypothetical crack-propagation resistance curve of a duplex microstructure $(A' + B)$ and that of its individual components (A), (A'), and (B) is shown in Fig. 14. A B1-type material (component (A) in Fig. 14) would have a more or less constant crack-propagation resistance, a B1-version surface-strengthened by grinding $((A')$ in Fig. 14) would exhibit a strong initial resistance to cracking decreasing to the level of the bulk resistance (A) after penetration of the residual compressive surface layer, and (B), a low-strength microcracked B2 material, would show the characteristic increase in crack-propagation resistance. The purpose of the microstructure is to combine both reasonable strength with R-curve behavior. Clearly, curve $(A' + B)$ is not obtained by just superpositioning (A') and (B).

The hypothetical crack propagation in a duplex structure is also shown in Fig. 13 where the initial flaw (c_0) propagates into zone B_0 which is under isostatic compression with a tensile hoop stress (σ_t) which is maximum at the boundary matrix A/zone B. This stress state occurs because the ZrO_2 particles in zones B

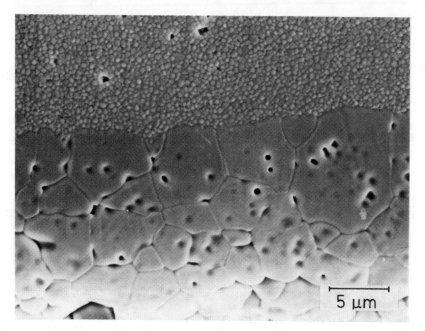

Fig. 12. Scanning electron micrograph of a spray-reacted 2Y-TZP (see Ref. 81) sintered for 2 h at 1500°C in a Y_2O_3 powder bed. The 15 µm cubic surface layer prevents the $t \to m$ transformation under autoclave conditions (250°C, 1 h, 6 bar water vapor).

have transformed to m symmetry and hence have expanded, while those in the matrix remain in t form at ambient temperature after cooling from the sintering temperature. Before c_0 can cut through B_0 cracks, c_1 and c_2 will form as a result of the stress field around B_1 and B_2. This process will continue to a certain extent, leading to crack branching as a consequence of a "crack-cascade" effect.

Al_2O_3-based C5 materials were prepared by dry-mixing agglomerates A (forming-matrix A in Fig. 13) with 10 to 20 vol% of agglomerates B, and pressing, sintering, or isostatically hot-pressing the mixture. Agglomerates A with diameters 10 to 20 µm were fabricated by spray-drying Al_2O_3-15 vol% ZrO_2 (microstructure type B1, see also Fig. 9(b)) and larger agglomerates B by either spray-drying (diameter 20 to 40 µm) or dry tumbling (50 to 150 µm) Al_2O_3 with 20 to 50 vol% ZrO_2 (microstructure type B2). The SEM photographs in Fig. 15 demonstrate that the crack propagation initiated from Vickers indents in a C5 composite with 20 vol% B zones. The arrows mark the cracks which always run into the compressive B zones. The load-deflection curve in Fig. 16 shows the controlled fracture of a C5 Al_2O_3 (16 vol% of B in matrix A'), matrix A' containing 15 vol% t-ZrO_2 and zones B 50 vol% m-ZrO_2. This is in sharp contrast to the catastrophic fracture of the respective matrix A' alone.

Al_2TiO_5 (B2, high strength and low expansion): Aluminum titanate exhibits almost zero thermal expansion (up to ≈700°C) if the grain size exceeds a critical diameter (≈10 µm).[84] In such case, microcracks develop due to the strong thermal expansion anisotropy of the lattice. Consequently, the strength is extremely low (20–40 MPa), which makes the material useful only in compressed-state engineering components. At smaller grain sizes, hence no microcracking, Al_2TiO_5

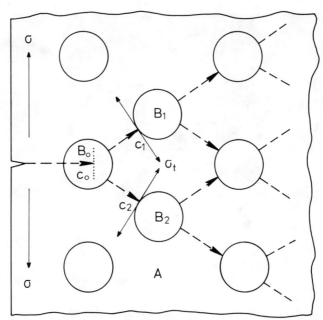

Fig. 13. Schematic crack propagation in a duplex material (C5) consisting of matrix A (e.g., Al_2O_3 + 15 vol% t-ZrO_2) and dispersed spherical zones B (e.g., Al_2O_3 + 30 vol% m-ZrO_2). Crack c_0 will be stopped in zone B_0 (which is under isostatic compression); on further increase of stress σ, cracks c_1 and c_2 may develop attracted by the tensile hoop stress field around B_1 and B_2.

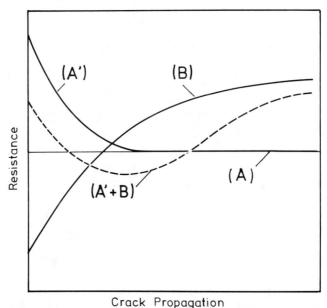

Fig. 14. Schematic R curves of a composite (A' + B) with duplex microstructure (C5) and that of its single components (A') and (B). For further details see text.

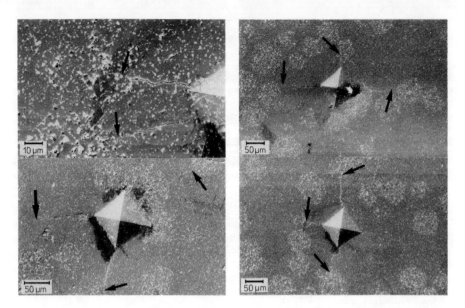

Fig. 15. 500 N Vickers indents in a duplex material consisting of 15 vol% compressive zones B (composed of Al_2O_3 + 20 vol% m-ZrO_2; B2 type) dispersed in matrix A (Al_2O_3 + 15 vol% t-ZrO_2; B1 type). Cracks starting from the indent corners are always attracted by zones B (see arrows).

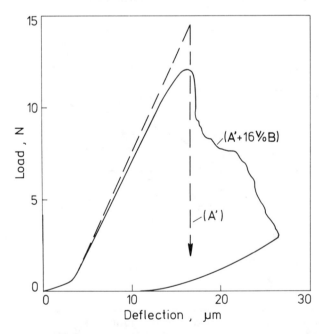

Fig. 16. Load-deflection curve of the duplex material shown in Fig. 15 compared with that of the matrix material A' alone.

Al-Titanate (Low-Expansion Ceramic)

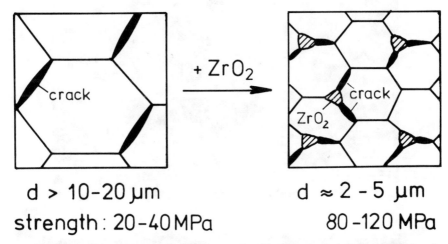

Fig. 17. Strength-enhancing effect of intergranular m-ZrO_2 particles in Al_2TiO_5 (microstructure B2). Left: Microcracks necessary for low thermal expansion form only at grain sizes $d > d_c \approx 10$ μm (Ref. 84). Right: m-ZrO_2 particles cause both reduction in grain size (<5 μm) and microcrack formation required for low expansion; the strength is consequently improved.

is correspondingly stronger, but the thermal expansion becomes high (7×10^{-6} K^{-1}).

Adding ZrO_2 particles allows the grain size to be reduced and still retain a low expansion. (The reason for this behavior is shown schematically in Fig. 17.) The microcracks required for the low expansion behavior are formed by the ZrO_2 particles when transforming to m symmetry. The reduced grain size also results in improved strengths (80–120 MPa). Al_2TiO_5 microstructures without (a) and with (b) 15 vol% ZrO_2 are seen in Fig. 18.[85] The grain-boundary microcracks are difficult to identify in the slightly overetched surfaces.

High-Temperature Design

Some ZTC will be used at elevated temperatures, i.e., at $T > T_0$ (700°–800°C, see Fig. 1). In this case, it may be necessary to take measures to maintain strength at higher application temperatures, even at the expense of a lowered room-temperature strength or toughness. Some potential methods for the respective microstructural development are listed in Table I, together with examples which are to be examined at present or in the future. Some of the examples will now be discussed.

Grain coarsening may become a problem due to concurrent ZrO_2 particle coarsening. B4 microstructures, although of possibly lower strength at room temperature than B1 types, may be made to meet some high-temperature requirements. The preparation and hot-forging of single-crystal PSZ have been presented elsewhere.[43] Numerous methods can be proposed whereby a reduction or even pre-

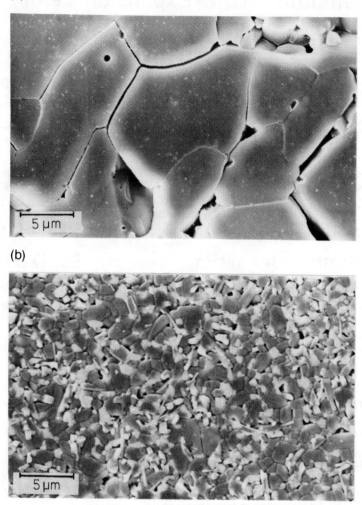

Fig. 18. Scanning electron micrographs of low-expansion Al_2TiO_5 sintered at 1400°C, 1 h: (a) α (0°–700°C) $\approx 0.5 \cdot 10^{-6}$/K, strength = 40 MPa; (b) with 12 vol% m-ZrO_2, α (0°–700°C) $\approx 2.5 \cdot 10^{-6}$/K, strength = 120 MPa.

vention of glassy grain-boundary phases is possible. Especially in the new TZP materials, which usually contain amorphous grain-boundary phases,[21] a need for some refining processing will be necessary. However, it is as yet unclear to what extent the amorphous phase contributes to the densification and toughness; i.e., the glassy phase may be an essential feature for the fabrication and enhanced mechanical properties of TZP materials. In most other ZTC for high-temperature application, annealing procedures, similar to those applied to Sialons,[86] leading to the crystallization of the grain-boundary phase, will probably be of advantage. Also additives which, for some reason, act as glassy phase getters, such as ZrO_2 in mullite[75] or SrO in Si_3N_4,[87] seem worth further investigation. Fiber or whisker reinforcement of ZTC, especially of TZP or PSZ ceramics, appears to be an attractive possibility of enhancing the high-temperature strength. The technological difficulty of incorporating the whiskers may be overcome by direct isostatic hot-pressing. Furthermore, the directional solidification of ZrO_2-containing eutectics[61] should be reinvestigated with respect to additional transformation toughening.

Some of the suggestions (Table I) for high-temperature design of ZTC are treated in more detail with first experimental results.

Al_2O_3-rich spinel (C4, precipitation-hardening): As found in PSZ single crystals (A6)[41] and in hot-pressed Al_2O_3-rich spinel,[88] second-phase precipitates within grains can result in considerable high-temperature strength. Thus, the achievement of a combination of such precipitation-hardening and ZrO_2 toughening would appear to be a desirable objective to enhance the mechanical properties at both room and high temperatures.

Previous experiments have shown that spinel ($Al_2O_3 \cdot MgO$) can be considerably strengthened by ZrO_2 additions.[12,68] It is yet unclear whether intragranular precipitation of α-Al_2O_3 is possible in polycrystalline Al_2O_3-rich spinel, as indicated in Ref. 88. However, nucleation and precipitation at free surfaces,[89] hence at the grain boundaries, could also lead to enhanced high-temperature mechanical properties. Furthermore, it can be assumed that the typical aging conditions for α-Al_2O_3 precipitations in spinel do not seriously change the phase composition and morphology of the dispersed ZrO_2 particles.

To study this potential C4 system, two spinel compositions, marked 1 and 2 in the phase diagram[90] in Fig. 19, were chosen.

(1) Prereacted MgO-1.5 Al_2O_3 spinel was attrition-milled without and with 15 vol% ZrO_2 and sintered for 2 h at 1600°C.[76] The room-temperature strength of the as-sintered composite was 560 MPa, as compared to 260 MPa for the ZrO_2-free spinel. The intergranular ZrO_2 particles were ≈70% in t symmetry. This phase state did not change for aging times up to 70 h at 1300°C. However, no α-Al_2O_3 precipitates were found by TEM, indicating that the Al_2O_3 solute content was too low.

Table I. Some Potential Methods for Improving the High-Temperature Mechanical Properties of ZrO_2-Toughened Ceramics

Methods	Examples
Grain coarsening	Spinel, Al_2O_3
Hot-forging of single crystals of ZrO_2/HfO_2	Y-PSZ
Prevention of glassy grain-boundary phase	Mullite, Al_2O_3, TZP, PSZ
Special grain-boundary design	Mg/Y-PSZ
Fiber reinforcement	TZP + Al_2O_3 or spinel whiskers
Directional solidification of eutectics	Al_2O_3-ZrO_2, ZrO_2-MgO, Al_2O_3-MgO-ZrO_2
Precipitation hardening	Al_2O_3-rich spinel
Retention of residual compressive surface stresses	Mullite, Al_2O_3
HfO_2 alloying (increase of A_s)	Al_2O_3, spinel

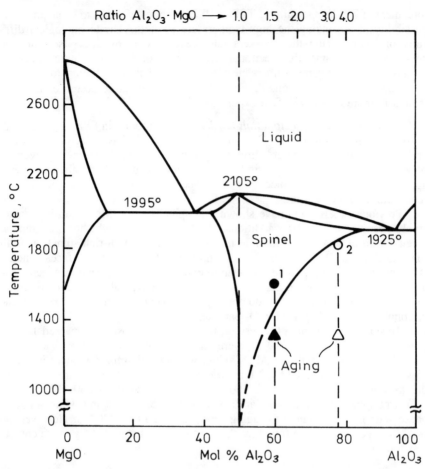

Fig. 19. MgO-Al$_2$O$_3$ phase diagram (Ref. 90) with spinel composition 1 and 2 used for ZrO$_2$ toughening. On aging at 1300°C, α-Al$_2$O$_3$ precipitation was possible only with composition 2.

(2) Rapidly solidified MgO-3.5 Al$_2$O$_3$ was attritor-milled for 16 h using Mg-PSZ milling media. The milling wear resulted in a ZrO$_2$ volume fraction of $\approx 13\%$. After sintering for 30 min in N$_2$ at 1800°C (position 2 in Fig. 19) the composites contained relatively large (5–10 μm) grains of composition 2 with m-ZrO$_2$ particles at the boundaries. Aging times >0.5 h caused α-Al$_2$O$_3$ platelets to precipitate within the grains. The strength has not yet been measured; however, due to the coarse m-ZrO$_2$ particles, resulting from the sintering conditions used, it is expected to be low. Thus, the sintering condition must be modified such that most ZrO$_2$ particles are retained in t symmetry, e.g., by short-time direct isostatic hot-pressing at $T < 1800°C$ or by reducing the Al$_2$O$_3$ content.

Al$_2$O$_3$ (B1, B4, retention of compressive surface stresses): Residual compressive surface stresses, introduced by grinding, for instance, will usually anneal out when the material is taken to $T > A_f$ ($\approx 1000°C$) and the strength is consequently

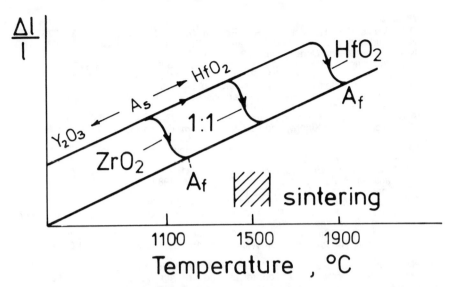

Fig. 20. Influence of HfO_2 alloying on transformation temperatures (A, M) of ZrO_2 particles. Temperatures given in the schematic diagram are based on Al_2O_3-matrix B1 microstructures containing 15 vol% ZrO_2, 1 ZrO_2:1 HfO_2, or HfO_2.

reduced.[12,68] One possibility of retaining the stress state after an anneal at $T > A_f$ is by alloying the ZrO_2 particles with HfO_2, which leads to increased M and A temperatures (Fig. 20). For instance, 50 mol% HfO_2 in a 15 vol% $(Zr, Hf)O_2 - Al_2O_3$ composite shifts A_f to 1460°C,[91] indicating that compressive stresses would be retained up to this temperature. It must, however, be assumed that stress annealing will take place, thereby reducing the residual stresses, the extent of which will depend on time and temperature. The technological problem associated with HfO_2 alloying is that the t phase becomes more difficult to retain to room temperature, i.e., d_c is reduced (e.g., for 15 vol% $(Zr_{0.5}Hf_{0.5})O_2$ in Al_2O_3, $d_c \approx 0,3$ μm).[91] For HfO_2 contents of >50 mol%, normal sintering at $T > A_f$ would probably result in increased grain sizes opposing the retention of the t modification.

Another possibility of retaining compressive stresses appears to be by incorporating round t-ZrO_2 particles, which are rather stable in the virgin state, but which, however, retain martensite nucleating defects when taken to $>A_f$ once they have previously transformed to m symmetry. This behavior has been shown to apply to in situ-reacted mullite-ZrO_2 which typically contains rounded ZrO_2 particles.[63]

Acknowledgments

The writer thanks R. H. J. Hannink for constructive criticism of the manuscript and M. V. Swain for R-curve measurements of duplex materials. Part of the work was supported by the German Science Foundation (DFG). The research on duplex structures (C5) is presently being supported by Stiftung Volkswagenwerk.

References

[1] R. C. Garvie, R. H. Hannink, and R. T. Pascoe, *Nature (London),* **258**, 703 (1975).
[2] H. H. Sturhahn, W. Dawihl, and G. Thamerus, *Ber. Dtsch. Keram. Ges.,* **52**, 703 (1975).
[3] N. Claussen, *J. Am. Ceram. Soc.,* **59** [1-2] 49-51 (1976).
[4] D. L. Porter and A. H. Heuer, *J. Am. Ceram. Soc.,* **60** [3-4] 183-84 (1977).
[5] T. K. Gupta, F. F. Lange, and J. H. Bechtold, *J. Mater. Sci.,* **13**, 1464 (1964).
[6] U. Dworak, H. Olapinski, and G. Thamerus; p. 543-50 in Science of Ceramics, Vol. 9. Edited by K. J. deVries. The Nederlandse Keramische Vereniging, Amsterdam, 1977.
[7] P. F. Becher, *J. Am. Ceram. Soc.,* **64** [1] 37-39 (1981).
[8] A. G. Evans and A. H. Heuer, *J. Am. Ceram. Soc.,* **63** [5-6] 241-48 (1980).
[9] R. W. Rice, K. R. Mc Kinney, and R. P. Ingel, *J. Am. Ceram. Soc.,* **64** [12] C-175-C-177 (1981).
[10] Advances in Ceramics, Vol. 3. Edited by A. H. Heuer and L. W. Hobbs. The American Ceramic Society, Columbus, OH, 1981.
[11] R. Stevens, *Trans. Br. Ceram. Soc.,* **80**, 81 (1981).
[12] N. Claussen, *Z. Werkstofftechn.,* **13**, 138 and 185 (1982).
[13] F. F. Lange, *J. Mater. Sci.,* **17**, 225 (1982).
[14] N. Claussen, *J. Am. Ceram. Soc.,* **61** [1-2] 85-86 (1978).
[15] N. Claussen, J. Steeb, and R. F. Pabst, *Am. Ceram. Soc. Bull.,* **56** [6] 559-62 (1977).
[16] K. Faber; this volume, pp. 293-305.
[17] W. Pompe, H.-A. Bahr, G. Gille, and W. Kreher, *J. Mater. Sci.,* **13**, 2720 (1978).
[18] N. Claussen, R. Cox, and J. S. Wallace, *J. Am. Ceram. Soc.,* **65** [11] C-190-C-191 (1982).
[19] N. Claussen and J. Jahn, *Ber. Dtsch. Keram. Ges.,* **55**, 487 (1978).
[20] M. Matsui, T. Soma, and I. Oda; for abstract see *Am. Ceram. Soc. Bull.,* **60** [3] 382 (1981).
[21] M. Rühle, N. Claussen, and A. H. Heuer, this volume, pp. 352-70.
[22] M. V. Swain, *J. Mater. Sci.,* **15**, 1577 (1980).
[23] F. F. Lange, *J. Am. Ceram. Soc.,* **63**, 38 (1980).
[24] A. N. Pilyankevich and N. Claussen, *Mater. Res. Bull.,* **13**, 413 (1978).
[25] W. K. Kriven, p. 1507 in Solid-Solid Phase Transformation. Edited by H. I. Aaronson, R. F. Sekerka, D. E. Laughlin, and C. M. Wayman. AIME, Warrendale, PA, 1982.
[26] R. C. Garvie, R. H. J. Hannink, and C. Urbani, *Ceramurgia Int.,* **6** [1] 19-24 (1980).
[27] R. H. J. Hannink, R. T. Pascoe, K. A. Johnson, and R. C. Garvie; pp. 116-36 in Advances in Ceramics, Vol. 3. Edited by A. H. Heuer and L. W. Hobbs. The American Ceramic Society, Columbus, OH, 1981.
[28] A. H. Heuer; pp. 98-115 in Advances in Ceramics, Vol. 3. (1981).
[29] R. R. Neurgaonkar, T. P. O'Holleran, and R. Roy; for abstract see *Am. Ceram. Soc. Bull.,* **56** [3] 289 (1977).
[30] N. Claussen, G. Lindemann, and G. Petzow, Proc. 5th CIMTEC, Lignano, Italy, 1982 and Ceram. Intern., **9**, 83 (1983).
[31] R. W. Trischuk, J. J. Scott, and J. E. Patchett, paper presented at ZIRCONIA '83, Stuttgart, 1983.
[32] H. Schubert, N. Claussen, and M. Rühle, this volume; pp. 766-73.
[33] M. Rühle; unpublished research.
[34] K. Kobayashi and T. Masaki, *J. Jpn. Ceram. Soc.,* **17**, 427 (1982).
[35] K. Nakajima, K. Kobayashi, and M. Murata, this volume; pp. 399-407.
[36] M. Yoshimura and S. Sōmiya, *Am. Ceram. Soc. Bull.,* **59** [2] 246 (1980).
[37] M. V. Swain, R. C. Garvie, and R. H. J. Hannink, *J. Am. Ceram. Soc.,* **66** [5] 358-62 (1983).
[38] R. H. J. Hannink and M. V. Swain, *J. Austr. Ceram. Soc.,* **18**, 53 (1982).
[39] R. H. J. Hannink and R. C. Garvie, *J. Mater. Sci.,* **17**, 2637 (1982).
[40] M. V. Swain and R. H. J. Hannink; this volume, pp. 225-39.
[41] R. P. Ingel; Ph. D. Thesis, Catholic University of Washington, 1982.
[42] R. P. Ingel, D. Lewis, B. A. Bender, and R. W. Rice, *J. Am. Ceram. Soc.,* **65** [9] C-150-C-152 (1982).
[43] R. P. Ingel, D. Lewis, B. A. Bender, and R. W. Rice, this volume, pp. 408-14.
[44] N. Claussen, *Ber. Dtsch. Keram. Ges.,* **54**, 420 (1977).
[45] N. Claussen and J. Jahn, *J. Am. Ceram. Soc.,* **61** [1-2] 94-95 (1978).
[46] R. W. Rice, Ceramic and Engineering Science Proceedings, p. 493 (July-August 1981).
[47] S. Hori, M. Yoshimura, S. Sōmiya, and R. Takahashi; this volume, pp. 794-805.
[48] R. C. Garvie, this volume, pp. 465-79.
[49] H. Kunz, R. Johannsen, and N. Claussen, *Z. f. wirtsch. Fertigg. (ZwF),* **78**, 529 (1983).
[50] M. Rühle and B. Kraus, p. 533 in Electron Microscopy, Vol. 10. Deut. Ges. f. Elektronenm., Frankfurt, 1982.
[51] M. Rühle, E. Bischoff, and N. Claussen, p. 1563 in Solid-Solid Phase Transformation. Edited by H. I. Aaronson, R. E. Sekerka, D. E. Laughlin, and C. M. Wayman. AIME, Warrendale, PA, 1982.

[52] A. H. Heuer, N. Claussen, W. M. Kriven, and M. Rühle, *J. Am. Ceram. Soc.*, **65** [12] 642–50 (1982).
[53] P. F. Becher, G. C. Culbertson, B. A. Mac Farlane, and C. Cm. Wu; for abstract see *Am. Ceram. Soc. Bull.*, **59** [3] 357 (1980).
[54] E. Di Rupo, E. Gilbert, T. G. Carruthers, and R. J. Brook, *J. Mater. Sci.*, **14**, 193 (1979).
[55] N. Claussen and J. Jahn, *J. Am. Ceram. Soc.*, **63** [3–4] 228–29 (1980).
[56] Sh. Yangyun and R. C. Brook, *Ceram. Int.*, **9**, 720 (1983).
[57] J. S. Wallace, M. Rühle, G. Petzow, and N. Claussen, pp. 155 in Ceramics and Ceramic-Metal Interfaces. Edited by T. Pask and A. G. Evans. Plenum, New York, 1981.
[58] J. S. Wallace; Ph. D. Thesis, Stuttgart, 1983.
[59] R. Mc Pherson, B. V. Shater, and Mi Ming Wong, *J. Am. Ceram. Soc.*, **65** [4] C-57–C-58 (1982).
[60] N. Claussen and J. Steeb, *J. Am. Ceram. Soc.*, **59** [9–10] 457–58 (1976).
[61] M. R. Jackson, J. L. Walter, F. D. Lemkey, and R. W. Hertzberg (eds.). Conference of In-Situ Composites-II, Xerox Publications, Lexington, MA, 1976.
[62] M. Rühle and W. M. Kriven, *Ber. Bunsenges. Phys. Chem.*, **87**, 222 (1983).
[63] J. S. Wallace, G. Petzow, and N. Claussen, this volume, pp. 436–42.
[64] M. V. Swain, R. C. Garvie, R. H. J. Hannink, R. Hughan, and M. Marmach, *Proc. Br. Ceram. Soc.*, **32**, 343 (1982).
[65] M. V. Swain and R. H. J. Hannink, Proceedings of an International Conference on Fracture V, Cannes, 1981.
[66] K. Haberko and R. Pampuch, presented at 5th CIMTEC, Lignano, Italy, 1982.
[67] M. V. Swain, *J. Mater. Sci.*, **15**, 1577 (1980).
[68] N. Claussen and M. Rühle; pp. 137–63 in Advances in Ceramics, Vol. 3. Edited by A. H. Heuer and L. W. Hobbs. The American Society, Columbus, OH, 1981.
[69] T. Kosmac, R. Wagner, and N. Claussen, *J. Am. Ceram. Soc.*, **64** [4] C-72–C-73 (1981).
[70] R. T. Pascoe and R. C. Garvie, p. 774 in Ceramic Microstructures '76'. Edited by R. Fulrath and J. Pask. Westview Press, Boulder, CO, 1977.
[71] R. Wagner; Ph. D. Thesis, University of Stuttgart, 1980.
[72] T. Kosmac, J. S. Wallace, and N. Claussen, *J. Am. Ceram. Soc.*, **65** [5] C-66–C-67 (1982).
[73] N. Claussen and D. P. H. Hasselman, p. 381 in Thermal Stresses in Severe Environments. Edited by D. P. H. Hasselman and R. H. Heller. Plenum, New York, 1980.
[74] M. V. Swain; pp. 355–69 in Fracture Mechanics of Ceramics, Vol. 6. Edited by R. C. Bradt, A. G. Evans, D. P. H. Hasselman, and F. F. Lange. Plenum, New York, 1982.
[75] S. Prochazka, J. S. Wallace, and N. Claussen, *J. Am. Ceram. Soc.*, **66** [8] C-125–C-127 (1983).
[76] K. Weisskopf and N. Claussen; unpublished research.
[77] P. F. Becher and V. J. Tennery, pp. 383–99 in Fracture Mechanics of Ceramics, Vol. 6. Edited by R. C. Bradt, A. G. Evans, D. P. H. Hasselman, and F. F. Lange. Plenum, New York, 1982.
[78] M. V. Swain and N. Claussen, *J. Am. Ceram. Soc.*, **66** [2] C-27–C-29 (1983).
[79] F. F. Lange and Davis; this volume, pp. 382–90.
[80] K. Tsukuma, Y. Kubota, and T. Tsukidate; this volume, pp. 382–90.
[81] H. Schubert, N. Claussen, and M. Rühle, paper presented at the Annual Meeting of the British Ceramic Society, London, Dec. 17, 1983.
[82] K. Tsukuma; personal communication.
[83] N. Claussen, J. S. Wallace, and M. Rühle; for abstract see *Am. Ceram. Soc. Bull.*, **63** [3] 453 (1984).
[84] J. J. Cleveland and R. C. Bradt; for abstract see *Am. Ceram. Soc. Bull.*, **55** [4] 396 (1976).
[85] K. Weisskopf; unpublished Ph. D. work results.
[86] J. Mukerji, P. Greil, and G. Petzow. *Sci. Sintering*, **15**, 43 (1983).
[87] N. Yamamoto; personal communication.
[88] S. Kanzaki, Z. Nakagawa, K. Hamano, and K. Saito, *Yogyo-Kyokai-Shi*, **88**, 59 (1980).
[89] W. T. Donlon, T. E. Mitchell, and A. H. Heuer, *J. Mater. Sci.*, **17**, 1389 (1982).
[90] D. M. Roy, R. Roy, and E. F. Osborn, *J. Am. Ceram. Soc.*, **36** [6] 185–90 (1953).
[91] N. Claussen, F. Sigulinski, and M. Rühle; pp. 164–67 in Advances in Ceramics, Vol. 3. Edited by A. H. Heuer and L. W. Hobbs. The American Ceramic Society, Columbus, OH, 1981.
[92] M. Shimada, K. Matsushita, S. Kuratani, T. Okamoto, K. Tsukuma, and T. Tsukidate, *J. Am. Ceram. Soc.*, **67** [2] C-23–C-24 (1984).

Microstructural Studies of Y_2O_3-Containing Tetragonal ZrO_2 Polycrystals (Y-TZP)

M. Rühle and N. Claussen

Max-Planck-Institut für Metallforschung
Institut für Werkstoffwissenschaften
Stuttgart, Federal Republic of Germany

A. H. Heuer

Case Western Reserve University
Department of Metallurgy and Materials Science
Cleveland, OH 44106

Strong, tough, Y_2O_3-containing 100% tetragonal ZrO_2 fine-grained polycrystals (Y-TZP) appear to have an unusually good combination of mechanical properties. We have now studied the microstructures of 11 different Y-TZPs containing 3–6 wt% Y_2O_3, most of which are commercially available. Although tetragonal (t)-ZrO_2 is the dominant phase in each material, cubic (c)-ZrO_2 and α-Al_2O_3 are present in some samples. All samples contain a continuous grain-boundary yttrium silicate phase which also usually contains Al_2O_3. The size distribution and volume fraction of each component of the microstructure, as well as variation in the local chemistry of the crystalline phases, have been determined by analytical electron microscopy, including EDS analysis.

Recently, Y_2O_3-containing, fine-grained tetragonal (t)-ZrO_2 polycrystals (Y-TZP) have become of interest for advanced structural applications. These ceramics are claimed to consist of t-ZrO_2 grains containing up to 6 wt% Y_2O_3 in solid solution. Such Y-TZP ceramics possess quite impressive mechanical properties — average flexural strengths as high as 1500 MPa and fracture toughness up to ≈ 10 MPa·m$^{1/2}$ are reported in the literature, see e.g., Claussen.[1] No microstructural analyses of these materials have been reported to date. In this paper TEM observations are summarized on the microstructures of 10 Y-TZP ceramics obtained from commercial sources, and 1 Y-TZP ceramic produced in our laboratory. The latter is henceforth referred to as material 1,[2] whereas the several commercial materials are identified as materials 2–11 in the following. All commercial materials were obtained during 1982–1983.

Experimental Details

Only a few of the manufacturers supplied information concerning the processing routes; all information available to us is included in Table I. The densities of all 11 materials were determined by the Archimedes technique. For those specimens available in large enough sizes, both the four-point bend strength and fracture toughness (measured using the indentation strength technique[3]) were determined.

As can be seen in Table I, strengths up to 1200 MPa were obtained for three materials, two of which (materials 2 and 7) had been hot-pressed and the third

Table I. Properties of Polycrystals

Material No.	Firing temp. (°C)	Density (g/cm³)	σ (MPa)	K_{Ic} MPa·m$^{1/2}$	Y_2O_3 X ray (wt%)	t-ZrO$_2$ Fraction (%)	t-ZrO$_2$ Grain size $\bar{d}$ (μm)	t-ZrO$_2$ Composition wt% Y_2O_3
1	S* 1550	5.9	890	9	3.5	≈100	0.8	3.6 (0.4)
2	HP 1400	6.15	1150	6	4.1	100	0.5	4.2 (0.4)
3	S 1600	6.01	800	11	4.0	≈85	0.9	3.5 (0.3)
4	S	6.00			4.0	70	2.0	3.7 (0.4)
5	S	5.82	910	11	5.4	63	0.2	3.3..4.2
6	S	5.99			8.7	60	0.7	4.6 (0.4)
7	HP	6.03	1180	6	6.2	58	0.5	3.2..4.5
8	S	5.02	870	10	6.2		0.5	4.0..5.4
9	S 1550	5.9	1200	5.5	5.9	80	0.4	4.4 (0.3)
10	S 1400	6.02	940§	9.5	5.5		0.4	1.8..4.2
11	S 1420	5.95			5.8	?	≈0.5	3.0..4.1

Table I. Properties of Polycrystals (continued)

Material No.	c-ZrO$_2$ Fraction (%)	c-ZrO$_2$ Grain size (μm)	c-ZrO$_2$ Composition wt% Y_2O_3	Transformability†	Al_2O_3 grains‡	Amorphous grain-boundary phase (nm)
1				N	+	≤10
2				D	+	≤4
3	≈15	3	9.3 (0.5)	E	+	≤7
4	≈30	5	9.0 (0.6)	VE	+	≤5
5	≈37	2.5	11.5 (0.6)	E	+	≤2
6	≈40	3	12.0 (0.4)	E	−	≤100
7	≈42	2.5	9.7 (0.5)	N	+	≤4
8		2.2	10.2 (0.5)	N	+	≤10
9	≈20	3	10.2 (0.4)	VD	+	≤14
10		0.8...1	8.0..9.6	VVE	−	≤2
11		0.6...2	6.7..9.4	D	+	≤10

*S: Sintered, HP: Hot-pressed.
†Transformability: VVE: very very easy; VE: very easy; E: easy; N: normal; D: difficult; VD: very difficult.
‡+, −: present, absent.
§These data were communicated by the manufacturer using the same techniques employed by us.

(material 9) sintered; however, the toughness of these materials was rather low, 5.5–6 MPa·m$^{1/2}$. Five sintered materials (1, 3, 5, 8, and 10) had strengths between 800 and 950 MPa and possessed fracture toughnesses between 9 and 11 MPa·m$^{1/2}$.

The overall chemical composition was analyzed by both calibrated X-ray fluorescence measurements and wet-chemical analysis. The Y_2O_3 contents varied between 3.5 and 8.7 wt% (Table I).

Specimens suitable for TEM were prepared in the conventional way, 3 mm diameter disks being carefully polished to a thickness of 40 to 50 μm prior to ion-thinning. The microstructural investigations were performed with a

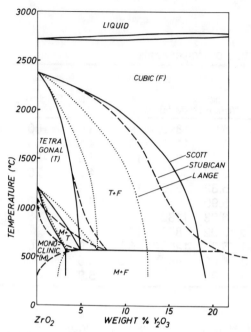

Fig. 1. Phase diagram in the ZrO_2-rich part of Y_2O_3 (Refs. 4–6).

JEOL JEM 200CX electron microscope fitted with a side-entry double-tilting stage and possessing microanalytical capability.

Three phase diagrams of the ZrO_2-rich part of the Y_2O_3-ZrO_2 phase exist in the literature[4-6] and are summarized in Fig. 1. Although the diagrams show qualitatively similar features, they differ on a quantitative basis. Nevertheless, they all predict that the materials investigated are either single-phase t-ZrO_2 or lie in the two-phase cubic (c) plus (t) field of the ZrO_2-Y_2O_3 diagram. Therefore, it was desirable to determine the local composition of the different materials, particularly when duplex microstructures were observed. Fortunately, a commercial EDS system (Tracor 2000) was linked to the microscope, and the local chemical composition could be determined with high spatial resolution ($\approx$20 nm) for all elements with $Z \geq 12$. For quantitative analysis, the EDS analysis utilized pure ZrO_2 and pure Y_2O_3 standards to facilitate data deconvolution and used theoretical values of k_{AB} for converting intensity ratios into composition ratios.[7] The accuracy of determination of the Y_2O_3 content of the Y-TZP ceramics was governed by statistical errors, since all systematic errors were excluded. The specimen drift was $\leq$0.1 nm/s, which results in a shift of the specimen $\leq$100 nm for counting times of 1000 s. When the chemical distribution within small grains was determined, we took care that the drift was very much smaller than 100 nm.

The buildup of a contamination spot during the analysis was mainly governed by surface contamination of the specimen, introduced during ion-thinning. The contamination could be drastically reduced if a liquid N_2 cooling trap was used in the ion thinner during final preparation of the electron-transparent TEM foils.

Microstructure

Low-magnification TEM study showed that materials 1 and 2 were composed only of t-ZrO_2 grains (Fig. 2) (and some minor components which will be dis-

Fig. 2. Typical microstructure of the 100% material t-ZrO_2. Grain size is in the range 0.8 μm (material 2).

cussed later). The other materials were duplex and contained small t-ZrO_2 grains and larger c-ZrO_2 grains. The size of the c-ZrO_2 grains was typically a factor of 3 to 10 larger than the t-ZrO_2 grains (Fig. 3). The mean grain sizes of all materials are listed in Table I.

t-ZrO_2 Grains

The t-ZrO_2 grains possess mean diameters between 0.2 and 2 μm (the diameter d is taken as $A^{1/2}$, where A is the projected area of the grain (Table I)). In some materials the grains showed pronounced faceting (Fig. 4(a)), while in others, the grains were rounded (Fig. 4(b)) and had large triple-point pockets between smooth grain faces. We speculate that rounding of the tetragonal grains occurred during liquid-phase sintering, but the presence of prominantly faceted grains in some materials that also contained a grain-boundary phase, and thus had presumably also densified by liquid-phase sintering, is not understood. The tetragonal symmetry of individual grains could be confirmed by electron diffraction (Fig. 5).

The mean grain size, $\bar{d}$, as well as the width of the distribution, $\overline{\Delta d}$, varied by a factor of 2 to 3 for materials 5, 7, and 11 for TEM specimen to TEM specimen and spatially within a single specimen. The microstructure was more uniform for the other materials.

The Y_2O_3 content of the t-ZrO_2 grains, and its spatial distribution within individual grains, differed from material to material. The t-ZrO_2 grains in materials 1, 2, 3, 4, 6, and 9 contained the same mean Y_2O_3 content in all grains. However, the sizable standard deviation may have resulted from real spatial variation of the composition within individual grains. This variation was certainly demonstrated for material 1 — within a single t-ZrO_2 grain, the Y_2O_3 content varied from the boundaries (3.7 ± 0.2 wt%) to the center of the grains (3.2 ± 0.2 wt%).

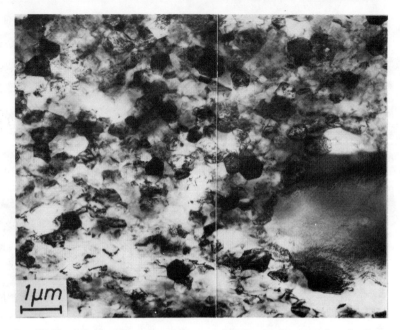

Fig. 3. Typical microstructure of material 9 which contains both t- and c-ZrO_2 (mean grain sizes are 0.4 and 3 μm, respectively). A large cubic grain (c) can be seen in the lower right.

In materials 5, 7, 8, 10, and 11, chemical equilibrium had not been reached during sintering. The concentration varied in the range noted in Table I, the accuracy of the extreme values being ±0.3 wt%. The variation is most pronounced in material 10, where the Y_2O_3 concentration differed by up to 2.4 wt% within one TEM specimen.

The t-ZrO_2 grains of some specimens (materials 2, 7, 9, and 11) frequently contained a high density of lattice defects, either as single dislocations or as arrays of dislocations (small-angle grain boundaries), as shown in Fig. 6.

c-ZrO₂ Grains

As already noted, the c grains were invariably larger than the t grains, although their number density was smaller. The c grains were not single phase, however, but themselves contained small t-ZrO_2 precipitates. A bright-field image of such a cubic grain is shown in Fig. 7, along with the corresponding diffraction pattern oriented to a [111] zone axis (Fig. 7(b)). Reflections unique to t-ZrO_2 (those with *odd, odd, even* indices) are visible on the diffraction pattern, and dark-field images taken with three different unique t-ZrO_2 reflections (Figs. 7(c)–(e)) show that the t-ZrO_2 precipitates are ≈10 nm in diameter. As discussed below, we believe that the t-ZrO_2 present in these large c-ZrO_2 grains must have precipitated during cooling following sintering.

Except for materials 10 and 11, the Y_2O_3 content of c-ZrO_2 grains varied between 9.3 and 12 wt%. The lower Y_2O_3 content of the c-ZrO_2 grains of materials 10 and 11 is further evidence that equilibrium had not been reached at the end of the sintering.

Fig. 4. Microstructure of Y-TZPs showing faceted (a) or rounded (b) grains. Bright-field electron micrographs.

Determination of Phase Equilibria

From the chemical analyses of the coexisting t-ZrO_2 and c-ZrO_2 grains, portions of the high-ZrO_2 end of the ZrO_2-Y_2O_3 phase diagram can be constructed (Fig. 8). Due to the lack of equilibration just cited, Fig. 8 includes only those data points for Y-TZP materials which were sintered and/or annealed for at least 6 h, and also contains information on other propriatory materials not discussed here for which the processing conditions are known; data obtained by Lanteri et al.[8] on single crystals annealed for up to 100 h are also included. The solvi separating the two-phase t-ZrO_2 plus c-ZrO_2 phase field from the single-phase t-ZrO_2 and c-ZrO_2 solid solution regions are believed accurate to ±0.3%. (Note the error bars of each datum in Fig. 8.) The volume fraction of t-ZrO_2 and c-ZrO_2 estimated from TEM (see Table I) agrees with the lever rule predictions of Fig. 8 for materials fired in the two-phase region.

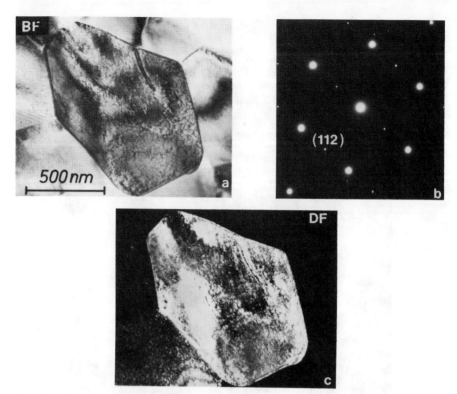

Fig. 5. Transmissions electron microscopy diffraction study of individual tetragonal grain. (a) Bright-field image; (b) selected area diffraction, [111] zone axis. The tetragonal reflection of a single variant is observed (*odd, odd, even* reflections are forbidden for c-ZrO_2); (c) dark-field image taken with the t-ZrO_2 reflection indicated (material 9).

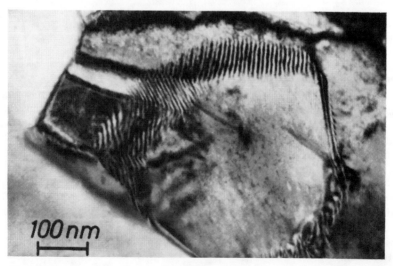

Fig. 6. Defects (small-angle grain boundary) in Y-TZP ceramic (material 9).

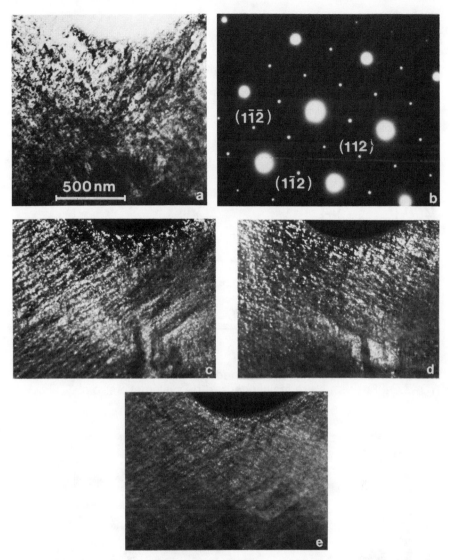

Fig. 7. TEM studies of a cubic grain (material 9). (a) Bright-field image; (b) diffraction pattern, [111] zone axis. On the diffraction pattern tetragonal reflections of all three variants are visible. (c) Dark-field image taken with variant 1, $g = (1\bar{1}2)$; (d) dark-field image taken with the tetragonal reflection of variant 2, $g = (1\bar{1}2)$; (e) dark-field image taken with the tetragonal reflection of variant 3, $g = (112)$.

Al_2O_3 Grains

Most materials contain grains which were unusually transparent to 200 kV electrons, one example being shown in Fig. 9. Evaluation of electron diffraction information, supplemented by microchemical analysis, showed that these grains were α-Al_2O_3. Up to 2 wt% Al_2O_3 was present in some materials; in fact, only two materials (6 and 10) were Al_2O_3-free.

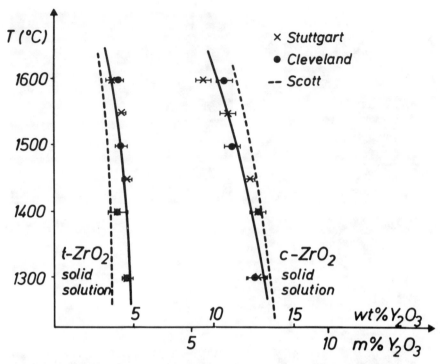

Fig. 8. Experimentally determined part of the phase diagram in Y_2O_3-ZrO_2. Error bars are mainly caused by systematic errors.

Fig. 9. Al_2O_3 grain embedded in TZP materials. Up to 2 wt% Al_2O_3 is present in some materials (material 9).

Amorphous Grain-Boundary Phase

One important result of the present study was that *all* grain boundaries of all 11 materials contained a continuous amorphous grain-boundary phase, whose width varied from material to material. The maximum thicknesses are noted in Table I. In some materials, sizable pockets of this glassy phase were observed at triple points. Figure 10 shows one example (material 2) where the maximum width of the grain-boundary phase is 5 nm, while Fig. 11 presents micrographs of material 6, which contains more of the glassy phase.

Analytical electron microscopy showed that the grain-boundary phase is rich in Y_2O_3 and SiO_2 in all materials, and contains Al_2O_3 in all materials except 6 and 10. Some ZrO_2 may also be present in this glassy phase but the ubiquitous ZrO_2 X rays present in EDS spectra could have arisen from matrix grains.

Transformation of t-ZrO_2

All t-ZrO_2 grains are metastable with respect to transformation to m symmetry, although a stress of some sort is required to nucleate the transformation (see below). Thus the relative "transformability" of various materials was of interest. Individual t-ZrO_2 grains could be induced to transform during examination in the 200-kV electron microscope or by *in situ* straining experiments in the high-voltage electron microscope. Both types of experiments are reported in this section.

Transformation in the Electron Beam

The transformability of t-ZrO_2 grains in the various materials was tested by simply exposing the specimen to a very intense electron beam. Foils are heated

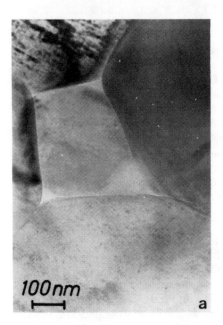 

Fig. 10. Amorphous grain-boundary phase in TZP (material 2). (a) Shows a bright-field image, in which the grains and pockets are clearly visible; (b) is a dark-field image taken with a part of the diffuse scattering caused by the amorphous ring. The amorphous areas show bright contrast.

Fig. 11. Amorphous grain-boundary phase (material 6). (a) Bright-field image; (b) dark-field image taken with the region of an amorphous ring. As in Fig. 10, the bright areas correspond to an amorphous region.

during such irradiation, and undergo buckling, resulting in stress-induced nucleation of the martensitic transformation. After a certain exposure time (during which we assume a minimum strain was reached), one grain in the specimen started to transform (Fig. 12); the m product is invariably twinned.

Table I includes information on the relative transformability of the several materials studied by this simple test, the transformability varying from "very, very easy" to "very difficult." The transformability depended on the grain size as well as on the Y_2O_3 content. Larger t-ZrO_2 grains of comparable Y_2O_3 content transformed more easily than smaller t-ZrO_2 grains, whereas for a given grain size a higher concentration of Y_2O_3 stabilized the t-ZrO_2 grains against transformation, i.e., it lowered the chemical driving force of the $t \rightarrow m$ transformation at low temperatures.

It was observed[9] that the nucleation of the transformation is inhomogeneous, in that the transformation starts (i) either in regions of high stress concentrations (close to the grain boundary but still within the grain) or (ii) at defects present within the grains, typically subgrain boundaries.

The transformation strain quite frequently caused neighboring grains to transform (an autocatalytic effect) (Fig. 13). On occasion, the course of transformation within a single grain could be followed; Fig. 14 shows an example. The velocity of the transformation front was surprisingly slow in materials with high Y_2O_3 content, and is a further indication of the effectiveness of Y_2O_3 in stabilizing t-ZrO_2. In those materials with variable Y_2O_3 contents (materials 5, 7, 8, 11, and especially 10), the transformation generally started at grains with the lowest Y_2O_3 content.

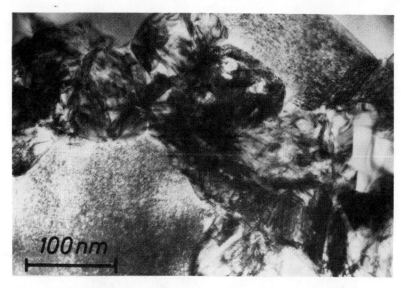

Fig. 12. t-ZrO_2 transformed under the electron beam. Typical twins can be observed.

In Situ Straining Experiments

In situ straining experiments[10] were performed on materials 2 and 3. Material 3 developed a continuous transformation zone in the wake of the crack, whereas "clusters" of t-ZrO_2 grains transformed adjacent the propagated crack in material 2.[10] Further experiments are clearly required to understand the mechanics of crack-tip stress-induced transformation in TZP.

Microcracking

In marked contrast to other transformation-toughened ZrO_2-containing ceramics, microcracks were present in as-fired specimens, even prior to any transformation (Fig. 15). The microcracks are probably due to the residual stresses present in Y-TZP due to thermal expansion anisotropy, and were frequently observed at those grain boundaries with the thinnest grain-boundary phase.

The materials were heavily microcracked after the martensitic transformation, and radial as well as tangential microcracks could be observed (Fig. 16). The fraction of microcracked grain boundaries adjacent transformed grains was ≈ 0.4.

Low-Temperature Degradation of Y-TZP

It is now well established[1,11–13] that Y-TZP ceramics degrade during long-time, low-temperature ($\approx 250°C$) annealing. The degradation can be accelerated by annealing the specimen at this temperature for 1 h in an autoclave containing high-pressure water at 2 Pa. Figure 17 shows the extent of this degradation in materials 2, 3, 5, and 7. A heavily cracked and transformed layer of m-ZrO_2 is formed near the surface after this autoclave test, whose extent varied from material to material. We suspect that high Y_2O_3 content and fine grain size stabilize the t-ZrO_2 against degradation, and that the width and composition of the amorphous grain-boundary phase may also be important variables. Interestingly, material 10 showed nearly no transformed layer after the 1-h autoclave test.

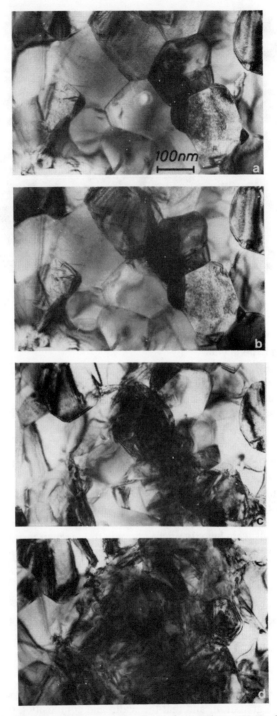

Fig. 13. Autocatalytic effects. (a) Transformation starts in one grain; (b)–(d) the transformation continues in neighboring grains during further electron irradiation (material 3).

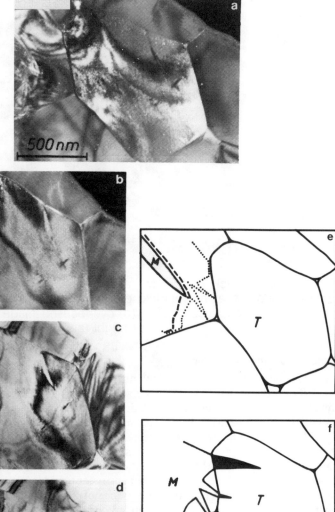
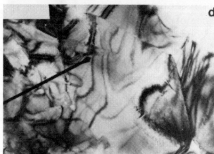
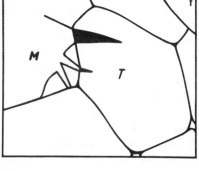

Fig. 14. Partially transformed t-ZrO_2 grain. (a) The transformation starts in the upper left grain, in which "martensite" (m-ZrO_2) needles are visible; (b) during further exposure, the transformation of the grain continues slowly; (c) the transformation reaches the grain boundary and causes elastic loading of the central grain; (d) the transformation starts in the grain indicated by the pointer; (e) and (f) schematic representations of the processes shown in Figs. (a)–(d).

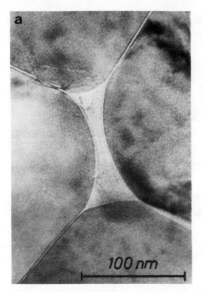 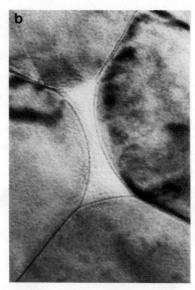

Fig. 15. Microcracking prior to transformation. In materials with a thin grain-boundary layer, microcracking is often observed. (a) Regular bright-field image (underfocused mode); (b) overfocused image. The amorphous grain-boundary phase is visible as bright (dark) lines.

Discussion and Conclusion

The range of microstructural and microchemical features present in this group of Y-TZP ceramics permits an understanding of microstructural evolution and resultant mechanical property correlation that would not have been possible had only one or two materials been available for detailed characterization.

The common microstructural feature for all materials was a continuous (i.e., perfectly wetting) grain-boundary phase containing Y_2O_3, SiO_2, possibly ZrO_2, and in most materials Al_2O_3. A eutectic exists between 1300° and 1400°C in the system Al_2O_3-SiO_2-Y_2O_3,[14] and the propensity for liquid formation during sintering will certainly be augmented by the presence of ZrO_2 as a fourth constituent. All materials are thus assumed to have undergone densification via liquid-phase sintering. The low temperature at which fully dense material can be realized, and the fine grain size of the t-ZrO_2, suggest that the liquid involved in this densification forms readily, cannot have a very high viscosity, and provides a classical medium for easy transport and rapid approach to chemical homogenization, while at the same time limiting grain-coarsening.

Silicon dioxide is a common impurity in ZrO_2 powders derived from $ZrSiO_4$, but may have been added intentionally in some materials. Likewise, Al_2O_3 may have been added intentionally (the presence of Al_2O_3 in Y-PSZ has already been reported.[15]) In some materials, however, we suspect that the Al_2O_3 and SiO_2 were present as wear debris from high (85–95 wt%) Al_2O_3 grinding balls, which are commonly used for particle and agglomerate comminution. We further surmise that those materials free of Al_2O_3 must have been processed using ZrO_2-based grinding media or via colloidal techniques that avoided ball-milling completely.

The extent of homogeneity was quite variable. Material 9 was the most homogeneous, both with regard to Y_2O_3 distribution within the t-ZrO_2 grains and

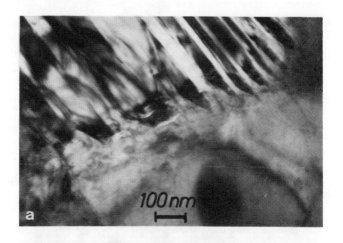

Fig. 16. Microcrack adjacent a transformed particle. (a) Tangential cracking; (b) radial cracking. The microcracks pass through the amorphous grain-boundary film.

to the width of the grain-size distribution. This may point to chemical homogenization of the starting powders via coprecipitation, for example. Material 1 (produced in our laboratories) was sintered from coprecipitated powders, but nevertheless showed gradients in Y_2O_3 content from the center to the edge of individual t-ZrO_2 grains, clearly a result of slower solid state interdiffusion in the system ZrO_2-Y_2O_3 than diffusion of these species in the silicate grain-boundary phase. The more dramatic chemical inhomogenity of materials 5, 7, 8, and 10 points to systems far from equilibrium at the end of sintering, and is thought to result from the use of separate sources of ZrO_2 and Y_2O_3 in the green body, which had not been adequately homogenized prior to and during sintering. Material 10 was unique in that it was Al_2O_3-free, but nevertheless had a very thin grain-boundary phase and was most inhomogeneous in regard to the Y_2O_3 distribution within the t-ZrO_2 (and c-ZrO_2) grains. Material 11 was also unique in that the starting powder is known to have been produced by quenching of a liquid ZrO_2-Y_2O_3 alloy followed by particle comminution. The resulting variation of

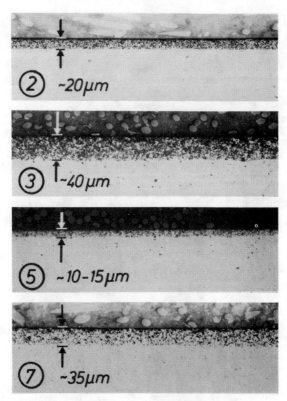

Fig. 17. Degradation of Y-TZP. Materials 2, 3, 4, and 7 were tested in an autoclave at 250°C for 1 h at 2 Pa under high-pressure water. A cross section of the different materials is shown, each containing a transformed near-surface layer.

Y_2O_3 content from grain to grain in both the t-ZrO_2 and c-ZrO_2 phases must reflect solute segregation during crystallization and rapid cooling of the molten material, with inadequate chemical homogenization during further processing.

The solvi defining the two-phase t-ZrO_2 plus c-ZrO_2 field in the ZrO_2-Y_2O_3 binary (Fig. 8) are close to those determined by Scott,[4] but the two-phase field is somewhat narrower than he determined. It is clear that only two materials, 1 and 2, were actually sintered in the single-phase t-ZrO_2 portion of the phase diagram; for the other materials which showed reasonable chemical homogenity for both t- and c-ZrO_2 (3, 4, 6, and 9), the compositions of the coexisting phases are determined by the sintering temperature and the solvi shown in Fig. 8 (the lack of equilibrium of materials 5, 7, 8, 10, and 11 has already been alluded to).

For the materials at equilibrium, the volume fraction of t- and c-ZrO_2 is determined by the starting composition, the sintering temperature, and the application of the lever rule to the phase diagram of Fig. 8. The fact that the c-ZrO_2 phase invariably showed a larger grain size than the t-ZrO_2 matrix is attributed to the difficulty of nucleating this phase during sintering. We believe that the material transport for the growth of c-ZrO_2 occurs more rapidly in the grain-boundary phase than does nucleation of new c-ZrO_2 grains, so that the few grains with c symmetry that did form tended to grow until the system achieved its equilibrium phase content. Note that this process can occur and yield chemically homogeneous c-ZrO_2 grains even when the t-ZrO_2 matrix is chemically inhomogeneous, as in materials 5, 7, and 8. Since we view the grain-boundary phase as a potent medium for chemical homogenization during sintering, the nonuniformity and low Y_2O_3 content of the c-ZrO_2 grains in materials 10 and 11 is surprising and may reflect the absence of Al_2O_3 in the glassy phase (material 10), and/or liquid-phase origin of the ZrO_2 particles (material 11).

As already noted, all c-ZrO_2 grains contained small t-ZrO_2 precipitates, which must have formed during cooling from the sintering temperature. In accordance with our view that the c-ZrO_2 grains grew from a silicate liquid phase, no other defects were present.

The t-ZrO_2 grains, on the other hand, frequently contained isolated dislocations and low-angle grain boundaries. This was especially pronounced for material 11, which must have undergone extensive plastic deformation during quenching of the molten alloy. The low sintering temperature was inadequate to permit annealing of these defects. In addition, strain contours were present in all materials, and are due to residual strains, almost certainly caused by the anisotropic thermal contraction of t-ZrO_2, which is also the cause of the observed microcracking.

Consider next the correlation between mechanical properties and microstructure. There is only imperfect correlation between strength and t-ZrO_2 grain size — the two coarsest-grained materials of the group whose strength was determined, materials 1 and 3, and the finest grain-sized ceramic, material 5, all had strengths below 900 MPa, whereas the three strongest materials (2, 7, and 9) had a medium grain size and the lowest toughness (5.5–6 $MPa \cdot m^{1/2}$). Materials 1, 3, 5, 8, and 10 all had toughness in the range 9 to 11 $MPa \cdot m^{1/2}$ and strengths between 800 and 1000 MPa, yet the grain size of t-ZrO_2 varied from 0.2 to 0.9 μm. The good combination of strength and toughness of material 10 may be due to the variable Y_2O_3 content of the t-ZrO_2, the low-Y_2O_3 grains being readily transformed and thus contributing to the high K_{Ic}. The *in situ* straining experiments[9] showed that the higher toughness of material 3 compared to material 2 could be correlated with a continuous transformation zone in the former material, combined with crack-branching, crack deflection, and extensive microcracking accompanying transformation; all these features were less marked in material 2.

The low-temperature degradation is not yet understood. Although the data are sparse, we suspect that higher Y_2O_3 content and thinner grain-boundary layers result in retarded degradation in the autoclave test. There are two possible explanations, a classical stress corrosion mechanism starting at the specimen surface, or a surface-nucleated isothermal martensitic transformation at these low temperatures. The autoclave observations show that the presence of water accelerates the degradation, but does not discriminate between these two explanations. Further experiments on specimens annealed at low temperatures under different moisture conditions should permit such discrimination.

Acknowledgments

A. H. Heuer's research on Y-TZP has been supported by DOE under Contract No. DE–AC02–83ER45006. He also acknowledges the receipt of a Humboldt Senior Scientist Award which made possible his stay at the Max-Planck-Institut, where the experimental work reported in this paper was performed. Partial support by the Deutsche Forschungsgemeinschaft (DFG) is acknowledged.

References

[1] N. Claussen, Proc. Europ. Coll. on Ceramics in Advanced Energy Technologies, Petton, Sept. 1982; to be published.
[2] H. Schubert, N. Claussen, and M. Rühle; this volume, pp. 766–73.
[3] M. V. Swain and N. Claussen, *J. Am. Ceram. Soc.*, **66** [2] C-27–C-29 (1983).
[4] G. G. Scott, *J. Mater. Sci.*, **10**, 1527 (1975).
[5] V. S. Stubican, R. C. Hink, and S. P. Ray, *J. Am. Ceram. Soc.*, **61** [1–2] 17–21 (1978).
[6] F. F. Lange; pp. 255–74 in Fracture Mechanics of Ceramics, Vol. 6. Edited by R. C. Bradt, A. G. Evans, D. P. H. Hasselmann, and F. F. Lange. Plenum, New York, 1983.
[7] J. J. Hren, J. I. Goldstein, and D. C. Joy, Introduction to Analytical Electron Microscopy. Plenum, New York, 1979.
[8] V. Lanteri, A. H. Heuer, and T. E. Mitchell; this volume, pp. 118–30.
[9] L. Ma and M. Rühle; unpublished work.
[10] M. Rühle, B. Kraus, A. Strecker, and D. Waidelich; this volume, pp. 256–74.
[11] O. T. Masaki and K. Kubayashi, Proceedings of the Annual Meeting of the Japanese Ceramic Society, 1981.
[12] M. Matsui, T. Soma, and I. Oda; this volume, pp. 371–81.
[13] M. Watanabe, S. Iio, and I. Fukuura; this volume, pp. 391–98.
[14] I. A. Bondar and F. Y. Galakhov, *Izv. Akad. Nauk USSR Ser. Krim.*, **7**, 1325 (1963).
[15] B. V. Narasimha Rao and T. P. Schreiber, *J. Am. Ceram. Soc.*, **65** [3] C-44–C-45 (1982).

Effect of Microstructure on the Strength of Y-TZP Components

M. Matsui, T. Soma, and I. Oda

NGK Insulators Ltd.
Research and Development Laboratory
Mizuho, Nagoya, Japan

The effects of microstructure on the strength and durability were studied for Y-TZP (yttria tetragonal zirconia polycrystals) with various microstructures prepared by both sintering and sintering-precipitation methods. In the case of the sintered-type Y-TZP, a maximum flexural strength of 980 MPa was obtained by selecting the Y_2O_3 content and the sintering temperature. In the case of the precipitated-type Y-TZP, the strength changed with the aging time, and the maximum strength was 400 MPa. When the sintered-type Y-TZP was used, fatigue was proved to be caused by the tetragonal (t)-to-monoclinic (m) phase transformation accompanied by a large volume expansion. The microstructure strongly affected the durability. The fatigue mechanism was discussed in terms of the transformation rate of the unstabilized t phase to the m phase. No degradation was observed in the Y-TZP with finely controlled microstructures.

Recently, many kinds of high-performance ceramics have been developed for high-temperature structural applications. Among these ceramics, the tetragonal zirconia polycrystal (TZP) is one of the most promising materials because of its excellent thermal and mechanical properties. Especially when used in a ceramic-metal composite system for thermal insulation such as adiabatic engine components,[1-3] low thermal conductivity and high thermal expansion are required to meet the insulating purpose and reduce difficulties with ceramic-metal attachment.

The mechanical properties of Y-TZP ceramics depend strongly on its microstructure, and the strength of Y-TZP can be increased up to about 1000 MPa by controlling the microstructure. On the other hand, some Y-TZPs show fatigue phenomena when used at high temperatures. The effect of microstructure on the strength and fatigue behavior in Y-TZP was studied; the fatigue mechanism is discussed in terms of the transformation rate of the tetragonal phase to the monoclinic phase.

Strength and Microstructure

As shown by the phase diagram for the system ZrO_2-Y_2O_3[4] (Fig. 1), Y-TZP can be prepared by two methods. One method is to sinter in the region where the tetragonal phase is stable, as shown by A, B, C, D, E, and F in Fig. 1. Y-TZP can be formed by sintering at high temperatures where a substantial portion is the cubic phase, and then aging at a lower temperature to precipitate more tetragonal phase in cubic grains, as shown by P in Fig. 1. Both Y-PSZ and Y-TZP were prepared, and their strengths were related to their microstructures.

Y-TZP

Mixed powders of ZrO_2 and Y_2O_3 were formed into disks (50 by 60 by 7 mm^3) followed by isostatic pressing at 2000 kg/cm^2. The compacts were sintered in air

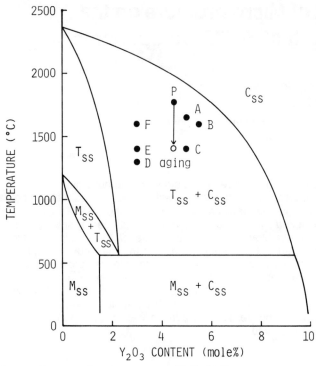

Fig. 1. Phase diagram for the system ZrO_2-Y_2O_3.

for 3 h and cooled in the furnace. The cooling rate was about 3°C/min. The samples (3 by 4 by 40 mm³) for flexural strength were prepared by machining with diamond tools according to standard JIS R 1609. The surface finishing was done with 800-grit diamond wheel. The flexural strength was measured by a 4-point bending fixture with 30-mm outer span and 10-mm inner span, where the crosshead speed of the testing machine was 0.5 mm/min. In Table I, the strengths for the

Table I. Flexural Strength of Various Sintered-Type Y-TZP.

Sample	Y_2O_3 content (mol%)	Fired temp (°C)	Flexural strength (MPa)
A	5	1650	430
B	5.5	1600	400
C	5	1400	650
D	3	1300	700
E	3	1400	920
F	3	1550	740
G	2	1400	500
H	2.5	1400	980
I	4	1400	820
J	6	1400	400
K	3	1200	400
L	3	1360	800
M	3	1450	880
N	3	1500	850

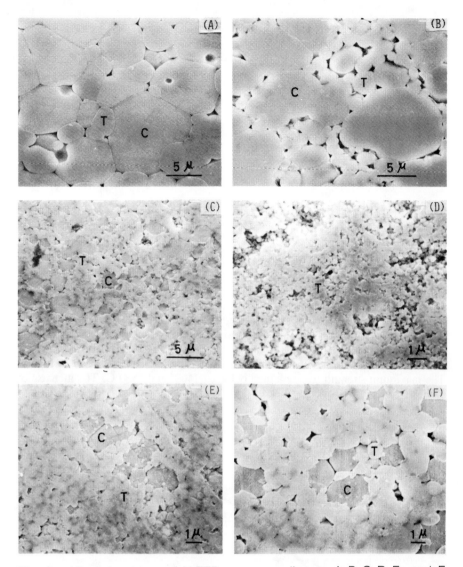

Fig. 2. Microstructures of Y-TZP corresponding to A, B, C, D, E, and F in Fig. 1. Sintering temperatures and Y_2O_3 contents are as follows: (A) 1650°C, 5 mol%; (B) 1600°C, 5.5 mol%; (C) 1400°C, 5 mol%; (D) 1300°C, 3 mol%; (E) 1400°C, 3 mol%; and (F) 1550°C, 3 mol%.

samples are shown along with the Y_2O_3 contents and sintering temperatures. Figure 2 shows typical microstructures for samples A, B, C, D, E, and F that were chemically etched using 46% HF at 20°C. These Y-TZP are composed of large cubic-phase grains and small tetragonal-phase grains. Sizes of the tetragonal grains in samples A, B, C, D, E, and F are approximately 5, 3, 0.5, 0.2, 0.5, and 1.5 μm, respectively. The ratios of cubic- and tetragonal-phase grains agree approximately with those predicted from the phase diagram. Sample E had the highest strength of these samples. The relationship between the Y_2O_3 content and the strength is shown in Fig. 3. The maximum strength of 980 MPa was obtained

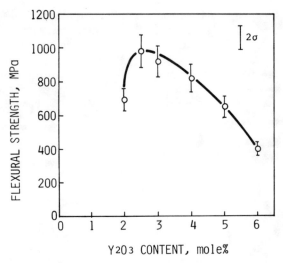

Fig. 3. Effect of Y_2O_3 content on flexural strength.

at 2.5 mol% Y_2O_3. The decrease in strength above 2.5 mol% Y_2O_3 is attributed to the decrease in the amount of the tetragonal phase. The strength degradation phenomenon below 2.5 mol% is attributed to the t-to-m transformation. The relationship between the firing temperature and the strength is shown in Fig. 4. The maximum strength was obtained between 1400° and 1500°C. The increase of

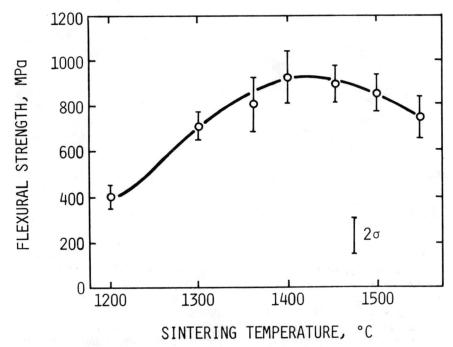

Fig. 4. Effect of sintering temperature on flexural strength.

strength with sintering temperatures up to 1400°C is explained by the increase in the density of the fired body. Above 1500°C, where the density is nearly constant, the decrease in strength is attributed to the growth of t and c grains. The grain sizes of these phases in the sample fired at 1550°C are about 1.5 and 4 μm, respectively. The grain growth may increase the size of the critical Griffith flaw or may cause the spontaneous transformation from t to m. These results indicate that the stabilizer content and the firing temperature must be finely controlled to obtain a high-strength Y-TZP.

Precipitated-Type PSZ

A sample containing 4.5 mol% Y_2O_3 was fired at 1770°C for 3 h and cooled to room temperature to obtain the cubic phase. This sample was aged for 168 h at 1400°C to precipitate the t phase. Figure 5 shows the microstructures before and after the aging. Tetragonal precipitates ($\approx$0.3 μm long) are seen in cubic grains, as shown in Fig. 5(C). A precipitate-free zone is observed near the boundaries of cubic-phase grains, as shown in Fig. 5(D). This boundary zone must be noted because it cannot be strengthened by aging, and it tends to become the origin of the fracture. The fracture surface of the sample before aging shows mainly a transgranular fracture mode and that after aging shows a mixed mode of intergranular and transgranular fractures. This implies that the transgranular strength was increased by precipitation of the tetragonal phase. The relationship between the aging time and the strength is shown in Fig. 6. The maximum strength of 400 MPa was obtained by aging for 140 h. The increasing strength up to 140 h is considered to be the result of strengthening of the c grains by the t-ZrO_2 precipitate phase. The decreasing strength after 140 h is considered to be caused by the low strength of the grain boundary. The strength of the Y-PSZ is lower than that of the Y-TZP. The strength of the Y-PSZ is understood to be limited by the existence of large c grains.

Durability of Y-TZP

Y-TZP is strengthened by the existence of t-ZrO_2. It is well known, however, that some Y-TZPs exhibit strength degradation by aging or thermal cycle tests at mediate temperatures. Figure 7 shows the cracks in the PSZ which indicated the strength degradation after heating/cooling cycles between room temperature and 800°C. Cracks started from contact points between t and c grains and propagated in c grains and along grain boundaries. The t-to-m transformation is accompanied by about 5% volume increase and is responsible for the strength degradation of Y-TZP. To study the durability of Y-TZP, the thermal expansion hysteresis, strength change after aging, and creep behavior were measured for samples D, E, and F that had the same amount of Y_2O_3 added and were sintered at different temperatures. The thermal expansion hysteresis was measured between room temperature and 900°C by using a fused quartz dilatometer. The heating and cooling rates were 1° and 5°C/min. The creep was measured between room temperature and 1000°C by a 4-point bending fixture with a 30-mm outer span and a 10-mm inner span, using a test bar of 2 by 3 by 40 mm^3. The creep strain was calculated from the deflection of the sample.[5]

Effect of Temperature

Thermal expansion curves for samples D, E, and F are shown in Fig. 8. The solid and dashed lines represent the experimental results for heating and cooling rates of 5° and 1°C/min, respectively. Thermal expansion curves for samples D

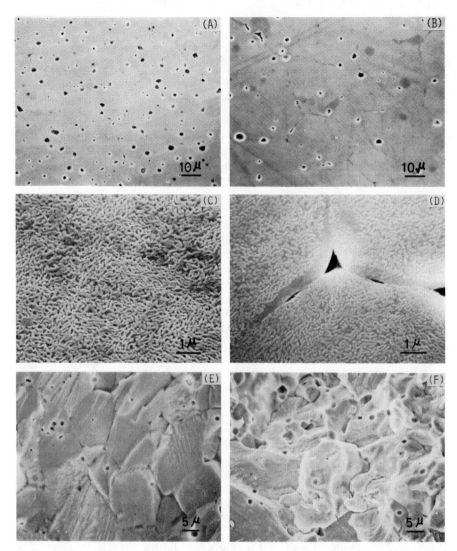

Fig. 5. SEM micrographs of aged Y-PSZ: (A) before aging, (B, C) after aging at 1400°C for 168 h, (D) grain-boundary portion after aging at 1400°C for 70 h, and (E, F) fracture surfaces before and after aging for 168 h, respectively. t-ZrO_2 precipitates are observed in c grain (C), and precipitate-free zones are present near some grain boundaries (D).

and E exhibited no hysteresis. The curve for sample F exhibited a large hysteresis, indicating that the t-to-m transformation occurred around 200°C and the reverse transformation occurred around 800°C. The effect of the heating and cooling rate on the hysteresis curve is noteworthy. The sample length recovered its initial state for the slower cooling rate of 1°C/min. For the faster cooling rate of 5°C/min, the sample contracted about 0.1% in length. The results show that the transformation has not fully progressed at the faster cooling rate, and this slow transformation is isothermal. Figure 9 shows the effect of the aging temperature (for the aging time

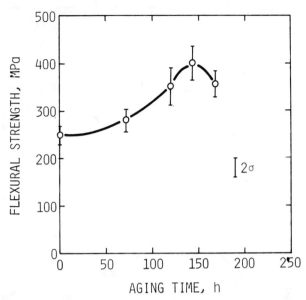

Fig. 6. Change in flexural strength during aging for tetragonal-phase precipitation.

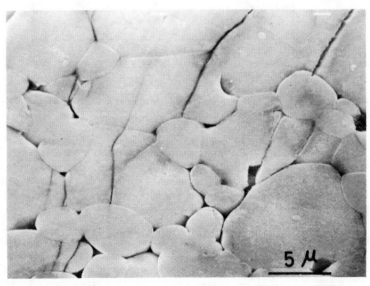

Fig. 7. Scanning electron micrograph showing cracks caused by thermal cycle in sample A.

of 2000 h) on the flexural strength of sample E. The strength degradation was observed only around 200°C. Sample D showed no strength degradation.

Effect of Stress

Figure 10 shows creep curves for sample E tested at 400 MPa and various temperatures. Creep phenomenon was observed around 250°C. Between 350° and

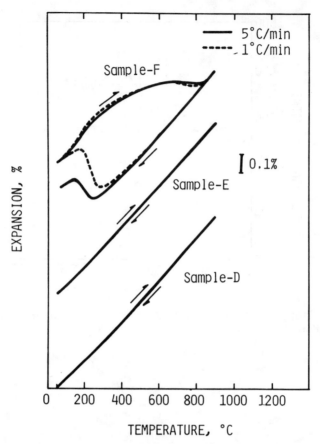

Fig. 8. Thermal expansion curves of samples D, E, and F. Hysteresis appears in sample F but not in samples D and E.

1000°C, no creep phenomenon was observed. The creep rate was highest at 250°C. Using the dye penetration test, no crack was observed at the surface of the sample after the creep test. The tensile and compressive surfaces of the creep sample were examined by X-ray diffraction analysis. The amount of m-ZrO_2 increased on both surfaces after the creep test. The results indicate that the stress enhanced the t-to-m transformation and caused the creep. The overall kinetics of martensitic transformations is dependent mainly on the nucleation process. The applied stress is considered to minimize the activation energy for nucleation of the m-ZrO_2. In Fig. 10, it is noteworthy that the creep rate changes with strain. Below 250°C, the creep rate increased with strain. Above 250°C, the creep rate decreased with strain. The behavior of the strain rate below 250°C is attributed to the autocatalytic effects, which are well known in the martensitic transformation of metals.[6,7] By this effect, the stress field around the m-ZrO_2 enhances the new nucleation of the monoclinic phase. The behavior of the creep strain rates above 250°C may be explained by the fact that the amount of transformation is determined by the degree of supercooling from the thermodynamic phase equilibrium temperature, as in the case of the martensitic transformation of metals. Figure 11 shows the Arrhenius plots of the strain rate at 0.3% strain under various applied stresses. The creep rate increased

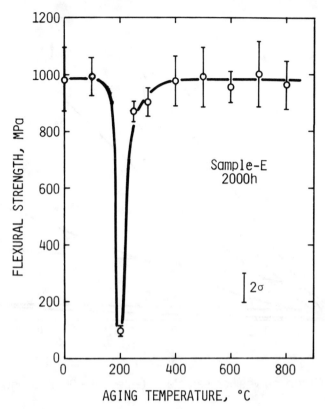

Fig. 9. Effect of aging temperature on flexural strength after 2000-h aging.

with stress. The temperature T_d, where the strain rate shows the maximum, was 275°C at 600 MPa and decreased with decreasing applied stress. T_d is estimated to be about 200°C when stress is extrapolated to 0. This extrapolated value shows a good agreement with the results of thermal expansion and aging tests.

The transformation rate without external stress was highest at about 200°C, as shown by the thermal expansion curve in Fig. 8 and by the strength aging temperature curve in Fig. 9. A similar behavior of the martensitic transformation is usually observed in metals.[6] This behavior appears because the activation energy of the martensitic transformation is a function of the degree of supercooling and the applied stress.

Sample D showed no creep phenomena. Sample F showed a more drastic creep phenomenon than sample E. Thus, the durability of the Y-TZP depends strongly on the microstructure. The Y-TZP with small tetragonal grains, sintered at low temperature, exhibited no fatigue. The reason for this has not been clarified. However, the following are possible causes for this phenomenon: (1) Succession of the transformation is interrupted by the grain boundaries. This effect is large in Y-TZP with small t grains. (2) Nucleation of the m phase is difficult to start in the case of small t grains. Since the t-to-m transformation is accompanied by shear deformation of the lattice due to cooperative movement of ions, a certain size is required for the transformation. When a grain is smaller than that critical size, the phase transformation does not occur. Since the lattice is discontinuous at the grain

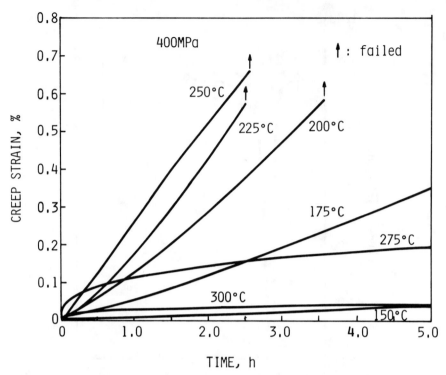

Fig. 10. Creep curves for sample E tested at 400 MPa and various temperatures.

boundary, even if a grain is a little larger than the critical size, the presence of grain boundaries tends to increase the probability for inhibition of the transformation, resulting in a slower transformation rate. (3) The effect of surface energy may change the thermodynamic equilibrium transformation temperature.[8]

Conclusion

In the case of the sintered-type Y-TZP, a maximum flexural strength of 980 MPa was obtained by selecting the Y_2O_3 content and the sintering temperature. In the case of Y-PSZ, the strength changed with aging time, and the maximum flexural strength was 400 MPa. Around 250°C, strength degradation and a creep phenomenon were observed in the Y-TZP with microstructures containing large grains. The strength degradation and creep phenomenon were shown to be caused by the t-to-m transformation. The rate of transformation from metastable t to m was dependent on both temperature and applied stress. An autocatalytic effect was observed in t-to-m transformation of Y-TZP below 250°C. No degradation was observed in the Y-TZP whose microstructure was finely controlled.

References

[1]W. Bryzik, "TACOM/Cummins Adiabatic Engine Program," Proceedings of the 20th Automotive Technology Development Contractor's Coordination Meeting, Dearborn, MI, Oct. 25–28 (1982).

[2]R. Kamo and W. Bryzik, "Cummins-TARADCOM Adiabatic Turbocompound Engine Program," 1981 SAE International Congress and Exposition, SAE paper 810070, Detroit, MI, Feb. 23–27 (1981).

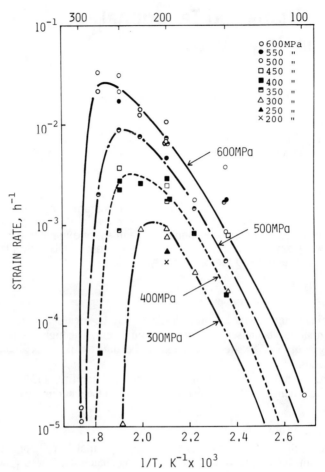

Fig. 11. Arrhenius plots of strain rate under various static 4-point bending stresses.

[3]M. E. Woods and I. Oda, "PSZ Ceramics for Adiabatic Engine Components," 1982 SAE International Congress & Exposition, SAE paper 820429, Detroit, MI, Feb. 22–26 (1982).
[4]H. G. Scott, "Phase Relationships in the Zirconia-Yttria System," *J. Mater. Sci.,* **10**, 1527–35 (1975).
[5]G. W. Hollenberg, G. R. Terwilliger, and R. S. Gordon, *J. Am. Ceram. Soc.,* **54** [4] 196–99 (1971).
[6]J. Burke; The Kinetics of Phase Transformation in Metals. Pergamon, Elmsford, NY (1965).
[7]J. W. Christian; p. 916 in The Theory of Transformations in Metals and Alloys, Pergamon, Oxford (1965).
[8]J. E. Bailey, D. Lewis, Z. M. Librant, and L. J. Porter, "Phase Transformations in Milled Zirconia," *Trans. J. Br. Ceram. Soc.,* **71** [1] 25–30 (1972).

Thermal and Mechanical Properties of Y_2O_3-Stabilized Tetragonal Zirconia Polycrystals

K. Tsukuma, Y. Kubota, and T. Tsukidate

Toyo Soda Manufacturing Co., Ltd.
Research Laboratory
Shinnanyo, Yamaguchi 746, Japan

Dense Y_2O_3-stabilized tetragonal zirconia (Y-TZP) ceramics containing 2–6 mol% Y_2O_3 were fabricated from fine powders by normal sintering and hot isostatic pressing. The microstructure, mechanical properties, and thermal stability of such materials were examined. Grain size and Y_2O_3 content significantly influenced the mechanical properties and thermal stability.

Y_2O_3-stabilized tetragonal zirconia (Y-TZP) may be regarded as one of the new structural materials because of its enhanced toughness. The high fracture toughness and strength of Y-TZP are considered to be due to the stress-induced phase transformation.[1,2]

We have studied the microstructure, thermal stability, and mechanical behavior of fully dense Y-TZP bodies. In this paper, a brief summary of these results is presented.

Experimental Procedure

Submicrometer ZrO_2 powders containing 2–6 mol% Y_2O_3 prepared by the coprecipitation technique were used as starting materials. These powders were pressed uniaxially at 40 MPa and then pressed isostatically at 300 MPa. The compacted bodies were sintered between 1400° and 1600°C for 2 h in air. The hot isostatic pressing was performed between 1400° and 1600°C under 150 MPa pressure in Ar gas using the presintered bodies.

The microstructure was observed by scanning electron microscopy (SEM) and scanning transmission electron microscopy (STEM). The three-point bending strength was measured under the conditions of 30-mm span length and 0.5-mm/min crosshead speed. The size of the test bar was 3 by 4 by 40 mm. The fraction of transformed monoclinic phase was measured by X-ray diffraction (XRD) of the broken surface of the test bar. The fracture toughness was measured by the microindentation method with a load of 50 kg. K_{Ic} values were calculated using the equation reported by Niihara et al.[3]

The thermal stability was examined by an aging test at 200°–300°C and a thermal cycle test between room temperature and 800°C. The aged samples were evaluated by the measurement of bend strength, X-ray phase identification, and light-microscopy observations.

Results and Discussion

Microstructure

The lattice constants and bulk densities of sintered Y-TZP with 2–6 mol% Y_2O_3 are shown in Table I. The XRD patterns revealed that the 2 mol% Y-TZP consisted of only the tetragonal phase and the 6 mol% Y-PSZ was almost completely cubic; 3 and 4 mol% Y-TZP consisted of a mixture of tetragonal and cubic phases. The lattice constants of both phases were constant in the composition range 2–6 mol% within experimental error. The Y_2O_3 content in the tetragonal and cubic phases determined from lattice constants using the relationship reported by Scott[4] were 2.0 and 6.2 mol%, respectively. The microstructures of bodies sintered at 1450°C are shown in Fig. 1.

Y-TZP with 2 mol% Y_2O_3 was composed of uniform fine grains (about 0.4 μm). The number of large grains (about 1 μm) increased with increasing Y_2O_3 content. Y_2O_3 solute partition and phase identification of particles in a 3 mol% Y-TZP body sintered at 1500°C were examined by STEM (Fig. 2).

The microstructure of the specimens was composed of two types of particles, relatively large grains giving no contrast and small grains having a twinned structure. The former was identified as the cubic phase from the electron diffraction pattern. The latter could not be identified, but it is assumed to be the monoclinic phase because of the twinned structure and small Y_2O_3 content. It is apparent that the monoclinic phase resulted from the stress-induced transformation of the tetragonal phase during preparation of the thin film.

Energy dispersive spectroscopy (EDS) microanalysis of the cubic and monoclinic phases revealed that the cubic phase contained 6.9 ± 0.5 mol% Y_2O_3 and the monoclinic phase contained 1.8 ± 0.5 mol% Y_2O_3. These results are almost in agreement with those determined by the XRD method and the prediction from the Y_2O_3–ZrO_2 phase diagram provided in Ref. 4.

Consequently, it is confirmed that Y-TZP with 2–6 mol% Y_2O_3 are mixtures of two end-members; one is the tetragonal phase with about 2 mol% Y_2O_3 and the other is the cubic phase with 6–7 mol% Y_2O_3.

Strength and Fracture Toughness

Fracture toughness: The relationship between the amount of stress-induced phase transformation and Y_2O_3 content is shown in Fig. 3. The amount of transformation increased with decreasing Y_2O_3 content. The 2 mol% Y-TZP revealed an extremely large amount of transformation.

Table I. Lattice Constants and Bulk Density of Y_2O_3–ZrO_2 Ceramics*

Sample	Lattice constants (Å)	Bulk density (g/cm³)
2 mol% Y_2O_3	a = 5.095, c = 5.180 (T)	6.08
2.5 mol% Y_2O_3	—	6.07
3 mol% Y_2O_3	a = 5.096, c = 5.180 (T)	6.08
3.5 mol% Y_2O_3	—	6.06
4 mol% Y_2O_3	a = 5.098, c = 5.180 (T), a = 5.134 (C)	6.05
6 mol% Y_2O_3	a = 5.133 (C)	6.03

*T = tetragonal and C = cubic.

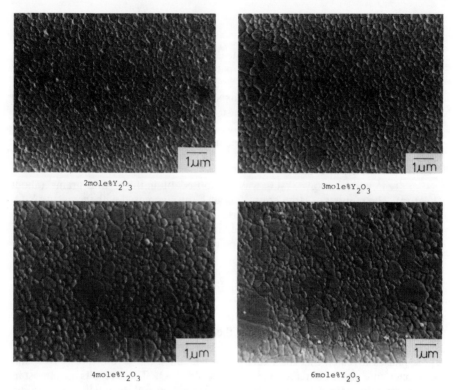

Fig. 1. Scanning electron micrographs of 2, 3, 4, and 6 mol% Y_2O_3–ZrO_2 ceramics show thermally etched surface of bodies sintered at 1450°C for 2 h.

The relationship between fracture toughness and Y_2O_3 content is shown in Fig. 4. K_{Ic} values increased with decreasing Y_2O_3 content. This relationship is similar to that of the stress-induced transformation. Fracture toughness depends strongly on the amount of stress-induced transformation, which is consistent with the experimental and theoretical results reported by Lange.[2]

Strength: The hot isostatic pressing produced almost perfectly dense Y-TZP bodies. Nevertheless, the strength of the isostatically hot-pressed body was strongly dependent on the presintering temperature. The relationship between bend strength and presintering temperature is shown in Fig. 5. This figure shows that a low presintering temperature (1400°C) is required to obtain high-strength isostatically hot-pressed bodies. Many large pores were observed on bodies presintered at high temperature (1600°C). These pores, which may result from grain growth, were not completely eliminated by hot isostatic pressing. The lower strength is due to the remains of these pores. Significant strength enhancement was obtained when presintering and hot isostatic pressing were performed under optimum conditions.

The dependence of bending strength on Y_2O_3 content is shown in Fig. 6. Y-TZP containing 3 mol% Y_2O_3 yielded the maximum strength, which is contrary to the expectation from K_{Ic} values and the amount of transformation. To explain the strengthening of Y-TZP with 2–3 mol% Y_2O_3, we may have to consider that transformation-toughening depends on the microstructure. The results suggest that

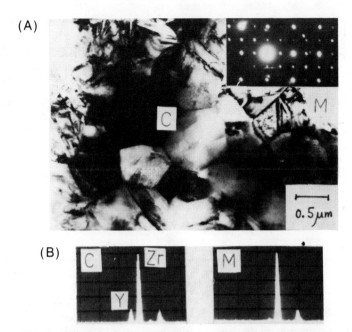

Fig. 2. Scanning transmission electron micrograph of TZP with 3 mol% Y_2O_3: (A) Electron diffraction pattern of (100) cubic plane where C is the cubic phase and M is the monoclinic phase; (B) EDS microanalysis of cubic and monoclinic grains.

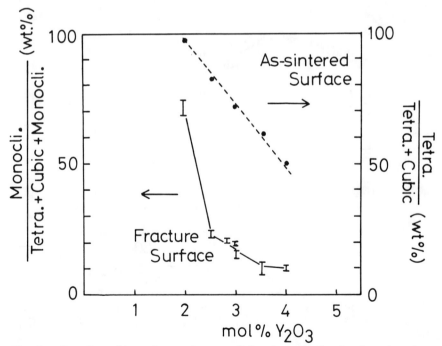

Fig. 3. Fraction of transformed monoclinic phase in the fractured surface plotted vs Y_2O_3 content.

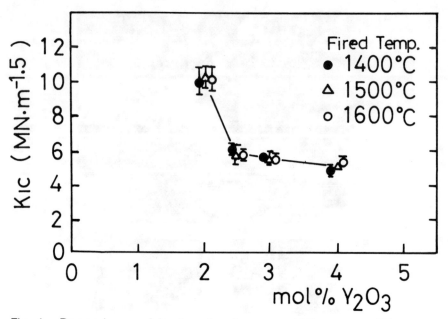

Fig. 4. Dependence of fracture toughness on Y_2O_3 content (firing temperature: ●, 1400°C; △, 1500°C; ○, 1600°C).

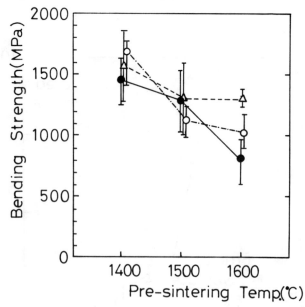

Fig. 5. Relationship between presintering temperature and three-point bending strength for isostatically hot-pressed TZP with 3 mol% Y_2O_3: (●) 1400°C, 150 MPa, 0.5 h; (△) 1500°C, 150 MPa, 0.5 h; (○) 1600°C, 150 MPa, 0.5 h.

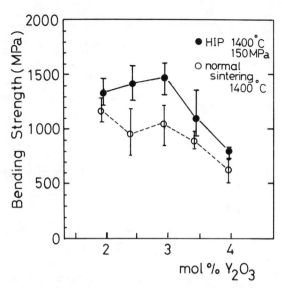

Fig. 6. Dependence of bending strength on Y_2O_3 content in normal sintered and isostatically hot-pressed Y-TZP (conditions of normal sintering: 1350°C, 2 h; conditions of hot isostatic pressing: 1400°C, 150 MPa, 0.5 h).

a small amount of cubic phase contributes to the strengthening, because 2 mol% Y-TZP with no cubic phase exhibited a lower strength than 3 mol% Y-TZP with a small amount of cubic phase.

Thermal Stability

Grain-size dependence: The relationship between the grain-size and the 230°C aging time is shown in Fig. 7. The sintered bodies which were composed of grains of size greater than 1 μm showed a large amount of transformation and a remarkable decrease in strength but, on the other hand, those bodies which contained grains of less than 0.4 μm showed no significant change of phase content and strength even after 1500 h.

Figure 8 represents the cross section of 3 mol% Y-TZP degraded by 230°C aging. The cracking, which occurred due to the volume expansion accompanying the transformation, propagated from the surface inward. This result suggests that the tetragonal particles on the surface are more unstable than those in the bulk. The propagation of cracking is responsible for the remarkable decrease in strength.

Thus, the aging test at low temperature (the unstable region of the tetragonal phase) revealed that the stability of the metastable tetragonal phase depends strongly on the grain size and the existence of surrounding particles.

Dependence on Y_2O_3 content: The dependence of thermal stability on Y_2O_3 content is shown in Fig. 9. The grain sizes of all specimens were almost the same, about 0.4 μm. TZP with 1.9 mol% Y_2O_3 degraded after only 24 h of aging, but TZP with more than 2.3 mol% Y_2O_3 showed no significant degradation.

The results of thermal cycle tests between room temperature and 800°C are shown in Fig. 10. TZP with 1.9 mol% Y_2O_3 showed rapid degradation, which is the same as the result of 300°C aging. In this test, the degraded sample showed

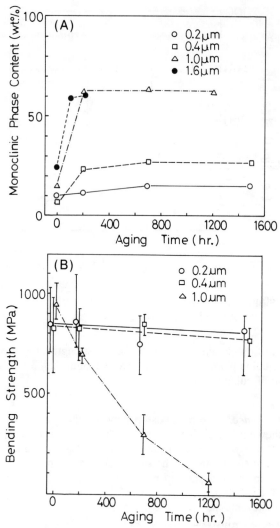

Fig. 7. Relationship between grain size and 230°C aging time for TZP with 3 mole Y_2O_3: (A) Grain-size dependence of phase transformation on the surface; (B) grain-size dependence of bending strength.

no visible cracking as observed in the 300°C aging test; instead, the material reduced to powder owing to the repetition of expansion and constriction accompanying the transformation. These results suggest that the stability of metastable tetragonal phase depends strongly on the Y_2O_3 content.

Conclusions

The strength of Y-TZP was enhanced significantly by hot isostatic pressing. The dependence of strength on Y_2O_3 content was different from that of fracture toughness and amount of transformation. This result is not consistent with that obtained from the concept of transformation-toughening. The thermal stability of Y-TZP depended on the grain size, Y_2O_3 content, and existence of surrounding grains.

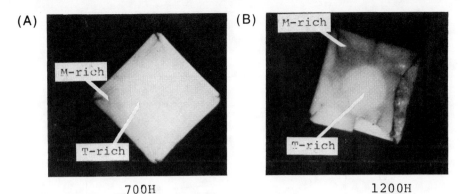

Fig. 8. Light micrographs of TZP with 3 mol% Y_2O_3 degraded by aging: (A) after 700 h; (B) after 1200 h.

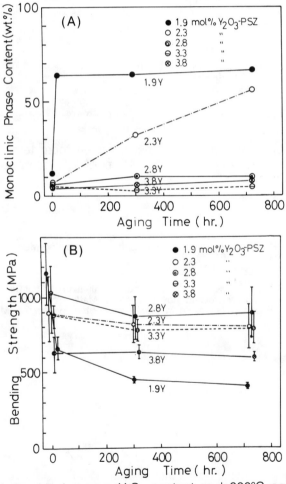

Fig. 9. Relationship between Y_2O_3 content and 300°C aging time: (A) phase transformation dependence on Y_2O_3 content; (B) bending strength dependence on Y_2O_3 content.

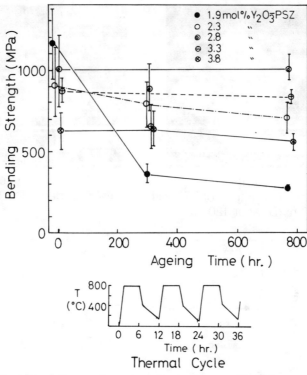

Fig. 10. Result of thermal cycle test between room temperature and 800°C. The samples are PSZ with 1.9–3.8 mol% Y_2O_3 composed of grains about 0.4 µm in size.

References

[1]T. K. Gupta, F. F. Lange, and J. H. Bechtold, "Effect of Stress-Induced Phase Transformation on the Properties of Polycrystalline Zirconia Containing Metastable Tetragonal Phase," *J. Mater. Sci.*, **13**, 1464–70 (1978).

[2]F. F. Lange, "Transformation Toughening Part 3. Experimental Observation in the ZrO_2–Y_2O_3 System," *J. Mater. Sci.*, **17**, 240–46 (1982).

[3]K. Niihara, R. Morena, and D. P. H. Hasselman, "Evaluation of K_{Ic} of Brittle Solids by the Indentation Method with Low Crack-to-Indent Ratios," *J. Mater. Sci. Lett.*, **1**, 13–6 (1982).

[4]H. G. Scott, "Phase Relationship in the Zirconia-Yttria System," *J. Mater. Sci.*, **10**, 1527–35 (1975).

Aging Behavior of Y-TZP

Masakazu Watanabe, Satoshi Iio, and Isamu Fukuura

Research Department
NGK Spark Plug Co., Ltd
Mizuho, Nagoya, Japan

The aging behavior of Y-TZP was studied, with emphasis on the effect of microstructure on the tetragonal (t)-to-monoclinic (m) transformation. The stability of t-ZrO_2 in Y-TZP materials was examined by measuring density, strength, and phase contents after they were aged for 100 h at temperatures between 100° and 800°C. Drastic decreases in both density and strength were observed for samples aged at 200°–300°C. XRD results indicated that the strength degradation and m-ZrO_2 content were linearly correlated. It was thus concluded that this change was due to the $t \rightarrow m$ transformation of ZrO_2. The above conclusion was in agreement with the results of the thermal expansion analysis, in which abnormal changes were observed at 200°–550°C on heating. Various Y-TZP materials with different mean grain size ranging from 0.1 to 1.0 μm were aged for 1000 h at 300°C to examine the effect of microstructure on the $t \rightarrow m$ transformation. It was found that a critical grain size (D_c) for the $t \rightarrow m$ transformation does exist, for example, 0.2 μm for Y-TZP with 2 mol% Y_2O_3; D_c is largely dependent on Y_2O_3 content.

Fine-grained tetragonal ZrO_2 (TZP, tetragonal zirconia polycrystals), consisting primarily of t-ZrO_2 grains, was reported recently.[1] It has been suggested that the toughness of this material is derived from the tetragonal (t)-to-monoclinic (m) transformation which occurs in the process zone around a crack front.[2] Thus, it is important to know how one can retain t-ZrO_2, so it can act as a toughening agent in the sintered body.

The ZrO_2-Y_2O_3 phase diagram[3,4] suggests that the phase composition of ZrO_2 around 1500°C, where Y-TZP may be sintered, changes from cubic (c)-ZrO_2 to t-ZrO_2 through a t–c two-phase region as the Y_2O_3 content decreases from 7 to 3 mol%. Gupta et al.[5] reported that the stability of t-ZrO_2 in the sintered body depends on the grain size, and t-ZrO_2 larger than the critical grain size ($D_c \approx 0.3$ μm) transforms to m-ZrO_2 when cooled after sintering. More recently, Lange[6] reported that D_c increases from 0.2 to 1.0 μm as the Y_2O_3 content increases from 2 to 3 mol%.

In this paper, the aging behavior of Y-TZP materials at temperatures below 1000°C is reported, with emphasis on the effect of microstructure on the $t \rightarrow m$ transformation.

Experimental Procedure

Sample Preparation and Aging

Samples containing 2–5 mol% Y_2O_3 were prepared from m-ZrO_2 powder and Y_2O_3 powder. The starting powders were mixed, ground by wet ball-milling, pressed at 150 MPa, and sintered at temperatures ranging from 1400° to 1600°C using an electric furnace. After sintering, surfaces of the samples were ground with 140-grit diamond wheels.

Y-TZP samples were aged for 100 and 1000 h at various temperatures between 100° and 800°C. The stability of t-ZrO_2 was examined by measuring density, strength, phase content, and thermal expansion.

Measurement

Strength was measured by a three-point bend method (sample size, 4 by 8 by 25 mm; span, 20 mm). X-ray diffractometry was used to estimate the phase content at the surface of the as-ground and 3 aged samples: {111} reflection intensities were measured to determine the ratio of m-ZrO_2 to total ZrO_2 $(c + t + m)$.[7] Scanning electron microscopy was used to examine the microstructure and estimate the mean grain size of the t-ZrO_2 in the sintered bodies. A dilatometer was used to examine thermal expansion behavior of some samples from room temperature to 1000°C, with a heating rate of 10°C/min.

Results and Discussion

Density and Strength Measurements

Figures 1 and 2 show density and strength, respectively, of samples containing 4 mol% Y_2O_3 sintered at 1600°C after they were aged in air at various temperatures for 100 h. Drastic decreases in both density and strength were observed for samples aged at 200° and 300°C. Macroscopic cracks were observed in these samples. On the other hand, for the samples aged at temperatures above 400°C, neither density nor strength showed significant change.

X-ray Diffractometry

The ratio of m-ZrO_2 to total ZrO_2 of the samples described in the previous section was measured quantitatively by X-ray analysis. Figures 3 and 4 show, respectively, the density and the strength plotted vs m-ZrO_2 content. XRD results indicated that the m-ZrO_2 content for aged samples increased more than 10% compared with the sample* before aging, and both strength and density decreased

*For the as-ground sample, m-ZrO_2 content was about 10%.

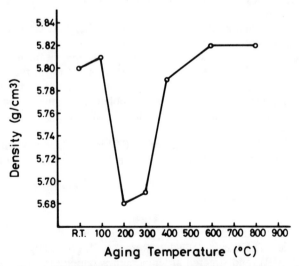

Fig. 1. Density vs aging temperature for Y-TZP containing 4 mol% Y_2O_3 (held for 100 h); sintering time, 1 h; atmosphere, air.

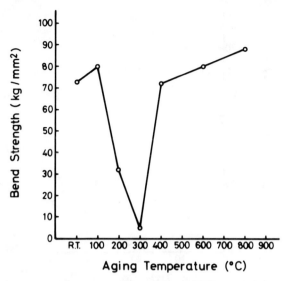

Fig. 2. Strength vs aging temperature for Y-TZP containing 4 mol% Y_2O_3 (held for 100 h).

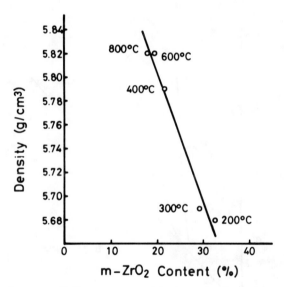

Fig. 3. Density vs m-ZrO_2 content, with the sample held at various temperatures for 100 h.

almost linearly with increasing m-ZrO_2 content. The m-ZrO_2 content of the samples aged at 200°–300°C increased drastically, up to about 30%, because of the $t \rightarrow m$ transformation. From these results, it was concluded that t-ZrO_2 retained in the Y-TZP samples is easily transformed to m-ZrO_2 by aging at temperatures between 200° and 300°C.

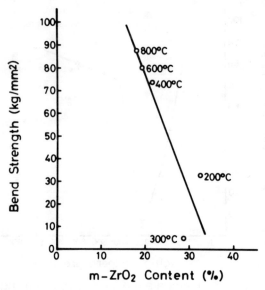

Fig. 4. Strength vs *m*-ZrO$_2$ content, with the sample held at various temperatures for 100 h.

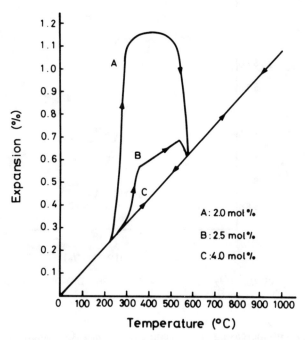

Fig. 5. Thermal expansion behavior of Y-TZP containing various Y$_2$O$_3$ contents. Sintering conditions for A and B: 1 h at 1500°C. Sintering conditions for C: 1 h at 1600°C.

Dilatometry

Thermal expansion measurements were made on samples containing 2, 2.5, and 4 mol% Y_2O_3. As shown in Fig. 5, the thermal expansion coefficients of the various TZP samples at high temperatures were almost the same (10.8×10^{-6}/°C). More importantly, abnormal thermal expansion behavior was observed in the samples containing 2 and 2.5 mol% Y_2O_3: on heating, a sharp expansion beginning at about 200° and a sharp shrinkage at about 550°C were observed. In contrast, no abnormal behavior was observed on cooling. It is possible that this phenomenon was caused by a volume change due to the $t \rightarrow m$ transformation. These results on 2 and 2.5 mol% Y_2O_3-doped samples are consistent with the results of the aging behavior described in the previous sections, whereas the results on the samples with 4 mol% Y_2O_3 are inconsistent. Therefore, thermal expansion measurements were carried out again by maintaining the 4 mol% Y_2O_3 sample for 100 h at 300°C. As shown in Fig. 6, expansion during holding at this temperature was observed as expected, and some remnant expansion was observed after cooling.

These results imply that the net free energy change on $t \rightarrow m$ transformation is negative at the aging temperature. Thus, it seems that the experimental results cannot be interpreted sufficiently using the "end-point" thermodynamics approach, but that it would be important to consider the nucleation problem to understand such transformation behavior, as suggested by Heuer et al.[8]

For the stability of t-ZrO_2 in Y-TZP, Andersson and Gupta[9] explicitly considered the nucleation argument by assuming that the martensitic $t \rightarrow m$ transformation occurs by the growth of preexisting embryos. Their argument is based on an end-point calculation, and thus it does not explain the present results; if the $t \rightarrow m$ transformation does not occur at room temperature, it does not occur at higher temperature, according to their theory. Let us now consider a situation where nuclei or embryos do not exist in the sintered bodies. As pointed out by Heuer et al.,[8] the nucleation barrier must be surmounted for the transformation to

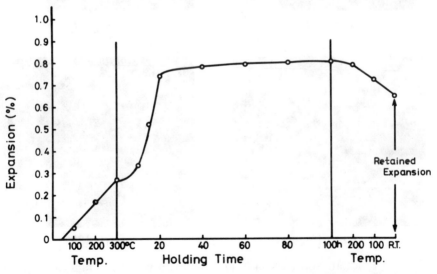

Fig. 6. Dimensional changes on aging Y-TZP at 300°C. See text for details.

occur. Following the notation of Heuer et al., the nucleation barrier, ΔF^*, will likely depend on the net driving force, decreasing with increasing $\Delta F(\text{chem})$, the free energy difference between the free energy of the constrained t-ZrO_2 particle and that of the free m-ZrO_2 particle. If the net free energy change on transformation, ΔF, is negative, the transformation can occur at temperatures at which $\Delta F^* < kT$. Applying the above arguments to the present results, it may be possible that the transformation condition, $\Delta F^* < kT$, was satisfied by the heat treatment between 200° and 300°C. At the higher temperatures, ΔF^* would increase drastically as $\Delta F(\text{chem})$ decreases and, consequently, kT may be smaller than ΔF^*. With increasing amounts of Y_2O_3, $\Delta F(\text{chem})$ decreases and ΔF^* increases. Thus, samples with more Y_2O_3 additions would be more resistant to transformation, consistent with the present results. Although the above discussion is quite speculative, it seems to explain the results qualitatively.

Srivastava et al.[3] stated that there is a eutectoid transformation (t-$ZrO_2 \rightarrow m$-$ZrO_2 + c$-ZrO_2) at about 565°C and 7.5 mol% $YO_{1.5}$ in the system ZrO_2-Y_2O_3. This implies that the $m \rightarrow t$ transformation on heating to about 550°C is possible even in a sintered body.

Fig. 7. Fracture surfaces of Y-TZP samples with various Y_2O_3 content; sintering time, 1 h.

Microstructure

Figure 7 shows microstructures of samples containing 2–5 mol% Y_2O_3, sintered at 1500°C. These SEM photographs reveal that c-ZrO_2 grain size and c-ZrO_2 content increase with increasing Y_2O_3 content, but the mean t-ZrO_2 grain size is almost independent of Y_2O_3 content. Detailed grain-size measurements were made on the samples containing 2 and 4 mol% Y_2O_3 as a function of sintering temperature. These results are shown in Fig. 8. Mean grain size increased with increasing sintering temperature, from 0.1 μm at 1400° to 1.0 μm at 1600°C.

Critical Grain Size

The various Y-TZP samples with different mean grain sizes ranging from 0.1 to 1.0 μm were aged for 1000 h at 300°C to examine the critical grain size (D_c) for the $t \rightarrow m$ transformation. Degradation in both density and strength for all aged samples was investigated to estimate D_c in the same way as already described. These results are shown in Fig. 9; D_c increased from 0.2 to 0.6 μm, when the Y_2O_3 content was increased from 2 to 5 mol%.

As discussed by Lange,[10] the net free energy change on transformation increases with increasing grain size. Extrapolation of the nucleation argument results in decreasing ΔF^* with increasing grain size if ΔF^* decreases with increasing ΔF. This grain-size dependence of ΔF^* may be the origin of D_c. Since ΔF^* increases with the Y_2O_3 content, D_c should also increase.

Conclusion

The aging behavior of Y-TZP was studied, with emphasis on the effect of the microstructure on the $t \rightarrow m$ transformation. The results may be summarized as follows: (1) Aging the Y-TZP materials at temperatures between 200° and 300°C caused drastic degradation in both strength and density due to the $t \rightarrow m$ trans-

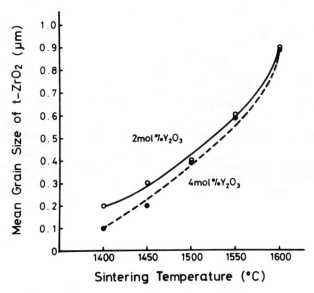

Fig. 8. Mean grain size of t-ZrO_2 as a function of sintering temperature; sintering time, 1 h.

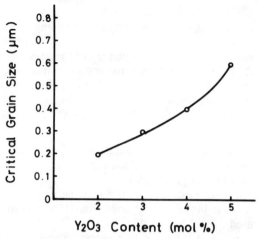

Fig. 9. Critical grain size as a function of Y_2O_3 content (when sample is held at 300°C for 1000 h).

formation. (2) The critical grain size for the $t \to m$ transformation increased from 0.2 to 0.6 μm between 2 and 5 mol% Y_2O_3. (3) The stability of the t-ZrO_2 retained in the sintered body was found to be affected by Y_2O_3 content, grain size, and aging time.

Acknowledgments

The authors deeply appreciated discussions with their colleague Yo Tajima.

References

[1] T. K. Gupta, J. H. Bechtold, R. C. Kuznichi, L. H. Adoff, and B. R. Rossing, *J. Mater. Sci.*, **12**, 2421 (1981).
[2] D. L. Porter, A. G. Evans, and A. H. Heuer, *Acta Metall.*, **27**, 1649 (1979).
[3] K. K. Srivastava, R. N. Patil, C. B. Chandary, K. V. G. K. Gokhale, and E. C. Subbarao, *Trans. Br. Ceram. Soc.*, **73**, 85 (1974).
[4] G. G. Scott, *J. Mater. Sci.*, **10**, 1527 (1975).
[5] T. K. Gupta, F. F. Lange, and J. H. Bechtold; *J. Mater. Sci.*, **13**, 1464 (1978).
[6] F. F. Lange, *J. Mater. Sci.*, **17**, 240 (1982).
[7] R. C. Garvie and P. S. Nicholson, *J. Am. Ceram. Soc.*, **55** [6] 303–305 (1972).
[8] A. H. Heuer, N. Claussen, W. M. Kriven, and M. Rühle, *J. Am. Ceram. Soc.*, **65** [12] 642–50 (1982).
[9] C. A. Andersson and T. K. Gupta; pp. 184–201 in Advances in Ceramics, Vol. 3. Edited by A. H. Heuer and L. W. Hobbs. The American Ceramic Society, Columbus, OH, 1981.
[10] F. F. Lange, *J. Mater. Sci.*, **17**, 225 (1982).

Phase Stability of Y-PSZ in Aqueous Solutions

K. Nakajima, K. Kobayashi, and Y. Murata

Toray Industries Inc.
Toray Research Center Inc.
Shiga-520, Japan

Phase stability in aqueous solutions was studied for Y-PSZ specimens containing 2–5 mol% Y_2O_3. The amount of monoclinic phase on the surface of each specimen increases with their dipping time in aqueous solutions, such as sulfuric acid, nitric acid, caustic soda, and distilled water, at elevated temperatures. In the case of specimens containing smaller amounts of Y_2O_3, cracks initiated at the surface after a few hours of dipping. Subsequently, the specimens disintegrated. It was observed by transmission electron microscopy that the cracks grow along the grain boundaries and that microstructural changes, such as twin formation induced by transformation from tetragonal to monoclinic phase, exist in the grains. From these observations it appears that a corrosion mechanism may be involved in cracking.

Recently many studies on ZrO_2-toughened ceramics have been conducted.[1-3] The toughening mechanism has been attributed to stress-induced transformation from metastable tetragonal phase to monoclinic phase under external stresses such as impact stress or crack-tip stresses.[4-8]

Increase in monoclinic phase in Y-PSZ containing metastable tetragonal phase after aging is accompanied by a decrease in the mechanical properties, such as toughness and flexural strength.[9,10]

In this paper, studies of the phase transformation of Y-PSZ induced at lower temperature in aqueous solutions and of the morphology with the use of transmission electron microscopy (TEM) and electron probe microanalyzer (EPMA) are described.

Experimental Procedure

Zirconia powders containing 2.0, 2.5, 3.5, 4.5 and 5.0 mol% Y_2O_3 were prepared as follows: Mixed aqueous solutions of zirconium oxychloride and yttrium chloride were heated until water evaporated thoroughly; the deposits were calcinated at 1000°C for 3 h in air, washed with water to remove Cl^- ions, and then ball-milled. The particle size appeared to be 0.05–0.1 μm (as observed by TEM).

The powder was pressed into rods at a pressure of 147 MPa and then sintered at 1550°C for 2 h in air. Impurities contained in specimens are shown in Table I.

About 0.2 mm of the specimen surface was removed carefully by grinding, and the roughness of the surface was regulated to be between 0.8 and 1.0 μm as measured using Talysurf IV,* which uses a sharply pointed stylus to trace the profile of the surface irregularities. The dimensions of the specimens for the dipping and flexural strength tests were 3 by 4 by 36 mm. The crystal structure of

---

*R. Taylor Hobson Co., Leicester, UK.

Table I. Impurity Level

	Impurity level (wt%)				
Ca	Si	Al	Ti	Mg	Na
0.13–0.18	0.02–0.09	<0.05	<0.05	<0.02	<0.01

the specimens was mainly tetragonal (and cubic); their monoclinic fractions are shown in Table II.

Dipping tests were performed in aqueous solutions with 30% sulfuric acid at temperatures between 41° and 107°C and in distilled water in an autoclave at temperatures between 100° and 180°C. Flexural stength was tested using the three-point bend test (span of 30 mm) at a crosshead speed of 0.5 mm/min. Fracture toughness was measured by the single-edge notched bar method at a crosshead speed of 0.05 mm/min. The notch depth was 2.0 mm, with its tip sharply edged.

Microstructure and phases were analyzed by X-ray diffraction (XRD),[3] TEM, and scanning electron microscope-EPMA (SEM-EPMA). Samples for SEM-EPMA were etched in 98% sulfuric acid between 160° and 170°C for 20–40 min.

Results and Discussion

Mechanical Properties

Mechanical and physical properties of specimens are summarized in Table III.

Phase Transformation in Aqueous Solutions

The relationship between transformation of Y-PSZ from tetragonal-to-monoclinic phase ($t \rightarrow m$ transformation) in 30% sulfuric acid and time and temperature is shown in Fig. 1. The intensities for two-phase components were measured by XRD, and then the monoclinic fraction was calculated according to the method of Porter and Heuer.[3] Dipping temperatures >63°C caused an increase

Table II. Fraction of Monoclinic Phase

	Specimen*				
	2.0Y*	2.5Y	3.5Y	4.5Y	5.0Y
Nominal content of yttria (mol%)	2.0	2.5	3.5	4.5	5.0
Monoclinic fraction (%)	23.3	13.7	8.8	5.3	3.3

*Notation for the samples.

Table III. Mechanical Properties

	Specimen				
	2.0Y	2.5Y	3.5Y	4.5Y	5.0Y
Density (g/cm^3)	5.86	6.02	5.96	5.96	5.95
Hardness (HV)	1170	1250	1290	1310	1330
Flexural strength (MN/m^2)	294	1024	833	568	475
Fracture toughness (MN/m$^{3/2}$)	4.9	9.7	6.9	5.4	4.3

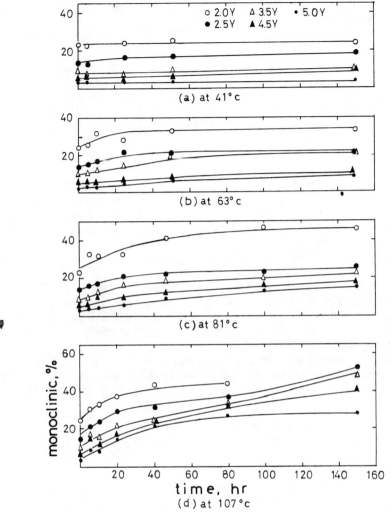

Fig. 1. Time dependence of the increase in monoclinic fraction in 30% H_2SO_4.

in the monoclinic fraction of all specimens. But the above-mentioned increase could not be observed at 41°C even after a dipping time of 150 h.

The rate of phase conversion from tetragonal-to-monoclinic phase can be given by Eqs. (1) or (2).

$$K = -\left(\frac{1}{V_t}\right)\frac{dV_t}{dt} \qquad (1)$$

$$\ln V_t = \text{constant} - Kt \qquad (2)$$

where V_t is the tetragonal fraction at time t.

If the specimen consists of metastable tetragonal and monoclinic phase (this assumption must be corrected as shown later in this section by considering the existence of the cubic phase), then Eqs. (1) and (2) are rewritten as

$$K = \left(\frac{1}{1-V_m}\right)\frac{dV_m}{dt} \qquad (3)$$

$$\ln(1-V_m) = \ln(1-V_{m_0}) - Kt \qquad (4)$$

where V_m is the monoclinic fraction and V_{m_0} is V_m at $t = 0$. Equation (4) did not hold for long dipping times (longer than about 50–100 h). But Eq. (4) held true for short dipping times. Therefore, activation energies of the $t \to m$ transformation in sulfuric acid for different specimens are estimated from $(dV_m/dt)_{t=0}$ vs $1/T$ as shown in Fig. 2, on the basis of the following equation derived from Eq. (3):

$$\left(\frac{dV_m}{dt}\right)_{t=0} = (1-V_{m_0})K \qquad (5)$$

It appears that the lower Y_2O_3 content corresponds to a lower activation energy.

The results of the dipping tests in distilled water in an autoclave are shown in Fig. 3. A similar tendency of the $t \to m$ transformation, as shown in Fig. 1, was observed at about 100°C, but at temperatures above 140°C, the saturation phenomena of monoclinic fractions were observed in distilled water.

The values of saturated monoclinic fractions in different specimens containing different Y_2O_3 content are shown in Table IV. In these specimens, tetragonal fractions in the equilibrium state at 1550°C, as obtained from the ZrO_2-$YO_{1.5}$ phase diagram by Scott,[11] are also given. However, because the tetragonal fractions are changeable with the cooling process, the actual tetragonal fractions of as-received specimens are somewhat different from those obtained in an equilibrium state at the sintering temperature of 1550°C.

The experimentally obtained saturated monoclinic fraction of each specimen can be assumed to be the transformed fraction of the "transformable" tetragonal phase[11,12] under the test conditions, whereas the other phase is cubic. Then in

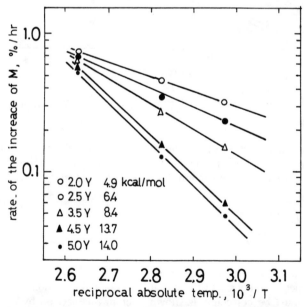

Fig. 2. Rate of increase of monoclinic phase vs reciprocal absolute temperature in 30% H_2SO_4.

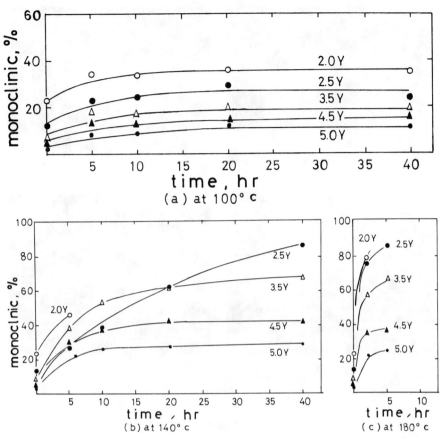

Fig. 3. Time dependence of the increase in monoclinic fraction in distilled water.

Table IV. Saturated Monoclinic Fraction for Specimens Dipped in Distilled Water (from Fig. 3b, c)

	Specimen				
	2.0Y	2.5Y	3.5Y	4.5Y	5.0Y
Observed (%)		~90	~70	~40	~30
Tetragonal phase* (%)	~92	~82	~63	~44	~35

*From phase diagram (Ref. 11).

consideration of the existence of cubic phase, the activation energies of transformable $t \rightarrow m$ transformation of 2.5, 3.5, 4.5 and 5.0 mol% Y_2O_3 are 5.8, 5.9, 5.5 and 4.2 kcal/mol, respectively.

Some specimens cracked or disintegrated after dipping tests.

Distribution of Yttrium and Structure Analysis

Figure 4 shows SEM photographs of specimens which were lapped and etched in 98% H_2SO_4, as indicated in the experimental procedure. Structures of all specimens consist of two types of regions, simple and relatively large crystalline grains (A, cubic phase) and white and fine grains (B). Region B is the

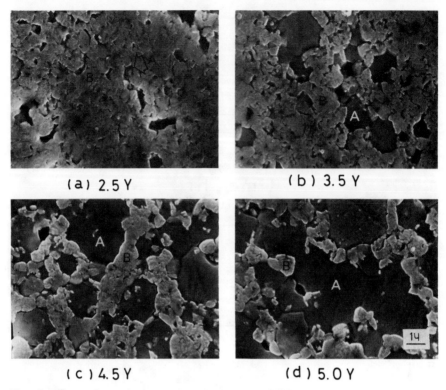

Fig. 4. Scanning micrographs of etched Y-PSZ.

main structure of 2.5 mol% Y_2O_3 and decreases as the Y_2O_3 content of a specimen increases.

As shown in Table V, region A is much richer in Y_2O_3 than region B, although the values were not strictly confirmed by EPMA. These results suggest that region B is the metastable tetragonal phase transformed to monoclinic through the etching process and is gradually transformed in sulfuric acid and in distilled water as mentioned earlier. On the other hand, the cubic region A did not undergo any phase changes nor did the Y_2O_3 distribution change before and after etching.

Morphology After Dipping Tests

Crack generation on certain specimens during dipping tests was found as mentioned above. As a typical case, Fig. 5 shows an SEM photograph of 2.0 mol% Y_2O_3 cracked after dipping in 30% sulfuric acid at 107°C for 80 h. Such cracks were also observed for some Y-PSZ specimens in boiling 20% hydrochloric acid, 30% nitric acid, 30% oxalic acid, and 30% caustic soda solutions.

Table V. Yttria Content in Regions A and B in Fig 4*

	Content (mol%)		
	2.5Y	4.5Y	5.0Y
Region A	6.8	7.1	7.5
Region B	2.4	3.0	3.5

*Typical values calculated from the intensity of Y Kα line.

Fig. 5. Scanning micrograph of 2.0 mol% Y_2O_3 after dipping in 30% H_2SO_4 at 107°C for 80 h.

Figure 6 shows a TEM photograph of 2.5 mol% Y_2O_3 before the dipping test. Micropores lie scattered at grain boundaries, but the inside of a grain is dense and relatively uniform, and the glassy phase could not be found using electron diffraction and TEM observation. This specimen was dipped in 30% sulfuric acid for 96 h at 107°C, and the surface was observed by TEM (Fig. 7). Microstructures such as twin formation appeared throughout the grains, and cracks were observed along the grain boundaries. They seemed to have initiated at triple points and grew along the

Fig. 6. Electron micrograph of as-received 2.5 mol% Y_2O_3.

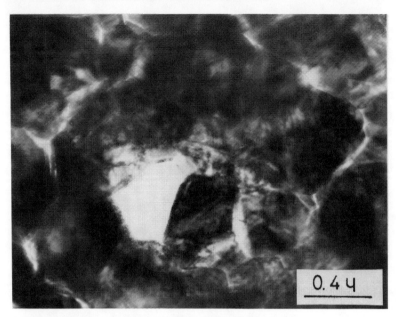

Fig. 7. Electron micrograph of 2.5 mol% Y_2O_3 after dipping in 30% H_2SO_4 at 107°C for 96 h.

grain boundaries (Fig. 8). A plain, translucent part appeared near the grain boundaries and along the edge of the sample. This part was identified as amorphous by electron-beam diffraction as shown in Fig. 8, and Zr and Y contents of this part were much lower than those of the specimen before the dipping test as studied by

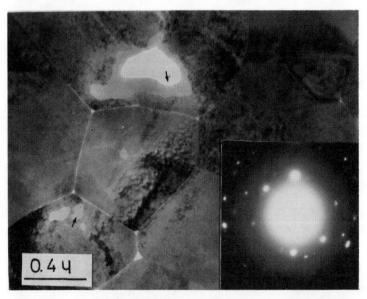

Fig. 8. Electron micrograph and microdiffraction pattern of 2.5 mol% Y_2O_3 thin film after dipping in 98% H_2SO_4 at 104°C for 0.5 h.

STEM analysis. These facts seem to indicate that the sample was attacked by sulfuric acid and was corroded.

Crack generation of Y-PSZ is observed at 300°–400°C during long aging tests in air,[9,10] but in aqueous solutions it takes place more catastrophically at lower temperatures in shorter times, possibly because of a multiple effect of the decrease in the surface energy and corrosion which promotes phase transformation.

Summary

Phase transformation of Y-PSZ from tetragonal-to-monoclinic phase was observed in sulfuric acid solution at temperatures above 63°C. The transformation was also observed in distilled water and acidic and basic aqueous solutions.

The transformation showed a saturation in the monoclinic fraction, and the saturated monoclinic fraction varied according to the Y_2O_3 content of the materials sintered under similar conditions.

In aqueous solutions, Y-PSZ showed microstructural changes in grains and cracking along grain boundaries at lower temperatures ($\approx 100°C$) and in shorter times than previous aging tests in air at 300–400°C. The cracks were attributed not only to the phase transformation but also to corrosion.

Acknowledgments

The authors thank T. Ogata for bending tests and fracture toughness measurements. The authors are grateful to M. Yoshikawa and K. Yamamoto for SEM-EPMA analysis.

References

[1] R. C. Garvie and P. S. Nicholson, "Structural and Thermomechanical Properties of Partially Stabilized Zirconia in the $CaO-ZrO_2$ System," *J. Am. Ceram. Soc.*, **55** [3] 152–57 (1972).
[2] N. Claussen, J. Steeb, and R. F. Pabst, "Effect of Induced Microcracking on the Fracture Toughness of Ceramics," *Am. Ceram. Soc. Bull.*, **56** [6] 559–62 (1977).
[3] D. L. Porter and A. H. Heuer, "Microstructural Development in MgO-Partially Stabilized Zirconia (Mg-PSZ)," *J. Am. Ceram. Soc.*, **62** [5–6] 298–305 (1979).
[4] R. C. Garvie, R. H. Hannink, and R. T. Pascoe, "Ceramic Steel?" *Nature (London)*, **258**, 703 (1975).
[5] T. K. Gupta, J. H. Bechtold, R. C. Kuznicki, and L. H. Cadoff, "Stabilization of Tetragonal Phase in Polycrystalline Zirconia," *J. Mater. Sci.*, **12**, 2421 (1977).
[6] T. K. Gupta, F. F. Lange, and J. H. Bechtold, "Effect of Stress Induced Phase Transformation on the Properties of Polycrystalline Zirconia Containing Tetragonal Phase," *J. Mater. Sci.*, **13**, 1464 (1978).
[7] F. F. Lange; E.C.M. 3, Vol. 3, Cambridge (1979).
[8] A. G. Evans, "Transformation Toughening in Partially Stabilized Zirconia," *Acta Metall.*, **29**, 447 (1981).
[9] K. Kobayashi, H. Kuwajima, and T. Masaki, "Phase Change and Mechanical Properties of ZrO_2-Y_2O_3 Solid Electrolyte after Aging," *Solid State Ion.* **3,4**, 489 (1981).
[10] N. Yamamoto; p. 375 in Ceramic Science and Technology at Present and in Future, 1981.
[11] H. G. Scott, "Phase Relationships in the Zirconia-Yttria System," *J. Mater. Sci.*, **10**, 1527 (1975).
[12] F. F. Lange and D. J. Green; pp. 217–25 in Advances in Ceramics, Vol. 3. Edited by A. H. Heuer and L. W. Hobbs, The American Ceramic Society, Columbus, OH, 1981.

Physical, Microstructural, and Thermomechanical Properties of ZrO₂ Single Crystals

R. P. INGEL, D. LEWIS, B. A. BENDER, AND R. W. RICE

U.S. Naval Research Laboratory
Washington, DC

The physical and thermomechanical properties of Y_2O_3-partially stabilized ZrO_2 (Y-PSZ) single crystals are reviewed and compared to crystals containing CaO and MgO. These Y-PSZ crystals show impressive strength and fracture toughness over the range 25°–1500°C. The temperature dependence of the toughness in Y-PSZ and Y_2O_3-cubic stabilized ZrO_2 (Y-CSZ) is discussed in relation to the different toughening mechanisms operative in these materials.

Most studies of partially stabilized zirconia (PSZ) have utilized polycrystalline materials fabricated by conventional powder processing techniques,[1-6] with the limitations of grain boundaries, porosity, and other processing defects. Conclusions regarding the temperature dependence of the toughening have been limited by the presence of grain boundaries which weaken at elevated temperatures. Some of the limitations noted above have been reduced or eliminated by the availability of relatively large single crystals of cubic stabilized ZrO_2 (CSZ) and, more recently, single crystals of Y-PSZ, both produced by directional solidification from the melt.[7,8] While this fabrication technique may be used to produce zirconia single crystals with a variety of stabilizers, it has proved most useful in the production of the Y-PSZ crystals.

The availability of these single crystals has made possible a comprehensive study of the physical and mechanical properties of Y-PSZ, some details of which are reported here. Comparisons with crystals containing CaO and MgO are also presented. Other results are reported elsewhere, including theoretical calculations of transformation-toughening.[9-12]

Experimental Procedure

ZrO_2 single crystals, typically 1–3 cm in diameter and 4–6 cm in length, containing 0–20 wt% Y_2O_3 were grown in skull melts 7–30 cm in diameter.* Crystals partially stabilized with CaO and MgO were similarly obtained for comparison with the Y-PSZ crystals. The Mg-PSZ and Ca-PSZ crystals contained 2.8 wt% MgO and 4 wt% CaO compositions, respectively, in the middle of the cubic-tetragonal phase field in each system. Fully stabilized Ca-CSZ crystals with 9 wt% CaO were also evaluated.

Density was measured by the Archimedes technique. Phase content and crystal lattice parameters were determined by X-ray diffraction analysis (XRD).

---
*Ceres Corp., North Billerica, MA.

Laue back-reflection analysis was used to determine the crystal growth directions and to orient specimens for mechanical property measurements. Thin sections, prepared by standard techniques, were examined by optical and transmission electron microscopy. Elastic constants of the single crystals were determined by the acoustic pulse-echo overlap technique.[13]

Specimens for flexural-strength and fracture-toughness testing were prepared by standard techniques, with the specimen length approximately parallel to the crystal-growth direction. Bars 2 by 3 mm in cross section were cut for strength and single-edge notched beam (SENB) tests, the latter having a 0.3-mm diamond sawn notch. These were tested in three-point flexure on a span of 12 mm at $-196°C$ in liquid N_2 and at $22-1500°C$ in air. Applied-moment double cantilever beam (AMDCB) tests at room temperature used precracked specimens, 1 by 5 by 12 mm, with a center groove 0.3 mm wide and ~0.6 mm deep.

Hardness was measured using a Vicker's diamond indentor with loads ranging from 0.5 to 800 N. Fracture toughness was also measured in this procedure at the higher loads using the standard indentation fracture toughness analysis.[14] Finally, fracture surface topography, fracture features, and fracture origins were examined by scanning electron microscopy (SEM), as were the large-scale features of the microstructure.

Results and Discussion

A summary of much of the physical and mechanical property data discussed below is given in Table I.

Physical Characterization

The crystals ranged from a transparent and pale yellow appearance at 12–20 wt% Y_2O_3 to a banded and cloudy appearance at 7–8 wt%, an opaque white at 4–5 wt%, and opaque with a greenish-gray tint at <4 wt% Y_2O_3. Compositions between 4 and 20 wt% Y_2O_3 had smooth facets between crystals, while at <4 wt% the crystals had rough surfaces. The Ca- and Mg-PSZ crystals, although smaller than the Y-PSZ crystals, were similar in appearance.

Transmission light microscopy of thin sections showed a "tweedlike" structure for <8 wt% Y-PSZ. This structure is presumably the effect of the overlapping

Table I. Comparison of Physical and Mechanical Properties of ZrO_2 Crystals

Crystal	Composition (wt%)	Precipitates (vol%)	Density (kg/m³)	Hardness (GPa)	Young's modulus (GPa)	Flexural strength (MPa)	Fracture toughness (MPa·m$^{1/2}$)
Mg-PSZ	2.8	48	5790	14.4	200	685 ± 48	4.82 ± 0.56
Ca-PSZ	4.0	38	5850	17.1	210	661 ± 33	3.97 ± 0.46
Ca-CSZ	9	0	5680	17.2	210	241 ± 28	2.54 ± 0.13
Y-PSZ	5	52	6080	13.6	233	1384 ± 80	6.92 ± 0.14
Y-CSZ	20	0	5910	16.1	233	346 ± 58	1.91 ± 0.16
Polycrystalline zirconia					Young's modulus (GPa)	Flexural strength (MPa)	Fracture toughness (MPa·m$^{1/2}$)
Mg-PSZ[3,4,22,26]					200	430–700	4.7–15
Ca-PSZ[1,2]					200–217	400–650	5.0–9.6
Y-PSZ[6,22]					210–238	696–980	5.8–9.0

strain fields of a high density of precipitates. X-ray diffraction showed only cubic zirconia for compositions with 15–20 wt% Y_2O_3, with increasing amounts of tetragonal zirconia with decreasing Y_2O_3 content, down to 5 wt%. Compositions between 0 and 4 wt% Y_2O_3 showed increasing amounts of monoclinic phase and decreasing amounts of tetragonal and/or cubic phases with decreasing stabilizer level, and the zirconia was single-phase monoclinic at 0 wt% Y_2O_3. The Mg-PSZ crystals contained cubic, tetragonal, and some monoclinic phases; the Ca-PSZ contained cubic and tetragonal phases, and the Ca-CSZ, of course, was all cubic. X-ray diffraction analysis gave cubic lattice parameters for the system Y_2O_3-ZrO_2 in good agreement with polycrystalline[15] and single-crystal[16] measurements. The lattice strains ($\approx$0.002–0.004) for specific compositions were determined by differences in lattice parameters between single crystals and finely powdered single-crystal samples and are summarized elsewhere.[9,17] Laue back-reflection analysis of 4–20 wt% Y_2O_3 crystals confirmed the $\langle 110 \rangle$ as the preferential growth direction, in agreement with earlier results.[7,16]

Transmission electron microscopy (TEM) and selected-area diffraction (SAD) on 5–12 wt% Y-PSZ clearly indicated that these compositions yielded a cubic solid-solution matrix, with increasing volume fraction of tetragonal second phase with decreasing Y_2O_3 content — 30 vol% at 8 wt% Y_2O_3 increasing to ~50 vol% at 5 wt% Y_2O_3. Transmission electron microscopy revealed a finely oriented dispersion of platelike precipitates, similar to those observed in polycrystalline Y-PSZ,[2,3,5] as shown in Fig. 1(C). Additional characterization of the precipitate dispersion is given elsewhere.[9] The Mg- and Ca-PSZ crystals showed different precipitate morphologies from the Y-PSZ, as shown in Figs. 1(A) and 1(B). Ca-PSZ crystals contained roughly equiaxial precipitates, while the precipitates in the Mg-PSZ were flattened ellipsoids.

Mechanical Properties

Elastic constants were relatively independent of composition for 4–20 wt% Y_2O_3 whereas the <4 wt% Y_2O_3 crystals containing the monoclinic phase showed an increase in the acoustic damping. The cubic single-crystal elastic constants were $C_{11} = 410$ GPa, $C_{12} = 110$ GPa, and $C_{44} = 60$ GPa, in agreement with prior

Fig. 1. Precipitate morphologies (TEM, DF) for (A) Ca-PSZ, (B) Mg-PSZ and (C) Y-PSZ single crystals, printed reversed for enhanced contrast.

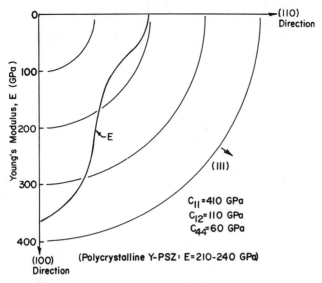

Fig. 2. Variation of calculated Young's modulus with direction in a ⟨110⟩ plane.

work.[18-20] Young's modulus as a function of orientation (Fig. 2), shows that ⟨111⟩ and ⟨100⟩ are the low and high modulus directions, respectively. The Zener anisotropy ratio, $2C_{44}/(C_{11} - C_{12})$, of 0.4 shows that zirconia is one of the most elastically anisotropic cubic ceramics. The Hashin-Shtrikman upper and lower bounds[21] for the Young's modulus and Poisson's ratio for polycrystalline zirconia are 226 and 233 GPa and 0.315 and 0.321, respectively, in excellent agreement with actual measurements on polycrystalline zirconia which range from 200–240 GPa and 0.300–0.315, respectively.[22]

Hardness for the various compositions decreased with increasing tetragonal phase, decreased greatly with increasing monoclinic content, and decreased moderately with increasing indentor load. The hardness reached a plateau for loads greater than 5 N of ≈16 GPa for the cubic crystals, 13–14 GPa for the cubic-tetragonal materials, and ≈6.6 GPa for the pure monoclinic material. The decreasing hardness with increasing tetragonal content suggests that the precipitates are relieving the stresses around the Vicker's indentor by transformation and twinning.[3,23,24] The hardness of the Mg-PSZ and Ca-PSZ crystals was 14.4 and 17.1 GPa, respectively, while that of the Y-PSZ crystals was 13.6 GPa.

Fracture toughness showed a pronounced maximum at ≈4 wt% Y_2O_3 (see Fig. 3). Toughness increases coincided with increases in volume fraction of the tetragonal precipitates as the stabilizer content decreased. The bulk of the fracture toughness results agree quite well with those obtained by other investigators and other techniques.[5,6] Only small changes in fracture toughness occurred for heat treatment at 1400°–1500°C for up to 170 h, indicating that this treatment does not effect sufficient microstructural changes to affect the mechanical properties. Indentation fracture toughness determined on a ⟨110⟩ surface showed moderate anisotropy in Y-CSZ crystals, ranging from 1.6 MPa·m$^{1/2}$ for cracks oriented in the ⟨111⟩ direction to 2.6 MPa·m$^{1/2}$ for cracks in the ⟨100⟩ direction. These results agree roughly with that expected from constant fracture surface energy and the

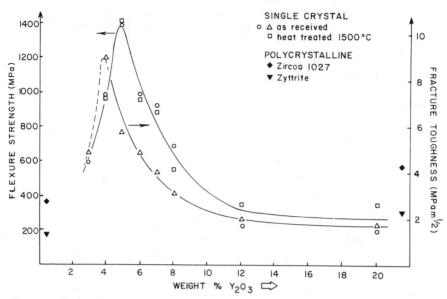

Fig. 3. Flexural strength and fracture toughness vs composition for Y-PSZ single crystals.

variation of Young's modulus with orientation. No significant orientation dependence of K_{Ic} was noted in a preliminary measurement of a PSZ crystal. This appears reasonable in view of the homogeneous, three-dimensional dispersion of precipitates. The indentation fracture toughness for the Mg-PSZ, Ca-PSZ, Ca-CSZ, and Y-CSZ crystals were 4.8, 4.0, 2.5, and 1.9 MPa·m$^{1/2}$, respectively.

Flexural Strength

As shown in Fig. 3, the flexural strength of the crystals increased from ≈200 MPa for the Y-CSZ to 1400 MPa at 5 wt% Y_2O_3, paralleling the increases in volume fraction of tetragonal precipitates in the cubic matrix, and was 144 MPa for the fully monoclinic material. The sensitivity to machining-induced transformation was evaluted by heating to eliminate the transformed surface layer, followed by remachining or polishing of the specimens. The compositions from 12–20 wt% showed significant improvements with heat treatment or polishing, e.g., from ≈200 to 350 MPa for the 20 wt% crystals. The flexural strength of Y-PSZ was not sensitive to heat treatment,[†] in agreement with XRD results which confirmed the absence of monoclinic zirconia on ground surfaces. Overall, it appears that the high strength of the PSZ crystals results from the bulk toughness of the material and does not depend on the presence of a compressive surface layer associated with machining damage.

The Mg-PSZ, Ca-PSZ, and Ca-CSZ crystals showed flexural strengths consistent with their K_{Ic}s: 685, 661, and 241 MPa, respectively.

[†]The 5 wt%-Y_2O_3 crystal had a strength of ≈1200 MPa when heated to 1800°C, probably due to precipitate growth during heat treatment and subsequent transformation to monoclinic symmetry on cooling. XRD confirmed the presence of the monoclinic phase in the heat-treated crystal, which had not been present in the as-received material. Very careful surface grinding, with small depth of feed, increased the strength to ≈2 GPa.

Examination of fracture surfaces showed typical fractographic features which permitted ready identification of the fracture origins, which were typically machining flaws. Fracture toughness values calculated from the flaw size, shape, and failure stress agreed quite well with those determined by other techniques, indicating similar micro- and macrotoughnesses. The fracture surfaces of the Y-CSZ specimens were very smooth except for distinctive zones of hackle, typical of single-crystal fracture markings.[25] The PSZ compositions exhibited the single-crystal fracture markings and a finer scale of microroughness. This microroughness, also observed on transgranular fracture surfaces in large grain-size polycrystalline Mg-PSZ,[22,26] results from the stress-induced transformation, impeding crack propagation.[27,28]

High-Temperature Mechanical Properties

The effect of test temperature on flexural strength, fracture toughness, and stress-strain behavior for the Y-PSZ and Y-CSZ crystals is presented elsewhere.[10] Only a brief summary is given here. The results for Y-PSZ show the expected decline in toughness and strength with increasing temperature, up to the transformation temperature, as would be expected. However, the toughness of the Y-PSZ material for temperatures >1100°C remains approximately twice that of the Y-CSZ. This result indicates the large role that crack–precipitate interaction (CPI) toughening plays in Y-PSZ. The flexural strength of the Y-PSZ crystals also remains about twice that of the Y-CSZ, up to the highest test temperature, ≈1500°C. In the high-temperature regime, 1000°–1500°C, the Y-CSZ crystals show a strain-rate-dependent yield stress, where the Y-PSZ crystals show little or no evidence of plasticity. This result suggests another benefit of the precipitate dispersion, i.e., limiting large scale plasticity at high temperature. Taken together with the room-temperature results, the high-temperature results suggest that the precipitate dispersion in Y-PSZ may provide toughening and strengthening over a wide temperature range by a combination of three mechanisms: transformation-toughening, CPI-toughening, and dispersion-strengthening.

Summary and Conclusions

Skull-melting technology can be utilized to obtain PSZ single crystals of quite a large size, with microstructures similar to those of polycrystalline materials, fracture toughness comparable to polycrystalline PSZ, and strength 2–3 times greater. The absence of grain boundaries and other defects associated with polycrystalline materials has simplified the interpretation of the measured mechanical properties, especially their temperature dependence. The high-temperature mechanical properties are significantly better than those of any polycrystalline PSZ.[10] The temperature dependence of the strength and fracture toughness of the Y-PSZ crystals, and comparison to the Y-CSZ crystals, shows that, while the major contribution to toughening at low temperatures is the transformation of tetragonal precipitates, other mechanisms, such as CPI-toughening, become important at moderate and high temperatures. The high-temperature flexural creep behavior of the Y-PSZ crystals also appears to benefit from the dispersion-strengthening effect of the dense array of precipitates. Thus, the precipitates in the Y-PSZ provide toughening and strengthening over a wide range of temperatures through the combination of transformation-toughening and CPI-toughening at low temperatures, CPI-toughening at moderate temperatures, and precipitate-dispersion-strengthening at high temperatures.

References

[1] R. C. Garvie, R. H. J. Hannink, and C. Urbanik, *Ceramurgia Int.*, **6** [1] 19–24 (1980).

[2] R. H. J. Hannink, K. A. Johnston, R. T. Pascoe, and R. C. Garvie; pp. 116–36 in Advances in Ceramics, Vol. 3. Edited by A. H. Heuer and L. W. Hobbs. The American Ceramic Society, Columbus, OH, 1981.

[3] D. L. Porter and A. J. Heuer, *J. Am. Ceram. Soc.*, **62** [5–6] 298–305 (1979); "Mechanisms of Toughening Partially Stabilized Zirconia (PSZ)," *ibid.*, **60** [3–4] 183–84 (1977).

[4] D. L. Porter, A. G. Evans, and A. H. Heuer, *Acta Metall.*, **27** [10] 1649–54 (1979).

[5] P. G. Valentine, R. D. Maier, and T. E. Mitchell; NASA Report NASA-CR-164438, 1981.

[6] F. F. Lange, "Transformation Toughening: Part 1; Size Effects Associated with the Thermodynamics of Constrained Transformations," *J. Mater. Sci.*, **17**, 225–34 (1982); "Transformation Toughening: Part 2; Contributions to Fracture Toughness," *ibid.*, **17**, 235–36 (1982); "Transformation Toughening: Part 3; Experimental Observations in the $ZrO_2-Y_2O_3$ System," *ibid.*, **17**, 240–46 (1982); "Transformation Toughening: Part 4; Fabrication, Fracture Toughness and Strength of $Al_2O_3-ZrO_2$ Composites," *ibid.*, **17**, 247–54 (1982); "Transformation Toughening: Part 5; Effect of Temperature and Alloy on Fracture Toughness," *ibid.*, **17**, 255–62 (1982).

[7] V. I. Aleksandrov et. al.; pp. 421–80 in Current Topics in Material Science, Vol. 1. Edited by E. Kaldis. North-Holland Publishing, Amsterdam, 1978.

[8] K. Nassau, *Lapidary J.*, **31**, 900 (1977); "Cubic Zirconia, An Update," *ibid.*, **35**, 1194 (1981).

[9] R. P. Ingel, Ph.D. Thesis, Catholic University, Washington, DC, 1982; Univ. Microfilms Int. #8302474.

[10] R. P. Ingel et. al., *J. Am. Ceram. Soc.*, **65** [9] C-150–C-151 (1982).

[11] R. P. Ingel et. al., *J. Am. Ceram. Soc.*, **65** [7] C-108–C-109 (1982).

[12] R. P. Ingel and D. Lewis; unpublished work.

[13] H. Kolsky; Stress Waves in Solids. Dover Publications, New York, 1963.

[14] A. G. Evans; pp. 112–35 in Fracture Mechanics Applied to Brittle Materials. Edited by S. W. Freiman. ASTM, Metals Park, OH, 1979.

[15] H. G. Scott, *J. Mater. Sci.*, **10** [9] 1527–35 (1975).

[16] V. I. Aleksandrov et. al., *Izv. Akad. Nauk SSSR, Neorg. Mater.*, **12** [2] 273–77 (1976).

[17] J. E. Thiebaud; B. S. Thesis, Alfred University, Alfred, NY, (1981).

[18] V. I. Aleksandrov et. al., *Sov. Phys. Solid State*, **16** [5] 1456–59 (1975).

[19] A. Feinberg and C. H. Perry, *J. Mater. Sci.*, **42**, 513–18 (1981).

[20] N. G. Pace et. al., *J. Mater. Sci.*, **4**, 1106–19 (1969).

[21] Z. Hashin and S. Shtrkiman, *J. Mech. Phys. Solids*, **10**, 343–52 (1962).

[22] D. Lewis, R. P. Ingel, and L. Schioler; unpublished work.

[23] T. W. Coyle, U.S. Naval Research Laboratory, Washington, DC, 1982, unpublished results.

[24] R. H. Hannink and M. V. Swain, *J. Mater. Sci.*, **16**, 1428–31 (1981).

[25] R. W. Rice; ASTM, Metals Park, OH, in press.

[26] R. H. J. Hannink and M. V. Swain, *J. Austr. Ceram. Soc.*, **18** [2] 59–62 (1982).

[27] K. T. Faber and A. G. Evans, *Acta Metall.*, **31** [4] 565–76 (1983).

[28] K. T. Faber and A. G. Evans, *Acta Metall.*, **31** [4] 577–84 (1983).

Ripening of Inter- and Intragranular ZrO$_2$ Particles in ZrO$_2$-Toughened Al$_2$O$_3$

B. W. Kibbel and A. H. Heuer

Case Western Reserve University
Department of Metallurgy and Materials Science
Case Institute of Technology
Cleveland, OH 44106

Ripening of intergranular and intragranular ZrO$_2$ particles in ZrO$_2$-toughened Al$_2$O$_3$ (ZTA) was studied at 1600°C. The increase in average ZrO$_2$ particle size and average Al$_2$O$_3$ grain size, as well as the change in ZrO$_2$ particle-size distributions, are given and analyzed in terms of classical Ostwald ripening theory, on the one hand, and coalescence due to particle drag, on the other.

The enhanced mechanical properties of ZrO$_2$-toughened Al$_2$O$_3$ (ZTA),[1] compared to single-phase Al$_2$O$_3$, depend sensitively on the ZrO$_2$ particle size, morphology, and distribution. Transformation-toughening requires that the tetragonal (t) symmetry of the ZrO$_2$ particles be retained at room temperature; this can be accomplished if the particle size is below a certain critical size, which varies from 0.5 to 2.0 μm,[2] depending, for isolated particles, on particle morphology and solute content.

Two kinds of ZrO$_2$ particles exist in ZTA—faceted intergranular ZrO$_2$ particles located between Al$_2$O$_3$ grains (mostly at grain corners) and spherical intragranular ZrO$_2$ particles, which are located within the Al$_2$O$_3$ grains (Fig. 1). The intergranular particles are generally larger than intragranular particles, because of the faster coarsening kinetics associated with the grain boundaries, and have a well-defined critical size. Intragranular particles are more resistant to transformation, presumably due to difficulty of nucleating the martensitic transformation in a spherical particle,[3] and do not show a well-defined critical size.

Second-phase particles in ceramic compacts can ripen by two distinct processes—Ostwald ripening or coalescence. During Ostwald ripening, small particles disappear while large ones grow because of the variation of solubility with particle size (the Thompson–Freundlich relation):

$$C_r = C_0 \exp(2\gamma\Omega/RTr) \qquad (1)$$

Here, C_r is the solute concentration in the matrix adjacent to a particle of radius r, C_0 the equilibrium solute concentration, γ the matrix-particle interfacial energy, and Ω the molar volume of the solute. For all but the smallest particles, $C_r \approx 1/r$. Thus, the resulting solute concentration in the matrix near small particles is larger than near large particles, and diffusion down this gradient causes larger particles to grow at the expense of smaller ones (Fig. 2(A)).

Alternatively, particles can grow by coalescence, if they are dragged by migrating matrix grain boundaries. In this process, grain growth and disappearance of small matrix grains causes particles to meet, coalesce, and thus form larger particles (Fig. 2(B)).

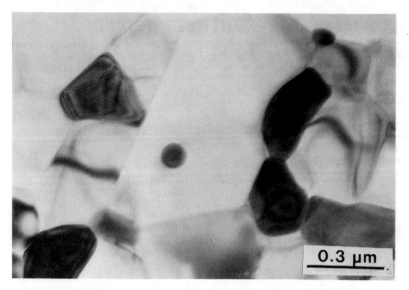

Fig. 1. TEM micrograph showing inter- and intragranular ZrO_2 particles.

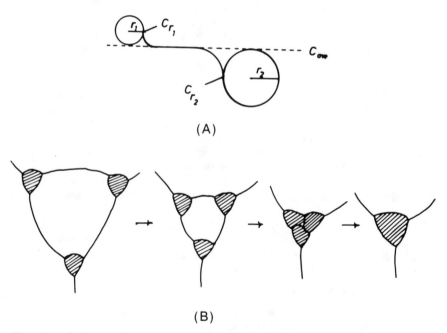

Fig. 2. Schematic representations of (A) Ostwald ripening and (B) coalescence.

Experimental Technique and Results

Ripening of inter- and intragranular ZrO_2 particles in two ZTAs of different character was studied. Both materials were of near theoretical density. Material I* contained 15 vol% ZrO_2; the t/m ratio was ≈4, and most of the ZrO_2 particles were

Fig. 3. TEM micrographs of intergranular material. Both micrographs show the same area with the objective aperture removed from the RHS to help discriminate between ZrO_2 particles and strongly diffracting Al_2O_3 grains.

intergranular (Fig. 3). Material II* contained 10 vol% ZrO_2 (with 1 mol% Y_2O_3 addition), and the t/m ratio was ≈19. This material has some intergranular ZrO_2, but most of the particles were intragranular (Fig. 4).

Samples of both materials were heated at 1600°C for up to 40 h. Material I was studied using scanning electron microscopy (SEM), but the fine intragranular particles in material II required transmission electron microscopy (TEM) examination.

Stereological analysis of intergranular ripening was performed by measuring areal fractions on polished sections automatically with a Cambridge image analyzer. The data were then deconvoluted to reflect sampling and sectioning bias by the Saltykov method.[4]

For the intragranular particles, particle size was measured directly by TEM, as the majority of particles were small compared to the foil thickness. Random lines were used to sample the intragranular ZrO_2 particles, and the data were deconvoluted to reflect sampling bias, as follows. Since the probability of a particle being chosen for measurement is proportional only to the particle radius, each size class was divided by its radius and the distribution renormalized. The Al_2O_3 grain size in both materials was measured by the usual linear intercept method, corrected for volume fraction.

Typical micrographic results for the intergranular and intragranular cases are shown in Figs. 5 and 6, respectively. Significant growth has occurred for both types of particles. Tables I and II show the mean ZrO_2 particle radii and the mean

---

*Materials I and II were kindly provided by Drs. N. Claussen (Max-Planck-Institut für Metallforschung, Stuttgart, Federal Republic of Germany) and P. F. Becher (Oak Ridge National Laboratory, Oak Ridge, TN), respectively.

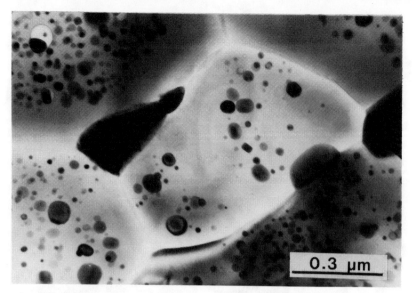

Fig. 4. TEM micrograph of intragranular material.

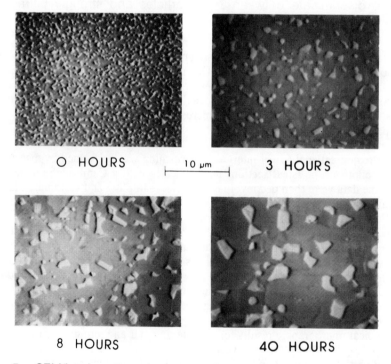

Fig. 5. SEM/backscattered micrograph results of intergranular ripening treatments at 1600°C.

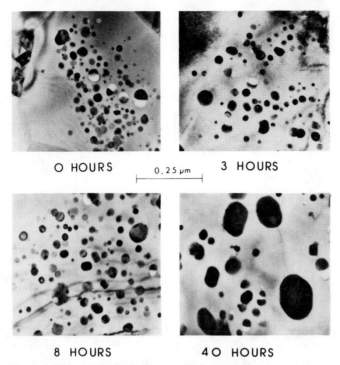

Fig. 6. TEM micrograph results of intragranular ripening treatments at 1600°C.

Al_2O_3 grain size for the several heat treatments employed. The ZrO_2 particle-size distributions are shown in Figs. 7 and 8; the intergranular particles assume a left-skewed distribution on a logarithmic scale after annealing, whereas the intragranular particles have an approximate logarithmic normal distribution.

Discussion

Tables I and II show that intergranular ZrO_2 particles grow faster than intragranular particles, despite their initially larger size; the kinetics of particle growth are clearly faster on the grain boundaries than within Al_2O_3 grains.

The data of Table I for intergranular particles (material I) were analyzed in various ways but a plot of $\bar{r}^4 - \bar{r}_0^4$ vs time, t (where the bars signify an average and r_0 is the starting size), appeared most satisfactory (Fig. 9). (Data of Green[5] of a similar ripening study in ZTA also fit this same time dependence.) Speight[6] showed that for Ostwald ripening via grain-boundary diffusion

$$\bar{r}^4 - \bar{r}_0^4 = \left(\frac{4\gamma\Omega CD_{gb}\delta}{3ABRT}\right)t \tag{2}$$

where D_{gb} is the grain-boundary diffusion coefficient, δ the grain-boundary "thickness," L the matrix grain size, A a constant dependent on surface and grain-boundary energies, and B a constant dependent on area fraction of a boundary occupied by particles.

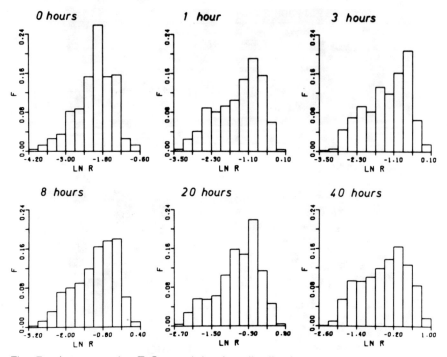

Fig. 7. Intergranular ZrO$_2$ particle-size distributions.

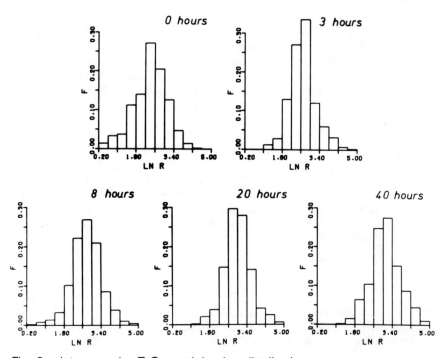

Fig. 8. Intragranular ZrO$_2$ particle-size distributions.

Table I. Intergranular Ripening

Duration at 1600°C (h)	Mean ZrO$_2$ particle size (μm) [σ]*	Mean Al$_2$O$_3$ radius ±10% error (μm)
0	0.13 [0.15]	0.19
1	0.23 [0.25]	0.42
3	0.27 [0.29]	0.53
8	0.43 [0.45]	0.66
20	0.50 [0.53]	0.99
40	0.59 [0.68]	1.34

*$\sigma = \exp(\sigma_{\log})$, where $\sigma_{\log}$ = standard deviation of the log distribution.

Table II. Intragranular Ripening

Duration at 1600°C (h)	Mean ZrO$_2$ particle size (nm) [σ]*	Mean Al$_2$O$_3$ radius ±10% error (μm)
0	8.6 [7.0]	0.51
3	13.1 [8.1]	0.85
8	13.2 [9.6]	0.93
20	17.9 [11.2]	1.16
40	19.0 [13.5]	1.35

*$\sigma = \exp(\sigma_{\log})$, where $\sigma_{\log}$ = standard deviation of the log distribution.

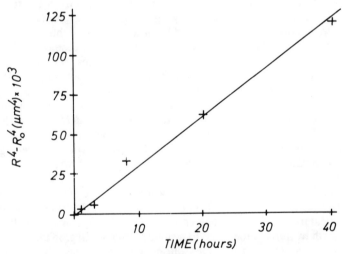

Fig. 9. $\bar{r}^{-4} - \bar{r}_0^{-4}$ vs t for intergranular ZrO$_2$.

Despite this "classical" time dependence, coalescence and not Ostwald ripening is thought to dominate intergranular ripening in our experiments. Many ZrO$_2$ particles are actually multiple particle clusters (Fig. 10); in fact, such clusters constituted 50 vol% of the ZrO$_2$ in material I but were treated as single particles in the data analysis. We believe that such particle clustering is good evidence for coarsening by coalescence, which as we show next also gives an $r^4 \approx t$ coarsening law. Furthermore, the diffusivity of ZrO$_2$ along Al$_2$O$_3$ grain boundaries implied by Eq. (2) at 1600°C is 5×10^{-6} cm^2/s—a figure which is high.

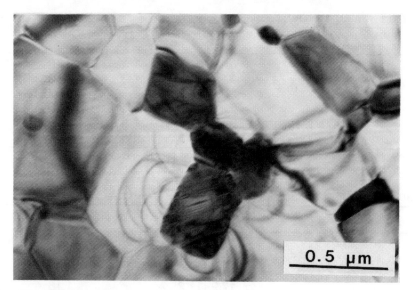

Fig. 10. TEM micrograph of ZrO$_2$ particle cluster.

As coalescence (which in fact starts during sintering) proceeds, all of the Al$_2$O$_3$ grain corners become "filled" with ZrO$_2$ particles. Given a fixed volume fraction and the absence of "breakaway" grain growth, a constant ratio exists between average Al$_2$O$_3$ grain size and ZrO$_2$ particle size, and thus both the matrix grains and the second-phase particles grow with the same time dependence. The particles exert a strong drag on the migrating boundaries, whose velocity v is controlled by the velocity at which the ZrO$_2$ particles can migrate. It is well established[7–9] that the velocity at which such a particle can migrate, assuming the necessary mass transport occurs by interfacial diffusion, is inversely proportional to the particle volume, i.e., $v \sim 1/r^3$. We thus expect $v \sim dG/dt \sim dr/dt \sim 1/r^3$ or $\bar{r}^4 - \bar{r}_0^4 = Kt$ where G is the grain size and K a constant.

Thus, a $t^{1/4}$ time dependence will be found even if particle coarsening occurs by coalescence. Coarsening of ZrO$_2$ by coalescence in ZTA was also observed by Lange and Hirlinger.[10]

Although it is clear from the microstructure of the intragranular material (material II) that exaggerated Al$_2$O$_3$ grain growth occurred during fabrication, the fabrication heat treatments were long enough for the Al$_2$O$_3$ grains to "exhaust" this stage of growth by impingement and to resume normal grain growth. No additional abnormal grain growth was observed during ripening. Thus, the population of intragranular grains was stable, and the particle-size distribution was not altered through breakaway.

Particle growth, due to the location of the intragranular ZrO$_2$ particles within the Al$_2$O$_3$ grains, could have occurred by Ostwald ripening involving bulk diffusion. Theory for such a process has been well developed by Wagner[11] and Lifshitz and Slyozov (L–S–W),[12] who predict a $t^{1/3}$ law. Our data do not follow such a law; the coarsening kinetics have a much smaller time exponent, the rate slowing significantly with time (Fig. 11). Furthermore, L–S–W theory predicts that particles undergoing Ostwald ripening by bulk diffusion will approach a size

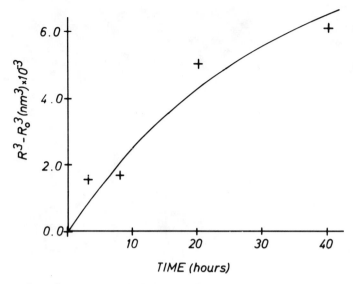

Fig. 11. $r^3 - r_0^3$ vs t for intragranular ZrO_2.

distribution that is skewed to the left and has a sharp cutoff at 1.5 $\bar{r}$. The intragranular ZrO_2 grain-size distributions (Fig. 8) do not show this form—the observed log-normal distributions extend far beyond 1.5 $\bar{r}$.

We believe that growth of intragranular ZrO_2 particles is primarily occurring by grain-boundary diffusion, because bulk diffusion is slow in this system. Assume that significant amounts of solute can be exchanged only when particles are on migrating Al_2O_3 grain boundaries. If two particles are both intersected by a migrating grain boundary, the longer can grow at the expense of the smaller. As grain growth proceeds, there will be a decrease in frequency of Al_2O_3 boundary/ZrO_2 particle contacts, and the growth rate will decrease with time. Strong evidence for this notion is provided by the many "denuded" or "particle-free" zones observed near Al_2O_3 grain boundaries (Fig. 12). We are presently attempting to model this type of coarsening, but it seems clear that its time dependence should be lower than that for either bulk or grain-boundary diffusion-controlled Ostwald ripening.

Conclusions

Ripening of intragranular and intergranular ZrO_2 particles in polycrystalline Al_2O_3 was studied at 1600°C. The intragranular particles appear to coarsen by coalescence, as particles are dragged by migrating Al_2O_3 grain boundaries, and obey an $r \approx t^{1/4}$ law. The ultimate rate-controlling process for such coarsening is the rate at which an individual ZrO_2 particle can migrate, i.e., the interfacial ZrO_2-Al_2O_3 diffusion kinetics.

The intragranular particles coarsen at a much slower rate but are still dominated by interfacial diffusion kinetics. Bulk diffusion is so slow in this system that large particles can coarsen (while smaller ones shrink) only if both types of particles are situated on grain boundaries. This leads to growth kinetics much slower than an $r \sim t^{1/2}$ or $t^{1/4}$ Ostwald ripening law and also to "denuded" zones adjacent Al_2O_3 grain boundaries.

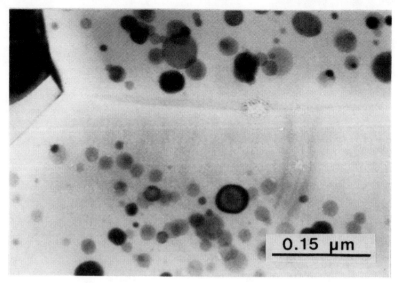

Fig. 12. Particle-free zone in intragranular material.

Acknowledgments

This research was supported by the DOE under Grant No. DEAS0277, ERR04217. A. H. Heuer acknowledges the Alexander von Humboldt Foundation for a Senior Scientist Award, which made possible his sabbatical leave at the Max-Planck-Institut für Metallforschung, Stuttgart, Federal Republic of Germany, where this paper was written. B. W. Kibbel acknowledges the Deutshe Akademische Austauschdienst for support while writing this paper in residence at MPI.

References

[1] N. Claussen, F. Sigulinski, and M. Rühle, "Design of Transformation-Toughened Ceramics"; pp. 137–63 in Advances in Ceramics, Vol. 3. Edited by A. H. Heuer and L. W. Hobbs. The American Ceramic Society, Columbus, OH, 1981.

[2] F. F. Lange and D. J. Green, "Effect of Inclusion Size on the Retention of Tetragonal ZrO_2; Theory and Experiments"; pp. 217–25 in Advances in Ceramics, Vol. 3. Edited by A. H. Heuer and L. W. Hobbs. The American Ceramic Society, Columbus, OH, 1981.

[3] A. H. Heuer, N. Claussen, W. M. Kriven, and M. Rühle, "Stability of Tetragonal ZrO_2 Particles in Ceramic Matrices," J. Am. Ceram. Soc., 65 [12] 642–50 (1982).

[4] S. A. Saltykov; p. 163 in Stereology. Edited by H. Elias. Springer-Verlag, New York, 1967.

[5] D. J. Green, "Critical Microstructure for Microcracking in Al_2O_3–ZrO_2 Composites," J. Am. Ceram. Soc., 65 [12] 610–14 (1982).

[6] M. V. Speight, "Growth Kinetics of Grain Boundary Precipitates," Acta Metall., 16 [1] 133–35 (1968).

[7] R. J. Brook, "Pore-Grain Boundary Interactions and Grain Growth," J. Am. Ceram. Soc., 52 [1] 56–57 (1969).

[8] F. M. A. Carpay, J. Am. Ceram. Soc., 60 [1–2] 82–83 (1977).

[9] C. H. Hsueh, A. G. Evans, and R. L. Coble, "Microstructural Development During Final/Intermediate Stage Sintering—I. Pore/Grain Boundary Separation," Acta Metall., 30, 1269 (1982).

[10] F. F. Lange and M. Hirlinger, "Hindrance of Grain Growth in Al_2O_3 by ZrO_2 Inclusions"; in Strengthening and Strength Uniformity of Structural Ceramics Annual Report, Vol. 2. Rockwell International Science Center, Thousand Oaks, CA, 1983.

[11] C. Wagner, Z. Elektrochem., 65, 581 (1961).

[12] I. M. Lifshitz and V. V. Slyozov, "Kinetics of Precipitation from Supersaturated Solid Solutions," J. Phys. Chem. Solids, 19 [1–2] 35–50 (1961).

Anomalous Thermal Expansion in Al$_2$O$_3$-15 Vol% (Zr$_{0.5}$Hf$_{0.5}$)O$_2$

W. M. Kriven* and E. Bischoff

Max-Planck-Institut für Metallforschung
Institut für Werkstoffwissenschaften
Stuttgart, Federal Republic of Germany

One possible origin of the anomalous thermal expansion of Al$_2$O$_3$-15 vol% (Zr$_{0.5}$Hf$_{0.5}$)O$_2$—diffusional phase separation on heating to a ZrO$_2$-rich t phase and a HfO$_2$-rich m phase—was investigated. STEM, EDS, and TEM techniques showed that virtually all particles were homogeneous and that such diffusional phase separation cannot explain the anomalous result.

In a previous study from this laboratory,[1] Al$_2$O$_3$ containing 15 vol% (Zr$_{0.5}$Hf$_{0.5}$)O$_2$ intergranular particles was fabricated by sintering at 1550°C. The linear thermal expansion was monitored in a dilatometer as the material was cycled at 10°C/min through its transformations. As illustrated in Fig. 1,[1] when heated, the dilatometer curve showed a deviation from smooth volume increase between 800° and 1460°C, i.e., below the M_S temperature. Since X-ray diffractometry indicated that 30% of the particles remained tetragonal (t) at room temperature, the volume increase was attributed to the reverse $t \rightarrow m$ transformation on heating.[1] This effect was not observed when Al$_2$O$_3$ containing a similar quantity of pure ZrO$_2$ was studied.[1,2]

The aim of this work was to investigate one possible origin of the thermal expansion anomaly. The as-sintered microstructure contained (Zr$_{0.5}$Hf$_{0.5}$)O$_2$ solid-solution particles.[1] If the $t \rightarrow m$ transformation in such a solid solution could occur by a diffusional reaction, a two-phase region would have to be present on the equilibrium phase diagram, as shown schematically in Fig. 2. Thus, a (Zr$_{0.5}$Hf$_{0.5}$)O$_2$ particle heated into the two-phase region should decompose to a ZrO$_2$-rich t phase and a HfO$_2$-rich m phase, with a net volume increase due to the lower density of the m component. Such phase separation of the 30% t particles present in the sintered samples could be the origin of the anomalous expansion. Therefore, specimens quenched from different stages in the expansion curve, i.e., with various histories (I to IV in Fig. 1), were examined for phase separation by TEM, STEM, and EDS analytical techniques.

Thin sections for TEM were prepared by standard methods and examined in a JEOL 200CX analytical microscope[†] fitted with an X-ray detector and a TN-2000 data analysis system. The investigation consisted of taking spectra from up to five parts of a solid-solution particle. In each of specimens I to IV, about 12 particles, including a few small intragranular crystals, were examined. From measured

---

*Now at the Materials Research Laboratory and the Department of Ceramic Engineering, University of Illinois at Urbana-Champaign.
[†]Japan Electron Optics Co., Tokyo, Japan.

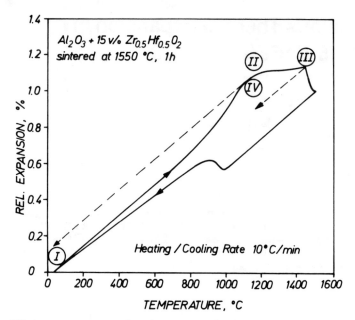

Fig. 1. Dilatometer curve of a sintered Al_2O_3 + 15 vol% $Zr_{0.5}Hf_{0.5}O_2$ composite. The specimens examined are labeled as (I) as-sintered at 1550°C for 1 h, (II) a specimen quenched from 1140°C in the temperature region of maximum expansion, (III) a specimen quenched from 1460°C at greater than maximum expansion, (IV) a cycled specimen annealed at 1150°C, the point of maximum expansion, for 14 days.

intensities, the composition was calculated using the Cliff-Lorimer method,[3] with K_{AB} values based on the standardless technique introduced by Goldstein et al.[4]

The microstructures of all specimens were similar, in that the particles were intergranular and usually monoclinic and twinned. Very few tetragonal particles were found in any specimens; we believe this is due to the martensitic transformation occurring in untransformed particles during foil preparation.

The EDS spectra for specimens I to IV were evaluated. For all specimens examined, the data generally fluctuated statistically about the theoretical composition of 50 mol% ZrO_2. No evidence for phase separation was found, other than occasional inhomogeneities which might be due to incomplete mixing during specimen fabrication.

Minor twin relaxation effects often occurred within particles of specimens II and III. For example, stepped $(110)_m$ boundary interfaces, domains of closure, $(100)_m$ twins of varying width, and internal $(110)_m$ twins were observed, indicating that post-transformational internal rearrangement had occurred. Such processes were identical to those occurring in Al_2O_3-ZrO_2[5] and mullite-ZrO_2.[6] No evidence was found for major twin coarsening, which may arise as an artifact when a thin TEM foil is annealed, as it was in the previous study.[1]

It is seen that experimental EDS analyses provided no evidence for ZrO_2-HfO_2 phase separation within the limits of experimental confidence. Twin annealing or relaxation was also unable to account for the additional volume increase at high temperatures. Another explanation for the observed effects must be sought.

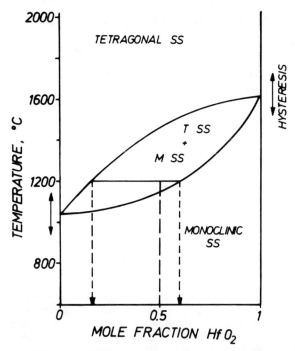

Fig. 2. Schematic version of the ZrO_2-HfO_2 phase diagram under equilibrium conditions, showing how solid-solution particles in the two-phase region may separate into Zr-rich tetragonal and Hf-rich monoclinic phases.

Acknowledgments

N. Claussen is gratefully thanked for providing the specimens and for valuable information concerning them. Thanks are due to H. Schubert for performing the heat treatments. L. H. Schoenlein and R. P. Ingel graciously donated the PSZ and PSH reference samples. The authors acknowledge thoughtful discussions with A. H. Heuer, W. Mader, and M. Rühle.

References

[1] N. Claussen, F. Sigulinski, and M. Rühle, "Phase Transformations of Solid Solutions of ZrO_2 and HfO_2 in Al_2O_3 Matrix"; pp. 164–67 in Advances in Ceramics, Vol. 3, Science and Technology of Zirconia. Edited by A. H. Heuer and L. W. Hobbs. The American Ceramic Society, Columbus, OH, 1981.
[2] A. H. Heuer, N. Claussen, W. M. Kriven, and M. Rühle, "Stability of Tetragonal ZrO_2 Particles in Ceramic Matrices," *J. Am. Ceram. Soc.*, **65** [12] 642–50 (1982).
[3] G. Cliff and G. W. Lorimer, "The Quantitative Analysis of Thin Specimens," *J. Microsc.*, **103**, 203–207 (1975).
[4] J. I. Goldstein, J. L. Costley, G. W. Lorimer, and S. J. B. Reed, SEM/1977/I, Edited by O. Johari, IITRI, 315–24 (1977).
[5] W. M. Kriven, "Martensite Theory and Twinning in Composite Zirconia Ceramics"; pp. 168–83 in Advances in Ceramics, Vol. 3. Edited by A. H. Heuer and L. W. Hobbs. The American Ceramic Society, Columbus, OH, 1981.
[6] E. Bischoff and M. Rühle, "Twin Boundaries in Monoclinic ZrO_2 Particles Confined in a Mullite Matrix," *J. Am. Ceram. Soc.*, **66** [2] 123–27 (1983).

Improvement in the Toughness of β''-Alumina by Incorporation of Unstabilized Zirconia Particles

J. G. P. Binner and R. Stevens

University of Leeds
Department of Ceramics
United Kingdom

S. R. Tan

Chloride Silent Power Ltd.
Runcorn
United Kingdom

Approximately 2-, 5-, 10-, and 15-wt% ZrO_2 particles were added to β''-alumina ceramic electrolyte in the form of tubes, using the chemical decomposition of sodium metazirconate in the presence of α-alumina and spinel, according to the reaction $xAl_2O_3 + Na_2ZrO_3 \rightarrow Na_2O \cdot xAl_2O_3 + ZrO_2$. An increase of $\approx 75\%$ in the toughness of the ceramic β''-alumina was achieved by the addition of ≈ 15 wt% ZrO_2, at the expense of a small reduction in strength caused by the formation of microcracks. With the ≈ 2 and ≈ 5 wt% additions of ZrO_2, the stress-induced transformation-toughening mechanism was observed to be predominant, giving an increase in both toughness and strength.

The use of β''-alumina as the solid electrolyte membrane for the sodium/sulfur battery is well established.[1-3] It is now recognized that electrolyte breakdown is one of the principal causes of cell failure, several mechanisms having been suggested. A particular failure mode is that of electrical breakdown due to short circuit by sodium dendrite penetration into the β''-alumina.

The dendrites are initiated at microcracks or imperfections at the Na$^+$ exit surface and cause current concentration due to the resulting lower local resistance. The effect is cumulative and can lead to catastrophic failure. Several models for this phenomenon have been proposed,[4-6] and one significant result of the theories has been the development of an expression for a critical current density, i_{cr}, below which sodium penetration cannot occur.[5]

Virkar[7] calculated that the critical current density is proportional to the fourth power of the stress-intensity factor, K_{Ic}. Thus any improvement which can be achieved in the K_{Ic} values for β''-alumina electrolyte ceramic will produce a commensurate increase in the critical current density. This should, in turn, result in both a longer battery life and allow a higher charging current density to be used with a reduced charging time.

In an attempt to achieve an increase in toughness, the much-documented technique which uses ZrO_2 to "transformation-toughen" ceramics[8-10] was exploited. Unlike the work performed by Viswanathan et al.,[11] a route which inserts the ZrO_2 particles into the host matrix by a chemical decomposition reaction was used.[12] Lange et al.[13] also studied the transformation strengthening of β''-alumina. However, in their work use was made of yttria partially stabilized zirconia (Y-PSZ) powders. Dependent on the ZrO_2 particle size and volume fraction present, both stress-induced transformation of metastable tetragonal ZrO_2 and microcracking are considered as toughening mechanisms.

Experimental Procedure

Appropriate quantities of the raw materials, listed in Table I, required to produce a β''-alumina composition of 8.6 wt% Na_2O, 1.0 wt% MgO, 0.5 wt% Li_2O, and x wt% ZrO_2 ($x \approx$ 0, 2, 5, 10, and 15) were vibromilled in water for 2 h prior to spray-drying in a spray drier* (inlet temperature of 190°C, outlet temperature of 130°C, and wheel speed of 18 000 rpm). The resulting powder was wet-bag isostatically pressed at 275 MPa to produce closed-end tubes 700-mm long by 38-mm outside diameter. The green shapes were bisque-fired at 900°C for 5 h to remove volatiles prior to zone sintering. Sintering was carried out in an rf induction-heated sintering furnace,[14] with a hot-zone temperature of 1625°C, at a traverse speed of 40 mm/min to produce ceramic β''-alumina tubes of greater than $\approx$98% theoretical density, according to the reaction:

$$aNa_2ZrO_3 + bNa_2CO_3 + cMgO \cdot Al_2O_3 + 2dLiOH \cdot H_2O + eAl_2O_3$$
$$\rightarrow (a + b)Na_2O \cdot cMgO \cdot dLi_2O \cdot (c + e)Al_2O_3 + aZrO_2 + bCO_2 \uparrow + 3dH_2O \uparrow \quad (1)$$

The strength of the β''-alumina ceramic was measured using "C" ring tests in both compression and tension modes. Prior to the testing the outside 1 mm of both cut faces of the rings was removed by grinding to reduce cutting damage and subsequently polished using 600-grit diamond paper.

Small sections of each of the tubes were mounted in epoxy resin and polished to a mirror finish[15] which enabled the toughness (K_{Ic}) to be measured by the indentation method of Anstis et al.[16] with loads of 20 and 30 N. One of the variables required for the calculation of K_{Ic} is Young's modulus, and hence small rectangular bars were cut from the tube walls, $\approx$6 mm wide and $\approx$25 mm long, and dynamically tested using the resonance technique of Astbury and Davis.[17] The experimental values of Young's modulus obtained were corrected for theoretical

---

*Niro Atomizer Inc., Columbia, MD.

Table I. Raw Materials Used in the Production of β''-Alumina Tubes

Raw materials	Source
Al_2O_3	A 16 SG (Alcoa)
$MgAl_2O_4$	Produced at Chloride Silent Power Ltd. (Runcorn)
$NaCO_3$	Analar (BDH)
$LiOH \cdot H_2O$	GPR (BDH)
Na_2ZrO_3	Produced at Chloride Silent Power Ltd. (Runcorn)
H_2O (deionized)	Produced at Chloride Silent Power Ltd. (Runcorn)

density, with the equation[18]

$$E_0 = E/(1 - bP) \qquad (2)$$

where E_0 is the value of the elastic modulus for a fully dense compact, E the experimental value, P the fractional porosity, and b a constant, ≈4.95 for β''-alumina (value determined from a plot of Young's modulus vs density, Fig. 1).

The amount of ZrO_2 retained in the tetragonal form in the β''-alumina matrix was determined using X-ray diffraction (XRD) techniques[19] on small pieces of the tubes. Scanning electron microscopy (SEM) was employed to examine the microstructures.

Results and Discussion

Scanning electron microscopy of the five compositions revealed that the microstructure of the β''-alumina matrices was essentially fine grained (1–3 μm) with little or no exaggerated grain growth. The size of the ZrO_2 particles developed by the reaction and subsequent fabrication process ranged from <1 to ≈10 μm, with the occurrence of particle agglomeration tending to increase with increasing ZrO_2 content.

The most notable difference between the microstructures was the occurrence of microcracking at the higher ZrO_2 compositions. The progress of a crack tip past ZrO_2 particles which have caused microcracking of the adjacent matrix will result in the growth and branching of the preexisting microcracks to form an extensive network. Figures 2 and 3 show SEM photographs of ≈2 and ≈15 wt% ZrO_2 additions to β''-alumina. There is no evidence of any microcracking with the ≈2 wt% level. However, extensive cracks were found running through the

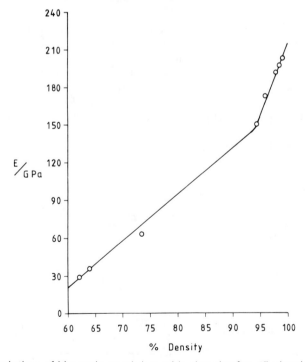

Fig. 1. Variation of Young's modulus with density for β''-alumina.

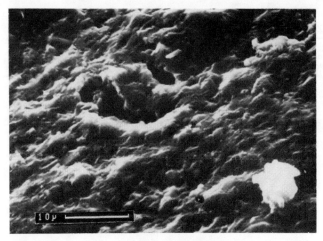

Fig. 2. Addition of ≈2 wt% ZrO_2 to β''-alumina.

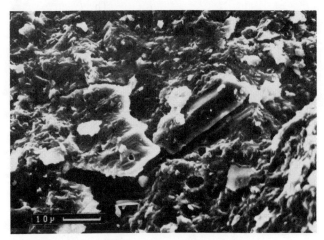

Fig. 3. Addition of ≈15 wt% ZrO_2 to β''-alumina.

≈15 wt% specimen. This is believed to be indicative of the microcracks being present in this specimen prior to the new fracture surface being formed. The existence of microcracks was noted to a lesser degree in both the ≈5 and ≈10 wt% ZrO_2-doped specimens.

Table II summarizes the results of the mechanical tests. The high tetragonal retention with ≈2 wt% additions can be ascribed to both the smaller average size of the particles and also, possibly, to a degree of stabilization from the magnesia content of the β''-alumina. Butler and Heuer[20] stated that they found no significant diffusion of MgO into ZrO_2 in an α-alumina system; however, the possibility of this occurring in β-alumina cannot yet be discounted. The increasing degree of agglomeration observed with the higher levels of ZrO_2 content explains the sharp reduction in tetragonal retention evident from Table II.

The increase in K_{Ic} (Fig. 4) indicates that toughening has occurred, particularly with incorporation of a high volume fraction of ZrO_2. When this observation

Table II. Mechanical Properties of β″-Alumina Tubes Incorporating Unstabilized Zirconia Particles

Experimental ZrO₂ (wt%)	Porosity (%)	Tetragonal retained (%)	Avg. K_{Ic} (MPa·m$^{-3/2}$)	Avg. σ_f (MPa) Compression	Tension	E_0 (GPa)
0.00	2.1		2.22	207	252	214
1.65	1.8	~80	2.23	236	264	213
3.71	0.8	~30	2.27	234	269	207
9.91	0.9	~20	2.56	223	254	205
12.40	1.3	~15	3.58	218	249	207

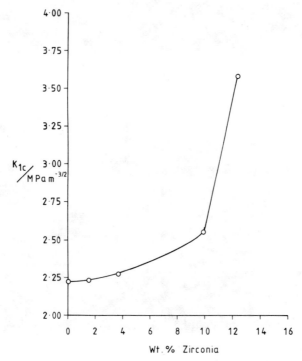

Fig. 4. Toughness (K_{Ic}) of β″-alumina tubes incorporating unstabilized ZrO₂ particles.

is combined with the results of the strength tests (Fig. 5), it becomes apparent that both the mechanisms referred to earlier are operational. With the low ZrO₂ levels the stress-induced transformation mechanism is believed to be predominant, resulting in an increase in both toughness and strength. However, as the ZrO₂ content increased and the particles became larger, the generation of microcracks occurred, and the strength therefore decreased until, ultimately, with the ≈15 wt% level the "microcracking" mechanism became predominant.

In comparison, Viswanathan et al.[11] reported strengths of 227–310 MPa for 15 wt% additions, thus demonstrating the loss of potential strength in the present work due to the growth and ensuing transformation of the zirconia particles. (The strengths of ≈152–172 MPa reported for zirconia-free β″-alumina are possibly

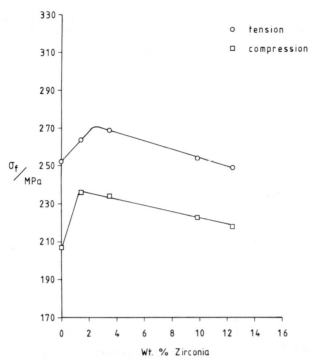

Fig. 5. Strength (σ_f) of β''-alumina tubes incorporating unstabilized ZrO_2 particles.

low. Virkar and Gordon[21] reported values of ≈200 MPa for 4-point bend strengths for fine-grained sintered material compared to 207 MPa for "C" rings under compression in the present work.)

The results of the toughness measurements are more difficult to compare, since Viswanathan et al.[11] elected to use the method of Evans and Charles[22] rather than that of Anstis et al.[16] The former method is highly dependent on the load employed, presumably owing to its empirical rather than theoretical basis.

The values measured for Young's modulus in the present work (Fig. 6) display a steady decrease with increasing ZrO_2 content, owing to a combination of the slightly lower value for zirconia compared with β''-alumina (207 GPa vs 214 GPa) and the occurrence of microcracking with the higher zirconia levels. Further experiments[23] on specimens of similar material to determine the thermal shock resistance have produced results which indicate that just over half the strength is retained when quenched in water at 0°–700°C with the ≈15 wt% zirconia additions, the considerable improvement in thermal shock resistance clearly demonstrating that microcracks are present.

Conclusions

The addition of unstabilized ZrO_2 second-phase particles to β''-alumina resulted in an increase in the toughness of ceramic tubes; an increase of ≈75% was achieved with ≈15 wt% additions. This increase, however, occurred via the generation of microcracks in the β''-alumina matrix, resulting in reduced strength. The presence of such microcracks would also aid the ingress of metallic sodium if the tubes were used as the electrolyte in a sodium/sulfur cell, causing an increased

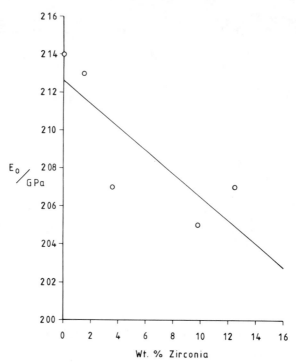

Fig. 6. Young's modulus (E) of β''-alumina tubes incorporating unstabilized ZrO_2 particles.

likelihood of dendrite formation in opposition to the original intention of reducing this phenomenon.

Incorporation of a stabilizing oxide into the ZrO_2 second-phase particle could result in an increase in both toughness and strength, producing the desired effect of allowing an increased critical current density, i_{cr}.

References

[1]G. J. May, "The Development of β-Alumina for Use in Electrochemical Cells: A Survey," *J. Power Sources*, **3**, 1 (1978).
[2]R. M. Dell and P. T. Moseley, "β-Alumina Electrolyte for Use in Sodium/Sulphur Batteries. Part 1. Fundamental Properties," *J. Power Sources*, **6**, 143 (1981).
[3]R. Stevens and J. G. P. Binner, "Structure, Properties and Production of β-Alumina," *J. Mater. Sci.* (in press).
[4]R. D. Armstrong, T. Dickinson, and J. Turner, "The Breakdown of β-Alumina Ceramic Electrolyte," *Electrochim. Acta*, **19**, 187 (1974).
[5]D. K. Shetty, A. V. Virkar, and R. S. Gordon, "Electrolytic Degradation of Lithia-Stabilised Polycrystalline β''-Alumina"; p. 651 in Fracture Mechanics of Ceramics. Edited by R. C. Bradt, D. P. H. Hasselman, and F. F. Lange. Plenum, New York, 1977.
[6]M. P. J. Brennan, "The Failure of β-Alumina Electrolyte by a Dendritic Penetration Mechanism," *Electrochim. Acta*, **25**, 621 (1980).
[7]A. V. Virkar, "On Some Aspects of the Breakdown of β''-Alumina Solid Electrolyte," *J. Mater. Sci.*, **16**, 1142 (1981).
[8]R. C. Garvie, R. H. Hannink, and R. T. Pascoe, "Ceramic Steel?" *Nature (London)*, **258**, 703 (1975).
[9]N. Claussen, "Fracture Toughness of Alumina with an Unstabilized Zirconia Dispersed Phase," *J. Am. Ceram. Soc.*, **59** [1–2] 49 (1976).

[10] A. G. Evans and A. H. Heuer, "Review—Transformation Toughening in Ceramics and Martensitic Transformations in Crack-Tip Stress Fields," *J. Am. Ceram. Soc.*, **63** [5–6] 241 (1980).
[11] L. Viswanathan, Y. Ikuma, and A. V. Virkar, "Transformation Toughening of β''-Alumina by Incorporation of Zirconia," *J. Mater. Sci.*, **18**, 109 (1983).
[12] J. G. P. Binner, R. Stevens, and S. R. Tan, "Preparation and High Temperature Stability of Sodium Zirconate," *Trans. J. Br. Ceram. Soc.*, **82**, 98 (1983).
[13] F. F. Lange, B. I. Davis, and D. O. Raleigh, "Transformation Strengthening of β''-Alumina with Tetragonal Zirconia," *J. Am. Ceram. Soc.*, **66** [3] C-50 (1983).
[14] I. W. Jones and L. J. Miles, "Production of β-Alumina Electrolyte," *Proc. Br. Ceram. Soc.*, **19**, 161 (1971).
[15] M. McNamee and J. G. Ashurst, "Ultrafast Ceramographic Preparation of Sodium β-Alumina," *Metallography*, **15**, 281 (1982).
[16] G. R. Anstis, P. Chantikul, B. R. Lawn, and D. B. Marshall, "A Critical Evaluation of Indentation Techniques for Measuring Fracture Toughness: I. Direct Crack Measurements," *J. Am. Ceram. Soc.*, **64** [9] 533 (1981).
[17] (a) N. F. Astbury and W. R. Davis, "An Introduction to Dynamic Testing"; p. 187 in The A. T. Green Book. Edited by N. F. Astbury and others. British Ceramic Research Association, Stoke-on-Trent, England, 1959.
(b) W. R. Davis, "Measurement of the Elastic Constants of Ceramics by Resonant Frequency Methods," *Trans. Br. Ceram. Soc.*, **67** [11] 515 (1968).
[18] E. A. Dean and J. A. Lopez, "Empirical Dependence of Elastic Moduli on Porosity for Ceramic Materials," *J. Am. Ceram. Soc.*, **66** [5] 366 (1983).
[19] P. A. Evans, R. Stevens, and J. G. P. Binner, "Quantitative X-ray Diffraction Analysis of Polymorphic Mixtures of Tetragonal and Monoclinic Zirconia," *Trans. J. Br. Ceram. Soc.*, **83** [2] 39 (1984).
[20] E. P. Butler and A. H. Heuer, "X-ray Microanalysis of Zirconia Particles in Zirconia-Toughened Alumina," *J. Am. Ceram. Soc.*, **65** [12] C-206 (1982).
[21] A. V. Virkar and R. S. Gordon, "Fracture Properties of Polycrystalline Lithia-Stabilized β''-Alumina," *J. Am. Ceram. Soc.*, **60** [1–2] 58 (1977).
[22] A. G. Evans and E. A. Charles, "Fracture Toughness Determination by Indentation," *J. Am. Ceram. Soc.*, **59** [7–8] 371 (1976).
[23] J. R. G. Evans and R. Stevens, "Thermal Shock of β''-Alumina with Zirconia Additions"; unpublished work.

Microstructure and Property Development of in Situ-Reacted Mullite-ZrO$_2$ Composites

JAY S. WALLACE, GÜNTER PETZOW, AND NILS CLAUSSEN

Max-Planck-Institut für Metallforschung
Pulvermetallurgisches Laboratorium
Stuttgart, Federal Republic of Germany

Dense mullite-ZrO$_2$ composites can be produced by in situ reaction of Al$_2$O$_3$ and ZrSiO$_4$ powders in a two-step sintering cycle. Unusual properties, particularly pertaining to tetragonal phase retention and transformation-toughening, are a direct result of the microstructural development route in these composites. The relationships found in this system help indicate which factors are responsible for tetragonal phase retention and transformation-toughening.

As a result of a combination of chemical inertness, low thermal conductivity, and thermal expansion coefficient, mullite (nominally 3Al$_2$O$_3 \cdot$ 2SiO$_2$) is an interesting candidate for advanced materials applications. The relatively poor mechanical properties, as well as difficulty in sintering high-purity mullite, have prevented its more widespread use.

One method of improving both the sintering behavior and mechanical properties of mullite is through ZrO$_2$ additions. In solid solution, ZrO$_2$ increases the sintering rate of mullite[1,2] and, as a second phase, acts as a grain-growth inhibitor[1] and transformation-toughening agent.[3-5] One method of producing mullite-ZrO$_2$ composites, in situ reaction of Al$_2$O$_3$ and ZrSiO$_4$ powders,[3,6] is discussed here, particularly with reference to microstructural development, tetragonal (t)-phase retention, and transformation-toughening.

Experimental

Mixtures of Al$_2$O$_3$* and ZrSiO$_4$† powders with Al$_2$O$_3$:SiO$_2$ molar ratios between 3:2 and 2:1 were milled for 2–8 h in an attritor mill using isopropyl alcohol. After the mixture was dried and sieved, the powder was isostatically pressed into bars at 630 MPa. These bars were sintered and reacted in a two-step heating cycle, first at 1400°C to densify the powders and then at 1575°C to react to mullite and ZrO$_2$. The bulk retained t fraction, X_t, was determined on as-sintered surfaces using integrated X-ray intensities.[7] The transformability of the t particles was estimated from the change in t fraction on diamond-machined surfaces, ΔX_t. Mechanical test bars of approximate dimensions 2 by 6 by 32 mm were notched to a depth, a, of 1–2 mm using a 50-μm thick diamond blade. The fracture toughness was measured by the SENB method, using four-point loading (28- and 14-mm spans) in a very "soft" testing machine and a loading rate of 0.25-mm min^{-1}. Microstructural development (e.g., critical ZrO$_2$ particle size) was observed in TEM.

---

*Alcoa A16, Aluminum Company of America, Pittsburgh, PA.
†Quarzwerke Priem,—325 mesh Zircon Sand, Bielefeld, Federal Republic of Germany.

Results and Discussion

Sintering and Reaction

Sintering of Al_2O_3 + $ZrSiO_4$ bodies at 1400°C to high densities (>95% theoretical density) in short times (30 min) offered no difficulties (Fig. 1). The rapid densification observed is, in part, due to fine starting powders and high green densities, but is also believed to be due to subsolidus reaction between Al_2O_3 and $ZrSiO_4$. No crystalline phases other than Al_2O_3 and $ZrSiO_4$ could be detected by XRD after sintering 60 min at 1400°C. After the mixture was sintered 300 min at 1400°C, however, ZrO_2 peaks appeared, indicating that decomposition of $ZrSiO_4$ was taking place. As the decomposition temperature of $ZrSiO_4$ was in excess of 1550°C,[8,9] the appearance of ZrO_2 indicated that a chemical intermixing of Al_2O_3 and SiO_2 was taking place. Such a subsolidus reaction between Al_2O_3 and SiO_2 to form noncrystalline "mullite" has been well documented[10,11] and described as "an essential part of the reaction"[10] because of the relatively high diffusion rates.

Very low diffusion rates have been found in crystalline mullite; EELS measurements[12] indicated that large Al_2O_3 concentration gradients existed within individual mullite grains. In addition, Al_2O_3 and $ZrSiO_4$ particles separated by ≈1 μm within a single mullite grain did not react even after 15 h at 1550°C, and intragranular ZrO_2 particles remained nearly constant in size.[12] Similar effects were found here; specimens that were heated to 1575° without holding at 1400°C had low densities (≈90%) and were incompletely reacted, even after extended times. A similar heating cycle with an intermediate 60-min hold at 1400°C produced specimens with high densities (>98%) which were completely reacted.

On the basis of these observations, the following sintering scheme is proposed. At low temperature, $T < 1440$°C, Al_2O_3 and $ZrSiO_4$ particles (Fig. 2(a))

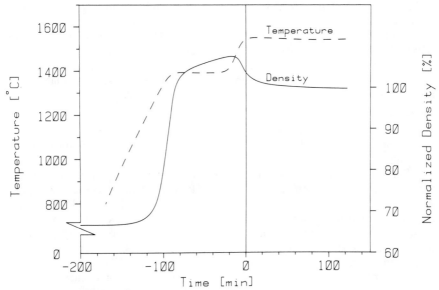

Fig. 1. Typical sintering reaction cycle for Al_2O_3 + $ZrSiO_4$ particles; nearly theoretical density is reached after 60 min at 1400°C. The reaction of Al_2O_3 + SiO_2 to form mullite at 1450°C is accompanied by a decrease in the specific density of the sample.

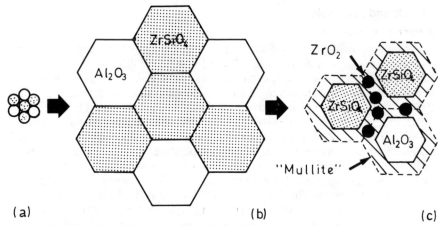

Fig. 2. Schematic of the sintering reaction cycle for Al_2O_3 + $ZrSiO_4$ powders (a). Sintering at 1400°C results in densification of the powder body without significant reaction to produce crystalline mullite (b). With increasing time at 1400°C and above, crystalline ZrO_2 nucleates and grows in an amorphous matrix (c). At some later time, the mullite matrix crystallizes around the ZrO_2 particles retaining the ZrO_2 particle morphology.

sinter to form dense Al_2O_3-$ZrSiO_4$ bodies with only limited reaction (Fig. 2(b)). When the temperature is raised, Al_2O_3 reacts rapidly with $ZrSiO_4$ to form a noncrystalline "mullite" matrix and ZrO_2 particles. Because the ZrO_2 particles are growing in a noncrystalline phase that is without grain boundaries, the resulting ZrO_2 particle morphology is rounded; i.e., no facets are formed (Fig. 2(c)). When crystalline mullite forms, the diffusion rates slow appreciably and changes in the compositional gradients in the mullite as well as ZrO_2 particle morphology occur slowly. Further change in compositional gradients of the mullite matrix, as well as in ZrO_2 particle morphology, occurs very slowly. Growth of intergranular ZrO_2 particles takes place primarily by grain-boundary diffusion and is particularly rapid in those materials that contain a glassy grain-boundary phase.

Tetragonal Phase Retention

The critical ZrO_2 particle size for spontaneous transformation on cooling to room temperature was ≈ 1.2 μm, a value far greater than that found in most other particulate systems.[13]

Surface grinding of composite with bulk t contents 30% < X_t < 60% caused transformation to m symmetry, reducing X_t to 17% in the near-surface region. Annealing for short times (12 min) at temperatures from 1200° to 1575°C, i.e., $T > A_f$, resulted in virtually no change in surface X_t on cooling to room temperature. Thus, all former t particles, which were transformed by grinding, returned again to m symmetry. This behavior was in contrast to other ZrO_2 systems where, after annealing at $T > A_f$, the t form was retained.[5,14]

The critical particle size in the fused mullite-ZrO_2 composites was only ≈ 0.6 μm[1] despite virtually identical bulk chemistry, elastic moduli, and thermal expansion coefficients in the two systems. This was probably due to the differences in ZrO_2 particle morphology (Fig. 3), the faceted ZrO_2 particles in the fused-mullite composites[1] acting as strain singularities, which, in combination with thermal expansion mismatch strains, produced stable m nuclei.[15] The ZrO_2 particles

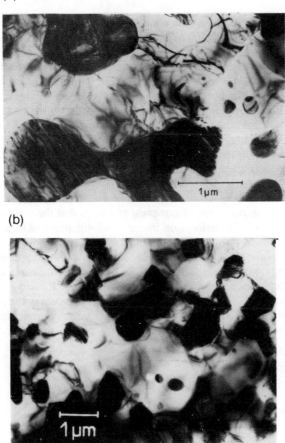

Fig. 3. Bright-field TEM micrographs of ZrO_2 particles embedded in a mullite matrix. Note the rounded ZrO_2 morphology in the in situ-reacted composites (a) and the faceted angular morphology (b) in the fused-mullite-ZrO_2 bodies.

in the in situ-reacted mullite-ZrO_2 were rounded (Fig. 3) and thus did not contain these nuclei producing strain singularities. Only large strain fields, such as those resulting from surface grinding, were able to trigger the m symmetry. A possible explanation is that, once these relatively large particles were made to transform by any means, they retained defects when annealed at $T > A_f$, which acted as nuclei for the transformation on cooling.

Fracture Toughness

The transformation behavior outlined above presents difficulties in preparing specimens without surface compressive stresses for mechanical testing. This compressive surface stress produces unrealistically high measured K_{Ic} and bend strength values, as evidenced in Al_2O_3-ZrO_2 composites.[14] As this stress, i.e., transformed particles, cannot be eliminated in these mullite-ZrO_2 composites by a short anneal at $T > A_f$, the fracture toughness measured here would be expected to be too high.

To determine the effect of the compressive stress, a series of specimens was sintered at 1575°C, each specimen for a total of 120 min. Before the sintering cycle was completed, however, some mechanical test bars were removed, machined, and then returned to complete the 120-min sintering cycle. As a result, the microstructures and bulk t fractions ($X_t \approx 50\%$) were the same for each of the specimens. The surface t fraction was $X_t = 17\%$ for the bars machined at the end of the 120-min sintering cycle, while X_t was 50% (equal to the bulk t content) for those sintered for 30 min at 1575°, machined, and then annealed for a further 90 min at 1575°C (Fig. 4). For these latter specimens, there was no surface compressive stress before testing because the surface t fraction was the same as that in the bulk. The measured fracture toughness did not reflect this change in surface compressive stress but remained essentially constant with increasing annealing time after machining (Fig. 4).

One possible explanation of our present data is that the crack-closure forces due to t-ZrO_2 particles transforming in the crack-tip strain field are of the same order of magnitude as those produced on machined surfaces. To test this hypothesis, specimens were machined after being sintered at 1400° and then reacted at 1575°C to form mullite and ZrO_2. As no ZrO_2 existed during the machining operation, no net surface compressive stresses due to transformed ZrO_2 could exist. X-ray diffraction analysis confirmed that X_t on the surface was the same as that in the bulk. These specimens, however, showed the same dependence on the transformable t fraction as those specimens which contained surface compressive

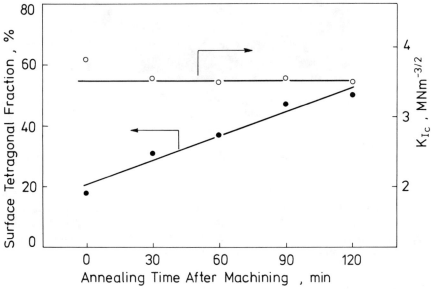

Fig. 4. Measured fracture toughness and surface tetragonal fraction as a function of annealing time at 1575°C after machining. Since all bodies had a total time at 1575°C (before and after machining) of 120 min, the bulk tetragonal fraction for all samples was 50%. Note that the measured fracture toughness is essentially independent of the surface tetragonal fraction.

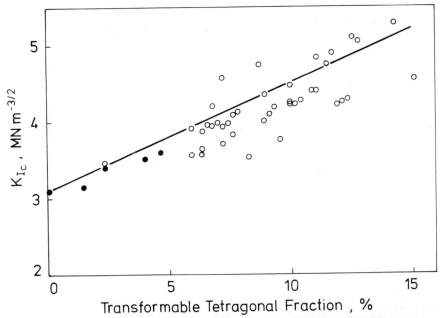

Fig. 5. Measured fracture toughness as a function of the transformable tetragonal fraction which was varied by changing the milling times and the sintering schedule. Solid circles are specimens without compressive zones.

stresses (Fig. 5). Since the measured fracture toughness scales with the transformable t fraction, we believe that crack-closure forces, resulting from either surface grinding or transformation in the crack-tip strain field, are responsible for the measured increase in fracture toughness. This corresponds closely with the mechanics approaches proposed by McMeeking and Evans[16] and Budiansky et al.[17]

Conclusion

High-density mullite-ZrO_2 composites can be fabricated at moderate temperatures (1575°C) using the in situ reaction between Al_2O_3 and $ZrSiO_4$. The unusual microstructural evolution route involved forms ZrO_2 particles with rounded morphology. Due to this morphology, nucleation of the monoclinic phase is difficult and metastably retained tetragonal particles result. These relatively large tetragonal particles easily transform in the strain field of an advancing crack, causing crack-closure forces which result in a higher measured fracture toughness.

Acknowledgments

The authors thank S. Prochazka for helpful discussions and E. Bischoff and M. Rühle for the TEM investigations.

References

[1]S. Prochazka, J. S. Wallace, and N. Claussen, "Microstructure of Sintered Mullite-Zirconia Composites," *J. Am. Ceram. Soc.*, **66** [8] C-125–C-127 (1983).
[2]J. Moya; private communication.

[3] N. Claussen and J. Jahn, "Mechanical Properties of Sintered in Situ-Reacted Mullite-Zirconia Composites," *J. Am. Ceram. Soc.*, **56** [3] 228–29 (1980).

[4] N. Claussen and J. S. Wallace, "Reply," *J. Am. Ceram. Soc.*, **64** [4] C-79–C-80 (1981).

[5] N. Claussen, "Transformation-Toughened Ceramics," *Z. Werkstofftechn.*, **13** [3] 138–47; **13** [4] 185–96 (1982).

[6] E. Di Rupo, E. Gilbert, T. G. Carruthers, and R. L. Brook, "Reaction Hot-Pressing of Zircon-Alumina Mixtures," *J. Mater. Sci.*, **14** [3] 705–11 (1979).

[7] R. C. Garvie and P. S. Nicholson, "Phase Analysis in Zirconia Systems," *J. Am. Ceram. Soc.*, **55** [6] 303–305 (1972).

[8] J. R. Anseau, J. P. Biloque, and P. Fierens, "Some Studies on the Thermal Solid State Stability of Zircon," *J. Mater. Sci.*, **11** [3] 578–82 (1976).

[9] W. C. Butterman and W. R. Foster, "Zircon Stability and the ZrO_2-SiO_2 Phase Diagram," *Am. Mineral.*, **52** [5–6] 880–85 (1967).

[10] W. G. Staley, Jr. and G. W. Brindley, "Development of Non-Crystalline Material in Subsolidus Reactions Between Silica and Alumina," *J. Am. Ceram. Soc.*, **52** [11] 616–19 (1969).

[11] R. F. Davis and J. A. Pask, "Diffusion and Reaction Studies in the System Al_2O_3-SiO_2," *J. Am. Ceram. Soc.*, **55** [10] 525–31 (1972).

[12] J. S. Wallace, N. Claussen, M. Rühle, and G. Petzow, "Development of Phases in In Situ-Reacted Mullite-Zirconia Composites"; pp. 155–65 in Surfaces and Interfaces in Ceramic and Ceramic-Metal Systems. Edited by J. A. Pask and A. G. Evans. Plenum, New York, 1981.

[13] N. Claussen and M. Rühle, "Design of Transformation Toughened Ceramics"; pp. 137–63 in Advances in Ceramics, Vol. 3. Edited by A. H. Heuer and L. W. Hobbs. The American Ceramic Society, Columbus, OH, 1981.

[14] N. Claussen and J. Jahn, "Transformation in ZrO_2 Particles in a Ceramic Matrix," *Ber. Dtsch. Keram. Ges.*, **55** [11] 487–91 (1978).

[15] M. Rühle and W. M. Kriven, "Stress Induced Transformation in Composite Zirconia Ceramics," *Ber. Bunsenges. Phys. Chem.*, **87**, 222–29 (1983).

[16] R. M. McMeeking and A. G. Evans, "Mechanics of Transformation Toughening in Brittle Materials," *J. Am. Ceram. Soc.*, **65** [5] 242–46 (1982).

[17] B. Budiansky, J. W. Hutchinson, and J. C. Lambropaulos, "Continuum Theory of Dilatant Transformation Toughening in Ceramics," *Int. J. Solids Struct.*, **19** [4] 337–55 (1983).

Size Effect on Transformation Temperature of Zirconia Powders and Inclusions

I. MÜLLER AND W. MÜLLER

FB 9, Hermann-Föttinger-Institut
Berlin, Federal Republic of Germany

The tetragonal-monoclinic phase transformation is delayed in ZrO_2 powders and ZrO_2-Al_2O_3 ceramics. The delay depends on the size of the ZrO_2 crystallites and it is different on heating and cooling. A model is presented which can describe several aspects of this behavior.

Models which explain the size effect in zirconia, which makes small crystallites transform at lower temperatures than big ones, are discussed. The effect occurs in ZrO_2 powders as well as in ZrO_2-Al_2O_3 ceramics. In the two cases the diameter of the crystallites is quite different, requiring different explanations for the size effect. Indeed, in powders, where the diameter is a few nanometers, the size effect — according to the model — is due to surface tension which inhibits the tetragonal-monoclinic phase change more effectively in small crystallites than in big ones. On the other hand, for ZrO_2 inclusions of micrometer size in the Al_2O_3 matrix, the phase change is inhibited by the elastic tension in the matrix; it is conjectured that small inclusions are surrounded by more matrix than big ones, and some evidence for the validity of this conjecture can be adduced.

Deformation-temperature curves of ZrO_2-Al_2O_3 ceramics are drawn on the basis of this conjecture, and they show the transition to occur in the same range of temperature as is observed.

The size effect in powders is different on heating and on cooling, and the model is capable of explaining this phenomenon.

Phenomenology

Size Effect in Powder

The expansion-temperature curve of a ZrO_2 bar specimen was measured by Fehrenbacher and Jacobson.[1] It is characterized by a big decrease in length during the tetragonal-monoclinic phase change on heating and an increase in length in the reverse transition on cooling. The corresponding transition temperatures depend on the size of the ZrO_2 crystallites: Small crystallites (≈ 4 nm) transform at lower temperatures than big ones (≈ 100 nm). Quantitative data are given in papers by Garvie[2] and Maiti et al.[3]

Size Effect in ZrO_2-Al_2O_3 Ceramics

The case of ZrO_2-Al_2O_3 ceramics is qualitatively analogous to that of ZrO_2 powders: There are length changes at different temperatures on heating and cooling so that the deformation temperature diagram has a hysteresis. Claussen[4] presented such curves, from which the influence of ZrO_2 particle size is clearly visible: The longer the milling time (i.e., the smaller the particles), the lower the transformation temperatures. Quantitative data can be obtained from Heuer et al.[5] It should be mentioned that the size of the ZrO_2 particle lies in the range of micrometers instead of nanometers, as in the case of the ZrO_2 powder.

Model

Transition in ZrO_2 Spheres

To facilitate analytical calculations, we regard the crystallites of the powder as being spherical. Figure 1 shows such a crystal during the transformation: A sphere containing monoclinic (m)-ZrO_2 grows in a sphere of tetragonal (t)-ZrO_2 in a spherically symmetric manner. The size effect will be explained as a consequence of surface tensions acting on the spheres. This idea originated with the work of Filipovich and Kalinina[6] and Garvie.[2,7]

Transition in ZrO_2-Al_2O_3 Ceramics

An m-ZrO_2 sphere grows inside a t-ZrO_2 sphere, but now both are surrounded by a spherical shell of Al_2O_3 (cf. Fig. 2). The surface energies as well as the elastic energy of the Al_2O_3 matrix will be taken into account. But the influence of the elastic matrix is much larger than the influence of the surface tensions. This is due to the ZrO_2 particle size, which is of the order of micrometers instead of nanometers, as in the case of the powder. Therefore, we conclude that surface tensions cannot be the reason for the size effect in ZrO_2-Al_2O_3 ceramics.

While it seems to be acceptable that the ZrO_2 particles can be treated as spheres (cf., for example, Fig. 3 in Ref. 4), the surrounding Al_2O_3 matrix is by no means spherical. Rather, it is a multiply connected three-dimensional body. However, we shall consider a "sphere of action" around each ZrO_2 particle (cf. Fig. 3) and assume that the spherical shell of Al_2O_3 in Fig. 2 represents the influence of the nonspherical matrix as far as the phase transition is concerned. This assumption is an important feature of our model.

Analysis and Results

Stability Condition

The Helmholtz free-energy F of a body in a volume V tends to a minimum in time t as equilibrium is approached, provided that the temperature is constant

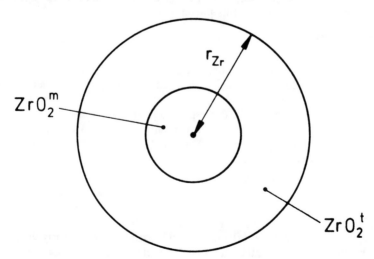

Fig. 1. Model for ZrO_2 crystallite.

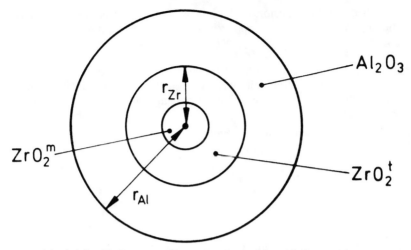

Fig. 2. Model for ZrO$_2$ crystallite surrounded by Al$_2$O$_3$ matrix.

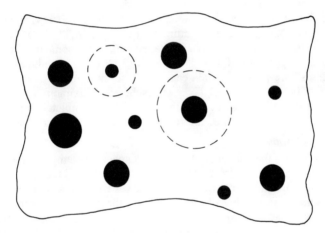

Fig. 3. Spheres of action around an inclusion.

and uniform in V and the pressure vanishes on the surface of V^8:

$$\frac{dF}{dt} \leq 0 \tag{1}$$

In a linear-elastic body one must distinguish three terms in F. The first term is an elastic energy which, according to the formulas of linear elasticity,[9] can be written as

$$F_{el} = \frac{1}{4\mu} \int_V \left\{ t_{ij}t_{ij} - \left[\frac{\lambda}{(3\lambda + 2\mu)}\right] t_{kk}^2 \right\} dV \tag{2}$$

where t_{ij} is the Cauchy stress tensor and λ and μ are the Lamé coefficients. The second is the interfacial energy F_s which for a surface O has the form

$$F_s = \sigma O \tag{3}$$

where σ is the surface tension.[10] The last term is the thermal free energy which, for the present case, has been approximately given by Filipovich and Kalinina.[6]

$$F_T = (4\pi/3)r_{Zr}^3 \rho_t q (1 - T/T_b) x \tag{4}$$

where r_{Zr} denotes the radius of the ZrO$_2$ particle, ρ_t and ρ_m are the mass densities of tetragonal and monoclinic zirconia, respectively, q is the specific heat of immersion, T_b the equilibrium transition temperature, and x the volume fraction of tetragonal zirconia.

Constitutive Function and Numerical Values

For the radial stress component, one finds in the linear theory of elasticity[9]

$$t_{rr} = (3\lambda + 2\mu)a - 4\mu\left(\frac{b}{r^3}\right)$$
$$- (3\lambda + 2\mu)\left[\alpha(T - T_R) + \left(\frac{\rho_t - \rho_m}{3\rho_m}\right)(1 - x)\right] \tag{5}$$

where a and b are two constants of integration which have to be determined from the boundary conditions on the interface between the spheres of Figs. 1 and 2. The boundary conditions express the requirement that stress and strain be continuous at the interfaces or else that a jump of stress is determined by the surface tension. α is the coefficient of thermal expansion, which enhances the stress unless the body is in its reference state, with temperature T_R. $(\rho_t - \rho_m/3\rho_m)(1 - x)$ describes the occurrence of stresses during the phase change which is accompanied by a volume expansion, since $\rho_t \neq \rho_m$. If $x = 1$, $(\rho_t - \rho_m/3\rho_m)(1 - x)$ reduces to zero; no "phase stresses" are present in this case, because the transformation has not yet begun. If $x = 0$, $(\rho_t - \rho_m/3\rho_m)(1 - x)$ reaches its maximum value; the phase stresses are now fully present, and the material is completely monoclinic. In Table I, the numerical values of the relevant constants[2,4,11-14] are listed.

Results for ZrO$_2$ Powder

Figure 4 shows the free energy of the two-sphere system described earlier. It is a function of the volume fraction x which is plotted for three temperatures and for one radius r_{Zr} of a powder particle. These curves are the result of a superposition of a linear function in x, due to F_T in Eq. (4), of the x-dependent surface energies and of the elastic part, proportional to $(1 - x)^2$. Actually, the surface terms have the essential influence on the shape of the curve. Note that, for $T = 1380$ K, the two minima at $x = 0$ and $x = 1$ have the same energy. This may therefore be considered as the temperature of the phase transformation. For each r_{Zr}, curves can be plotted, and in this manner the transformation temperature T can be deduced. The resulting relation r_{Zr} vs T is shown in Fig. 5, together with the data of Refs. 2 and 3.

As far as the consideration of surface tension goes, the above argument is akin to an argument given in Refs. 6 and 7. The difference lies in the fact that we consider elastic energies in addition to surface energies, and take account of the different densities of the two phases.

Table I. Data for ZrO$_2$ and Al$_2$O$_3$

λ_{ZrO_2} (N/m²)	μ_{ZrO_2} (N/m²)	$\lambda_{Al_2O_3}$ (N/m²)	$\mu_{Al_2O_3}$ (N/m²)	α_{ZrO_2} (K^{-1})	$\alpha_{Al_2O_3}$ (K^{-1})	σ_t (N/m)	σ_m (N/m)	ρ_m (g/cm³)	$\rho_t q$ (g/cm³)	$\rho_t V$ (J/m³)	T_b (K)
7×10¹⁰	8×10¹⁰	9×10⁹	9×10⁹	1.1×10⁻⁵	8.5×10⁻⁶	0.77	1.13	5.68	6.1	2.82×10⁸	1448

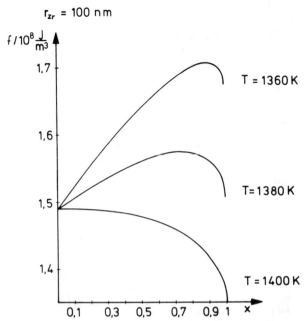

Fig. 4. Free energy of a crystallite as a function of the phase fraction in ZrO_2.

Results for ZrO_2-Al_2O_3 Ceramics

Figure 6 shows the free energy of a spherical ZrO_2 particle embedded in an Al_2O_3 sphere of action (cf. Fig. 2) plotted as a function of the tetragonal fraction x. As before, the curves result from a superposition of a linear function representing the thermal free energy and of surface- and elastic-energy terms, but now the elastic part, which is proportional to $(1 - x)^2$, plays the dominant role. This explains the drastic difference in the shape of the curves in Fig. 4 and Figs. 6 or 7.

Figure 6 shows that, with decreasing temperature, the minimum shifts from the right to the left. The phase change is complete when the minimum reaches $x = 0$.

Figure 7 shows the same situation as Fig. 6, except that the radius of the inclusion r_{Zr} has been decreased while r_{Al} has been kept constant so that there is more matrix to inhibit the phase transformation, and therefore the transition temperature decreases.

This is the key to an understanding of the size effect: Smaller particles are surrounded by more Al_2O_3 matrix than large particles; therefore, their transition is more effectively inhibited.

If this conjecture is valid, one should be able to determine the size r_{Al} of the sphere of action from the measured values of the transition temperatures for given sizes r_{Zr} of the inclusion. Table II shows the values of r_{Al} thus calculated. When the values of the table are used to plot r_{Al} vs r_{Zr} one obtains a linear dependence as shown by the dashed line in Fig. 8.

Such a linear dependence is supported by additional observations. Indeed, we have evaluated a bright-field HVEM micrograph of an Al_2O_3–4 vol% ZrO_2 speci-

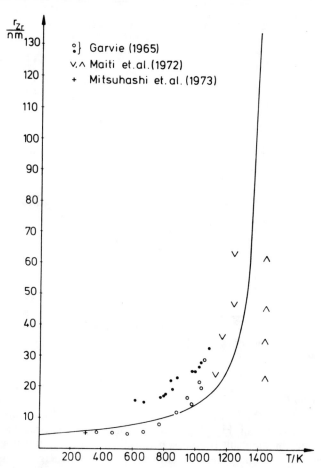

Fig. 5. Transition temperature as a function of crystallite size.

Table II. Prediction of Spheres of Action Radii for Several Transition Temperatures*

T_M (K)	r_{Zr} (µm)	r_{Al} (µm)
477	0.31	0.442
588	0.34	0.478
707	0.38	0.524
842	0.47	0.632
908	0.56	0.743
966	0.69	0.903
1013	0.87	1.126

*T_M and r_{Zr} are according to measurements made by Heuer et al. (Ref. 5) in Al_2O_3 16 vol% ZrO_2.

men that has reached us by courtesy of Dr. W. M. Kriven and Dr. S. Schmauder (MPI für Metallforschung, Stuttgart) and obtained the plot shown in Fig. 8. The dots in that plot show the radius of the sphere of action r_{Al} as a function of the inclusion radius r_{Zr}, where r_{Al} was taken to be the distance of the center of an

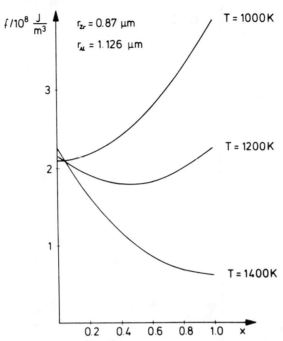

Fig. 6. Free energy of matrix and inclusion as a function of phase fraction for large inclusion.

inclusion to the surface of its next neighbor. Although there is some scatter of data, the overall linear character of the function r_{Al} (r_{Zr}) is clearly visible. The discrepancy between the two straight lines of Fig. 8 may be due to the fact that the dashed line was drawn for 16 vol% ZrO_2, whereas the micrograph has only 4 vol% ZrO_2.

Expansion-Temperature Curves

To compute an expansion temperature curve using our spherical model (Fig. 2), we construct a "one-dimensional ceramic body" by stacking ZrO_2-Al_2O_3 blocks of various size, (cf. Fig. 9). For computing expansion, it is necessary to know the size distribution of the ZrO_2 particles. Small particles occur more frequently than big ones. A typical case is shown in Fig. 10. This figure has served as a motivation for the assumed distribution shown in Fig. 11, which is used to compute the expansion-temperature diagram (Fig. 12). This curve must be compared with the experimental curves in Ref. 4.

It is clear that the curve of Fig. 12 makes a gradual transition between the two parallel dashed lines as the phase transformation proceeds. There is no hysteresis in our curve, because all arguments refer to the thermodynamic equilibrium. We proceed to discuss this point.

Hysteresis

Hysteresis in ZrO_2 Powder

The hysteresis observed in Ref. 1 implies that, on cooling, the phase transition occurs at a lower temperature than on heating. Inspection of the data[3] confirms this. The corresponding points in Fig. 5 have been marked as $\wedge$ for heating and $\vee$ for cooling. Note that, on heating, there is virtually no influence of the particle

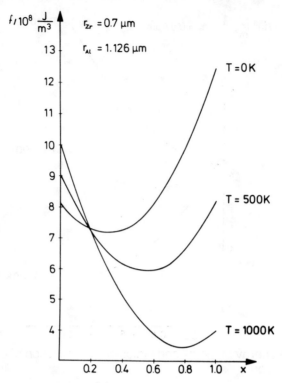

Fig. 7. Free energy of matrix and inclusion as a function of phase fraction for small inclusion.

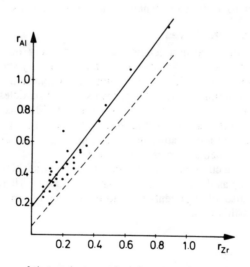

Fig. 8. Radius r_{Al} of the spheres of action as a function of the radius r_{Zr} of the embedded ZrO_2 particles.

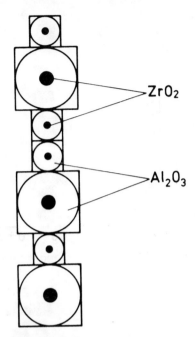

Fig. 9. Model for a ZrO_2-Al_2O_3 bar.

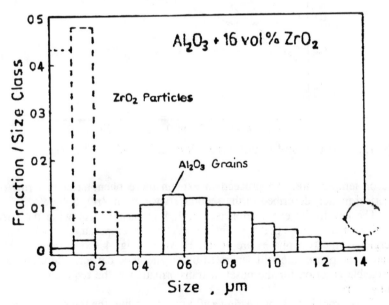

Fig. 10. Distribution of ZrO_2 inclusions and Al_2O_3 grains according to size (Ref. 5).

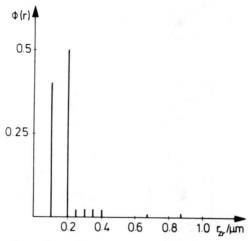

Fig. 11. Distribution of the ZrO₂ inclusions according to size (assumed).

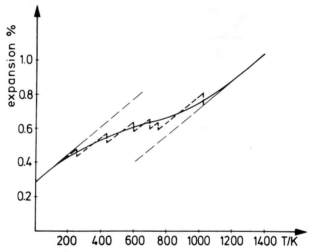

Fig. 12. Expansion-temperature curve of ZrO₂-Al₂O₃ (predicted).

size on temperature. We proceed to explain these phenomena in terms of the spherical model, described in the section Transition in ZrO_2 Spheres.

The key to the explanation is Fig. 4. Previously, in drawing the curve of Fig. 5 we had assumed that the transition occurs at a temperature for which the free energies of the two phases are equal. However, Fig. 4 shows that at that temperature the phases are still separated by an energetic barrier. Therefore, a more reasonable criterion for the onset of transformation is the requirement of a vanishing barrier.

Thus, from Fig. 4, on heating a 100-nm crystallite, the transformation should occur only at 1400 K where the left barrier has vanished. On cooling, the transformation should occur only at 1320 K where the right barrier has virtually vanished. On that basis, one can draw a curve for $r_{Zr}(T)$, both on heating and on

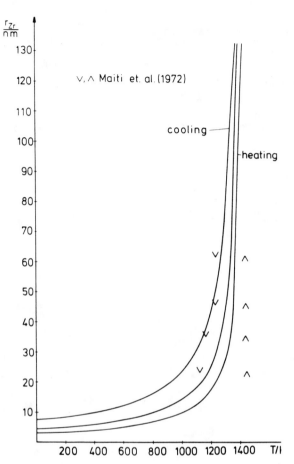

Fig. 13. Transition temperature as a function of crystallite size on heating and cooling.

cooling. These curves are shown in Fig. 13. Note that the heating curve is very steep so that there is little influence of the particle size on the transition temperature, just as was observed by Maiti et al.[3]

Hysteresis in ZrO_2-Al_2O_3 Ceramics

So far our model does not describe the hysteresis effect for ZrO_2-Al_2O_3 ceramics. An argument akin to the one in the previous subsection is impossible here, because the free-energy curves of Fig. 6 do not show any barriers.

References

[1] L. L. Fehrenbacher and L. A. Jacobson, "Metallographic Observation of the Monoclinic-Tetragonal Phase Transformation," *J. Am. Ceram. Soc.*, **48** [3] 157–61 (1965).
[2] R. C. Garvie, "The Occurrence of Metastable Zirconia as a Crystallite Size Effect," *J. Phys. Chem.*, **69**, 1238 (1965).
[3] H. S. Maiti, K. V. G. K. Gokhale, and E. G. Subbarao, "Kinetics and Burst Phenomenon in ZrO_2 Transformation," *J. Am. Ceram. Soc.*, **55** [6] 317–22 (1972).
[4] N. Claussen, "Umwandlungsverstärkte keramische Werkstoffe," *Z. Werkstofftech.*, **13**, 138 (1982).

[5]A. H. Heuer, N. Claussen, W. Kriven, and M. Rühle, "Stability of Tetragonal ZrO_2-Particles in Ceramic Matrices," *J. Am. Ceram. Soc.*, **65** [12] 642–50 (1982).
[6]V. N. Filipovich and A. M. Kalinina, "Critical Amorphization Radius of Crystals," *Struct. Glass*, **5**, 34 (1965).
[7]R. C. Garvie, "Stabilization of the Tetragonal Structure in Zirconia Microcrystals," *J. Phys. Chem.*, **82**, 218 (1978).
[8]W. Dreyer, I. Müller, and P. Strehlow, "A Study of Equilibrium of Interconnected Balloons," *Q. J. Mech. Appl. Math.*, **35**, 423 (1982).
[9]I. N. Sneddon, "The Classical Theory of Elasticity"; in Handbuch der Physik, Vol. 6. Springer-Verlag, Berlin, 1958.
[10]A. Sommerfeld; Lectures on Theoretical Physics, Vol. 2; Ch. 17. Academic Press, New York, 1950.
[11]E. D. Whitney, "Electrical Resistivity and Diffusionless Phase Transformations of Zirconia at High Temperatures and Ultrahigh Pressures," *J. Electromech. Soc.*, **112**, 91 (1965).
[12]D'Ans Lax; Taschenbuch für Chemiker und Physiker, Vol. 1, 3rd ed. Springer-Verlag, New York, 1967.
[13]W. H. Gitzen, Ed.; Alumina as a Ceramic Material. The American Ceramic Society, Columbus, OH, 1970.
[14]R. C. Garvie; Zirconium Dioxide and Some of its Binary Systems in High Temperature Oxides, Part II, Oxides of Rare Earths, Titanium, Zirconium, Hafnium, Niobium and Tantalum. Edited by A. M. Alper. Academic Press, New York, 1970.

Relationship Between Morphology and Structure for Stabilized Zirconia Crystals

D. MICHEL

CNRS LA 302, C.E.C.M.
94400 Vitry, France

Crystals with cubic phases were grown by "skull-melting" in systems containing zirconia and yttrium or lanthanide sesquioxides. For highly defective fluoritelike phases (rate of oxygen vacancies ranging from 0.08 to 0.14), the crystal habit was not that usually displayed by crystals with the CaF_2 structure. Observed faces and growth directions were found to correspond to forms with low Miller indices for the rhombohedral fluorite superstructure (Y_6UO_{12} type, space group $R\bar{3}$, $Z = 1$). Results indicated that "short-range" order interactions, involved in the crystallization process from the melt, induce the morphology of the crystals.

The fluorite structure is known to accommodate grossly nonstoichiometric compositions. Stabilized zirconias are thus fluoritelike phases $Zr_{1-x}M_xO_{2-y}$ with extended defects (oxygen vacancies and substitutional cations M) on both anionic and cationic sublattices.

In systems involving ZrO_2 and a stabilizing oxide (MgO, CaO, Y_2O_3, or Ln_2O_3 with Ln = lanthanide), such oxygen-deficient phases are stable over wide ranges of temperature and composition. Defects are more or less randomly distributed, depending on the vacancy and substitution rates, the stabilizing element, and temperature.

For specific compositions, fluorite superstructures have been found in different systems[1–3]: $CaZr_4O_9$ (monoclinic) and $Ca_6Zr_{19}O_{44}$ (rhombohedral)[4–8]; $Mg_2Zr_5O_{12}$[9] and $M_4Zr_3O_{12}$ (M = Sc, Y; Ln = Er→Lu)[10–19] with a Y_6UO_{12}-type rhombohedral structure[20]; and $Ln_2Zr_2O_7$ (cubic, pyrochlore-type) with Ln = La→Gd.[11,21]

These ordered phases are generally stable only at low temperatures and undergo order→disorder transitions at high temperatures into cubic defective fluorites. Nevertheless, completely random distributions of defects are not achieved for heavily doped zirconia with a high rate of oxygen vacancies. Short-range order phenomena are revealed by "diffuse scattering" features on patterns of X-ray, neutron, or electron diffusion. For instance, Fig. 1 shows the (111) back-reflection Laue photograph of an $(Er_{0.57}Zr_{0.43})O_{1.71}$ crystal. Diffuse spots are obtained in addition to fluorite fundamental spots.

The present paper reports observations on the crystal habit of "disordered" phases, indicating that the morphology of crystals obtained by solidification from the melt is strongly influenced by short-range order phenomena. Results have been interpreted by considering structural features of the ordered rhombohedral phase (Y_6UO_{12}-type) and using the analysis of "periodic bond chains" (pbc), according to the theory of Hartman and Perdok[22] and Hartman.[23]

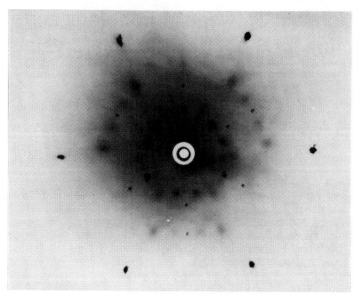

Fig. 1. As-grown $Er_4Zr_3O_{12}$ crystal with a disordered cubic structure. Back-reflection Laüe pattern on (111).

Experimental Procedure

The studied samples are single crystals of ZrO_2 stabilized with M_2O_3 oxides (M = Y, Er, or Yb) in the composition range $M_xZr_{1-x}O_{2-x/2}$ with $0.3 < x < 0.57$.

Starting materials were high-purity powders (99.9 grade) of ZrO_2* and yttrium, erbium, or ytterbium sesquioxide.[†] Crystals were grown by directional solidification from the melt using the skull-melting technique developed in our laboratory.[24–27] Large crystals (15 by 5 by 5 mm) were obtained with a solidification rate of about 5 mm/h from ingots of 500 g. Crystals displayed a prismatic morphology elongated along the growth direction. Crystal habits were determined from back-reflection Laue patterns on faces oriented by goniometry using autocollimation.

Results

Crystal Structure

Powder X-ray diagrams indicated that the structure of as-grown crystals was cubic (fluorite $Fm3m, Z = 4$) for all the considered compositions ($0.3 < x < 0.57$). At the $M_4Zr_3O_{12}$ composition, crystals would display the rhombohedral fluorite superstructure, which is the stable form in the phase diagram at room temperature. The temperature for the order-disorder transitions from the rhombohedral to the cubic structure are $T = 1620°C$ for M = Yb,[11] $T = 1500°C$ for M = Er,[2] and $T = 1250°C$ for M = Y,[12] respectively. In fact, as the ordering process is very sluggish, crystals grown from the melt exhibited the high-temperature disordered form retained at room temperature.

The transformation into the rhombohedral phase requires atom diffusion and can be induced only by thermal treatments below the transition temperature.

---

*Produits Chimiques Ugine Kuhlmann, Paris, France.
†Rhône-Poulenc, Paris, France.

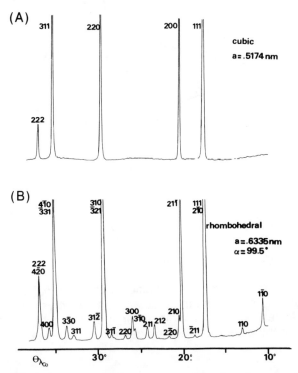

Fig. 2. Powder diffraction diagram for $Yb_4Zr_3O_{12}$ crystals: (A) as-grown, disordered cubic; (B) annealed 1 week at 1200°C, ordered rhombohedral.

Ordering can be achieved for the ytterbium compound by annealing it for a few days at 1200°C (Fig. 2), but much longer treatments (at least two weeks) are needed for the erbium or yttrium compounds to give rise to the extra lines characteristic of the rhombohedral superstructure.

Crystal Habit

Crystals with a fluorite-type structure generally exhibit habits defined by {001} and {111} forms of natural faces and ⟨110⟩ or ⟨100⟩ preferred growth directions. The crystals of stabilized ZrO_2 did not display crystal habits usually found with fluorite crystals. On the contrary, different forms of faces with high Miller indices were systematically observed. The most frequent and extended faces found in the study of crystals with different compositions between $(M,Zr)O_{1.85}$ and $(M,Zr)O_{1.714}$ were defined by the following forms: {321}, {531}, {139}, {031}, or {311}. Most crystal faces were found in the zone with the crystallographic directions ⟨211⟩, ⟨321⟩, or ⟨310⟩ corresponding to the growth axes. Typical examples are shown in Fig. 3 and Table I, which gives experimental results for various crystals.

Discussion

Results indicated that crystallization from the melt of highly defective phases was conditioned by factors leading to preferential growth along determined crystallographic directions which are not principal axes of the fluorite structure. Interpretation of the observed habit can be found by considering the relationship between the rhombohedral Y_6UO_{12} type and the parent fluorite type. The corre-

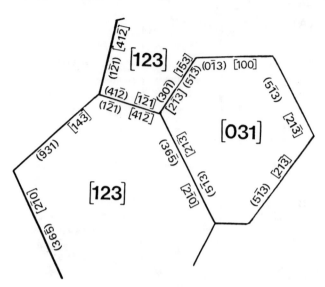

Fig. 3. Transverse section of a solidified ingot (normal to the growth direction). Indexing is given for three adjacent single crystals of $Er_4Zr_3O_{12}$. Figures in brackets are Miller indices for the edges between crystals and those in parentheses are Miller indices for the faces perpendicular to the section.

Table I. Crystal Habit Observed for $Y_xZr_{1-x}O_{2-x/2}$ Crystals with Different Compositions

Growth axis	Faces
$x = 0.3$	
1$\bar{3}$9	$\bar{3}$21 310 031
$x = 0.3$	
1$\bar{4}$3	111 $\bar{3}$01 931 $\bar{1}$23
$x = 0.36$	
310	1$\bar{3}$0 1$\bar{3}$1 1$\bar{3}$2 1$\bar{3}$3 $\bar{1}$33 $\bar{1}$31
$x = 0.3$	
531	$\bar{2}$31 1$\bar{3}$4

spondence between the rhombohedral and the cubic unit cells in direct and reciprocal space is given in Table II. In addition, the indexing with reference to cubic axes is reported for directions and planes with low Miller indices.

The data in Table II suggest that the growth directions and faces observed on the studied crystals correspond to the simplest directions and planes of the Y_6UO_{12}-type structure. In a recent paper, we interpreted the habit of In_6WO_{12} crystals (which display the same structure) in terms of a pbc analysis.[28] The Hartman theory predicts a preferential growth along chains consisting of the strongest bonds of the structure. The crystal habit is defined by F faces which contain at least two coplanar pbc vectors.[22,23] In In_6WO_{12}, the strongest bonds are the tungsten-oxygen ones, forming isolated octahedral WO_6 groups. Periodic bond chains were searched in directions connecting WO_6 groups by short In-O bonds. The analysis of the structure led to two pbc directions $\langle 001 \rangle_R$ and $\langle 110 \rangle_R$. Only the three forms of F faces $\{001\}_R$, $\{110\}_R$, and $\{111\}_R$ can be obtained with these pbc vectors. In_6WO_{12} crystals grown from the vapor phase actually had natural faces corresponding to the expected forms. These faces are represented on Fig. 4 by a stereographic projection normal to the 3-fold axis. Taking into account the relationship of the In_6WO_{12} structure with fluorite, the indices of the faces and growth directions are $\{531\}_c$, $\{312\}_c$, $\{931\}_c$, and $\langle 211 \rangle_c$ and $\langle 130 \rangle_c$, respectively. It appears then that highly defective $(M, Zr)O_{2-x}$ phases have the same crystal habit as crystals with the Y_6UO_{12}-type structure.

For $M_4Zr_3O_{12}$ compounds, the habit of crystals obtained by solidification is formed at temperatures in the stability range of the cubic disordered phase. Nevertheless, because of the high rate of oxygen vacancies (one vacancy for seven available anionic sites), it is likely that strong repulsive interactions between vacancies are present even at the high temperature at which the crystal habit is defined.

Ordering phenomena are, of course, limited to a short range in cubic disordered phases. For crystals displaying crystal habits related to those of In_6WO_{12} crystals, the range of interactions between "clusters" (octahedral groups MO_6 formed by paired oxygen vacancies) is at least 0.6–0.8 nm, which are the lengths of the observed pbc vectors.

For compositions other than $(M, Zr)O_{12}$, a similar interpretation can be proposed, assuming that, in a series of compositions between $(M, Zr)O_{1.85}$ and $(M, Zr)O_{1.714}$, the local organization of defects is related to the arrangement found in the ordered phase. Results concerning the structure of intermediate phases in

Table II. Relationships Between the Fluorite Structure and the Y_6UO_{12} Type (Rhombohedral, Space Group $R\bar{3}$, $Z = 1$)*

Axes	Planes
$a_R = (1/2)[21\bar{1}]_F$	$a_R^* = (1/7)[53\bar{1}]_F^*$
$\alpha_R = 99°59'$	$\alpha_R^* = 78°46'$
$\langle 100 \rangle_R \rightarrow \langle 211 \rangle_C$	$\{100\}_R \rightarrow \{513\}_C$
$\langle 110 \rangle_R \rightarrow \langle 013 \rangle_C$	$\{1\bar{1}0\}_R \rightarrow \{321\}_C$
$\langle 211 \rangle_R \rightarrow \langle 431 \rangle_C$	$\{11\bar{1}\}_R \rightarrow \{931\}_C$
$\langle 11\bar{1} \rangle_R \rightarrow \langle 210 \rangle_C$	$\{110\}_R \rightarrow \{421\}_C$
$\langle 1\bar{1}0 \rangle_R \rightarrow \langle 321 \rangle_C$	$\{211\}_R \rightarrow \{635\}_C$

*The correspondence is given for selected directions and faces.

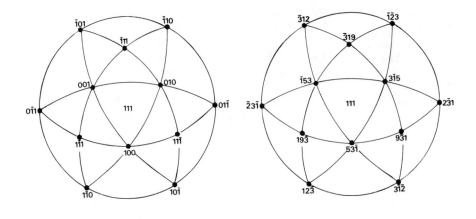

RHOMBOHEDRAL INDEXATION   CUBIC CORRESPONDENCE

Fig. 4. Crystal habit for In_6WO_{12} crystals grown from the vapor phase (stereographic projection III). Faces are indexed with reference to the rhombohedral (left) and cubic (right) lattices.

binary rare-earth oxides indicate that the ordered arrangements M'_nO_{2n-2} (M' = Ce, Pr, and Tb) are also closely related to the M_7O_{12} type[29] and can be described as formed by alternate intergrowth of MO_2 and M_7O_{12}.

In conclusion, our results provide experimental evidence that short-range order phenomena occur during the crystallization process of stabilized ZrO_2 with high vacancy rates. The habit of crystals appeared to be influenced by the same interactions that lead to the ordered rhombohedral arrangement at low temperatures.

References

[1]R. Collongues, A. Kahn, and D. Michel, "Superionic Conducting Oxides," *Annu. Rev. Mater. Sci.,* **9**, 123–50 (1979).
[2]H. J. Rossell, "Ordering in Anion-Deficient Fluorite-Related Oxides"; pp. 47–63 in Advances in Ceramics, Vol. 3. Edited by A. H. Heuer and L. W. Hobbs. The American Ceramic Society, Columbus, OH, 1981.
[3]D. J. M. Bevan and E. Summerville, "Mixed Rare-Earth Oxides"; pp. 401–521 in *Handbook on Physics and Chemistry of Rare Earths,* Vol. 3. Edited by K. A. Gschneidner, Jr. and Le Roy Eyring. North-Holland, Amsterdam, 1979.
[4]D. Michel, "Ordered States in Fluorite Phases ZrO_2-CaO at the $4ZrO_2$-CaO Composition." (in Fr.), *Mater. Res. Bull.,* **8**, 943–50 (1973).
[5]S. G. Allpress and H. J. Rossell, "A Microdomain Description of Defective Fluorite-Type Phases $Ca_xM_{1-x}O_{2-x}$ (M = Zr, Hf; x = 0.1–0.2)," *J. Solid State Chem.,* **15**, 68–78 (1975).
[6]V. S. Stubican and J. R. Hellman, "Phase Equilibria in Some Zirconia Systems"; pp. 25–36 in Advances in Ceramics, Vol. 3. Edited by A. H. Heuer and L. W. Hobbs. The American Ceramic Society, Columbus, OH, 1981.
[7]S. P. Ray and V. S. Stubican, "Fluorite-Related Ordered Compounds in the ZrO_2-CaO and ZrO_2-Y_2O_3 Systems," *Mater. Res. Bull.,* **12**, 549–56 (1977).
[8]J. R. Hellman and V. S. Stubican, "The Existence and Stability of $Ca_6Zr_{19}O_{44}$ Compound in the System ZrO_2-CaO," *Mater. Res. Bull.,* **17**, 459–65 (1982).
[9]C. Delamarre, "The Existence and Structure of a New M_7O_{12} Compound in Systems ZrO_2 (or HfO_2)-MgO" (in Fr.), *C. R. Hebd. Seances Acad. Sci., Ser. C* **269**, 113–15 (1969).
[10]J. Lefevre, "Structural modifications of Fluorite-Related Phases Involving ZrO_2 or HfO_2" (in Fr.), *Ann. Chim.,* **8**, 117–49 (1963).
[11]M. Perez Y Jorba, "Study of Zirconia–Lanthanide Sesquioxide Systems" (in Fr.), *Ann. Chim.,* **7**, 459–511 (1962).

[12] H. G. Scott, "Phase Relationships in the Yttria-Rich Part of the Yttria–Zirconia System," *J. Mater. Sci.*, **12**, 311–16 (1977).

[13] D. Michel, A. Kahn, and M. Perez Y Jorba, "Cation Ordering in A_6BO_{12} Compounds with the Y_6UO_{12} Type" (in Fr.), *Mater. Res. Bull.*, **11**, 857–66 (1970).

[14] M. R. Thornber, D. J. M. Bevan, and J. Graham, "Crystal Structures of the Intermediate Phases $Zr_5Sc_2O_{13}$ and $Zr_3Sc_4O_{12}$," *Acta Crystallogr., Sect. B*, **24**, 1183–90 (1968).

[15] M. R. Thornber and D. J. M. Bevan, "Crystal Structures of the High and Low Temperature Forms of $Yb_4Zr_3O_{12}$," *J. Solid State Chem.*, **1**, 536–44 (1970).

[16] M. R. Thornber, D. J. M. Bevan, and E. Summerville, "Phase Studies in the Systems ZrO_2-M_2O_3 (M = Sc, Yb, Er, Dy)," *J. Solid State Chem.*, **1**, 545–53 (1970).

[17] V. S. Stubican, R. C. Hink, and S. P. Ray, "Phase Equilibria and Ordering in the System ZrO_2–Y_2O_3," *J. Am. Ceram. Soc.*, **61** [1] 17–21 (1978).

[18] C. Pascual and P. Duran, "Subsolidus Phase Equilibria and Ordering in the System ZrO_2-Y_2O_3," *J. Am. Ceram. Soc.*, **66** [1] 23–27 (1983).

[19] S. P. Ray, V. S. Stubican, and D. E. Cox, "Neutron Diffraction Investigation of $Zr_3Y_4O_{12}$," *Mater. Res. Bull.*, **15**, 1419–23 (1980).

[20] S. F. Bartram, "Crystal Structure of the Rhombohedral $MO_3 \cdot 3R_2O_3$ Compounds (M = U, W, or Mo) and Their Relation to Ordered M_7O_{12} Phases," *Inorg. Chem.*, **5**, 749–54 (1966).

[21] D. Michel, M. Perez Y Jorba, and R. Collongues, "Fluorite→Pyrochlore Transitions and Order-Disorder Phenomena in Phases $(1 - x)ZrO_2$–xLn_2O_3," (in Fr.), *Mater. Res. Bull.*, **9**, 1457–68 (1974).

[22] P. Hartman and W. G. Perdok, "On the Relations between Structure and Morphology of Crystals," *Acta Crystallogr.*, **8**, 49–52 (1955); **8**, 521–29 (1955).

[23] P. Hartman, "Structure and Morphology"; pp. 367–402 in *Crystal Growth: An Introduction*. Edited by P. Hartman. North-Holland, Amsterdam, 1973.

[24] D. Michel, M. Perez Y Jorba, and R. Collongues, "Growth of Stabilized Zirconia Crystals and Properties of Cubic Phases ZrO_2-CaO," (in Fr.), *C. R. Hebd. Seances Acad. Sci., Ser. C*, **266**, 1602–1604 (1968).

[25] A. M. Anthony and R. Collongues, "Modern Methods of Growing Single Crystals of High-Melting-Point Oxides"; pp. 147–249 in *Preparative Methods in Solid State Chemistry*. Edited by P. Hagenmuller, Academic Press, New York, 1972.

[26] D. Michel, "Defect Ordering in Crystals of Refractory Oxides Involving ZrO_2 or Al_2O_3" (in Fr.), *Rev. Int. Hautes Temp. Refract.*, **9**, 225–42 (1972).

[27] D. Michel, M. Perez Y Jorba, and R. Collongues, "Growth from Skull-Melting of Zirconia-Rare Earth Oxide Crystals," *J. Cryst. Growth*, **43**, 546–48 (1978).

[28] D. Michel and A. Kahn, "The Structure of Indium Tungstate In_6WO_{12}: Its Relation with the Fluorite Structure," *Acta Crystallogr., Sect. B*, **38**, 1437–41 (1982).

[29] Le Roy Eyring, "The Binary Rare-Earth Oxides"; pp. 337–99 in *Handbook of the Physics and Chemistry of Rare-Earths*. Edited by K. A. Gschneidner, Jr. and L. Eyring: North-Holland, Amsterdam, 1979.

Section IV
Structural and Other Applications

Structural Applications of ZrO_2-Bearing Materials 465
R. C. Garvie

ZrO_2 Ceramics for Internal Combustion Engines 480
U. Dworak, H. Olapinski, D. Fingerle, and U. Krohn

Plasma-Sprayed Zirconia Coatings 488
P. Boch, P. Fauchais, D. Lombard, B. Rogeaux, and M. Vardelle

Microstructure and Durability of Zirconia Thermal Barrier Coatings 503
D. S. Suhr, T. E. Mitchell, and R. J. Keller

Thermal Diffusivity of Zirconia Partially and Fully Stabilized with Magnesia 518
W. J. Buykx and M. V. Swain

Mechanical, Thermal, and Electrical Properties in the System of Stabilized $ZrO_2(Y_2O_3)/\alpha\text{-}Al_2O_3$ 528
F. J. Esper, K. H. Friese, and H. Geier

The Reaction-Bonded Zirconia Oxygen Sensor: An Application for Solid-State Metal-Ceramic Reaction-Bonding 537
R. V. Allen, W. E. Borbidge, and P. T. Whelan

Properties of Metal-Modified and Nonstoichiometric ZrO_2 544
R. Ruh

Diffusion Processes and Solid-State Reactions in the Systems Al_2O_3-ZrO_2(Stabilizing Oxide)(Y_2O_3, CaO, MgO) 546
T. Kosmac, D. Kolar, and M. Trontelj

Structural Applications of ZrO$_2$-Bearing Materials

R. C. GARVIE

Advanced Materials Laboratory
CSIRO Division of Materials Science
Melbourne, Victoria, Australia 3001

Products made of new composite ceramics and refractories containing ZrO$_2$ particles as a dispersed phase outperform those made of conventional materials by several hundred percent. Mg-PSZ (magnesia partially stabilized ZrO$_2$) materials strengthened by transformation-toughening and surface compressive stresses have been tested successfully in such hard-wearing applications as tappet facing materials, powder compaction dies, and dry bearings. Suitable heat treatments induce thermal shock resistance in these ceramics for application as hot extrusion dies and components for vehicular engines. Al$_2$O$_3$-ZrO$_2$ materials have enhanced toughness, strength, and wear resistance due to a combination of microcracking and surface strengthening mechanisms. Consequently, the efficiency of industrial grinding wheels and metal cutting tool bits made of these materials compared to those using Al$_2$O$_3$ alone has increased 8- and 3-fold, respectively. A different type of microcracking toughening mechanism is utilized in ZrSiO$_4$-ZrO$_2$ materials, which are strong, dense, and thermal shock-resistant. This new class of "refractories" has been tested successfully as low-cost, high-performance tundish nozzles for the continuous casting of steel. It is concluded that ZrO$_2$-based materials comprise an authentic revolution in the development of brittle materials.

"Classical" attempts to strengthen or toughen ceramics involved reducing the flaw size, generating surface compressive stresses, or introducing a distribution of second-phase particles. Modest success was enjoyed by these methods such as strengthening alumina for cutting tool bits by reducing the grain size, hardening spinel watch bearings by precipitating Al$_2$O$_3$, and toughening dental porcelain by dispersing particles of the same oxide.[1,2] Structural applications of materials strengthened by these techniques were inhibited because severe environments could generate large surface flaws. Such flaws could penetrate surface compressive stresses with disastrous consequences, and any increase in toughness due to second-phase particles was small.[3,4] It is now possible for the first time to produce large increases in fracture energy and to generate, in effect, impenetrable surface compressive layers by dispersing zirconia particles of a critical size and quantity in a suitable matrix.[5-8] The success of the new ceramics stems from the ≈4% increase in volume that accompanies the tetragonal-to-monoclinic ($t \rightarrow m$) transformation in ZrO$_2$ microcrystals. The expanding, transforming inclusion interacts with the surrounding matrix (e.g., cubic stabilized zirconia, CSZ) to produce in the matrix a tensile hoop stress and a radial compressive stress. These intensive but highly localized stresses interact with the tensile stress field, driving the main crack to generate the most powerful toughening mechanisms known to date that operate in ceramic oxides.

A variety of toughening/strengthening mechanisms is generated by the dispersed phase, depending on whether the $t \rightarrow m$ reaction is induced dynamically during crack propagation or is made to happen a priori by mechanical or thermal processes. These mechanisms appear either singly or in combination in several particle/matrix composites, so that analysis is difficult. The composites can be divided into the categories of precipitate-toughened (e.g., t- and m-ZrO_2 precipitates within a CSZ matrix) and particulate-toughened (e.g., t- and m-ZrO_2 particles in CSZ, alumina, mullite, and zircon). In addition there is the single-phase, all t-ZrO_2 material, made from the system Y_2O_3-ZrO_2, known as tetragonal stabilized yttria zirconia (TSYZ). In the main, materials discussed in the present work are limited to those now used as industrial products and include precipitate-toughened magnesia partially stabilized zirconia (Mg-PSZ) and particulate-toughened alumina and zircon composites containing intergranular ZrO_2 particles.

Two related toughening/strengthening mechanisms are generated by the compressive transformational stresses. The first occurs concurrently with crack propagation and is known as transformation-toughening. In this author's opinion, transformation-toughening is effective in the bulk only in the precipitate-toughened high-zirconia materials alloyed with CaO or MgO and also TSYZ. Possibly transformation-toughening occurs in Y-PSZ materials, but there is no published evidence on this point. X-ray studies of strong polycrystalline Y-PSZ in the author's laboratory showed no evidence of transformation-toughening (R. T. Pascoe; private communication). The microstructures of Ca- and Mg-PSZ materials feature tetragonal precipitates of ZrO_2, ≈ 0.1–0.15 μm in the longest dimension, distributed in matrix grains of CSZ, 50–70 μm in diameter. The microstructure of TSYZ comprises $\approx 100\%$ t-ZrO_2 grains that are in the range 0.2–1.0 μm, the optimum size depending on the yttria content.[9] The tensile stress field at the tip of the main crack triggers the $t \rightarrow m$ transformation in precipitates in the alkaline earth-PSZ materials and in grains in the TSYZ alloy within a critical size range located in a "process zone" at the crack tip. The compressive stresses generated in the zone reduce the applied stress field below its critical value. To maintain propagation of the crack, extra energy must be supplied to the crack tip, which appears as an enhanced fracture energy. The second mechanism is known as surface-strengthening and occurs when the $t \rightarrow m$ transformation is induced, a priori, by grinding or other mechanical processes in surface layers of any composite that contains stress-transformable t-ZrO_2 particles.[10] The compressive stresses produced in this way extend to several micrometers or even tens of micrometers below the surface and are effective in inhibiting propagation of surface flaws.[11]

If the $t \rightarrow m$ transformation temperature is arranged to occur between the firing temperature of the composite and room temperature, then this reaction occurs a priori in the bulk of the material, thereby generating a different set of toughening mechanisms. These include crack deflection and/or branching and also microcracking. It is likely that the first two mechanisms occur when the m-ZrO_2 particles are close to the critical size required to preserve the tetragonal structure. Microcracking can occur when the m-ZrO_2 particles are somewhat larger than the critical size. The hoop stress of these particles can combine with the tensile stress field at the crack tip to initiate and extend microcracks. It seems that yet another toughening mechanism is invoked when the m-ZrO_2 particles are considerably larger than the critical size, as discussed later.

The various toughening mechanisms are associated with critical sizes of the ZrO_2 microcrystals. Knowledge of these size effects allows the design of microstructures with optimum thermomechanical properties for a particular application. For example, to fully utilize transformation-toughening in PSZ, the mean size of the precipitates should be close to the upper limit of a critical size range, defined as that value beyond which the precipitates would spontaneously undergo the $t \rightarrow m$ reaction during cooling of the material from the firing temperature to room temperature. In practice, this desired microstructure is obtained in Mg-PSZ materials by aging products at $\approx 1420°C$ that were previously solution-fired at 1700°C so as to obtain the peak strength (≈ 600 MPa), which coincides with the peak toughness (≈ 9 MPa·m$^{1/2}$), as shown in Fig. 1. At peak strength, the maximum number of precipitates possible has grown by Ostwald ripening to fall within the critical size range. The parallel trend shown by the MOR and K_{Ic} curves is consistent with the proposed toughening mechanism and shows that transformation-toughened PSZ obeys the Griffith equation. Surface strengthening occurs in PSZ to the extent that the fracture strength of a polished sample is increased by $\approx 20\%$ when the surfaces are ground.[10]

The critical size at room temperature for the $t \rightarrow m$ transformation of intergranular particles in Al$_2$O$_3$-ZrO$_2$ composites is in the range ≈ 0.5–0.8 μm.[12,13] Commercial materials are made by simply mixing ZrO$_2$ particles of the appropriate size with alumina powder (usually containing a small amount of MgO) and sintering or hot-pressing until densification is nearly complete. Ostwald ripening can be used also to "fine-tune" the final particle size distribution. Also, alloying the ZrO$_2$ particles with such oxides as Y$_2$O$_3$ can increase the critical size. Transformation-toughening in the bulk in Al$_2$O$_3$-ZrO$_2$ alloys can be only about

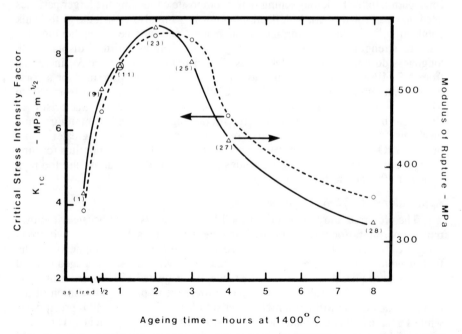

Fig. 1. Aging time dependence of the strength (MOR) and toughness (K_{Ic}) of an Mg-PSZ alloy aged at 1420°C. The numbers in parentheses are the monoclinic contents in a ground surface.

one-fourth as effective as it is in PSZ because the volume fraction of particles and the effective dilatational transformation strain in the former are only $\approx 1/3$ and $1/2$, respectively, relative to the latter. The reduction in the transformation strain is due to the low mean value of the thermal expansion of alumina (8×10^{-6} °C^{-1}) compared to that of t-ZrO$_2$ (14×10^{-6} °C^{-1}).[12] Surface strengthening, however, is important in Al$_2$O$_3$-ZrO$_2$ composites. The reason for this apparent anomaly is the fact that the depth of the surface-transformed zone is very much larger than that of the typical surface flaw size, as discussed by Swain.[14] This mechanism can increase the strength and fracture toughness by ≈ 2.5 times relative to the matrix in alloys containing 16 vol% ZrO$_2$ particles that are nearly all tetragonal.[11] That the observed increase in properties in such materials is due to surface strengthening is confirmed by the fact that the mechanical properties were reduced nearly to those of the matrix alone, after being annealed.[11] As before, in the case of the PSZ alloys, the MOR and K_{Ic} of surface-strengthened Al$_2$O$_3$-ZrO$_2$ alloys follow a parallel trend. It is interesting to note that Ruf and Evans also observed that there was no measurable toughening in the bulk when intergranular t-ZrO$_2$ particles were distributed in ZnO.[15]

The maximum toughening, in the bulk, in Al$_2$O$_3$-ZrO$_2$ composites occurs when the intergranular m-ZrO$_2$ particles are ≈ 1.5–3.0 μm in diameter, considerably larger than the critical size.[6] At present, the relative contribution of each of the possible mechanisms associated with m-ZrO$_2$ particles has not been clarified. Green[13] observed a marked decrease in the Young's modulus of Al$_2$O$_3$ composites when the $t \rightarrow m$ transformation occurred in ZrO$_2$ particles ≈ 0.8 μm in diameter. It is tempting to infer from this observation that microcracking makes an important contribution to the bulk toughening in the composite containing the larger particles cited above. Probably crack deflection and/or bowing also contribute to the bulk toughening.[15] Unlike transformation-toughening and surface strengthening, the fracture strength and toughness of Al$_2$O$_3$-ZrO$_2$ composites that display bulk toughness do not follow parallel trends. For example, whereas the K_{Ic} increased from ≈ 5 MPa·m$^{1/2}$ for the matrix to a peak value of ≈ 10 MPa·m$^{1/2}$ for a composite containing about 16 vol% ZrO$_2$ particles, the MOR decreased from 580 to 510 MPa for the same materials.[16] This behavior could be characteristic of ZrO$_2$-bearing composites in which microcracking is a significant toughening mechanism. As the volume fraction of particles increases, there is a greater probability of ZrO$_2$ particle agglomeration causing microcracks to link up, thus forming a larger critical flaw size. (The K_{Ic} values quoted immediately above are too high by about 30% due to the measuring technique.[17])

Applications of PSZ Materials

The preferred alloying oxide for the PSZ ceramics is MgO because it allows more opportunity for microstructural engineering than CaO or Y$_2$O$_3$. Two broad categories of material are distinguished according to the required application. The first, aged to maximum strength, is designated MS while the second, aged beyond this point to acquire thermal shock resistance, is designated TS. Properties of these materials are listed in Table I.[18] The main difference in phase composition of the two materials is that MS has a higher content of transformable t-ZrO$_2$ precipitates while TS has a higher concentration of transformed m-ZrO$_2$ precipitates. The major effect of this microstructural engineering on the properties is to increase the ratio of the integral work-to-fracture energy (γ_{wof}) to the fracture initiation energy (γ_I) from ≈ 1 for MS to ≈ 2–3 for TS ceramics. This means that MS excels in applica-

Table I. Typical Phase Composition and Properties of Two Mg–PSZ Types (MS and TS)

	MS	TS
Tetragonal precipitates (%)	24	5
Monoclinic precipitates (%)	9	25
gbm-ZrO_2 (%)	7	10
MOR (MPa)	630	600
K_{Ic} (MPa·m$^{1/2}$)	9	8–15
γ_{wof}/γ_I	~1	~2–3
Weibull modulus	22	22

tions requiring maximum strength, toughness, and wear resistance, whereas the high value of γ_{wof}/γ_I for TS indicates it is the first choice for applications involving thermal shock. Note that the enhanced thermal shock resistance of TS was acquired at the expense of only ≈5% of the peak strength.

Microstructural engineering of Mg-PSZ ceramics stems from the recent discovery that aging at 1100°C allows the development of a series of microstructures which display a continuous spectrum of thermomechanical properties ranging from high strength and wear resistance at one end to high thermal shock resistance at the other.[19] An example of these phenomena is shown by the data in Fig. 2, which is a plot of the initial strength (MOR_i) and also the strength retained after damage by a standard thermal shock test (MOR_r) as a function of aging time at 1100°C for an Mg-PSZ alloy.[19] The thermal shock test simulated conditions occurring during the hot extrusion process and consisted of quenching samples preheated to 450°C in a bath of molten aluminum at 900°C. At ≈4 h of aging, the samples had maximum strength but no thermal shock resistance. When the aging time was increased to ≈9 h, the MOR_i decreased somewhat, but the MOR_r increased suddenly from 0 MPa at 8 h to ≈390 MPa. Thermal shock resistance is induced by the formation of monoclinic precipitates that generate a new type of toughening mechanism (as yet undefined) in addition to transformation-toughening. The strength-aging curve noted in Fig. 2 is not due to particle growth as is the case for "classical" aging experiments with PSZ but rather due to an anion defect ordering process in the solid that causes the requisite loss of coherency.[20] Although a range of properties can be developed also by aging at ≈1400° rather than at 1100°C, the latter temperature is preferred because of the larger effective fracture energy that can be obtained, as shown in Fig. 3. Also, quality control is easier when products are aged at 1100° rather than at 1400°C because the peak properties are maintained over a broader time interval.

An established application of TS materials is in dies for the hot extrusion process, as depicted schematically in Fig. 4. Brass or copper billets weighing about 300 kg and heated to ≈800° and 900°C, respectively, are extruded under a pressure of ≈15 MPa through the die that was heated to ≈450°C, to form rods or tubes with diameters in the range 0.5–12 cm. For steel, the billets are heated to ≈1100°C. Thus, the ceramic die is exposed to high thermal-shock and mechanical-impact stresses, amounting to ≈190 and ≈350 MPa, respectively. These combined stresses are close to or exceed the fracture stress of Mg-PSZ materials in the range 400°–800°C.[21] To enable the die to withstand such severe conditions, it is inserted in a tool steel case by shrink-fitting, a process that generates compres-

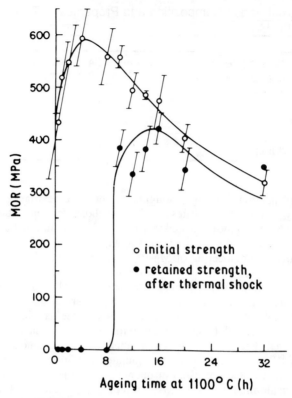

Fig. 2. Initial (MOR$_i$) and retained strength (MOR$_r$) of an Mg-PSZ alloy aged at 1100°C (Ref. 19).

sive stresses in the die amounting to several hundred MPa. A finite element stress analysis has shown that generation of these shrinkage stresses is of overriding importance for the successful use of ceramic dies.[22] The economic advantage of using PSZ dies is their very long, maintenance-free life compared to that of the competing high-alloy metal dies, such as Stellite.* Further advantages of ceramic dies are the superior surface finish and dimensional control of the extruded product. Metal dies must be removed from production after every 12–15 extrusions to have the surface dressed, which is expensive. In service, although the ceramic dies are twice the cost of conventional metallic versions, the 4-fold increase in die life gives considerable economic advantage.

A potential application of enormous technological and economic impact is the use of ceramics in vehicular engines.[23-25] In this regard, PSZ ceramics are leading candidate materials because of their exceptional strength, toughness, low coefficient of friction, and thermal expansions that are similar to those of steel. The adiabatic diesel engine with no cooling system is an attractive concept.[24] The engine runs hot and is therefore fuel-efficient; the savings in fuel and increase in

---

*Stellite, Cabot Corp., Kokomo, IN.

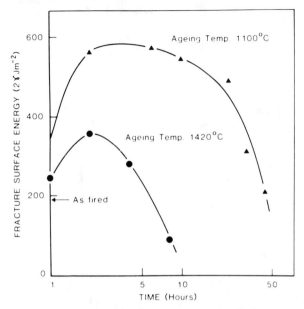

Fig. 3. Comparison of the fracture surface energy of an Mg-PSZ alloy aged at 1420° and 1100°C (Ref. 21).

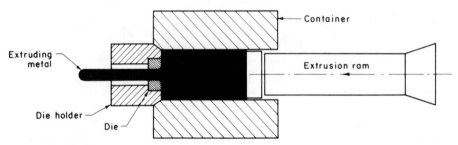

Fig. 4. Schematic diagram of the hot-extrusion process.

power to be expected from this design are 22 and 37%, respectively. Fuel savings of 30–50% were achieved in a recent 13-month field test in which a truck was powered by an adiabatic engine with components protected by a layer of plasma-sprayed CSZ.[26] Figure 5 is a schematic drawing of a more advanced design that will have engine components made of Mg-PSZ, including cylinder liners, piston caps, valve guides, valve seats, a valve, and a hot plate. Mg-PSZ (MS) ceramics have been used successfully as cam follower faces (tappets) in conventional diesel engines. The data are presented in Fig. 6 as plots of percent dimensional change of the component as a function of running time in the test facility and show the importance of matching the microstructure of PSZ to the application. Peak-aged (MS) material showed negligible wear during the test, whereas overaged (TS) material showed unacceptable expansion. Any monoclinic material in the latter sample was repeatedly cycled through the inversion due to the periodic thermal

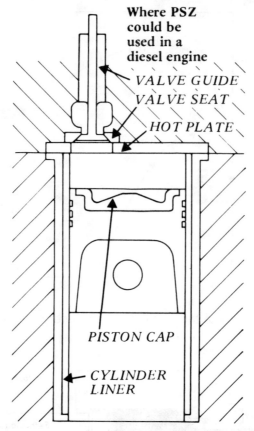

Fig. 5. Schematic diagram of the adiabatic diesel engine showing ceramic components.

and mechanical stresses generated by the test. The alternating, periodic transformational stresses initiated and extended microcracks, which caused the expansion, in the process known as "ratchetting."[27]

PSZ ceramics must be considered as structural bioceramic materials. The presently used materials, stainless steel and high-density polyethylene (HDPE), have deficiencies in that the metal suffers fatigue and corrosion while the HDPE creeps and has an unacceptably high rate of wear in knee prostheses. In recent years, alumina has been used successfully as an alternative prosthetic material. However, PSZ materials are superior to commercial alumina ceramics in all the important mechanical and wear properties. For example, PSZ possesses a higher Weibull modulus,[28] superior wear resistance, and lower susceptibility to stress corrosion than does alumina.[29] Recent in vivo tests of six months duration showed that PSZ ceramics were biocompatible and did not suffer degradation of properties.[29]

Mg-PSZ (MS) also shows great resistance to wear and is a candidate material for applications involving abrasive slurries, such as linings for hydrocyclones used to fractionate aqueous slurries of minerals in the mining industry and also as

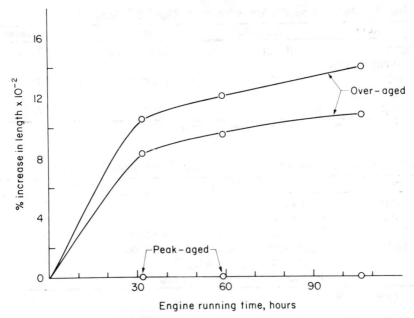

Fig. 6. Wear data for peak- and overaged Mg-PSZ cam follower faces (tappets) for a conventional diesel engine.

components and linings for pumps. The results of a standard wear test in which cylinders of different ceramics were tumbled with aqueous slurries of various abrasive grits are given in Table II, which lists the rate of volume decrease of the cylinder in the "steady-state" region.[30] PSZ materials were found to be much more wear-resistant than alumina, KT-silicon carbide, and porcelain in these experiments. Other examples in which the superior wear resistance has been exploited successfully are listed in Table III.

Applications of Al_2O_3-ZrO_2 Materials

Historically, toughened abrasive grain for industrial grinding wheels was the first application of alumina-zirconia alloy material.[31] The powerful impact of zirconia science and technology in the abrasive grain industry is shown by the data in Table IV, which lists each innovation in microstructural development, the year in which it occurred, and the grinding ratio.[32] The latter quantity is a measure of

Table II. Slurry Abrasion Wear of Ceramics*

Slurry	Ceramics[†]			
	Al_2O_3	KT–SiC	Mg–PSZ	Porcelain
Al_2O_3	95.8	16.4	5.2	29.2
Quartzite	4.7	3.2	1.4	5.5
Limestone	3.3	1.4	0.7	5.8

*Reference 30.
[†]Steady-state rate of volume (%) decrease $\times 10^3$, (s^{-1}).

Table III. Miscellaneous Applications of Mg–PSZ (MS)

Industry	Application	Present material	Improvement*
Paper manufacturing	Autoclave nozzle	Stainless steel	30-fold
Powder metallurgy	Powder compaction die	Tungsten carbide	4-fold
Mining	Dry bearing	Steel	6-fold

*Defined as the increase in component lifetime.

Table IV. Improvement in the Efficiency of Industrial Grinding*

Year	Material	Grinding ratio
1950	Fused Al_2O_3	100
1956	Fused, cast Al_2O_3	186
1962	Fused, cast Al_2O_3-ZrO_2 (slow cooled)	256
1977	Fused, cast Al_2O_3-ZrO_2 (fast cooled)	586
1978	Fused, cast Al_2O_3-(CeO_2/ZrO_2) (fast cooled and aged ~1000°C/3 h)	750

*Reference 32.

the efficiency of the grinding process and is defined as the ratio of the weight of metal removed to the weight of wheel lost. The first exploitation of the enhanced properties of a dispersion of ZrO_2 particles probably was done unconsciously in 1962. The grain was made by electrofusing the alumina-zirconia batch, casting it into molds, slow-cooling, and finally crushing the material to the appropriate grit size. By 1977 the abrasive grain industry used fast cooling, which produced the requisite t-ZrO_2 phase.[33] The marked enhancement of the grinding ratio associated with fast cooling is likely due to the presence of both m- and t-ZrO_2 particles. It is interesting to note that recent attempts in the literature to characterize abrasive grain use only stereological concepts to relate the toughness of the grains to the microstructure rather than the modern fracture mechanics approach.[34]

The most recent refinement of microstructural control occurred in 1978 and comprised the alloying of the alumina-zirconia batch with CeO_2, followed by aging the material to optimize the toughness.[35] The grinding ratio of this sophisticated grain constitutes nearly an 8-fold improvement on the fused-alumina grain first used in 1950.

Fine-grained alumina-metal cutting tool bits made by hot-pressing began to be used extensively during the 1950s.[36] It is a natural development to use alumina toughened by a dispersion of ZrO_2 particles in this application.[37] To appreciate the impact of this technology, it is necessary to distinguish between continuous and interrupted cutting. During interrupted cutting, each revolution imposes repeated severe mechanical impacts on the tool bits. To withstand this punishing environment, it was discovered that both t- and m-ZrO_2 particles were necessary, as shown by the data listed in Table V.[38] Although adding m-ZrO_2 particles to alumina enhances the bulk toughness and therefore the tool bit life (i.e., the number of passes) in interrupted cutting, the flank wear increases to unacceptably high levels in continuous cutting. The microcracking-toughening mechanism in the latter cutting mode causes enhanced wear due to surface chipping. However, when

Table V. Lifetime and Wear Data for Al$_2$O$_3$ and ZrO$_2$ Tool Bits*

	Cutting Mode					
	Continuous[†]			Interrupted[†]		
	A	A + m−Z	A + (m,t−Z)	A	A + m−Z	A + (m,t−Z)
No. of passes	25			25	53	70
Flank wear (mm)	0.20	0.78	0.23	0.20		

*Reference 38.
[†]A = Al$_2$O$_3$; A + m−Z = Al$_2$O$_3$ + 16 vol% m−ZrO$_2$; A + (m,t−Z) = Al$_2$O$_3$ + 16 vol% (m−+t−ZrO$_2$); m/t, ≈1.

≈50% of the particles are tetragonal, the surface strengthening mechanism is invoked, which simultaneously increases the lifetime of the bit in interrupted cutting by a factor of 3, while reducing the flank wear to that of the pure alumina bit in continuous cutting.

Another interesting application related to the toughening phenomena in Al$_2$O$_3$-ZrO$_2$ composites is the toughening of the fast-ion conductor, β''-alumina, by a dispersion of ZrO$_2$ particles. The toughness of these composites was increased from 2.5–3.0 MPa·m$^{1/2}$ to 5–8 MPa·m$^{1/2}$ by distributing ≈16 wt% particles of zirconia ≈1.0 μm in diameter.[39,40] The surface of this solid electrolyte tends to degrade rapidly under the stress of a large current density, which inhibits the development of batteries employing this material. If microcracking can be avoided, the surface strengthening effect and bulk transformation-toughening mechanisms of the composite allow the safe use of large current densities, which could have considerable technological and economic impact in this area.

Refractory Applications of Zirconia Composite Materials

Traditional refractories have a useful degree of thermal shock resistance at the expense of their inherent strength. These materials work only because the microstructure is porous, with many coarse grains. Such materials do not store significant amounts of elastic energy during thermal shock. However, they are susceptible to enhanced slag attack and erosion by flowing molten metal.[41] This situation is improved dramatically by dispersing m-ZrO$_2$ particles of the appropriate size and amount in a suitable matrix phase. It is now possible to design refractories that are dense, strong, and thermal-shock-resistant. The system ZrSiO$_4$-ZrO$_2$ provides an excellent model for creating high-performance refractories able to handle molten metals. Typical products include foundry crucibles, tundish pouring nozzles, and flow control systems (sliding gates) for the continuous casting of steel.

Figure 7 is a schematic drawing of the continuous casting process used in the steel industry. The rate of flow of metal from the tundish vessel into the mold must be constant for several heats if the process is to be successful. This condition requires that the tundish pouring nozzle withstand the severe thermal shock that occurs with every heat. In addition, the nozzle bore must have a high resistance to erosion by the flowing metal stream at ≈1650°C. The current material of choice is a refractory grade of PSZ which, although expensive, is not a particularly suitable refractory for its intended purpose. The strength of this material is only 20–30 MPa, with values of the open porosity up to ≈20%. The siliceous grain

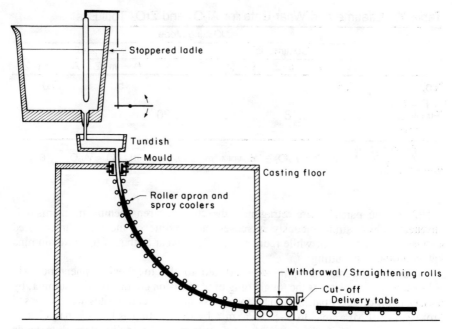

Fig. 7. Schematic diagram of the continuous casting process.

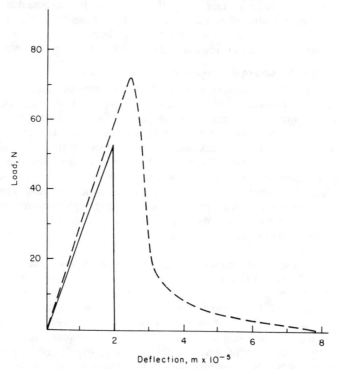

Fig. 8. Load-deflection curves for DZ (solid line) and DZ10 (dashed line) materials.

boundaries soften at the working temperature, accelerating erosion of the bore. If any slag is present, the material is destabilized, which causes even more rapid deterioration of the nozzle.[42]

Dissociated zircon containing about 10 wt% m-ZrO_2 particles ≈ 13 μm in diameter (DZ10) is a promising low-cost, high-performance candidate nozzle material.[43] Zircon itself is not easy to sinter, but when it is dropped through a plasma furnace, it dissociates to form spheres of reactive silica containing microcrystals of ZrO_2 about 0.2-μm long. The milled powder reaction-sinters readily to form a dense, fine-grained (2–3 μm) product. The addition of m-ZrO_2 particles induces a high-integral work-to-fracture energy, as shown by the load deflection curves plotted in Fig. 8. The (yet to be defined) toughening mechanism renders the composite material thermal-shock-resistant. Figure 9 is an optical micrograph of DZ10 samples that were damaged by first heating them to 600°C and then quenching them into water at room temperature. The micrograph shows many microcracks, which seem to cut the ZrO_2 particles and then are arrested. The properties of DZ10 are compared with those of a commercial clay-bonded zircon material in Table VI. Recent field tests confirmed the superior properties of DZ10. A small experimental tundish was fitted with a DZ10 nozzle at an Australian foundry. After 37 heats of molten stainless steel had been poured through the nozzle at a mean temperature of 1650°C, it was still in use, although the nozzle had never been preheated. Field tests of nozzles at a major US steel company showed that there was no significant cracking due to thermal shock, and less erosion occurred compared to refractory-grade PSZ nozzles.

Summary

Data presented in this brief review show that the ZrO_2-bearing materials have superior properties and hence performance during use, several hundred percent greater than those of conventional ceramics or refractories. The versatile matrix/particle combinations can be selected so as to produce composites that are

Fig. 9. Optical micrograph of DZ10 thermally shocked from 600°C, × 160.

Table VI. Properties of Dissociated Zircon + 10 Wt% ZrO_2 (DZ 10) and Clay-Bonded Zircon*

Properties	Material	
	DZ 10	Clay-bonded zircon
MOR (MPa)	149	22
K_{Ic} (MPa·m$^{1/2}$)	2.9	1.5
γ_{wof}/γ_I	3.3	1.0
R_{st} (m$^{1/2}$·K^{-1})	3.6	3.5
Open porosity (%)	2	21

*Reference 43.

tough for the cutting, grinding, and extrusion of solid metals, hard-wearing for abrasive environments or those characterized by large frictional forces, and extremely thermal-shock-resistant (with no loss of strength) for handling molten metals. In the light of these remarks, "revolutionary" does not seem too strong a word to describe the present and future impact of the science and technology of zirconia materials on the industrial world.

Acknowledgments

I thank my colleagues N. Gane, R. H. J. Hannink, M. J. Murray, and M. V. Swain for assistance in obtaining information and also for helpful discussion.

References

[1] G. K. Bansal and A. A. Heuer, "Precipitation in Nonstoichiometric Magnesium Aluminate Spinel," *Philos. Mag.*, **29** [4] 709–22 (1974).
[2] E. H. Greener, J. K. Harcourt, and E. P. Lautenschlages; Materials Science in Dentistry. Williams and Wilkins, Baltimore, MD, 1972.
[3] H. P. Kirchner; Strengthening of Ceramics. Marcel Dekker, New York, 1979.
[4] A. G. Evans and T. G. Langdon, "Structural Ceramics," *Prog. Mater. Sci.*, **21** [2–4] 171–442 (1976).
[5] R. C. Garvie, R. H. J. Hannink, and R. T. Pascoe, "Ceramic Steel?," *Nature (London)*, **258** [5337] 703–704 (1975).
[6] H. H. Sturhahn, W. Dawihl, and G. Thamerus, "Applications and Properties of Sintered ZrO_2," *Ber. Dtsch. Keram. Ges.*, **52** [3] 59–62 (1975).
[7] N. Claussen, "Fracture Toughness of Al_2O_3 with an Unstabilized ZrO_2 Dispersed Phase," *J. Am. Ceram. Soc.*, **59** [1–2] 49–51 (1976).
[8] D. L. Porter and A. H. Heuer, "Mechanisms of Toughening Partially Stabilized Zirconia (PSZ)," *J. Am. Ceram. Soc.*, **60** [3–4] 183–84 (1977).
[9] F. F. Lange, "Transformation Toughening" (Parts I–V), *J. Mater. Sci.*, **17** [1] 225–63 (1982).
[10] R. T. Pascoe and R. C. Garvie, "Surface Strengthening of Transformation Toughened Zirconia"; Ceramic Microstructures '76. Edited by R. M. Fulrath and J. A. Pask. Westview, Boulder, CO, 1977.
[11] N. Claussen and M. Rühle, "Design of Transformation Toughened Ceramics"; pp. 137–63 in Advances in Ceramics, Vol. 3. Edited by A. H. Heuer and L. W. Hobbs. The American Ceramic Society, Columbus, OH, 1981.
[12] A. H. Heuer, N. Claussen, W. M. Kriven, and M. Rühle, "Stability of Tetragonal ZrO_2 Particles in Ceramic Matrices," *J. Am. Ceram. Soc.*, **65** [12] 642–50 (1982).
[13] D. J. Green, "Critical Microstructures for Microcracking in Al_2O_3-ZrO_2 Composites," *J. Am. Ceram. Soc.*, **65** [12] 610–14 (1982).
[14] M. V. Swain, "Grinding-Induced Tempering of Ceramics Containing Metastable Tetragonal Zirconia," *J. Mater. Sci. Lett.*, **15**, 1577–79 (1980).
[15] H. Ruf and A. G. Evans, "Toughening by Monoclinic Zirconia," *J. Am. Ceram. Soc.*, **66** [5] 328–32 (1983).
[16] N. Claussen, "Fracture Toughness of Al_2O_3 with an Unstabilized ZrO_2 Dispersed Phase," *J. Am. Ceram. Soc.*, **59** [1–2] 49–51 (1976).

[17] B. Mussler, M. V. Swain, and N. Claussen, "Dependence of Fracture Toughness of Alumina on Grain Size and Test Technique," *J. Am. Ceram. Soc.*, **65** [11] 566–72 (1982).

[18] R. R. Hughan, R. H. J. Hannink, R. C. Garvie, M. J. Murray, M. V. Swain, and R. K. Stringer; International Patent Application, No. PCT/AY83/00.

[19] R. H. J. Hannink and R. C. Garvie, "Sub-Eutectoid Aged, Mg-PSZ Alloy with Enhanced Thermal Up-Shock Resistance," *J. Mater. Sci.*, **17** [9] 2637–43 (1982).

[20] H. J. Rossell and R. H. J. Hannink, "Microstructures of $MgO\text{-}ZrO_2$ Alloys"; this volume, pp. 139–51.

[21] R. H. J. Hannink and M. V. Swain, "Magnesia-Partially Stabilized Zirconia: The Influence of Heat Treatment on Thermomechanical Properties," *J. Aust. Ceram. Soc.*, **18** [2] 53–62 (1982).

[22] S. T. Gulati, J. D. Helfinstine, and A. D. Davis, "Determination of Some Useful Properties of Partially Stabilized Zirconia and the Application to Extrusion Dies," *Am. Ceram. Soc. Bull.*, **59** [2] 211–19 (1980).

[23] R. N. Katz, "Ceramics for Vehicular Engines: State-of-the-Art"; pp. 449–67 in Materials Science Monograph—Energy and Ceramics, Vol. 6. Edited by P. Vincenzini, Elsevier, Amsterdam, 1980.

[24] R. Kamo, M. E. Woods, and W. C. Geary, "Ceramics for Adiabatic Diesel Engine"; pp. 408–87.

[25] U. Dworak, D. Fingerle, H. Olapinski, and U. Krohn, "Application of ZrO_2 Ceramics in Internal Combustion Engines"; this volume, pp. 480–87.

[26] S. Robb, "Cummins Successfully Test Adiabatic Engine," *Am. Ceram. Soc. Bull.*, **62** [7] 755–56 (1983).

[27] R. C. Garvie and P. S. Nicholson, "Structure and Thermomechanical Properties of Partially Stabilized Zirconia in the $CaO\text{-}ZrO_2$ System," *J. Am. Ceram. Soc.*, **55** [3] 152–57 (1972).

[28] R. C. Garvie, M. F. Goss, and C. Urbani, "Weibull Modulus Studies of Magnesia-Partially Stabilized Zirconia Ceramics"; Science of Ceramics, Vol. 12. Saint-Vincent, Italy, June 1983.

[29] R. C. Garvie, C. Urbani, D. K. Kennedy, and J. C. McNeuer, "Bioceramic Applications of Mg-PSZ Ceramics"; to be published in *Journal of Materials Science*.

[30] M. V. Swain and M. F. Goss, "Slurry Abrasion Wear of Ceramics"; CSIRO. Adv. Mater. Lab. Rept. No. AML–81–10 (1981).

[31] L. Coes, Jr.; pp. 61–67 in Abrasives. Springer-Verlag, New York, 1971.

[32] R. A. Rowse and J. E. Patchett, "Innovations in Grinding Materials"; pp. 215–28 in Sagamore Army Materials Research Conference Proceedings, 1978.

[33] A. K. Kuriakose and L. J. Beaudin, "Tetragonal Zirconia in Chilled Cast Alumina-Zirconia," *J. Am. Ceram. Soc.*, **46**, 45–50 (1977).

[34] R. E. Sheplet and E. D. Whitney, "Effect of Microstructure Friability in the System Al_2O_3," *Wear*, **46**, 281–94 (1978).

[35] G. R. Watson, "Alumina-Zirconia Alloy"; Can. Pat. No. 1 031 169, 1978.

[36] A. G. King and W. M. Wheildon, Ceramics in Machining Processes. Academic Press, New York, 1966.

[37] U. Dworak and H. Olapinski, "Cutting Plate for Free-Cutting Machine"; Ger. Offen. 2 923 213, 1980.

[38] N. Reiter, "A New Ceramic Cutting Material with Superior Toughness," *Eng. Digest*, **40** [9] 17–23 (1979).

[39] L. Viswanathan, Y. Ikoma, and A. V. Virkar, "Transformation Toughening of β-Alumina by Incorporation of Zirconia," *J. Mater. Sci.*, **18** [1] 109–13 (1983).

[40] J. Binner, R. Stevens, and S. Tan, "ZrO_2 Toughening of $\beta\text{-}Al_2O_3$"; this volume, pp. 428–35.

[41] W. S. Trettner, "Refractories Technology," *Am. Ceram. Soc. Bull.*, **58** [7] 715–18 (1979).

[42] D. F. Beal, "Bore Erosion of Zirconia Tundish Nozzles"; for abstract see *Am. Ceram. Soc. Bull.*, **59** [8] 860 (1980).

[43] R. C. Garvie, "Improved Thermal Shock Resistant Refractories from Plasma-Dissociated Zircon," *J. Mater. Sci.*, **14**, 817–22 (1979).

ZrO₂ Ceramics for Internal Combustion Engines

ULF DWORAK, HANS OLAPINSKI, DIETER FINGERLE, AND ULRICH KROHN

Feldmühle Aktiengesellschaft
Fabrikstr. 23–29
D-7310 Plochingen
Federal Republic of Germany

This paper discusses material aspects of various zirconia grades, property design, and joining techniques of various structural ZrO_2 parts suitable for insulation, as well as for friction and wear applications in gasoline and diesel engines.

Over the last ten years, the automotive industry has initiated a number of research and development programs to improve fuel economy, reliability and maintenance costs, environmental conditions (e.g., exhaust gases and noise), and comfort of engine systems. Because ceramic parts may impact these requirements positively, the use of ceramics for future diesel and gasoline engines for both passenger cars and industrial vehicles is currently being evaluated (Table I).

Material Selection

The properties of the ceramic material required for the various fields of application are obviously quite different. Further differentiations in the required material properties depend on the design of parts and the joining technique. It seems necessary to select the best material part-by-part to meet the required needs, which means that currently there is no ideal material available which covers the whole range of applications. There are many ceramic materials and manufacturing processes with corresponding properties and, consequently, with performance characteristics, which are most suitable for specific engine parts.

The influence of insulation on engine efficiency has been discussed for cooled, uncooled, and adiabatic diesel engines.[1] Reducing the energy losses to the cooling system and using the high exhaust energy of turbocharged engines to drive a second turbine, coupled to the engine output shaft via the transmission (the so-called turbocompound engine) resulted in a gain of up to 25% in the specific fuel consumption. Partially stabilized zirconia (PSZ), which offers high strength, low thermal conductivity, and high toughness, is currently of great interest for insulated engine components. The high thermal expansion coefficient of PSZ is close to that of cast iron, which simplifies the ceramic-metal attachment. The small mismatch in the thermal expansion between metal and PSZ reduces the assembly interference requirements when they are combined and used at high temperatures. The use of an interference fit is a suitable joining technique for PSZ. By superimposing compressive forces, this technique leads to a reduction in the high thermal stresses which appear during operation.

Table I. Requirements for Ceramic Diesel and Gasoline Engine Components

Thermal insulating components (fuel economy)
　Piston heads
　Cylinder liners
　Cylinder heads
　Portliners
　Manifold liners
　Turbocharger housings
　Turbocharger heat shields

High-temperature strength components (fuel economy)
　Turbocharger hot wheels

Weight-reduction components (fuel economy)
　Piston pins
　Turbocharger hot wheels

Wear- and corrosion-resistant components (reliability improvements)
　Cylinder liner segments
　Valve seats
　Valve guides
　Bypass valve guides
　Tappet inserts
　Rocker arm inserts
　Bearings
　Seal rings

Exhaust-gas improvement components (environment)
　Precombustion chambers
　Glow plugs
　Portliners

Characteristics of PSZ Grades

PSZ-ceramics have been commercialized in engineering applications since 1970.[2,3] Table II shows the grades and properties of ZrO_2 ceramics.

Ca/Mg-PSZ (Zt 35), with approximately 20% t precipitates in a c matrix, consists of large (60–70 μm) grains. This material is used in wire drawing technology.[2]

Mg-PSZ (ZN 40) possesses high strength and fracture toughness due to the high proportion ($\approx$40–50%) of t precipitates.

Mg-PSZ (ZN 50) is a modified-grade material with a reduced thermal expansion coefficient. Figure 1 illustrates the different dilatation characteristics of ZN 40 and ZN 50. These materials consist of grains typically 60–70 μm in size surrounded by a thin grain-boundary film of m phase (Fig. 2).

Y-TZP (ZN 100) (Fig. 3) is a ZrO_2 with high strength and toughness. This is a very-fine-grained material ($\approx$0.3 μm) with nearly 100% t phase.

Mg-PSZ (ZN 20) is designed for critical thermal shock conditions. Thermal stresses are reduced by a low thermal expansion. The expansion characteristics as

Table II. PSZ Grades and Properties of ZrO$_2$ Ceramics

Material	Designation	E (GPa)	σ^*_{BRT} (MPa)	σ^*_{TRT} (MPa)	K_{Ic} (MPa·m$^{1/2}$)	m	HV$_{RT}$	ρ (g·cm^3)	λ (W/m·K)	c (J/g·K)	$\alpha_{RT-1000°C}$ (10^{-6}/K × 10)
Standard Grades											
Ca/Mg-PSZ	Zt 35	200	300	215	4.8	>20	1300	≥5.70	2.1	0.4	9.8
Mg-PSZ	ZN 40	200	500	400	8.1	>20	1200	≥5.73	2.1	0.4	9.8
Modified Grades											
Mg-PSZ	ZN 50	200	580		9.0		900	≥5.68	2.5	0.4	~7.0
Y-PSZ	ZN 100	190	1050		9.7	>12		≥5.97		0.4	9.3
Mg-PSZ	ZN 20	180	350		3.5			≥5.6	3.8	0.4	≈5.5

*B = bending,
T = tensile,
RT = room temperature.

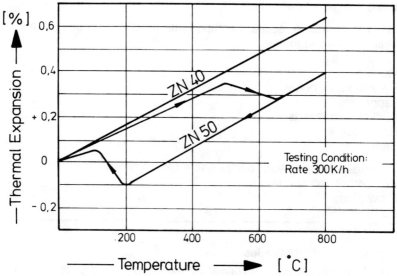

Fig. 1. Dilatation curves of Mg-PSZ (ZN 40 and ZN 50).

a function of temperature exhibit a pronounced hysteresis, due to the high amount of *m* phase present.

Properties

Strength and Fracture Toughness

Figure 4 shows the strength and fracture toughness of ZN 40 as functions of temperature. These strength values were obtained on test bars (3.5 by 7 by 40 mm) ground with a 120-mesh diamond wheel and broken in 4-point bending in air at a stress rate of 1.75 MPa/s (Fig. 4(A)). The fracture toughness,[4] obtained with single-edge-notched beam tests (Fig. 4(B)), corresponds closely to the strength dependence.

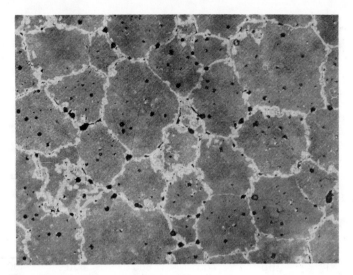

Fig. 2. Microstructure of Mg-PSZ (ZN 50).

Fig. 3. Microstructure of Y-TZP (ZN 100).

Dynamic Fatigue

The dynamic fatigue of ZN 40 was tested in cyclic loading experiments using tensile test rods.[5] As shown in Fig. 5, the tensile strength is reduced from the initial 400 MPa to ≈200 MPa after 10^7 load cycles.

Thermal Stability

Isothermal heat treatment up to 3000 h at 600° and 800°C in air did not degrade PSZ grades ZN 40 and ZN 50 (Fig. 6). The physical characteristics of

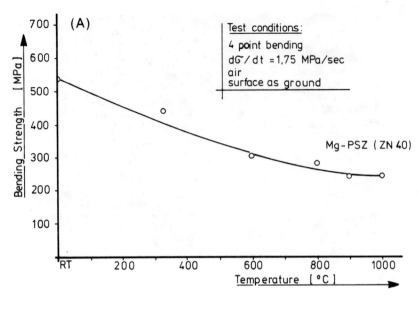
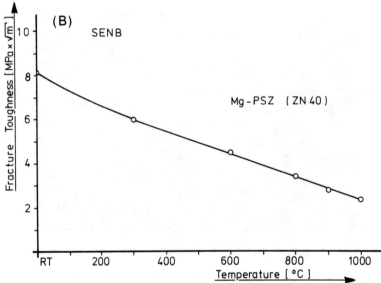

Fig. 4(A). Bending strength of Mg-PSZ (ZN 40) vs temperature.
(B). Fracture toughness of Mg-PSZ (ZN 40) vs temperature.

ZN 40 change over time during heat treatment at 1000°C. This effect is related to a subeutectoid aging,[6–8] where growth of m-ZrO_2, mainly in the grain-boundary region, can be observed. Simultaneously the t-phase content is reduced, leading to a reduction in the transformation-toughening effect.

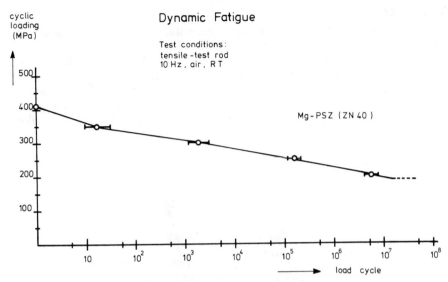

Fig. 5. Dynamic fatigue of Mg-PSZ (ZN 40).

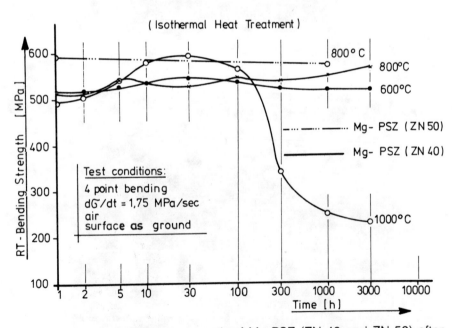

Fig. 6. Room-temperature strength of Mg-PSZ (ZN 40 and ZN 50) after heat treatments indicated.

Applications

A number of essential components for internal combustion engines have been tested successfully during the last two years. Some of these parts are shown in Figs. 7–9. Figure 7 illustrates shrink-fitted PSZ cylinder liners in a cast iron block. A ZN 40 bypass valve guide for a turbocharger device (Fig. 8) was press-fitted into a cast iron housing. A permanent temperature gradient of approximately 800°C was applied over the length of the valve guide. One remarkable test result was a drastic reduction in wear, to one-tenth that of the older design using superalloy valve guides. A further example, which has been successfully tested in a small

Fig. 7. PSZ cylinder liners.

Fig. 8. PSZ bypass valve guides for a turbocharger.

Fig. 9. PSZ valve seats.

diesel engine, is a valve seat (Fig. 9). Since the ZrO_2 ceramic (ZN 40) has a low elastic modulus, it can withstand high cyclic mechanical loading and severe thermal shock conditions.

Conclusion

Due to their low thermal conductivity, high strength, and high fracture toughness, ZrO_2 ceramics are candidate materials for insulation of internal combustion engines and for wear parts under severe environmental conditions. Further potentials of these materials depend on specific variations of material properties required for engine components.

References

[1] P. Walzer, *Brennst.-Wärme-Kraft*, **35**, 46–50 (1983).
[2] H. H. Sturhahn, G. Thamerus, and H. C. Eichas, *Draht*, **9**, 487–90 (1974).
[3] H. H. Sturhahn and P. Schorr, *Bänder Bleche Rohre*, **9**, 347–49 (1975).
[4] Lishing Li and R. F. Pabst; pp. 371–82 in Fracture Mechanics of Ceramics, Vol. 6. Plenum, New York, 1983.
[5] R. F. Pabst, I. Bognar, and J. B. Zwissler, *Ber. Dtsch. Keram. Ges.*, **57** [2] 13–16 (1980).
[6] R. C. Garvie, R. R. Hughan, and R. T. Pascoe, *Mater. Sci. Res.*, **11**, 263–74 (1978).
[7] R. P. Ingel, D. Lewis, B. A. Bender, and R. W. Rice, *J. Am. Ceram. Soc.*, **65** [9] C-150–C-152 (1982).
[8] R. H. I. Hannink, *J. Mater. Sci.*, **18**, 457–70 (1983).

Plasma-Sprayed Zirconia Coatings

P. Boch, P. Fauchais, D. Lombard, B. Rogeaux, and M. Vardelle
La CNRS 320
F. 87065 Limoges, France

Particle trajectories in the plasma jet and gas-particle heat transfer were studied. Correlation between spraying parameters and crystallographic, microstructural, and mechanical properties of zirconia-sprayed coatings were carried out on zirconia-calcia and zirconia-yttria materials. The annealed states were studied, chiefly for zirconia-yttria. The annealing temperatures were 550°, 900°, and 1400°C. The crystallographic phases were determined. The mechanical properties of the coatings (hardness, thermal expansion, Young's modulus, strength, and thermal shock resistance) were evaluated.

Plasma spraying is a well-established means of forming thermal-barrier coatings (TBCs). In this process, the powder particles injected into a plasma jet are quickly melted and propelled onto a substrate where they are quenched ($\approx 10^5$ $K \cdot s^{-1}$). TBCs have to withstand severe conditions for a sufficiently long time: zirconia-based coatings have been widely studied[1] due to their low thermal conductivity (≈ 1.5 $W \cdot m^{-1} \cdot K^{-1}$) and their medium thermal expansion ($\approx 10^{-5}$ K^{-1}), which permit a good match with metal substrates. Numerous parameters, such as plasma-torch characteristics and spraying conditions, determine the structure and the properties of the TBCs. However, for a given material, the structure of the deposit is essentially determined by the size, kinetic energy, and viscosity of the particle on impact. Hence, the influence of spraying conditions must be considered through their influence on particle velocity and temperature.

Many studies have been devoted to the case of gas turbines, which operate at high temperatures. The inlet temperature can be in excess of 1200°C, and the superalloys which are used can reach about 900°C at the ceramic-metal interface.[2] On the other hand, diesel engines are made out of materials which cannot sustain such temperatures: cast iron or aluminum alloys have to remain below about 350°C and stellitic steels (valves) below 600°C. Other differences between the requirements of diesel engines and turbines may be found.[3] For instance, controversial data are given for stabilizing additions of zirconia-based ceramics, suggesting that diesel engines could require smaller amounts than turbines. This is the reason for the choice of partially stabilized zirconia (PSZ). Moreover, it is known that some PSZs can exhibit a mechanical toughening.[4]

Plasma-sprayed coatings which appear to be the best for turbines are not necessarily adaptable for diesel engines. Therefore, this paper describes an experimental investigation of relationships between plasma-spraying conditions and the resulting properties of zirconia coatings, in order to give useful information for diesel uses.

Experimental Procedure

Zirconia powders were sprayed in air, with a conventional dc plasma torch working with an Ar-H$_2$ gas mixture (Ar, 75 L·min^{-1} at 3 bars; H$_2$, 15 L·min^{-1} at 3 bars). The particles were radially injected into the plasma jet with argon as the powder-carrier gas. The substrate material was 0.38 wt% carbon steel, and the grit-blasting powder was alumina. An air jet was blown onto the coating while spraying in order to cool it. The characteristics of zirconia powders are presented in Table I.

The phase determination was performed by X-ray diffraction (transmission diffraction on a Guinier chamber) in the (111) and (400) regions. The (111) region allows the separation of the (111) diffraction of the cubic and tetragonal phases from the (111) and (11$\bar{1}$) diffraction of the monoclinic zirconia; the (400) region helps for an accurate separation of the cubic phase, with one (400) peak only from the tetragonal one, with two (400) peaks, (040) and (004). Quantitative evaluation of the phases was made by measuring the area under each peak.[5]

A diesel engine works in thermal cycles. Such cycles correspond to annealings of the ceramic deposits; they can also provoke thermal shock damage. Hence, three annealing temperatures were chosen, 550°, 900°, and 1400°C. The first temperature corresponds to a maximum possible gradient of 200°C across the layer protecting the cast iron- or the aluminum-based pieces ($T_{max} \approx 350°C$); the second one corresponds to a gradient of 300°C in the case of valves ($T_{max} \approx 600°C$); the last does not correspond to the working temperature range but was chosen high enough to lead to equilibrium phases.

The thermal shock properties of the composite pieces (substrate + coatings) were studied using an automated bench and an ultrasonic control method and have been published elsewhere.[6,7]

The thermal expansion (α) was measured on prismatic beams (1.5 by 4 by 10 mm^3) (electronic differential dilatometer ADAMEL); the strength (σ_f) was measured by the biaxial flexure of disks (0.7–1.5 mm thick, 20 mm in diameter); Young's modulus (E) was determined from the resonant frequencies of these disks[8]; the critical value (ΔT_c) of the thermal shock resistance was determined by quenching the samples from a furnace at various temperatures with boiling water at 100°C. Measurement of the resonant frequencies of the disks allows determination of the damage curve, using one sample only.

Results and Discussion

Plasma-Particle Energy Transfer

The quality of a coating depends, to a large extent, on the velocity and temperature of the particles prior to their impact onto the substrate. These particle

Table I. Main Characteristics of Sprayed Powders

Material	Size range (μm)	Density	Main phase
ZrO$_2$–8 wt% Y$_2$O$_3$	1		
ZrO$_2$–6 wt% Y$_2$O$_3$	10–25	6.0±0.1	Tetragonal
ZrO$_2$–4 wt% Y$_2$O$_3$			
ZrO$_2$–8 wt% Y$_2$O$_3$	20–75	6.0±0.1	Tetragonal
ZrO$_2$–7.5 wt% CaO	5–15	5.7±0.1	Cubic

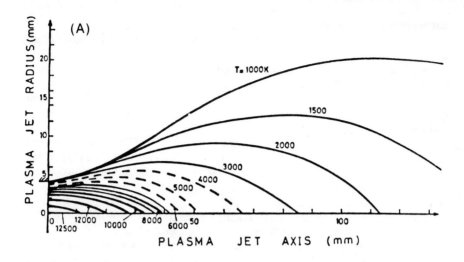

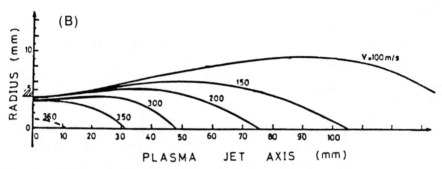

Fig. 1. (A) Temperature and (B) velocity isocontours of an Ar-H_2 plasma jet as function of the radial and the axial distance from the nozzle exit.

parameters are essentially governed by the characteristics of the plasma jet (flow and temperature fields), the particle trajectory (depending on the momentum transfer), and the gas-particle heat transfer.

Experimental temperatures and axial velocity isocontours of the Ar-H_2 plasma jet are given in Fig. 1. The plasma jet can be divided into three regions[9]: (1) The core region, where the plasma temperature is high (10 000–12 000 K) and relatively constant, which extends 10–12 mm from the nozzle exit; (2) the transition region, where the plasma temperature falls quickly to less than 3000 K, which is 100 mm from the nozzle exit; and (3) the last region, where the temperature drops gradually as the plasma jet mixes with the ambient atmosphere.

The radial temperature gradients and the velocity gradients are particularly steep around the boundaries of the core region (up to 4000 K·mm^{-1} and 200 m·s^{-1}·mm^{-1}, respectively). Such high gradients are to be considered in the study of the thermal treatment of particles in the plasma jet. Therefore, the experiment requires a close control of the particle injection and trajectory.

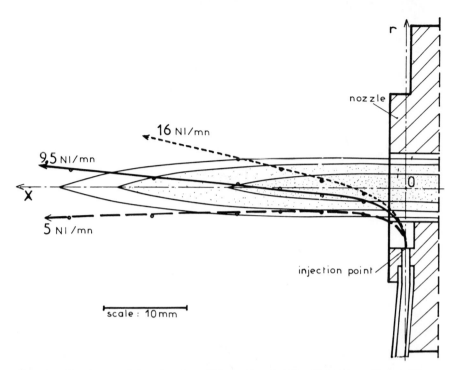

Fig. 2. Particle trajectories for different powder-carrier gas flow rates.

Particle Trajectories: To determine the particle trajectories, we measured the number of particles traveling at different points in the jet by counting, during a given time, the pulses resulting from the light scattered by the particles passing through a focused laser beam.[9] Then the trajectories were determined using the maximum of the radial distribution of the particles.

The results obtained for yttria-containing powders (10–25 μm) are given in Figs. 2 and 3. Figure 2 shows the particle trajectories for different flow rates of the carrier gas for a given flow rate of powder. Figure 3 shows, for the optimal gas flow rate, the particle radial distribution at different distances from the nozzle exit. It can be seen that, for the optimal flow rate, most of the particles (more than 90%) pass through the hot region of the plasma. The remaining particles travel in the periphery of the jet. Consequently, they are not melted and have a low velocity. Furthermore, we observed that the number of molten particles collected onto the substrate is a maximum when the mean particle jet has a deviation of 10°–15° from the torch axis in the opposite direction of the injector.

Particle Evaporation: We studied particle evaporation by emission spectroscopy. For calcia-containing powders, for instance, we observed the intensity evolution of the most intense lines of Zr (383.5 nm) (3835 Å)) and Ca (393.3 nm (3933 Å)). It is worth noting that the particle-surface temperature quickly reaches the evaporation temperature (2 or 3 mm from the injection point) and that the evaporation rates are different for Ca and Zr. This shows that the composition of the deposit may be slightly different from the composition of the initial powder.

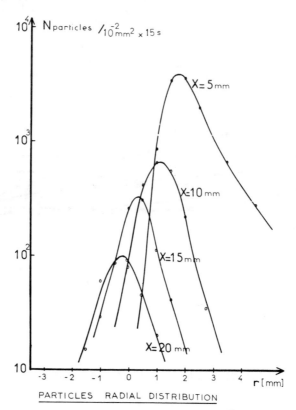

Fig. 3. Particle radial distribution for different distances from the nozzle exit.

Relationship between Zirconia Coating Properties and Spraying Conditions

Besides the particle size and the nature of the stabilizer (Ca or Y), we studied the influence of the following process parameters: electrical power input, in the range 15–40 kW, torch-to-substrate distance (D_t) in the range 50–150 mm, powder-injector location (D), with internal injection ($D = 0$) and external injection ($D = 10$ mm); flow rate of the powder-carrier gas, in the range 5–17 L·min^{-1}, and flow rate of the cooling gas, in the range 160–640 L·min^{-1}. Coatings were stripped off the substrate before porosity, microstructural texture, nature and quantity of phases, and mechanical properties were determined.

Microstructural Study

Porosity: Plasma-sprayed coatings exhibit mostly interconnected open porosity so that closed porosity does not exceed 10% of total porosity. Thus, the following results, which are related only to open pores, can be extended to total porosity.

Experimental measurement on velocity and surface temperature of particles in the plasma jet shows that the velocity increases with power input but that particle temperature changes only a little, because the dwelling time in the hot zone of the plasma jet remains the same.[10]

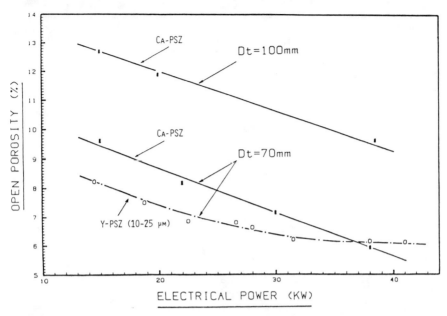

Fig. 4. Open porosity of the Y_2O_3-ZrO_2 deposits vs torch-to-substrate distance.

This leads to a small decrease of porosity (about 2% for yttria-containing coatings and 4% for calcia-containing coatings) when the power increases (Fig. 4). Identical variations of porosity vs power are obtained for the two yttria-zirconia size ranges, but coarse powders give a higher porosity by about 2%. This small difference comes from the large size range of the 20–75 μm powder.

To obtain dense coatings, the torch-to-substrate distance should enable the particles to impinge on the target in the liquid state with the greatest kinetic energy. However, these requirements are difficult to meet: A high velocity reduces the residence time and thus the particle temperature; since the particles have to be melted, the substrate must be located a few millimeters downstream from the point where the particles reach their maximum temperature. Therefore, this location does not correspond to their maximum of velocity.[10] Referring to Fig. 5, dealing with yttria-containing powders, the best location of the substrate is 75 mm from the nozzle exit (open porosity of about 6.5% for the 10–25-μm powder). The flow rate of the carrier gas determines the initial velocity of particles and thus their trajectory in the plasma jet (Fig. 2). Of course, the optimal trajectory depends on the size range and, for yttria-containing powders in the 10–25 μm range, minimum porosity is found for a 9.5-L·min^{-1} gas flow rate, whereas, for the 20–75 μm range, an 8-L·min^{-1} rate is necessary. Another injection parameter which could modify the coating properties is the location of the powder injector (D). It determines the temperature zones through which the particles pass. In fact, its influence on porosity is very slight and can be neglected.

It is difficult to choose the optimal porosity for TBCs because of contradictory effects. A high porosity decreases thermal conductivity, which improves the insulation, and limits crack propagation during thermal shocks; a small porosity im-

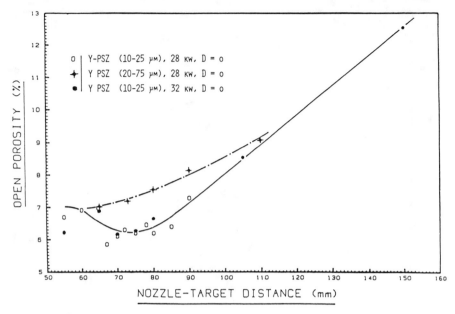

Fig. 5. Open porosity vs power input.

proves corrosion resistance and increases hardness and strength. Hence, the best value depends on the specific TBC use.

Hardness: For porous materials, hardness is directly related to porosity: comparisons between the changes in porosity and hardness vs various spraying parameters confirmed this correlation. For minimal porosity, Vicker's hardness is about 700, even if it can be lower than 300 for more porous coatings.

Microstructure: As usual for plasma-sprayed coatings, the ceramic deposits have an anisotropic layered structure, with a dense array of microcracks.[6] Examples of microcracks are shown in Fig. 6.

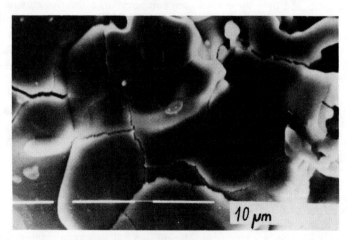

Fig. 6. Microcracks in as-sprayed deposits.

Table II. Relative Amount (mol%) of Phases in As-Sprayed States

Phase	Relative amount of phases (mol%)				
	7.5 wt% CaO	10 wt% CaO	4 wt% Y_2O_3	6 wt% Y_2O_3	8 wt% Y_2O_3
Cubic	75	90			
Tetragonal			>90	99	99
Monoclinic	25	10	<10	1	1

Crystallographic Study

 The As-Sprayed State: Table II shows that the as-sprayed state is a two-phase state. For CaO-ZrO_2 materials, the major phase is cubic; for Y_2O_3-ZrO_2 the major phase is the "nontransformable tetragonal" one, described by Miller et al.[5] This tetragonal phase is not the equilibrium one: the splat cooling has not allowed the phase separation (in the domain above the eutectoid temperature, Fig. 7), which means that the tetragonal phase has the nominal composition of the compound. The tetragonal phase largely predominates in 6 and 8 wt% Y_2O_3 materials (monoclinic content of about 1%).

 Y_2O_3-ZrO_2 coatings exhibit good phase stability with respect to spraying parameters. Indeed, increasing power input slightly increases the tetragonal content, but this concentration does not change with any other parameter studied.

 On the other hand, destabilization is observed in CaO-ZrO_2 coatings when the power input and the powder-carrier gas flow rate increase or when the injector is in the $D = 0$ position (Fig. 8). We found that the monoclinic phase content can be in excess of 40% if plasma temperature and dwelling time in the plasma core are

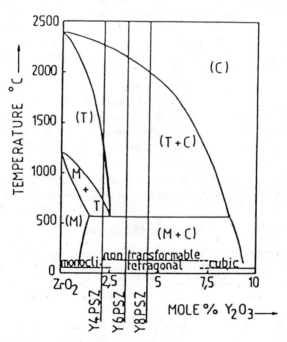

Fig. 7. Partial view of the ZrO_2-Y_2O_3 diagram.

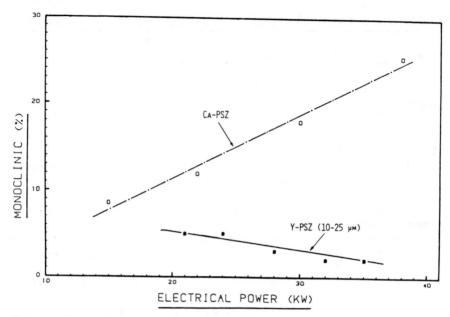

Fig. 8. Effect of power input on monoclinic content of CaO-ZrO_2 deposits.

sufficient. The flow rate of the cooling gas also influences destabilization and, with an elevated power input, the monoclinic phase can decrease from 35 to 25% when the cooling air flow rate is increased from 160 to 640 $L \cdot min^{-1}$. If the hypothesis of Ca evaporation explains destabilization in most cases, the influence of cooling indicates another cause of destabilization which is not clearly understood.

The Annealed States: The soak time was always 100 h. Table III shows that the monoclinic phase content is important in the low-yttria (4%) material, chiefly after annealing at 550°C, close to the eutectoid temperature. For the ZrO_2–6 wt% Y_2O_3 and ZrO_2–8 wt% Y_2O_3 compounds the monoclinic phase does not develop. However, the ZrO_2–8 wt% Y_2O_3 material leads to the phase separation [tetragonal (medium yttria) → tetragonal (low yttria) + cubic (high yttria)]: after annealing at 1400°C, the relative proportions roughly correspond to the phase diagram. The metastability of the tetragonal phase agrees with data of Miller et al.[5]

Thermal and Mechanical Properties

Thermal Expansion: The $\Delta l/l$ curves were obtained for a temperature variation rate of 250 $K \cdot h^{-1}$. All the samples had been previously annealed for 100 h

Table III. Relative Amount (mol%) of Monoclinic Phase in 100-h Annealed States

State	Relative amount of monoclinic phases (mol%)				
	7.5 wt% CaO	20 wt% CaO	4 wt% Y_2O_3	6 wt% Y_2O_3	8 wt% Y_2O_3
As-sprayed	25	10	<10	1	1
100 h, 550°C	25	10	70–75	2–5	2–5
100 h, 900°C	30–40	25–30	30–35	2–5	2–5
100 h, 1400°C	—	—	35–40	4–8	7

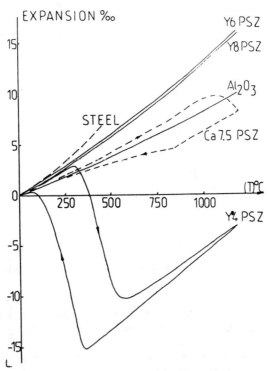

Fig. 9. Thermal expansion/temperature variations.

at 500°C to obtain a stable state. Figure 9 shows that Y6 PSZ and Y8 PSZ have regular thermal expansion behavior, the mean coefficient ($\alpha_{20}^{1000} \approx 13 \times 10^{-6}$ K^{-1}) being higher than alumina but lower than 0.38 wt% carbon steel. On the other hand, Y4 PSZ exhibits a strong contraction during heating, due to the monoclinic → tetragonal transformation.

This transformation begins close to 300° and ends close to 575°C (eutectoid temperature). During cooling, these temperatures decrease by about 200°C for the inverse transformation (tetragonal → monoclinic). This strong dilatometric anomaly seems to eliminate Y4 PSZ as a suitable coating because of the thermal fatigue effects at the substrate-deposit interface and in the deposit itself. The X-ray diffraction measurement performed at various temperatures on similar samples gave data in good agreement with the dilatometry (Fig. 10). The insensitivity to the heating rate corresponds to the martensitic nature of the transformation, with no thermal activation (as long as other phase-separation mechanisms do not occur).

For the zirconia-calcia materials the mean thermal expansion is less than the Y6 PSZ and Y8 PSZ one. The dilatometric anomaly occurs at a relatively high temperature (>850°C), which does not forbid use of these materials in diesel engines.

Strength and Young's Modulus: The strength and the modulus of elasticity of materials in the as-sprayed state are low in comparison to sintered ceramics of the same composition. This is a common fact for plasma-sprayed ceramic deposits,[11] due to the dense microcrack array (Fig. 6) and the residual stresses.

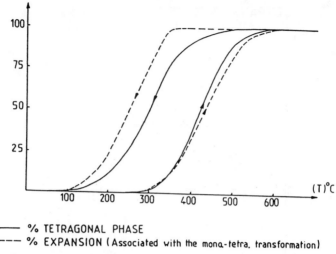

— % TETRAGONAL PHASE
--- % EXPANSION (Associated with the mono-tetra. transformation)

Fig. 10. Dilatometric anomaly/phase amount vs temperature.

The values for the three yttria-containing materials (4, 6, and 8 wt% of Y_2O_3) appeared to be similar ($\sigma_f \approx$ 100 MPa, $E \approx$ 50 GPa). These values were not very sensitive to spraying parameters.

However, when the torch-to-substrate distance is too short, microcracks develop and cause a dramatic reduction of mechanical properties (σ_f lower than 20 MPa). In contrast, the strength of ZrO_2–7.5 wt% CaO deposits is dependent on power input, as can be seen in Fig. 11. The influence of the monoclinic phase content on the mechanical properties of CaO-ZrO_2 coatings has to be determined by further experiments.

The influence of annealing was studied for the Y_2O_3-ZrO_2 materials only. The changes in mechanical properties appeared to be similar for the three compositions (Fig. 12). The low-temperature annealing (550°C) leads to a decrease of σ_f and E, which seems to be associated with a propagation of cracks, releasing the residual stresses. The annealing at higher temperatures (900° and chiefly 1400°C) leads to a strong increase in the mechanical properties. Similar results were obtained for calcia-stabilized zirconia.

Changes in mechanical properties cannot be simply related to the phase transformations because of the similar behavior for Y4 PSZ (which develops up to 70% monoclinic phase content) and for Y6 PSZ and Y8 PSZ (where the monoclinic phase content does not increase). As in the case of alumina,[12] the increase in strength and elastic modulus should be related to the healing of microcracks. This can be compared to the presintering of powder compacts, but here no shrinkage occurs. Besides, it is worth noting that the peculiar microstructure of the deposits does not appear to be damaged by the dilatometric anomaly in Y4 PSZ. This interesting behavior is currently being studied.

Thermal Shock Resistance: TBCs are used in applications involving thermal cycling. The main reason for spalling is the differential thermal expansivity between substrate and deposit.[13] Expansion coefficient ranks in the increasing order: zirconia, cast iron and carbon steels, superalloys, and aluminum alloys. Therefore, the case of turbines (superalloys) is between the two cases for diesel (cast iron and

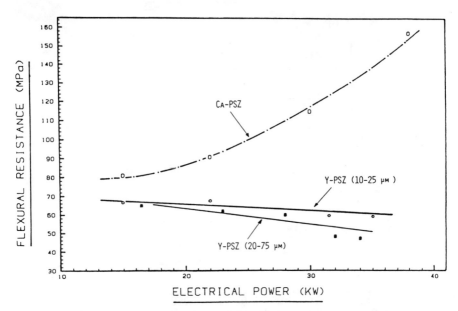

Fig. 11. Flexural strength of Ca-stabilized deposits vs power input.

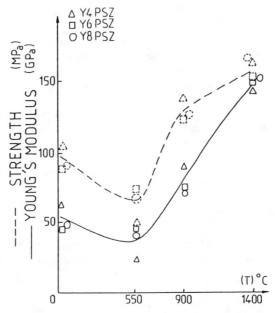

Fig. 12. Mean values of strength and of Young's modulus at 20°C vs annealing temperature (ZrO_2-Y_2O_3 compounds).

aluminum alloys). It is possible to decrease the stresses at the interface by a "segmentation" of the coating, perpendicular to the substrate.[1] However, the intrinsic thermal shock resistance of the coating must be taken into account, prior to all subsequent investigations on TBCs.

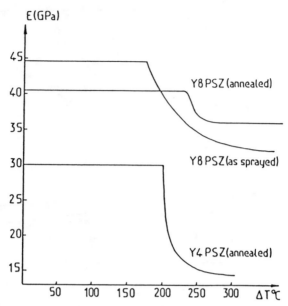

Fig. 13. Resonant frequencies of samples quenched to boiling water vs ΔT.

For pieces subjected to thermal shocks of increasing severity, the critical temperature gradient (ΔT_c) corresponds to the initiation of new cracks (thermoelastic theory)[14] or the propagation of preexisting ones (energetic theory).[15] In the case of thermal fatigue, slow crack growth occurs, which leads to a slightly smaller value of ΔT_c than in the case of a single shock.[16]

The thermal shock test used is a nondestructive method,[8] which relates the frequency (or Young's modulus) decrease of a vibrating sample to the extent of the thermal damage suffered. The ceramic coatings (20-mm diam. disks) had been previously separated from the substrates. The choice of boiling water as a quenching medium helps to avoid uncertainties[17] in the determination of ΔT_c.

Figure 13 shows the variations of Young's modulus with temperature drop (ΔT) for three specimens: an annealed ZrO_2–$4Y_2O_3$, an as-sprayed ZrO_2–$8Y_2O_3$, and an annealed ZrO_2–$8Y_2O_3$ (annealing 100 h at 550°C). The ΔT_c values were 200°, 180°, and 230°C, respectively. For ZrO_2–8 wt% Y_2O_3, the annealing improves both the critical gradient (ΔT_c) and the thermal damage ($\approx \Delta E/E$). The mean value of ΔT_c is about 200°C, which must be compared to the R parameter[15]:

$$R \approx \sigma_f[(1 - \nu)/E\alpha] \tag{1}$$

By taking $\sigma_f \approx 80$ MPa, ν (Poisson's ratio) ≈ 0.2, $E \approx 40$ GPa, $\alpha \approx 10^{-5}$ K^{-1}, R is about 160°C, that is, slightly under ΔT_c. This small difference can be attributed to the fact that Biot's number is not infinite, in the case of water quench, which reduces the thermal stress level and therefore increases ΔT_c. Thermal cycling at $\Delta T > \Delta T_c$ does not lead to a continuously decreasing strength for ZrO_2–6 wt% Y_2O_3 and ZrO_2–8 wt% Y_2O_3. In contrast, such cycling does lead to progressive weakening of the ZrO_2–4 wt% Y_2O_3 coating, the propagation of cracks being assisted by the tetragonal/monoclinic transition.

Conclusions

The main conclusions drawn from this study are the following:

(1) Plasma-particle momentum and heat transfer take place principally in the plasma core.

(2) The torch-to-substrate distance is the major factor influencing the zirconia coating microstructure.

(3) Porosity and hardness are not dependent on the same spraying parameters as flexural strength and monoclinic phase content.

(4) Significant disparities in the behavior of $CaO-ZrO_2$ and $Y_2O_3-ZrO_2$ coatings are observed. For Y8 PSZ coatings, phases and flexural strength remain constant with spraying parameters, whereas, for Ca7.5 PSZ, monoclinic content and mechanical properties are influenced by parameters such as power input or carrier-gas flow rate.

(5) The peculiar microstructure of the as-sprayed state (pores, cracks, and lamellar grains) leads to low values of mechanical properties of the coatings (hardness, strength, and Young's modulus). These values are of the same order for Ca PSZ and Y PSZ and are not strongly dependent on the amount of stabilizer, even after annealing.

(6) Y6 and Y8 PSZ have a very low amount of monoclinic phase, in the as-sprayed state and the annealed states. In contrast, a significant monoclinic phase can develop in Ca PSZ and Y4 PSZ.

(7) The tetragonal/monoclinic transition induces a large dilatometric anomaly for Y4 PSZ at temperatures located in the diesel working temperature range. This leads to a poor resistance to thermal shock damage.

(8) The strong difference in behavior between 4 wt% Y_2O_3 and 6 wt% Y_2O_3 PSZ calls for a study of the intermediate compositions.

References

[1] R. J. Bratton and S. K. Lau, "Zirconia Thermal Barrier Coatings"; pp. 229–39 in Advances in Ceramics, Vol. 3. Edited by A. H. Heuer and L. W. Hobbs. The American Ceramic Society, Columbus, OH, 1981.

[2] A. S. Grot and J. K. Martyn, "Behavior of Plasma-Sprayed Ceramic Thermal Barrier Coatings for Gas Turbine Applications," *Am. Ceram. Soc. Bull.*, **60** [8] 807–11 (1981).

[3] I. Kvernes, "Coating of Diesel Engine Components," in Seminar on Coatings for H.T. Applications, Petten, France, 8–11 March, 1982.

[4] F. F. Lange, "Transformation Toughening," *J. Mater. Sci.*, **17**, 225–63 (1982).

[5] R. A. Miller et al., "Phase Stability in Plasma Sprayed, Partially Stabilized Zirconia Yttria"; pp. 241–53 in Advances in Ceramics, Vol. 3. Edited by A. H. Heuer and L. W. Hobbs. The American Ceramic Society, Columbus, OH, 1981.

[6] P. Boch et al., "Production and Properties of ZrO_2 Based Thermal Barriers," *Sci. Ceram.*, in press.

[7] P. Boch et al., "Study of the Thermal Fatigue Resistance of Plasma-Sprayed Zirconia Coatings for Diesel Engine Components," International Symposium on Ceramic Components for Engine, Hakone, 17–21 Oct. 1983.

[8] J. C. Glandus and P. Boch, "Utilisation d'éprouvettes en forme de disques pour la mesure des caractéristiques mécaniques de matériaux fragiles," *Mem. Sci. Rev. Metall.*, **1**, 27–32 (1980).

[9] A. Vardelle, J. M. Baronnet, M. Vardelle, and P. Fauchaus, *IEEE Trans. Plasma Sci.*, **PS-8** [4] 417–24 (1980).

[10] A. Vardelle, M. Vardelle, and P. Fauchais, *Plasma Chem. Plasma Process.*, **2** [3] 255–91 (1982).

[11] G. F. Hurley and F. D. Gac, "Structure and Thermal Diffusivity of Plasma-Sprayed Al_2O_3," *Am. Ceram. Soc. Bull.*, **58** [5] 509–11 (1979).

[12] P. Boch et al., "Sintering of Plasma-Sprayed Alumina Deposits"; in Ceramic Powders. Elsevier Publishing Co., New York, in press.

[13] S. Rangaswamy and H. Herman, "Thermal Expansion Study of Plasma-Sprayed Oxide Coatings," *Thin Solid Films*, **73**, 43–52 (1980).

[14]W. D. Kingery, "Factors Affecting Thermal Stress Resistance of Ceramic Materials," *J. Am. Ceram. Soc.*, **38** [1] 3–15 (1955).
[15]D. P. H. Hasselman, "Unified Theory of Thermal Shock Fracture Initiation and Crack Propagation in Brittle Ceramics," *J. Am. Ceram. Soc.*, **52** [11] 600–604 (1969).
[16]N. Kamiya and O. Kamigaito, "Prediction of Thermal Fatigue Life of Ceramics," *J. Mater. Sci.*, **14**, 573–82 (1979).
[17]P. F. Becher, "Effect of Water Bath Temperature on the Thermal Shock of Al_2O_3," *J. Am. Ceram. Soc.*, **64** [1] C-17–C-18 (1981).

Microstructure and Durability of Zirconia Thermal Barrier Coatings

D. S. Suhr and T. E. Mitchell

Case Western Reserve University
Department of Metallurgy and Materials Science
Cleveland, OH 44106

R. J. Keller

TRW Inc.
Materials and Manufacturing Technology Center
Cleveland, OH 44117

Various combinations of plasma-sprayed bond coatings and zirconia ceramic coatings on a nickel-based superalloy substrate were tested by static thermal exposure at 1200°C and cyclic thermal exposure to 1000°C. The bond coats were based on Ni-Cr-Al alloys with additions of rare earth elements and Si. The ceramic coats were various ZrO_2-Y_2O_3 compositions, of which the optimum was found to be ZrO_2-8.9 wt% Y_2O_3. Microstructural analysis showed that resistance to cracking during thermal exposure is strongly related to deleterious phase changes. Zones depleted of Al formed at the bond coat/ceramic coat interface due to oxidation and at the bond coat/substrate interface due to interdiffusion, leading eventually to breakdown of the bond coat. The 8.9% Y_2O_3 coating performed best because the as-sprayed metastable (high-Y_2O_3) tetragonal phase converted slowly into the low-Y_2O_3 tetragonal plus high-Y_2O_3 cubic-phase mixture, so that the deleterious monoclinic phase was inhibited from forming. Failure appeared to start with the formation of circumferential cracks in the zirconia, probably due to compressive stresses during cooling, followed by the formation of radial cracks due to tensile stresses during heating. Cracks appeared to initiate at the Al_2O_3 scale/bond coat interface and propagate through the zirconia coating.

Two-layer thermal barrier coatings consisting of a NiCrAlY inner metallic bond coat layer and an outer ZrO_2-Y_2O_3 ceramic coat layer were investigated as a means to increase the durability of air-cooled gas turbine airfoils[1-4] by preventing oxidation and corrosion of the Ni- or Co-based superalloy and by reduction of the metal temperature. Zirconia has suitable properties for the thermal barrier coat because it has a low thermal conductivity, a relatively high coefficient of thermal expansion, and good chemical stability. However, it has poor mechanical stability because of the large volume change (about 4%) during the transformation from the tetragonal (t) to the monoclinic (m) phase. The largest improvements in mechanical properties can be obtained by partially stabilizing the ZrO_2 with additions of CaO, MgO, or Y_2O_3.[5] Of these, Y_2O_3 performs best, in part because CaO and MgO tend to vaporize.

Thermal barrier coatings are conventionally applied by plasma-spraying, where the molten ZrO_2-Y_2O_3 particles impact the cold substrate and solidify into the high-temperature cubic (c) phase. The cooling rate is fast enough that much of the c phase transforms directly into a t phase of the same composition. (This has been termed a "diffusionless" transformation[6] or, more accurately, a "displacive" transformation.[7]) The nonequilibrium t phase is retained to room temperature on rapid cooling; it has a high Y_2O_3 content,[8] and has also been observed in skull-grown crystals.[7] This nonequilibrium t phase is not stable at high temperatures and transforms to an equilibrium mixture of c and low-Y_2O_3 t phase. The equilibrium t phase may further transform to m symmetry on cooling to room temperature. The transformation is accompanied by a large volume change, which may cause failure of the ceramic coating by delamination and spallation.

The bond-coat composition is also important, since the addition of rare earth elements can inhibit spallation of the protective Al_2O_3 scale. Furthermore, interdiffusion between the bond coat and substrate alloying elements has been shown to have a significant effect on the durability of the two-layer thermal barrier coatings.[9,10]

Plasma-spraying operating parameters are also important in improving the durability of coatings[11] because they control aspects of the as-sprayed microstructure such as pore morphology and distribution within the ceramic coat and oxide stringers within the bond coat. In addition, spraying parameters have strong effects on residual stress states in the coating. In the present work, various combinations of bond coats and ceramic coats of different compositions have been investigated and tested under static and cyclic thermal exposure conditions. Particular attention has been paid to the correlation between durability and microstructure in the coatings and at the various interfaces in the sandwich structures.

Experimental Procedures

Plasma-Spraying Techniques

The metallic bond coats (0.2-mm thick) were applied by a low-pressure plasma spray (LPPS) process onto solid PWA 1422 Ni-based superalloy rods (0.5 in. in diameter) which were sand-blasted and degreased before spraying. After the bond coat was applied, the specimens were annealed in argon at 1080°C for 4 h and then the ZrO_2-Y_2O_3 ceramic coat (0.5-mm thick) was overlaid by air-plasma-spraying (APS).

The microstructures of the ceramic coat and bond coat are strongly influenced by the various plasma-spraying parameters. Some of the most important parameters are (1) gun-to-work distance, (2) particle size, (3) particle-size distribution, (4) plasma gun power, (5) flow-gas compositions, (6) substrate temperature (i.e., preheat), and (7) powder feed rate. Spraying parameters for the ceramic coat were optimized by inspecting transverse sections of bars sprayed with various parametric combinations and by subjecting bars to accelerated cyclic thermal tests to 1100°C. In general, parameters were chosen which produced coatings with uniformly distributed small porosity and which minimized substrate heating. LPPS bond coatings were used to avoid the problem of oxide films that form between splats in APS bond coatings.[12]

Coating Compositions

The composition of the Ni-17Cr-14Al bond coat was varied in order to study the effects of small additions of alloying elements (Y, Zr, Hf, Si) on the behavior of the thermal barrier coatings, as described elsewhere.[2,13,14] The behavior of the bond coatings was evaluated by high-temperature exposure, using the same ZrO_2-8.9 wt% Y_2O_3 ceramic coatings, and the best bond coating was selected for further testing in which three additional ZrO_2-Y_2O_3 compositions were applied. (All the compositions are given in weight percent unless otherwise specified.) These systems were then subjected to high-temperature exposure and microstructural observation in order to find the optimum thermal barrier coating system.

Thermal Treatment

Three specimens from each coating system were heat-treated by static thermal exposure at 1200°C for 1, 10, and 100 h and by cyclic thermal exposure — 1000°C for 50 min followed by forced-air-cooling for 10 min. Cyclic thermal exposure tests were performed for up to 500 cycles or until failure was observed by visual detection of cracks in the ceramic coat.

Physical Examinations

Microstructural and chemical changes were examined by optical microscopy, scanning electron microscopy (SEM), energy dispersive X-ray analysis (EDAX), X-ray diffractometry (XRD), and transmission electron microscopy (TEM). The SEM/EDAX work was done with a Cambridge Stereoscan.* The ZrO_2-Y_2O_3 coatings were removed mechanically and ground to powder using a mortar and pestle for phase analysis by XRD. The XRD trace around the {111} peaks of the c, t, and m phases and around the {400} peaks of the c and t phases were analyzed carefully. The {111} peaks of the m phase are well separated, but the c and t phases are coincident because the {111} interplanar spacings of these two phases are almost the same. However, the {400} peaks from these two phases can be deconvoluted from the diffractometer trace and the intensities from each phase were used for phase analysis.

The mole fractions were calculated from the measured X-ray intensities as described by Miller et al.[8]:

$$\frac{M_m}{M_{c,t}} = 0.82 \frac{I_m(11\bar{1}) + I_m(111)}{I_{c,t}(111)} \quad (1)$$

$$\frac{M_c}{M_t} = 0.88 \frac{I_c(400)}{I_t(400) + I_t(004)} \quad (2)$$

The morphology of the c, t, and m phases of the ceramic coat and of the oxide film on the bond coat was analyzed by TEM imaging and diffraction techniques, using an analytical electron microscope equipped with EDAX and scanning attachments.†

Results

The general microstructure will be described before discussing the cyclic and static thermal tests on the various thermal barrier coating systems in Table II. More detailed discussion follows on phase analysis of the ceramic coating, phase changes due to oxidation and interdiffusion in the bond coat, and preliminary TEM analysis.

General Microstructures

Microstructural changes in the 8.9% Y_2O_3 system after static thermal exposure for 1, 10, and 100 h at 1200°C are shown in the optical micrographs in Fig. 1. The general microstructure after 100 h static thermal exposure is shown

---

*Cambridge Stereoscan, Cambridge Scientific Industries, Cambridge, MD.
†Phillips EM-400T, Phillips Electronic Instruments, Inc., Mahwah, NJ.

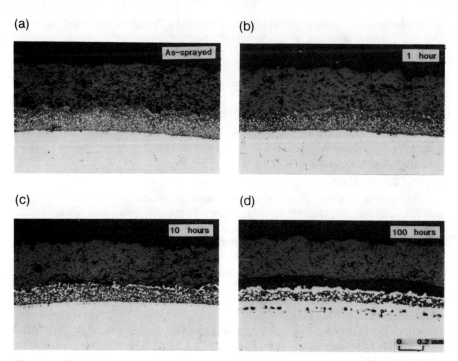

Fig. 1. Optical microstructures of an 8.9% Y_2O_3 system, as-sprayed and after static thermal exposure for 1, 10, and 100 h at 1200°C.

in Fig. 2, where the various microstructural features are also labeled. There is little observable change in the ceramic coat after 1-h exposure; it is still adherent to the bond coat and no cracks have developed within the ceramic coat. After 10-h exposure, a network of cracks has developed at and/or near the bond coat/ceramic coat interface; other ZrO_2-8.9% Y_2O_3 coatings exhibited cracks similar to but slightly more extensive than those in Fig. 1(c). After 100-h exposure, all of the systems had failed by delaminating near the ceramic coat/bond coat interface, leaving behind small amounts of ZrO_2 at valleys in the bond coat surface. The network of axial cracks and circumferential cracks became more extensive, whereas the ceramic coat became substantially more dense than in the as-sprayed condition.

In Figs. 1 and 2, the sprayed bond coat consists of Al-rich β-NiAl phase (dark) and Cr-rich γ/γ' (white) phase mixture. (γ is the fcc Ni solid solution and γ' is the Ni_3Al ordered precipitate.) There was a large increase in the grain sizes of the β and γ/γ' phases during the high-temperature thermal exposure and zones depleted of β phase developed at both of the bond coat interfaces. The depleted zone at the ceramic coat/bond coat interface has a low Al content due to oxidation, and the depleted zone at the bond coat/substrate interface has a low Al content due to interdiffusion.

The microstructures of the thermally cycled specimens were different from those of the static thermally exposed specimens. The depleted zones, especially those at the ceramic coat/bond coat interface, were smaller than those of the static thermally exposed specimens. This is probably due to the higher temperature (1200°C) of the static thermal tests, which results in more rapid diffusive changes.

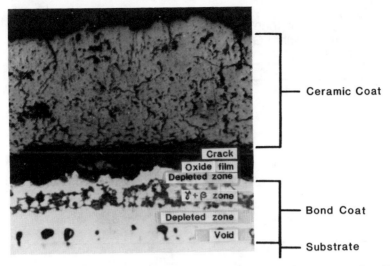

Fig. 2. General optical microstructure of an 8.9% Y_2O_3 system after 100 h static thermal exposure at 1200°C.

Typical microstructures of thermally cycled specimens are shown in Fig. 3. No porosity at the bond coat/substrate interface is evident in Fig. 3, in comparison with the prominent porosity developed after static thermal exposure (Fig. 1). However, precipitation of extraneous phases occurred both in the bond coat and in the substrate due to interdiffusion. The system in Fig. 3(b) failed after 290 cycles, but that in Fig. 3(a) did not fail even after 500 cycles.

Cyclic Thermal Exposure Failure Analysis

As described elsewhere,[15] thermal cycling showed that Y is highly beneficial to Ni-Cr-Al bond coats and that addition of Si gives further improvement. To

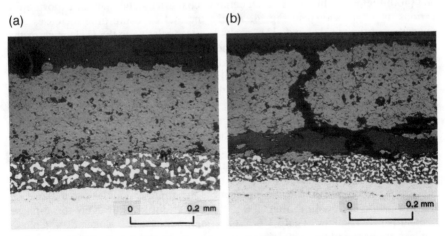

Fig. 3. Optical microstructures of specimens from an 8.9% Y_2O_3 system with (a) a good bond coat after 500 cycles and (b) a poorer bond coat after 290 cycles (50 min 1000°C, 10 min air-cool).

Table I. Effect of Y_2O_3 Content in the Ceramic Coat on the Number of 1000°C Cycles to Failure for Individual Specimens with a Given Bond Coat (Ni–Cr–Al with Y and Si Additions)

Y_2O_3 (wt%)	No. of cycles to failure
4.3	7
	15
6.1	233
	233
8.9	>500
	>500
	>500
19.6	321
	>500

investigate the dependence of Y_2O_3 content in the ceramic coat on the behavior of thermal barrier coatings, four ZrO_2-Y_2O_3 compositions of ceramic coats were applied onto the best bond coat. The effects of Y_2O_3 content on cyclic thermal exposure behavior are shown in Table I. The average number of cycles to failure of the 4.3% Y_2O_3 system is only 15 cycles because of the formation of the m phase in low-Y_2O_3 content partially stabilized zirconia (PSZ) during cyclic thermal exposure. Phase transformations during high-temperature thermal exposure will be discussed in the next section. The poor performance of the 6.1 and 19.8% Y_2O_3 systems compared to that of the 8.9% system confirms previously reported data.[2,3] The 8.9% Y_2O_3 system did not fail even after 500 cycles.

Phase Analysis

Phase analysis of the ZrO_2-Y_2O_3 ceramic coat after static and cyclic thermal exposure was performed by X-ray diffractometry, as described earlier. The {111} and {400} regions of the 8.9% Y_2O_3 ceramic coat after static thermal exposure for various times are shown in Fig. 4. As the thermal exposure time increases, the intensities of the m $(11\bar{1})$ and (111) peaks increase relative to the (111) peak of the c and t phases in the {111} region and the intensities of the c (400) peak increase with respect to the t (004) and (400) peaks in the {400} region. The difference in 2θ angle between the (004) and (400) peaks of the t phase increases with increasing thermal exposure time. This means that the c/a ratio of the t phase increases due to a decrease in the Y_2O_3 content.[6]

Mole fractions of the phase were calculated from Eqs. (1) and (2) and the results are summarized in Table II. The large amounts of the t phase in the as-sprayed material are due to the "nontransformable" t phase formed during rapid cooling, as described by Miller et al.[8] and discussed further in Discussion.

Table II shows that the 4.3% Y_2O_3 material has a large amount of m phase in the as-sprayed condition. The mole percent of the m phase doubles due to the $t \rightarrow m$ transformation on cooling after 100-h thermal exposure; however, the c-phase content hardly changes. The 6.1% Y_2O_3 system initially has almost 80% t phase but less m phase and the transformation rate to the m and c phases is much slower than in the 4.3% Y_2O_3 system. In the as-sprayed condition the amounts of t and

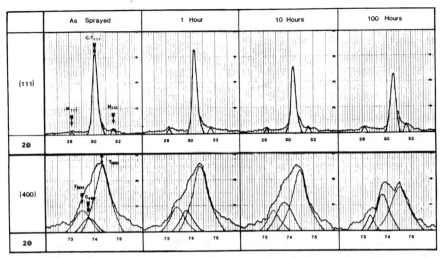

Fig. 4. Experimental X-ray diffraction patterns and deconvoluted peaks in the {111} and {400} regions for plasma-sprayed ZrO_2-8.9% Y_2O_3 after 1, 10, and 100 h static thermal exposure at 1200°C.

Table II. Phase Analyses (mol%) after Thermal Exposure at 1200°C

System (%Y_2O_3)	Phase	As-sprayed	1	10	100
4.3	Monoclinic	22	37	41	44
	Cubic	4	5	4	3
	Tetragonal	74	58	55	53
6.1	Monoclinic	16	17	18	20
	Cubic	6	9	10	11
	Tetragonal	78	74	72	69
8.9	Monoclinic	8	9	9	12
	Cubic	13	15	22	31
	Tetragonal	79	76	69	57
19.6	Monoclinic	3	3	2	2
	Cubic	70	76	84	88
	Tetragonal	27	21	14	10

c phases increase as the Y_2O_3 content increases, whereas the amount of m phase decreases.

The 8.9% Y_2O_3 system also has a large amount ($\approx$80%) of t phase in the as-sprayed condition and the c-phase content increases significantly (to $\approx$30%) at the expense of the t phase after 100 h at 1200°C. The 19.6% Y_2O_3 system consists mostly of c phase (70%) but it also has some t phase and a very small amount of m phase (perhaps due to unreacted powder.) Annealing causes the c-phase content to increase at the expense of the t phase as expected from the equilibrium phase diagram.

Phase changes after cyclic thermal exposure show the same trends as for static thermal exposure, except that the phase changes are generally slower, due to the lower temperature (1000°C) used. The 4.3% Y_2O_3 system initially contained 22% m phase and a specimen which failed after only 7 cycles showed an increase in m-phase content to 32%. In contrast, the 8.9% Y_2O_3 system did not fail even after 500 cycles and the structure was correspondingly stable, the t phase decreasing from 79 to 68% and the m phase increasing from 8 to 15%. The 6.1 and 19.0% Y_2O_3 systems likewise had relatively small changes in phase content.

Microchemical Analysis

The average Y_2O_3 content in the ceramic coat did not change even after 100 h of static thermal exposure, although it varied by a few percent with position in the as-sprayed specimens. There were, however, large changes in local chemical compositions within the bond coat and substrate as a function of exposure time. Results from point analyses of the chemical composition in the bond coat after 1-, 10-, and 100-h static thermal exposure are summarized in Table III and will be discussed below. The microstructure of the as-sprayed bond coat consists of a mixture of β-NiAl grains and grains of a γ fcc Ni solid-solution matrix containing fine γ'-Ni$_3$Al precipitates. The β phase is fairly constant in composition and contains relatively high Al and low Cr. The γ/γ' grains are also constant in composition and contain relatively low Al and high Cr. The Si, Co, and Cr contents

Table III. Chemical Compositions (wt%) in the Bond Coat after Static Thermal Exposure at 1200°C

Time	Al	Si	Cr	Co	Ni	Ti	Phase	Remark
As-sprayed	4.5	2.8	26.9	2.2	63.6		γ/γ'	
	12.7	2.0	15.8	1.7	67.8		β	
1 h	2.5	1.1	19.7	2.8	73.9		γ/γ'	Depleted zone near ceramic coat
	6.1	1.0	11.6	1.9	79.4		β	Below depleted zone
	3.0	1.0	18.5	2.6	74.8		γ/γ'	Below depleted zone
	7.2	0.9	9.1	1.9	80.9		β	Near substrate
	9.9	1.6	17.6	2.7	67.8	0.4	γ/γ'	Near substrate
	2.1	1.4	19.2	3.5	73.3	0.5	γ	Depleted zone near substrate
10 h	2.4	1.2	17.6	4.6	74.2		γ/γ'	Depleted zone near ceramic coat
	3.6	1.8	17.4	4.2	73.0		γ/γ'	Middle of bond coat
	8.5	1.4	9.8	3.8	75.6	0.9	β	Middle of bond coat
	3.4	2.2	17.4	4.8	71.5	0.7	γ/γ'	Depleted zone near substrate
100 h	6.4	2.8	11.4	7.7	70.3	1.5	γ/γ'	Depleted zone near ceramic coat
	7.5	2.4	5.0	4.9	78.4	1.8	β	Middle of bond coat
	5.1	3.6	11.4	8.4	70.8	0.7	γ/γ'	Depleted zone near substrate

in the γ/γ' regions are larger than in the β phase, since these elements tend to stabilize the γ/γ' phases.

After 1-h exposure, depleted zones are already evident on both sides of the bond coat (Table III). The depleted zone near the interface between the ceramic coat and the bond coat has a lower Al and Si content but a higher Cr content than in the as-sprayed condition, due to the preferential oxidation of the Al and Si. The Al and Si contents in the β and γ/γ' phases near the ceramic coat/bond coat interface are lower than those far from the interface. In fact, the loss of Al is probably sufficient to convert the γ/γ' to single-phase γ near the interface. The depleted zone near the interface between the bond coat and substrate has a lower Al content, similar to the depleted zone near the ceramic coat. In addition, it is apparent from Table III that Ti and Co diffuse out from the substrate into the bond coat due to interdiffusion.

These trends continue with increasing exposure time and the depleted zones near both interfaces become wider. The only substantial change is a general loss of Cr. Large Kirkendall voids form in the bond coat just above the original bond coat/substrate interface, while small voids were observed at the original interface, as seen in Fig. 1.

Depletion at the ceramic coat/bond coat interface is due to preferential oxidation of Al. The oxide film formed on the bond coat is formed by diffusion of the bond coat alloying elements outward and interaction with oxygen diffusing rapidly through the ZrO_2. Chemical compositions of the oxide as a function of exposure times are summarized in Table IV. After short time exposure, many types of transient oxides are observed, especially NiO and spinels ($NiAl_2O_4$, $NiCr_2O_4$). After long time exposure, the oxide is mainly the most stable oxide, Al_2O_3, with about 10% Cr_2O_3.

Other Microstructural Aspects

In this section we present some preliminary observations of the microstructure, particularly the ceramic coating and its interface with the bond coat. The microstructure is highly variable because of the varying morphologies and cooling rates of the splats sprayed on the rough surface. Figure 5(a) shows a region of fine equiaxed grains in the ZrO_2-4.3% Y_2O_3, which is 100% t phase and has an actual composition of 8.3% Y_2O_3. By contrast Fig. 5(b) shows a neighboring region which is also fine grained but has transformed to m symmetry with resultant grain-boundary cracking; the composition in this region is 1.6% Y_2O_3. Sometimes large-grained regions were observed, especially in the middle of the ceramic coat where the cooling rate is slower. In the example shown in Fig. 6, a colony of t precipitates is seen which consists of t variants in a twin relationship and which probably formed during cool-down from plasma spraying (see Discussion). In regions where the splat morphology is clear, as in Fig. 7, columnar t grains are

Table IV. Chemical Composition (wt%) of the Oxide Film Formed at the Ceramic Coat/Bond Coat Interface

Time at 1200°C (h)	Al_2O_3	Cr_2O_3	CoO	NiO
1	72.6	10.7	0.9	15.8
10	21.8	27.6	2.8	47.8
100	93.1	6.9		

(a) (b)

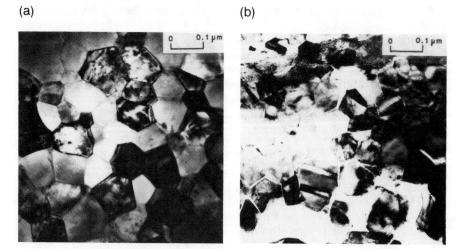

Fig. 5. TEM bright-field images of the as-sprayed 4.3% Y_2O_3 ceramic coat showing regions of fine, equiaxed grains of (a) tetragonal phase and (b) monoclinic phase. The actual Y_2O_3 content of the tetragonal phase region is 8.3% Y_2O_3 and that of the monoclinic region is 1.6% Y_2O_3.

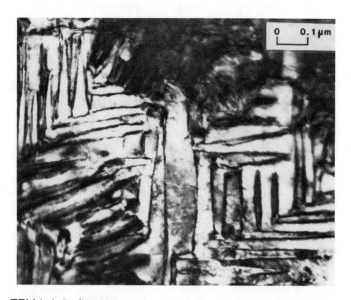

Fig. 6. TEM bright-field image of the as-sprayed 8.9% Y_2O_3 ceramic coat showing the tetragonal colony structure which was developed within a large-grained region due to slow cooling in the middle section of the ceramic coat.

observed, indicating a high cooling rate parallel to the columns. Note that the splats are as small as a few tenths of a micrometer thick while their widths are probably several hundred micrometers.

Fig. 7. TEM bright-field image showing the ZrO_2-Y_2O_3 splat morphology with columnar tetragonal grains which have grown perpendicular to the splat boundary (6.1% Y_2O_3) ceramic coat after 100 h at 1200°C.

Changes in microstructure on aging or cycling are difficult to ascertain because of the variable starting structure. The fine-grained t grains appear to be remarkably stable; in fact, Fig. 7 is from a specimen aged at 1200°C for 100 h. Figure 8 is from a 8.9% Y_2O_3 specimen cycled 503 times to 1000°C and shows

Fig. 8. TEM bright-field image of the 8.9% Y_2O_3 ceramic coat after 503 cycles. The tetragonal phase has partly transformed to monoclinic in a large-grained region while the surrounding fine-grained material has remained tetragonal.

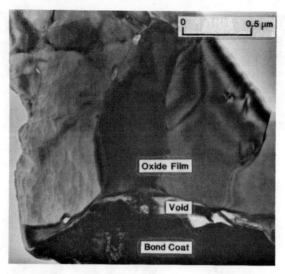

Fig. 9. TEM bright-field image of Al_2O_3 oxide film formed at the ceramic coat/bond coat interface after 10 h at 1200°C.

a large-grained region which has partly transformed from t to m symmetry, while the surrounding fine-grained material has remained. Fine porosity is also evident in the large grains; this may possibly be due to entrapped argon from the plasma-spraying process.

Attempts have also been made to examine the bond coat/ceramic coat interface. Figure 9 shows a transverse section through the Al_2O_3 oxide film on the bond coat. The columnar grains of Al_2O_3 are ≈ 2 μm thick and are compact and protective. However, voids have formed at the bond coat interface, leading to the suspicion that this may be a source of weakness. In fact, SEM analysis of the underside of ceramic coatings that have failed and spalled (Fig. 10) shows that Al_2O_3 has adhered to a large area of ZrO_2. The Al_2O_3 shown in Fig. 10 is smooth relative to the flaky appearance of the ZrO_2 caused by the splat morphology, and there are dimples on the Al_2O_3, presumably due to the voids formed at the interface. Finally, the Al_2O_3 is concave, indicating that it has broken away from the tops of hills on the bond coat and that cracks then propagate through the ZrO_2.

Discussion

The microstructures produced by the plasma-spraying process are first to be discussed and then the mechanisms responsible for failure of the coatings. The initial amounts of the t, c, and m phases are summarized in Fig. 11 and the stability of each phase to thermal exposure is shown in Table II. In general, the most durable coatings have the highest t contents and the least m material. The as-sprayed t-ZrO_2 phase is not the equilibrium phase but is formed during rapid cooling of the molten ZrO_2 particles after they are sprayed onto the cold substrate. This t phase has been called "nontransformable" by Miller et al.,[8] who suggested that it is formed directly from the high-temperature cubic phase by a "diffusionless" transformation. Lanteri et al.[7] refer to it as the t' phase with a high Y_2O_3 content to distinguish it from the equilibrium t phase; they suggest further that t' is formed from the cubic phase by a "displacive" transformation.

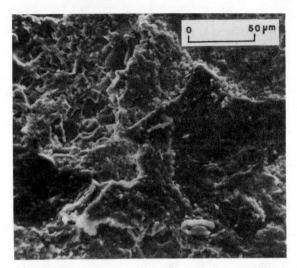

Fig. 10. SEM image of the underside of a ceramic coat that had failed and spalled showing alumina scale which has adhered to a large area of zirconia (8.0% Y_2O_3 ceramic coat after 100 h at 1150°C).

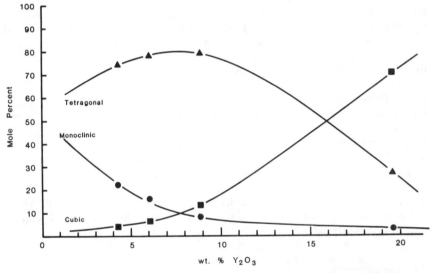

Fig. 11. Percent of each phase in as-sprayed ZrO_2-Y_2O_3 coatings as a function of powder composition.

The TEM results given in the last section show that the fine-grained t' regions are remarkably stable to aging at 1200°C. Large-grained regions contain t colonies in a cubic matrix which are presumably formed during the relatively slow cooling of the outer splats. These colonies are observed in single crystals aged in the two-phase $c + t$ region in the temperature range 1400°–1500°C.[7] The increase in c phase of the 6.1 and 8.9% Y_2O_3 compositions during aging must be due to the reaction $t' \rightarrow c + t$. However, the TEM results also show that the composition within the ceramic coating is highly variable. This could be due either to variations in the Y_2O_3 content of the original powder from a lack of ZrO_2/Y_2O_3 homogenization or to a loss of Y_2O_3 by evaporation from overheated liquid drops. Whatever the reason, the result is a much higher m phase content in the 4.3 and 6.1% Y_2O_3 compositions than might be expected if the compositions were uniform. Certainly these two compositions are much less durable than the 8.9% Y_2O_3 coating, and this correlates with the m-phase content. On the other hand, the inferior performance of the 19.6% Y_2O_3 coating, which contains very little monoclinic phase, must be due either to poor mechanical properties of the c phase or possibly a high oxygen diffusivity, leading to accelerated oxidation of the bond coat.

There is no single cause of failure in the different thermal barrier coating systems; however, there are requirements which, if not met, will tend to lead to premature failure. For example, if the Y_2O_3 content is too low, failure will occur from poor thermal and mechanical stability due to the presence of the m phase. The relatively poor performance of the high-Y_2O_3 material is probably due to the inferior fracture toughness of fully stabilized zirconias compared with PSZs. The 6–8 wt% Y_2O_3 material represents an optimum composition with respect to the good mechanical properties of the metastable t' phase and its slow conversion into the (also metastable) c plus t phase mixture, which also has good mechanical properties. The ceramic coating then lasts only as long as the bond coat can retain its oxidation resistance by the formation of an adherent Al_2O_3 scale. The presence of Y is vital to the adherence of the scale (by a mechanism not yet agreed upon), and the bond coat must be thick enough to avoid depletion of Al and Y by interdiffusion with the substrate and by thickening of the oxide layer.

The failure mechanism of the best systems is not clear. The residual stresses from plasma-spraying are presumably compressive in the ceramic coat on cooling. During heating they tend to become tensile and again compressive during cooldown. Transverse sections (e.g., Figs. 1 and 2) show that circumferential cracks form first, probably due to compressive stresses acting on the weakened bond coat/ceramic coat interface as oxidation progresses. The radial cracks which lead to final failure must then develop from tensile stresses in the delaminated regions during heating. It is important to note that the circumferential cracks appear to propagate through the ceramic coating but may nucleate at the Al_2O_3/bond coat interface rather than the ZrO_2/Al_2O_3 interface, indicating the importance of oxidation. In fact, cracks may initiate under the oxide scale, as suggested by the voids in this region, as shown in Fig. 9, and the adherence of Al_2O_3 to spalled ceramic coatings (Fig. 10).

Acknowledgments

This research was supported by the U.S. Air Force, Wright-Patterson Air Force Base, Contract No. F33615–81–C–5063. Helpful discussions with Robert Ruh are acknowledged.

References

[1] R. J. Bratton and S. K. Lau; pp. 226–40 in Advances in Ceramics, Vol. 3. Edited by A. H. Heuer and L. W. Hobbs, The American Ceramics Society, Columbus, OH, 1982.
[2] S. Stecura, *Am. Ceram. Soc. Bull.*, **61** [2] 256–62 (1982).
[3] S. Stecura, *Thin Solid Films*, **73**, 481–89 (1980).

[4]R. D. Maier, C. M. Scheuermann, and C. W. Andrews, *Am. Ceram. Soc. Bull.,* **60** [5] 555–60 (1981).
[5]E. C. Subbarao; pp. 1–24 in Advances in Ceramics, Vol. 3. Edited by A. H. Heuer and L. W. Hobbs, The American Ceramics Society, Columbus, OH, 1982.
[6]H. G. Scott, *J. Mater. Sci.,* **12** [2] 311–16 (1977).
[7]V. Lanteri, A. H. Heuer, and T. E. Mitchell; this volume, pp. 118–30.
[8]R. A. Miller, J. L. Smialek, and R. G. Garlick; pp. 241–53 in Advances in Ceramics, Vol. 3. Edited by A. H. Heuer and L. W. Hobbs, The American Ceramics Society, Columbus, OH, 1982.
[9]A. R. Nicoll, W. Kleemann, and R. Engel, *Thin Solid Films,* **95**, 245–54 (1982).
[10]P. A. Siemers and W. B. Hillig, NASA N82–10040.
[11]S. Stecura; NASA Tech. Mem. TM–81724 (1981).
[12]R. W. Smith, *Thin Solid Films,* **84**, 59–72 (1981).
[13]M. A. Gedwill; NASA TM–81567.
[14]H. W. Grumling and R. Bauer, *Thin Solid Films,* **95**, 3–20 (1982).
[15]R. J. Keller; TRW Final Report to U.S. Air Force under Contract No. F33615–81–C–5063, 1983.

Thermal Diffusivity of Zirconia Partially and Fully Stabilized with Magnesia

W. J. Buykx

AAEC Lucas Heights Research Laboratories
Sutherland, New South Wales, 2232, Australia

M. V. Swain

Advanced Materials Laboratory
CSIRO Division of Materials Science
Melbourne, Victoria, 3001, Australia

The thermal diffusivity of a range of zirconias partially stabilized with MgO (Mg-PSZ) and of fully MgO-stabilized zirconia (Mg-CSZ) has been determined using the laser flash technique. The Mg-PSZ range included variations in phase distribution and microstructure induced by fabrication heat treatments. Using a theoretical model, the thermal diffusivity of tetragonal zirconia (t-ZrO_2) precipitates was calculated and was found to be significantly higher than that of cubic zirconia (c-ZrO_2) and monoclinic zirconia (m-ZrO_2). The thermal diffusivity of subeutectoid aged Mg-PSZ increased with the thickness of the continuous m-ZrO_2 grain-boundary layer. The temperature dependence of the thermal diffusivity of Mg-PSZ containing m-ZrO_2 precipitates at room temperature was also measured and revealed a hysteresis effect similar to that observed in thermal expansion. The thermal diffusivity of Mg-PSZ aged at 1420°C was found to be dependent on the laser pulse energy used.

The thermal properties of ZrO_2-based ceramics are of considerable scientific and technological significance. Fully stabilized zirconia (CSZ), with its defect fluorite structure, is one of the poorest crystalline thermal conductors known, because its random oxygen vacancies act as phonon scattering sites. Monoclinic zirconia (m-ZrO_2), the other phase of zirconia readily available at room temperature, lacks the point defects of the cubic phase and is a much better thermal conductor. Little work has been performed on the thermal properties of tetragonal zirconia (t-ZrO_2), because in a pure state it exists only as very fine powders or small precipitates. In Mg-PSZ, a combination of all three phases of ZrO_2 can be present, the amount of each phase being controlled by composition, firing conditions, heat treatment, and temperature.

Although the main technological interest in ZrO_2 is undoubtedly its high fracture toughness due to transformation-toughening,[1,2] recent developments in engines have also stressed the importance of thermal properties. In particular, the concept of the adiabatic or insulated diesel engine, with the prospect of considerably improved fuel efficiency and reduction in component parts, has excited the interest of many engine makers.[3] Other areas where the thermal properties of ZrO_2 play a role is where thermal shock resistance is important. Materials with higher thermal conductivity reduce the severity of the thermal stresses developed when the ambient temperature of the body is changed rapidly.

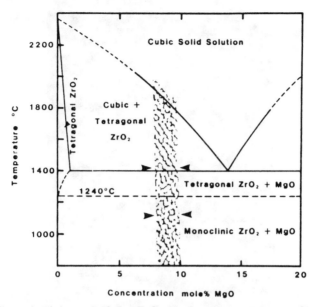

Fig. 1. Zirconia-rich end of the ZrO_2-MgO phase diagram. Shaded area indicates composition range used for commercial alloys. Aging treatments are carried out in the cubic tetragonal region at 1420°C or the subeutectoid monoclinic ZrO_2 + MgO region at 1100°C.

Characterization of the microstructure and theoretical models have been found to be powerful tools in the interpretation of the measured thermal diffusivity of multiphase materials.[4,5] These methods were applied to the present study of the thermal diffusivity of a wide range of phase compositions and microstructures of MgO-stabilized ZrO_2.

Materials

The Mg-PSZ materials used in this study contained 10 mol% MgO. They were prepared by simultaneously sintering and solution heat-treating in the cubic, single-phase region of the phase diagram (Fig. 1). They were then cooled in a controlled manner[6] such that all the precipitates remained in the tetragonal form at room temperature. Finally, they were heat-treated above the eutectoid temperature at 1420°C in the cubic tetragonal phase field or below the eutectoid temperature at 1100°C in the m-ZrO_2 + MgO region of the phase diagram. As-fired Mg-PSZ contains a fine precipitate of oblate spheroid particles of t-ZrO_2 in the cubic matrix phase (Fig. 2).* The axial ratio of the precipitates is approximately 0.15. With heat treatment at 1420°C (cubic tetragonal region), the precipitates retain their shape but decrease in number and increase in size. After 3–4 h at 1420°C, they lose coherency with the matrix and transform to m-ZrO_2 on cooling. The microstructures of Mg-PSZ heat-treated at 1420°C for 2 h (containing both t- and m-ZrO_2 precipitates) and 8 h (containing mostly m-ZrO_2 precipitates) are shown in Figs. 3 and 4,

---

*Figures 2–4 were taken with a scanning electron microscope on specimens which were polished and etched in hot phosphoric acid (150°C, 15 s).

Fig. 2. Tetragonal precipitate in as-fired Mg-PSZ, bar length 0.3 μm.

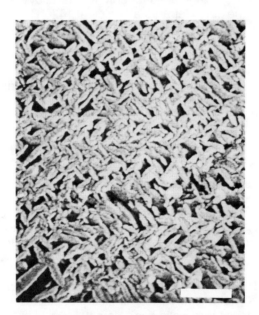

Fig. 3. Tetragonal + monoclinic precipitate in Mg-PSZ aged for 2 h at 1420°C, bar length 0.75 μm.

respectively. Subeutectoid heat treatment at 1100°C results in the decomposition of the cubic phase into MgO-rich m-ZrO_2 at grain boundaries and intragranular pores. The thickness of the continuous monoclinic phase at the grain boundaries increases with time. After 32 h at 1100°C, all the precipitates have been destabilized and

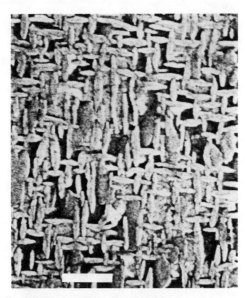

Fig. 4. Monoclinic precipitate in Mg-PSZ aged for 8 h at 1420°C, bar length 0.75 μm.

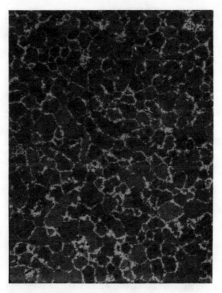

Fig. 5. Monoclinic zirconia grain-boundary material in Mg-PSZ aged for 24 h at 1100°C, average grain size 50 μm.

transformed to m-ZrO_2 on cooling. This destabilization, which is not due to coarsening of the precipitates, is discussed in some detail elsewhere.[7] The microstructures of subeutectoid aged Mg-PSZ are shown in Figs. 5–7 for aging times of 36, 66, and 96 h, respectively.

Fig. 6. Monoclinic zirconia grain-boundary material in Mg-PSZ aged for 66 h at 1100°C, average grain size 50 μm.

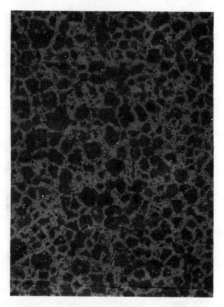

Fig. 7. Monoclinic zirconia grain-boundary material in Mg-PSZ aged for 96 h at 1100°C, average grain size 50 μm.

Mg-CSZ containing 14 mol% MgO was 100% c-ZrO_2. Part of this material was completely decomposed at 1100°C to m-ZrO_2 and MgO.

The phase compositions of all the materials used in this study are shown in Table I. Disks, 10-mm diameter by 1-mm thick, were machined from these mate-

Table I. Phase Composition and Thermal Diffusivity

Chemical composition	Heat treatment	Phase composition (%)					Thermal diffusivity (10^{-6} m²·s⁻¹)		
		Cubic	t precipitate	m precipitate	Grain boundary m phase	Calculated	Measured		
								3 J	6 J
14% MgO-CSZ		100						0.49	0.48
14% MgO-CSZ	8 h/1100°C				100			1.29	1.28
10% MgO-PSZ	As-fired	60	30	<5	<5			0.83	0.81
	2 h/1420°C	65	25	10	<2	0.775		0.67	0.61
								0.70(G)	0.67(G)
	8 h/1420°C	70	<5	24	<2	0.675		0.76	0.66
								0.74(G)*	0.65(G)
	24 h/1100°C	55		27	18	0.77		0.77	0.75
	66 h/1100°C	35		25	40	0.93		0.93	0.91
	96 h/1100°C	20		25	55	1.07		1.05	1.02

*G is for ground specimen; all other specimens were polished.

rials and polished on both faces. In the case of Mg-PSZ aged at 1420°C, the effect of surface treatment was investigated briefly.

Thermal-Diffusivity Measurement

The flash technique[8] was used to measure thermal diffusivity. An energy pulse of 0.3-ms duration from a neodymium-glass laser was deposited on one face of the 10-mm diameter by 1-mm thick disk-shaped specimens. The temperature history of the opposite face was measured by using a Chromel-Alumel† thermocouple (0.1-mm wire diameter) and displayed on a storage oscilloscope. The thermal diffusivity, α, of the material was calculated from

$$\alpha = 0.139 \, L^2/t_{1/2} \tag{1}$$

where L is the specimen thickness and $t_{1/2}$ is the time required for the rear-face temperature to reach one-half of its maximum value. The front face of the specimens was coated with a thin sputtered layer (<1 μm) of platinum to prevent laser radiation reaching the rear-face thermocouple by transmission through the translucent specimens.

Two specimens were measured for each composition. On each specimen three measurements at a pulse energy of 3 J and three at a pulse energy of 6 J were obtained. Consecutive measurements were made alternately at 3 and 6 J. The means and standard deviation of the six thermal-diffusivity determinations at each pulse-energy level were calculated.

Calculation of the Thermal Diffusivity of Multiphase Materials

The theory of the conductivity of multiphase materials employs two model microstructures: the variable dispersion model, in which at least one phase consists of isolated particles (the other phase(s) may be discrete or continuous), and the penetration model, in which all phases are continuous. The microstructures of the materials used in this work were judged to be of the variable dispersion type.

The general formula for the thermal conductivity of a two-phase material with variable dispersion microstructure is[9]:

$$(1 - V_D) = \left(\frac{k_M}{k}\right)^a \left(\frac{k_D - k}{k_D - k_M}\right) \left(\frac{k + bk_D}{k_M + bk_D}\right)^d \tag{2}$$

†Hoskins Mfg. Co., Detroit, MI.

where

$$a = \frac{F(1-2F)}{1-\cos^2\theta(1-F)-2F(1-\cos^2\theta)} \qquad (2a)$$

$$b = \frac{1-\cos^2\theta(1-F)-2F(1-\cos^2\theta)}{2F(1-\cos^2\theta)+\cos^2\theta(1-F)} \qquad (2b)$$

$$d = \left(\frac{F(1-2F)}{1-\cos^2\theta(1-F)-2F(1-\cos^2\theta)}\right.$$
$$\left. + \frac{2F(1-F)}{2F(1-\cos^2\theta)+\cos^2\theta(1-F)}\right) - 1 \qquad (2c)$$

The variables k, k_M, and k_D are the thermal conductivities of the two-phase material, the matrix phase, and the dispersed phase, respectively; V_D, F, and $\cos^2\theta$ are, respectively, the volume fraction, form factor, and orientation factor of the dispersed phase. Particle shapes are approximated by ellipsoids of revolution, with form factor F a function of axial ratio z/x, where z is the axis of revolution and x the radius. Values of F range from $F = 0$ for an infinitely thin platelet to $F = 1$ for an infinitely long needle-shaped particle. The orientation factor $\cos^2\theta = 0$ (axis of revolution at right angles to direction to heat flow) to $\cos^2\theta = 1$ (axis parallel to heat flow). The theoretical models are applicable to thermal conductivity and the other field properties of multiphase materials. In this work, thermal diffusivity was measured. Thermal conductivity, k, and thermal diffusivity, α, are related by

$$\alpha = k/(\rho C_p) \qquad (3)$$

where ρ is the density and C_p is the specific heat. Since ρC_p (the heat capacity per unit volume) is relatively insensitive to variations in composition and structure,[10] the thermal conductivity model is applicable also to thermal diffusivity.

Calculations were carried out in two steps. First, the thermal diffusivity of c-ZrO_2 (matrix) containing precipitates (dispersed phase) was calculated. The precipitates were approximated by ellipsoids of revolution with an axial ratio $z/x = 0.15$, for which $F = 0.1$. There was no preferred orientation, so the random orientation factor $\cos^2\theta = 1/3$ was adopted. In those cases where both m- and t-ZrO_2 precipitates were present, the volume-fraction-weighted average of the measured thermal diffusivity of m-ZrO_2 and the calculated thermal diffusivity of t-ZrO_2 was used for the thermal diffusivity of the dispersed phase (the calculation of the thermal diffusivity of t-ZrO_2 will be discussed later).

In step 2, the overall thermal diffusivity of the material was calculated by regarding the grain-boundary m-ZrO_2 as the matrix in which the equiaxed ($F = 1/3$), randomly oriented ($\cos^2\theta = 1/3$) precipitate containing c-ZrO_2 grains, with thermal diffusivity as calculated in step 1, are the dispersed phase.

Results

Table I summarizes the phase compositions, the mean room temperature values of thermal diffusivity measured at laser pulse energies of 3 and 6 J, and, where applicable, the calculated values of thermal diffusivity of the materials investigated. The mean room temperature values of thermal diffusivity measured at laser pulse energies of 3 and 6 J on polished specimens of the Mg-PSZ compositions are shown plotted against the corresponding calculated thermal-diffusivity values in Fig. 8. The error bars represent ±2 standard deviations.

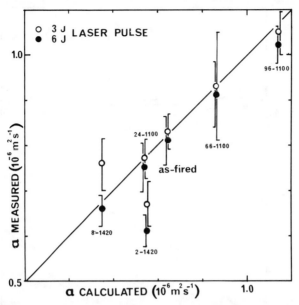

Fig. 8. Correlation between measured and calculated thermal diffusivity values of Mg-PSZ.

The thermal diffusivity of Mg-PSZ aged for 8 h at 1420°C was measured as a function of temperature up to 900°C during both heating and cooling by using a 6-J pulse. The results are shown in the top part of Fig. 9.

The room-temperature thermal diffusivity of t-ZrO_2 precipitates was calculated to be 1.74×10^{-6} $m^2 \cdot s^{-1}$, as discussed below.

Discussion

The thermal diffusivity of subeutectoid aged Mg-PSZ is approximately proportional to the total volume fraction of m-ZrO_2 present (precipitates + grain-boundary material).

The measured thermal-diffusivity values of the Mg-PSZ compositions containing only c- and m-ZrO_2 (i.e., the subeutectoid aged materials) are in good agreement with the values calculated on the basis of the measured thermal diffusivities of c- and m-ZrO_2 which demonstrates the validity of the model used (Fig. 8). There is a small difference (0.02×10^{-6} $m^2 \cdot s^{-1}$) between the thermal diffusivities measured at laser pulse energies of 3 and 6 J. This probably represents a systematic error.

The measured thermal diffusivity of as-fired Mg-PSZ is similarly insensitive to laser pulse energy. It is, therefore, reasonable to assume that the theoretical model is valid in this case. The theoretical expression for the thermal diffusivity of as-fired material, in which only the thermal diffusivity of t-ZrO_2 is unknown, was set equal to the mean of the pooled thermal-diffusivity values measured at 3- and 6-J laser pulse energy. This yielded a value of 1.74×10^{-6} $m^2 \cdot s^{-1}$ for the thermal diffusivity of t-ZrO_2.

Using this value, the theoretical thermal diffusivities of the Mg-PSZ compositions aged for 2 and 8 h at 1420°C could now be calculated and compared with the measured values (Fig. 8). Both these compositions exhibit a large pulse-energy

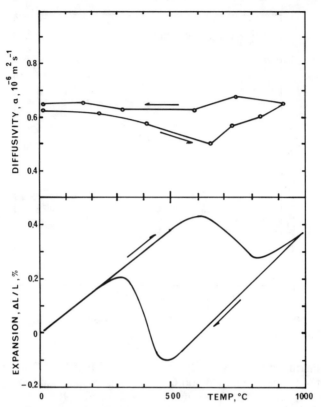

Fig. 9. Thermal diffusivity and thermal expansion of Mg-PSZ aged for 8 h at 1420°C as function of temperature.

effect on measured thermal diffusivity. The effect is reversible and, therefore, cannot be due to microcracking. The probability that the difference in thermal diffusivity measured at the two pulse-energy levels is due to chance is less than 1%. For Mg-PSZ aged for 8 h at 1420°C the thermal-diffusivity value measured at 6 J agrees well with the calculated value, but the value measured at 3 J is much higher.

For Mg-PSZ aged for 2 h at 1420°C the thermal-diffusivity values measured at 3 and 6 J fall far short of the calculated values for this composition.

These phenomena cannot be explained fully at this time, but the following comments are believed to be relevant. In as-fired Mg-PSZ, the t-ZrO_2 precipitates are small, with a largest dimension of 35 nm. The thermal diffusivity is not affected by laser pulse energy. In Mg-PSZ aged for 2 and 8 h at 1420°C, the t-ZrO_2 precipitates have grown to a largest dimension of 250 nm or greater and a smallest dimension of 45 nm. They are at the limit of coherency with the matrix.[11] An increase/decrease in laser pulse energy leads to a decrease/increase in thermal diffusivity. This suggests that a small amount of energy causes t-ZrO_2 precipitates of critical size to transform to m-ZrO_2 and that there is no unique, stable value of thermal diffusivity for these materials.

The measured thermal diffusivities of 2 h/1420°C Mg-PSZ specimens with as-machined, unpolished surfaces were a little higher than those of polished specimens, and the pulse-energy dependence was much less (Table I). This may be due to the high concentration of m-ZrO_2 precipitates in the machined surfaces. Rubbing the surfaces of polished 8 h/1420°C Mg-PSZ specimens on No. 280-grit abrasive paper caused a very small change in the measured thermal diffusivities and did not affect the pulse-energy dependence (Table I).

The temperature dependence of the thermal diffusivity of Mg-PSZ aged for 8 h at 1420°C, containing mostly m-ZrO_2 precipitates, is shown in the upper half of Fig. 9. The temperature dependence of the thermal conductivity of dielectric solids is represented by $k = (A + BT)^{-1} + k_e$, where the constant A is the thermal resistance due to imperfections, BT the lattice resistance, and k_e the electronic contribution which remains insignificant until fairly high temperatures are reached. The thermal conductivity of ZrO_2-4 wt% MgO decreases steadily from room temperature to approximately 1150°C.[12] On heating, the thermal diffusivity of Mg-PSZ decreased with temperature up to 650°C, as expected, and then rose sharply to 900°C, reflecting the $m \rightarrow t$ ZrO_2 transition of the precipitate. This is in good agreement with the thermal expansion curve for the same material,[11] shown in the bottom half of Fig. 9. The thermal contraction curve shows the reverse $m \rightarrow t$ ZrO_2 transition more clearly than the thermal-diffusivity curve on cooling.

Conclusions

(1) The room-temperature thermal diffusivity of t-ZrO_2 precipitate in Mg-PSZ is approximately 35% higher than that of monoclinic zirconia.

(2) The thermal diffusivity of subeutectoid aged Mg-PSZ increases with the amount of m-ZrO_2 present. The continuous m-ZrO_2 grain-boundary phase enhances the thermal shock resistance of this material.

(3) The thermal diffusivity of Mg-PSZ containing t-ZrO_2 precipitates of near-critical size is strongly influenced by laser pulse energy.

(4) The temperature dependence of the thermal diffusivity of Mg-PSZ aged for 8 h at 1420°C is subject to hysteresis similar to that observed for thermal expansion.

References

[1] R. C. Garvie, R. T. Pascoe, and R. H. J. Hannink, *Nature (London)*, **258**, 703–705 (1975).
[2] M. V. Swain and R. H. J. Hannink; pp. 1559–70 in Advances in Fracture, Vol. 4. Pergamon Press, Elmsford, NY, 1981.
[3] R. Kamo, M. Woods, and W. Geary, "Ceramics for Adiabatic Diesel Engines"; p. 468 in CIMTEC, Proceedings of the 4th Conference on Energy and Ceramics, 1979.
[4] W. J. Buykx, "The Effect of Microstructure and Microcracking on the Thermal Conductivity of UO_2-U_4O_9," *J. Am. Ceram. Soc.*, **62** [7–8] 326–32 (1979).
[5] W. J. Buykx, "Specific Heat, Thermal Diffusivity and Thermal Conductivity of SYNROC, Perovskite, Zirconolite and Barium Hollandite," *J. Nucl. Mater.*, **107**, 78–82 (1982).
[6] R. T. Pascoe, R. H. J. Hannink, and R. C. Garvie; pp. 447–54 in Science of Ceramics 9. Edited by K. J. de Vries, Nederlandse Keramische Vereniging, The Netherlands, 1977.
[7] R. H. J. Hannink, "Microstructural Development of Sub-eutectoid Aged MgO-ZrO_2 Alloys," *J. Mater. Sci.*, **18**, 457–70 (1983).
[8] W. J. Parker, R. J. Jenkins, C. P. Butler, and G. L. Abbott, "Flash Method of Determining Thermal Diffusivity, Heat Capacity and Thermal Conductivity," *J. Appl. Phys.*, **32** [9] 1679–84 (1961).
[9] B. Schulz, "Die Abhängigkeit der Feldeigenschaften Zweiphasiger Werkstoffe von ihrem Gefügeaufbau," *Kernforschungszent. Karlsruhe, [Ber.] KFK*, **KFK-1988** (1974).
[10] D. R. Flynn, "Thermal Conductivity of Ceramics"; in Mechanical and Thermal Properties of Ceramics. Edited by J. B. Wachtman, Jr., NBS Spec. Publ. (U.S.), No. **303** (1969).
[11] R. H. J. Hannink and M. V. Swain, "Magnesia-Partially Stabilised Zirconia: The Influence of Heat Treatment on Thermo-mechanical Properties," *J. Aust. Ceram. Soc.*, **18** [2] 53–62 (1982).
[12] Y. S. Touloukian, R. W. Powell, C. Y. Ho, and M. C. Nicolaou, "Thermal Diffusivity"; pp. 451–53 in Thermophysical Properties of Matter, Vol. 10. IFI/Plenum, New York/Washington, 1970.

Mechanical, Thermal, and Electrical Properties in the System of Stabilized $ZrO_2(Y_2O_3)/\alpha\text{-}Al_2O_3$

F. J. Esper, K. H. Friese, and H. Geier
Robert Bosch GmbH
Department of Materials Science
Stuttgart, Federal Republic of Germany

In the system of stabilized $ZrO_2(Y_2O_3)/\alpha\text{-}Al_2O_3$, the mechanical strength of the ZrO_2-rich materials is determined by strains in the microstructure. These are induced by distortions of the lattice caused by different thermal expansion coefficients of the components of the microstructure. Different densification of the various compositions in the sintering process also influences the mechanical strength. The thermal shock resistance depends on the mechanical and thermal properties of the compositions and is remarkably improved with increasing alumina content.

The coefficient of thermal expansion of stabilized ZrO_2 ceramic is greater than that of partially stabilized ZrO_2 (PSZ). For several reasons the thermal shock resistance of the fully stabilized ZrO_2 (FSZ) is lower than that of the partially stabilized one.
Banister et al.[1] reported that the thermal expansion coefficient of a Y_2O_3-stabilized ZrO_2 (YSZ) body can be decreased by addition of $\alpha\text{-}Al_2O_3$. Furthermore, it is known that other properties of FSZ can be altered by the addition of $\alpha\text{-}Al_2O_3$.[2-4]
The aim of this paper is to examine the mechanical and thermal properties of the binary system YSZ and $\alpha\text{-}Al_2O_3$.

Experimental Procedure

The following raw materials were used: ZrO_2 with 2 mass% HfO_2 and a maximum of 0.5 mass% impurities, >99 mass% Y_2O_3, and >99.5 mass% Al_2O_3. The compositions of zirconia and yttria were in the molar ratio 92.5:7.5. Varying amounts of alumina were added to this mixture.
The powder mixtures of ZrO_2, Y_2O_3, and Al_2O_3 were milled in a vibratory mill. Al_2O_3 mills and grinding media were used. Specific surface area of the resulting powders was 15 m^2/g. The powders were compacted to test specimens of different shape and size with compacting pressure of 300 bars (30 MPa). The test specimens were sintered and simultaneously stabilized in an electric furnace at a temperature of 1530°C in air.
The density of the sintered samples was determined by water displacement. Radial crushing strength was measured using ring-shaped test specimens. Test bars were used to determine the dynamic Young's modulus. The same samples were used to measure the thermal expansion coefficient up to a temperature of 1200°C using a high-temperature dilatometer. Cylindrical test samples were used for determining the thermal conductivity by means of the well-known thermal flux method,

where the axial temperature distribution is compared with that of reference specimens. Thermal shock resistance was obtained by the acoustic emission analysis described in Ref. 5. Electrical conductivity was measured using cylindrical specimens, provided with Pt thick-film electrodes.

The samples used for the X-ray analysis were also utilized for defining the microstructure after thermal etching at a temperature of 1400°C for 1 h in air. Strains in the microstructure were determined by analyzing the profile width of the X-ray diffraction lines of cubic zirconia or α-alumina as a function of the diffraction angle.

Results

With respect to microstructure and mechanical strength, three ranges of the composition can be distinguished. In the range 100–65 mass% ZrO_2 (composition range 1) the properties of the ZrO_2 matrix are dominant. The 65–20 mass% ZrO_2 range (composition range 2) can be described as a transitional range, and the range from 20 to 0 mass% ZrO_2 (composition range 3) is determined by the properties of the α-Al_2O_3. These three ranges are less pronounced for the Young's modulus.

Sintered Density and Microstructural Analysis

After the specimens were sintered at 1530°C for 6 h, they had a sintered density greater than 93% of the density of the solid body computed from the phase composition of each specimen. The sintered density decreases continuously from 5.8 g/cm^3 for 100 mass% ZrO_2 to 3.94 g/cm^3 for 0 mass% ZrO_2 (equal to

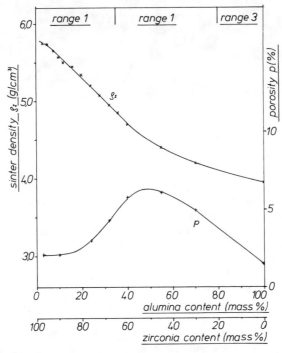

Fig. 1. Influence of alumina content on the sintered density and the porosity of YSZ/Al_2O_3 ceramics.

(A)

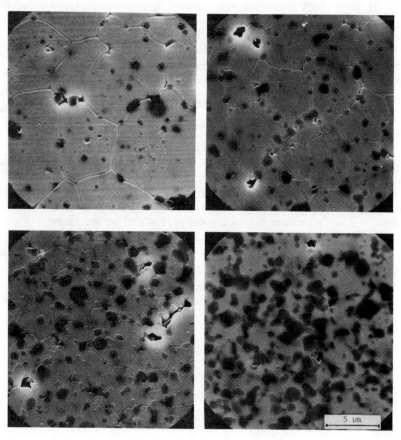

100 mass% α-Al_2O_3). In the 60:40 ZrO_2-Al_2O_3 composition range, an inflection in the curve is observed. The porosity as function of the composition shows a maximum in this range (Fig. 1). By X-ray analysis, only cubic ZrO_2 and α-Al_2O_3 are detected. Secondary phases resulting from possible reactions between ZrO_2 and Al_2O_3 or between Y_2O_3 and Al_2O_3 were not observed. Also, the Auger electron analysis gave no indication of such reactions.

Figure 2(A) shows the microstructures of compositions with 97, 92, 84, and 68 mass% ZrO_2 (3, 8, 16, and 32 mass% Al_2O_3, respectively) as determined by scanning electron microscopy (SEM) and Fig. 2(B) the microstructures of compositions with 60, 45, 30, and 0 mass% ZrO_2 (40, 55, 70, and 100 mass% Al_2O_3, respectively).

In composition range 1, where the ZrO_2 properties are predominant, the size of the ZrO_2 grains decreases with decreasing ZrO_2 content, whereas the size of the Al_2O_3 grains increases. Al_2O_3 particles are found within ZrO_2 grains and in the grain boundaries. It is seen that the Al_2O_3 grains are pinched between the ZrO_2

(B)

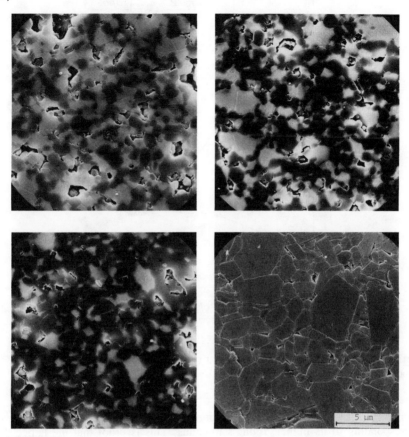

Fig. 2. Scanning electron microscopy micrographs from YSZ/Al$_2$O$_3$ ceramics.

grains (Fig. 2(A), composition with 97 mass% ZrO$_2$). The ZrO$_2$ grains in the vicinity of the Al$_2$O$_3$ grains are severely deformed. Butler and Drennan[6] observed similar results in mixtures of fully Y$_2$O$_3$-stabilized ZrO$_2$ and Al$_2$O$_3$.

Composition range 2 is characterized essentially by large pores and cracks. The grain size of both components of the microstructure is almost equal. The number of pores decreases markedly with decreasing ZrO$_2$ content (composition range 3). The microstructure of the samples with 0 mass% ZrO$_2$ is that of well-known alumina ceramics.

The microstructures show the same dependence of porosity on composition as determined (see Fig. 1) by the correlation between the sintered density and the density of the solid body.

The analysis of the profile width of the X-ray lines points at lattice distortions not only of the ZrO$_2$ but also of the Al$_2$O$_3$ grains. The lattice distortion of the composition with 97 mass% ZrO$_2$ was determined to be 0.4%, that of the composition with 68 mass% ZrO$_2$ to be 1.5%, and that for a composition with

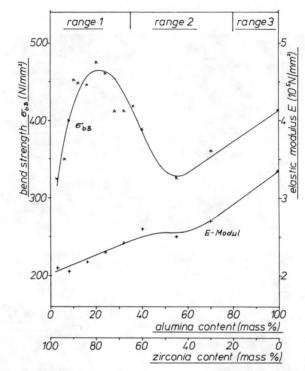

Fig. 3. Influence of alumina content on the bend strength and the elastic modulus of YSZ/Al$_2$O$_3$ ceramics.

59 mass% ZrO$_2$ to be 1%. The distortion of the Al$_2$O$_3$ lattice was always less, corresponding to its higher Young's modulus. This is proof that the lattice distortions are caused by stresses.

The determined diffraction spectra do not correspond to grain-size effects. The yttria was homogeneously distributed in the zirconia grains as shown by transmission electron microscopy (TEM) analysis.

Mechanical, Thermal, and Electrical Properties

Figure 3 shows that the mechanical strength has a maximum in composition range 1, in composition range 2 it has a minimum, and it reaches the strength of Al$_2$O$_3$ in composition range 3.

Young's modulus increases with decreasing ZrO$_2$ content, remains practically constant in the range of 60–30 mass% ZrO$_2$, and then increases to the Young's modulus of alumina. The coefficient of thermal expansion of these materials decreases continuously with decreasing ZrO$_2$ content (Fig. 4). The thermal conductivity, on the contrary, increases with decreasing ZrO$_2$ content and decreases with increase in temperature (Fig. 5). In the range 65–20 mass% zirconia, the slope of the curves for the thermal expansion coefficient and the thermal conductivity is less than in the other composition ranges. The thermal shock resistance increases markedly in composition range 1 and changes little in the other two composition ranges (Fig. 6). The electrical conductivity is described by the temperature at which the ceramic in question has an electrical resistance of 10^6 $\Omega \cdot$ cm,

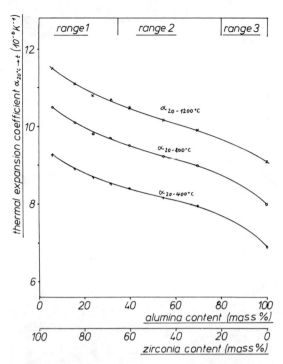

Fig. 4. Correlation between alumina content and thermal expansion coefficient of YSZ/Al$_2$O$_3$ ceramics in different temperature ranges.

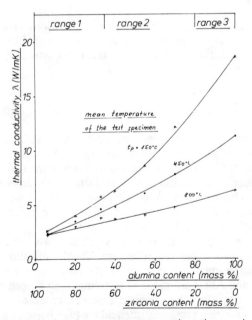

Fig. 5. Influence of alumina content on the thermal conductivity of YSZ/Al$_2$O$_3$ ceramics.

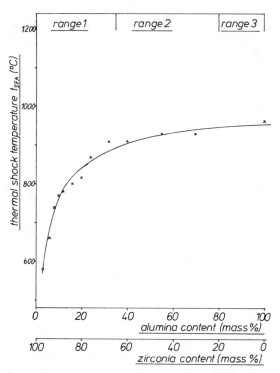

Fig. 6. Thermal shock temperature as a function of the alumina content of the YSZ/Al$_2$O$_3$ ceramics.

which usually is an upper limit for electrical applications of zirconia ceramics. This temperature increases continuously with decreasing ZrO$_2$ content (Fig. 7). Practically, the compositions with a ZrO$_2$ content less than 20 mass% can be considered as insulators.

Discussion

The properties of the compositions in the investigated system can be explained as follows:

Lattice distortions of both components, cubic ZrO$_2$ and α-Al$_2$O$_3$, were determined. The distortions increased with increasing Al$_2$O$_3$ content. This increase is a result of the different thermal expansion coefficients of the two components. Due to these lattice distortions, strains occur. These strains are responsible for the steep increase of the mechanical strength in composition range 1 with decreasing ZrO$_2$ content. Simultaneously, the grain size of ZrO$_2$ decreases, improving the mechanical strength by a smaller amount.

In composition range 2, the strains, which increase with increasing Al$_2$O$_3$ content, produce cracks in the microstructure. The densification is hindered more and more with increasing Al$_2$O$_3$ content. In the case where approximately equal amounts of ZrO$_2$ and Al$_2$O$_3$ are present, sintering is no longer governed by ZrO$_2$ and still not by Al$_2$O$_3$. Thus, low shrinkage occurs, resulting in a relatively high porosity. The cracks in the microstructure and the high porosity are responsible for the decreasing strength.

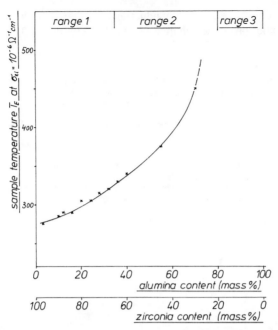

Fig. 7. Influence of alumina content on the electrical conductivity of YSZ/Al$_2$O$_3$ ceramics.

With decreasing ZrO$_2$ content in composition range 3, sintering is governed by the Al$_2$O$_3$ characteristics so that the lattice distortions and the porosity decrease, whereas the mechanical strength increases.

The Young's modulus increases continuously from composition range 1 through composition range 3. The shallower slope of the curve in the composition range 2 is probably caused by the higher porosity. The same statement can be made for the thermal expansion coefficient and thermal conductivity. The change of the slope of the two curves is not so pronounced as for the Young's modulus.

The remarkable increase in the thermal shock resistance in composition range 1 is due to the improved mechanical strength. Further, it results from the change in the coefficient of thermal expansion and thermal conductivity in this composition range. That the thermal shock resistance does not diminish with decreasing mechanical strength depends on the increased porosity for the materials of composition range 2. Both effects can compensate each other. In this case the change of the thermal conductivity and the thermal expansion coefficient influence the thermal shock resistance in a positive way. The electrical conductivity is determined by the composition.

Conclusions

The mechanical strength determined by the radial crushing method with as-fired samples in the system of fully stabilized ZrO$_2$(Y$_2$O$_3$)/α-Al$_2$O$_3$ is essentially influenced by strains occurring in the different materials and by the different sintering behavior. By adding about 20 mass% alumina, the mechanical strength can be increased by 50%. The thermal shock resistance of the same composition is improved from 550° to 850°C. Due to relatively high content of the non-conductive Al_2O_3, the electrical conductivity is altered by a remarkably small amount. The other properties change practically with the ZrO_2 content.

References

[1] M. J. Bannister, N. A. McKinnon, and R. R. Hughan, Australian Patent 8395-76, August 18, 1977; German Pat. Application DOS 2 754 522, June 8, 1978.

[2] F. J. Rohr, "Solid Electrolyte for Fuel Cells," German Pat. Application DAS 1 671 704, September 23, 1971.

[3] K. C. Radford and R. J. Bratton, "Zirconia Electrolyte Cells," *J. Mater. Sci.*, **14**, 59–65 (1979).

[4] F. J. Esper, K. H. Friese, and H. Geier, "Solid Electrolyte Oxygen Sensors," U.S. Pat. 4 221 650, September 9, 1980.

[5] F. J. Esper and H. M. Wiedenmann, "Thermal Shock Resistance of Ceramics Determined by Acoustic Emission Analysis," *Ber. DKG*, **55** [12] 507–10 (1978).

[6] E. P. Butler and J. Drennan, "Microstructural Analysis of Sintered High-Conductivity Zirconia with Al_2O_3 Additions," *J. Am. Ceram. Soc.*, **65** [10] 474–78 (1982).

The Reaction-Bonded Zirconia Oxygen Sensor: An Application for Solid-State Metal-Ceramic Reaction-Bonding

R. V. ALLEN, W. E. BORBIDGE, AND P. T. WHELAN

CSIRO, Division of Chemical Physics
Clayton, Victoria, Australia 3168

The process of solid-state metal-ceramic reaction-bonding can be used to advantage in the construction of a zirconia in situ type oxygen sensor, in which a ZrO_2-Pt-Al_2O_3 reaction bond forms a seal between a zirconia solid electrolyte sensing tip and an alumina shank of a length which varies to suit the installation. Results of tests on the mechanical properties of the sensor are presented, in particular the strength and vacuum tightness of the reaction bond and the effects of thermal cycling and high operating temperatures on these properties. The special electrical characteristics of the sensor are described, and data are presented on the response time and accuracy of measurement of p_{O_2} over the operating temperature range.

Metal-ceramic reaction-bonding is a patented technique for the direct, solid-state bonding of metals to ceramic oxides, which can be used to form strong, hermetic seals between a wide range of materials.[1,2] Both noble metals and transition metals can be bonded to single-oxide and mixed-oxide ceramics. Materials commonly used for bonding are Pt, Pd, Au, Ni, and Cu to Al_2O_3, ZrO_2, MgO, SiO_2, and BeO ceramics.

The process involves heating the metal and ceramic under light pressure to improve contact, to a temperature about 90% of the melting point of the lowest melting component (usually the metal). Time at temperature to achieve a strong bond can vary from a few minutes to a few hours. The atmosphere for bonding is usually oxidizing, i.e., air, but vacuum or inert atmospheres such as argon are sometimes used to prevent excessive oxidation when non-noble metals are used. To achieve optimum bond strength, the surfaces of the ceramics (and in some cases also the metals) are polished to optical flatness prior to bonding.

A feature of the reaction-bonding process is that the bonds so formed generally retain their properties of high strength and vacuum tightness at high operating temperatures. The technique is appropriate for a variety of applications, but it is particularly suited to those involving high temperatures, for which the commonly used metallizing and brazing techniques for joining metals to ceramics are usually unsuitable.

Reaction-bonding involving ZrO_2 as the ceramic oxide has been investigated extensively. Most of the common types of ZrO_2 (i.e., Y-, Ca-, or Mg-stabilized and also partially stabilized zirconia) have been tested and found to bond satisfactorily to a wide range of metals. Table I shows some of the combinations which can be reaction-bonded using ZrO_2, with their typical bond strengths. From Table I it can be seen that materials differing widely in thermal expansion can be success-

Table I. Combinations of Materials Which Have Been Reaction-Bonded and Typical Bond Strengths Attained*

Bond configuration	Bond strength[†] (MPa)	Temperature used for bonding (°C)
ZrO_2(CaO stab)-Pt-Al_2O_3	154	1450
ZrO_2(Y_2O_3 stab)-Pt-Al_2O_3	>110[‡]	1450
ZrO_2(MgO stab)-Pt-Al_2O_3	170	1450
ZrO_2-Au-Al_2O_3	99	1040
ZrO_2-Au-mullite[§]	42	1040
ZrO_2-Au-MACOR[¶]	51	900
ZrO_2-Au-stainless steel	24[§]	950
ZrO_2-Pt-stainless steel	16	1130
ZrO_2-nicrosil[§*]-Al_2O_3	18	1200
ZrO_2-copper	87	1000
ZrO_2-Ni-ZrO_2	14[§]	1020
PSZ[††]-Cu-steel	52	1000

*ZrO_2 refers to CaO-stabilized zirconia, except otherwise indicated. Metals can be bonded in the bulk form, e.g., solid rod or tubing, but most test bonds are made by using metal foils, sandwiched between ceramic rods or tubes. The bonding of steels normally requires an intermediate metal foil.
[†]4-point bend test.
[‡]Fails within ceramic—bond remains intact.
[§]Mullite is an aluminosilicate.
[¶]MACOR is a brand name for machinable glass.
[**]Nicrosil is a Ni-Cr-Si alloy.
[††]PSZ is partially stabilized zirconia.
[§]Shear tested.

fully reaction-bonded, e.g., Ca-stabilized ZrO_2-Au-$3Al_2O_3 \cdot 2SiO_2$ (mullite), for which the linear coefficients of thermal expansion (in units of 10^{-6} cm$^{-1} \cdot$ °C^{-1}) are as follows: ZrO_2, $\alpha = 10.0$; Au, $\alpha = 16.9$; mullite, $\alpha = 5.3$.[3] Typical bond strength for this combination is 42 MPa.

Recent research on reaction-bonding has been directed mainly toward exploiting the process in solving technological problems.[4] To this end, an oxygen sensor has been developed in which a ZrO_2-Pt-Al_2O_3 reaction bond, with its properties of high strength and leak tightness at high temperatures, is used to advantage as the basis of the construction.

The Reaction-Bonded Oxygen Sensor

General Description

The reaction-bonded oxygen sensor is of the direct, in situ ZrO_2 solid electrolyte type, suitable for providing continuous measurement of oxygen concentration in hot gaseous environments.

The sensor is constructed from a ZrO_2 thimble (Ca- or Y-stabilized) reaction-bonded via a platinum foil washer to an Al_2O_3 shank of a length to suit the installation (see Fig. 1). The ZrO_2 thimble is platinized using platinum paste on both the inner and outer surfaces to form the electrodes of the oxygen sensor. The outer electrode is connected by the platinum foil bond to a platinum wire inside the sensor, which connects to the terminal head as the outer electrode wire. A thermocouple element (Pt–Pt13Rh) is fitted within the body of the sensor so as to contact the inner platinized surface of the ZrO_2; the platinum lead of the thermocouple also acts as the inner electrode wire.

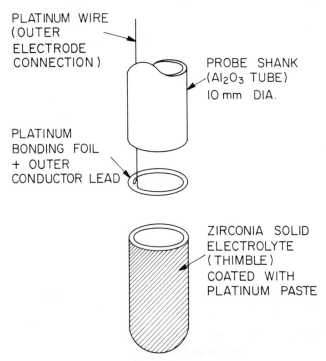

Fig. 1. Exploded view of the reaction-bonded sensor. Thermocouple pair not shown.

The main advantage of this type of construction is that there are no wires external to the sensor, where they may be subject to mechanical damage or chemical attack. The lack of external wires also eliminates problems with sealing and gas tightness at the terminal head of the probe. Also, with the reaction-bonded sensor, the outer electrode is firmly anchored (the platinum reaction bond in effect is the outer electrode), which compares favorably with sensors where the outer electrode wire is simply wrapped around the ZrO_2 and can easily become detached.

The use of a standard, short thimble of ZrO_2 facilitates application of the platinum paste to form the inner electrode (the platinum paste is applied to the inner electrode and fired before bonding the sensor). This means that the inner electrode can easily be prepared to a consistent, standard thickness, which results in improved and predictable electrode performance.

Mechanical Properties

The mechanical performance of the ZrO_2-Pt-Al_2O_3 reaction bond forming the basis of the sensor has been studied using a series of test bonds, formed under identical conditions to those used for oxygen sensors (i.e., 4 h at 1450°C bonding temperature and 1 MPa clamping pressure). The base strength of these bonds was found experimentally to be very high, averaging 154 MPa (modulus of rupture, 4-point bend test). The effect of subsequent high operating temperatures on bond strength was determined for a selection of the bonds by testing them to failure in a hot bend rig at temperatures between ambient and 1000°C. The results are plotted in Fig. 2, where it can be seen that the bonds retain high strength at elevated

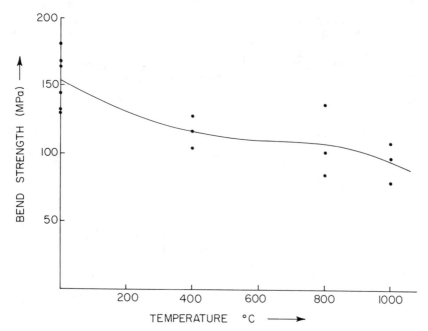

Fig. 2. Relationship between strength of ZrO_2-Pt-Al_2O_3 bonds and their subsequent operating temperature.

temperatures (95 MPa at 1000°C). This is a slight reduction from the ambient strength; however, some fall in strength with temperature is normal for all ceramic materials; ceramic-metal bonds could therefore be expected to show a similar decrease in strength with temperature.

The effect of thermal cycling on bond strength has also been evaluated. After 100 cycles between 25° and 1000°C, at 90°C/min, bond strength had deteriorated slightly from an average of 154 to 117 MPa. Vacuum tightness of the bonds had generally remained intact throughout the cycling, with the worst-case leak rate after cycling being $\approx 10^{-5}$ STD $cm^3 \cdot s^{-1}$. Overall, these results indicate a satisfactory ability to withstand thermal cycling over a wide temperature range.

Extending these results to reaction-bonded oxygen sensors, it can be stated that the bond sealing the ZrO_2 sensing element to the alumina shank is strong and vacuum-tight and will retain these properties at high operating temperatures and after repeated thermal cycling.

These properties make this type of sensor most suitable for direct in situ applications (e.g., in boilers, furnaces, flues, and kilns), where the sensor is required to operate at temperatures above 600°C and up to about 1300°C.

Electrical Characteristics

When the two sides of the ZrO_2 sensor are exposed to different oxygen partial pressures p'_{O_2} and p''_{O_2}, a voltage is developed between the sensor's electrodes, which is given by the Nernst equation.

$$E_N = (RT/4F) \ln (p'_{O_2}/p''_{O_2}) \qquad (1)$$

where R and F are the gas and Faraday constants and T is the temperature. The

Nernst voltage arises as a result of equilibrium electrochemical reactions occurring at each electrode:

$$O_2(\text{gas}) + 4e^-(\text{metal}) = 2O^{2-}(\text{electrolyte}) \tag{2}$$

When the oxygen partial pressure differential between the two electrodes is zero, reaction (2) will proceed at an equal rate at each electrode, and no voltage will be developed between the sensor's output leads. In practice, there are inequalities and electrothermal offsets which can generate small voltages, referred to as the cell constant potential; for the reaction-bonded sensor this is typically <0.2 mV.

In normal operation the sensor's inner electrode is held at a fixed reference oxygen partial pressure p'_{O_2}, usually air, 0.2095 atm (2.0545×10^4 Pa). The outer electrode senses oxygen partial pressure variations, generally at a lower oxygen concentration than the inner; thus, it becomes the anode of the cell by producing a negative output voltage with respect to the inner electrode which becomes the cathode.

The oxygen partial pressure dependence of the electrode reactions causes ionic concentration gradients to be formed in the electrolyte, which can be represented by equivalent electrical components. Figure 3 shows these components in relation to the total equivalent circuit and the physical configuration of the sensor.

Interface Impedances: The components R_A, C_A and R_K, C_K of Fig. 3 represent the anode and cathode interface impedances and are a direct result of the electrode reactions (2) occurring at the double layer, a region extending a few angstroms into

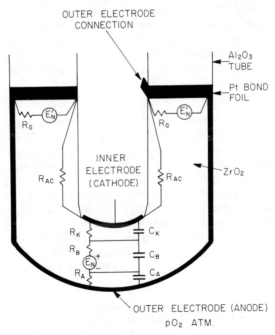

Fig. 3. Total equivalent circuit shown diagrammatically in relation to the physical configuration of the sensor.

the electrolyte immediately in contact with the platinum electrode. The capacitances formed here are large, typically 500–1000 μF and must be taken into consideration when interfacing to control instrumentation. The resistive components R_A and R_K vary with oxygen partial pressure and also electrode condition; this latter characteristic can be used as a means of detecting sensor aging. For instance, if the electrodes have deteriorated due to abrasion or chemical attack (more likely to occur at the outer sensing electrode) and have lost material to the point where there may no longer be a continuous electrical contact over the surface area, the interface resistance will increase. This problem can be reduced by the use of a filter or coating over the sensor tip to give electrode protection.[5]

Bulk Impedance: The bulk resistance, shown as R_B in Fig. 3, and its associated small capacitance, C_B, arise from the ionic conductivity of the zirconia electrolyte bulk material; the capacitance has a typical value 0.01 μF.[6] R_B varies logarithmically with temperature and causes the total sensor resistance (i.e., $R_T = R_B + R_K + R_A$) to vary typically between about 15 Ω at 1000° and 500 kΩ at 600°C. The bulk electrolyte resistance also has an "aging" property in that it will increase in value with time at operating temperature due to the gradual decline in ionic conductivity of the electrolyte.[7] Ca-stabilized ZrO$_2$ is worse in this respect than Y-stabilized ZrO$_2$. These aging effects are not peculiar to the reaction-bonded sensor but are common to all ZrO$_2$ oxygen sensors. In practice, the resistance of a sensor can increase considerably before there is any significant loss of Nernst voltage, causing inaccuracy in measurement of oxygen concentration.

Internal Resistance: The circuit element shown as R_{AC} in Fig. 3 is peculiar to the reaction-bonded sensor and arises because the outer electrode is connected to the platinum reaction bond. This, in effect, creates a second cell between anode and cathode on the inside of the sensor; however, both electrodes are exposed to the same air reference atmosphere, and no Nernst voltage can be generated from this cell. Also the area of the cathode is designed to be small compared to the anode (approximate ratio 1:8); thus, the amount of unplatinized electrolyte between the cathode and the inner bond connection is large, making R_{AC} a high resistance. Its shunting effect on the sensor's output voltage is therefore negligible.

The Bond Cell: Another potential source of Nernst voltage in the reaction-bonded sensor is the cell formed across the ZrO$_2$–Pt bond, shown as E'_N and R_G in Fig. 3. This cell does experience an oxygen partial pressure differential $p'_{O_2} - p''_{O_2}$, but its potential is effectively short-circuited by the platinum bond washer and has no influence on the sensor output voltage E_N. Any tendency for E'_N, R_G, and also R_{AC} to produce ionic circulating currents within the electrolyte, and as a result cause oxygen semipermeability or leakage through the sensor, has not been found to be significant in terms of sensor accuracy. Also, because the reaction bond forms a vacuum-tight seal, separating the gas atmosphere at the inner from that at the outer electrode, high absolute accuracy is obtained over the working range of the sensor; the deviation from Nernstian response is within ±0.2 mV over the operating temperature range.

Response Time: The response time of the sensor to gas composition changes has been measured as ≈150 ms under laboratory conditions; it is considered that the experimental arrangement is a limiting factor and that the actual sensor response time is much faster. Measurements under industrial conditions, in a coal-fired power station environment, have shown response times to be of the order of a few milliseconds.

Probe Assembly

For installation in industrial sites, the sensor requires physical protection in the form of a metal or ceramic sheath and a suitable head shell for termination of the electrical connections. The thermocouple element is supported in a length of 3-bore ceramic tubing of such diameter as to fit into the inside of the sensor. At the head shell the thermocouple is spring-loaded so as to hold the thermocouple bead against the sensor's inner electrode with a light pressure. The thermocouple in conjunction with the outer electrode wire forms a three-wire sensor connector arrangement, whereby the platinum leg of the Pt–Pt13Rh couple is the common connection between the sensor and thermocouple output voltages. The remaining bore in the thermocouple tube is used as a pathway for supply of an air reference atmosphere at a flow rate of approximately 0.5 L/min to the inner electrode.

Conclusion

Reaction-bonding is a versatile technique. Its use in the construction of a ZrO_2 oxygen sensor is an application which utilizes the properties of reaction-bonding to best advantage. These are high strength at high operating temperatures, vacuum tightness of the bonded seal, the ability to effectively join materials of differing thermal expansion, and its application to a wide range of metals and ceramics.

References

[1] CSIRO and The Flinders University, South Australia, "Chemical Bonding of Metals to Ceramic Materials," Australian Pat. 452 651, Italian Pat. 920 003, British Pat. 1 352 775, U.S. Pat. 4 050 956.

[2] H. J. de Bruin, A. F. Moodie, and C. E. Warble, "Ceramic-Metal Reaction Welding," *J. Mater. Sci.*, **7**, 909–18 (1972).

[3] W. H. Kohl; Handbook of Materials and Techniques for Vacuum Devices. Reinhold Publishing Co., New York, 1967.

[4] F. P. Bailey and W. E. Borbidge; pp. 525–33 in Surfaces and Interfaces in Ceramic and Ceramic-Metal Systems. Edited by J. A. Pask and A. G. Evans. Plenum, New York, 1981.

[5] B. C. H. Steele, J. Drennan, R. K. Slotwinski, N. Bonanos, and E. P. Butler, "Factors Influencing the Performance of Zirconia-Based Oxygen Monitors"; pp. 286–97 in Advances in Ceramics, Vol. 3. Edited by A. H. Heuer and L. W. Hobbs. The American Ceramic Society, Columbus, OH, 1981.

[6] J. E. Bauerle, "Study of Solid Electrolyte Polarization by a Complex Admittance Method," *Phys. Chem. Solids,* **30**, 2657–70 (1969).

[7] R. M. Dell and A. Hooper; pp. 291–312 in Solid Electrolytes. Edited by P. Hagenmuller and W. Van Gool. Academic Press, New York, 1978.

Properties of Metal-Modified and Nonstoichiometric ZrO$_2$

ROBERT RUH

Air Force Wright Aeronautical Laboratories
Wright-Patterson Air Force Base, OH 45433

Metal additions of Ti or Cr above the solubility limit significantly improve the microstructure and properties of polycrystalline ZrO$_2$ through liquid-phase-enhanced sintering. For Zr additions, the same improvement is observed, the material becomes nonstoichiometric, and the temperature of the cubic(c)–tetragonal(t) transformation is lowered to 1490°C at 63 at.% O.

Metal additions significantly affect the stoichiometry and properties of polycrystalline ZrO$_2$ sintered under vacuum. Early work by Weber et al.[1] revealed that Ti-modified ZrO$_2$ was the most inert and most suitable ceramic crucible material for melting Ti. In addition to being nonreactive, this material possessed excellent thermal shock resistance. High-temperature XRD studies showed that Ti-modified ZrO$_2$ traversed the monoclinic(m)-tetragonal(t) transformation similarly to unalloyed ZrO$_2$. It was postulated that the improved properties might be due to increased thermal conductivity, plastic deformation of the metal phase, or Ti solid solution in ZrO$_2$. To gain a better understanding of metal-modified ZrO$_2$ and determine the mechanisms for property improvement, more detailed investigations were accomplished on several zirconia-metal systems that exhibited the improved properties.

A comprehensive study of the reactions in the ZrO$_2$-Ti vertical section was accomplished.[2] Specimens were fired under vacuum and examined using microstructural analysis, microhardness determinations, lattice parameter determinations, electron probe analysis, and thermal conductivity measurements. Results revealed that up to 4 at.% Ti was retained in solid solution in ZrO$_2$ and ≈10 mol% ZrO$_2$ was retained in solid solution in Ti. For the latter case, the Zr entered the Ti lattice substitutionally and the oxygen entered interstitially.

Compressive strength measurements on samples of ZrO$_2$ plus 5 and 15 at.% Ti compositions revealed room-temperature values of 965 MPa and values decreasing to ≈690 MPa at 500°–800°C.[3] An aging phenomenon dependent on time, temperature, and oxygen pressure was found to increase the room-temperature and high-temperature strengths. Compositions above the solubility limit were able to traverse the $t \rightarrow m$ transformation repeatedly without degradation, whereas compositions below the solubility limit retained little strength after traversing this transformation.

From the above results, it appeared that the abrupt change in microstructure caused by titanium additions was responsible for the sharp change in properties, rather than the solid solubility affecting the inherent properties of the ZrO$_2$ itself. Consequently, the mechanism by which this dense, highly sintered structure was formed was studied in the system Ti-ZrO$_2$[4] using diffusion couples. Results revealed that in the immediate vicinity of the Ti source a two-phase region existed containing appreciable amounts of liquid. In the normally prepared compositions in which Ti powder is distributed as uniformly as possible throughout zirconia, it appears that the highly sintered body that results is produced by the formation and overlap of similar regions in which, at least initially, enough liquid was present to allow liquid-phase-enhanced sintering.

The final study of the system Ti-ZrO_2 was a determination of the pseudo-binary phase diagram.[5] Results confirmed the Ti solubility limit in ZrO_2 and established the existence of $(Ti, Zr)_3O$ as a third phase at low temperatures.

The system ZrO_2-Cr was investigated[6] since Cr-modified ZrO_2 possessed the same ability to traverse the $t \rightarrow m$ transformation as Ti-modified ZrO_2. Results revealed that 1 at.% Cr went into solid solution in ZrO_2 and that dense sound samples were produced via liquid-phase-enhanced sintering.

A portion of the system Zr-ZrO_2 was investigated in detail[7] after the work of Gebhardt et al.[8] revealed a sharp change in the solubility limit of α-Zr in ZrO_2 at 1577°C, as evidenced by the precipitation of α-Zr striations. The phase relations were studied by metallographic analysis, high-temperature XRD analysis, and room-temperature lattice parameter studies. Results revealed that the $c \rightarrow t$ transformation for oxygen-deficient ZrO_2 (63 at.% O) is lowered to 1490°C. Alpha Zr has greater solubility in the c phase than in the t phase, and the transformation is characterized by the exsolution of α-Zr, which is microstructurally characterized as striations. The presence of the c phase and $c \rightarrow t$ transformation was verified by high-temperature XRD analysis.

A technique was developed for the preparation of dense microstructurally sound bodies of stoichiometric ZrO_2.[9] Specimens were sintered under vacuum to 2300°C for 3 h and then buried in raw ZrO_2 powder in an Al_2O_3 crucible and fired to 1000°C in air. Complete reoxidation occurred within 18 h, and densities of 94–96% of theoretical were obtained.

References

[1](a) B.C. Weber, H.J. Garrett, F.A. Mauer, and M.A. Schwartz, *J. Am. Ceram. Soc.*, **39** [6] 197–207 (1956).
 (b) B.C. Weber, W.M. Thompson, H.O. Bielstein, and M.A. Schwartz, *J. Am. Ceram. Soc.*, **40** [11] 363–73 (1957).
[2] R. Ruh, *J. Am. Ceram. Soc.*, **46** [7] 301–307 (1963).
[3] H.A. Lipsitt and R. Ruh, *J. Am. Ceram. Soc.*, **47** [12] 645–46 (1964).
[4] R. Ruh, N.M. Tallan, and H.A. Lipsitt, *J. Am. Ceram. Soc.*, **47** [12] 632–35 (1964).
[5] R.F. Domagala, S.R. Lyon, and R. Ruh, *J. Am. Ceram. Soc.*, **56** [11] 584–87 (1973).
[6] R. Ruh and H.J. Garrett, *J. Am. Ceram. Soc.*, **47** [12] 627–29 (1964).
[7] R. Ruh and H.J. Garrett, *J. Am. Ceram. Soc.*, **50** [5] 257–61 (1967).
[8] E. Gebhardt, H.D. Seghezzi, and W. Duerrschnabel, *J. Nucl. Mater.*, **4** [3] "I," 241–54, "II," 255–68, "III," 269–71 (1961).
[9] H.J. Garrett and R. Ruh, *Am. Ceram. Soc. Bull*, **47** [6] 578–79 (1968).

Diffusion Processes and Solid-State Reactions in the Systems Al_2O_3-ZrO_2 (Stabilizing Oxide) (Y_2O_3, CaO, MgO)

T. Kosmač, D. Kolar, and M. Trontelj

"E. Kardelj" University
"J. Stefan" Institute
Ljubljana, Yugoslavia

Diffusion processes and solid-state reactions between Al_2O_3 and stabilized ZrO_2 containing Y_2O_3, CaO, or MgO in solid solutions were investigated using diffusion couples. In the system Al_2O_3-$ZrO_2(Y_2O_3)$, diffusion of Y_2O_3 into Al_2O_3 does not take place and no YAG formation could be observed. In the systems Al_2O_3-ZrO_2(CaO) and Al_2O_3-ZrO_2(MgO), the stabilizing oxides react with Al_2O_3, resulting in the formation of Ca aluminates and Mg-Al spinel, respectively.

Stabilized ZrO_2 ceramics usually contain small amounts of Al_2O_3, either as an impurity or as a sintering aid, which reportedly promotes enhanced densification of ZrO_2 at lower temperatures.[1,2] A number of reports concerning ZrO_2-toughened Al_2O_3 ceramics (ZTA) doped with Y_2O_3 confirmed that Y_2O_3-stabilized ZrO_2 (YSZ) cannot be substantially destabilized by Al_2O_3.[3-5] CaO-stabilized ZrO_2 (CSZ), on the other hand, reacts with Al_2O_3, resulting in destabilization of c-ZrO_2.[6] Some experimental observations[7] also indicate a similar influence of Al_2O_3 on MgO-stabilized ZrO_2 (MSZ), as described for CSZ. Binary phase equilibria in the systems Al_2O_3-ZrO_2 (stabilizing oxide) are reasonably well established, whereas for the ternary systems only a limited amount of information appears to be available.[6,8]

The present study is concerned with diffusion processes and solid-state reactions in the systems Al_2O_3-YSZ, Al_2O_3-CSZ, and Al_2O_3-MSZ, in the temperature range 1300°–1600°C. For these purposes, diffusion couples between stabilized ZrO_2 and Al_2O_3 were examined by electron microprobe and X-ray analysis. The results obtained could also be applied in the preparation of ZTA ceramics, where, by the addition of an appropriate stabilizing agent, the relative amount of dispersed t-ZrO_2 particles within the Al_2O_3 matrix can be significantly increased.

Materials and Experimental Procedure

Starting materials for the preparation of stabilized ZrO_2 were commercial high-purity ZrO_2* with SiO_2, TiO_2, and Fe_2O_3 impurity levels below 0.01%, $CaCO_3$,[†] and Mg and Y acetates.[‡] After homogenization in a ball mill, powder mixtures were dried and calcined at 1400°C for 6 h. The crushed powders were

---

*G 10 grade, Magnesium Elektron Ltd., Twickenham, UK.
[†]pa grade, E. Merck, Darmstadt, FRG.
[‡]99.9% purity, Ventron Gmbh, Karlsruhe, FRG.

pressed into disks (16 mm in diameter by 5-mm thick) and sintered at 1700°C for 4 h in air. The stabilizing oxide concentrations chosen were 3.5 and 8% Y_2O_3, 4.5 and 8% MgO, and 5.5 and 8% CaO by weight, respectively.

Diffusion couples were prepared either by pressing small crushed ZrO_2 pieces into Al_2O_3[§] pellets or by annealing flame-melted Al_2O_3 spheres (50–200 μm in diameter) on polished ZrO_2 surfaces, which enables a well-defined contact area to be formed between the two components. After heat treatment, diffusion couples were sectioned and polished before electron microprobe analysis. X-ray analysis was conducted on composite materials, sintered from Al_2O_3 and 20 wt% stabilized ZrO_2 powder mixtures. For these experiments commercial ZrO_2 solid solution powders[¶] were used with the same composition as that used for diffusion couples, but with slightly higher impurity contents.

Results and Discussion

After being sintered, the stabilized ZrO_2 disks used for diffusion couples were 80–85% dense, with a grain size ranging from about 2–10 μm. Phase compositions were qualitatively examined by X-ray analysis and varied with the kind and amount of stabilizing oxide according to the binary phase equilibria.

The System Al_2O_3-YSZ

It was shown that Y_2O_3 addition increases the t/m ratio of dispersed ZrO_2 particles in an Al_2O_3 matrix.[3-5] It was also reported that higher Y_2O_3 addition to ZTA results in a composite material with c-ZrO_2 particles within the matrix.[3] Our results show no changes in the stability of YSZ in an Al_2O_3 matrix and are therefore in accord with previously reported observations. Furthermore, quantitative analysis of diffusion couples did not reveal the presence of any Y_2O_3 in the Al_2O_3 (Fig. 1), despite the fact that, in the Al_2O_3-Y_2O_3 binary system, the solubility of Y_2O_3 was reported to be in the range of 1 mol%.[9] On the other hand, small amounts of ZrO_2 were detected in Al_2O_3, which in turn is also slightly soluble in YSZ. Both solute concentrations were estimated to be in the range of 0.1 wt%, but we were not in a position to determine whether the solutes are concentrated on the grain boundaries or randomly distributed in the bulk of the material. This result supports the work of Bernard[10] and Butler and Drennan,[2] who observed a slight Al_2O_3 solubility in YSZ.

The Systems Al_2O_3-MSZ and Al_2O_3-CSZ

MSZ and CSZ display quite different behaviors when they are in contact with Al_2O_3. X-ray analysis of sintered Al_2O_3 composites with 20 wt% MSZ or CSZ confirmed that both stabilizing oxides react with Al_2O_3, resulting in destabilization of ZrO_2. In the system Al_2O_3-MSZ, $MgAl_2O_4$ spinel starts to form at temperatures well below 1300°C. With increased annealing temperatures, the $MgAl_2O_4$ lattice parameter decreases, indicating an increasing amount of Al_2O_3 being dissolved in $MgAl_2O_4$ solid solution. It should also be noted that the solubility of ZrO_2 in the reaction product was found to be negligible. The distribution of Al_2O_3 and MgO in the diffusion couple between an Al_2O_3 sphere and an MSZ plate after a 4-h annealing time at 1600°C shows that the reaction starts on the surface of the Al_2O_3 sphere, whereas no MgO could be detected in the core of the sphere (Fig. 2). The predominant mass transport in the initial reaction stage is therefore Mg^{2+} surface

---

[§]99.98%, A6 Extrapure, Ugine Kuhlmann, Jarie, France.
[¶]S grade, Magnesium Elektron.

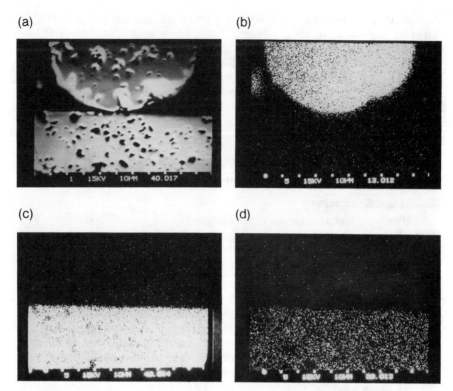

Fig. 1. Diffusion couple between Al_2O_3 sphere and YSZ plate, annealed for 4 h at 1600°C: (a) electron image; (b) distribution of Al; (c) distribution of Zr; (d) distribution of Y.

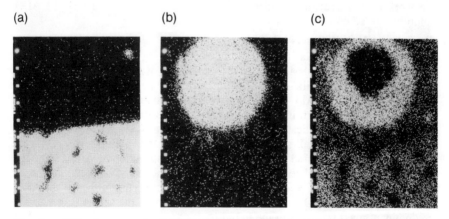

Fig. 2. Diffusion couple between Al_2O_3 sphere and MSZ plate, annealed for 4 h at 1600°C: (a) distribution of Zr; (b) distribution of Al; (c) distribution of Mg.

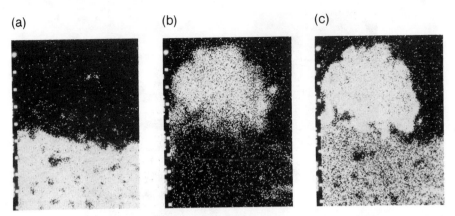

Fig. 3. Diffusion couple between Al_2O_3 sphere and CSZ plate, annealed for 4 h at 1600°C: (a) distribution of Zr; (b) distribution of Al; (c) distribution of Ca.

diffusion, followed by penetration into the bulk by grain-boundary and/or bulk diffusion through the layer of the reaction product. Another possible mechanism could be transport of MgO via the gas phase, a possibility already considered to account for the observed increase in the amount of m-ZrO_2 on the surface of sintered Mg–PSZ ceramics.[11] To check this possibility, Al_2O_3 spheres on Pt foil were covered by a small MSZ crucible and annealed at 1600°C for 24 h. Without direct contact between Al_2O_3 and MSZ, the presence of MgO on the surface of the Al_2O_3 spheres could not be detected. This experiment indicates that gas-phase transport of MgO does not play a significant role in the reaction between Al_2O_3 and MSZ.

In the system Al_2O_3-CaO, a number of aluminates exist, ranging from $CaO \cdot 6Al_2O_3$ (CA_6) to $3CaO \cdot Al_2O_3$ (C_3A). No solubility regions were reported to exist between the various compounds.[12] Microprobe analysis of the diffusion couple showed that CA_6 forms below 1300°C; however, the reaction layer after a 24-h annealing time at 1300°C was only a few micrometers thick. Qualitative analysis of the reaction layer formed during the 4-h heat treatment at 1600°C revealed the presence of 8% CaO, 3% ZrO_2, and 89% Al_2O_3 by weight, indicating that a considerable amount of ZrO_2 can be dissolved in CA_6. More insight into the phase equilibria was obtained by annealing Al_2O_3 spheres on a CSZ plate (Fig. 3). After 4 h at 1600°C, the Al_2O_3 sphere became irregularly shaped, showing a large contact area with the CSZ plate. The Ca concentration within the sphere close to the contact area is higher than in the upper part, indicating the presence of two different aluminates. Quantitative analysis confirmed that Ca concentrations correspond to CA_2 and CA_6, respectively. This observation is in accordance with the binary phase diagram, i.e., once the total amount of the Al_2O_3 available is consumed in CA_6 formation, the phase equilibrium is shifted toward CA_2. Such a situation appears during sintering of CSZ doped with a small amount of Al_2O_3. According to the Al_2O_3-CaO phase diagram, formation of lower Ca aluminates results in the lowering of the solidus line (modified by the presence of ZrO_2), such that liquid can be formed below 1600°C if CA_2 is present. The wettability

Fig. 4. Distribution of Mg and Al in diffusion couple between Al_2O_3 and MSZ, containing 4.5 wt% MgO, annealed at 1600°C for 4 h.

of CSZ by the liquid phase seems to be very good, as it penetrates along the CSZ grain quickly.

Both Al_2O_3-MSZ and Al_2O_3-CSZ diffusion couples also showed a characteristic concentration gap of the stabilizing oxide within the ZrO_2 phase at the contact area, as predicted by the equilibrium diagrams. This gap starts to form only after the stabilizing oxide reaches the lowest solubility limit in c-ZrO_2.

For the diffusion couple between Al_2O_3 and MSZ (containing 4.5 wt% MgO), it was found by quantitative EPMA that, after a short annealing time (1 h) at 1600°C, the MgO concentration within the concentration gap was 0.2 wt%, whereas in the bulk of the MSZ, this concentration was 3.3–3.5 wt% (Fig. 4). It is interesting to note that in both regions an almost constant concentration level was observed. The MgO level in the concentration gap corresponds well to the limiting amount of MgO that can be incorporated into t-ZrO_2 solid solution,[13] whereas the concentration in the bulk of MSZ can be compared with the limiting amount of MgO in the c-ZrO_2 solid solution. Similar analysis of the Al_2O_3-MSZ couple, containing 8 wt% MgO, revealed a constant concentration level containing 3.5% MgO throughout the interfacial region, whereas the MgO concentration outside this region increased toward the center of the embedded MSZ piece. The nonuniform distribution of MgO in the central region of the MSZ piece confirms the presence of free MgO in MgO-rich MSZ. When these results are compared with the phase equilibria in the system ZrO_2-MgO at 1600°C, it appears that, at the beginning of the reaction, MgO is consumed from the c-ZrO_2 solid solution. So long as Mg-ion diffusion through the cubic phase can follow the rate of reaction, $MgAl_2O_4$ is in equilibrium with c-ZrO_2 containing a lower limiting amount of MgO. Once the diffusion path becomes too long or once the MgO in c-ZrO_2 solid solution reaches its lowest solubility level (i.e., 3.3–3.5% MgO), a layer of t-ZrO_2 containing 0.2% MgO starts widening, eventually becoming detectable as the MgO concentration gap registered by microprobe analysis.

A similar phenomenon was observed within the CSZ piece in an Al_2O_3 matrix (Fig. 5). The CaO concentrations determined within the gap and in the center of the CSZ piece were 0.6 and 5.5 wt%, respectively. The latter level again corre-

(a)

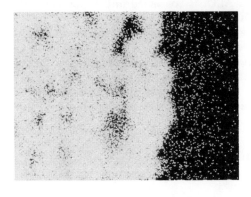

(b)

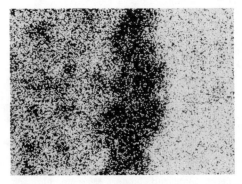

Fig. 5. Diffusion couple between Al_2O_3 and CSZ, containing 8 wt% CaO, annealed for 4 h at 1600°C: (a) distribution of Zr; (b) distribution of Ca.

sponds to the solubility limit of CaO in c-CSZ, whereas the former is less than the reported solubility limit of CaO in t-ZrO_2 (i.e., 2.2 wt%) by Stubican and Ray.[14] The systematically appearing intermediate step between the two concentration levels indicates, however, that, at the c/t interface, an equilibrium between CaO-poor c-ZrO_2 and CaO-rich t-ZrO_2 is established. Therefore, a mechanism of mass transport and CaO concentration gap formation identical with that described for the system Al_2O_3-MSZ can be hypothesized.

Conclusions

(1) The major aim of this work was to compare diffusion processes and solid-state reactions in the Al_2O_3-stabilized ZrO_2 composites.

(2) Within the Al_2O_3 matrix, Y_2O_3 remains in a stable ZrO_2 solid solution; no reaction and no solubility of Y_2O_3 in Al_2O_3 could be observed at temperatures up to 1600°C. The solubility of ZrO_2 in Al_2O_3, as well as that of Al_2O_3 in YSZ, was estimated to be in the range of 0.1 wt%.

(3) MSZ reacts with Al_2O_3, forming $MgAl_2O_4$ spinel. The solubility of ZrO_2 in the reaction product is negligible, whereas about 0.4 wt% of Al_2O_3 was found in the MSZ interphase region.

(4) The first reaction product in the system Al_2O_3-CSZ is $CaO \cdot 6Al_2O_3$. Once the total amount of Al_2O_3 is consumed in CA_6 formation, CA_2 is formed, both aluminates containing considerable amounts of ZrO_2.

(5) At the beginning of the reaction between Al_2O_3 and MSZ or CSZ, the stabilizing oxide is consumed from c-ZrO_2 solid solution. So long as Mg- or Ca-ion diffusion can follow the rate of reaction, the reaction product is in equilibrium with c-ZrO_2 containing a lower limiting amount of stabilizing oxide. Once the mass transport becomes too slow, a layer of t-ZrO_2 containing a low MgO or CaO content starts spreading from the contact area, resulting in the stabilizing oxide concentration gap.

Acknowledgment

This work was supported by Research Councile of Slovenia.

References

[1] K. C. Radford and R. J. Bratton, "Zirconia Electrolyte Cells, Part 1: Sintering Studies," *J. Mater. Sci.*, **14** [1] 59–65 (1979).

[2] E. P. Butler and J. Drennan, "Microstructural Analysis of Sintered High-Conductivity Zirconia with Al_2O_3 Additions," *J. Am. Ceram. Soc.*, **65** [10] 474–78 (1982).

[3] F. F. Lange, "Transformation Toughening, Parts I–V," *J. Mater. Sci.*, **17** [1] 225–62 (1982).

[4] P. F. Becher and V. J. Tennery, "Fracture Toughness of Al_2O_3-ZrO_2 Composites," Rept. ORNL WS-16366, 1981.

[5] N. Burlingame, "Toughening of Zirconia Composites"; M. S. Thesis, University of California, Berkeley, 1980.

[6] M. J. Bannister, "Development of the $SIRO_2$ Oxygen Sensor: Ternary Phase Equilibria in the System ZrO_2-Al_2O_3-CaO," *J. Austral. Ceram. Soc.*, **17** [1] 21–24 (1981).

[7] N. Claussen, "Transformation-Toughened Ceramics," *Z. Werkstofftechn.*, **13**, 138–47 and 185–96 (1982).

[8] W. D. Tuohig and T. Y. Tien, "Subsolidus Phase Equilibria in the System ZrO_2-Y_2O_3-Al_2O_3," *J. Am. Ceram. Soc.*, **63** [9–10] 595–96 (1980).

[9] E. M. Levin and H. F. McMurdie; Fig. 4370 in Phase Diagrams for Ceramists, 1975 Supplement. The American Ceramic Society, Columbus, OH, 1975.

[10] H. Bernard, "Sintered Stabilized Zirconia Microstructure and Conductivity," Rept. CEA-R-5090, CEN Saclay, France, 1981.

[11] T. S. Witkowski, L. Schoenlein, and A. H. Heuer, "Surface Transformation in Optimally Aged Mg-PSZ," for abstract see *Am. Ceram. Soc. Bull.*, **57** [9] 831 (1978).

[12] E. M. Levin and H. F. McMurdie; Fig. 4308 in Phase Diagrams for Ceramists, 1975 Supplement. The American Ceramic Society, Columbus, OH, 1975.

[13] C. F. Grain, "Phase Relations in the ZrO_2-MgO System," *J. Am. Ceram. Soc.*, **50** [6] 288–90 (1967).

[14] V. S. Stubican and S. P. Ray, "Phase Equilibria and Ordering in the System ZrO_2-CaO," *J. Am. Ceram. Soc.*, **60** [11–12] 534–37 (1977).

Section V
Electrolytic Properties and Applications

Defect Structure and Transport Properties of ZrO_2-Based Solid Electrolytes... 555
    J. F. Baumard and P. Abelard

Microstructural-Electrical Property Relationships in High-Conductivity Zirconias............................... 572
    E. P. Butler, R. K. Slotwinski, N. Bonanos, J. Drennan, and B. C. H. Steele

Low-Temperature Properties of Samaria-Stabilized Zirconia... 585
    M. Goge, G. Letisse, and M. Gouet

Influence of Impurities in Solid Electrolytes on the Voltage Response of Solid Electrolyte Galvanic Cells......... 591
    T. Reetz, H. Näfe, and D. Rettig

Low-Temperature Behavior of ZrO_2 Oxygen Sensors.......... 598
    S. P. S. Badwal, M. J. Bannister, and W. G. Garrett

Life and Performance of ZrO_2-Based Oxygen Sensors......... 607
    B. Krafthefer, P. Bohrer, P. Meonkhaus, D. Zook, L. Pertl, and U. Bonne

Accurate Monitoring of Low Oxygen Activity in Gases with Conventional Oxygen Gauges and Pumps................ 618
    J. Fouletier, E. Siebert, and A. Caneiro

Collaborative Study on ZrO_2 Oxygen Gauges................. 627
    A. M. Anthony, J. F. Baumard, and J. Corish

Computer-Controlled Adjustment of Oxygen Partial Pressure... 631
    F. Vizethum, G. Bauer, and G. Tomandl

Oxygen Sensing in Iron- and Steelmaking.................... 636
    D. Janke

Mixed Ionic and Electronic Conduction in Zirconia and Its Application in Metallurgy................................ 646
   M. Iwase, K. T. Jacob, and I. Ichise

ZrO_2 Oxygen and Hydrogen Sensors: A Geologic Perspective... 660
   G. C. Ulmer

Application of Zirconia Membranes as High-Temperature pH Sensors.. 672
   L. W. Neidrach

Preparation and Operation of Zirconia High-Temperature Electrolysis Cells for Hydrogen Production................. 685
   E. Erdle, A. Koch, W. Schäfer, F. J. Esper, and K. H. Friese

Defect Structure and Transport Properties of ZrO$_2$-Based Solid Electrolytes

J. F. BAUMARD AND P. ABELARD

Centre de Recherches sur la Physique des Hautes Températures C.N.R.S.
45045 Orleans-Cedex, France

Zirconia-based solid electrolytes, which are among the best known oxygen-ion conductors, exhibit features which are similar to other doped fluorite-structured oxides with respect to oxygen transport. In stabilized zirconia and heavily doped ceria or thoria, the electrical conductivity is sensitive to the nature of the dopant introduced, and it goes through a maximum when the concentration of dopant is varied. Recent investigations indicate that defect interactions, mainly dopant(s)-vacancy attractive effects, play an important role in the transport properties of grossly defective solid solutions. Frequency dependence of the conductivity observed in impedance spectroscopy is explained on the basis of variable barrier heights for the diffusion process, due to differences in the local defect interactions.

Interest in the fluorite-structured oxides, of which stabilized zirconia is one, has been stimulated by the high anionic conductivity of these materials when they are doped with low-valence cations, such as Ca^{2+}, Y^{3+}, or rare-earth cations. It is recognized that electrical conductivity takes place through the movement of oxygen ions via vacancies that compensate the lower valence of the di- or trivalent impurities which enter the lattice at cation sites.[1] The additives are generally present in the mixed oxide in the proportion of 5–15 mol%, so as to give solid solutions with formulas $Zr^{4+}_{1-x}M^{2+}_xO_{2-x}$ and $Zr^{4+}_{1-2x}M^{3+}_{2x}O_{2-x}$, respectively, in the case of zirconia. The exceptional conductivities of these materials are attributed to two main factors. First, fluorite structure is suitable for anionic migration, with relatively low activation energies.[2] Second, a large concentration of oxygen vacancies, equivalent to a large concentration of potential carriers, is introduced within the anion sublattice for the charge compensation of the dopant. In the last two decades, a lot of detailed studies concerning defect equilibria and ionic conductivity of zirconia have been conducted with respect to the nature and concentration of dopant, temperature, and oxygen partial pressure. Several recent reviews on these topics are available.[3–7] At the same time, development of experimental techniques, such as impedance spectroscopy which allows accurate determination of conductivity,[7] has provided sets of reliable data on the bulk behavior, with a negligible or small influence of other limiting steps in the oxygen transport, for instance at the grain boundaries or at the electrode-electrolyte interfaces. The numerous experimental works on stabilized zirconia have established several essential features: (1) Ionic conductivities follow more or less approximately an Arrhenius law over a wide temperature range, with activation energies of ≈ 1 eV; (2) the variation of electrical conductivity with dopant content exhibits a maximum which occurs around 12–13 mol% in CaO-stabilized zirconia and around 8–9 mol% in Y_2O_3-stabilized zirconia; (3) the decrease of the conductivity with increasing dopant content is accompanied by an increase in the activation energy; (4) for a given concentration of aliovalent species required to stabilize the fluorite structure, the conductivity depends on the nature of the dopant.

During the past few years, substantial improvement has been achieved in the understanding of the transport properties of these grossly defective solid solutions. It has become possible to handle, more than qualitatively, effects of defect interactions that were suspected to play an important role in the carrier motion. In addition, reliable data have now been obtained for other compounds belonging to the same series of fluorite-structured oxides, such as doped ceria[6,8] or thoria.[9] Pure ceria and thoria adopt the fluorite structure, a fact that enables the investigation of the effects of doping over a wider concentration range. With respect to zirconia, this represents a clear advantage from a basic point of view, since difficulties inherent in the complex phase diagrams of ZrO_2-based compounds[10] can be avoided.

Effect of Dopant Concentration

It has been known for some time that isothermal plots of ionic conductivity of doped zirconia exhibit maxima for a dopant concentration of MO or M_2O_3. These maxima lie in the vicinity of the zirconia-rich boundary of the cubic-phase field.[1] Some examples taken from the literature are given in Fig. 1. That the conductivity goes through a maximum as a function of concentration is generally

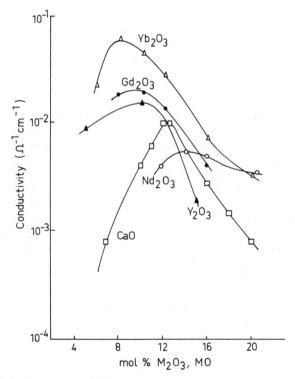

Fig. 1. Variation of conductivity with dopant concentration for various doped zirconias (T = 1080 K).

attributed to defect interactions in the grossly defective solid solution. Using the well-known symbolism of Kröger and Vink, one can write the incorporation reaction of CaO and Y_2O_3 as

$$CaO \xrightarrow{ZrO_2} Ca_{Zr}'' + V_O^{\cdot\cdot} + O_O^x \tag{1}$$

$$Y_2O_3 \xrightarrow{ZrO_2} 2Y_{Zr}' + V_O^{\cdot\cdot} + 3O_O^x \tag{2}$$

Starting with the hypothesis of a more or less random distribution of dopant ions in the cation sublattice, it is expected that the mobile oxygen vacancies should experience two types of interaction, namely, repulsive with respect to other positively charged carriers and attractive with respect to aliovalent cations. Several attempts have been made to explain these effects in the literature. Attractive interactions could result first in the formation of dopant-vacancy complexes, the formation of which is described by pseudochemical equilibria as

$$Ca_{Zr}'' + V_O^{\cdot\cdot} \rightleftharpoons (Ca_{Zr}, V_O)^x \tag{3}$$

$$Y_{Zr}' + V_O^{\cdot\cdot} \rightleftharpoons (Y_{Zr}, V_O)^{\cdot} \tag{4}$$

$$2Y_{Zr}' + V_O^{\cdot\cdot} \rightleftharpoons (2Y_{Zr}, V_O)^x \tag{5}$$

Even if the mobility of vacancies trapped into associates is strongly diminished when compared with that of "free" vacancies, this argument does not explain the decrease of conductivity with the dopant content observed for stabilized zirconias, since only the relative and not the absolute number of free defects decreases. Furthermore, as pointed out by Catlow,[11] such an approach rests on two important assumptions, viz., that the structure may be described in terms of free defects and localized associates which are separated by regions of perfect lattice and that the concentrations of these species may be treated by the mass action law formalism within the framework of thermodynamics. It is doubtful if, for the defect concentrations of interest here, the concept of defect associate is still valid when the dopant concentration is so large that, on an average, every vacancy has one such aliovalent cation in its immediate neighborhood.

Schmalzried[12] attempted to handle electrostatic interactions of oppositely charged defects with the Debye-Hückel approach at low dopant content ($x \lesssim 10^{-4}$) and perfect ordering for larger concentrations ($x \gtrsim 10^{-2}$). Results of his calculations justify a continuous increase in the activation energy of the ionic conductivity with the dopant concentration. Even qualitatively, however, these models cannot explain the behavior observed over the whole set of compositions since, whereas the conductivity goes through a maximum, the activation energy first decreases and then increases only beyond the maximum in question (Fig. 2). Possible effects of structural and microstructural changes at the phase boundary between the phase fields of cubic and cubic + tetragonal forms, lying close to the composition with maximum conductivity, were also conjectured.[1] Although such effects cannot be ignored a priori for zirconia-based electrolytes, it is likely that other reasons must exist, as similar trends are also observed in the variation of transport properties for some fluorite-structured solid solutions, for instance, doped ceria, in the domain of a single-phase, cubic oxide.[6,8]

Many times the tendency to ordering in these highly defective solids has been assumed to play an important role in transport properties.[6] It is understood here that oxygen vacancies entering as building units into ordered phases, e.g., microdomains, would be almost perfectly trapped and thus would not contribute significantly to the anionic mobility. Unfortunately, for doped oxides that adopt the

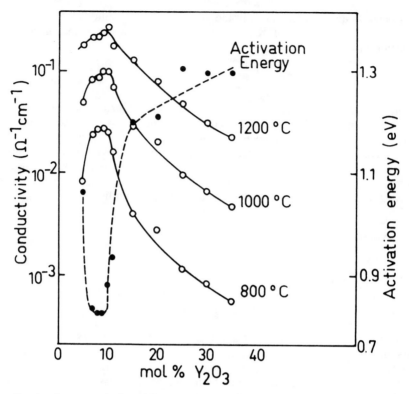

Fig. 2. Isotherms of electrical conductivity and the variation of activation energy for Y_2O_3-doped zirconia.

fluorite structure, our knowledge of the details of the defect structure is much less than for nonstoichiometric pure compounds, such as CeO_{2-x} or PrO_{2-x}, where long-range order of vacancies is demonstrated by the existence of superstructures.[13] Basically, the evolution toward equilibrium, at least below 1000°C, is complicated by the very low mobilities of the cations, which prevent complete formation of the thermodynamically preferred structures.[2] In the case of zirconia, ordered phases such as $CaZr_4O_9$ are now known to appear on the CaO-rich side of the cubic-phase field, during aging in the vicinity of 1000°C, but for most other systems, including Y_2O_3-doped ZrO_2, details of the microstructure are still controversial.[2,14,15] It must be recognized that a tendency to ordering is compatible with the observation that the anionic conductivity, for instance in CaO-doped ZrO_2, falls off at high CaO content. If the resistivity of the small, ordered domains is large, the greater proportion of the current flowing in the sample would be carried in the cubic matrix. As the composition becomes closer to $CaZr_4O_9$, increasing proportion of microdomains will block the paths in the fluorite matrix, thus causing a rapid decrease in the conductivity. While it seems likely, from aging experiments, that ordering causes a net decrease in the conductivity, there is no clear evidence that the large variation of conductivity with composition (Figs. 1–3) results from this tendency to order. In a recent paper, Butler et al.[16] used atomistic calculations to differentiate the energetic stability of microdomains with respect to isolated de-

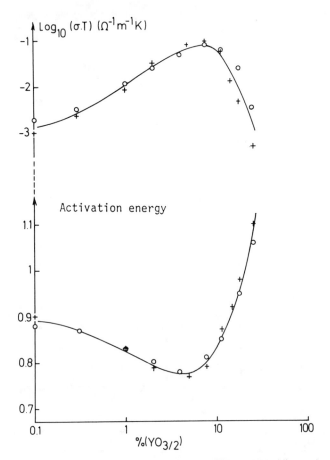

Fig. 3. Variations of electrical conductivity ($T = 450$ K) and activation energy for Y_2O_3-doped ceria.

fects, but the results were inconclusive. Thus, at this stage, it seems reasonable to choose, for a quantitative treatment, the simplest model for the distribution of aliovalent cations, namely, a random distribution of dopants frozen in from the high temperature at which the materials are prepared. It will be shown that some insight may indeed be obtained with this simple assumption, which clearly would be revised to enter the details of transport in aged materials.

Activation Energy of Transport

Dopant-vacancy interactions lead to the typical variation of electrical conductivity with composition. In the cation sublattice, for low dopant concentrations, most aliovalent ions will be isolated in the matrix, and very few of them will have a common oxygen site as a nearest neighbor. Thus any attractive interaction between oxygen vacancies and dopant ions will result in the formation of complexes involving single vacancies bound to the isolated cations in question. Taking the example of a Y_2O_3-containing material, at sufficiently low temperature, essentially one-half of the dopant will be under the associated form $(Y_{Zr}, V_O)\cdot$, while the other

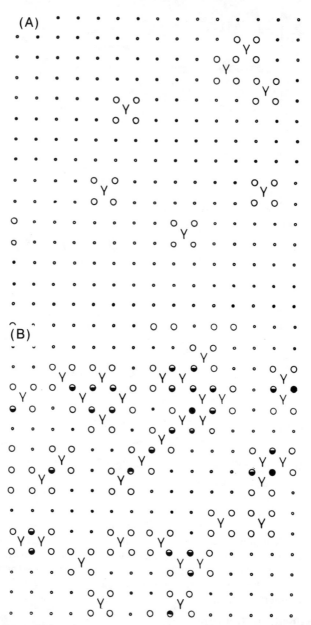

Fig. 4. (A) Two-dimensional representation of a random distribution of Y ions in a square-centered cubic lattice. For clarity, only cationic sites occupied by yttriums are depicted. Oxygen sites that enter the coordination polyhedra of dopant are represented by large circles (O) and those that are surrounded by four normal tetravalent cations by smaller circles (O). Yttriums occupy 4% of cation sites. (B) The representation is the same when yttriums occupy 16% of the cation sites. Oxygen sites surrounded by four tetravalent cations are denoted by (O) and those that are nearest neighbours of one, two, and three yttriums are denoted by (O), (◐), and (●), respectively.

will be present as the Y'_{Zr} species (remember that two yttriums are necessary to create one oxygen vacancy); in addition, very few vacancies resulting from the dissociation of the associates will be free in the lattice. Experimental evidence in favor of the existence of these complexes has been given by Nowick and his group from dielectric and inelastic relaxation phenomena on Y_2O_3-doped ceria.[17,18] When the defect concentration increases, there is a larger probability that a given anionic site becomes the nearest neighbor of two, and maybe three or even four, dopant cations. These sites are expected to act as deeper traps, at least from simple Coulombic considerations, for the oxygen vacancies, the mean mobility of which should then decrease. A two-dimensional representation of a random distribution of Y dopant in a square-centered lattice is depicted in Fig. 4. In the case of a small content of yttria (Fig. 4(A)), very few anionic sites belong to more than one coordination polyhedra of Y^{3+} ions. On the other hand, for a larger content (16 at.% Y^{3+}), many oxygen sites (Fig. 4(B)) are nearest neighbors of two and three Y^{3+}. During long-range motion, the oxygen vacancies will thus have to jump over higher barriers to reach the saddle point.

Atomistic calculations performed by Butler et al.[16] indirectly support the idea of variable activation energies for the jump processes. The binding energies of associates, defined with respect to the total energy of isolated defects that enter the associates, are given in Table I for zirconia. From the calculations it turns out that complex defects such as $(Ca_{Zr}, V_O)^\times$ and $(Y_{Zr}, V_O)^.$ are considerably stabilized by binding energies of 0.7 and 0.3 eV, respectively. This is in agreement with the fact reported previously that in doped CeO_2 aliovalent cations and vacancies form dipoles diluted in the lattice. Complex defects such as $(2Ca_{Zr}, V_O)''$ and $(2Y_{Zr}, V_O)^\times$ appear to be even more stable by ≈ 0.35 eV. It is, then, not unrealistic to infer that activation energies for the individual jumps will differ according to the local configurations around the mobile vacancies during their long-range motion.

We recently attempted to treat the problem of interactions semiquantitatively. The *path probability method*, developed by Kikuchi,[20] formerly used to investigate diffusion in alloys[21] or beta alumina[22] was adapted to the case of doped fluorite-structured oxides.[8] For reasons that are outside the scope of this paper, it is simpler to handle the problem by considering oxygen ions as the mobile species rather than vacancies. Both kinds of interactions mentioned ahead have been included, but owing to otherwise inherent difficulties in the mathematical treatment, only short-range interactions, limited to the nearest cation and anion neighbors, were considered. Attractive interaction between dopant and oxygen vacancies was treated as a repulsive interaction between yttriums and oxygen ions when the latter entered the coordination polyhedra of the dopant. Two energetic states of the oxygen ion were defined, namely, α and β (Fig. 5). In the state α, the anion is surrounded by four tetravalent ions, such as Zr^{4+} or Ce^{4+}, while in the state β one or several impurities are nearest neighbors. State α is assumed to possess an energy reduced

Table I. Binding Energies of Various Defect Clusters in Doped Zirconia*

Defect clusters	Binding energies (eV)			
	Mg^{2+} (0.89 Å)	Ca^{2+} (1.12 Å)	Y^{3+} (1.015 Å)	Gd^{3+} (1.08 Å)
MV_O	−1.19	−0.69	−0.28	−0.17
$(M_2)V_O$	−2.1	−1.04	−0.63	−0.44
$M(V_O)_2$	−1.01	−0.78		

*The effective radii for eight-coordinated cations are given in parentheses, with $r_{Zn^{4+}} = 0.84$ Å. (Ref. 19).

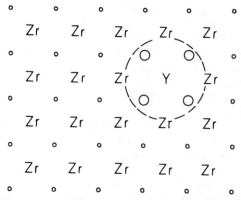

Fig. 5. Oxygen ions which occupy a site (O) close to a dopant cation possess an energy increased by W (β sites) compared to the ions that are nearest neighbors of four tetravalent cations (α sites).

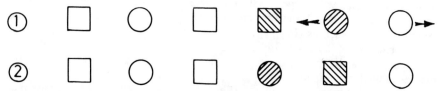

Fig. 6. Repulsive interaction between carriers favors ordering on the oxygen sublattice.

by W compared to a site β. We also included repulsive interaction between carriers. Within the framework of this model, such an interaction reflects the tendency to order in the mobile sublattice[22] (Fig. 6). Whatever the exact nature of vacancy arrays may be in heavily doped fluorites, a long-range order is precisely observed for nonstoichiometric systems, such as CeO_{2-x}, on the basis of the periodicity of a cluster comprising two vacancies at the opposite end of the [111] diagonal in one of the eight cubes of the mother fluorite structure.[13] Repulsive interaction between two oxygen ions increases their individual energy by a quantity ε.

The model of Sato and Kikuchi[22] allows the calculation of the probability of occupancy for α and β sites as well as the probability of occurrence of various pairs O_i-O_j, V_i-O_j, O_i-V_j, V_i-V_j ($i,j = \alpha, \beta$), where O and V denote an oxygen ion and a vacancy, respectively. Thus it is possible to weight the influence of the local environment of a vacancy on its transition rate. An example of a configuration is given in Fig. 7, where the barrier height is diminished by a quantity $W + 4\varepsilon$, since the oxygen ion is located on a β site before jumping and possesses four next-nearest neighbors that destabilize it by 4ε. From the calculations, it turns out that the conductivity is the product of three factors

$$\sigma = \sigma_0 F_\varepsilon F_W \qquad (6)$$

where σ_0 is the electrical conductivity in the absence of any interaction, and F_ε and F_W represent the effects of repulsive and attractive interactions, respectively.[8,23] The contribution of the factor F_ε, which remains close to unity as long as the level of

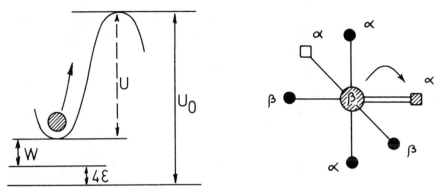

Fig. 7. Configuration for which the environment of an ion modifies the barrier height for diffusion.

cation substitution is smaller than 30–40%, does not play a significant role in the domain of compositions of interest. The variation of F_ε is in good agreement with the results obtained by Barker and Knop[24] with a blocking site model, in which no vacancies, owing to their same effective charge, can occupy nearest-neighbor anion sites. In Eq. (6), the conductivity σ_0 is given by the classical formula

$$\sigma_0 \propto \rho(1 - \rho) \exp(-U_0/kT) \tag{7}$$

where $\rho = 1 - x/2$ is the fraction of empty anionic sites and U_0 the uniform barrier height for diffusion in an otherwise perfect lattice.

$$F_W = \sum_{i,j} \left[\frac{p_O^i p_V^j}{\rho(1 - \rho)} \right] \exp(-E_{ij}/kT) \tag{8}$$

where p_O^i and p_V^j denote the probability of finding an oxygen ion and a vacancy in site i, respectively, and $\exp(-E_{ij})$ is a mobility factor related to individual processes $\alpha\alpha$, $\alpha\beta$, $\beta\alpha$, and $\beta\beta$. The total conductivity is then the sum of four contributions $\sigma_{\alpha\alpha}$, $\sigma_{\alpha\beta}$, $\sigma_{\beta\alpha}$ ($=\sigma_{\alpha\beta}$), and $\sigma_{\beta\beta}$.[8] Coulombic repulsion W between an anion and a dopant was chosen as 0.4 eV, from a rough calculation corresponding to distances of 2.5 Å (0.25 nm) between charges and a dielectric permittivity of 25. The activation energy U_0 for diffusion of the vacancy in the perfect lattice is taken as 0.6 eV, in agreement with data published in the literature.[17] According to the plots of each contribution to the conductivity (Fig. 8), the interval of concentration may be divided into three domains. At low dopant concentration, charge motion takes place essentially in the α positions. Partial conductivity $\sigma_{\alpha\alpha}$ is independent of concentration. This result is in agreement with the simple model of associates for which the equilibrium (Eq. (4)) is fully displaced to the right. In this case

$$[(Y'_{Zr}, V_O)^{\cdot}] = [Y'_{Zr}] \tag{9}$$

where the quantities between brackets denote concentrations. Application of the mass action law to Eq. (4) leads to $[V_O^{\cdot\cdot}] = C(T)$, $C(T)$ being a constant independent of dopant concentration. The activation energy of the conductivity is simply $(U_0 + W)$, the total energy needed to dissociate the pairs and move the vacancy in the α sublattice. For larger concentrations of dopant, influence of β sites builds up, and the dominant mechanism becomes the exchange between α and β positions.

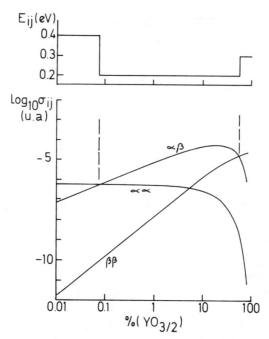

Fig. 8. Variation of various contributions to the total electrical conductivity. Mobility terms in Eq. (8) were chosen as $E\alpha\beta = W/2$, $E\alpha\alpha = 0$, $E\beta\alpha = -W/2$, and $E\beta\beta = 0.3$ eV. This choice has no direct consequence on the variation of each individual contribution owing to the form of Eq. (8).

The electrical conductivity then goes through a maximum for a theoretical concentration of about 10 mol% Y_2O_3. At higher concentrations of dopant, the probability of finding V_β-O_β pairs is an increasing function of dopant content. Strictly speaking, it would be necessary to define a larger number of energetic states in the model, especially when the probability that two (or more) dopant ions share a common site in their coordination polyhedra becomes significant. Deep trapping is expected to contribute significantly to the rapid decrease in conductivity beyond its maximum. Under these conditions, however, the model becomes more or less intractable, unless drastic approximations are made in the mathematical treatment. We preferred to simulate the effect of trapping near two or more impurities by taking a low mobility factor $E_{\beta\beta}$, assumed as 0.3 eV in Eq. (8), in agreement with the results of atomistic calculations.[16]

It is noteworthy that this model could reproduce the maximum observed in conductivity experiments as well as the variation in the activation energy found for the doped fluorite oxides (Figs. 2, 3, and 9). Conceptually, the treatment does not differ very much from that of Nakamura and Wagner,[25] who used a quasi-chemical approach, also with two energetic states 1 and 2, assumed to be relatively free and relatively bound, respectively, and for which different mobilities are assigned to various 1-1, 1-2 (or 2-1), and 2-2 transitions. However, the basic reasons are explicit and appear more clearly in the present development. It must also be noticed that, using the similar idea of variable barrier heights, a computer simulation was able to derive the remarkable feature of a maximum value in the electrical conductivity.[11]

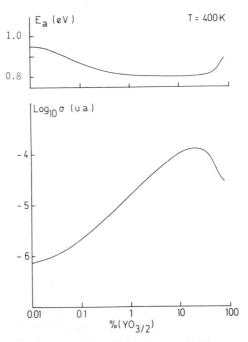

Fig. 9. Calculated variations of electrical conductivity and activation energy within the framework of the model.

Frequency-Dependent Conductivity and the Distribution of Jump Probabilities

Impedance spectroscopy is now classically used to determine the respective contributions to the impedance of a cell containing an ionic conductor.[7] Its development, after the original work of Bauerle,[26] who applied this technique for the first time to stabilized zirconia, has allowed the acquisition of considerable data on the limiting steps to the transport, e.g., bulk resistivity, transfer through the grain boundaries, and mass and charge transfer at the electrode-electrolyte interfaces. As a first, but often good, approximation, data points due to the bulk contribution are distributed in the complex impedance representation over semicircles going through the origin of the axes (Fig. 10). Thus an equivalent circuit, consisting of a parallel assembly of a resistance and a capacitance, which affords a semicircle in the complex impedance representation when the frequency of the applied voltage varies, is often used to simulate the behavior of the bulk. However, experience proves that the semicircles obtained with ionic conductors as stabilized zirconia are depressed by an angle α below the real axis (Fig. 10). This implies that the electrical conductivity and/or the dielectric properties are dispersive. Assuming a homogeneous medium, a total conductance σ_T may be defined as

$$\sigma_T(\omega) = \sigma(\omega) + j\varepsilon_0\varepsilon(\omega)\omega \tag{10}$$

where all symbols have evident significance. The real and imaginary parts of the conductivity are interrelated by the relations of Kramers-Kronig.[27] The dc conductivity is now expressed as

$$\sigma_0 = \left(\frac{Nq^2\langle r^2 \rangle}{6kT}\right)\left(\frac{1}{\bar{t}}\right) \tag{11}$$

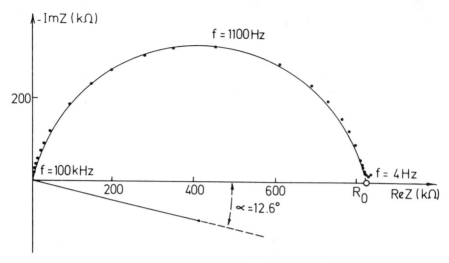

Fig. 10. A semicircle obtained in the complex impedance representation for a stabilized zirconia single crystal (T = 519 K).

where N and q are the concentration and charge of the oxygen vacancies, respectively, and $\langle r^2 \rangle$ stands for their mean-square displacement after elementary jumps. In this formula, the jump frequency has been replaced by the inverse of the mean residence time $\bar{t}$ on a site. The parameter $\bar{t}$ was calculated from Eq. (11) for the shortest jump distances in the lattice, and then a normalized curve $\sigma(\omega)/\sigma_0$ was plotted vs $\omega \bar{t}$ in Fig. 11. Points define a single curve within experimental error. It is then clear that the dispersion of total conductivity $\sigma_T(\omega)$ must be related to long-range motion of vacancies and not to frequency-dependent dielectric properties, due, for instance, to reorientation of dipoles formed by associates such as $(Y'_{Zr}, V_O)^{\cdot}$.

This is not entirely unexpected if one considers that, during long-range motion, carriers will experience barriers with variable heights to jump from one site to another. A distribution of jump frequencies will then cause a dispersion in the conductivity.[8] To interpret the frequency dependence, we adapted the formalism of Scher and Lax,[28] which is based on a generalized theory of the mobility in the approach of continuous random walk of carriers on a lattice. Basically, the disorder is contained in a single function $Q(t)$ which represents the probability that the particle remains on its site for a time t before it moves. The conductivity variation is related to the frequency of the applied field as follows:[27]

$$\tilde{Q}(\omega)/\bar{t} = [\sigma(\omega)/\sigma_0 + j\omega\bar{t}]^{-1} \tag{12}$$

where $\tilde{Q}(\omega)$ is the Fourier transform of $Q(t)$. The distribution $Q(t)$ results from a distribution in jump frequencies $F(W)$:

$$Q(t) = \int_0^\infty F(W) \exp(-Wt) \, dW \tag{13}$$

The dynamics of the motion is incorporated into $Q(t)$, or equivalently $\tilde{Q}(\omega)$, from which one can deduce $F(W)$ by computational techniques. It is found that a logarithm of normal distribution of jump frequencies fairly fits the variation of $\tilde{Q}(\omega)/\bar{t}$, as shown in Fig. 12. The distribution of barriers is then nothing but a

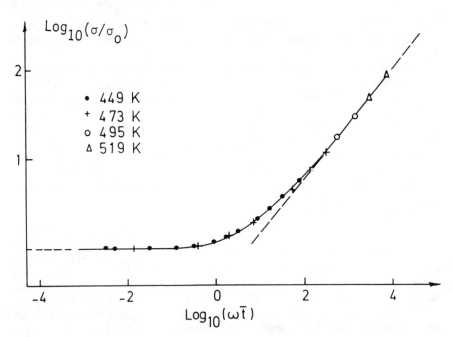

Fig. 11. Variation of the real part of total conductivity with frequency.

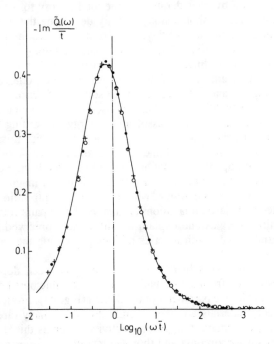

Fig. 12. A plot of the imaginary part of $Q(\omega)$ vs frequency. Data obtained at several temperature ($\bigcirc$, 473 K; $\bullet$, 495 K; +, 519 K) are located on a single curve within experimental error when coordinates are normalized by factors $1/\bar{t}$ and $\bar{t}$ (see Eqs. (11) and (12)).

Gaussian curve, with half-width of 0.13 eV. This value is in good agreement with that proposed by Baudry et al.[29] from ^{181}Ta gamma-gamma angular correlation studies.

This distribution of energy barriers justifies the previous statements about the effects of the environment of a vacancy on its transition rate toward neighboring sites. It explains why, in the complex impedance representation, data points are distributed over semicircles that are depressed below the real axis. In agreement with the discussion in the previous section, it shows that a complete description of the dynamics of vacancies in heavily doped materials probably necessitates a large set of activation energies rather than a spectrum containing a few discrete values. For this purpose, it would clearly be of interest, in the future, to determine the spatial extent of screened interactions between the defects contained in the lattice.

Influence of the Nature of Dopant

Qualitatively, similar trends are observed for each dopant regarding the variation of electrical conductivity with oxygen vacancy concentration for oxide fluorites. However, for a given host lattice, the conductivity varies largely according to the nature of the dopant cation. For instance, Y_2O_3-stabilized zirconia is a better ion conductor than a CaO-stabilized material. It has also been established that the "best" dopant differs according to the host matrix, e.g., ZrO_2, CeO_2, or ThO_2. For instance, scandium oxide, Sc_2O_3, yields the highest conductivities for ZrO_2-based electrolytes,[30] whereas the performances of Sc_2O_3-doped ceria are rather poor.[31]

Kilner and Brook[32] suggested that the differences between the fluorite materials could be explained by consideration of the total energy for oxygen diffusion. The preexponential factor should not primarily determine the differences. They took the total energy as the sum of the energy of migration of an isolated vacancy and the energy needed to dissociate a defect pair containing a vacancy and an impurity. This holds, according to previous considerations, for low dopant concentrations, when the concept of associates is valid. However, one can compare the changes in this regime and extrapolate it with some confidence to heavily doped specimens (Fig. 1). The differences that are observed in the total activation energy would reflect mostly changes in the association energy. According to the results of atomistic calculations, the latter was found to vary as the size mismatch between the host and dopant cations (Table I and Figs. 13 and 14). Variation of the binding energy is steeper for dopant radii that are smaller than the host. Similarity of this curve with that obtained for the binding energies between magnesium vacancies and various isovalent cations in MgO[33] where there is no contribution of Coulombic effects, suggested that elastic relaxations around the defect pairs could play a major role in the binding energies and explain the differences observed in the conductivities of materials doped with a series of ions possessing the same valency as rare-earth cations.

Finally, according to Kilner,[33] good matching between dopant oxide and matrix can be estimated from a plot that compares the cubic host lattice parameter and the pseudocubic lattice parameter of the dopant (Fig. 14). From this plot, it can be seen why CeO_2 materials are conveniently doped by many rare-earth oxides, preferably Gd_2O_3, to achieve a large conductivity, whereas this doping becomes much more difficult for zirconia and thoria compounds.

Conclusion

An attempt was made to explain the complex variations of ionic conductivity of doped fluorite oxides, including stabilized zirconia, with respect to the oxygen

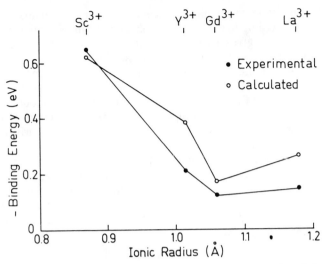

Fig. 13. Experimental and calculated binding energies for defect pairs in doped CeO_2 (taken from Ref. 32).

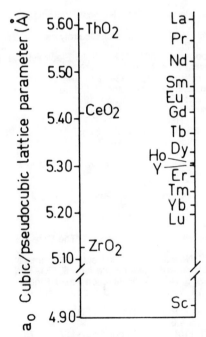

Fig. 14. Schematics of the lattice parameters for the fluorite oxides (extrapolated in the case of ZrO_2) and pseudocubic lattice parameters for rare-earth oxides (taken from Ref. 32).

vacancy concentration and the nature of dopant. The models used here extend the conventional concept of impurity-vacancy associates which, presumably, holds only at low dopant content. Electrical conductivity of these solid electrolytes proceeds by a multimode mechanism, since the height of potential barriers for diffusion varies according to the local environment of a mobile anion. Distribution of activation energies gives rise to a distribution of transition rates, which is responsible for the frequency dependence of the conductivity.

Although there is limited hope that quite a substantial improvement in the electrical performances of these materials may indeed be achieved during the next few years, this group of fluorite oxides will probably receive a lot of attention due to its specific interest as model compounds for ionic conduction. For instance, limitation of transport by grain boundaries, which appears clearly in complex impedance measurements, deserves future investigations, in conjunction with structural and microanalytical studies of the intergranular area. Also, regarding the bulk properties, no quantitative, or at least semiquantitative, attempt has been made to take account of order, especially in the cation distribution, within a treatment of transport properties. Obviously, this calls for complementary work, the results of which would undoubtedly be useful for modeling of the behavior of many other superionic conductors.

References

[1] E. C. Subbarao; pp. 1–24 in Advances in Ceramics, Vol. 3. Edited by A. H. Heuer and L. W. Hobbs. The American Ceramic Society, Columbus, OH, 1981.

[2] C. R. A. Catlow; pp. 61–99 in NonStoichiometric Oxides. Edited by O. Toft Sorensen. Academic Press, New York, 1981.

[3] T. H. Etsell and S. N. Flengas, "The Electrical Properties of Solid Oxide Electrolytes," Chem. Rev., 70, 339–76 (1970).

[4] T. Takahashi; pp. 989–1051 in Physics of Electrolytes, Vol. 2. Edited by J. Hladik. Academic Press, New York, 1972.

[5] R. M. Dell and A. Hooper; pp. 291–312 in Solid Electrolytes. Edited by P. Hagenmuller and W. Van Gool. Academic Press, New York, 1978.

[6] J. A. Kilner and B. C. H. Steele; pp. 233–69 in Ref. 2.

[7] M. Kleitz, H. Bernard, E. Fernandez, and E. Schouler; pp. 310–36 in Ref. 1.

[8] P. Abelard; Ph.D. Thesis, Orléans, France, 1983.

[9] A. Hammou, J. Chim. Phys., 72, 439–47 (1975).

[10] V. S. Stubican and J. R. Hellmann; pp. 25–36 in Ref. 1.

[11] C. R. A. Catlow; pp. 36–44 in Computer Simulation in the Physics and Chemistry of Solids. Edited by C. R. A. Catlow, W. C. Mackrodt, and V. R. Saunders. The Science Research Council, Daresbury, England, 1980.

[12] H. Schmalzried, "On Correlation Effects of Vacancies in Ionic Crystals," Z. Phys. Chem. (Wiesbaden), 105, 47–62 (1977).

[13] O. Toft Sorensen; pp. 1–59 in Ref. 2.

[14] J. B. Cohen, F. Faber, Jr., and M. Morinaga; pp. 37–46 in Ref. 1.

[15] H. J. Rossell; pp. 47–63 in Ref. 1. See also: L. H. Schoenlein, L. W. Hobbs, and A. H. Heuer, "Precipitation and Ordering in Calcia and Yttria-Stabilized Zirconia," J. Appl. Crystallogr., 13, 375–79 (1980).

[16] V. Butler, C. R. A. Catlow, and B. E. F. Fender, "The Defect Structure of Anion Deficient ZrO_2," Solid State Ion., 5, 539–42 (1981).

[17] Da Yu Wang and A. S. Nowick, "Dielectric Relaxation from a Network of Charged Defects in Dilute $CeO_2:Y_2O_3$ Solid Solutions," Solid State Ion., 5, 551–54 (1981).

[18] M. P. Anderson and A. S. Nowick, "Relaxation Peaks Produced by Defect Complexes in Cerium Dioxide Doped with Trivalent Cations," J. Phys. (Paris), 42, Suppl. No. 10, C5, 823–28 (1981).

[19] R. D. Shannon and C. T. Prewitt, "Effective Ionic Radii in Oxides and Fluorides," Acta Crystallogr., Sect. B, 25, 925–46 (1969).

[20] R. Kikuchi, "The Path Probability Method," Suppl. Prog. Theor. Phys., 35, 1–64 (1966).

[21] R. Kikuchi and H. Sato, "Substitutional Diffusion in an Ordered System," J. Chem. Phys., 51, 161–82 (1969).

[22] H. Sato and R. Kikuchi, "Cation Diffusion and Conductivity in Solid Electrolytes," J. Chem. Phys., 55, 677–701 (1971).

[23] P. Abelard and J. F. Baumard; unpublished work.

[24] W. W. Barker and O. Knop, "Statistics with Constraint for Simple and Face-centred Cubic Arrays: Application to Ionic Conductivity and Non-Stoichiometry in Fluorite Oxides," Proc. Br. Ceram. Soc., 19, 15–27 (1971).

[25] A. Nakamura, and J. B. Wagner, Jr., "Defect Structure, Ionic Conductivity, and Diffusion in Calcia-Stabilized Zirconia," *J. Electrochem. Soc.*, **127**, 2325–33 (1980).

[26] J. E. Bauerle, "Study of Solid Electrolyte Polarization by a Complex Admittance Method," *J. Phys. Chem. Solids*, **30**, 2657–70 (1969).

[27] P. Abelard and J. F. Baumard, "Study of the d.c. and a.c. Electrical Properties of an Yttria-Stabilized Zirconia Single Crystal," *Phys. Rev. B*, **26**, 1005–17 (1982).

[28] H. Scher and M. Lax, "Stochastic Transport in a Disordered Solid. I. Theory," *Phys. Rev. B*, **7**, 4491–4519 (1973).

[29] A. Baudry, P. Boyer, and A. L. De Oliveira, "^{181}Ta Gamma-Gamma Angular Correlation Study of Oxygen Self-Diffusion in Heavily Defective Solids," *J. Phys. Chem. Solids*, **43**, 871–79 (1982).

[30] T. M. Gur, I. D. Raistrick, and R. A. Huggins, "Ionic Conductivity of 8 mol% Sc_2O_3-ZrO_2 Measured by Use of Both A.C. and D.C. Techniques," *Mater. Sci. Eng.*, **46**, 53–62 (1980).

[31] R. Gerhardt-Anderson and A. S. Nowick, "Ionic Conductivity of CeO_2 with Trivalent Dopants of Different Ionic Radii," *Solid State Ion.*, **5**, 547–50 (1981).

[32] J. A. Kilner and R. J. Brook, *Solid State Ion.*, **6**, 237–52 (1982).

[33] J. A. Kilner, "The Role of Dopant Size in Determining Oxygen Ion Conductivity in the Fluorite Structure Oxides," Proceedings of the International Conference on Non-Stoichiometric Compounds, Perpignan-Alenya, France, 1982.

Microstructural–Electrical Property Relationships in High-Conductivity Zirconias

E. P. BUTLER, R. K. SLOTWINSKI, N. BONANOS, J. DRENNAN,* AND B. C. H. STEELE

Imperial College
Department of Metallurgy and Materials Science
London SW7 2BP, England

The relationship between electrical properties and microstructure has been examined for calcia and calcia + magnesia partially stabilized zirconia as a function of thermal history at a typical two-phase aging temperature and for yttria fully stabilized zirconia as a function of small additions of SiO_2 and Al_2O_3. The evolution of the experimental complex resistivity plots has been correlated with grain-interior and grain-boundary microstructures as determined by optical, scanning, and transmission electron microscopy (at 100 and 1000 kV potentials), X-ray diffraction, and X-ray microanalysis.

The addition of aliovalent oxides, such as Y_2O_3, CaO, or MgO to ZrO_2 creates a whole range of ceramic "alloys" possessing significant ionic conductivity and good mechanical properties when composition and fabrication parameters are suitably optimized. In fully stabilized ZrO_2 (FSZ), the normal high-temperature cubic (c) phase is retained to room temperature; such materials are extensively used as solid-state electrolytes in oxygen monitors, although their mechanical strength and thermal-shock resistance can be quite poor. Mobile oxygen-ion vacancies are generated by cation dopant substitution for Zr^{4+} in the host lattice, and maximum ionic conductivity generally occurs at a solute level corresponding to the minimum quantity necessary to achieve full stabilization.[1] Conductivity increases with type of dopant: $Yb_2O_3 > Y_2O_3 > CaO > MgO$.

By contrast, partially stabilized ZrO_2 (PSZ), with a lower concentration of dopant, has superior mechanical strength, fracture toughness, and thermal-shock resistance, but possesses a lower electrical conductivity, at least in the as-fired state. Optimum mechanical properties are achieved by aging PSZ following high-temperature firing to create a microstructure consisting of a cubic matrix containing a uniform dispersion of fine metastable tetragonal (t) particles. Longer aging coarsens the t precipitates to such a size that they martensitically transform to monoclinic (m) symmetry on cooling. Thus, the microstructural progression of the grain interiors with aging time, but following cooling to room temperature, is $c \rightarrow c + t \rightarrow c + t + m \rightarrow c + m$. At grain boundaries, particles of m phase are produced at an earlier stage in this sequence, as a result of easier nucleation and enhanced growth in boundary localities.

We have demonstrated that the ac complex impedance technique,[2] employed at relatively low temperatures ($\approx 300°C$) to achieve good separation of interior and

---

*Now with CSIRO Division of Materials Science, Melbourne, Victoria 3001, Australia.

boundary components of conductivity, is particularly sensitive to microstructural changes on aging.[3] Since c, t, and m phases do not have identical electrical properties, conductivity measurements provide useful information about the aging and transformation sequences. Overaging PSZ samples results in an increase in grain-interior resistivity as the $t \rightarrow m$ transformation takes place[3,4]; therefore, large quantities of m phase are clearly undesirable in high-conductivity ZrO_2 ceramics. It has been suggested[5,6] that the t phase is also less conductive than the c phase so that the individual phase conductivities (σ) may follow the order $\sigma_c > \sigma_t > \sigma_m$. If this is correct, any PSZ should always have inferior electrical properties to its more highly doped FSZ partner. Recent indications are, however,[7] that this situation holds only for either as-fired or overaged materials and that short-term aging can induce conductivity improvements resulting in PSZ material having bulk conductivity comparable to FSZ.

Impedance analysis is particularly sensitive to the presence of second-phase particles or segregant layers at grain boundaries, and early applications of the technique to ZrO_2 ceramics demonstrated the strong grain-boundary blocking effect of SiO_2 impurities.[8] In Y-FSZ, SiO_2 has a dramatic effect on conductivity,[9] effecting decreases both in the grain-interior (bulk) component (σ_{gi}) and the grain-boundary component (σ_{gb}). The fall in σ_{gb} is generally attributed to the existence of high-resistivity grain-boundary phases inherited from liquid-phase sintering, which cause the process of conduction across the boundaries to be partially blocked. More modest decreases in total conductivity have been reported with additions of TiO_2 to Ca-FSZ[10] and Fe_2O_3 and Bi_2O_3 to Y-FSZ.[11-13] The effect of Al_2O_3 additions, however, remains contentious: Increases in σ_{gb} were reported by Bernard,[14] but decreases in both σ_{gb} and σ_{gi} noted by other investigators.[13,15] It has been suggested that Al_2O_3 may interact with SiO_2 during sintering,[16] and conductivity improvements may take place, therefore, only in slightly impure (SiO_2-containing) material.[17]

In this contribution, the relationship between electrical properties, as determined by impedance analysis, and microstructure, as characterized by transmission electron microscopy (TEM), scanning electron microscopy (SEM), and X-ray diffraction (XRD) techniques, is explored for PSZ material of CaO and (CaO + MgO) compositions and for Y-FSZ material containing deliberate additions of Al_2O_3 and SiO_2. We begin by briefly examining the ac technique and associated electrical modeling, to provide a basis for the interpretation of the results obtained.

Modeling and Measurement of ac Behavior

The ac impedance of an ionic conductor measured in a two-terminal configuration contains contributions from the interiors of the grains, the grain boundaries, and the electrode–electrolyte interfaces, which can be resolved, as shown by Bauerle.[2] Figure 1(A) shows a schematic plot of the complex impedance for a polycrystalline sample of stabilized ZrO_2. The circuit chosen by Bauerle to model the electrical behavior is shown in Fig. 1(B). It expresses the idea that migrating oxygen ions are sequentially blocked by grain interiors, grain boundaries, and electrodes and that there is no appreciable conduction along grain boundaries. It is found empirically that, in ZrO_2 ceramics, the activation energies for R_{gi} and R_{gb} are identical, although segregated grain-boundary phases are known to be poor ionic conductors and might be expected to display a higher activation energy. This effect lead Bauerle to put forward the concept of easy paths, i.e., regions of the grain boundary where intergranular contact is established. Thus R_{gb} is caused by

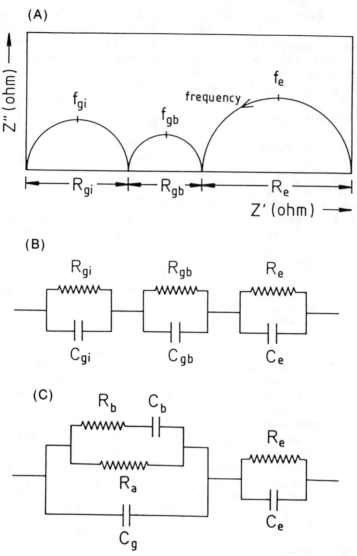

Fig. 1. Electrical response of polycrystalline zirconia: (A) schematic complex impedance plot; (B) equivalent circuit diagram proposed by Bauerle, (Ref. 2) showing components of the impedance arising from the grain interiors (r_{gi}), grain boundaries (r_{gb}), and electrodes (e); (C) equivalent circuit proposed by Schouler, (Ref. 9) showing parallel connection of geometric capacitance (C_g), nonblocked (R_a), and blocked (R_b, C_b) ionic paths.

constricted ionic paths parallel to the grain-boundary capacitance C_{gb}. A measure of the degree of blocking may be obtained from the ratio R_{gb}/R_{gi}.

A slightly different model, proposed by Schouler,[9] divides the ionic current into two parallel paths, one of which (R_b, C_b) is capacitatively blocked at the grain boundary, while the other (R_a) is not. This situation is represented by the circuit of Fig. 1(C); the geometric capacitance C_g and interfacial impedance R_e, C_e are also included. The grain-interior resistance does not appear here as one component but as the combination of R_a and R_b in parallel. The proportion of the ionic current blocked is then given by $\beta = R_b^{-1}/(R_a^{-1} + R_b^{-1})$. The same ratio β may be expressed in terms of the series model as $\beta = R_{gb}/(R_{gi} + R_{gb})$.

Both models successfully explain the identical activation energies of R_{gi} and R_{gb} and can be shown to be equivalent. Bernard and coworkers[14,18] found that the parallel model gives a more consistent description of the variation of blocking with microstructural variables, such as grain size and porosity. In many cases the choice depends on physical intuition, and either model may be used to describe ZrO_2 ceramics. In our opinion, the series model is perfectly adequate for situations in which there is a substantial amount of second phase at grain boundaries, i.e., in all but the purest materials. It has been widely applied and has the additional advantage that the R-C elements are related to the complex impedance plot in a one-to-one manner, thus simplifying the analysis.

Working with very pure materials, Verkerk et al.[19] suggested that grain-boundary impedances are caused by a thin layer of material depleted of vacancies by the action of a positive space charge and segregation of solute to the grain boundaries; they calculated values for the resistivity of this region. Such calculations are inappropriate where there are significant amounts of grain-boundary phase, since the effective area of intergranular contact is unknown.

Electrical properties can also be affected by the presence of a second phase within the grains, such as the tetragonal particles occurring in PSZ. A model derived by Fricke,[20,21] which is applicable to two-phase mixtures, treats the mixture as a dispersion of randomly oriented ellipsoids of conductivity σ_2 in a continuous medium of conductivity σ_1. The total conductivity σ_0 is then given by the relation

$$\sigma_0 = \sigma_2 + \frac{(\sigma_1 - \sigma_2)(1 - x_2)}{1 + (x_2/3)\sum_{n=1}^{3}(\sigma_1 - \sigma_2)/(\phi_n \sigma_1 + \sigma_2)}$$

where σ_0 is the complex conductivity of the dispersion, σ_1 and σ_2 are the complex conductivities of the matrix and the dispersed phase, respectively, x_2 is the volume fraction of the dispersed phase, and $\phi_{1,2,3}$ are form factors which depend on the axial ratios of the ellipsoidal particles defined by the semiaxes a, b, c, where $a \geq b \geq c$.

By introducing complex conductivities, the ac behavior of PSZ may be modeled over all frequencies for the grain interiors. More important, this equation may be used to infer the conductivity of the dispersed phase from measurements of the total conductivity.[7]

Impedance measurements were performed on a system with a Solartron 1174 frequency response analyzer over the frequency range 10^{-3}–10^6 Hz at an applied potential of 0.2 V. A measurement temperature of 300°C was chosen because at this temperature the available frequency range spans the gi, gb, and e arcs, allowing a graphical estimation of the quantities R_{gi} and R_{gb}. The impedance data were converted and plotted as the complex resistivity $\rho^* = Z^*A/d$, where A is the cross-sectional area and d the thickness of the samples. Circuit elements denoted by lower-case letters are corrected for sample shape. There was no evidence of significant electronic conductivity at the measurement temperature.

Microstructure–Electrical Property Relationships
Effect of Aging on Ca- and (Ca, Mg)-PSZ

Samples of PSZ ceramics were prepared by Anderman & Ryder Ltd., from MEL grade "S" zirconia powder mixed with a required quantity of stabilizers to give 8 mol% Ca-PSZ and 6.6 mol% plus 3.2 mol% (Ca, Mg)-PSZ. These compositions were chosen to minimize the solution treatment temperature while providing a reasonably high volume fraction of the tetragonal phase. The samples were prefired at 1770°C for 4 h, solution-treated in the c single-phase field[22] at 1830°C, and aged to various extents in the $c + t$ field at 1400°C. Further preparation details are given elsewhere.[3] The resultant ceramic bodies were consistently ≈97% dense, with no open porosity; typical final impurity content was 0.2% SiO_2 and 0.01% Al_2O_3.

Optical microscopy, SEM, TEM, energy-dispersive X-ray microanalysis (EDX), and XRD were all carried out in a standard way, as described previously.[3] Ac electrical measurements were carried out on samples studied by XRD and then treated in the usual manner.[7]

X-ray studies, on ground and 1-μm diamond-polished surfaces, showed the presence of c and t phases in the as-fired samples of both systems. The t peaks were poorly defined but on aging became sharper and remained so until aging times of typically 20 h for Ca-PSZ and 50 h for (Ca, Mg)-PSZ. Beyond these times, t peaks were no longer observed and m reflections, which first appeared after about 5 h of aging, became pronounced. The polymorph method of phase-content analysis[23] gave the precipitate phase content as ≈30% in Ca-PSZ and ≈18% in (Ca, Mg)-PSZ. Examination of peak shapes indicated less strain in the ternary system.

Optical microscopy (Fig. 2) revealed that the ceramics were large-grained (80–100 μm), with a uniform distribution of closed porosity. No significant grain growth was observed on aging. The appearance of the grain boundaries differed in the two systems, the (Ca, Mg)-PSZ ceramic exhibiting much straighter boundaries, with appreciably smaller quantities of grain-boundary phase.

Transmission electron microscopic examination of thin foils allowed study of t precipitate morphology and growth (Fig. 3) and the conditions under which the t precipitates transformed to the m phase on cooling. In Ca-PSZ, the majority of the precipitates transformed after ≈20 h of aging when their average size was ≈90 nm. By comparison, transformation took place in (Ca, Mg)-PSZ when the precipitates were ≈400 nm in size; it began at an earlier time (after only ≈10 h) and was still incomplete after 50-h aging. The most likely explanation for this result is the effect of the lower volume fraction of t precipitates in the ternary system. It is anticipated that with high t precipitate volume fractions, the interacting stress fields around individual transforming particles could cause a "domino effect" and result in transformation at quite a well-defined aging time.

Examination of the grain-boundary localities revealed large twinned m precipitates in both systems, shown by EDX to be deficient in stabilizer. EDX also confirmed that segregation of stabilizer and silica impurity to the other grain-boundary phases occurs.[3] These phases in Ca-PSZ were predominantly crystalline and evenly distributed along the whole length of the grain boundaries, while those in (Ca, Mg)-PSZ were mainly glassy and more localized, resulting in some sections of "clean" grain boundary. A feature unique to the system (Ca, Mg)-PSZ was zones free of precipitates occurring along certain grain boundaries (Fig. 4). These are cubic regions where precipitates seem to have dissolved in the path of a moving

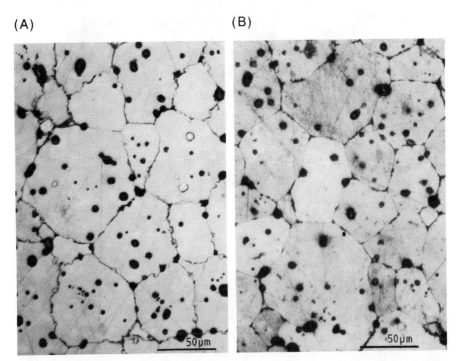

Fig. 2. Optical micrographs of as-fired, polished, and thermally etched surfaces of (A) Ca-PSZ and (B) (Ca, Mg)-PSZ.

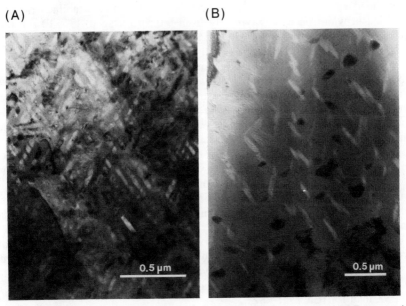

Fig. 3. Transmission electron micrographs of peak-aged PSZ ceramics (1000 kV): (A) Ca-PSZ aged for 15 h at 1400°C and (B) (Ca, Mg)-PSZ aged for 10 h at 1400°C.

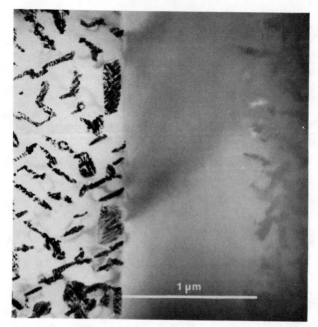

Fig. 4. Transmission electron micrograph of a precipitate-free zone in (Ca, Mg)-PSZ aged for 20 h at 1400°C. The vertical boundary to the left side of the micrograph delineates the migrated grain-boundary position.

grain boundary. Preliminary EDX analysis showed them to be stabilizer-rich; they increased in width on aging, typically from ≈0.5 (5-h aged) to ≈10 μm (50-h aged). Their mechanism of formation and effect on grain-boundary resistivity is currently under investigation.

Ac impedance measurements obtained for samples aged for various times produced complex resistivity plots, shown in Fig. 5 for Ca-PSZ and in Fig. 6 for (Ca, Mg)-PSZ. Significant changes with aging are clearly observable for both materials. Chord lengths r_{gi} and r_{gb} (Fig. 1) were estimated graphically; the electrode arc, not being a property of PSZ, was not considered. The variations of these quantities appear in Fig. 7.

The grain-boundary component of resistivity, r_{gb}, shows a gradual rise with aging in the binary system but does not vary as widely as r_{gi}. The increase can be interpreted[3] as being due to growth of monoclinic and other resistive grain-boundary phases, as evidenced by TEM. By contrast, there is no systematic increase in r_{gb} in the ternary system which has a different initial grain-boundary morphology (Fig. 2).

The resistivity of grain interiors, r_{gi} (Fig. 7) is of most interest and shows three well-defined stages of behavior: Over the first few minutes of aging there is a sharp fall (stage I), after which the value of r_{gi} stays approximately constant (stage II), until finally it rises again (stage III). Stage III is more pronounced for Ca-PSZ than for (Ca, Mg)-PSZ.

Stage I can be rationalized in terms of equilibration processes. Before beginning the 1400°C aging, the as-fired samples have a nonequilibrium structure,

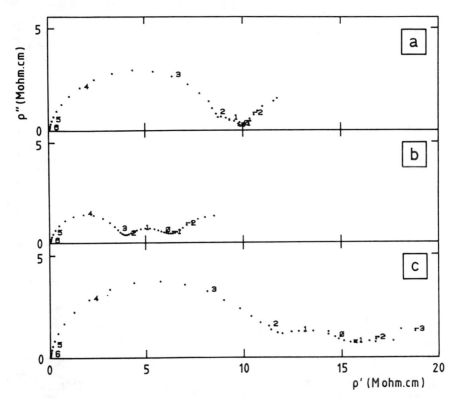

Fig. 5. Complex resistivity plots obtained at 300°C for Ca-PSZ samples: (a) as-fired, (b) aged 15 h, and (c) aged 30 h. Numbers on plots denote logarithm of frequency.

quenched-in by the relatively fast furnace cool. This implies a lower than equilibrium solute content in both c and t phases relative to that pertaining at 1400°C. Thus, in the early part of aging diffusion results in stabilizer enrichment and an increase in the volume fraction of the t precipitates. Since the conductivity of a dispersed phase, σ_0, is insensitive to size and only slightly sensitive to shape of the precipitates, as discussed in a previous publication, the fall in r_{gi} during stage I cannot be explained by particle growth or morphological changes. Compositional changes, however, may be responsible, since in the process of equilibration c becomes enriched in stabilizer, and the conductivity of c is known to increase with stabilizer content, reaching a maximum at ≈ 12.5 mol%.[24] The increase in volume fraction of t phase provides an alternative explanation only if this phase has a higher conductivity than the cubic matrix. Quantitative analysis of the stage II → stage III transition presented elsewhere[7] directly supports this suggestion, i.e., at 300°C, $\sigma_t > \sigma_c$.

Stage II may be regarded simply as a period of coarsening at constant volume fraction of t phase. The approximate constancy of r_{gi} during this stage is consistent with this interpretation, since σ_0 should be invariant with the precipitate particle size. Since the t strain energy is increasing during this period, this would imply that σ_0 is also insensitive to localized interfacial stresses in a two-phase mixture.

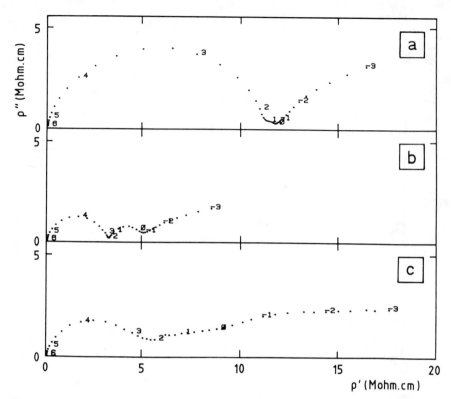

Fig. 6. Complex resistivity plots obtained at 300°C for (Ca, Mg)-PSZ samples: (a) as-fired, (b) aged 10 h, and (c) aged 60 h. Numbers on plots denote logarithm of frequency.

The large increase accompanying the transition from stage II to stage III, well defined in the binary system but also present, although less pronounced, in the ternary system, corresponds to the $t \to m$ transformation of the precipitates, as shown by XRD and TEM (see Fig. 4) and demonstrates conclusively that $\sigma_t > \sigma_m$.

The extraction of the individual conductivities of the cubic, tetragonal, and monoclinic phases has been reported[7] and indicates that $\sigma_t > \sigma_c > \sigma_m$. The implication of this finding is that the presence of a high volume fraction of t phase should be beneficial in high-conductivity zirconia ceramics. Preliminary experimental measurements on $\approx 100\%$ t material[25] confirm this relationship.

Effect of Additives

Samples of 6 mol% Y-FSZ containing additions of Al_2O_3 and SiO_2 were prepared from MEL grade SC16Y12 powder, Gallenkamp Griffin γ-grade Al_2O_3, and finely ground Angolan quartz. The Al_2O_3 and SiO_2 quantities were varied from 1 to 3 and 0.4 to 3 mol%, respectively, covering an Al_2O_3/SiO_2 range of 0.33–7.5. After extensive mixing and ball-milling, the powders were isostatically pressed into pellets and sintered at 1600°C. Theoretical densities for all samples were between 85 and 90%, and grain sizes were 10–20 μm.

Selective complex resistivity plots obtained at 300°C showing the influence of these additives on the component cord lengths r_{gi} and r_{gb} are presented in Fig. 8:

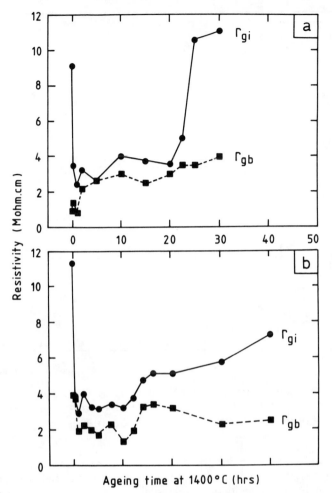

Fig. 7. Variation with aging of resistivities of grain interior (r_{gi}) and grain boundary (r_{gb}), corrected for sample shape for (a) Ca-PSZ and (b) (Ca, Mg)-PSZ.

(a) With 1% Al_2O_3 + 3% SiO_2 the ac response is characterized by a large grain-boundary arc ($r_{gb}/r_{gi} \approx 4.2$). (b) As the Al_2O_3 level is increased, this component of resistivity falls ($r_{gb}/r_{gi} \approx 1.9$), and the total conductivity of the ceramic improves by some 37%. The grain interior resistivity r_{gi} is almost unaffected. (c) Reduction in the SiO_2 level ($A/S \approx 7.5$) results in a further reduction in r_{gb} ($r_{gb}/r_{gi} \approx 1.1$) and improvement in total conductivity. Again, little change takes place in r_{gi}.

The results demonstrate that grain boundaries in Y-FSZ containing SiO_2 are initially made more conductive to oxygen ions as Al_2O_3 is added. Al_2O_3 is sparingly soluble in Y_2O_3–ZrO_2,[13,14] so discrete second-phase particles are present[16] which play a contributory role in enhancing sinterability by their ability to pin boundaries and hence reduce growth in the later stages of sintering. The conductivity im-

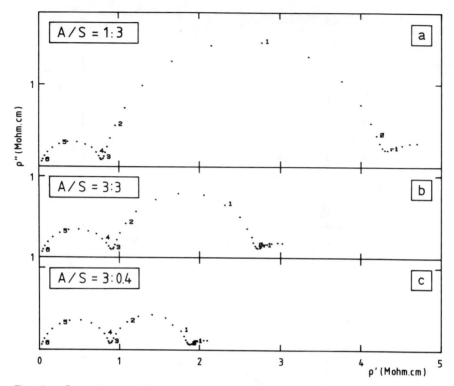

Fig. 8. Complex resistivity plots for Y_2O_3–ZrO_2 containing varying levels of Al_2O_3 and SiO_2. As the Al_2O_3/SiO_2 ratio (A/S) increases, (a) → (b) → (c), the grain-boundary component of resistivity decreases.

provements with Al_2O_3 additions demonstrated in Fig. 8 can be taken as supportive evidence for the grain-boundary scavenging model of Butler and Drennan.[16]

Although a small quantity of SiO_2 is soluble in ZrO_2,[18] it is envisaged that at the impurity levels used in this investigation, SiO_2-rich liquid phases form and wet the ZrO_2 grains at the sintering temperature. Initial consolidation is accompanied by a high driving force for grain growth. Chemical interactions between the second-phase Al_2O_3 particles and SiO_2-rich layers occur, but since boundary motion is rapid, it is expected that these encounters are relatively brief. In the later stages of densification, however, boundary motion is slower and the Al_2O_3 particles are, therefore, more effective pinners. In the time scale of pinning, it is postulated that selective partitioning of the SiO_2 to the Al_2O_3 particle interfaces takes place, a process assisted by rapid grain-boundary diffusion and wetting forces. This attraction leads to SiO_2-rich cusps adjacent the Al_2O_3 particles[26] and, in certain circumstances, to interfacial reactions producing particulate products. A high-voltage electron micrograph showing this stage of interaction appears in Fig. 9. The grain boundary runs vertically to the left side of the Al_2O_3 particle and makes an inclined angle to the surface of the thin foil. The particulate products remain to be positively identified, but mullite and zirconia, and possibly zircon, would appear to be likely candidates. Such areas are relatively uncommon in thin

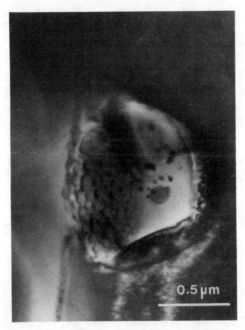

Fig. 9. An α-Al_2O_3 particle pinning and interacting chemically with a grain boundary in Y_2O_3–ZrO_2. Note inclusions within the particle and interfacial products (1000 kV) (Ref. 16).

foil samples examined at 100 kV, exacerbating the problem of microchemical identification with EDX.

The A/S value for mullite formation under equilibrium conditions is 1.5; at lower values mullite + SiO_2-rich liquid are predicted to form from the Al_2O_3-SiO_2 phase diagram at typical sintering temperatures (1600°C). It has been suggested[17] that the conflicting results of the effects of Al_2O_3 on the conductivity can be reconciled by reference to the A/S ratio. Significant improvements are expected only in relatively impure (SiO_2-containing) material when sufficient Al_2O_3 is added to avoid mullite + SiO_2-rich liquid phases, i.e., $A/S \geq 1.5$.

Acknowledgments

The financial assistance of Anderman & Ryder, Ltd. and the Science and Engineering Research Council is gratefully acknowledged.

References

[1] J. A. Kilner and B. C. H. Steele; pp. 233–69 in Nonstoichiometric Oxides. Edited by O. Toft Sorensen. Academic Press, New York, 1981.
[2] J. E. Bauerle, *J. Phys. Chem. Solids*, **30** [12] 2657–70 (1969).
[3] R. K. Slotwinski, N. Bonanos, B. C. H. Steele, and E. P. Butler; pp. 41–53 in Engineering with Ceramics, Vol. 32. Edited by R. W. Davidge. British Ceramic Society, Stoke-on-Trent, England, 1982.
[4] S. P. S. Badwal, *J. Aust. Ceram. Soc.*, **18**, 35–37 (1982).

[5]F. K. Moghadam, T. Yamashita, and D. A. Stevenson; pp. 364–79, Advances in Ceramics, Vol. 3. Edited by A. H. Heuer and L. W. Hobbs. The American Ceramic Society, Columbus, OH, 1981.
[6]F. K. Moghadam and D. A. Stevenson, *J. Am. Ceram. Soc.*, **65** [4] 213–16 (1982).
[7]N. Bonanos, R. K. Slotwinski, B. C. H. Steele, and E. P. Butler, *J. Mater. Sci.*, **19**, 785–93 (1984).
[8]N. M. Beekmans and L. Heyne, *Electrochem. Acta*, **21** [4] 303–10 (1976).
[9]E. Schouler; Ph. D. Thesis, National Polytechnic Institute of Grenoble, France, 1979.
[10]K. C. Radford and R. J. Bratton, *J. Mater. Sci.*, **14** [1] 66–69, (1979).
[11]K. Keizer, A. J. Burggraaf, and G. DeWith, *J. Mater. Sci.*, **17**, 1095–1102 (1982).
[12]R. V. Wilhelm and D. S. Howarth, *Am. Ceram. Soc. Bull.*, **58** [2] 228–32 (1979).
[13]M. J. Verkerk, A. J. A. Winnubst, and A. J. Burggraaf, *J. Mater. Sci.*, **17**, 3113–22 (1982).
[14]H. Bernard, Report CEA-R-5090, Commissariat a l'Energie Atomique, CEN-Saclay, France, (1981).
[15]M. V. Inozenitsev and M. V. Perfil'ev, *Elektrokhimiya*, **11**, 1031–36 (1975).
[16]E. P. Butler and J. Drennan, *J. Am. Ceram. Soc.*, **65** [10] 474–78 (1982).
[17]J. Drennan and E. P. Butler, *Sci. Ceram.*, **12**, 267–72 (1984).
[18]M. Kleitz, H. Bernard, E. Fernandez, and E. Schouler; pp. 310–36 in Advances in Ceramics, Vol. 3. Edited by A. H. Heuer and L. W. Hobbs. The American Ceramic Society, Columbus, OH, 1981.
[19]M. J. Verkerk, B. J. Middelhuis, and A. J. Burggraaf, *Solid State Ion.*, **6**, 159–70 (1982).
[20]H. Fricke, *Phys. Rev.*, **24**, 575 (1924).
[21]H. Fricke, *J. Phys. Chem.*, **57** 934–7 (1953).
[22]V. S. Stubican and J. R. Hellmann; pp. 25–36 in Advances in Ceramics, Vol. 3. Edited by A. H. Heuer and L. W. Hobbs. The American Ceramic Society, Columbus, OH, 1981.
[23]R. C. Garvie and P. S. Nicholson, *J. Am. Ceram. Soc.*, **55** [6] 303–305, (1972).
[24]A. Nakamura and J. B. Wagner, *J. Electrochem. Soc.*, **127**, 2325 (1980).
[25]N. Bonanos, R. K. Slotwinski, B. C. H. Steele, and E. P. Butler, *J. Mater. Sci. Lett.*, **3**, 245–48 (1984).
[26]J. Drennan and E. P. Butler, *J. Am. Ceram. Soc.*, **65** [11] C-194–C-195 (1982).

Low-Temperature Properties of Samaria-Stabilized Zirconia

M. Goge, G. Letisse, and M. Gouet

Université Paris Val de Marne
Laboratoire de Thermodynamique et d'Electrochimie des Matériaux
94010 Creteil Cedex, France

Samaria-stabilized zirconia was studied from 400° to 800°C and oxygen pressures of over 10^{-5}–1 atm (9.8×10^{-1} –9.8×10^4 Pa). Such materials show considerable electronic conductivity above 800°C but were considered in the past to be of no interest in the manufacture of potentiometric oxygen sensors and fuel cells in spite of their ionic conductivity being similar to that of calcium-stabilized zirconia (CSZ). Ionic transference numbers below 800°C using emf measurements of concentration cells and complex impedance and admittance spectra have now been determined. It appears that the conductivity is essentially ionic below 600°C.

Potentiometric sensors based on stabilized ZrO_2 are currently used between 600° and 900°C. Decreasing the working temperature would be of interest for many uses. But the lower the temperature, the greater the ZrO_2 membrane resistivity and the slower the redox reactions at the electrode. It would, therefore, be judicious to replace the three-fold point (gas-metal-oxide) by a double one (gas-mixed conductor electrolyte). We already know of such an oxide–Sm-stabilized ZrO_2, previously studied at high temperature by Gouet et al.[1] and now considered between 400° and 800°C. In this low-temperature range, we measured the emf of concentration cells. These electrical values are profoundly influenced by the non-reversibility of the reactions at the electrodes. The interface phenomena were observed from complex impedance and admittance diagrams.

Experimental Procedure

Three samples of ZrO_2 stabilized with 8, 12, and 15 mol% Sm_2O_3 were prepared for this study from conventional mixing of commercial powders. They were isostatically pressed at 250 MPa and sintered in air for 20 h at 1873 K and cooled at the rate of 70°C/h to room temperature. Then they were machined to form pellets (1-mm thick with a 12-mm diameter) and cylinders (30-mm long with a 4-mm diameter). Platinum electrodes were painted on both sides and heated at 1200°C for 6 h.

Figure 1 is a photomicrograph of an 8 mol% Sm_2O_3-doped ZrO_2 sample showing a grain size of about 5 μm.

A concentration cell was built using alumina tubes for the part carrying the pellet and copper material for the other parts. The sample was contacted mechanically by a spring-loaded device.

Two oxygen pressures were obtained by mixing oxygen and nitrogen and were established on each side of the sample and controlled at their entrance and exit in the apparatus by a ZrO_2 gage. Partial oxygen pressures varied from 10^{-5} to 1 atm

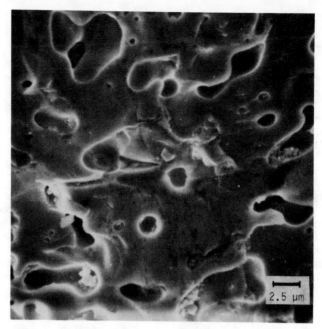

Fig. 1. Photomicrograph of a 8-mol% Sm_2O_3-doped ZrO_2.

(9.8066×10^{-1} – 9.8066×10^4 Pa). Electromotive force measurements were performed with a millivoltmeter of input impedance higher than 10^{12} Ω. AC conductivity measurements are made from 900° to 1500°C on long samples at a fixed frequency of 1000 Hz. Below 800°C, we used a frequency-response analyzer (Solartron 1174) from 10^{-1} to 10^6 Hz and at 50-mV constant voltage.

Results and Discussion

The ratio of the measured emf and the theoretical value vs temperature is shown in Fig. 2 and compared to Gouet et al.'s[1] results obtained in 1975. No dependance on the partial oxygen pressure was observed, and a good reproducibility with time was found.

Similar decreases of cell emf were obtained by Matsui[2] and Arakawa et al.[3] on yttrium-stabilized zirconia (YSZ) and calcium-stabilized zirconia (CSZ) at low temperature. There is no doubt about the reversibility of the redox reactions below 900°C, so we studied the interface phenomena using complex impedance and admittance diagrams.

For the existing experimental conditions, the impedance and admittance diagrams are generally composed of four depressed semicircles, as shown in Fig. 3, for the cells O_2, $Pt/ZrO_2 + Sm_2O_3/Pt$, and O_2.

Individual phenomena can be attributed to each arc from the variations of their intersections on the real axis vs parameters such as temperature, partial oxygen pressure, measurement voltage, nature and surface of electrode, etc. The relaxation time deduced from the frequency of the imaginary maxima can also be compared with the values obtained for similar cells.[4-8] The distinctive features of the impedance and admittance diagrams plots are (1) two semicircles in the high-frequency range, somewhat depressed (≈0°–10°) under the real axis, one being

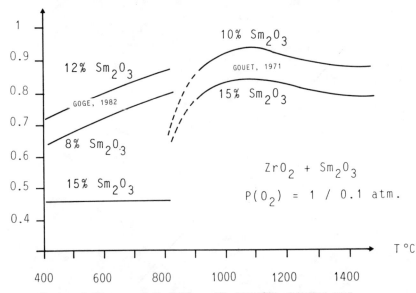

Fig. 2. Ratio of measured and theoretical emf vs temperature.

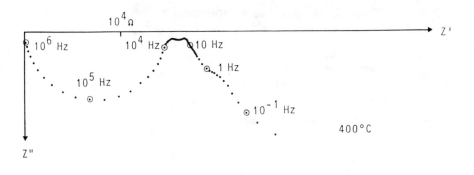

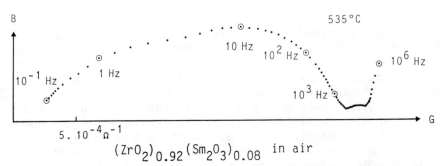

Fig. 3. Impedance and admittance diagrams for the cells O_2, Pt/ZrO_2 + Sm_2O_3/Pt, O_2.

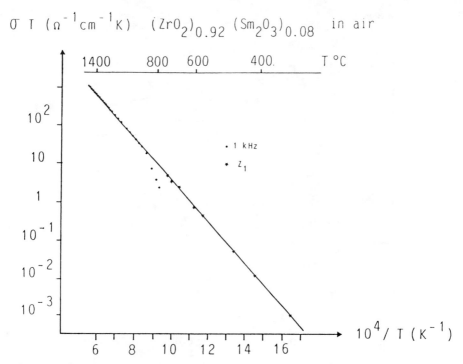

Fig. 4. Conductivity vs reciprocal temperature.

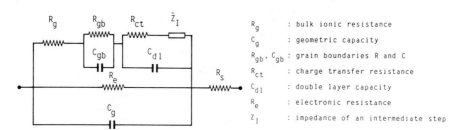

Fig. 5. Electrical circuit equivalent to one with mixed conductors.

attributed to the grain resistance and the other to the grain boundaries and (2) two more-depressed semicircles ($\approx 10°$–$30°$) in the low-frequency range attributed to the electrode reaction. The points "Z_1" or "G_1" permit the conductivity of the grains to be calculated. Conductivity vs the reciprocal temperature is represented in Fig. 4. "High"-temperature conductivity is also shown in Fig. 4.

An electrical circuit (Fig. 5) with an electronic resistance equivalent to the one with mixed conductors studied by Gur et al.[9] can be proposed as a convenient way to analyze the complex spectra. It is evident that such a circuit does not take into account the depression of the semicircles but is sufficient to allow exploitation of the extrapolated points on the real axis.

For the prevailing galvanostatic conditions, the ratio A between the measured and the theoretical emf can be written as

$$A = \frac{\text{measured emf}}{\frac{RT}{4F} \ln(P_1/P_2)} = \frac{R_e}{R_g + R_{gb} + R_{ct} + R_l + R_e} \qquad (1)$$

Extrapolated points on the real axis at very low frequency (G_4) and high frequency (G_1) are used to determinate the ionic transference number, which is equal to the ratio $R_e/(R_e + R_g)$ and is easily deduced from $\sigma_i/(\sigma_i + \sigma_e)$ when the form factor is known.

G_1 and G_4 are related to the resistances of the equivalent circuit:

$$G_1 = \left(\frac{1}{R_e} + \frac{1}{R_g}\right)(1 - R_s G_1)$$

$$G_4 = \left(\frac{1}{R_e} + \frac{1}{R_g + R_{gb} + R_{tc} + R_l}\right)(1 - R_s G_4) \qquad (2)$$

R_s is a reference resistance used in the electrical set.

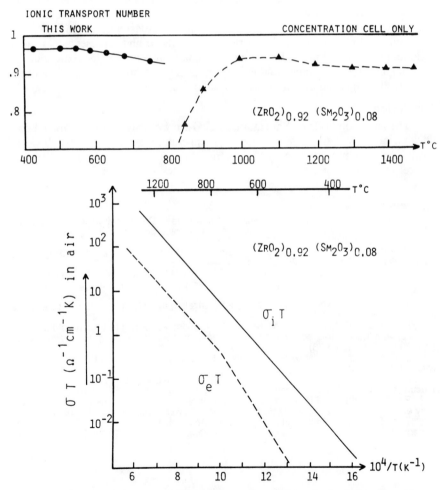

Fig. 6. Ionic transference number vs temperature for a ZrO_2 sample.

Table I. Migration Enthalpies of Anions

Sm_2O_3 (mol%)	ΔH_i (eV)
8	1.01
12	1.34
15	1.33

Thus we calculate the ionic transference number:

$$t_i = \frac{R_e}{R_e + R_g}$$
$$= 1 - \left[\frac{G_4(1 - R_sG_4)}{G_1(1 - R_sG_1)}\right](1 - A) \tag{3}$$

Figure 6 depicts the ionic transference number vs the temperature for a ZrO_2 sample. When compared to values obtained at higher temperatures, it appears that the decrease previously observed near 900°C was an artifact coming from the large interfacial resistances and that such solid solutions are essentially ionic.

Ionic and electronic conductivities are calculated and represented vs reciprocal temperature in Fig. 6. Because of the small value of the electronic transference number, the electronic conductivity σ_e cannot be obtained with certainty. In Table I, the migration enthalpies ΔH_i of anions are given.

Conclusion

The solid solutions of Sm-stabilized ZrO_2, previously known as mixed conductors, are essentially ionic in the low-temperature range of 400°–800°C and partial oxygen pressures of 10^{-5}–1 atm (9.8066×10^{-1}–9.8066×10^4 Pa). Hence, they can be used as electrolytic ionic materials in oxygen sensors.

References

[1]M. Gouet, B. Chappey, and M. Guillou, "Nature de la Conductivité Électrique à Haute Température dans le Système Oxyde de Zirconium-Oxyde de Samarium," *C. R. Hebd. Seances Acad. Sci.*, **280**, 117–19 (1975).

[2]N. Matsui, "Complex Impedance Analysis for the Development of Zirconia Oxygen Sensors," *Solid State Ion.*, **3/4**, 525–29 (1981).

[3]T. Arakawa, A. Saito, and J. Shiokawa, "Efficiency of Noble Metal Electrodes for Zironcia Oxygen Sensors in Detecting Oxygen at Lower Temperatures," *Bull. Chem. Soc. Jpn.*, **55**, 2273–74 (1982).

[4]J. E. Bauerle, "Study of Solid Electrolyte Polarization by a Complex Admittance Method," *J. Phys. Chem. Solids*, **30**, 2657–70 (1969).

[5]E. Schouler, G. Giraud, and M. Kleitz, "Applications According to Bauerle of the Plot of Complex Admittance Diagrams in the Electrochemistry of Solids," *J. Chim. Phys. Physiochim. Biol.*, **70** [9] 1309–16 (1973).

[6]D. Y. Wang and A. S. Nowick, "The Grain Boundary Effect in Doped Ceria Solid Electrolytes," *J. Solid State Chem.*, **35**, 325–33 (1980).

[7]M. J. Verkerk and A. J. Burggraaf, "Oxygen Transfer on Substituted ZrO_2, Bi_2O_3, and CeO_2 Electrolytes with Platinum Electrodes," *J. Electrochem. Soc.*, **130** [1] 78–84 (1983).

[8]F. K. Moghadam and D. A. Stevenson, "Influence of Annealing of the Electrical Conductivity of Polycrystalline ZrO_2 + 8 wt% Y_2O_3," *J. Am. Ceram. Soc.*, **65** [4] 213–16 (1982).

[9]T. M. Gur, I. D. Raistrick, and R. A. Huggins, "Ac Admittance Measurements on Stabilized Zirconia with Porous Platinum Electrodes," *Solid State Ion.*, **1**, 251–71 (1980).

Influence of Impurities in Solid Electrolytes on the Voltage Response of Solid Electrolyte Galvanic Cells

T. REETZ, H. NÄFE, AND D. RETTIG

Akademie der Wissenschaften der DDR
Zentralinstitut für Kernforschung
Rossendorf, DDR-8051 Dresden, German Democratic Republic

Solid electrolyte specimens were studied by both the polarization technique and the solid electrolyte titration method. The total polarization current consists of a time-dependent ionic and electronic branch as well as of a time-independent electronic branch. The time-dependent current is a result of the presence of aliovalent impurity cations. The oxygen exchange between the solid electrolyte material and the ambient atmosphere has been correlated to the content of those impurities, which alter their oxidation level in the corresponding oxygen pressure range. The results are capable of explaining many measuring effects observable at low temperatures.

At low temperatures ($T < 1000$ K) and high voltages ($|V| > 1000$ mV), solid electrolyte sensors respond slowly to changes in measuring conditions. This is especially the case when gas sensors are used to measure weakly buffered gas mixtures. Furthermore, liquid sodium oxygen meters working between 400° and 450°C can show a time-dependent voltage over a period of some thousand hours. Moreover, this behavior is characteristic of the electrolyte material used in the sensor. Under certain conditions, there are also problems in coulometric applications which express themselves in time-dependent deviations from Faraday's law.

The reasons for these properties and the importance of electrolyte impurities with respect to this behavior are discussed.

Experimental Procedure

The investigations were performed with electrolyte samples manufactured by different producers (for compositions and purities, see Table I). Using the polarized cell technique (Hebb-Wagner method[1,2] see Fig. 1), the polarization current was measured as a function of time, polarization voltage, temperature, and oxygen pressure of the reversible electrode. The oxygen content of the gas repre-

Table I. Compositions of the Solid Electrolyte Materials

		Impurity content (at. ppm)								
Material	Formula	Fe	Ti	Ce	Mn	Ni	Cr	Al	Si	Mg
YDT	$Th_{0.85}Y_{0.15}O_{1.925}$	64	50	8	2	150	100	12000	1500	1200
CSZ 1	$Zr_{0.85}Ca_{0.10}Mg_{0.05}O_{1.85}$	1080	1040	90	26	50	40	2100	3000	
CSZ 2	$Zr_{0.85}Ca_{0.15}O_{1.85}$	160	560	6	16	7	14	1800	4500	1200
CSZ 3	$Zr_{0.85}Ca_{0.13}Mg_{0.02}O_{1.85}$	104	160	5	8	2	20	1500	4500	

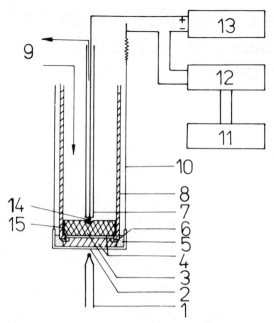

Fig. 1. Schematic diagram of setup for the polarization cell: (1) thermocouple, (2) steel cup, (3) electrolyte pellet (lower surface completely platinized with platinum paste; upper surface platinized only in the middle of the pellet), (4) steel disk pressing a platinum plate on the platinized electrolyte surface, (5, 6) glasses with various softening points, (7) alumina tube, (8) metallic tube, (9) reference gas inlet/outlet, (10) wire with springs, (11) recorder, (12) potentiostat, (13) ammeter, (14) ball of platinum wire pressed on the platinized area of the electrolyte, and (15) glass seal.

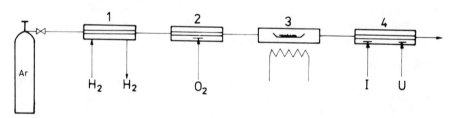

Fig. 2. Experimental arrangement: (1) hydrogen diffusion unit (silicon resin hose in a thermostat), (2) electrochemical oxygen addition, (3) reaction tube, (4) solid electrolyte titration cell.

senting the reversible electrode was continuously analyzed during polarization. For this purpose, an Ar/O_2 mixture of known partial pressure and flow rate was led through the reference electrode compartment. After passing the porous platinum film of the electrolyte surface, the gas mixture was conducted through a solid electrolyte titration cell. This cell (unit 4 in Fig. 2) consists of a ZrO_2 tube with two separately arranged pairs of platinum electrodes: one for potentiometric measure-

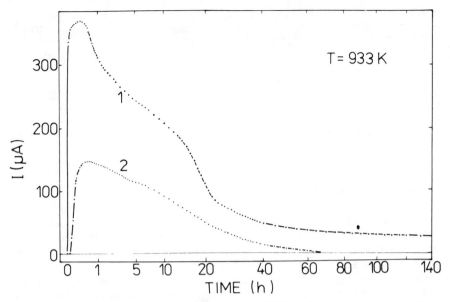

Fig. 3. Polarization current through the YDT specimen as a function of time (individual data points with extrapolated curves from recorder): (1) total current and (2) oxygen ion current.

ments and the other for coulometric purposes.[3] Using this arrangement and an electronic device,* potentiostatic measurements can be performed. Hence, alterations in the oxygen current of the titration cell correspond to alterations in the oxygen content of the gas flowing through the ZrO_2 tube.

In addition, powdered electrolyte material previously annealed in an atmosphere of a known oxygen pressure p'_{O_2} was exposed to an Ar-H_2-H_2O mixture of known oxygen pressure p''_{O_2}, which could be changed, in steps, between 10^{-2} and 10^{-11} Pa using the equipment depicted in Fig. 2. After the gas mixture of known flow rate had passed the electrolyte sample, its oxygen potential was again measured with a titration cell (unit 4 in Fig. 2). These investigations were complemented by electron spin resonance measurements (EPR) of the electrolyte material annealed under different oxygen partial pressures.

Results and Discussion

Figure 3 shows the time dependence of the polarization current through the ThO_2 specimen (933 K, V_{pol}(air) = -2100 mV, curve 1). The oxygen current of the Ar/O_2-reversible electrode changes simultaneously with alteration of the total current (curve 2). Curve 2 corresponds to a release of oxygen. The oxygen flux disappears after a finite time. This means that the source of the oxygen is the electrolyte itself.

The difference between the total and the oxygen ion current is an electronic current which slowly decreases before reaching a stationary level after more than 1000 h at 933 K. The magnitude of the steady state current corresponds to the

---

*OXYLYT, ZfK Rossendorf, Dresden, German Democratic Republic.

Table II. Content of Fe and Ti in the CSZ Specimens and Oxygen Exchange After Changing the Oxygen Pressure from p'_{O_2} to p''_{O_2} at 1100°C

Material	Impurity content (at. ppm)		Oxygen exchange Δc_0 (molecules of O/10^6 g)	
	Fe	Ti	A*	B†
CSZ 1	1080	1040	−9.4	+8.2
CSZ 2	160	560	−1.34	+4.4
CSZ 3	104	160	−0.88	+1.34

*At $p'_{O_2} = 2 \times 10^4$ Pa amd $p''_{O_2} = 1.6 \times 10^{-6}$ Pa.
†At $p'_{O_2} = 3.2 \times 10^{-16}$ Pa and $p''_{O_2} = 2.24 \times 10^{-10}$ Pa.

n and p conductivities hitherto known from extrapolating high-temperature data. Therefore the time dependence of the electronic current must be interpreted as a consequence of the total conductivity, which is the sum of the well-known steady state conductivities and an additional time-dependent branch exceeding the stationary level for a very long time. The occurrence of this branch is probably due to the presence of impurities dissolved in the electrolyte. The time for reaching the steady state is longer at lower temperatures. The amount of oxygen released from the electrolyte corresponds to the current-time integral of curve 2 in Fig. 3. If this integral is determined at various polarization voltages, a relation is obtained between the oxygen exchange (release or capture), Δc_0 of the electrolyte, and the oxygen potential in equilibrium with the electrolyte. The derivative of such a curve, expressed as $dc_0/d(\log p''_{O_2}{}^{1/4})$ approximately results from measuring the oxygen release after changing the oxygen pressure of the Ar-H$_2$-H$_2$O-mixture in steps and reequilibrating the sample. The results obtained for the ZrO$_2$ sample CSZ-1 are given in Fig. 4. The data in Table II reveal that the amount of oxygen exchanged relates directly to the impurity content of the electrolyte material.

The redox equilibrium of certain impurity cations is established in favor of the formation of oxygen vacancies, depending on the oxygen potential,

$$2xO_O + 4Me^{y'} \rightleftharpoons xO_2 + 2xV_O^{\cdot\cdot} + 4Me^{(y-x)'} \tag{1}$$

(Kröger–Vink notation, Me$^{y'}$ = impurity cation on a lattice or interstitial site or at the grain boundary.) From the mass action law it follows that

$$K = \left(\frac{p_{O_2}}{p_0}\right)^{x/4} \frac{[V_O^{\cdot\cdot}]^{x/2} c_{red}}{[O_O]^{x/2} c_{ox}} \tag{2}$$

where p_0 = reference pressure = 101 325 Pa.

The equilibrium constant K is a quantitative measure of the redox stability of the impurity defect under consideration. Taking into account the mass and charge balances

$$c_{red} + c_{ox} = c_{Me} \tag{3}$$

$$[V_O^{\cdot\cdot}] = (1/2)[D_H'] + (x/2)c_{red} \tag{4}$$

as well as the facts that $[D_H']$ (concentration of dopant cations on host lattice sites) and $[O_O]$ are practically constant and $xc_{red}/[D_H'] \ll 1$, one obtains

$$K' = \left(\frac{p_{O_2}}{p_0}\right)^{x/4}\left(\frac{c_{red}}{c_{Me} - c_{red}}\right) \quad \text{with} \quad K' = K\left(\frac{2[O_O]}{[D_H']}\right)^{x/2} \tag{5}$$

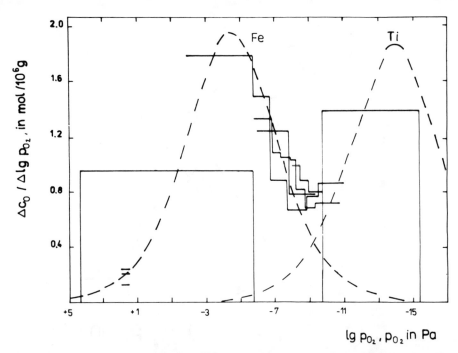

Fig. 4. Dependence of the oxygen exchange at a certain pressure range ($\Delta c_O/\log p_{O_2}$) from the logarithm of the oxygen pressure: (———) mean oxygen exchange experimentally determined at the corresponding oxygen pressure range; (– – – –) calculated curves according to Eq. (8).

According to Eq. (4)

$$d[V_O^{\cdot\cdot}] = (x/2)\, dc_{red} = -dc_O \qquad (6)$$

is obtained. Hence, the oxygen released corresponds to the content of the impurity cations of the electrolyte which are present in a reduced form.

As a result, the measurement of the oxygen exchanged can serve as a characterization method for the electrolyte material.[4] From Eq. (5) it follows that the magnitude of K' can be obtained from the maximum of the $dc_O/d(\log p_{O_2}^{x/4})$ vs $\log p_{O_2}^{x/4}$ curve. Besides, x can be assumed to be equal to 1:

$$K' = p_{O_2}^{1/4}(max)/p_0^{1/4} \qquad (7)$$

Furthermore, a calculation of K' is possible according to the equation

$$\Delta c_O = \left(\frac{K' c_{Me}}{2}\right)\left(\frac{1}{K' + (p'_{O_2}/p_0)^{1/4}} - \frac{1}{K' + (p''_{O_2}/p_0)^{1/4}}\right) \qquad (8)$$

with the assumption that in the range p'_{O_2}-p''_{O_2}, different redox systems do not disturb one another. Equation (8) follows from Eqs. (5) and (6).

Figure 4 reveals that there are two maxima at 1373 K, corresponding to two equilibria of form (1), one above and the other below 10^{-9} Pa. In accordance with Table II, we assumed that the two maxima were caused by redox equilibria

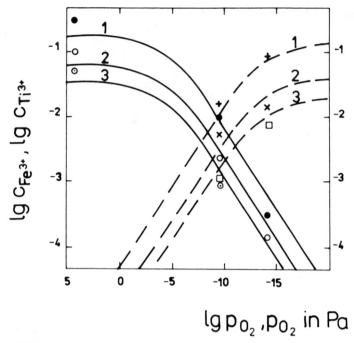

Fig. 5. Dependence of the Fe^{3+} and Ti^{3+} concentrations in the CSZ specimens 1–3 (Table I) from the oxygen pressure at 1100 °C: (———) Fe^{3+} concentrations according to Eq. (9); (– – – –) Ti^{3+} concentrations according to Eq. (10); (●1, ○2, ⊙3) intensities of the Fe^{3+} EPR spectra in relative units; (+1, ×2, □3) intensities of the Ti^{3+} EPR spectra in relative units.

Fe^{3+}/Fe^{2+} (above 10^{-9} Pa) and Ti^{3+}/Ti^{4+} (below 10^{-9} Pa), both dissolved in the electrolyte. From the data in Table II and with the help of Eq. (8) the following values result at 1373 K: $\ln K'_{Fe} = -5.5 \pm 0.3$ and $\ln K'_{Ti} = -10.8 \pm 0.5$. An additional temperature dependence of the oxygen exchanged in the range 1173–1373 K leads to the following expressions (with T in Kelvins):

$$\ln K'_{Fe} = 5.5 - 15000/T \pm 0.3 \tag{9}$$

$$\ln K'_{Ti} = -0.25 - 14500/T \pm 0.5 \tag{10}$$

The equilibration of the redox pairs mentioned above could be confirmed by means of electron spin resonance measurements.[5] In Fig. 5, the relative intensities of the Fe^{3+} and Ti^{3+} spectra are compared with the concentrations of the respective ions calculated from the constants K'_{Fe} and K'_{Ti} and the total Me concentration of the electrolyte.

Conclusions

The relationships described here provide a reliable method for characterizing electrolyte materials. This method is suitable only for those impurities which influence the measured properties of the electrolyte. On the other hand, the results are capable of explaining many measured properties observed at low temperatures, where the processes described are very slow. For instance, the necessity of oxygen exchange in order to achieve the redox equilibria requires well-buffered electrode substances and good electrode kinetics. If these prerequisites are not met, the oxygen exchange will be slow and will depend on the temperature; significant changes of the oxygen content in the electrode material will also occur.

Furthermore, a time-dependent branch of the electronic conductivity accounts for time-dependent, low transference numbers in special potentiometric and coulometric applications. The additional conductivity branch, which may be effective over very long times, is the reason for the voltage drifts, e.g., in liquid sodium oxygen meters.

The results indicate that the unwanted phenomena may be reduced by reducing the content of those impurities whose K' values lie in the oxygen potential range expected to be covered by the practical application.

References

[1] M. H. Hebb, *J. Chem. Phys.*, **20**, 185 (1952).
[2] C. Wagner; p. 361 in Proceedings 7th Meeting of the International Committee on Electrochemical Thermodynamics and Kinetics, Lindau 1955. Butterworth Scientific Publications, London, 1957.
[3] K. Teske and W. Gläser, *Microchim. Acta*, **I/S-6**, 653 (1975).
[4] H. Näfe and D. Rettig, ZfK Report 422, 1980.
[5] D. Rettig, K. Teske, and I. Ebert, ZfK Report 510, 1983.

Low-Temperature Behavior of ZrO_2 Oxygen Sensors

S. P. S. BADWAL, M. J. BANNISTER, AND W. G. GARRETT

CSIRO, Division of Materials Science
Advanced Materials Laboratory
Melbourne, Victoria, Australia 3001

The relative importance of the solid electrolyte and the electrodes in determining the low-temperature behavior of stabilized zirconia oxygen sensors is considered. Contrary to general belief, the electrodes play the more important role at low temperatures. The performance may be greatly improved by using, instead of porous platinum, oxide electrodes comprising solid solutions based on UO_2. Laboratory tests and plant trials show that ideal behavior in oxygen-excess gases can be achieved below 400°C.

Below about 600°C, stabilized ZrO_2 oxygen sensors generally develop emf errors that increase with decreasing temperature.[1] Their impedance and response time also rapidly increase,[2] and they can become excessively sensitive to the gas flow rate and to trace levels of combustibles such as CO.[3]

Moderate temperature applications for oxygen sensors exist in molten metals such as lead and sodium,[4] in automotive engine exhausts,[5] and in boiler flue gases.[6] In the last case, operation below about 450°C would eliminate the explosion hazard that exists with present-day heated probes or analyzers.

It is widely believed that electrolytes with higher ionic conductivity are required to improve the low-temperature performance.[7] However, despite the use of better conductors such as doped CeO_2[8] or Bi_2O_3,[9] thin films of stabilized ZrO_2,[10] or complex solid electrolytes such as $SrCl_2$–KCl–SrO[11] and β/β''–Al_2O_3,[12] no major improvements in low-temperature performance have been achieved.

In this paper, evidence is presented that it is the electrode rather than the electrolyte that plays the major role in determining sensor behavior at low temperatures. Supporting information already exists in the literature. For example, with conventional porous Pt electrodes, the O_2/O^{2-} exchange reaction occurs at or near the gas/electrode/electrolyte three-phase interface,[13] and the low-temperature performance depends critically on the electrode microstructure.[8] Compared with Pt, Ag electrodes give better behavior at low temperatures,[14] possibly because of easier transport of oxygen through the latter metal.

The nonstoichiometric solid solutions based on UO_2 exhibit many properties[15,16] similar to those required for electrodes on oxygen sensors: (i) They are electronic conductors with high oxygen ion diffusion coefficients and high solubility for oxygen. (ii) They are structurally stable over wide ranges of temperature and oxygen partial pressure. (iii) Being isostructural with stabilized ZrO_2, they should have very similar thermomechanical properties to the solid electrolyte. It is shown here that electrodes based on these materials provide very good sensor performance at low temperatures.

Electrode Kinetics

Experiments were performed to assess the relative contributions of the electrode and the electrolyte in a cell at low temperatures. Complex impedance spectroscopy over the frequency range from 1 mHz to 1 MHz was used to study electrode kinetics, particular attention being paid to the effect of heat treatment on the electrode microstructure and thus on the kinetic parameters.

Three types of electrodes were investigated: (1) Porous Pt electrodes were comprised of (a) 0.9-μm thick sputtered coatings or (b) and (c) commercial Pt pastes.* (2) Fluorite solid solution electrodes with the general formula

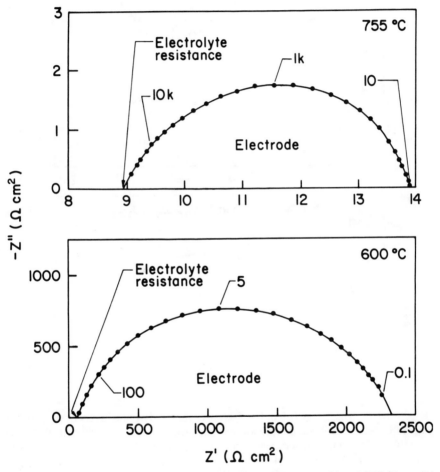

Fig. 1. Complex impedance spectra for a Pt paste No. 6082/7 mol% YSZ/Pt paste No. 6082 cell in 100% oxygen, showing the effect of temperature on the relative roles of the electrode and electrolyte resistance. The data have been normalized to 1 cm^2 of electrode/electrolyte contact area.

---

*Hanovia Liquid Gold No. 6082 and 8907, respectively.

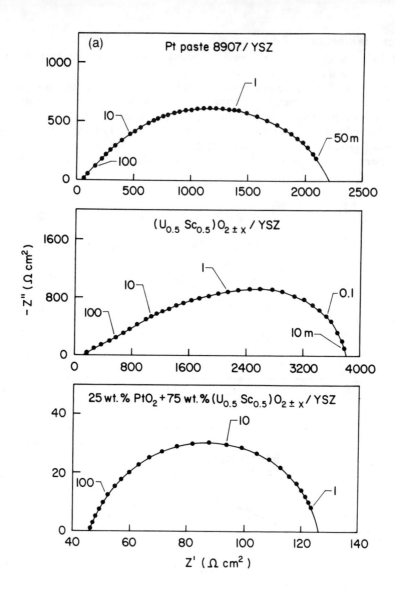

$(U_zSc_{1-z})O_{2\pm x}$, where $z = 0.5$ or 0.38, were comprised of thin layers of the powder painted on as a fine paste in ethanol. (3) Electrode mixtures of 25 wt% PtO_2 and 75 wt% $(U_zSc_{1-z})O_{2\pm x}$, where $z = 0.5$ (PtU1) or 0.38 (PtU2), were painted on as fine paste in triethylene glycol. X-ray diffraction showed that heating after application of the electrodes converted the PtO_2 to Pt metal without reaction with the $(U_zSc_{1-z})O_{2\pm x}$.

The electrolytes used for these experiments were disks of 7 or 10 mol% YSZ with 92–94% of theoretical density. After assembly, the various cells were given in situ heat treatments in stages at 600°, 750°, and 900°C. After each heat treatment, impedance spectra were recorded on cooling. In the first two figures, only the electrode arcs in the complex impedance plane are shown. The electrode behavior represented by these arcs was often complicated; nevertheless, the left

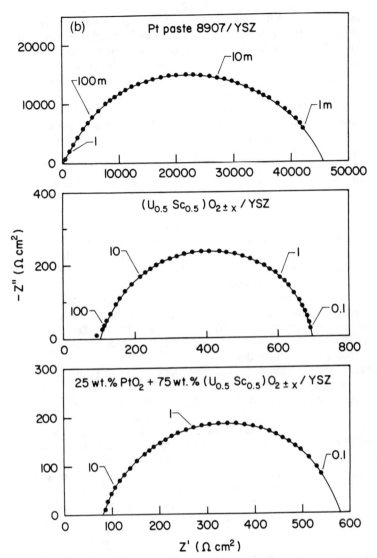

Fig. 2. Complex impedance spectra at 600°C for three electrodes in 100% oxygen after heat treatment at (a) 600°C and (b) 900°C. The data have been normalized to 1 cm² of electrode/electrolyte contact area.

intercept on the real axis represents the total electrolyte resistance (grain boundary plus volume), and the difference between the right and left intercepts is the total electrode resistance. In the remainder of this section, we summarize the more important findings of these experiments.

First, because the activation energy associated with the charge-transfer process at Pt electrodes is higher than that due to conduction through the solid electrolyte, the relative contribution of the electrodes to the total impedance increases with decreasing temperature. For example, Fig. 1 shows for Pt paste electrodes on 7 mol% YSZ that, even though the electrode resistance was comparable with that of the electrolyte at 755°C, it was by far the greater of the two at 600°C. The relative dominance of the electrode resistance is even greater at lower temperatures. It is also greater at lower oxygen partial pressures, since, as the oxygen partial pressure is decreased, the electrode resistance increases while the electrolyte resistance remains constant.

Second, the low-temperature behavior of porous Pt electrodes deteriorates after exposure to higher temperatures. Comparison of Figs. 2(a) and 2(b) shows that heating at 900°C markedly increased the resistance (R_0) and relaxation time (τ_0) of the electrode process at 600°C for porous Pt on YSZ. This deterioration was due to sintering and grain growth of the Pt (Fig. 3), which caused reductions in the fine electrode porosity, in the total electrode/electrolyte contact area, and in the extent of the three-phase contact region, together with the isolation of some Pt particles.

By contrast, the microstructure of $(U_zSc_{1-z})O_{2\pm x}$ electrodes did not change on heating (Fig. 3). In fact, R_0 and τ_0 decreased markedly due to improved electrode/electrolyte contact (Fig. 2). In the case of PtU1 and PtU2, R_0 and τ_0 were initially substantially less than for porous Pt, and they increased only slightly after heating at 900°C (Figs. 2(a) and 2(b)).

Additional experiments with PtU2 electrodes heat-treated only at 600°C showed that, with careful control of the surface roughness of the electrolyte, an electrode resistivity as low as 10–20 $\Omega \cdot cm^2$ and a time constant of ≈ 10 ms could be achieved at 600°C. These values are at least an order of magnitude below the best obtained with Pt electrodes.

These results demonstrate that electrodes based on $(U_zSc_{1-z})O_{2\pm x}$ materials have greater microstructural stability than porous Pt and are capable of showing much lower electrode resistivities and time constants on stabilized ZrO_2 at low temperatures. Detailed kinetic studies[17] suggest that such materials actively participate in the Faradaic redox reactions, allowing oxygen charge transfer to proceed by means of oxidation/reduction reactions at the gas/electrode interface, followed by the diffusion of oxygen ions to the electrode/electrolyte interface and their transfer across that interface to the solid electrolyte. Thus, the limitations inherent in preferred reaction at the three-phase interface, as occurs with porous platinum electrodes, are to a large extent overcome. This undoubtedly is the reason for the better kinetic behavior of electrodes based on $(U_zSc_{1-z})O_{2\pm x}$ materials.

Sensor Tests

Sensors were constructed with electrodes of porous Pt, PtU1, and PtU2. These materials were each applied to a closed-end tube made by welding a solid electrolyte disk across the open end of an Al_2O_3 tube. A similar design, using pellets of solid electrolyte instead of disks, was described previously.[18] To make a strong leak-free join, it is necessary to achieve a close thermal expansion match between the solid electrolyte and the Al_2O_3 tube.[19] In the present case this was achieved by incorporating 50 wt% of Al_2O_3 in the disks[20] and by closely controlling the stabilizer concentration to ensure slight understabilization. Both measures degrade the ionic conductivity.[21] To partly compensate, Sc_2O_3 rather than Y_2O_3 was used as the stabilizer; nevertheless, the conductivity was only about one-twentieth of the value found in the Al_2O_3-free, fully stabilized material (Fig. 4).

All sensors were heated to 600°C in air prior to testing in order to eliminate organics and, in the case of PtU1 and PtU2, to decompose PtO_2 to Pt metal.

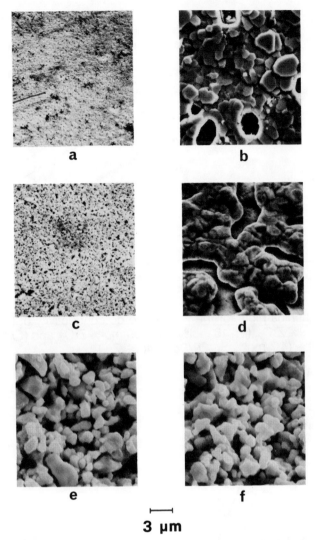

Fig. 3. Scanning electron micrographs of sputtered Pt ((a), (b)), Pt paste No. 6082 ((c), (d)), and $(U_{0.5}Sc_{0.5})O_{2\pm x}$ ((e), (f)) after heat treatment at 600°C ((a), (c), and (e)) and 900°C (b), (d), and (f).

Performance tests were carried out at 25°C intervals between 300° and 600°C. Experiments included the determination of the cell emf with air at both electrodes or with air at the inner electrode and 1–100% O_2 in N_2 at the outer electrode and the effect on cell emf of a factor of 10 variation in the flow rate of the internal gas. All tests were repeated after the sensors had been reheated in air, first to 750° and then to 900°C.

Results after firing to 600°C are shown in Fig. 5 (air vs 1.15% O_2 in N_2) for PtU2 and Pt paste No. 6082. In each case the sensor with PtU2 electrodes showed

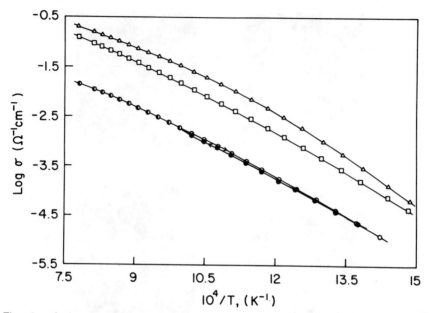

Fig. 4. Arrhenius plots for the four-probe dc conductivity of (△) 8 mol% Sc_2O_3/92 mol% ZrO_2, (□) 10 mol% Y_2O_3/90 mol% ZrO_2, and (○ ●) 50 wt% Al_2O_3/50 wt% (4.7 mol% Sc_2O_3/95.3 mol% ZrO_2).

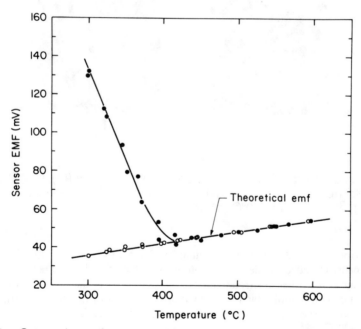

Fig. 5. Comparison of sensors with electrodes of either (●) Pt paste No. 6082 or (○) PtU2. Air vs 1.15% O_2 in N_2. Both types of electrode were baked on at 600°C.

negligible error down to 300°C, whereas with Pt paste there were significant errors below a "breakdown temperature" of 450°C. Other forms of porous Pt electrodes gave similar results to paste No. 6082, whereas PtU1 was similar to, although not quite as good as, PtU2. Results in other atmospheres were similar to those shown in Fig. 5.

Heating to 900°C increased the "breakdown temperature" of all sensors. Nevertheless, sensors provided with PtU1 or PtU2 electrodes performed satisfactorily down to 360° after a 900°C treatment. In all cases the cell emf became dependent on the internal air flow rate only below the "breakdown temperature." For porous Pt electrodes, similar behavior to that shown in Fig. 5 was found using lower-impedance Ca-stabilized ZrO_2 closed-end tubes,[†] demonstrating that the electrolyte impedance does not play a major role in determining the low-temperature behavior.

Similar sensors with PtU2 electrodes also performed well in the exhaust duct of a natural-gas-fired furnace at 400°–485°C and in the flue of a small oil-fired boiler at 350°–400°C. In the latter case it was essential to provide a separate catalyst to oxidize traces of unburnt fuel; otherwise abnormally high emfs were observed.

These results provide strong evidence that the low-temperature behavior of ZrO_2 oxygen sensors is determined more by the electrodes than by the electrolyte. Perhaps the most powerful argument for the importance of the electrodes is the fact that, despite use of a solid electrolyte whose conductivity is a factor of 20 below the best that might have been achieved, sensors with electrodes based on urania/scandia gave Nernstian behavior in the laboratory to approximately 300°C, well below the lowest temperature achieved with porous Pt electrodes even on higher conductivity electrolytes.

Conclusions

Although it is widely believed that the oxygen ion conductivity of the solid electrolyte determines the low-temperature limit of oxygen sensors, there is literature evidence to suggest that the nonideal emfs and slow response rates observed at low temperatures depend more on the physical and chemical nature of the electrodes. This hypothesis is strongly supported by the results presented here. Complex impedance studies on stabilized ZrO_2 electrolytes with various electrodes show that the resistance and time constant of the oxygen-transfer reaction at the electrodes depend strongly on the electrode morphology and composition. Compared with porous Pt electrodes, mixtures of Pt and urania/scandia solid solutions give electrode resistances and time constants lower by an order of magnitude. These improvements are consistent with the ability of urania/scandia solid solutions to exchange oxygen with the surrounding atmosphere and with the solid electrolyte and to allow the diffusion of both electrons and oxygen ions. These properties in turn enable the oxygen-exchange reaction to be less dependent on the availability of three-phase gas/electrode/electrolyte contacts. A further advantage of urania/scandia electrodes over porous Pt is their ability to resist microstructural changes on exposure to high temperatures, enabling a fine porous microstructure to be retained. Sensors incorporating such electrodes are undergoing plant trials with a view to their eventual use by industry.[22]

---

†Corning Glass Works, Corning, NY.

Acknowledgments

We thank F. T. Ciacchi for his experimental assistance. Partial support for this work was provided under the National Energy Research Development and Demonstration Program, which is administered by the Commonwealth Department of National Development and Energy of Australia.

References

[1] A. M. Anthony, J. F. Baumard, and J. Corish; Collaborative Study Group on Zirconia-Based Oxygen Gauges, Second Rept. Commission on High Temperature and Solid State Chemistry, I.U.P.A.C., 1983.
[2] M. Kleitz and J. Fouletier; p. 103 in Measurement of Oxygen. Edited by H. Degn, I. Balslev, and R. Brook. Elsevier, Amsterdam, 1976.
[3] H. Okamoto, H. Obayashi, and T. Kudo, Solid State Ionics 3/4, 453 (1981).
[4] C. C. H. Wheatley, F. Leach, B. Hudson, R. Thompson, K. J. Claxton, and R. C. Asher; p. 556 in Liquid Metals. Edited by R. Evans and D. A. Greenwood. Conference Series No. 30, The Institute of Physics, 1976.
[5] P. McGeehin, Trans. J. Br. Ceram. Soc., 80, 37 (1981).
[6] B. C. H. Steele, J. Drennan, R. K. Slotwinski, N. Bonanos, and E. P. Butler; p. 286 in Advances in Ceramics, Vol. 3. Edited by A. H. Heuer and L. W. Hobbs. The American Ceramic Society, Columbus, OH, 1981.
[7] J. A. Kilner and R. J. Brook, Solid State Ionics, 6, 237 (1982).
[8] R. T. Dirstine, W. O. Gentry, R. N. Blumenthal, and W. Hammetter, Am. Ceram. Soc. Bull., 58, 778 (1979).
[9] T. Takahashi, T. Esaka, and H. Iwahara, J. Appl. Electrochem., 7, 303 (1977).
[10] M. Croset, J. P. Schnell, G. Velasco, and J. Siejka, J. Appl. Phys., 48, 775 (1977).
[11] A. Pelloux, J. P. Quessada, J. Fouletier, P. Fabry, and M. Kleitz, Solid State Ionics, 1, 343 (1980).
[12] J. L. Lundsgaard and R. J. Brook; p. 159 in Measurement of Oxygen. Edited by H. Degn, I. Balslev, and R. Brook. Elsevier, Amsterdam, 1976.
[13] S. Karpachov and A. Filjajev, Z. Phys. Chem., 238, 284 (1968).
[14] T. Arakawa, A. Saito, and J. Shiokawa, Bull. Chem. Soc. Jpn. 55, 2273 (1982).
[15] S. P. S. Badwal and D. J. M. Bevan, J. Mater. Sci., 14, 2353 (1979).
[16] S. P. S. Badwal, D. J. M. Bevan, and J. O'M Bockris, Electrochim. Acta, 25, 1115 (1980).
[17] S. P. S. Badwal; J. Electroanal. Chem., 161, 75 (1984).
[18] R. K. Stringer and K. A. Johnston: U.S. Pat. No. 4 046 661, Sept. 6, 1977.
[19] M. J. Bannister, W. G. Garrett, K. A. Johnston, N. A. McKinnon, R. K. Stringer, and H. S. Kanost, Mater. Sci. Monogr., 6, 211 (1980).
[20] M. J. Bannister, N. A. McKinnon, and R. R. Hughan: U.S. Pat. No. 4 193 857, Dec. 23, 1980.
[21] S. P. S. Badwal; J. Mater. Sci., 18, 3230 (1983).
[22] S. P. S. Badwal and M. J. Bannister; Australian Patent Application No. PF7857/83.

Life and Performance of ZrO_2-Based Oxygen Sensors

B. Krafthefer, P. Bohrer, P. Moenkhaus, D. Zook, L. Pertl, and U. Bonne
Honeywell Inc.
Bloomington, MN

We have fabricated both differential (concentration-cell-type) and absolute (electrode-active, self-referenced-type) sensors, with sputtered and thick-film electrodes and have investigated the effects of long-term exposures to various environments (up to 4000 h at 700°C) by monitoring sensor output and impedance. Physical changes of the Pt electrode and ZrO_2-electrolyte structure were investigated by scanning electron microscopy. The life-test results of differential sensors show decreases in sensor conductivities and output with age. These aging effects are reduced as grain size decreases and as Pt-electrode adherence is improved. With aging we observed an increase in the Pt-electrode pore size by more than a factor of 10 and an increase in the concentration of Si impurities in the Pt. We also fabricated absolute sensors on the basis of the reversible oxidation of Pd, with both the two- and the three-electrode configurations. We confirmed that large-impedance changes (50 ×) as well as substantial hysteresis occurred when the oxygen was varied to reduce or oxidize the Pd working electrode. The output of the three-electrode sensors decays with time, without establishing a definite plateau. About 1 s after the reduction is interrupted, the output voltage corresponds to the published value of Gibbs free energy. We conclude that more work needs to be done to shed light on the operating mechanism.

The simplicity of fabrication and use of oxygen sensors based on the Nernstian concentration cell has led to the wide application of such sensors for control of automotive and stationary combustion.[1-3] The insensitivity to CO_2 and SO_2, the lack of calibration requirements (at least in principle), solid-state ruggedness, and reasonable cost have made it the preferred method for oxygen measurement. However, the need for an on-line reference requires a rather large, costly, and sizeable power-consuming sensor if ambient air is used as reference. An encapsulated reference[7,8] requires a more complex cell and closer temperature control. A third possibility is based on an absolute, electrode-active sensor,[4-6] which requires no reference chamber but more complex signal processing. With these three ZrO_2-based approaches, there have been some life or aging problems and signal drift which, if solved or at least better understood, would lead to increased user satisfaction.

The purpose of this paper is to report on the performance and life-test results obtained with several types of sensors and to attempt to interpret them in order to derive fabrication or user methods that would reduce these problems. As shown below, this attempt has led to some progress, but several observed effects still remain that are insufficiently understood.

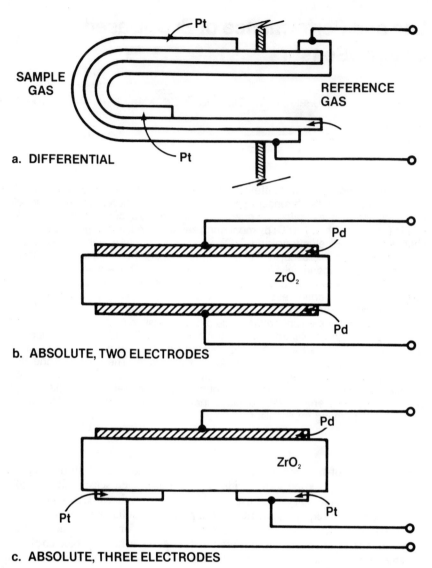

Fig. 1. Zirconia-based oxygen sensors fabricated and tested in this study.

Experimental Procedure

We have fabricated and tested oxygen sensors of both the differential (O_2-concentration chain or conventional Nernstian, purchased and from our production facilities) and absolute[6] (metal redox or electrode-active) types. We describe in this section fabrication and test methods.

Differential sensors made with different materials and platinum-electrode processing techniques were set up for life test at constant temperature cycling between 1 and 21% O_2 (room air). We are reporting here on the results obtained with three batches. The zirconia cells were test-tube shaped (see Fig. 1(*a*)), about

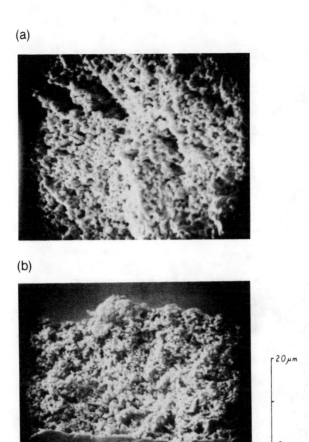

Fig. 2. Micrograph of the inner electrode of the new sensor. Views show (a) electrode outer surface and (b) electrode–electrolyte interface.

1 cm in diameter, and fabricated by slip casting. Most sensors were doped with 8 mol% Y_2O_3 and fired at a temperature of 1600°–1620°C. Batch 1 was made by an outside vendor. Batches Nos. 2 and 3 are representative of two consecutive generations of sensors made by the Honeywell Ceramics Center. They were found to have a much smaller grain size and better electrode adherence, as determined from scanning electron microscopy studies.

Two types of absolute sensors were made, using two or three electrodes, as shown in Fig. 1(b) and (c). These flat sensors also were "slip" cast and cut to size (about 1 by 2 by 0.1 cm^3). Both sputtered and thick-film electrodes of Pt and Pd were used. Initial adhesion problems with the Pd electrodes were eliminated after increasing the electrode "firing" temperature to 1380°C.

Operating temperatures of 600°C were used unless otherwise indicated. The experimental system set up for the three-electrode system consists of a current supply that gets periodically switched on to cathodically reduce the PdO to Pd. A MACSYM II programmable data logger was used to electronically record the working electrode potential relative to the Pt reference electrode.

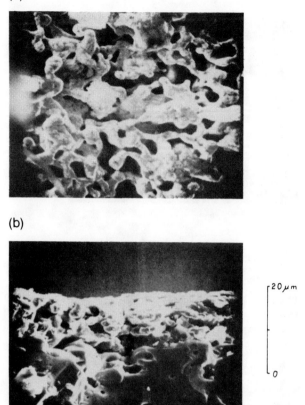

Fig. 3. Micrograph of the outer electrode of aged sensor run for 4000+ h: (a) electrode outer surface and (b) electrode–electrolyte interface.

Results and Discussion

The scanning electron micrographs of Figs. 2–4 show the changes that occurred in the electrode structure of the differential O_2 sensors before and after more than 4000 h of life test. The pore size within the platinum electrode and related dimensions were found to increase by over a factor of 10. Although not determined quantitatively, it appeared that the amount of Pt also decreased (i.e., part of the Pt had sublimed or reacted), as also observed by others.[9]

We believe that this structural change in the Pt electrodes is the primary cause of the observed changes in sensor impedance and output, as shown in Fig. 5 for the three batches mentioned in the previous section. Figure 5 shows the result of plotting impedance and output of the differential sensors vs operating time. The sensors were maintained at a constant temperature of 700°C and cycled from room air to an ambient containing 1% O_2 three times every hour. The measurements of batches 2 and 3 indicate significant improvements in performance with respect to batch 1 and were found to be associated with ZrO_2 of smaller grain size and better

(a)

(b)

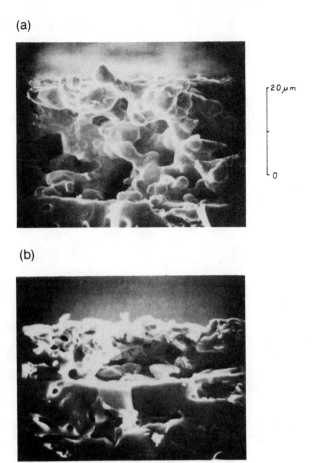

Fig. 4. Micrograph of electrodes of sensor after 4000+ h of operation.

adherence of the electrodes, as observed via scanning electron micrographs. This may be consistent with earlier findings regarding the influence of grain and grain boundaries on conductivity changes.[10] If we define sensor operating life as the time when the impedance has increased by a factor of three, Fig. 5 reveals that the sensor life ranges from 700 to 2000 h (for batches 1 and 3, respectively). The associated signal output decreased by 16 and 3% for batches 1 and 3, respectively.

For an evaluation of the absolute, electrode-active approach, sensors were fabricated with one or more Pd-paste electrodes.[6] These sensors showed an increase in impedance of about a factor of 50 as the oxygen concentration was increased from 2 to 10% at 730°C (see Fig. 6). In spite of the slow change in oxygen concentration (3–5 min per 1% O_2 change), we observed considerable hysteresis, i.e., we found differences in sensor impedance depending on whether the oxygen concentration was increased or decreased. We interpreted this as an indication that the Pd $\rightleftharpoons$ PdO phase change involves a nucleation process analogous to supercooling or superheating.[13]

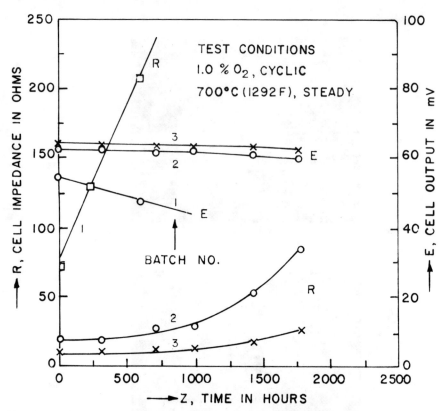

Fig. 5. Life-test results of oxygen measuring cells. Impedance (R) and output (E) vs time for three generations of sensor cells.

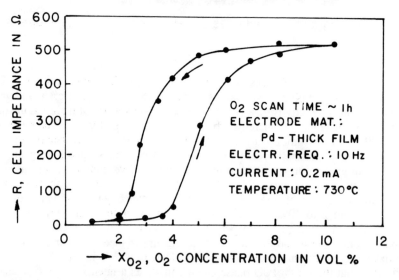

Fig. 6. Slow scan of two-electrode cell impedance vs oxygen concentration showing hysteresis.

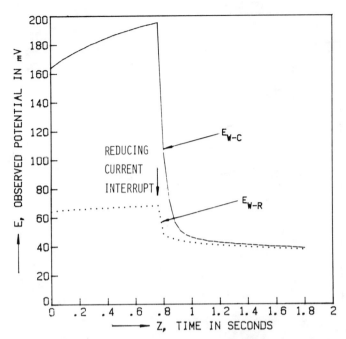

Fig. 7. Typical potential decay after interrupting current vs time.

The characteristics of the Pd/PdO electrode were evaluated using a three-terminal electrode configuration to measure the transient behavior of the electric potential before and after current interruption.[5,11] The voltage output of the absolute, electrode-active oxygen sensor should be given as

$$E = [\Delta G° - RT \ln (P_S^{1/2})]/nF \qquad (1)$$

where R is the universal gas constant, F, the Faraday constant, n, the charge-transfer number, with a value of 2 in this case, T, the temperature in K, P_S, the oxygen concentration of the unknown sample gas, and $\Delta G° = -113900 + 99.9T$ J/mol^{-1}, the Gibbs free energy of formation of PdO[5,12] at 1 atm of O_2.

Figure 7 shows an example of voltage measurements; the two curves represent the observed time dependence of the working-reference voltage (W-R) and working-counter voltage (W-C), respectively, with reducing current applied between working and counter electrodes before the current interrupt, as indicated. Note that E_{W-C} is nearly equal to E_{W-R}, within 0.5 s after interrupt.

The difficulty with the E_{W-R} curve of Fig. 7 is that the voltage continues to decay without reaching a definite plateau corresponding to $\Delta G°$, as expected from Eq. (1). This problem is illustrated by plotting the E_{W-R} curve with different time scales, as shown in Fig. 8. The fast (almost vertical) initial decay and the subsequent slower change suggest three decay processes. The first is the fast decay of the voltage drop across the electrolyte, and the second corresponds to the decay of voltage across the working electrode double layer,[5,6] as the amount of available oxide-free metal decreases. The third we believe to be the formation of PdO due to the direct interaction between atmospheric O_2 and Pd, which effectively shunts the process of most interest represented by Eq. (1).

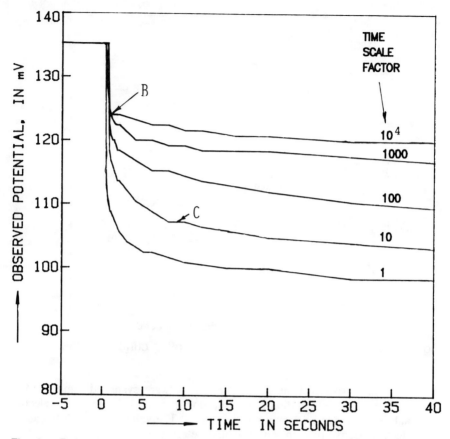

Fig. 8. Rate of voltage decay for a three-electrode absolute sensor.

In order to shed more light on these findings, we repeated these experiments for several oxygen concentrations (Fig. 9). These data show that the O_2 *response* is the same at different times after interrupt. There is close agreement between the values read at point C (see Fig. 8) and the theoretical values corresponding to Eq. (1). We therefore selected this point as the definition of the value for tests at this time.

Preliminary results of a short (3 weeks) life test are plotted in Fig. 10. Although the sensor output voltages vary by an unacceptably large amount, it is premature to draw negative conclusions about sensor performance without further studies on the sensor operating mechanism.

The temperature dependence of the output voltage for the three types of sensors is compared in Fig. 11. The temperature dependence of the absolute sensor is similar to that measured for encapsulated PdO[8] but stronger than that of the differential sensors. We feel that further studies of cell-polarization effects using the three-electrode configuration may lead to a better understanding of the operating mechanism of this promising sensor.

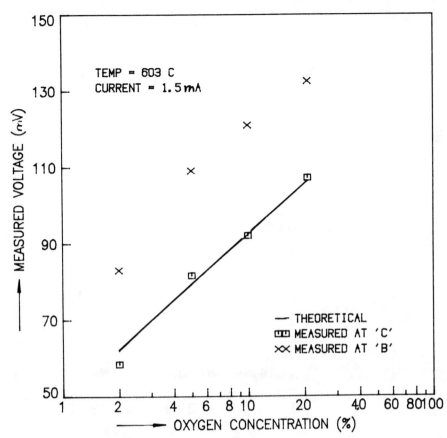

Fig. 9. Comparison between theoretically expected sensor output and two sets of measured values.

Conclusions

(1) The type of materials and processing used in the fabrication of the oxygen sensors significantly influences sensor life and performance. Defining "life" as the operating time during which the sensor *impedance* changes by less than a factor of 3, life spans from 700 (batch 1) to 2000 h (batch 3) were found, depending on ZrO_2 grain size and electrode adherence. The sensor signal output changed from 3 to 16% for sensor batches 3 and 1, respectively.

(2) Both types of Pt electrodes become increasingly porous with operating time and probably contribute to the observed long-term increase in cell impedance.

(3) To achieve more predictable, drift-free operation, we should aim for a small-grain-size electrolyte and uniform, relatively thin, and well-adhering Pt electrodes.

(4) Although experimental evidence points toward establishment of a fast chemical equilibrium of the redox electrode process, the initial cell values are higher than those expected from Gibbs free energies[8] by about 30 mV at 600°C.

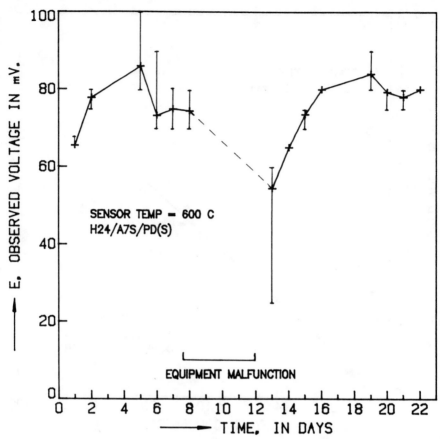

Fig. 10. Life test of self-referenced oxygen sensor in air.

(5) A better understanding of the various mechanisms involved in electrode degradation, impedance increase, hysteresis of O_2 response (two-electrode cell), and sensor dynamics is expected to lead to means to minimize the drift in the sensor output signal.

References
[1] E. Hamman, H. Manger, and L. Steinke; SAE Paper 770401, International Automotive Engineering Congress, Detroit, MI, Feb. 1977.
[2] R. E. Hetrick, W. A. Fate, and W. C. Vassell, *IEEE Trans. Electron Devices,* **ED-29**, 129 (1982).
[3] R. H. Torborg and U. Bonne; Paper No. 78-49.6, APCA 71st Annual Meeting, Houston, TX, 25–30, June 1978.
[4] S. P. S. Badwal and H. J. deBruin, *Austr. Chem. Eng.,* **30**, 9 (1979).
[5] H. J. deBruin; Proceedings of the International Meeting on Chemical Sensors, Fukuoka, Japan, Sept. 1983. Also H. J. deBruin and S. P. S. Badwal, "Electrode Active Oxygen Monitor," U.S. Pat. No. 4 326 318, April 27, 1982.
[6] S. P. S. Badwal and H. J. deBruin, *J. Electrochem. Soc.,* **129**, 1921 (1982).
[7] C. F. Bauer, L. B. Welch, K. J. Youtsey, and F. R. Szofran; U.S. Pat. No. 4 040 929, Aug 9, 1977.

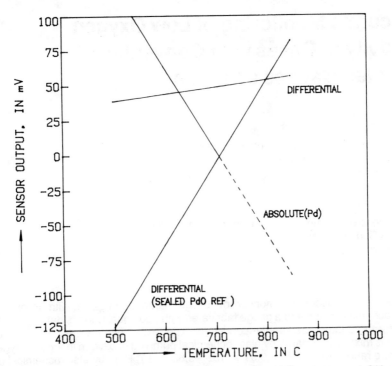

Fig. 11. Temperature dependence of ceramic oxide sensors at 2% oxygen partial pressure.

[8] F. J. Gutierrez and G. Vitter, *J. Phys. E.*, **16**, 361 (1983).
[9] S. Pizzini, et. al., *J. Appl. Electrochem.*, **3**, 153 (1975).
[10] F. K. Moghadam and D. A. Stevenson, *J. Am. Ceram. Soc.*, **65**, [4] 213–16 (1982).
[11] Da Yu Wang and A. S. Nowick, *J. Electrochem. Soc.*, **126**, [7] 1155–65 (1979).
[12] J. S. Wagner, *J. Electrochem. Soc.*, **119**, 68 (1967).
[13] S. D. Bader, L. Richter, and T. W. Orent, *Surf. Sci.*, **115**, 501 (1982).

Accurate Monitoring of Low Oxygen Activity in Gases with Conventional Oxygen Gauges and Pumps

J. FOULETIER AND E. SIEBERT

Laboratoire d'Energétique Electrochimique, Enseeg-BP 75
38402 Saint Martin d'Heres
Cedex, France

A. CANEIRO

Laboratorio de Termodinamica, Centro Atomico de Bariloche
8400-S.C. de Bariloche
Rio Negro, Argentina

It is possible to accurately monitor low oxygen activities in gases with an oxygen electrochemical pump and a gauge fabricated with good quality industrial ZrO_2 tubes under appropriate experimental conditions. Trial experiments were performed on gaseous mixtures prepared by electrochemical pumping of oxygen in argon (oxygen pressure range: 10^{-7}–1 atm (9.81×10^{-3}–9.80×10^4 Pa)), by electrochemical reduction of pure carbon dioxide (oxygen pressure range: 5×10^{-11}–5×10^{-19} atm (4.90×10^{-6}–4.90×10^{-14} Pa)), and by electrochemical oxidation of pure hydrogen and argon–hydrogen mixtures (oxygen pressure range: 10^{-19}–10^{-25} atm (9.80×10^{-10}–9.81×10^{-21} Pa)). An accuracy of 2% on the oxygen activity measurement can now be obtained over the whole range of pressures investigated.

Oxygen gauges incorporating ZrO_2-based electrolytes have been widely developed for applications in both laboratory and industrial environments.

The monitoring of oxygen pressure in inert gases and vacuum is one of the main laboratory applications. Over the last 10 years, the joint efforts of instrument and ceramic manufacturers have led to a noticeable improvement in ZrO_2 tube quality and to the production of oxygen sensors. Recently, the Commission of High Temperatures and Refractory Materials of the International Union of Pure and Applied Chemistry (IUPAC) organized an international program to compare the characteristics of oxygen gauges formed on commercial ZrO_2 tubes. The results of the electrochemical tests carried out in our laboratory have been published,[1] and the cooperative study is described in another chapter of this volume.[2] One of the main conclusions was that the performances of the gauges do not vary significantly with the origin and composition of the tubes in the range 10^{-7}–1 atm (9.81×10^{-3}–9.80×10^4 Pa) of oxygen.

The aim of this paper is to show that it is also possible to accurately monitor low oxygen activities in gases with conventional ZrO_2 gauges and pumps under appropriate experimental conditions.

Description of the Gauges

The gauges were formed on closed-ended Y or Ca-doped ZrO_2 tubes purchased from various suppliers.[2] The gauge assembly was described in detail previously[1] and was intentionally chosen to be simple and inexpensive. Platinum electrodes were used (Pt paste Degussa 308 A*). These coatings were fired at 900°C in air for 1 h. Pure platinum wires, fixed to the coatings by a second layer of platinum paint, were used as current leads. Air, in contact with the external electrode, was used as a reference gas. A glass part, sealed on the top of the ZrO_2 tube, carried the analyzed gas to the measuring electrode. The temperature of this electrode was measured with a Pt–Pt10Rh thermocouple. The cell was placed in an electric furnace; the temperature difference between the two electrodes was less than 1°C. The optimal working temperature of the gauge was found to depend on the gaseous mixture used; systematic tests were performed between 500° and 900°C.

Gas Circuit

As in previous experiments,[1,3–5] we coupled an electrochemical oxygen pump and a gauge. The electrochemical pump consists of a tube of 9 mol% YSZ with a metal coating on both the inner and outer surfaces. Vitrifiable platinum paint (Pt Degussa M 8005) was used to increase the adherence of the electrodes on the tube surface. The oxygen content of the gas flowing inside the tube was controlled by adjusting the continuous direct current passing between the two electrodes. Due to the rather high currents required to obtain certain experimental conditions, a rather large ZrO_2 tube was selected (2.5 cm in diameter) and the electrodes covered large parts of the inner and outer surfaces. The current densities were kept below 50 $mA \cdot cm^{-2}$. The operating temperature of the pump was in the range 700°–1000°C, depending on the nature of the flowing gas and the oxygen pressure range.

The gas circuit consisted of a gas tank (nominally pure argon, pure CO_2, pure H_2, or Ar–H_2(5%) mixture), the electrochemical pump, the oxygen gauge, and an accurate flowmeter. Another gauge was generally inserted in the gas circuit. It was used to check that the gas composition arriving in the investigated gauge remained constant, for instance, when the temperature of this gauge was varied. The additional gauge was therefore connected in the gas circuit immediately upstream of the investigated gauge. It was also used to detect the onset of oxygen permeation through the ZrO_2 tube of the investigated gauge. In this case, it was of course connected downstream.

The results quoted in this paper were obtained during long-term tests (more than 6 months). A complete analysis of the results obtained was given in Ref. 1.

Theoretical Relations: $P_{O_2} = f(I)$

The equations relating the composition of the gas circulating in the pump and the current passing through it were established previously.[1,6–9] The principle of the derivation and the basic equations are the following.

Part of this work was carried out under the auspices of the International Union of Pure and Applied Chemistry (IUPAC).
*Degussa Corp., Teterboro, NJ.

The oxygen flux, J, passing through the electrolyte wall of the pump is related to the current, I, by Faraday's law:

$$J = I/4F \tag{1}$$

Accordingly, the oxygen partial pressure, P_{O_2}, in the gas streaming in the pump tube, varies as a linear function of the current:

$$P_{O_2} = P_{O_2}^\circ + 0.209 Ip/D \tag{2}$$

where $P_{O_2}^\circ$ is the oxygen partial pressure upstream of the pump, I the current (in A), D the gas flow rate (in $L \cdot h^{-1}$ NTP), and p the total pressure in the pump (P_{O_2}, $P_{O_2}^\circ$, and p are in atm). This formula is valid only when the oxygen flux added or extracted from the carrier gas (i.e., $I/4F$) is negligible compared to the inert gas flux in the pump (i.e., $D/22.4$).[4]

When oxygenated molecules such as CO_2 are flowing inside the pump, direct electrochemical reduction can be performed. The electrode reaction is

$$CO_2(g) + 2e^-(\text{metal}) \rightarrow CO(g) + O^{2-}(\text{electrolyte}) \tag{I}$$

and the balance reaction is

$$CO_2(\text{inside}) = CO(\text{inside}) + 1/2 O_2(\text{outside}) \tag{II}$$

The gas component partial pressures can be calculated from the three basic equations[8,9]:

$$\frac{P_{CO} + 2P_{CO_2}}{P_{CO} + P_{CO_2}} = 2 - \frac{22.4}{2FD} I \tag{3}$$

$$P_{CO} + P_{CO_2} = p \tag{4}$$

$$\frac{P_{CO_2}}{P_{CO} P_{O_2}^{1/2}} = K_c(T) = \exp\left(-\frac{\Delta G^\circ(T)}{RT}\right) \tag{5}$$

where p is the total pressure and $K_c(T)$ and $\Delta G^\circ(T)$ are the equilibrium constant and the standard Gibbs energy change of reaction (II) at T_K. A recent determination of $\Delta G^\circ(T)$ gives[10]

$$\Delta G^\circ(T) = -283.328 + 0.08753 T_K \quad (\text{kJ} \cdot \text{mol}^{-1}) \tag{6}$$

with a standard deviation of 75 $J \cdot \text{mol}^{-1}$.[10]

Equations (3)–(6) give

$$P_{O_2} = \frac{1}{K_c^2(T)} \left(2.392 \frac{D}{I} - 1\right)^2 \tag{7}$$

$$P_{O_2} = \left(2.392 \frac{D}{I} - 1\right)^2 \exp\left(21.05 - \frac{68150}{T}\right) \quad (\text{atm}) \tag{8}$$

$$P_{CO} = 0.418 p \frac{I}{D} \tag{9}$$

$$P_{CO_2} = p\left(1 - 0.418 \frac{I}{D}\right) \tag{10}$$

When an Ar-H_2 mixture (hydrogen mole fraction, q) is supplied to the pump, the P_{H_2O}/P_{H_2} ratio can be controlled by direct electrochemical oxidation according to

$$H_2(g) + O^{2-}(\text{electrolyte}) \rightarrow H_2O(g) + 2e^-(\text{metal}) \tag{III}$$

The gas composition can be calculated from the equations[9]:

$$P_{H_2O} + P_{H_2} = qp \tag{11}$$

$$\frac{P_{H_2O}}{P_{H_2} + P_{H_2O}} = 0.418\frac{I}{qD} \tag{12}$$

$$\frac{P_{H_2O}}{P_{H_2}P_{O_2}^{1/2}} = K_H(T) = \exp\left(-\frac{\Delta G°(T)}{RT}\right) \tag{13}$$

The variation of the standard Gibbs energy of reaction

$$H_2 + 1/2 O_2 = H_2O \tag{IV}$$

is[11]

$$\Delta G°(T) = -247.657 + 0.05520 T_K \quad (kJ \cdot mol^{-1}) \tag{14}$$

These equations give

$$P_{O_2} = \frac{1}{K_H^2(T)}\left(2.392\frac{qD}{I} - 1\right)^{-2} \tag{15}$$

$$P_{O_2} = \left(2.392\frac{qD}{I} - 1\right)^{-2} \exp\left(13.278 - \frac{59571}{T}\right) \quad (atm) \tag{16}$$

$$P_{H_2O} = 0.418p\left(\frac{I}{D}\right) \tag{17}$$

$$P_{H_2} = p\left(q - 0.418\frac{I}{D}\right) \tag{18}$$

Monitoring of P_{O_2} in the Range 10^{-7}–1 atm (9.81×10^{-3}–9.80×10^4 Pa)

The possibility of such monitoring is well established. Results are given here for reference only. They were analyzed in detail in Ref. 1. Ideally, the gauge voltage obeys Nernst's law:

$$E = 4.958 \times 10^{-5} T \log\left(\frac{P_{O_2}}{0.209p}\right) \tag{19}$$

In Fig. 1(a), the corresponding variation of gauge voltage as a function of temperature is plotted for various constant oxygen pressures. (The oxygen pressure was fixed with the oxygen pump; i.e., a constant current was passed through the pump and the gas flow rate was maintained strictly constant.) Deviations from the theoretical straight lines were observed at low temperatures (due to excessive impedance) and at high temperatures (due to the polarizing oxygen semipermeability flux[3]). These two temperatures delineate the temperature interval of "ideal response" of the gauge at the analyzed oxygen pressure. Limits are quantitatively fixed at the points where the resulting measurement error in P_{O_2} is equal to 10%. The variations of these limits as a function of the measured oxygen pressure are given in Fig. 1(b). For example, with the tested gauge, it ranges from 600° to 800°C at 10^{-7} atm (9.80×10^{-3} Pa) of oxygen and from 500° to 950°C at 10^{-6} atm (9.80×10^{-2} Pa) of oxygen.

In this domain of ideal response, the possibility of gas monitoring can be assessed from various types of measurements. For instance, the reproducibility can be evaluated by comparing measurements by two oxygen gauges connected in series. Trial measurements were carried out with various zirconia tubes. The oxygen pressure in flowing argon was varied by the oxygen pump from

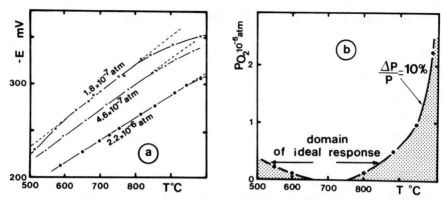

Fig. 1. Verification of Nernst's law: (a) E vs T for three oxygen partial pressures; (b) domain of ideal response corresponding to an error on the measured oxygen partial pressure <10% ($\Delta P_{O_2}/P_{O_2} < 0.1$).

Table I. Reproducibility of Oxygen Partial Pressure Measurements by Two Gauges Connected in Series

P_{O_2} (10^{-6} atm) (9.80×10^{-2} Pa) upstream gauge ($T = 730°C$)	P_{O_2} (10^{-6} atm) (9.80×10^{-2} Pa) downstream gauge ($T = 806°C$)	$\Delta P_{O_2}/P_{O_2}$ (%)
0.51	0.56	9.5
1.34	1.34	0
0.89	0.91	+2.2
1.93	1.89	−2.1
3.02	2.87	−5.2
4.87	4.58	−6.3
7.77	7.30	−6.4
11.4	10.6	−7.5
19.4	17.9	−8.4
29.4	27.1	−8.5
48.7	44.7	−8.5
98.5	89.9	−8.9
195	179	−8.9
388	353	−8.9
683	623	−9.6
939	902	−4.1

5×10^{-7} atm (4.90×10^{-2} Pa) up to 10^{-3} atm (9.8×10^1 Pa). Table I compares results obtained with two gauges. The average reproducibility deviation is 4.4% with a maximum deviation of 9.5%. Results obtained with all the tubes tested show good reproducibility of measurement.

Another way to estimate the accuracy of the measurement is based on the verification of Eq. (2), i.e., the "Faraday's law test." This test consists in verifying the linear relationship between the current passing through the electrochemical pump and the oxygen partial pressure measured with the downstream oxygen

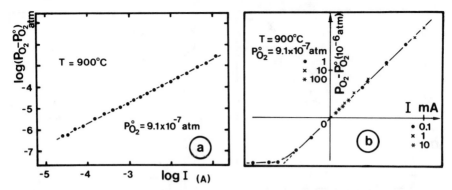

Fig. 2. Variation of the oxygen pressure in flowing argon as a function of the current passing through the upstream electrochemical pump: gauge temperature, 900°C; gas flow rate, 11.1 L·h^{-1} NTP; oxygen pressure upstream of the pump, 9.1 × 10^{-7} atm (8.92 × 10^{-2} Pa); (a) log $(P_{O_2}-P^°_{O_2})$ vs log (I) variation; (b) $(P_{O_2}-P^°_{O_2})$ vs I variation (experimental slope, 1.96 × 10^{-2} atm·A^{-1} (1.92 × 10^3 Pa·A^{-1}), calculated slope, 1.88 × 10^{-2} atm·A^{-1} (1.84 × 10^3 Pa·A^{-1})) (three scales indicated).

gauge. Figure 2(a) gives an example of this test over four orders of magnitude. The accuracy of the oxygen pressure monitoring is better illustrated by plotting $(P_{O_2} - P^°_{O_2})$ as a linear function of I on a multiscale diagram (Fig. 2(b)). The results fit the theoretical straight line with an average deviation of 1% and a maximum deviation of 8%. The slope of the experimental straight line, m_{exp} (determined by assigning an appropriate weight to each experimental point[1]), agrees well with the calculated slope, m_{calcd}, deduced from Eq. (2) ($m_{calcd} = 0.209I/D$). For the results reported in Fig. 2(b), experimental and calculated slopes agree to within 4% (m_{exp} = 1.96 × 10^{-2} atm·A^{-1} (1.92 × 10^3 Pa·A^{-1}) and m_{calcd} = 1.88 × 10^{-2} atm·A^{-1} (1.84 × 10^3 Pa·A^{-1})). As a counter example, Fig. 2(b) also shows the behavior of the gauge outside the domain of ideal response. The gauge voltage appears constant in a certain pumping current interval. A further increase in the reducing current produces a sharp jump toward more cathodic voltages, indicating a reduction of the traces of CO_2 and H_2O present in the gas and even an electrochemical reduction of the pump tube itself.[4] It should be pointed out that the gauge temperature during the test reported in Fig. 2 was intentionally selected high for demonstration. At temperatures around 700°C, an even greater accuracy is obtained. For example, at 693°C, the average deviation from the straight line is only 0.8%, the maximum deviation is 3.7%, and the experimental and calculated slopes agree to within 2%.[1]

Monitoring of P_{O_2} in the Range 10^{-11}–10^{-25} atm (9.80 × 10^{-7}–9.80 × 10^{-21} Pa)

In this oxygen-pressure range, H_2O-H_2 and CO_2-CO mixtures are generally used, due to their high buffer capacity. CO-CO_2 mixtures are conventionally prepared by mixing flowing gases with titrating pumps[12] and H_2-H_2O mixtures by bubbling hydrogen through water at a controlled temperature.[13] Although the feasibility of direct electrochemical reduction of CO_2 in a zirconia pump was demonstrated some time ago,[14] few investigations devoted to the control of CO-CO_2 ratio with an electrochemical pump have been reported.[4,15] The results

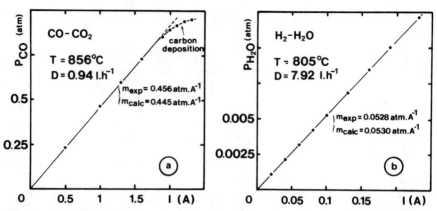

Fig. 3. (a) Variation of CO partial pressure as a function of the current passing through the pump (experimental slope, 0.456 atm·A^{-1} (4.47 × 10^4 Pa·A^{-1}), calculated slope, 0.445 atm·A^{-1} (4.36 × 10^4 Pa·A^{-1})). (b) Variation of the water-vapor partial pressure in pure hydrogen as a function of the current passing through the pump (experimental slope, 0.0528 atm·A^{-1} (5.17 × 10^3 Pa·A^{-1}), calculated slope, 0.0530 atm·A^{-1} (5.19 × 10^3 Pa·A^{-1})).

of a recent investigation demonstrated that the simple pump gauge device, described in the previous section, can be used to accurately monitor oxygen activity in the range 10^{-11}–10^{-25} atm (9.80 × 10^{-7}–9.80 × 10^{-21} Pa) at 800°C.

According to Eq. (9), the CO partial pressure in the flowing gas is proportional to the current passing through the pump. Quantitatively, the CO partial pressure was calculated from the gauge emf using Eqs. (8)–(10) and (19).

$$P_{CO} = \left[1 + 0.457 p^{1/2} \exp\left(2.32 \times 10^4 \frac{E}{T} + \frac{3.407 \times 10^4}{T} - 10.53\right)\right]^{-1}$$

(atm)   (20)

A check on this linear relationship is given in Fig. 3(a). The slope calculated from the gas flow rate ($m_{calc} = 0.418\ (p/D)$) equals the experimental slope calculated by linear regression to within 0.3%. With sufficiently buffered mixtures ($5 \times 10^{-3} < P_{CO}/P_{CO_2} < 5 \times 10^3$), the response time of the system is in the order of 1 min.

With a gas flow rate in the order of 10 L·h^{-1}, the linear relationship was verified over three orders of magnitude ($10^{-4} < P_{CO}/P_{CO_2} < 10^{-1}$). At a flow rate lower than 1 L·h^{-1} (see Fig. 3(a)), a noticeable deviation from the linear relationship is observed at high current density (corresponding to a carbon monoxide content higher than 80%). This deviation was attributed to carbon deposition on the platinum electrode[16] and a resulting P_{CO}/P_{CO_2} ratio controlled by the Boudouard reaction

C + CO$_2$ = 2CO   (V)

Likewise, water-vapor partial pressure was calculated from the gauge voltage and Eqs. (11)–(14) and (19):

$$P_{H_2O} = qp\left[1 + 2.187p^{-1/2}\exp\left(6.639 - \frac{29785}{T} - 23212\frac{E}{T}\right)\right]^{-1} \quad \text{(atm)} \tag{21}$$

An example to check on the linear relationship is given in Fig. 3(b). Agreement between the experimental and calculated slopes is better than 0.4%.

Two limiting factors should be considered with H_2-H_2O mixtures. An upper limit for P_{H_2O} is fixed by the dew point of the coldest part of the pump gauge device. Obviously, this limit can be raised by heating the metallic parts of the setup with wound resistors and/or by using Ar-H_2 mixtures instead of pure hydrogen. Another limit is set for very reducing mixtures obtained with pure hydrogen by an increase of the n-type electronic conductivity of the electrolyte.[17] For example, at temperatures around 900°C, the measurement error due to this phenomenon is noticeable even for water-vapor mole fractions of a few percent, whereas at 800°C (see Fig. 3(b)) the theoretical relationship is verified for water-vapor mole fractions <0.1%.

Examples of the corresponding variation of oxygen partial pressure, calculated from Nernst's law (Eq. (19)), are given in Fig. 4. The experimental results are compared to the theoretical curves deduced from Eqs. (8) (CO-CO_2 mixtures) and (16) (H_2-H_2O mixtures). These results illustrate that, by an appropriate choice of the monitoring method (reduction of CO_2, oxidation of H_2 or of H_2-Ar mixtures) of the gas flow rate and of the cell temperature, it is possible to fix the oxygen pressure in the range 10^{-11}–10^{-25} atm (9.80×10^{-7}–9.80×10^{-21} Pa).

Conclusion

An accuracy of 2% in oxygen pressure determinations with gauges formed on commercially available tubes can be obtained over the oxygen pressure range 1–10^{-25} atm (9.80×10^{4}–9.80×10^{-2} Pa). Such an accuracy can be obtained

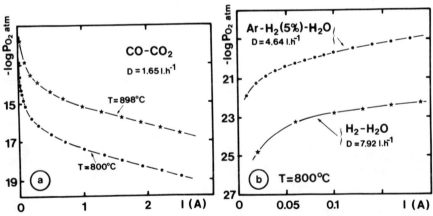

Fig. 4. Oxygen pressure measured with the gauge as a function of the direct current passing through the upstream electrochemical pump: (a) reduction of flowing carbon dioxide for two temperatures of the gauge (D = 1.65 L·h^{-1}); (b) oxidation of Ar-H_2(5%) mixture (D = 4.64 L·h^{-1}) and of pure hydrogen (D = 7.92 L·h^{-1}).

using conventional designs (i.e., with platinum electrodes, air reference electrodes, and working temperatures around 650°C for inert gases-oxygen mixtures and around 800°C for CO-CO_2 or H_2-H_2O mixtures). The association of an electrochemical oxygen pump proves to be a very useful and accurate method to monitor the gas composition.

However, it should be pointed out that, for the analysis of nonbuffered gaseous mixtures (oxygen in vacuum[5]) or at very high temperatures (higher than 900°C[1]), these gauges are not accurate enough and another type of design, such as the "zirconia-tip electrode,"[3] must be used.

References

[1] J. Fouletier, E. Mantel, and M. Kleitz, *Solid State Ionics*, **6**, 1 (1982).
[2] A. M. Anthony, J. F. Baumard, and J. Corish; this volume, pp. 627–30.
[3] J. Fouletier, P. Fabry, and M. Kleitz, *J. Electrochem. Soc.*, **123**, 204 (1976).
[4] J. Fouletier, G. Vitter, and M. Kleitz, *J. Appl. Electrochem.*, **5**, 111 (1975).
[5] Y. Meas, J. Fouletier, D. Passelaigue, and M. Kleitz, *J. Chimie Phys.*, **75**, 826 (1978).
[6] J. Fouletier, H. Seinera, and M. Kleitz, *J. Appl. Electrochem.*, **5**, 177 (1975).
[7] N. Fukatsu, I. Osawa, and Z. Kozuka, *J. Jpn. Inst. Metals*, **40**, 1263 (1976).
[8] N. Fukatsu, I. Osawa, and Z. Kozuka, *Trans. JIM*, **9**, 25 (1978).
[9] A. Caneiro, M. Bonnat, and J. Fouletier, *J. Appl. Electrochem.*, **11**, 83 (1981).
[10] A. Caneiro, J. Fouletier, and M. Kleitz, *J. Chem. Therm.*, **13**, 823 (1981).
[11] J. F. Elliot, M. Gleiser, and V. Ramakrishma, Thermochemistry for Steelmaking. Addison-Wesley, New York, 1963.
[12] J. Campserveux and P. Gerdanian, *J. Solid State Chem.*, **23**, 73 (1978).
[13] H. Ullmann, D. Naumann, and W. Burk, *Z. Phys. Chem.*, **237**, 337 (1968).
[14] M. Kleitz, J. Besson, and C. Déportes; p. 354 in Proceed. 2ème Journées Int. Etudes des Piles à Combustibles. Edited by Serai-Comasi. Bruxelles, 1967.
[15] N. Fukatsu, I. Osuki, and Z. Kozuka, *J. Jpn. Inst. Metals*, **40**, 1263 (1976).
[16] T. H. Etsell and S. N. Flengas, *Met. Trans.*, **3**, 27 (1972).
[17] W. A. Fisher and D. Janke, *Archiv. für Eisenhuettenw.*, **39**, 89 (1968).

Collaborative Study on ZrO$_2$ Oxygen Gauges

A. M. ANTHONY AND J. F. BAUMARD

Centre de Recherches sur la Physique des Hautes Temperatures, CNRS
45045 Orleans-Cedex, France

J. CORISH

Trinity College
Department of Chemistry
Dublin 2, Ireland

The Commission on High Temperature and Solid State Chemistry of the International Union of Pure and Applied Chemistry (IUPAC) has directed an international collaborative study of zirconia-based oxygen gauges. The purpose has been to examine the microstructures of the materials, evaluate the techniques used in their incorporation into gauges, and assess the response of these devices to standard tests. Commercially available materials were used and were supplied in the form of tubes made from five types of stabilized zirconia by the manufacturers listed in Appendix A. The names of the laboratories which participated in the study are given in Appendix B.

The results of the microstructural examination of the materials were presented at the previous meeting in this series: full details of the compositions, impurity contents, and physical characteristics of the electrolytes are also available in that report.[1] In a second series of tests within the collaborative study, the extent of the electronic contributions to the conductivities of these electrolytes at higher temperatures was investigated and also the effect of increases in their oxygen semipermeabilities. The results of these tests, in which a zirconia-tip electrode was used, have also been presented.[2] In the final phase of the collaboration, the tubes were distributed by the Commission to be used in the fabrication of oxygen gauges by the participants. These gauges were then subjected to standard procedures to verify the Nernst law, determine the symmetry of their response when the gases were interchanged, and measure any emf which developed when the same gas was allowed to flow on both sides of the cells. The data from these measurements were analyzed centrally, and we present here a brief summary of the principal results of these electrochemical tests. A full report, which includes details of all three aspects of the collaborative study, will be published under the auspices of the Commission.[3]

Electrochemical Tests

So that the gauges might reflect current working practices in the participating laboratories, no instructions were given as to their preparation from the tubes: The participants were requested only to carry out the standard tests listed above. All the laboratories chose platinum as the electrode material with Pt–PtRh thermocouples to measure the cell temperature. The electrodes were applied to small areas on the inside and outside of the closed end of the tubes as a paste or paint, which was then fired at a temperature well above the anticipated working range of the device. The gases chosen for verification of the Nernst law and investigation of the symmetry of the gauges were air and oxygen. Their use avoids the uncertainties and difficulties inherent in providing reference atmospheres from $CO-CO_2$ mixtures or metal-metal oxide couples. The zero emf was measured with air on both sides of the cell, and the effects of changes in the flow rates of the gases on the emf of the cells were also investigated. The full report[3] contains discussion of the errors which may arise in this kind of measurement because of physical permeability of the electrolyte to the gases, uncertainties in measurements of cell voltage, temperature or reference pressure of oxygen, nonuniform cell temperatures, and electronic leakage in the electrolyte.

The Nernst law was verified in the temperature range <800 to >1300 K, and Table I lists the results of a standard least-squares regression analysis of these data. The calculated values of the slopes of the lines, their intercepts at 0 K, the correlation factors of the linear regressions, and the mean departures from theoretical voltage are given. It has been assumed here that the temperature measurements were without error, since the uncertainties which may arise as a result of the deviations from standards expected in the thermocouple materials are considerably less than those from other sources. As is evident, the data for all experiments except three lay along or very close to straight lines, with values of their correlation factors close to unity. The data from IUPAC 3 in laboratory 5 and IUPAC 5 in laboratories 3 and 5, which have correlation factors <0.9, show that the gauges concerned had almost certainly not functioned properly. If the analysis in Table I was confined to measurements made at temperatures >880 K, there was a strong tendency to move toward the theoretical slope, find smaller intercepts at 0 K, and evaluate correlation factors even closer to unity, and it is apparent that a minimum operating temperature of 880 K should be recommended.[3] The data from laboratory 4 on IUPAC 1 using normal air/O_2 in both configurations represent the level of accuracy which can be attained with these gauges. For data above 880 K the value of the experimental slope is 3.360×10^{-2} mV·K^{-1} which is in error by 0.21% compared to the theoretical value. The intercept at 0 K is 0.025 mV, the correlation factor has a value of 0.9999, and the mean departure from the theoretical Nernst voltage was 0.059 mV, which represents a typical error of <0.3% in determining an unknown partial pressure of oxygen.

The values of the zero emf with air on both sides of the gauges were measured in the same temperature interval as was covered during verification of the Nernst law. The average absolute values of the zero emf lay within the range 0.1–0.2 mV for three laboratories; for the others, they were somewhat larger. Although an effort to find a correlation between the observed zero emf values and the performance of the same gauge during the Nernst test was partly successful, it clearly failed to establish the zero emf test as an alternative to the test based on the Nernst relationship.

Ideal behavior was not observed with respect to the symmetry of the cells. In spite of this, data obtained during either one or both modes of operation generally led to the same conclusions in the Nernst law testing of any device. Their symmetry tended to improve with increasing temperature, but such improvement did not necessarily correspond to better accuracy.

Extensive data on the effect of changes in the gas flow rates through the cells have also been reported.[3] The observed variations in cell voltages have been rationalized in one laboratory on the basis of changes in the total local pressure in the cell compartments caused by the different rates of flow of the gases.

Table I. Parameters from the Fitting of Nernst Law Data to Straight Lines*

Tube material		Laboratory 1	Laboratory 2	Laboratory 3	Laboratory 4	Laboratory 5	Laboratory 6
IUPAC 1	(i)	3.512	3.374	3.524	3.179†	3.480	3.323
	(ii)	−1.754	0.314	−1.815	0.978	−1.664	1.357
	(iii)	0.995	1.000	0.998	0.999	0.935	0.997
	(iv)	0.223	0.403	0.357	0.225	0.545	0.901
IUPAC 2	(i)	3.560	3.296	3.503	3.471†	3.229	
	(ii)	−2.318	0.298	−1.819	−2.562	1.694	
	(iii)	0.994	0.994	0.984	0.988	0.990	
	(iv)	0.314	0.352	0.545	0.701	0.450	
IUPAC 3	(i)	3.501	3.355	3.140	3.203†	(2.936)	
	(ii)	−1.692	0.167	1.745	0.290	(4.485)	
	(iii)	0.979	1.000	0.988	0.994	(0.705)	
	(iv)	0.474	0.059	0.837	0.634	(1.733)	
IUPAC 4	(i)	3.407		3.198	3.416†	2.933	3.336
	(ii)	−0.521		2.155	−2.000	4.716	1.114
	(iii)	0.998		0.996	0.992	0.992	0.995
	(iv)	0.167		0.396	0.671	0.534	0.785
IUPAC 5	(i)	3.375	3.431	(2.534)	2.996†	(3.900)	3.163
	(ii)	−0.191	−0.620	(5.904)	2.514	(−7.064)	3.226
	(iii)	0.999	0.996	(0.573)	0.998	(0.897)	0.997
	(iv)	0.125	0.213	(3.999)	0.635	(1.640)	1.090
IUPAC 1	(i)				3.357‡		
	(ii)				0.069		
	(iii)				1.000		
	(iv)				0.061		
Mode of operation		Air/O$_2$ and O$_2$/air	IUPAC 2, 3; Air/O$_2$ and O$_2$/air IUPAC 1, 5: O$_2$/air	Air/O$_2$ 1 measurement O$_2$/air	‡Air/O$_2$ and O$_2$/air; otherwise air/O$_2$	IUPAC 1: O$_2$/air IUPAC 2 to 5 air/O$_2$ and O$_2$/air	O$_2$/air

*The parameters in each case are as follows: (i) 10^2 slope in mV·K^{-1}; (ii) intercept at 0 K in mV; (iii) correlation factor; (iv) mean absolute value of the departure from the theoretical voltage in mV, assuming no error in the temperature measurements.

†Measurements made using synthetic air with a nominal composition of 21.70% oxygen; all other data refer to natural air.

Summary

The central analysis of the data from these electrochemical tests allows some general trends to be established despite the individual procedures and materials, other than the tubes, used to construct the gauges in the various laboratories. Nearly all the cells worked very well, and in the high oxygen chemical potentials obtained with air-oxygen mixtures, the Nernst law was obeyed, usually to within 0.2–0.3 mV of the theoretical voltage. The corresponding error in the partial pressure of oxygen is 1–2%. When carefully fabricated and operated, the devices can be much more accurate, and it is possible to obtain results which are within ≤0.2% of the theoretical emf. No significant differences were found in the performances of the various tubes which could be correlated with the different materials from which they were made. However, there were a few gauges which failed seriously and others which exhibited small but significant asymmetry in their emf values when the gases in their compartments were interchanged. It would be of considerable scientific interest to establish the reasons for these inadequacies. Finally, the quality and reproducibility of those gauges which appeared to be the best on the basis of the Nernst law test was in all cases confirmed by their performances in the other electrochemical tests.

Acknowledgment

We are grateful to the suppliers of the tubes and the participating laboratories whose combined goodwill made this collaboration possible.

Appendix A

Manufacturers who supplied tubes especially for this study were the following: Corning Glass Works, Ceramic Products Division, Solon, OH 44139, USA; Degussit, Postfach 7, Steinzeugstrasse, D-6800 Mannheim-71, Federal Republic of Germany; Nuclear Research Institute, 25068 Rez, Czechoslovakia; Nippon Kagaku Togyo Co Ltd, Wako Shoken Bldg. No. 3, Kitahama 3-chome, Higashi-ku, Osaka 541, Japan.

Appendix B

The people and the respective laboratories which participated in the project were the following: M. J. Bannister, CSIRO Division of Materials Science, Advanced Materials Laboratory, P. O. Box 4331, Melbourne, Australia; H. Rickert, M. Lange, and J. Lohmar, Universität Dortmund, Lehrstuhl fur Physikalische Chemie I, Otto Hahn Strasse, 4600 Dortmund, Federal Republic of Germany; A. M. Anthony and J. F. Baumard, CR due la Phisique des Hautes Temperatures, CNRS, 45045 Orleans Cedex, France; J. Fouletier, E. Mantel, and M. Kleitz, Laboratoire d'Energétique Electrochimique, La 265 ENSEEG, 38401 Saint Martin d'Hères, France; G. Petot and P. Ochin, Université de Paris-Nord, Laboratoire des Proprietes Méchaniques et Thermodynamiques des Materiaux, Avenue J. B. Clement, 93430 Villetaneuse, France; J. P. Bonnet, D. Benjelloun, and M. Onillon, Laboratoire de Chimie du Solide du CNRS, 351 Cours de la Liberation, 33405 Talence Cedex, France; J. Corish,* Department of Chemistry, University College, Dublin, 4, Ireland; Y. Saito, Tokyo Institute of Technology, Research Laboratory of Engineering Materials, 4259 Nagatsuta-Midori, Yokohama 227, Japan; B. C. Steele, J. Drennan, R. K. Slotwinski, N. Bonanos, and E. P. Butler, Imperial College, Wolfson Unit for Solid State Ionics, London SW7 2BP, United Kingdom; W. L. Worrell and G. M. Mehrotra, University of Pennsylvania, School of Metallurgy and Materials Science, Philadelphia, PA 19104, USA.

References

[1]B. C. H. Steele, J. Drennan, R. K. Slotwinski, N. Bonanos, and E. P. Butler; pp. 286–309 in Advances in Ceramics, Vol. 3. Edited by A. H. Heuer and L. W. Hobbs. The American Ceramic Society, Columbus, OH, 1981.
[2]J. Fouletier, E. Mantel, and M. Kleitz, *Solid State Ion.*, **6**, 1–13 (1982).
[3]*Pure Appl. Chem.*, in press.

*Present address: Department of Chemistry, Trinity College, Dublin, 2, Ireland.

Computer-Controlled Adjustment of Oxygen Partial Pressure

F. VIZETHUM, G. BAUER, AND G. TOMANDL

Universität Erlangen-Nürnberg
Institut für Werkstoffwissenschaften III (Glas und Keramik)
Federal Republic of Germany

An apparatus is described for adjusting the oxygen partial pressure of a flowing gas by a ZrO_2 solid-state electrolyte cell. The main components are a heated, metallized, calcium-stabilized ZrO_2 tube and a p_{O_2} sensor of the same material. The temperature and the p_{O_2} are controlled by a microcomputer. The nonlinear characteristic pump current vs p_{O_2} is derived theoretically and agrees well with experiment. The influence of the four gas parameters, initial p_{O_2}, initial p_{H_2O}, gas flow, and temperature, is analyzed.

Ceramics made of stabilized ZrO_2 are applied as oxygen sensors in a wide technical field, since they have a good ionic conductivity and an electrical transport number t close to 1.

Moreover, ZrO_2 ceramics are used as electrolytes for gas titrations.[1-3] The current through a ZrO_2 electrolyte according to the equation

$$O_2 + 4e^- \rightleftharpoons 2O^{2-} \tag{1}$$

indicates an interaction of a sample and the surrounding atmosphere. It corresponds to an equivalent amount of oxygen being released or consumed by the sample. On the other hand it should be possible to alter the oxygen partial pressure (p_{O_2}) of an enclosed gas volume by "pumping" O_2 through a ZrO_2 ceramic which is in direct contact with the gas. If a voltage is applied, a current of O^{2-} ions flows through the ceramic and thus the electrochemical potential of the oxygen at the electrodes is changed.

It is possible to use the ZrO_2 ceramics as a semipermeable diaphragm, the permeability of which may be controlled by an applied voltage. This application is mentioned in earlier papers but not described satisfactorily.[4-6] This paper describes an experimental arrangement for automatic control of oxygen partial pressure by means of a ZrO_2 ceramic "oxygen pump" and derives the fundamental equations.[6]

Theory

If Faraday's law is valid, i.e., the electric transport number t is equal to 1, gas reactions like

$$H_2O \rightleftharpoons H_2 + \frac{1}{2}O_2$$

or

$$CO_2 \rightleftharpoons CO + \frac{1}{2}O_2$$

must be taken into account. Whatever the initial O_2 content, a decrease in O_2 content will lead to an increase of H_2 content.

The following gas parameters must be considered: initial p_{O_2} ($p_{O_2}^0$), initial p_{H_2O} ($p_{H_2O}^0$), gas flow (f), and temperature (T). The current through the ZrO_2 is determined by the decrease in O_2 and H_2O contents, corresponding to an increase in H_2 content (p'_{H_2}).

$$\Delta p_{O_2} = \Delta p_{O_2}^0 - \Delta p_{O_2}$$
$$\Delta p_{H_2O} = \Delta p_{H_2O}^0 - \Delta p_{H_2O} \tag{2}$$

or

$$\Delta p_{H_2} = \Delta p_{H_2} - \Delta p_{H_2}^0 \tag{3}$$

Expressed in mole fractions, the total rate is:

$$\Delta n = \Delta n_{O_2} + \Delta n_{H_2} \tag{4}$$

The p_{H_2} and $p_{H_2O}^0$ must be calculated from the mass action equation of the dissociation of water:

$$p_{H_2} = p_{H_2O}/K_{H_2O} p_{O_2}^{1/2} \tag{5}$$

Because $p_{H_2O} = p_{H_2O}^0 - p'_{H_2}$, the actual p'_{H_2} is

$$p'_{H_2} = p_{H_2O}^0/(1 + K_{H_2O} p_{O_2}^{1/2}) \tag{6}$$

If Faraday's law is written with correction of the mole volume from normal conditions to high temperatures, the pump current, I_p, is

$$I_p = \sum_{i=1}^{r} n_i z_i F f T / V_M T_R \tag{7}$$

where $T_R = 298$ K, n_i = mole fractions of oxygen and dissociated molecules, z_i = charge number/mole, F = Faraday constant, f = gas flow, and V_M = mole volume. Characteristics calculated with these equations show good agreement with experimental values.

Experimental Equipment

In consideration of the results of previous papers,[1,2] a solid-state electrolyte (SSE) cell was constructed as shown in Fig. 1. The cell was placed into a glass tube ($d = 25$ mm, $l = 500$ mm), which was heated by a horizontal tube furnace. The two open ends of the glass tube were sealed with stoppers. The electrical connections, the main inlet of the reference gas and the Ca-stabilized ZrO_2 tube ($d = 12$ mm, $l = 600$ mm), were pushed axially into the glass tube and led outside through holes. Cooling coils at the end of the quartz glass tube provided sufficient cooling of the stoppers. Within the zone of constant temperature, the ZrO_2 was metallized with a porous Pt electrode layer by means of a Pt paste, which was heated in air to a temperature of 900°C.

Anode and cathode were contacted by Pt wires wound helically. A Pt–RhPt thermocouple was attached to the outside of the ZrO_2 tube and controlled the furnace temperature to exactly 900°C, the working temperature of the cell. The electrode wires were connected with a dc source. The applied pump voltage and the current flowing through the SSE were determined with two digital voltmeters and by means of the voltage drop at a measuring resistance. A reference gas flowed between the outside of the ZrO_2 tube and the quartz glass tube. The reaction gas

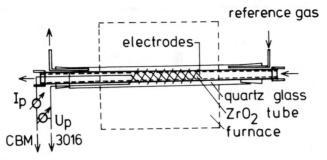

Fig. 1. Experimental arrangement.

was led through the ZrO_2 tube where the oxygen activity was adjusted. From there it was led to a separate measuring furnace by a capillary pipe. There the p_{O_2} was continually measured by a ZrO_2 O_2 sensor. With this arrangement it was possible to adjust the stationary states of p_{O_2} of a flowing gas in the region 10^{-4} to 10^{-20} bar at 900°C.

Gases

Pure nitrogen with a remaining $p^0_{O_2}$ of 3×10^{-5} bar was used as reaction gas. It was taken from the gas cylinder with precision control valves and was dried by concentrated sulfuric acid to a remaining water partial pressure of 7×10^{-5} bar. One part of the gas flow was saturated with water at room temperature and then led back to vary the water partial pressure of the reaction gas from 7×10^{-5} to 2×10^{-2} bar.

When CO_2 was added ($p^0_{O_2} = 1 \times 10^{-3}$ bar), the initial $p^0_{O_2}$ of the reaction gas increased, but the gas equilibrium CO/CO_2 must be taken into account. The gas flow was controlled by a precision measuring tube; the temperature of the gas was measured by a thermocouple at the top of the sensor.

Experimental Procedure

At constant temperature and constant gas flow, it is expected that in the measuring furnace a stationary equilibrium of O_2 activity can be adjusted after a corresponding waiting period. The relation of pump current and p_{O_2} is non-linear (Fig. 2).

At the beginning of the experiments, the p_{O_2} of the reaction gas was measured by the O_2 sensor while the electric circuit of the SSE cell was open; the water partial pressure and the gas flow were adjusted. Subsequently, a constant pump current through the electrolyte was adjusted and the change of the p_{O_2} in the measuring furnace was observed.

The constancy of the current was controlled by a microcomputer, which fitted the voltage of the dc source with a relative accuracy of 1×10^{-3}. The time for reaching a constant p_{O_2} in the measuring furnace depended strongly on the gas parameters p_{O_2}, $p^0_{H_2O}$, f, and T. It was impossible to lower the p_{O_2} below 1×10^{-6} bar at a water partial pressure $<10^{-4}$ bar. This behavior was interpreted as a hint that the electrolysis of gas molecules like H_2O or CO_2 plays an important role and must therefore be taken into account in the calculated O^{2-} current through the SSE. Even when the mixture was dried with concentrated sulfuric acid to $p^0_{H_2O} \approx 7 \times 10^{-5}$ bar, the water partial pressure and the oxygen partial pressure of the reaction gas were in the same range.

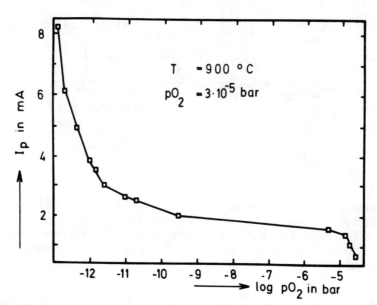

Fig. 2. Pump current (I_p) vs oxygen partial pressure (p_{O_2}).

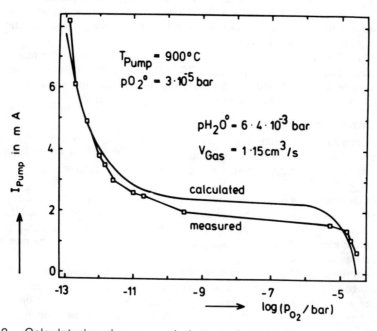

Fig. 3. Calculated and measured characteristics.

Results

The characteristic measured by the described arrangement is in excellent agreement with the characteristic calculated from the described equations (Fig. 3). Gas equilibria must be considered even when the p_{O_2} is adjusted by ZrO_2 cells, but it is possible to control these cells very accurately to equilibrate the temperature dependence of p_{O_2} during cooling or heating. In the characteristic I_p vs p_{O_2}, three ranges may be distinguished as follows:

(1) There is a range with 'direct' pumping of ionized O_2 molecules of the gas, where the p_{O_2} is decreased to 10^{-6} bar. A further decrease of p_{O_2} without the presence of dissociating molecules (H_2O, CO_2) is not possible.

(2) The oxygen content has about the same value as the hydrogen or carbon monoxide content, originating from dissociation (10^{-6} to 10^{-10} bar).

(3) The p_{O_2} decreases only due to gas reactions (indirect pumping). H_2O or CO_2 molecules are electrolyzed, and the O_2 is removed through the ZrO_2 ceramic. The proportion p_{H_2O}/p_{H_2} decreases and so does the p_{O_2}.

According to Ref. 3, the characteristic may be interpreted as a titration reaction of oxygen by hydrogen, originating at the electrode by electrolysis of water. This corresponds both to experiments, where the p_{O_2} could not be decreased below 10^{-6} bar when the gas was dried, and to the activation energy of 265 kJ/mol for the p_{O_2} controlling mechanism, which agrees with the formation enthalpy of water (255 kJ/mol). The use of ZrO_2 cells does not lead to a temperature-independent control of p_{O_2} in reducing atmospheres. The most important advantages are the fast response and the ability to accurately control the cell. This enables the adjustment of p_{O_2} independently of temperature.

References

[1] W. Weppner, Chen Li Chuan, and A. Rabenau, "Solid State Electrochemical Study of the Phase Diagram and Thermodynamics of the Ternary System Cu-Ge-O," *J. Solid. State Chem.*, **31**, 257–64 (1980).

[2] K. Teske and U. Gläser, "Verfahren zur coulometrischen Bestimmung geringer Wasserstoffgehalte in Gasen," *Mikrochem. Acta*, **I**, 653 (1975).

[3] K. Teske, "Bestimmung und Einstellung von O/U-Verhältnissen in Uranoxiden durch Festelektrolytchemie," *Kernenergie*, **24** [II] 1 (1981).

[4] C. B. Alcock and S. Zador, "Electrolytic Removal of Oxygen from Gases by Means of Solid Electrolyte," *J. Appl. Electrochem.*, **2**, 289 (1972).

[5] K. Agrawal, D. W. Short, R. Gruenke, and R. A. Rapp, "Control of Oxygen by Coulometric Titration," *J. Metall. Eng.*, **3**, 354 (1974).

[6] F. Vizethum, G. Bauer, and G. Tomandl; Science of Ceramics, Vol. 12, 1984; p. 757.

Oxygen Sensing in Iron- and Steelmaking

D. JANKE

Max-Planck-Institut für Eisenforschung GmbH
4000 Düsseldorf, Federal Republic of Germany

The properties of solid oxide electrolytes, the design of oxygen probes, and the benefits of the existing applications of these probes in the various stages of iron- and steelmaking are briefly reviewed. Furthermore, the potential future applications of oxygen probes in iron- and steelmaking technology are outlined in the present survey. In this context, the possibilities and limits of long-term measurements in steel melts are discussed.

Use of Oxygen Probes in the Various Stages of Iron- and Steelmaking

Single-reading probes for the in situ determination of dissolved oxygen in molten steel have achieved a high industrial standard and are regularly used in steelmaking operations. The probes are based on thin-walled tubes, closed at the lower end, consisting of partially stabilized ZrO_2, and on $Cr-Cr_2O_3$ powder mixtures representing the oxygen reference material. The measured emfs show a satisfactory accuracy and reproducibility within measuring intervals between 10 and 30 s after the rapid immersion of the sensors into the steel bath. The measuring devices are normally combined with a thermocouple sensor to simultaneously control the temperature of the steel bath.

Figure 1 is a schematic of the conventional production lines of iron- and steelmaking. Selected locations for useful operation of oxygen probes are marked in this scheme: (1) oxygen measurement in blast furnace hot metal before and after slag pretreatment in a transfer ladle; (2) end-point control for oxygen and carbon in the basic oxygen furnace (converter steelmaking); (3) control of oxidation and refining in electrofurnace steelmaking; (4) control of oxygen in casting ladles before and after homogenization and deoxidation of the steel bath; (5) control of carbon in the vacuum treatment of steel; (6) control of reoxidation of steel; (7) control of rimming action and deoxidation in the ingot casting of steel.

The probes must be operated in a wide range of oxygen activities,* according to the varying degrees of oxidation of the iron and steel melts in the stages of production (Fig. 2). The oxygen activities will be as high as 0.02–0.1 in the case of converter steel (2) or rimming steel (3) but as low as 0.0002–0.0008 in Al-deoxidized steel (5). Extremely low a_O (between 0.00005 and 0.0001) are attributed to blast furnace hot metal (1).

High-Temperature Properties of the Solid Electrolyte Materials
Composition and Chemical Stability

For industrial oxygen probes, ZrO_2-based electrolytes have been exclusively used so far. The commonly used partially stabilized zirconia (PSZ) is a two-phase

---

*Activity a_O of dissolved oxygen. On the basis of Henry's law, for a hypothetical 1 wt% solution, $a_O \equiv$ wt%.

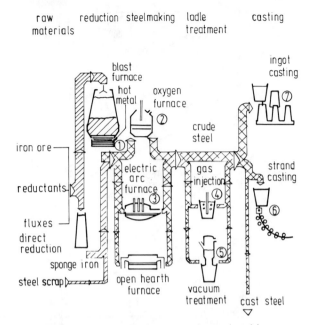

Fig. 1. Use of oxygen sensors in iron- and steelmaking.

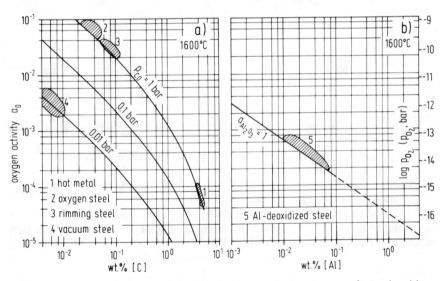

Fig. 2. Ranges of oxygen activity in the various stages of steelmaking.

(cubic + tetragonal) material with an excellent thermal shock resistivity, which enables rapid immersion of the probe into the molten steel. However, to secure a high oxygen ion conductivity, the cubic phase prevails in the structure. A typical composition is 7 mol% MgO. In some cases, fully stabilized zirconia (FSZ) was also used with lime additions from 10 to 20 mol%, which represents the stability

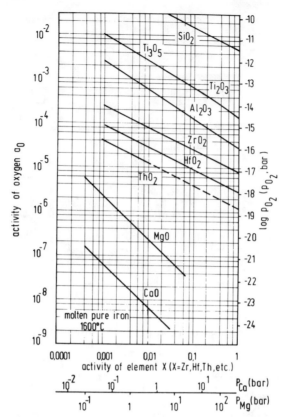

Fig. 3. Refractory oxide stabilities in contact with pure iron melts at 1600°C.

range of the cubic phase at higher temperatures. This paper refers to a zirconia electrolyte with 14 mol% CaO.

With special regard to ionic conductivity and chemical stability at low oxygen potentials, it is of interest to consider the use of hafnia- or thoria-based electrolyte materials. Reference is made in this report to solid electrolyte compositions of HfO_2 with 17 mol% CaO and ThO_2 with 8 mol% Y_2O_3.

The free energies of formation of ZrO_2, HfO_2, and ThO_2 as well as those of the dopant oxides MgO, CaO, and Y_2O_3[4] indicate that the ThO_2-Y_2O_3 combination exhibits the highest chemical stability. The refractory oxide stabilities in contact with pure iron melts at 1600°C are represented by the metal-oxygen reaction equilibria given in Fig. 3.

$$x[\text{Me}]_{\text{Fe, 1 wt\%}} + y[\text{O}]_{\text{Fe, 1 wt\%}} \leftrightarrows \text{Me}_x\text{O}_y(s) \tag{1}$$

Note that zirconia is essentially stable when exposed to an Al-deoxidized steel melt (0.05% Al, $a_O = 0.0002$). But decomposition and dissolution into molten Fe-C alloys (hot metal) at a_O between 10^{-5} and 10^{-4} must be considered to some extent.

Ionic and Electronic Conductivity

The solid oxide electrolytes must also be specified in view of their ionic and electronic conductivity. Partial electronic conductivity lowers the emf of an oxygen concentration cell:

oxygen reference ‖ oxide electrolyte ‖ metal melt

by partially short-circuiting the cell. According to Schmalzried's analysis,[1] the emf for oxide electrolytes with a mixed ionic and electronic conduction at low oxygen partial pressures ($p_{h'} \gg p_{O'_2,ref} > p_{O'_2,met} > p_{e'}$) is expressed by the equation

$$E = \frac{RT}{F} \ln \frac{p_{e'}^{1/4} + p_{O'_2,ref}^{1/4}}{p_{e'}^{1/4} + p_{O'_2,met}^{1/4}} \tag{2}$$

where the parameter $p_{e'}$ is defined as the oxygen partial pressure at equal ionic and excess electron conductivity ($t_{ion} = 0.5$), depending on the temperature and the type and composition of the electrolyte; $p_{e'}$ can be experimentally determined using a polarization technique. Its values for four typical electrolyte compositions based on zirconia, hafnia, or thoria are compiled in Table I.[2,3,5] In Fig. 4 the ionic transference number[1] of the various oxide electrolytes

$$t_{ion} = \left[1 + \left(\frac{p_{O_2}}{p_{e'}}\right)^{-1/4}\right]^{-1} \tag{3}$$

is shown as a function of the oxygen potential p_{O_2}. It is obvious from Fig. 4 that the electrolytes based on hafnia or thoria exhibit considerably lower partial electronic conductivity than ZrO_2-based electrolytes.

Oxygen Permeation

Accurate oxygen activities in molten steel are obtained from thin-walled single-reading probes at measuring intervals from 10 to 30 s after immersion. However, the long-term operation of these probes is impeded by oxygen transfer across the solid electrolyte.[6] The oxygen transport is induced by an electrical current due to the gradient of the oxygen potential between the molten metal and the oxygen reference electrode. The oxygen potential at both solid electrolyte interfaces is affected by this current. In particular, the electrolyte reference interface is polarized, inasmuch as the chemical Cr-Cr_2O_3 equilibrium is disturbed,[7] which causes a variation of the measured emf with time. The oxygen transport by ionic and electronic current

Table I. Parameters $p_{e'}$ ($t_{ion} = 0.5$) for Solid Oxide Electrolytes at 1200°–1600°C

Electrolyte	Composition (mol%)	Composition (wt%)	ln $p_{e'}$	Ref.	$p_{e'}$ at 1600°C (bar)
ZrO_2 (CaO, MgO) FSZ	84.0 ZrO_2 11.3 CaO 4.7 MgO	92.9 ZrO_2 5.5 CaO 1.6 MgO	$-68400/T + 21.59$	2	1.2×10^{-15}
ZrO_2 (MgO) PSZ	93.1 ZrO_2 6.9 MgO	97.6 ZrO_2 2.4 MgO	$-74370/T + 24.42$	5	5.1×10^{-16}
HfO_2 (CaO) FSH	83.5 HfO_2 16.5 CaO	95.0 HfO_2 5.0 CaO	$-70260/T + 20.35$	3	9.1×10^{-18}
ThO_2 (Y_2O_3)	92.0 ThO_2 8.0 Y_2O_3	93.0 ThO_2 7.0 Y_2O_3	$-82970/T + 26.38$	2	1.2×10^{-18}

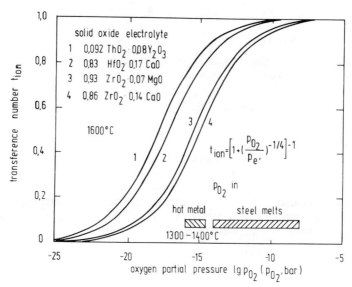

Fig. 4. Ionic transference numbers of various solid oxide electrolyte materials (Refs. 2, 3, 5).

$$j_{O^{2-}} = -\frac{\kappa_{O^{2-}}}{2Fl}(E - E^*)$$

$$= -\frac{\kappa_{O^{2-}}RT}{2F^2l}\left[\ln\left(\frac{p_{O_2,ref}^{1/4}}{p_{O_2,met}^{1/4}}\right) - \ln\left(\frac{p_{e'}^{1/4} + p_{O_2,ref}^{1/4}}{p_{e'}^{1/4} + p_{O_2,met}^{1/4}}\right)\right] \quad (4)$$

can be minimized by[8-10] (1) the application of an electrolyte with low electronic conductivity, (2) approach of the oxygen reference potential to the oxygen potential of the steel melt, and (3) increase of the electrolyte thickness in the oxygen probe.

Moreover, transfer of molecular oxygen is possible in polycrystalline ceramic materials, depending on the porosity (micropores and cracks), the thickness of the electrolyte, and the oxygen potential difference

$$j_{O_2} = -\left(\frac{1}{\tau}\right)\left(\frac{P}{l}\right)D_{O_2}\left(\frac{1}{RT}\right)(p_{O_2,ref} - p_{O_2,met}) \quad (5)$$

where j_{O_2} is the oxygen flux (mol·cm^{-2}·s^{-1}), τ tortuosity, ~0.1, p apparent porosity (volume fraction), and D_{O_2} gas diffusion coefficient, ~4 cm^2·s^{-1}.[11]

It has been shown that oxygen transfer due to ionic current prevails over oxygen gas diffusion at a thickness of 0.1 cm and apparent porosities from 0.5 to 2.0 vol%.

Oxygen Sensors

Construction and Response Time

Four essential designs of oxygen sensors are presented in Fig. 5. The tubular sensor of type A is preferably applied as a single-reading probe for short-term measurements in iron and steel melts under operational conditions. The needle sensor of type C represents a more recent development as a single-reading probe with possibly lower costs of manufacture and a quicker response after im-

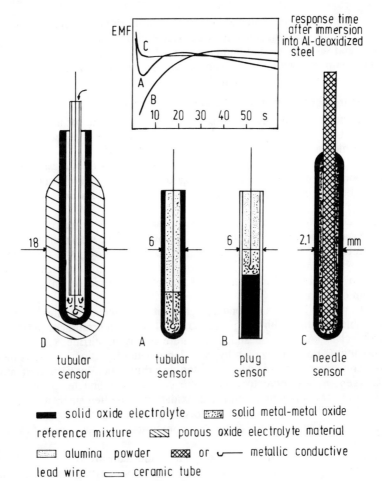

Fig. 5. Typical design of oxygen sensors for measurements in iron and steel melts.

mersion.[12,13] Sensor B was developed earlier for short-term measurements. In a modified form, it has recently been suggested for long-term measurements in steel melts. Type D represents a useful probe for continuous monitoring of oxygen, particularly in laboratory experiments.[3,5] Typical emf display curves are also included in Fig. 5 for various types of sensors after immersion in Al-deoxidized steel melts. It is shown that the accurate emf is attained after 5–8 s for needle sensors (type C), after 15–20 s for tubular sensors (type A), and 35–40 s for modified plug-type sensors (type B).

Calibration

Calibration of the oxygen sensors is possible in laboratory tests by using the following methods: (1) emf measurements in pure iron melts with sampling and subsequent oxygen determination by vacuum fusion analysis; (2) emf measurements in carbon-containing iron melts under controlled p_{CO} with sampling and subsequent carbon analysis where the measured oxygen activities are compared with the oxygen activities calculated from the equilibrium constant of the C-O-CO reaction; (3) emf measurements in Al-deoxidized iron melts at constant $a_{Al_2O_3}$ with sampling and subsequent aluminum analysis where the measured oxygen activities must be compared with the oxygen activities calculated from the equilibrium constant of the Al-O-Al$_2$O$_3$ reaction.

Emf Measurements in Iron- and Steelmaking Practice

It has been demonstrated that oxygen probes developed for molten steel may also be used for sensing the extremely low oxygen activities between 0.00001 and 0.0005 in molten pig iron at temperatures around 1400°C.

In the process of oxygen converter steelmaking, it is of interest to determine the end point of oxygen blowing with regard to the final carbon content. Knowledge of the degree of oxidation of the steel melt enables adequate additions of alloys.

The control of oxygen in the electric arc furnace is important from the blocking stage where the oxidation slag is removed and the steel melt is deoxidized using a refining slag with additions such as C, Fe-Si alloys, and CaC$_2$. Oxygen probe measurements serve to adjust the amount of deoxidizers to prevent either an increase in impurities or an extension of the refining period which favors H$_2$ and N$_2$ pickup of the steel.

In the past 5–10 years, secondary treatment in big teeming ladles has attracted considerable attention with various purposes: (1) homogenization of the steel bath with respect to composition and temperature by argon stirring; (2) deoxidation and alloying; (3) desulfurization by gas–slag powder injection; (4) adjustment of casting temperature; and (5) degassing and decarburization by vacuum treatment.

Oxygen probes serve to control the composition and temperature of the steel bath in these operations. In the deoxidation of molten steel, it is necessary to largely reduce the content of dissolved oxygen with appropriate additions of aluminum. Thereby, the amount of precipitated oxide impurities should be minimized.

The low oxygen activities established in the ladle treatment can easily be annulled by reoxidation during casting. The continuous casting system, consisting of teeming ladle, tundish, and mold, is shown schematically in Fig. 6.[14] Reoxidation may occur by reactions of the molten steel with air, fluxes, and refractories. The various reoxidation sites are marked. Oxygen probes, preferably in continuous operation, could be helpful tools to indicate reoxidation of the steel, e.g., in the tundish of the casting system. To meet these requirements, suitable long-term probes must be developed. A modified plug-type sensor, as shown in Fig. 7(a), has been suggested for this purpose.[7] The earlier developed standard plug-type sensor for short-term measurements is sketched in Fig. 7(b) for comparison. In Fig. 8 the behavior of modified plug-type sensors on immersion is shown in comparison with the commercial tubular sensors; both types are based on a Cr-Cr$_2$O$_3$ reference. It can be realized at two levels of oxygen that the plug-type sensors yield stable emf readings, whereas the emfs of the tubular sensors are correct in a short initial interval, with a subsequent time-dependent decay.

So far, reliable long-term measurements have been obtained over periods of 2 h.[7,15]

Conclusions

Oxygen probes based on partially stabilized ZrO$_2$ have been developed and established as a world-wide measuring routine in steelmaking processes. The

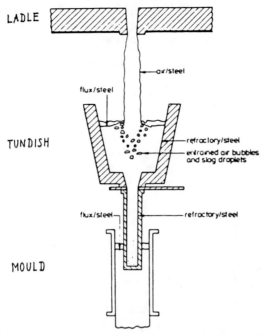

Fig. 6. Reoxidation sites in continuous casting of steel (Ref. 14).

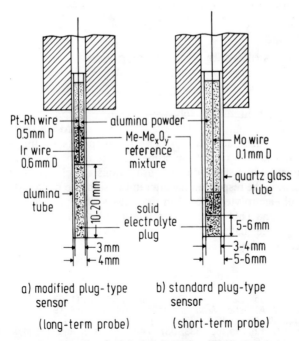

Fig. 7. Schematic sketch of plug-type sensors.

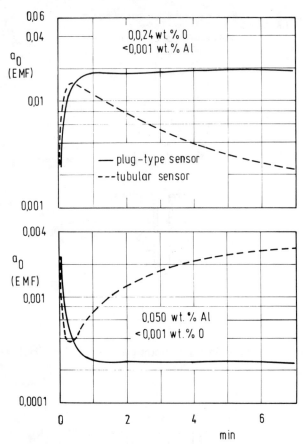

Fig. 8. Readings of emf by modified plug-type sensors and commercial tubular sensors in pure iron melts (Ref. 15).

oxygen devices are being used exclusively as single-reading probes with, so far, measuring intervals of only 10–20 s after immersion into the steel bath. Recent developments are dedicated to further improvements of the accuracy, reproducibility, and success rate of the probe measurements, to the reduction of the costs of manufacture by using simpler constructions and cheaper materials, and to a further decrease in the response time after immersion into molten steel. Stabilized ZrO_2 is a useful electrolyte material in most cases of application, but it is worth noting that HfO_2- and ThO_2-based electrolytes exhibit advantages at lower oxygen potentials in the steel bath due to their lower excess electron conductivity and higher chemical stability.

References
[1]H. Schmalzried, *Ber. Bunsenges. Phys. Chem.*, **66**, 572 (1962).
[2]D. Janke and W. A. Fisher, *Arch. Eisenhüttenwes.*, **46**, 683 (1975).
[3]D. Janke, *Metall. Trans.*, **13B**, 227 (1982).
[4]J. Barin and O. Knacke; pp. 333, 337, 742, 746, 888, and 916 in Thermochemical Properties of Inorganic Substances. Springer-Verlag, Berlin, 1973.

[5] D. Janke and W. A. Fisher, *Arch. Eisenhüttenwes.*, **46**, 755 (1975).
[6] D. Janke and H. Richter, *Arch. Eisenhüttenwes.*, **50**, 93 (1979).
[7] D. Janke, *Arch. Eisenhüttenwes.*, **54**, 259 (1983).
[8] H. H. Möbius and R. Hartung, *Silikattechnik*, **16**, 276 (1965); H. Ullmann, *Z. Phys. Chem.*, **237**, 71 (1968); T. Reetz, *ibid.*, **249**, 369 (1972).
[9] W. Pluschkell, *Arch. Eisenhüttenwes.*, **46**, 11 (1975); P. J. Kreyger, B. Slangen, and H. W. den Hartog, *Stahl Eisen*, **95**, 393 (1975).
[10] M. Iwase and T. Mori, *Trans. Iron Steel Inst. Jpn.*, **19**, 126 (1979).
[11] G. H. Geiger and D. R. Poirier; pp. 467–72 in Transport Phenomena in Metallurgy. Addison-Wesley, Reading, MA, 1973.
[12] D. Janke and K. Schwerdtfeger, *Stahl Eisen*, **98**, 825 (1978).
[13] D. Janke, K. Schwerdtfeger, J. Mach, and G. Bamberg, *Stahl Eisen*, **99**, 1211 (1979).
[14] K. Schwerdtfeger, *Arch. Eisenhüttenwes.*, **54**, 87 (1983).
[15] D. Janke, *Stahl Eisen*, **103**, 29 (1983).

Mixed Ionic and Electronic Conduction in Zirconia and Its Application in Metallurgy

M. Iwase and E. Ichise

Department of Metallurgy
Kyoto University
Kyoto, 606, Japan

K. T. Jacob

Department of Metallurgy
Indian Institute of Science
Bangalore 560012, India

Mixed ionic and electronic conduction in ZrO_2-based solid electrolytes was studied. The effect of impurities and second-phase particles on the mixed conduction parameter, P_n, was measured for different types of ZrO_2 electrolytes. The performance of solid-state sensors incorporating ZrO_2 electrolytes is sometimes limited by electronic conduction in ZrO_2, especially at temperatures >1800 K. Methods for eliminating or minimizing errors in measured emf due to electronically driven transport of oxygen anions are discussed. Examples include probes for monitoring oxygen content in liquid steel as well as the newly developed sulfur sensor based on a $ZrO_2(CaO)$ + CaS electrolyte. The use of mixed conducting ZrO_2 as a semipermeable membrane or chemically selective sieve for oxygen at high temperatures is discussed. Oxygen transport from liquid iron to CO + CO_2 gas mixtures through a ZrO_2 membrane driven by a chemical potential gradient, in the absence of electrical leads or imposed potentials, was experimentally observed.

Since the pioneering work of Kiukkola and Wagner[1] on solid oxide electrolytes, stabilized ZrO_2 has been used extensively for both thermodynamic and kinetic studies of high-temperature metallurgical systems. Equilibrium studies involving such electrolytes are based on open-circuit measurements of emf of electrochemical cells. In kinetic studies, either a constant dc voltage is applied across electrochemical cells and the current is measured as a function of time (potentiometric method) or a constant current is drawn through the cell and the voltage is measured as a function of time (galvanostatic method). Faraday's law gives the relation between current and the rate of oxygen transport through the electrolyte when operated in the ionic domain. These applications, however, have often been limited by the presence of electronic conduction in the electrolyte.

The transference number of ions in stabilized ZrO_2, in the oxygen pressure-temperature domain where both ionic and electronic conduction prevail, is given by[2]

$$t_{ion} = \left[1 + \left(\frac{P_{O_2}}{P_n}\right)^{-1/4} + \left(\frac{P_{O_2}}{P_p}\right)^{1/4}\right]^{-1} \qquad (1)$$

The two parameters, P_p and P_n, are defined as oxygen partial pressures at which the ionic conductivity, σ_{ion}, is equal to the p-type and n-type electronic conductivities, respectively. The proposed $P_{O_2}^{1/4}$ and $P_{O_2}^{-1/4}$ dependencies for the p-type and n-type electronic conductivities, respectively, arise from the defect equilibria

$$O_2 + 2V_O^{\cdot\cdot} = 2O_O^\times + 4h^{\cdot} \tag{2}$$

$$2O_O^\times = O_2 + 2V_O^{\cdot\cdot} + 4e' \tag{3}$$

where Kröger–Vink notation[3] is used. The open-circuit cell voltage, E^*, for an electrolyte that exhibits mixed ionic and electronic conduction is given by[2]

$$E^* = \frac{RT}{F}\left[\ln\left(\frac{P_{O_2}^{'1/4} + P_n^{1/4}}{P_{O_2}^{''1/4} + P_n^{1/4}}\right) + \ln\left(\frac{P_{O_2}^{''1/4} + P_p^{1/4}}{P_{O_2}^{'1/4} + P_p^{1/4}}\right)\right] \tag{4}$$

where R is the gas constant, F the Faraday constant, T temperature, and P'_{O_2} and P''_{O_2} are the oxygen partial pressures at each electrode.

Mixed ionic and electronic conduction in ZrO_2-based electrolyte was studied. The parameter, P_n, was measured for different types of ZrO_2 electrolytes. Methods for eliminating or minimizing errors in measured emf due to electrochemical semipermeability flux are discussed. Examples include probes for monitoring oxygen content in liquid steel as well as newly developed sulfur sensors based on a $ZrO_2(CaO) + CaS$ two-phase electrolyte.[4]

Measurements of the Parameter, P_n

The chemical compositions of three commercial ZrO_2 specimens (ZR-11, ZR-15, and ZR-15C) used for the P_n measurements are given in Table I, as well as ZR-S, which was produced at the author's laboratory. The latter was fabricated as follows. A fine mixture of $ZrO_2 + 15$ mol% CaO was prepared from high-purity nitrate solution by evaporation to dryness and subsequent decomposition of the nitrate. This material was mechanically mixed with 8 mol% CaS. The resulting material was cold-pressed into a pellet 15 mm in diameter by 5 mm thick in a steel die. The pellets were painted with a slurry of CaS. After being dried, they were

Table I. Characteristics of the ZrO_2 Samples Used for P_n Measurements

Sample code	Experimental method used for P_n measurements	Degree of stabilization (%)	Major impurities		Remark
			SiO_2	Al_2O_3	
ZR-11	Oxygen permeability Coulometric titration	95	1.50*	0.70*	11 mol% CaO
ZR-15	Coulometric titration	97	1.50*	0.70*	15 mol% CaO
ZR-15C	Coulometric titration	100	0.30*	0.30*	15 mol% CaO less impurities
ZR-S	Open-circuit emf measurement	100	(High purity)		Two-phase electrolyte $0.92[(ZrO_2)_{0.85}(CaO)_{0.15}] + 0.08$ CaS

*Percent by weight.

sintered for 20 h at 2073 K and 4 h at 2273 K under purified argon. The microstructure of the sintered product showed a fine dispersion of CaS in $ZrO_2(CaO)$.

The parameter, P_n, was measured using three techniques: the permeability flux method,[5-7] Swinkels' coulometric titration technique,[8-10] and open-circuit emf measurements.

Permeability Flux Method

The oxygen permeability of ZrO_2 electrolyte was measured as follows. A ZrO_2 tube was heated at 1673–1823 K using an SiC resistance furnace. The outside of the tube was flushed with a stream of either $N_2 + O_2$ or $CO + CO_2$ gas mixture, while the inside was filled with pure oxygen at 1 atm (1.01×10^5 Pa) pressure. The amount of oxygen inside the ZrO_2 tube was found to decrease gradually due to the permeation of oxygen through the electrolyte. The mercury leveler connected to the ZrO_2 tube was adjusted, therefore, to keep the pressure of pure oxygen at 1 atm (1.01×10^5 Pa) pressure, and the change of mercury level was measured at intervals of 1–3 min for 20–30 min. The ZrO_2 tubes were found to be almost impermeable when a pressure gradient of an inert gas was imposed across the tube. Hence, oxygen permeability occurs by ionic transport, not through physical pores or microcracks. Since the oxygen pressure gradient across the tube was kept constant during these experiments, the oxygen permeability through the ZrO_2 tube can be calculated using equations for steady-state diffusion.

The permeability measurement cells can be represented schematically by cell A

$$O_2 \ (P'_{O_2} = 101\,325 \text{ Pa})/ZrO_2(CaO)/N_2 + O_2 \ (P''_{O_2} = 100\text{–}40\,000 \text{ Pa})$$

and cell B

$$O_2 \ (P'_{O_2} = 101\,325 \text{ Pa})/ZrO_2(CaO)/CO + CO_2 \ (P''_{O_2} = 10^{-2}\text{–}10^{-7} \text{ Pa})$$

The oxygen permeability due to p-type and n-type electronic conduction is given by[5]

$$J_{O_2} = \frac{RT}{4F^2 L} [\sigma_p^\circ (P'^{1/4}_{O_2} - P''^{1/4}_{O_2}) - \sigma_n^\circ (P'^{-1/4}_{O_2} - P''^{-1/4}_{O_2})] \tag{5}$$

where J_{O_2} is the oxygen flux (mol·cm^{-2}·s^{-1}), L is the thickness of the electrolyte, and σ_p° and σ_n° are the p-type and n-type electronic conductivities, respectively, at $P_{O_2} = 1$ Pa. The introduction of the new parameter, $P_{p,n}$, facilitates the discussion of the specific cases which arise from Eq. (5).

$$P_{p,n} = (\sigma_p^\circ / \sigma_n^\circ)^{-2} \tag{6}$$

This parameter shows the oxygen partial pressure at which the p-type and n-type electronic conductivities are equal.[5] The special cases of Eq. (5) are given below.

(A) If the sequence of oxygen partial pressures is

$$P_p \gg P'_{O_2} > P''_{O_2} \gg P_{p,n} \tag{7}$$

Eq. (5) can be simplified to

$$J_{O_2} = \frac{RT}{4F^2 L} \sigma_p^\circ (P'^{1/4}_{O_2} - P''^{1/4}_{O_2}) \tag{8}$$

(B) When the oxygen partial pressures satisfy the condition

$$P_p \gg P'_{O_2} \gg P_{p,n} \gg P''_{O_2} \gg P_n \tag{9}$$

then

$$J_{O_2} = \frac{RT}{4F^2L}(\sigma_p^\circ P_{O_2}^{\prime 1/4} + \sigma_n^\circ P_{O_2}^{\prime\prime -1/4}) \tag{10}$$

(C) The condition

$$P_{p,n} \gg P_{O_2}' > P_{O_2}'' \gg P_n \tag{11}$$

leads to

$$J_{O_2} = \frac{-RT}{4F^2L}\sigma_n^\circ(P_{O_2}^{\prime -1/4} - P_{O_2}^{\prime\prime -1/4}) \tag{12}$$

The sequence of oxygen partial pressures in cells A and B satisfy the conditions given by Eqs. (7) and (9), respectively. The relationships between J_{O_2} and P_{O_2}'' obtained with cells A and B were in agreement with those predicted from Eqs. (8) and (10), respectively. The values of σ_p° and σ_n° obtained from the permeability measurements were combined with the available ionic conductivity data[11] to derive the values for P_n and P_p.

Coulometric Titration Technique

The experimental apparatus for this technique is shown in Fig. 1. An external dc current was applied to cell C

Pt/O_2 (101 325 Pa)/ZrO_2/Ag + O(l)/Mo

until the following condition is satisfied

$$P_{O_2}(Ag) \ll P_n \tag{13}$$

The cell voltages were measured on a strip chart recorder of 2 MΩ impedance; a sharp decay in voltage indicated that an open-circuit condition had been achieved. The open-circuit emf obtained immediately after the current was interrupted is given by

$$E = -\frac{RT}{4F}(\ln P_n - \ln 101\,325) \tag{14}$$

The Coulometric titration technique has the advantage of being simple and rapid. Hence, the majority of the ZrO_2 samples were tested with this technique.

Open-Circuit Emf Measurements

When p-type electronic conduction in ZrO_2 electrolyte is negligible in comparison with n-type electronic conduction, Eq. (4) becomes

$$E = \frac{RT}{F} \ln \left(\frac{P_{O_2}^{\prime 1/4} + P_n^{1/4}}{P_{O_2}^{\prime\prime 1/4} + P_n^{1/4}} \right) \tag{15}$$

Consequently, if oxygen partial pressures at each electrode, P_{O_2}' and P_{O_2}'', are known precisely, open-circuit measurements of the cell emf can be used to derive values of P_n. This technique was applied to two-phase ZrO_2(CaO) + CaS solid electrolyte using cell D:

Pt(Au)/Ar' + H_2S' + H_2'/ZrO_2(CaO) + CaS/Ar" + H_2S'' + H_2''/(Au)Pt

Since calcium sulfide coexists with dissolved calcium oxide in the ZrO_2 electrolyte, the sulfur-oxygen exchange reaction must be considered

$$2CaS + O_2 + 2H_2 = 2CaO(\text{in } ZrO_2) + 2H_2S \tag{16}$$

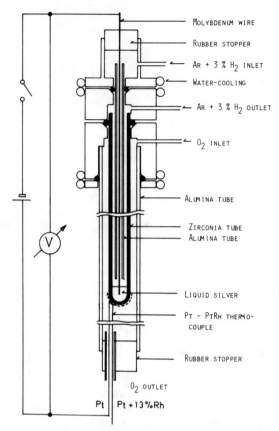

Fig. 1. Experimental apparatus for the Coulometric titration technique.

The gas phase establishes the sulfur potential at the electrolyte-electrode interface, and this is converted to an equivalent oxygen potential by reaction (16). The oxygen partial pressures can be precisely calculated by using available thermochemical data. The values of $\log (P_{H_2S}/P_{H_2})$ at each electrode were kept constant at -4.7609 and -1.8348, respectively. Special precautions had to be taken to ensure that there was no oxygen potential gradient at the electrolyte-gas mixture interface.[4] This will be described in a later section. Figure 2 shows the relation between open-circuit emf and temperature; the measured emfs were found to deviate significantly from the theoretical values calculated from the simple Nernst's equation ($E = RT/4F \ln (P'_{O_2}/P''_{O_2})$) because of the presence of n-type electronic conduction in the electrolyte. The values of P_n were derived using Eq. (15). It should be remembered that, when the difference between the measured emf and that given by the Nernst equation is small, the accuracy of derived P_n is poor.

Results and Discussion

For the $ZrO_2 + 11$ mol% CaO electrolyte (ZR-11), both the oxygen permeability flux technique and the Coulometric titration method were used to determine the P_n parameter. The results are given in Fig. 3. As shown in this figure, the oxygen permeability technique was less accurate than the other. This is attributed

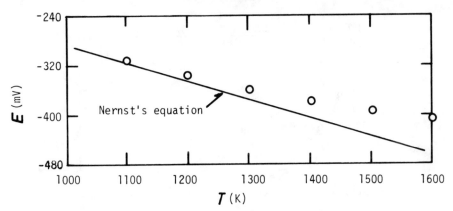

Fig. 2. Relation between open-circuit emfs and temperature obtained with cell D.

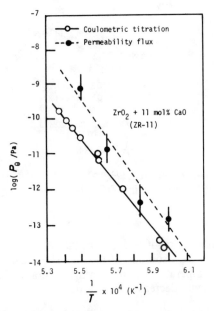

Fig. 3. Relation between log P_n and $1/T$ obtained by using the oxygen permeability flux method and the Coulometric titration technique for ZrO_2 + 11 mol% CaO sample (ZR-11).

to the uncertainties in the effective permeation area.[5] The P_n values obtained from permeability are, however, in satisfactory agreement with those by the Coulometric titration method.

The values of P_n for the ZrO_2 + 15 mol% CaO electrolytes obtained are compared in Fig. 4 with those for high-purity ZrO_2 of the same dopant concentration reported by Scaife et al.[10] The effect of impurities on electronic conduction can be readily seen; the P_n values increase with increasing concentrations of

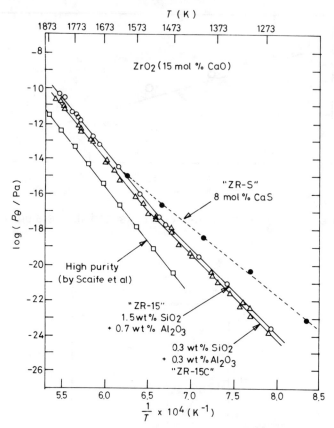

Fig. 4. Relation between log P_n and $1/T$ obtained by using the Coulometric titration technique for ZrO_2 + 15 mol% CaO samples (ZR-15, ZR-15C, and ZR-S).

SiO_2 and Al_2O_3. Furthermore, it is apparent that the dispersion of CaS in cubic ZrO_2 + 15 mol% CaO solid solution results in a significant increase in electronic conduction at lower temperatures, although the accuracy of P_n obtained by open-circuit emf measurements decreases with a decrease in temperature for reasons already discussed.

Method for Eliminating or Minimizing Errors in Emf of Electrochemical Cells due to Electronic Conduction

The oxygen potential at the electrolyte-electrode interface is often different from that in the bulk of the electrode. This difference is usually referred to as overvoltage or polarization. All polarization phenomena finally stem from the ionic "leakage" current through the electrolyte due to trace concentration of electronic defects. This flux is given by Eq. (5). Although in principle many types of polarization may occur, this paper will focus on diffusion polarization that can result in significant errors in measured open-circuit emf. While fundamental aspects of diffusion polarization in ZrO_2-based electrochemical cells were reviewed by Schmalzried[12] at the first zirconia conference, practical solutions of this prob-

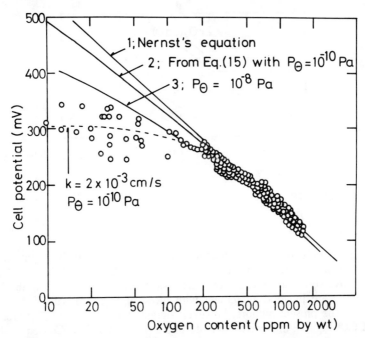

Fig. 5. Relation between measured emfs and oxygen content in liquid steel (Ref. 13).

lem, i.e., methods for eliminating or minimizing errors in measured cell emf due to such a diffusion polarization or a method for correcting measured emf for polarization effects, have been investigated only recently. In this study, two such methods are suggested.

Correction for Diffusion Polarization at Electrolyte-Liquid Steel Interface

An electrochemical cell incorporating a ZrO_2-based solid electrolyte and an appropriate reference electrode such as $Mo + MoO_2$ and $Cr + Cr_2O_3$ can be used for direct and instantaneous measurements of oxygen activity or concentration in liquid steel. Such electrolyte probes have been extensively used in steelmaking processes. However, industrial application of electrochemical oxygen probes has not been without problems. It has been reported, for example, that different types of commercial oxygen probes indicate different oxygen activities under industrial conditions. Such a difference in measured cell potential is attributed to the difference in the extent of the electronic conduction in specific ZrO_2 used in individual commercial oxygen probes. The diffusion polarization must be taken into account in order to relate the measured emfs and the bulk oxygen concentration in liquid steel at low concentrations.

Figure 5 shows the relation between measured cell emf of a galvanic sensor for oxygen in liquid steel (cell E described below) and the oxygen concentration obtained by chemical analysis.[13]

$$Mo/Mo + MoO_2/ZrO_2(15 \text{ mol\% } CaO)/O(\text{in Fe})/Mo$$

The measured cell voltages between 500 and 1800 ppm by weight of oxygen in

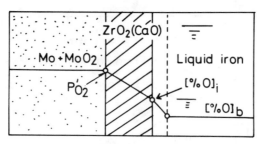

Fig. 6. Oxygen potential gradient at the liquid steel-electrolyte interface.

liquid steel agree essentially with those predicted from the Nernst equation:

$$E = \frac{RT}{4F} \ln \left(\frac{P'_{O_2}}{P''_{O_2}}\right) \qquad (17)$$

The cell potentials for oxygen concentration in the range 200–500 ppm are close to those given by the Nernst equation (curve 1). Between 100 and 200 ppm oxygen, however, cell voltages are much lower than those exhibiting the Nernst behavior but are in good agreement with those of Eq. (15) with $P_n = 10^{-8}$ Pa (curve 3). The present authors, however, obtained a value of 10^{-10} Pa for P_n at 1873 K (see Fig. 4); this corresponds to line 2 on the diagram. Consequently, the apparent agreement between the emf and those in curve 3 does not prove the applicability of Eq. (15). This view is supported by the results obtained below 100 ppm oxygen. The measured emfs in this range are significantly lower than those calculated in line 3 and very much lower than those calculated in line 2. These low emfs can be explained by the oxygen potential gradient near the electrolyte–liquid-steel interface.

A comprehensive interpretation of the results shown in Fig. 5 can be given by considering the oxygen flux through the electrolyte and the resulting diffusion polarization. From Eq. (5) it can be deduced that,

$$-\frac{dN(O^{2-})}{dt} = \frac{RT\sigma_{ion}}{2F^2L}\{\ln(1 + P_n^{1/4}K^{1/2}[\%O]_i^{-1/2}) - \ln(1 + P_n^{1/4}P_{O_2}^{'-1/4})\} \qquad (18)$$

where $dN(O^{2-})/dt$ is the number of moles of oxygen anion transported per unit time, $[\%O]_i$ the oxygen concentration in wt% in liquid steel at the liquid steel-electrolyte interface, and K the equilibrium constant for the reaction $\frac{1}{2}O_2 = O$ (1 wt% in liquid iron).

By assuming that there exists no oxygen potential gradient at the reference electrode-electrolyte interface, one may consider oxygen transport within the concentration gradient at the liquid steel-electrolyte interface, as shown in Fig. 6. The rate of oxygen transport across this flow boundary layer is given by

$$-\frac{dN(O)}{dt} = \frac{k\rho}{1600}([\%O]_b - [\%O]_i) \qquad (19)$$

where $dN(O)/dt$ is the number of moles of atomic oxygen transported through the flow boundary layer at unit time, k the mass transport coefficient for oxygen, ρ the density of liquid steel, and $[\%O]_b$ the bulk oxygen concentration in wt%. By combining Eqs. (18) and (19), the values of $[\%O]_i$ can be calculated as a function of $[\%O]_b$, and subsequently the relation between measured emf and $[\%O]_b$ can be

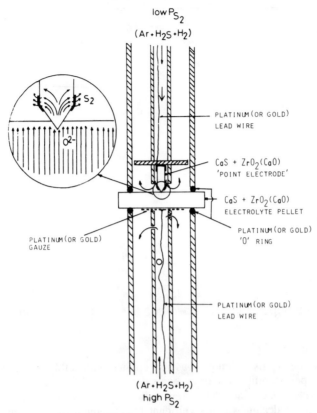

Fig. 7. "Point electrode" design used by the authors. The concept of point electrode was introduced by Fouletier et al (Ref. 14).

obtained as shown by the dashed line in Fig. 5. Although fairly large scatter is associated with potential measurements at oxygen levels below 100 ppm, the dashed line well represents the experimental results. As can be seen from Eqs. (18) and (19), the emf–$[\%O]_b$ relationship depends significantly on the mass transport coefficient, k. With respect to the experimental data shown in Fig. 5, the most appropriate values for k ranged from 1×10^{-3} to 7×10^{-3} cm/s, with a mean value of 2×10^{-3} cm/s. Equation (15) is not sufficient to determine oxygen levels less than about 200 ppm, and diffusion polarization at the liquid metal electrode must be considered.

Diffusion Polarization at the Pt/Ar + H$_2$S + H$_2$ Electrode

The two-phase electrolyte, ZrO$_2$(CaO) + CaS, can be used for monitoring sulfur potentials in the Ar + H$_2$S + H$_2$ gas mixtures, as described in the preceding section. At lower temperatures, the P_n values for this electrolyte are higher than the commercial-grade ZrO$_2$ and very much higher than the high-purity ZrO$_2$ electrolyte (see Fig. 4). Consequently, one should anticipate mass transport through the electrolyte and a concentration gradient in the gas phase in the vicinity of the electrolyte-gas mixture interface. A useful method for minimizing such concentration polarization was suggested by Fouletier et al.[14] Figure 7 illustrates their

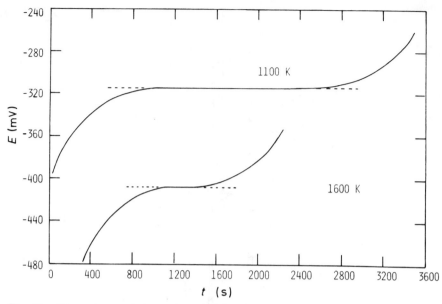

Fig. 8. Time dependence of open-circuit emfs of cell D after Coulometric titration.

"point electrode." This technique was also applied with cell D by the authors. The oxygen semipermeability flux or an equivalent sulfur flux is dissipated at the tip of the point electrode, since the current will take the lowest resistance path. This special electrode design was a significant improvement in the open-circuit emf measurements with cell D. Nevertheless, the cell emfs showed a significant decrease with time at higher temperatures and lower sulfur potentials; additional procedures were required to obtain stable cell potentials. An external current was applied to cell D to titrate uphill against the chemical potential gradient. Traces of time dependence of the emf at 1100 and 1600 K after a current is passed in the range 5 to 100 μA for 15–60 min are shown in Fig. 8. The emfs of the "plateau" region are considered to be free from the diffusion polarization effect at the electrode. The sequence of the oxygen potentials or equivalent sulfur potentials before and after such an uphill titration is illustrated schematically in Fig. 9.

Application of Mixed Conducting Zirconia in Metallurgy

As discussed above, the application of ZrO_2-based solid electrolytes at high temperatures has often been limited because of electronic conduction. Consequently, an effort has been made to obtain highly ionically conducting ZrO_2. Mixed conductors have been considered to be undesirable. However, mixed conductors have some useful applications in metallurgy. ZrO_2-based mixed conductors have been used for transporting oxygen from one phase to another in the absence of an external current,[15] e.g., for removing oxygen from liquid iron to CO + CO_2 gas mixtures. An alumina crucible was charged with about 500 g of pure iron at 1823 K under a stream of purified argon inside an SiC resistance furnace. A calcia-stabilized ZrO_2 tube (ZR-11) was immersed in liquid iron. The inside of the ZrO_2 tube was flushed with a stream of either CO + 112 ppm CO_2 or

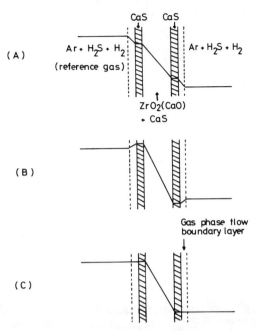

Fig. 9. Schematic diagram showing a gradient of chemical potential of oxygen or equivalent sulfur potential at the electrolyte-electrode interface: (A) before passing external current, (B) during the uphill titration, (C) during the period corresponding to the "plateau" region of emf shown in Fig. 8.

$CO + 896$ ppm CO_2 gas mixture. The initial oxygen concentration in liquid iron was adjusted to 580 ± 10 ppm by weight by virtue of the equilibrium reaction between three condensed phases

$$Fe(l) + O(\text{in Fe}) + (1 + x)Al_2O_3 = FeAl_{2+2x}O_{4+3x} \tag{20}$$

The rate of oxygen removal from liquid iron to the $CO + CO_2$ gas mixtures through the ZrO_2 tube, without the application of external voltages, was determined by oxygen analysis of a sample obtained by suction.

Figure 10 shows the time dependence of oxygen concentration in liquid iron. The initial oxygen content, 580 ± 10 ppm, decreased to 350 ppm after 200 min and to 200 ppm after 500 min. When the oxygen content in the melt became lower than 580 ppm, the decomposition of $FeAl_{2+2x}O_{4+3x}$ must take place

$$FeAl_{2+2x}O_{4+3x} = Fe(l) + O(\text{in Fe}) + (1 + x)Al_2O_3 \tag{21}$$

The change in the oxygen content of liquid iron suggests that oxygen transport through the electrolyte is faster than the decomposition of iron aluminate. The dashed curve in Fig. 10 represents the results of kinetic analysis based on the assumption that the rate of oxygen transport is limited by solid-state diffusion of oxygen anion through the ZrO_2 tube. This theoretical line is not in agreement with the experimental results. To fit the experimental data, it had to be assumed that the rate of oxygen transport is limited by the liquid-state diffusion of oxygen in the boundary layer near the ZrO_2-liquid-iron interface. The results of this

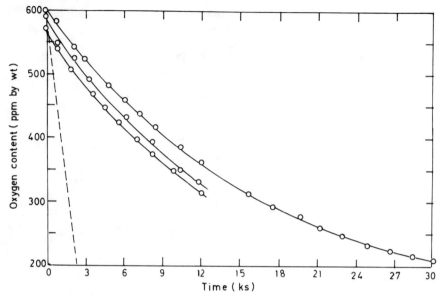

Fig. 10. Removal of oxygen from liquid iron to CO + CO_2 gas mixture through calcia-stabilized ZrO_2. Three sets of results were obtained with different flow rates of CO + CO_2 gas mixtures, indicating that the transport rate was not controlled by the gas-phase diffusion. Dashed line shows the results of kinetic analysis made assuming that the rate of oxygen removal is limited by the solid-state diffusion of oxygen anion through a ZrO_2 tube (Ref. 15).

kinetic analysis are particularly encouraging, since the rate of oxygen transport can be accelerated by stirring the liquid iron. A mixed conductor can thus be used for transporting oxygen from one phase to another without imposed potentials. Similarly mixed conducting ZrO_2 may be used for the thermolysis of water, as suggested by Anthony et al.[16]

Acknowledgment

The authors express their sincere appreciation to M. Takeuchi, T. Kawanami, and T. Yamasaki, Nippon Kagaku Togyo Co. Ltd., for their constructive comments and discussion. This study was supported financially by the Hyuga Hosai Scholarship of The Iron and Steel Institute of Japan (granted to M. I.), and this is gratefully acknowledged.

References

[1] K. Kiukkola and C. Wagner, "Measurements on Galvanic Cells Involving Solid Electrolyte," *J. Electrochem. Soc.*, **104**, 379 (1957).
[2] H. Schmalzried, "Ionen-und Elektronenleitung in Binaren Oxiden und ihre Untersuchung Mittels EMK-Messungen," *Z. Phys. Chem., Neue Folge*, **34**, 87 (1963).
[3] F. A. Kröger and H. J. Vink; p. 310 in Solid State Physics, Vol. 3. Edited by F. Seitz and D. Turnbull. Academic Press, New York, 1956.
[4] K. T. Jacob, M. Iwase, and Y. Waseda, "Sulfur Potential Measurements with a Two-Phase Sulphide-Oxide Electrolyte," *J. Appl. Electrochem.*, **12**, 55 (1982).

[5] M. Iwase and T. Mori, "Oxygen Permeability of Calcia-Stabilized ZrO_2," *Met. Trans., B,* **9B**, 365 (1978).

[6] M. Iwase and T. Mori, "Oxygen Permeability of Calcia-Stabilized ZrO_2 at Low Oxygen Partial Pressures," *Met. Trans., B,* **9B**, 653 (1978).

[7] W. A. Fischer; pp. 503–12 in Fast Ion Transport in Solid–Solid State Batteries and Devices. Edited by W. V. Gool. North-Holland, Amsterdam, 1972.

[8] D. A. J. Swinkels, "Rapid Determination of Electronic Conductivity Limits of Solid Electrolyte," *J. Electrochem. Soc.,* **117**, 1267 (1970).

[9] T. H. Etsell and S. N. Flengas, "N-Type Conductivity in Stabilized ZrO_2 Solid Electrolyte," *J. Electrochem. Soc.,* **119**, 1 (1972).

[10] P. H. Scaife, D. A. J. Swinkels, and S. R. Richards, "Characterisation of ZrO_2 Electrolyte for Oxygen Probes Used in Steelmaking," *High Temp. Sci.,* **8**, 31 (1976).

[11] K. S. Goto, "A Survey on Industrial Application of Oxygen Concentration Cells with Solid Electrolyte," *Trans. Iron. Steel. Inst. Jpn.,* **16**, 469 (1976).

[12] H. Schmalzried; pp. 254–71 in Advances in Ceramics, Vol. 3. Edited by A. H. Heuer and L. W. Hobbs. The American Ceramic Society, Columbus, OH, 1981.

[13] M. Iwase and A. McLean, "Evaluation of Electrochemical Oxygen Probes for Use in Steelmaking," *Solid State Ionics,* **5**, 571 (1981).

[14] J. Fouletier, P. Fabry, and M. Kleitz, "Electrochemical Semipermeability and Electrode Microstructure in Solid Oxide Electrolyte Cells," *J. Electrochem. Soc.,* **123**, 204 (1976).

[15] M. Iwase, M. Tanida, A. McLean, and T. Mori, "Electronically Driven Transport of Oxygen from Liquid Iron to $CO + CO_2$ Gas Mixtures Through Stabilized ZrO_2," *Met. Trans., B,* **12B**, 517 (1981).

[16] A. M. Anthony; pp. 437–54 in Advances in Ceramics, Vol. 3. Edited by A. H. Heuer and L. W. Hobbs. The American Ceramic Society, Columbus, OH, 1981.

ZrO_2 Oxygen and Hydrogen Sensors: A Geologic Perspective

GENE C. ULMER

Temple University
Geology Department
Philadelphia, PA   19122

The geosciences have been attracted to the high accuracy of ZrO_2 cells for both $f(O_2)$ and pH sensors. That the very same ZrO_2 membrane can be used above 600°C to sense $f(O_2)$ and used between 25° and 300°C (maybe higher) to sense pH has been demonstrated. Specific resistivity measurements for such cells follow the equation $\log R = -2.20 + 4000/T$ (for T(K) from 298–1573 K) (for Y_2O_3 levels of 4–8 mol%). In the lower-temperature regime, i.e., pH sensing, the ZrO_2 cell does not respond to changes in molecular O_2 or H_2 in its environment. Geochemical raw material impurities and ZrO_2 membrane fabrication techniques that affect $f(O_2)$ and pH sensing are discussed. The application of ZrO_2 cells to various geologic redox equilibria are demonstrated by a few selected examples.

In the geosciences, accurate pressure-temperature data for oxidation-reduction equilibria are scarce; the utility of ZrO_2 as a high-temperature oxygen sensor (>600°C currently) has been realized in geosciences for about two decades,[1] but the utility of ZrO_2 as a low-temperature (<300°C currently) hydrogen ion sensor has been possible for only the last 4 years.[2] In Fig. 1 are shown portions of oxygen fugacity–T space ($\approx$oxygen partial pressure) and portions of pH-Eh* space of interest in earth sciences. As delineated in Fig. 1(A), a few of the better-known redox equilibria are represented at 1 atm (9.80×10^4 Pa) — solid lines show the $f(O_2)$-T relationships for selected mineral-vapor equilibria: (CCO) represents the assemblage graphite-CO_2-CO-O_2; (WI) represents the assemblage wustite(nonstoichiometric FeO)-iron metal-O_2; (WM) represents the assemblage wustite-magnetite(Fe_3O_4)-O_2; (QFM) represents the assemblage quartz(SiO_2)-fayalite(Fe_2SiO_4)-magnetite-O_2. In Fig. 1(B), solid lines show the thermodynamic stability of water at 25°C and 1 atm (9.80×10^4 Pa), while the dashed boxed area shows typical environments for terrestrial waters. In both (A) and (B), word positions indicate generalized locations of geoscience environments. When suitable ZrO_2 electrolytic cells are used, these ranges of environments are opened to accurate investigation. Furthermore, efforts to extend ZrO_2-$f(O_2)$-T experiments to high pressure ($\approx$30 kbars) ($\sim 5.87 \times 10^1$ Pa) have been attempted[7] and efforts to extend ZrO_2-pH-T hardware to 300 bars (0.587 Pa) are also reviewed herein. Given that typical existing redox equilibria thermodynamic data bases[8–10] often disagree by orders of magnitude, the reproducibility generally claimed for ZrO_2

---

*Eh is the electron concentration in aqueous environments and is controlled by various redox equilibria.

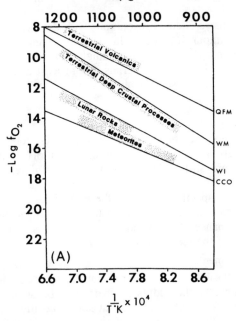

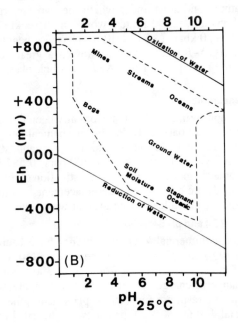

Fig. 1. Geoscience redox environments (see text for discussion): (A) QFM, WM, and WI, buffer assemblages whose redox values are known (Ref. 3) CCO data were first calculated (Ref. 4) and later experimentally determined (Ref. 5); (B) adapted from the data of Becking et al. (Ref. 6).

sensors of ±0.15 log units of $f(O_2)$ and ±0.1 pH units is very attractive to experimentalists.

Unfortunately, these claimed resolutions with ZrO_2 cells are easily forfeited when raw material impurities and/or fabrication techniques interfere with the sensing ability for either O^{2-} or H^+. Accordingly, it is necessary to understand the general geochemistry of the raw materials and at least a few of the fabrication parameters that can influence sensing abilities of ZrO_2 cells to be used in geologic redox research.

Geologic Methods

The specific geologic methods of utilizing ZrO_2 for $f(O_2)$ determination have been extensively treated in the literature.[7,11-13] In pH-Eh measurements with ZrO_2 cells, which are still under development, useful references are already available.[2,14,15]

Accordingly, the specific details of the techniques are not recounted here. However, to understand the role of raw material impurities and fabrication techniques on the performance of $f(O_2)$ and Eh-pH cells, it is necessary to understand the broader aspects of the techniques.

Intrinsic $f(O_2)$ Measurements

When a polyphasic rock or material can be separated into its constituent minerals or components, the individual parts can be subjected to an $f(O_2)$–T determination with a ZrO_2 $f(O_2)$ cell. If, at some specific temperature, two (or more) separates from the same host material have the same intrinsic $f(O_2)$, this temperature and $f(O_2)$ are thought to have genetic significance, i.e., these are the redox and temperature conditions under which the two minerals or components last communicated via oxygen. Limited work on these kinds of experiments at pressures as great as 20 kbars (3.91 × 10^1 Pa) has also begun.[5] As explained in a later section, the $f(O_2)$–T technique has allowed a great deal of research on the genesis of ancient or samples far removed from their place of origin (see data of Fig. 1(A)).

Hydrothermal pH Measurements

Control and/or monitoring of pH has historically been limited to temperatures <100°C and pressures ≲60 bars (0.12 Pa) by mechanical design difficulties and the chemical instabilities of glass electrodes. The availability of pH circuits at temperatures up to 300°C and pressures up to 300 bars (0.587 Pa) utilizing ZrO_2 pH sensors (potentiometric mode) allows the investigation of diverse hydrothermal problems such as sulfide ore migration and emplacement,[16] evaluation of geothermal brine corrosivity,[17] and nuclear waste ground water interactions.[15]

Relationship between $f(O_2)$ and pH Sensing

Some individual commercial ZrO_2 membranes have been demonstrated to respond in both the $f(O_2)$ and pH sensing modes, i.e., $f(O_2)$ response at >600°C and pH response of the same membrane from 25° to 300°C (and maybe higher). To evaluate any change in charge carrier in the two temperature regimes and to evaluate ZrO_2 membrane response for $f(O_2)$ or pH sensing, the specific resistivity of several commercial 4–8 mol% Y_2O_3 products was tested. In a nonclassical but simple method, the ZrO_2 tube to be measured was filled with 0.1N KCl and immersed in 0.1N KCl, and resistivities were measured with a pair of platinum electrodes with a high impedance electrometer.† The variability of path length

---

†Model 610C high impedance electrometer, Keithley Instruments, Inc., Cleveland, OH.

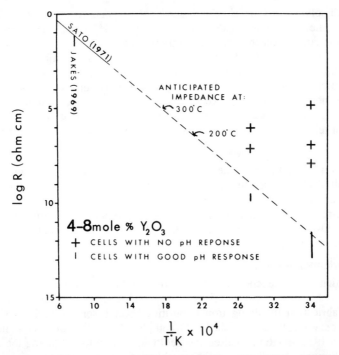

Fig. 2. The log R vs reciprocal T for ZrO_2 sensors. The solid line represents high-temperature values for the ZrO_2 cells showing accurate $f(O_2)$ sensing ability. The temperature extrapolation fits the equation log R = $-2.20 + 4000/T$ (T in K). Those cells with acceptable pH response are indicated; pH data for more than 100 cells are included.

between the two Pt wires was controlled but is not critical in that the resistivity of the small path length in $0.1N$ KCl is negligible compared to the resistivities being measured. Figure 2 shows results of these measurements.

A literature review[3] of resistivities for very accurate high-temperature $f(O_2)$ responses with ZrO_2 cells is shown by the solid line in Fig. 2, while a linear extrapolation of these data is shown by the dashed line. One set of data at high temperature[18] showed that membranes with a range of resistivities at 1000°C still performed accurately as $f(O_2)$ sensors. When tested at 25° and 90°C, it was shown[19] that over 100 ZrO_2 membranes from four suppliers were successfully "predicted" to have good pH responses. When log R at some T is 1 order of magnitude or more higher than the linear equation, poor or no pH response is obtained. Some 30 cells whose resistivities are represented by "+" data points in Fig. 2 were not suitable as pH sensors. High Y_2O_3 content, high porosity, microfractures, and chemical impurities (e.g., iron) are among the factors thought to be associated with poor pH response. Chemical impurities (e.g., iron) are not thought to be as important in pH response as in $f(O_2)$ response.

At high temperatures, it is well understood that ZrO_2 conductivity is related to the O^{2-} charge carrier. While Niedrach has claimed[20] that the pH sensing is related to oxygen ion conductivity, he makes clear[17] that his ZrO_2 cell circuit for pH does not respond at 285°C to changes in dissolved oxygen in the solution; i.e., it is not an $f(O_2)$ sensor. At 90°C we have also shown that the circuit

Ag/AgCl‖solution‖ZrO_2/Cu^0–Cu_2O does not respond to changes in dissolved hydrogen, i.e., it is not a sensor for H_2, only for H^+ ions. These observations lead to the conclusion that the conductivity mechanism of ZrO_2 as a pH sensor is still to be established. Classically, if the charge carrier changes in a semiconductor, the log R vs T plot shows a slope change at the temperature where the conductivity mechanism changes. Unfortunately, the resolution of data in Fig. 2 is not sufficient to prove or disprove such a slope change. Impedance spectroscopy may be necessary to establish the mechanism of pH sensing and the transition temperature between pH and $f(O_2)$ sensing in ZrO_2 membranes.

Geochemistry of ZrO_2 Raw Materials

Zircon, $ZrSiO_4$, or baddeleyite, ZrO_2, are the major source materials for nearly all ZrO_2 production. Among the common geochemical contaminants of these natural precursors are entrapped ilmenite, $FeTiO_3$, and ferrous oxide in solid solution, respectively. Specific examples of the problems with iron contamination in ZrO_2 $f(O_2)$ sensors are presented in the next section. See the appendix for further details of geochemical concerns.

Cell Fabrication Problems

Particularly in geochemical application, cell fabrication problems can quickly lead to troublesome or even useless $f(O_2)$ or pH readouts. For example, troubles in cell fabrication involving iron impurities became serious in 1978 when one fabricator saw one type of finished ZrO_2 cell jump from 50 ppm Fe_2O_3 to more than 700. This 1978 situation focused the geologic users into producing the *Intrinsic Oxygen Fugacity Newsletter,*[‡] an informal bulletin board-format letter exchanging problems and solutions in the use of ZrO_2 electrolytic cells. All of the fabrication problems summarized below are data presented in previous newsletters.

Testing of $f(O_2)$ Cells

With a gas mixing board[21] and with the JANAF-based $f(O_2)$ calculations for C-H-O gases at one bar (1.95 × 10^{-3} Pa),[22] a test matrix of CO_2-H_2 gas mixtures is used in the double-cell method.[13] In this technique, the two ZrO_2-platinum contacts on each cell are maintained by friction and spring-loading. Two ZrO_2 cells can be checked in the same furnace; usually a "vintage" or previously calibrated cell is kept in place to monitor daily accuracy, while the new cell to be tested forms the second half of the double-cell design. In the figures that follow, log MR is the log mixing ratio of CO_2/H_2; the solid lines are calculated from literature values,[22] and the indicated data points are those measured.

Figure 3 shows a test on an excellent ZrO_2 cell with 4.5 mol% Y_2O_3 fabricated in North America in 1972 with less than 50 ppm Fe_2O_3. The ability of the cell to reproduce expected $f(O_2)$ values is clearly demonstrated from 900° to 1200°C at log MR from -1 to $+2$. Historically, the nonideality at log $MR = -2$ has been interpreted as arising from electronic conductivity at these extreme reducing conditions. By comparison, Fig. 4 shows an identical test on a cell of ZrO_2-4.5 mol% Y_2O_3 with >700 ppm Fe_2O_3 content. The readings would drift to more reduced values, i.e., closer to theoretical values, if several days were allowed for each data point. This behavior correlated with the oxidation state of the iron

---

[‡]This newsletter is now circulated to over 100 labs around the world. The letter has no preconceived schedule—it appears only as there is news and usually not more than twice a year. It can be solicited by request from the author at his listed address.

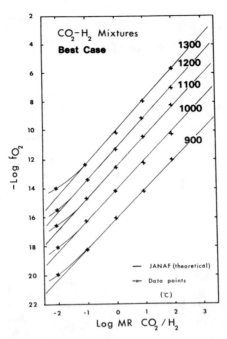

Fig. 3. $f(O_2)-T$ behavior of a ZrO_2-4.5 mol% Y_2O_3 cell with <50 ppm Fe_2O_3. See text for nonideality at log $MR = -2$.

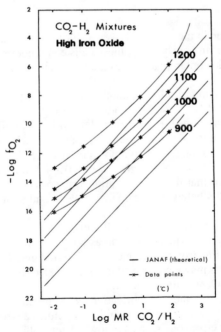

Fig. 4. $f(O_2)-T$ behavior of a ZrO_2-4.5 mol% Y_2O_3 cell with >700 ppm Fe_2O_3. See text for discussion.

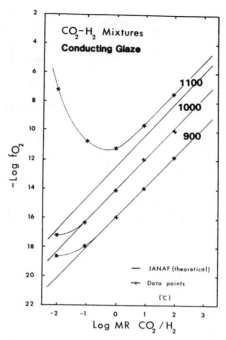

Fig. 5. $f(O_2)-T$ behavior of a ZrO_2-4.5 mol% Y_2O_3 cell with an accidental alkali silicate glaze. This glaze resulted from fuel impurities during manufacturing firing of the cell.

oxide in solid solution in the lattice. However, for work in which $f(O_2)$-T data were to be accumulated on an hourly basis, these cells could not be used. The oxidation state of the iron oxide contamination was a major factor in the cell's redox sensing ability.

Figure 5 shows the results of another manufacturing pitfall, i.e., a high alkaline ash content of fuels used in fabrication firing. In this cell's manufacture, it became coated with a potash-rich silicate glass which at high temperature shorted the cell's signal by acting as an ionic melt that coated the cell's exterior. At 1050°C, the cell's signal was no longer reliable nor reproducible. For cells from the same batch, a subsequent removal of the glaze with 40% HF for 1 h and testing of the cell at 1050°C proved that this interpretation was correct, i.e., with the removal of the glaze, the cell's behavior became acceptable and similar to that shown in Fig. 3.

Figure 6 shows a fabricated ZrO_2 cell with 8 mol% Y_2O_3 and with <50 ppm Fe_2O_3, which displayed n-type conduction as judged by its resistivity (Fig. 2) and by its $f(O_2)$ sensing. Electron probe analysis suggested that diopside ($CaMgSi_2O_6$) in the grain boundaries may have caused this type of electronic conduction because of traces of $NaFeSi_2O_6$ in solid solution.

Figure 7 shows yet another fabrication problem, i.e., a microtexture with fractures. Volume instabilities within the microstructure caused microfractures that leak gas, thus giving erroneous signals at lower temperatures. However, at higher temperatures where the fractures "heal," the ideal behavior is restored.

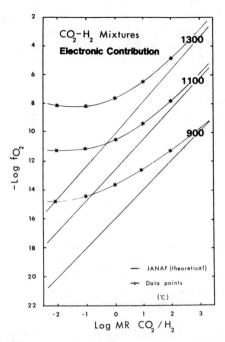

Fig. 6. $f(O_2)$–T behavior of a ZrO_2–8 mol% Y_2O_3 cell with n-type semiconductor behavior. See text for possible reasons for this behavior.

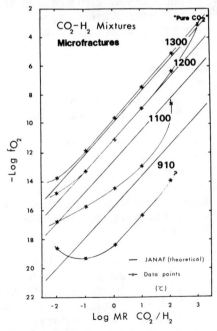

Fig. 7. $f(O_2)$–T behavior of a ZrO_2–$f(O_2)$ cell with microfractures; at higher temperatures, the fractures apparently "heal."

This is not an exhaustive list, but it demonstrates the problems that may arise. Of particular note is the observation that problems such as those shown in Figs. 6 and 7 lead to excellent reproducibility without any real accuracy. In other types of problems (Figs. 4 and 5), neither reproducibility nor accuracy could be achieved.

Testing of pH Cells

We have found that Y_2O_3 content does affect pH sensing in ZrO_2. For ZrO_2 with 4–8 mol% Y_2O_3, the relationship of resistivity to pH (and $f(O_2)$) sensing has already been discussed. While Niedrach[20] claims that "little difference (conduction) has been observed over a concentration range from 4.5–10.0 wt% Y_2O_3," a personal communication discloses that his results are actually best at 4.5 mol% Y_2O_3. This is different than our conclusion[16] that 8 mol% is optimum because this corresponds to a fully stabilized (i.e., single-phase cubic) ZrO_2.[23]

Microfracturing usually renders a ZrO_2 membrane useless for pH sensing, as shown in Fig. 2. If a ZrO_2 membrane filled with $0.1N$ KCl is exposed to a pressure differential and shows a decrease in resistivity, it is likely that the pressure differential has forced the electrolyte deeper into the microtexture, thus causing the resistivity decrease. This type of testing is more sensitive at room temperature for detecting microfractures than is the classic He-leak test. Furthermore, this type of testing is useful in the selection of porous ZrO_2 membranes necessary for the reference electrodes (salt bridges) needed in designing the hydrothermal ZrO_2 pH sensors.

In addition to the correlation of high porosity and microfractures with poor pH sensing, several labs have found variability even with a successful supplier of ZrO_2 cells—not every cell supplied will function. In some extreme cases, one in twenty cells will have pH sensing ability, even though all 20 cells are presumably of the same fabrication lot. A better understanding and prediction of pH sensing for ZrO_2 membranes may have to await impedance spectroscopy investigations into charge-carrying mechanisms.

Selected Geologic Applications

Three relevant areas of work with ZrO_2 $f(O_2)$ and ZrO_2 pH sensors are briefly discussed in this section. As early as 1966, it was shown[24] that ZrO_2 sensors emplanted in a drill hole in the crust of Makapuhi Lava Lake in the Hawaiian Islands could be used to directly measure the redox potential ($f(O_2)$) of gases evolving from a cooling volcanic magma. Subsequently, this technique was refined, and ZrO_2 $f(O_2)$ probes have been used to monitor Volcano, one of the volcanic Aeolian Islands north of Sicily, to aid in volcanic eruption prediction.[25] It has been found that as steam builds up in any volcanic magma chamber, the $f(O_2)$ values of the vapor exhalation of the volcano also change. Hence, implantation of redox sensors at many of the world's dangerous volcanoes seems desirable and has in fact already been done in over half a dozen volcanoes including Mt. Etna in Sicily.[26]

With geopolitically induced crises in many of the world's ore reserves, there is an ever-growing need to understand ore genesis to be able to predict ore locations. Studies along this line have begun. The role of $f(O_2)$ in the petrogenesis of the critical zone of the Bushveld Complex in the Transvaal, South Africa has been studied[27] in order to better understand chrome ore genesis. Because of its importance in the production of steel alloys, the $f(O_2)$–T relationships were determined for one of the world's larger niobium ore deposits.[28] Very recently the lateral

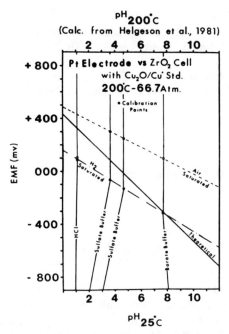

Fig. 8. Pt‖solution‖ZrO_2/Cu^0–Cu_2O sensor calibration. This circuit is sensitive to both Eh and pH as is shown. See text for further details.

$f(O_2)$ changes were established as important in the petrogenesis of the Merensky Reef, a platinum-enriched ore horizon of the Bushveld Complex of South Africa.[29]

The evaluation of geothermal brines is an obvious utilization for ZrO_2 pH sensors.[17] Many of the world's potential geothermal fields have such corrosive steam that expensive heat exchangers with recycled organics must be installed to isolate the steam from the turbine blades. Similarly, stability testing of waste-package isolation barriers to ground water in nuclear-waste subterranean repositories is an obvious utility of a pH sensor that could withstand burial at 1 km ($P \cong 300$ bars, 0.587 Pa) and heat buildup to 300°C.[15]

Figure 8 shows initial work in our lab on a pH sensor involving a Pt electrode and a ZrO_2/Cu^0-Cu_2O electrode. This circuit was calibrated at 200°C and 67.6 bars (0.132 Pa) in a pumpable titanium-lined autoclave. By comparison with the 25°C data of Fig. 1(B), it is clear that such a circuit would be useful in investigating a wide range of natural hydrothermal environments in which glass electrodes would quickly fail.

Conclusions

The applicability of ZrO_2 cells to $f(O_2)$ or pH sensing for technically and geologically important problems appears to be an expanding field of research. Increased demand seems likely for high-purity, high-performance Y_2O_3-stabilized ZrO_2 cells. The interrelationship between charge carrier and $f(O_2)$ and pH responses in ZrO_2 electrolytic cells awaits elucidation, perhaps by impedance spectroscopy. At fixed Y_2O_3 ranges, pH and $f(O_2)$ sensing seem to correlate with linear resistivity-temperature relationships.

Acknowledgments

Through the last decade, several sources of grant money have kept up the momentum on these studies. Earth Sciences NSF, Research Corp. of New York, Bethlehem Steel Corp., and DOE/Rockwell Hanford Operations are specifically thanked for their continued assistance. Contributors of data such as Owens Corning Fiberglass, Zircoa/Corning, Coors Porcelain, and Ceramic Oxide Fabricators are also to be thanked for their generosities.

Appendix

Geochemistry of ZrO_2 Raw Materials

Zircon, $ZrSiO_4$, is an accessory mineral associated with the late stages of crystallization of granites, syenites, and nepheline syenites, i.e., very SiO_2-rich rocks. In the weathering of these rocks, quartz (SiO_2) grains, ilmenite ($FeTiO_3$) grains, magnetite (Fe_3O_4) grains, and zircon grains are among those minerals resistant to mechanical and chemical disintegration. These minerals therefore form the bulk of detrital material in river and beach sands. Wave-sorting and wind-winnowing tend to concentrate in particular the higher-density magnetite, ilmenite, and zircon phases, in stringers and/or pockets of heavy mineral sands. In some zircon crystals, ilmenite and magnetite are even entrapped. Burial of these deposits by storm-emplaced sand can lead eventually to fossil beaches that may or may not subsequently be cemented into sandstones. Extraction from these sands by specific gravity techniques constitutes the bulk (95%) of the source of zirconium production.

Because of the similarity in ionic radii between Zr^{4+} (0.79 Å) (0.079 nm) and Hf^{4+} (0.78 Å) (0.078 nm), hafnium proxies for zirconium in solid solution usually between 1 and 4% by weight. Furthermore, while thorium is larger, Th^{4+} (1.02 Å) (0.102 nm), it too is in solid solution, often with sufficient thorium-induced radiation damage being present to have rendered the zircon crystals amorphous or *metamictized*. Unfortunately, ilmenite crystals of microscopic size are often entrapped (poikilitic texture) as numerous inclusions within the natural zircon crystals at the time of igneous origin. This ilmenite, while usually weakly magnetic, is so finely dispersed as to be inseparable by mechanical means and therefore produces iron and titanium contamination to the ultimate end-product, zirconium oxide. Entrapped magnetite is usually magnetic enough to be separable. This iron contamination is intolerable in many solid electrolyte applications of zirconia. Furthermore, the presence of thoria and hafnia enhances the neutron absorption cross section of zirconia. Thus, zirconia to be used in the nuclear industry is now typically extracted of its hafnia and thoria content.

Zirconium oxide, ZrO_2, is known mineralogically as baddeleyite, but it is quite rare. Baddeleyite extracted from Palabora, an extinct carbonate volcano in South Africa, contains in solid solution more than 2 wt% iron oxide, which again is difficult to impossible to extract.

The world mineral economic resources during the 1970s utilized 35 times (on average) more zirconium in nonmetallic usage than in metallic usage. The combined annual total metallic and nonmetallic usages in 1981 were more than 700 000 short tons (zirconium concentrate) with Australia, South Africa, U.S., U.S.S.R.,

China, and India being the primary producers.[30] Data for the U.S. in 1980–81 indicate that the glass and refractories industries utilized about 20% of the total consumption of zirconium. As usage of ZrO_2 increases, purity will become an ever more important problem.

References

[1] M. Sato, Econ. Geology, **60**, 812–18 (1965).
[2] L. Niedrach, J. Electrochem. Soc., **127**, 2122–30 (1980).
[3] H. P. Eugster and D. R. Wones, J. Petrol., **3**, 82–125 (1962).
[4] B. M. French, Rev. Geophys., **4**, 223–53 (1966).
[5] E. Woermann, B. Knecht, M. Rosenhauer, and G. C. Ulmer, Fortschr. Miner., **56**, 144–45 (1978).
[6] L. G. M. Baas Becking, I. R. Kaplan, and D. Moore, J. Geol., **65**, 243–84 (1960).
[7] G. C. Ulmer, M. Rosenhauer, E. Woermann, J. Ginder, A. Drory-Wolff, and P. Wasilewski, Am. Mineral., **61**, 653–60 (1976).
[8] JANAF; Joint Army, Navy, Air Force Thermochemical Tables, 2d ed. Natl. Bur. Stds: Ref. Data Ser., **No. 37**, 114 pp., 1971.
[9] H. C. Helgeson, D. H. Kirkham, and G. C. Flowers, Am. J. Sci., **281**, 1249–516 (1981).
[10] G. B. Naumov, B. N. Ryzhenko, and J. L. Khodakovsky, "Handbook of Thermochemical Data (translation available as NTIS, PB)," 226–722, 1971.
[11] M. Sato, pp. 43–99 in Research Techniques for High Pressure and High Temperature. Edited by G. C. Ulmer. Springer-Verlag, New York, 1971.
[12] M. Sato and M. Valenza, Am. J. Sci., **280–A**, 134–58 (1980).
[13] W. C. Elliott, D. E. Grandstaff, G. C. Ulmer, and T. Buntin, Econ. Geology, **77**, 1493–510 (1982).
[14] T. Tsurata and D. D. MacDonald, J. Electrochem. Soc., **129**, 1221–25 (1982).
[15] J. Myers, G. C. Ulmer, D. E. Grandstaff, R. Brozdowski, M. J. Danielson, and O. H. Koski, American Chemical Society Symposium; pp. 197–215 in Series 246. Edited by G. S. Barney, J. D. Navratil, and W. W. Schulz. The American Chemical Society, Washington, DC, 1984. Geochemical Behavior of Disposed Radioactive Waste; pp. 59–61 in Extended Abstracts Volume, 4th International Symposium on Water-Rock Interactions. International Association for Geochemistry and Cosmochemistry, 1983.
[16] W. L. Bourcier, M. A. McKibben, H. L. Barnes, G. C. Ulmer, R. A. Brozdowski, and D. E. Grandstaff.
[17] L. W. Niedrach; this volume, pp. **672–84**.
[18] D. Jakes, Chem. listy, **63**, 1073–91 (1969).
[19] G. C. Ulmer, D. E. Grandstaff, R. A. Brozdowski, H. L. Barnes, and W. L. Bourcier, Geol. Soc. Am., **14** [7] 635 (1982).
[20] L. W. Niedrach; U. S. Pat. No. 4 264 424 (Assignee: General Electric Co. — 16 claims).
[21] R. H. Nafziger, G. C. Ulmer, and E. Woermann; pp. 9–14 in Research Techniques for High Pressure and High Temperature. Edited by G. C. Ulmer, Springer-Verlag, New York, (1971).
[22] P. Deines, R. H. Nafziger, G. C. Ulmer, and E. Woermann, Bull. Earth Mineral Sciences Exp. Station, **88**, (1974); 129 pp.
[23] R. C. Hink; M. S. Thesis, The Pennsylvania State University, 1977.
[24] M. Sato and T. L. Wright, Science (Washington, D.C.), **153**, 1103–1105 (1966).
[25] M. Carapezza, S. Hauser, P. M. Nuccio, and M. Valenza, EOS, **61**, 402 (1980).
[26] M. Sato and J. G. Moore, R. Soc. London Philos. Trans., **A274**, 137–46 (1973).
[27] R. T. Flynn, C. Sutphen, and G. C. Ulmer, J. Petrology, **19**, 136–52 (1978).
[28] J. J. Friel and G. C. Ulmer, Am. Mineral, **59**, 314–18 (1974).
[29] T. J. Buntin, G. C. Ulmer, and D. P. Gold, Geol. Soc. Am., **14** [7] 456 (1983).
[30] W. S. Kirk; pp. 927–37 in U.S. Bureau of Mines: Minerals Yearbook, Vol. 1, 1981.

Application of Zirconia Membranes as High-Temperature pH Sensors

Leonard W. Niedrach

General Electric Corporate Research and Development
Schenectady, NY 12301

The zirconia pH sensor behaves much like the classical glass electrode, but it extends the range of measurement to much higher temperatures—about 300° vs 120°C. It also has virtues over the glass electrode at lower temperatures because of the absence of an "alkaline error." Like the glass electrode, it is insensitive to changes in the redox potential of the environment and, in turn, it exerts no influence on the environment. Such sensors have been finding application in the direct measurement of the pH of geothermal brines, of water in nuclear reactors, and in high-temperature corrosion studies. The sensors can also be used as "pseudoreference" electrodes for the measurement of redox and corrosion potentials in high-temperature media. Although available ceramics permit operation for at least 1–3 months, some evidence of slow degradation has been found. Ceramics having higher conductivities would be desirable.

The use of zirconia ceramics in the fabrication of fuel cells and oxygen sensors has been well known for years. More recently it has been shown that membrane electrodes fabricated from stabilized zirconia can be used for the measurement of pH.[1] This behavior, which resembles that of the classical glass electrode, has subsequently been investigated over a range of temperatures from ambient to about 285°C with pressurized water[2,3] and simulated geothermal brines.[4] It is the purpose of this paper to review the principles of the sensor, its present status, and some applications.

Description of the Sensor

The present preferred structure is shown in Fig. 1. In the initial work, sensors were fabricated in a fashion somewhat akin to that of the conventional glass electrode in that a buffered saline solution in contact with a chlorided silver wire served as the internal element within the membrane sheath.[1,2] While this type of internal element continues to be used,[3] it is now felt that a dry internal element similar to those often employed in zirconia oxygen sensors — viz., a dry mix of a metal and its oxide or two oxides of the same metal in different valence states — is preferable.[5] The copper-cuprous oxide couple has proved highly satisfactory for this purpose.

The dry mix of copper and cuprous oxide powders (1:1 by weight) is packed around a copper wire in a 1–2 in. long section near the closed end of the ¼ to ⅜ in. OD zirconia tube. This defines the active region of the sensor. A backing of glass wool is then added to help confine the powder. This arrangement offers several advantages over the earlier aqueous system: (i) it can be readily prepared by simple mixing of the powders and packing into the tube; (ii) when prepared in this fashion

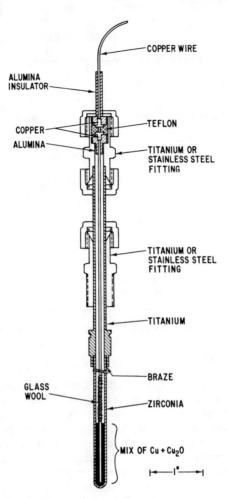

Fig. 1. Schematic diagram of the sensor.

it has been found to be extremely stable and reproducible; (iii) in contrast to aqueous internals, it permits ready designation of the active region of the sensor because it does not wet the wall with a conducting film; (iv) in the absence of an internal aqueous phase, seal fabrication is simplified.

The seal design employed in this sensor replaces compression seals that were less satisfactory. This new design should be more generally useful, particularly for field applications. It has been found that the thermal expansion coefficients of yttria-stabilized zirconia and chemically pure titanium are sufficiently well matched to tolerate the brazing conditions. Ticusil* and Ticuni* (active metal brazing alloys) were used as brazes. The former, while less corrosion-resistant than the latter, can be formed at 830°–850°C in an argon atmosphere. The latter has

---

*Manufactured by GTE Wesgo, Belmont, CA.

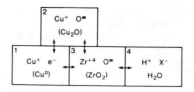

PHASES   **THERMODYNAMIC BALANCES**   **INTERPHASE POTENTIALS**

1-2 $_1\mu_{Cu^o} - (_1\mu_{e^-} - F_1\phi) - (_2\mu_{Cu^+} + F_2\phi) = 0$ $\Delta\Phi_{1-2} = 1/F(_2\mu_{Cu^+} + _1\mu_{e^-} - _1\mu_{Cu^o})$

2-3 $(_2\mu_{O^=} - 2F_2\phi) - (_3\mu_{O^=} - 2F_3\phi) = 0$ $\Delta\Phi_{2-3} = 1/F(\tfrac{1}{2}{_2\mu_{O^=}} - \tfrac{1}{2}{_3\mu_{O^=}})$

1-3 $\Delta\Phi_{1-3} = 1/F(_2\mu_{Cu^+} + \tfrac{1}{2}{_2\mu_{O^=}} + _1\mu_{e^-} - \tfrac{1}{2}{_3\mu_{O^=}} - _1\mu_{Cu^o})$

$= 1/F(\tfrac{1}{2}{_2\mu_{Cu_2O}} + _1\mu_{e^-} - \tfrac{1}{2}{_3\mu_{O^=}} - _1\mu_{Cu^o})$

3-4 $_4\mu_{H_2O} - (_3\mu_{O^=} - 2F_3\phi) - 2(_4\mu_{H^+} + F_4\phi) = 0$ $\Delta\Phi_{3-4} = 1/F(-\tfrac{1}{2}{_4\mu_{H_2O}} + \tfrac{1}{2}{_3\mu_{O^=}} + _4\mu_{H^+})$

1-4 $\Delta\Phi_{1-4} = 1/F(\tfrac{1}{2}{_2\mu_{Cu_2O}} + _4\mu_{H^+} - _4\mu_{Cu^o} - \tfrac{1}{2}{_4\mu_{H_2O}} + _1\mu_{e^-})$

Which is equivalent to the potential of the half-reaction:

$$Cu^o + \tfrac{1}{2}H_2O = \tfrac{1}{2}Cu_2O + H^+ + e^-$$

Fig. 2. Interphase potentials and related thermodynamic data for the sensor.

required temperatures of 950°–975°C, where considerable discoloration and degradation of the ceramic occurs. Although both seem acceptable at present, lower-melting, corrosion-resistant brazing materials would be preferable.

For measurements at the lower temperatures, when pressurization is not required, sensors have been prepared by sealing the ceramic to lime or lead glass.[6] This can be readily accomplished if care is taken to use a diffuse flame and to protect the ceramic with the aid of an intervening fused-silica shield. Mercury is then added to a depth of 1–2 in., and contact is made with a platinum wire. Attempts to establish a stable potential by adding mercuric oxide have been only moderately successful. A better approach has been to immerse the sensor and a platinum counterelectrode in a hot aqueous solution of dilute acid or base and electrolyze for several hours with the internal electrode of the sensor 30–100 V positive to the counterelectrode. Under these conditions, a film of mercuric oxide evidently forms at the ceramic-mercury interface to aid in the establishment of stable potentials. Alternatively, one may simply add the mercury and accept a slow drift in rest potential as oxide slowly forms over a longer period of time. Under these conditions, the drift in the rest potential is slow compared with the responses of interest.

Principle of Operation

For the sensor as shown in Fig. 1, several interphase potentials are involved in series between the external terminal (copper wire) and the solution being monitored. These and the thermodynamic balances involved in the potential determining reactions at the various interfaces are summarized in Fig. 2 after the approach of Vetter.[7] The balances are derived from the fact that at equilibrium the algebraic sum of the electrochemical potentials, $_i\eta_s$, of the reacting species is zero

$$\sum_i \nu_{si} \eta_s = 0 \tag{1}$$

where

$$_i\eta_s = {_i}\mu_s + {_i}z_s F_i\phi \tag{2}$$

and where $_i\mu_s$ is the chemical potential of reacting species s in phase i, $_iz_s$ the charge on the species, $_i\phi$ the Galvani (inner) potential in phase i, $_i\nu_s$ the stoichiometric factor of the species involved in the potential determining reaction, and F Faraday's constant.

In summing up the interphase potentials between phases 1 and 4 it is seen in Fig. 2 that the net change is equivalent to that for the simple copper-cuprous oxide couple:

$$2Cu + H_2O \rightleftarrows Cu_2O + 2H^+ + 2e^- \tag{3}$$

the thermodynamics for which relative to the SHE are well established.

To make an actual measurement a reference electrode is required, as shown in Fig. 3, which also indicates the potential changes that occur at the various interfaces included in the complete measuring system. Every effort must be made to maintain all of these interface potentials constant except that at the interface between the pH sensor and the solution to be measured. For work at elevated

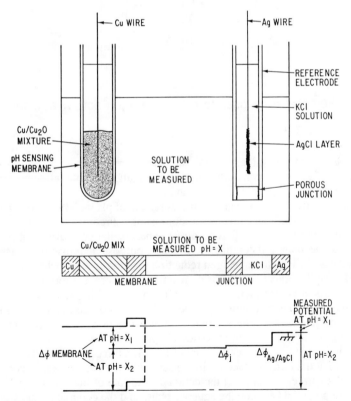

Fig. 3. Potential changes at various interfaces around measuring circuit.

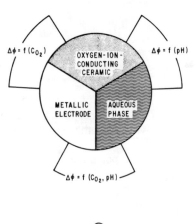

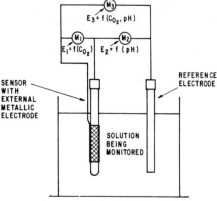

Fig. 4. Potential responses to pH and oxygen activity at different interfaces.

temperatures and pressures, several satisfactory reference electrodes have been described.[3,8,9] At 95°C, conventional, commercially available reference electrodes are satisfactory; e.g., $3M$ KCl/Ag/AgCl.

It is informative to compare the measurement of pH with that of oxygen partial pressure (or activity in an aqueous solution) using a zirconia-type sensor. Although detailed data are not available for a direct experimental comparison, it is possible to address the matter theoretically. For this purpose it is useful to assume an idealized case in which the oxygen reduction reaction is electrochemically reversible and complications from corrosion reactions and peroxide formation are absent. The comparison is also facilitated by the diagrams in Fig. 4, where the sensor is shown with a porous platinum electrode on its outer surface as typically employed with an oxygen sensor. The internal reference junction may be continued to be visualized as the copper-cuprous oxide couple.

As indicated in the upper diagram of Fig. 4, a three-phase interface is thereby established at the outer surface of the sensor. This involves the aqueous phase, the porous platinum electrode, and the ceramic. The role of the solution pH in establishing the potential between the aqueous phase and the ceramic remains as above. Through similar reasoning, and assuming the ideal behavior of oxygen at the platinum electrode, it can be shown that, in principle, the potential between the metal and the ceramic is a function only of the *molecular* oxygen activity (concentration in dilute solutions when "salting out" effects are negligible) in the water. In such a case the potential corresponds to that for the reaction:

$$O_2(aq) + 4e^-(Pt) \rightleftarrows 2O^{2-}(ceramic) \qquad (4)$$

Similarly the potential between the metal and the aqueous phase is a function of both the molecular oxygen activity and the pH and corresponds to that for the reaction

$$O_2(aq) + 4e^-(Pt) + 4H^+(aq) \rightleftarrows 2H_2O \qquad (5)$$

From this it follows that the potential between the internal electrode of the sensor and the external reference varies with the pH of the solution. The potential between the internal electrode and the outer, porous electrode reflects the molecular oxygen activity and that between the external electrode and the external reference is affected by both the pH and the molecular oxygen activity. When used in conjunction with an aqueous medium in this fashion, it is of interest to note that the platinum electrode could be completely separated from the ceramic; i.e., be introduced as a flag, and the above relationships would remain unaffected. It should also be noted, that if the oxygen activity (concentration in dilute solutions) in the aqueous phase is maintained constant as the pH is changed, the platinum electrode can, in principle, serve as a pH sensor.

In actual practice the behavior of the oxygen electrode on platinum deviates from ideality, in part because of electrochemical irreversibility of the primary reduction reaction, but also because of mixed potentials involving corrosion of the platinum and the formation of peroxide. As a result the potential between the platinum electrode and the inner electrode of the sensor will show some response to pH changes in addition to the main response to changes in molecular oxygen concentration. Similarly, some deviations from the ideal Nernstian behavior will be encountered in measurements of the potential of the platinum electrode against an external reference.

Characteristics of the Ceramic

Although sufficient information has not yet been accumulated to fully specify the characteristics of an acceptable ceramic, some general observations are in order. Early investigations with magnesia-, calcia-, and yttria-stabilized zirconias clearly indicated the superiority of the yttria-containing materials,[2] possibly because of their generally higher conductivities.[10] As a result all subsequent work has been done with yttria-stabilized ceramic, and little difference has been observed over a concentration range from 8.0 to 16.9 wt%. Conductivities of these ceramics determined over the temperature range from 25° to 285°C have been found to be reasonably consistent with earlier measurements at higher temperatures.[2] While quite acceptable, particularly for measurements above 95°C, the yttria-stabilized ceramic still imposes high input impedance requirements on the instrumentation that would be lessened with a more conducting ceramic. The literature[10,11] indicates that scandia as a stabilizer should result in lower impedance sensors, but this has not been investigated.

A variety of alternative oxygen ion-conducting ceramics should also be applicable as long as their oxygen ion transport numbers are close to 1.0 and that for electron and/or hole transport is at least an order of magnitude, and preferably several orders of magnitude, less. In this connection, just as in the case of an oxygen sensor, consideration must be given to the presence of oxidizing and reducing agents in the solution and their effect on the ceramic.

Suitable tubes fabricated by slip casting and isostatic pressing are available commercially. We have also fabricated some of our own tubes using slip casting. Thoroughly "laundered" zirconia-yttria powders[12] were used for this purpose, and the final firing was performed at 1550°–1600°C in air.

While complete analytical data are not available, it is felt that high-purity materials are mandatory for preparing satisfactory sensors. In particular, the concentrations of alumina and silica should be minimized. When present as impurities or additives, we have observed that they tend to accumulate at grain boundaries and promote intergranular attack. Their presence is also often manifested in fracture sections by intergranular fracture. It has been our experience that higher-purity, acceptable ceramics invariably fracture transgranularly.

Also of importance is the final microstructure of the ceramic. Although closed pores of any size *within* the ceramic should be innocuous, we have found that our best ceramics have the finest pores. More important, however, seems to be the minimization of open pores on the outer surface of the tubes. Some of these features may be recognized in the micrographs in Fig. 5 which are representative of a superior, an acceptable, and an inferior ceramic. The large subsurface pore with a restricted opening seen in the light micrograph of sample *b* would be expected to contribute to sluggish response.

Experimental Procedures

Much of the work to date has been aimed simply at demonstrating the capability of the sensors to respond to pH changes. For this purpose, relatively simple equipment has been required for tests below 100°C. For example, Teflon[†] vessels with reflux and provision for stirring and periodic additions of acid or base have been used.[6] At higher temperatures, "refreshed" autoclaves with provision for the addition of acid and base are required. All are straightforward and many applicable systems have been described.[2,3]

In all cases a reference electrode is required. Many alternatives have been employed. They may be divided into two categories: (1) those in which the reference couple—silver-silver chloride—is maintained at the operating temperature of the system[8] and (2) those in which the reference couple is maintained at ambient temperature, in which case a thermal potential is also involved.[3,9]

Because of the high impedance of the sensor, particularly at lower temperatures, a high input impedance electrometer is required for measurements.

Often a comparison standard is desired as a check on the response of a sensor. At temperatures up to about 100°C, a commercial glass electrode may be used for this purpose.[2] At higher temperatures, an alternative is required. In some cases the oxygen electrode has been used[1,2] and in others the hydrogen electrode.[3] All have proved satisfactory, when used with caution, and confirm the response of the sensors.

Performance Data

Behavior at 285°C

The primary interest in the zirconia sensors has been to exploit them at relatively high temperatures and pressures; e.g., 285°C at 1200 psi. Under these

---

[†]Trade name for tetrafluoroethene. Manufactured by E.I. du Pont de Nemours & Co., Wilmington, DE.

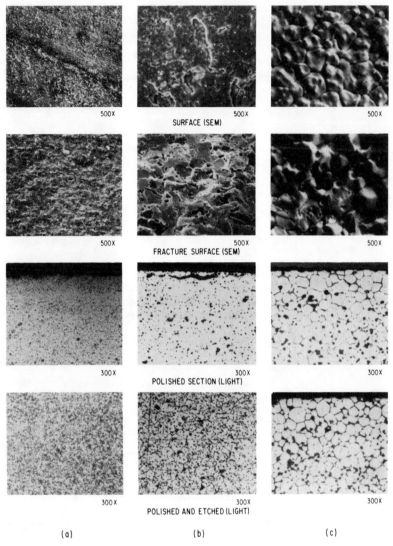

Fig. 5. Micrographs of three ceramics: (a) superior; (b) acceptable; (c) unacceptable.

conditions they have invariably performed well in both pure water and simulated brines for as long as 40 days. Such performance has been thoroughly documented elsewhere,[2,4] but representative response data are shown in Fig. 6. Although the one plot in Fig. 6 is that of the zirconia sensor vs an oxygen electrode serving as a comparison pH sensor, it has been demonstrated that both have responses to pH that are 90–97% of theoretical values at 285°C.[2] It should be noted that the apparent slow response of the sensor and the oxygen electrode to pH changes reflects the turnover time of the system rather than the time constants of the electrodes. The true response times of the electrodes are probably well under 1 s

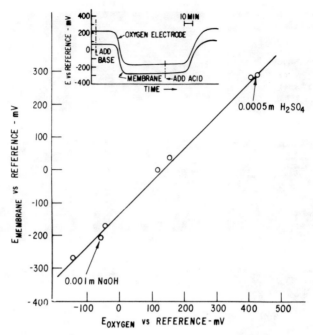

Fig. 6. Responses of a pH sensor and an oxygen electrode to pH changes at 285°C.

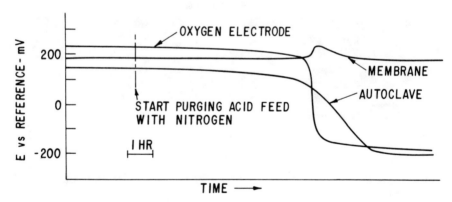

Fig. 7. Behavior of a sensor, the oxygen electrode, and the autoclave as the redox potential of the environment is changed at 285°C.

at 285°C. A similar correlation of response of a zirconia sensor with a hydrogen electrode has been obtained at 275°C.[3]

One point that must be reemphasized is that, like the glass electrode, the zirconia membrane electrode does *not* respond to changes in the redox potential of the solution, e.g., the dissolved molecular oxygen concentration. This is illustrated by the data in Fig. 7, which was obtained while $0.0005M$ sulfuric acid, initially aerated, was being fed to the autoclave.[2] On switching to a nitrogen purge, the

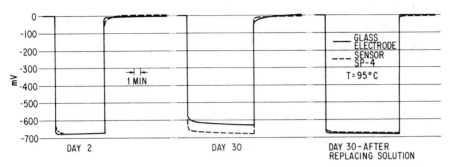

Fig. 8. Comparison of responses of a zirconia sensor and a glass electrode to rapid changes in pH at 95°C.

potentials of the autoclave and the platinum flag (oxygen electrode) immediately started to fall while that of the zirconia pH sensor remained constant. After several hours a more rapid drop occurred in the potentials of the autoclave and the platinum electrode. This is attributed to a change in the passivation reaction for the stainless steel autoclave. Instead of the passive film being maintained by the reduction of oxygen to water, water is reduced to liberate hydrogen. The platinum flag then clearly showed its sensitivity to this change in the redox environment. The oxygen-ion-conducting ceramic membrane, however, retained an essentially constant potential. The small, transient increase in potential is believed to have been a response to a real transient in pH associated with the change in the passivation reaction on the stainless steel. Looked at in another light, if the potential of the platinum electrode were referred to that of the pH sensor, the response would have been identical with that of an oxygen sensor having a platinum electrode applied directly to the outside of the zirconia tube.

Behavior at 95°C

Because original interest in the new sensor stemmed from the desire to measure the pH of water in nuclear reactors, initial attention was focused on performance at 285°C. More recently, extensive comparisons with commercial glass electrodes in the temperature range of overlapping capability have been made.[6,13] For this purpose, 95°C has been a convenient temperature because it permits operation at ambient pressure without interference from boiling, even when dilute solutions are used. This work demonstrated that the new zirconia sensor shows little, if any, sensitivity to alkali ions in basic solutions and, therefore, is free from the "alkaline error" associated with the glass electrode.[13]

It was also demonstrated that long life can be obtained with zirconia electrodes at 95°C. For this purpose sensors were alternately exposed to acid or base for several days at 95°C, and then an aliquot of base or acid was added in increments of sufficient size to swing from acidic to basic pH and vice-versa.[6] The transient responses of a glass electrode and a zirconia sensor on the first and thirtieth days of such an experiment are compared in Fig. 8. Here it is seen that the time constant of the zirconia electrode is comparable with that of the HA glass electrode. It is also evident that, as NaCl accumulated in the solution, the glass electrode lost some of its response. When the solution was replaced with fresh acid or base, the full response of the glass electrode was again obtained.

Over still longer periods of operation at 95°C, some loss of response *rate* was encountered, even with the best sensors. This is illustrated by the response data in

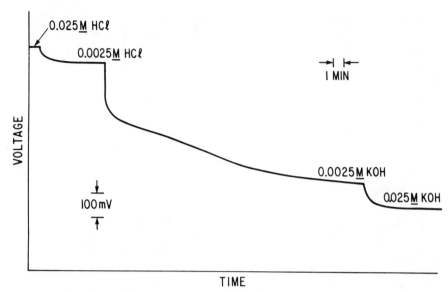

Fig. 9. Response of a three-month-old sensor to small increments of 5M KOH and 5M HCl; T = 95°C.

Fig. 9. In contrast to rapid responses from a new sensor over the entire pH range, the aged sensor has become more sluggish. In the more acidic and more basic regions, additions of acid or base result in responses that are only slightly slower than the original. In the mid-pH region, however, the response is considerably more sluggish, and a sigmoidal break is seen in the potential-time plot. Similar behavior has been observed with sensors that have been operated at 285°C for as long as 2 weeks. When the solution is cooled to 95°C and tested at this lower temperature, the theoretical voltage response to pH changes is still obtained, but even longer response times are observed.

This behavior suggests that the solutions are slowly penetrating into grain boundaries or other discontinuities in the structure and that functional groups are being titrated in the intermediate pH range. These functional groups may be derived from the slow hydrolysis of second-phase materials resulting from impurities in the ceramic. For example, alumina could slowly hydrolyze to AlOOH,[14] which would probably be titratable in the mid-pH region. Silica should also hydrolyze, but it would be expected that silicic acid would titrate in a more acidic region.

Even in the absence of impurities, segregation of yttria in grain boundaries could have a similar effect since $Y(OH)_3$ is stable relative to Y_2O_3 in water under the conditions employed.[15] In contrast to the above, ZrO_2 is stable relative to the hydrolyzed forms.[16,17]

While we are apparently dealing with a real phenomenon here, it does not seem to be a critical one at present with our best ceramics. Similar behavior has, however, been encountered with other ceramics when the pause associated with the "titration" has been over 1 h in duration.

Applications

In most of the work to date, emphasis has been placed on studying the performance of the sensors, per se. In a few cases, however, the sensor has been used effectively for other purposes. Taylor and Caramihas,[18] for example, employed the sensor effectively to follow pH changes in simulated crevices to enable interpretation of chemical effects. Zirconia sensors have also been operated successfully in the cooling water from a nuclear reactor,[19] and they are presently being used to follow the pH of geothermal brines being fed to a 10-MW power generating plant in El Centro, CA.[20] In this connection, extensive preliminary measurements were made with simulated brines containing as much as 20 wt% sodium chloride and 100 ppm hydrogen sulfide.[4] An excellent correlation was obtained between measured and calculated pH values over a pH range of 3–9 at 285°C. In this case agreement was well within a 0.5 pH unit, and at this time it is not certain whether the discrepancies were associated with the measured or calculated values of the pH.

Finally, the use of the sensor as a pseudoreference electrode for the monitoring of redox and corrosion potentials in high-temperature systems has been suggested.[5] The use of sensors in this mode is presently underway on a routine basis.[21]

Conclusions

The use of zirconia membrane sensors to determine the pH of high-temperature solutions is a viable application. Such sensors may also be useful for lower-temperature measurements in highly alkaline media because of the absence of the alkaline error associated with glass electrodes.

Although the characteristics of selected ceramics are adequate for present applications, slow degradation of response time is observed with even the best. It appears likely that further improvement will be possible with additional attention to purity of materials and detail of fabrication.

It would also be desirable to achieve higher conductivities. This might be achieved through the substitution of alternative stabilizers for the present yttria, perhaps scandia. It appears less likely that alternative systems based on ceria or bismuth oxide will prove attractive because of the aggressiveness of high-temperature aqueous environments.

Acknowledgments

S. Prochazka of General Electric Corporate Research and Development was extremely helpful by fabricating several high-density zirconia tubes for use in some of this work. Highly purified, yttria-stabilized zirconia powder was kindly prepared by C. Scott of the General Electric Lighting Research Laboratory, Cleveland, Ohio, for this purpose.

References

[1] L. W. Niedrach, *Science (Washington, D.C.)*, **207**, 1200 (1980).
[2] L. W. Niedrach, *J. Electrochem. Soc.*, **127**, 2122 (1980).
[3] T. Tsuruta and D. D. Macdonald, *J. Electrochem. Soc.*, **129**, 1221 (1982).
[4] L. W. Niedrach and W. H. Stoddard; Report No. PNL 4651, Pacific Northwest Laboratory, Richmond, WA, Feb. 1983. PNL is operated by Batelle Memorial Institute under prime contract No. DE-AC-06-76-RLO-1830 for the United States Department of Energy, Division of Geothermal Energy; *J. Electrochem. Soc.*, **131**, 1017 (1984).
[5] L. W. Niedrach, *J. Electrochem. Soc.*, **129**, 1445 (1982).
[6] L. W. Niedrach and W. H. Stoddard, *Ind. Eng. Chem. Prod. Res. Dev.*, **22**, 594 (1983).
[7] K. J. Vetter; pp. 17 ff in Electrochemical Kinetics. Translated by S. Bruckenstein and B. Howard. Academic Press, New York, 1967.
[8] M. E. Indig and A. R. McIlree, *Corrosion*, **35**, 288 (1979).
[9] M. J. Danielson, *Corrosion*, **35**, 201 (1979).
[10] T. H. Etsell and S. N. Flengas, *Chem. Rev.*, **70**, 339 (1970).

[11] R. M. Dell and A. Hooper; p. 291 in Solid Electrolytes. Edited by P. Hagenmuller and W. Vangool. Academic Press, New York, 1978.
[12] C. E. Scott and J. S. Reed, *Am. Ceram. Soc. Bull.*, **58**, 587–90 (1979).
[13] L. W. Niedrach, *Anal. Chem.*, **55**, 2426 (1983).
[14] G. C. Kennedy, *Am. J. Sci.*, **257**, 568 (1959).
[15] M. W. Shafer and R. Roy, *J. Am. Ceram. Soc.*, **42**, [11] 563–70 (1959).
[16] K. Nakamura, S. Hirano, and S. Somiya, *Am. Ceram. Soc. Bull.*, **56** [5] 513–515 (1977).
[17] E. Tani, M. Yoshimura, and S. Somiya, *J. Am. Ceram. Soc.*, **66** [1] 11–14 (1983).
[18] D. F. Taylor and C. A. Caramihas, *J. Electrochem. Soc.*, **129**, 2458 (1982).
[19] M. Indig, J. Weber, and W. H. Stoddard; unpublished work.
[20] G. Jensen and R. Robertus; unpublished work.
[21] L. W. Niedrach, et al.; unpublished work.

Preparation and Operation of Zirconia High-Temperature Electrolysis Cells for Hydrogen Production

E. ERDLE, A. KOCH, AND W. SCHAEFER

Dornier System GmbH
D-7990 Friedrichshafen
Federal Republic of Germany

F. J. ESPER AND K. H. FRIESE

Robert Bosch GmbH
D-7000 Stuttgart
Federal Republic of Germany

Zirconia (ZrO_2) with 9 or 10 mol% Y_2O_3 and 4 mol% Al_2O_3 is particularly well suited for use as a solid electrolyte in the high-temperature electrolysis of water vapor, especially in the form of small cylinders with a wall thickness of 0.3 mm and coated on both sides with porous electrodes. At an operating temperature of 1000°C and a voltage of 1.33 V, these electrolysis cells achieved a current density of more than 0.6 A/cm². This represents a specific electrical energy consumption of only 3.2 kW·h/m³ of H_2 (as compared with a minimum of 4.4 kW·h/m³ of H_2 for conventional electrolysis techniques).

Stabilized ZrO_2 is a good conductor of oxygen ions at high temperature. For this reason, zirconia has been tested for approximately 20 years as a solid electrolyte in oxygen sensors, high-temperature fuel cells, and electrolysis cells.[1-7]

Due to its advantages in terms of thermodynamics and reaction kinetics, the high-temperature electrolysis provides a method of producing hydrogen from water vapor with an improved efficiency, compared to conventional electrolysis techniques. The German development activities in the field of this new electrolysis technique are represented by the project HOT ELLY (*h*igh *o*perating *t*emperature *el*ectro*ly*sis).

An electrolysis membrane operates as follows: H_2O molecules are split at the porous cathode by an external electric field. Hydrogen remains at the cathode side while oxygen ions are transported through the gastight electrolyte and subsequently discharged at the anode as molecular O_2.

The simplest separation of the product gases hydrogen and oxygen can be attained if the solid electrolyte cells are produced in the form of hollow cylinders. This geometry provides many advantages, particularly in terms of connecting numerous cells in series and simplified gas transport in "electrolysis tubes."[1,8]

Composition of the Electrolyte Material

ZrO_2 can be used as a solid electrolyte only if a high ionic conductivity and the stabilization of the cubic fluorite phase are achieved by the addition of suitable

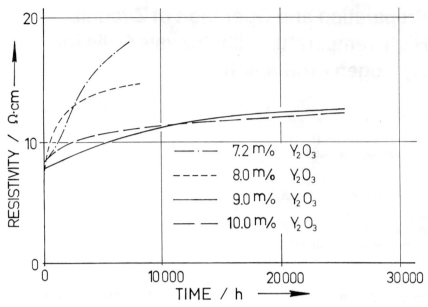

Fig. 1. Resistivities of electrolyte materials with varying Y_2O_3 content at 1000°C as a function of time.

divalent or trivalent oxides. The resulting ionic conductivities of ZrO_2 electrolytes vary from 0.055 to 0.25 $\Omega^{-1} \cdot cm^{-1}$ at 1000°C.[9]

An adequate conductivity at an acceptable cost can be achieved only by the addition of Y_2O_3 as the other rare earth oxides, e.g., Sc_2O_3, are too expensive. For the production of electrolytes, ZrO_2 samples containing 4 mol% Al_2O_3 and from 7.2 to 10 mol% Y_2O_3 were investigated.

Our studies showed that sufficiently low resistivity could be attained if the ceramic electrolyte was stabilized with 8–10 mol% Y_2O_3 (Fig. 1). However, if the electrolyte materials were annealed for a long time at 1000°C, the resistivity of samples with lower-Y_2O_3 content increased. Therefore, for operating periods greater than 20 000 h, the ceramic electrolyte must be stabilized with at least 9 mol% Y_2O_3 in order to minimize the aging effect.

Additionally, it was demonstrated that the purity of the Y_2O_3-stabilized zirconia (YSZ) did not influence the resistivity and aging behavior of the ceramic electrolytes (materials with overall 99% and 99.99% purity were compared).

The cost of the yttria (Y_2O_3) comprises more than half the cost of the electrolyte material. Therefore, electrolytes were produced which were stabilized with an inexpensive Y_2O_3 concentrate rather than with pure (99%) Y_2O_3. The Y_2O_3 concentrates had approximately the following composition: 60 wt% Y_2O_3, 30 wt% of the heavy rare earth oxides, principally Yb_2O_3, Er_2O_3, Dy_2O_3, and Gd_2O_3, and approximately 10 wt% of the light rare earth oxides. The raw material also contained less than 1% of other impurities, e.g., SiO_2, TiO_2, and Fe_2O_3. Fortunately, the rare earth oxides, particularly the heavy rare earths with their favorable ionic radii, can also stabilize the cubic fluorite phase.[10,11]

An addition of 17.9 wt% of the Y_2O_3 concentrate to the ceramic corresponded to a stabilization of ≈9 mol% Y_2O_3. The resistivity was nearly the same as that of the electrolyte with 9 mol% Y_2O_3.

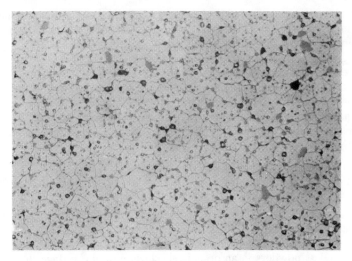

Fig. 2. Microstructure of an electrolyte material fully stabilized with Y_2O_3 concentrate (thermally etched cross section).

Production of Electrolyte Cylinders

The electrolyte must satisfy the following requirements: lowest resistance possible (i.e., smallest possible wall thickness) and sufficient mechanical stability. These somewhat contradictory requirements can be largely fulfilled by hollow cylinders, which are 10 mm long and have a wall thickness of 0.3 mm. These cylinders have a 1-mm thick rim at each end. The diameter of the continuous cylinder bore is 13.8 mm.

The material from which the electrolyte cylinders were pressed was produced by normal ceramic pulverization and granulating processes from a mixture of ZrO_2, Y_2O_3, and Al_2O_3 raw materials. The cylindrical shape was produced in the following way. First, components with a wall thickness greater than 1 mm were pressed, and the external profile of the cylinder was ground before sintering. By appropriate dimensioning of the cylinders, it was possible to take into account the shrinkage which occurs during sintering. Finally, the electrolyte cylinders were sintered in air at a temperature of 1550°C; the cylinders were placed on a zirconia substrate with a dense and smooth surface in order to reduce the distortion during sintering to a minimum.

The density and bending strength of the materials made with Y_2O_3 and the Y_2O_3 concentrate were 5.60 and 5.72 g/cm^3 and 340 and 315 MPa, respectively. The slightly higher density of the electrolyte material stabilized with Y_2O_3 concentrate can be attributed to the higher density of the rare earth oxides. A bending strength of 300–350 MPa is reasonable, considering the dense but relatively coarse structure of the stabilized zirconia (Fig. 2).

In addition to cubic zirconia, the electrolyte material also contained a small amount of Al_2O_3 as a second crystalline phase, which corresponded to the Al_2O_3 added to the raw material mixture. The phase composition of the ZrO_2 electrolyte did not depend on the use of pure Y_2O_3 (99%) or the Y_2O_3 concentrate.

The end surfaces of the electrolyte cylinders must be extremely flat and smooth ($R_a < 0.5$ mm) to allow a gastight junction of the electrolysis cells in long electrolysis tubes. Therefore, the electrolyte cylinders were so dimensioned that, after sintering, they were slightly longer than necessary. The required end tolerances (in terms of smoothness and flatness) were then attained by subsequent machining of both cylinder ends in a saw with a diamond blade and by a final polishing process.

In contrast, the wall surfaces of the electrolyte cylinders must be rough enough to ensure adequate adhesion of the electrode layers and an extensive three-phase boundary. Sufficient roughness was achieved by the selection of suitable grinding parameters during the shaping of the external surfaces. The internal cylinder surface was roughened by a separate grinding process that took place prior to sintering.

Production of Electrolysis Cells

For their use as high-temperature electrolysis cells, the electrolyte cylinders were coated with porous electrodes. Both the cathode (hydrogen/water vapor electrode) on the internal cylinder surface and the anode (oxygen electrode) were applied with simple spraying and sintering techniques. The electrode materials were sprayed as powders in an organic suspension on the surfaces to be covered and were subsequently fired.

The cathode was formed by a nickel-cermet layer approximately 100 μm thick; the ceramic component was a mixture of ceria and zirconia. The specific electrical conductivity of the cathode in a mixture of H_2/H_2O was approximately 800 $\Omega^{-1} \cdot cm^{-1}$ at 1000°C. The anode consisted of doped lanthanum-manganese-oxide and was approximately 300 μm thick. Its specific electrical conductivity in air was about 50 $\Omega^{-1} \cdot cm^1$ at a temperature of 1000°C. The porosity of both electrodes was about 50% in volume.

Operation of Electrolysis Cells

For laboratory test operation, the electrolysis cells were pressure-welded to a zirconia tube using gold rings and sealed with a lid (see Fig. 3). The gold rings also served as electrical connections. The electrodes were joined to the gold rings using a platinum paste.

Typical conditions for test operation were temperatures of 1000°C and gas flow rates of 100 cm^3/min of H_2O and 50 cm^3/min of H_2. The gas flow rates were measured at the cell inlet and outlet; the voltage was measured as a function of the current density.

Figure 4 presents the current-voltage characteristic of such a cell. The operating voltage of the cells is designed to be 1.33 V.[12] At this voltage a current density of about 0.6 A/cm^2 is reached (which will be slightly smaller for cells connected in series in an electrolysis tube due to losses in the interconnections). The cell voltage of 1.33 V corresponds (also in the case of an electrolysis plant) to a specific electrical energy consumption of 3.2 $kW \cdot h/m^3$ of H_2, which is considerably lower than the consumption for conventional electrolysis techniques.[13] These results clearly show that the high-temperature electrolysis of water vapor leads to substantial energy savings in comparison to conventional electrolysis techniques.

The long-term performance (1000 h) was checked at constant current density (0.3 $A \cdot cm^{-2}$) and a cathodic H_2 concentration of 33% with the anode in air. The cell voltage remained constant at 1.07 V during this period.

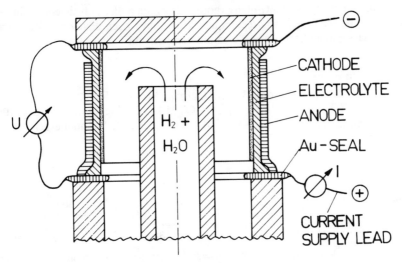

Fig. 3. Sample geometry for electrolysis tests.

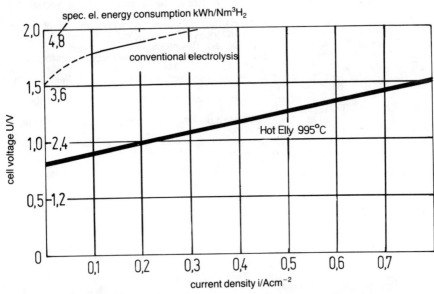

Fig. 4. Current-voltage characteristics and energy consumption of a zirconia high-temperature cell (HOT ELLY) and a conventional cell.

Conclusion

These measurements clearly demonstrate the great potential of high-temperature electrolysis, on the basis of ZrO_2 electrolytes, for future hydrogen production. In comparison with conventional electrolysis techniques, electrical energy savings of at least 28% are possible.

In an electrolysis plant, many thousands of such electrolysis cells would be required. The manufacturing techniques which have been developed are suitable for the mass production of such ceramic components.

Acknowledgments

This work was sponsored by the German Ministry for Research and Technology (BMFT) within the project HOT ELLY. The authors are grateful to W. Doenitz, H. Geier, G. Knoll, B. Santer, and R. Schmidberger for their kind support.

References

[1] W. Doenitz and R. Schmidberger, "Concepts and Design for Scaling Up High-Temperature Water Vapor Electrolysis," *Int. J. Hydrogen Energy*, **7**, 321–30 (1982).

[2] W. Doenitz, R. Schmidberger, and E. Steinheil, "Hydrogen Production by High Temperature Electrolysis of Water Vapor," *Int. J. Hydrogen Energy*, **5**, 55–63 (1980).

[3] F. J. Rohr, "High Temperature Fuel Cells"; pp. 431–50 in Solid Electrolytes. Academic Press, New York, 1978.

[4] H. Dueker, K. H. Friese, and W. D. Haecher, "Ceramic Aspects of the Bosch Lambda Sensor," SAE-Report 750223, presented at Automotive Engineering Congress and Exposition, Detroit, MI, Feb. 24–28, 1975.

[5] H. S. Isaacs, "Zirconia Fuel Cells and Electrolyzers"; pp. 406–18 in Advances in Ceramics, Vol. 3. Edited by A. H. Heuer and L. W. Hobbs. The American Ceramic Society, Columbus, OH, 1981.

[6] D. Janke, "Zirconia-, Hafnia- and Thoria-Based Electrolytes for Oxygen Control Devices in Metallurgical Processes"; pp. 419–36 in Advances in Ceramics, Vol. 3. Edited by A. H. Heuer and L. W. Hobbs. The American Ceramic Society, Columbus, OH, 1981.

[7] A. O. Isenberg, "Energy Conversion via Solid Oxide Electrolyte Electrochemical Cells at High Temperatures," *Solid State Ion.*, **3/4**, 431–37 (1981).

[8] W. Schaefer and R. Schmidberger, "Diffusionsschweissen von ZrO_2-Festelektrolyt-Zellen," *DVS Beri.*, **66**, 82–85 (1980).

[9] T. H. Etsell and S. N. Flengas, "Ionic Conductivity of Various Solid Electrolytes at 1000°C," *Chem. Rev.*, **70**, 339 (1970).

[10] A. K. Kuznetsov et al., "Ceramics Based on Zirconium Dioxide Stabilized with Yttrium Concentrate," *Ogneupory*, **6**, 45–48 (1971).

[11] D. W. Strickler and W. G. Carlson, Electrical Conductivity in the ZrO_2-Rich Region of Several M_2O_3–ZrO_2 Systems," *J. Am. Ceram. Soc.*, **48** [6] 286–89 (1965).

[12] W. Dönitz, G. Dietrich, and E. Erdle, *DECHEMA Monogr.*, **92**, 323–34, (1982).

[13] LURGI Express Information, T 1084/11.80, 1980.

Section VI
Processing

Processing Techniques for ZrO$_2$ Ceramics.................. 693
S. Wu and R. J. Brook

Sinterability of ZrO$_2$ and Al$_2$O$_3$ Powders: The Role of Pore Coordination Number Distribution........................ 699
F. F. Lange and B. I. Davis

Sintering Kinetics of ZrO$_2$ Powders........................ 714
A. Roosen and H. Hausner

Sintering of a Freeze-Dried 10 Mol% Y$_2$O$_3$-Stabilized Zirconia... 727
L. Rakotoson and M. Paulus

ZrO$_2$ Micropowders as Model Systems for the Study of Sintering 733
R. Pampuch

Wet-Chemical Preparation of Zirconia Powders: Their Microstructure and Behavior............................. 744
M. A. C. G. van de Graaf and A. J. Burggraaf

Preparation of Y$_2$O$_3$-Stabilized Tetragonal Polycrystals (Y-TZP) from Different Powders........................ 766
H. Schubert, N. Claussen, and M. Rühle

Preparation of Ca-Stabilized ZrO$_2$ Micropowders by a Hydrothermal Method................................ 774
K. Haberko and W. Pyda

Growth and Coarsening of Pure and Doped ZrO$_2$: A Study of Microstructure by Small-Angle Neutron Scattering.... 784
A. F. Wright, N. H. Brett, and S. Nunn

Al$_2$O$_3$-ZrO$_2$ Ceramics Prepared from CVD Powders............. 794
S. Hori, M. Yoshimura, S. Sōmiya, and R. Takahashi

Preparation of Mixed Fine Al$_2$O$_3$-HfO$_2$ Powders by Hydrothermal Oxidation................................ 806
H. Toraya, M. Yoshimura, and S. Sōmiya

Applications of Rapid Solidification Theory and Practice to Al_2O_3-ZrO_2 Ceramics 816
   G. Kalonji, J. McKittrick, and L. W. Hobbs

Synthesis and Sintering of ZrO_2 Fibers 826
   I. N. Yermolenko, T. M. Ulyanova, P. A. Vityaz, and
   I. L. Fyodorova

Epilogue 833
   R. J. Brook

Processing Techniques for ZrO$_2$ Ceramics

SUXING WU AND R. J. BROOK
University of Leeds
Department of Ceramics
Leeds LS2 9JT, United Kingdom

The search for optimum properties in zirconia and zirconia-containing ceramics has led to much work on fabrication procedures capable of providing homogeneous and reproducible materials with closely defined microstructures. The work has emphasized two aspects: first, the preparation of fine, agglomerate-free powders capable of yielding dense, flaw-free materials and, second, the identification of firing procedures and compositions specifically selected to enhance densification and microstructure development during heat treatment. The avenues for powder quality improvement are briefly introduced, and the targets for powder properties are reviewed. The possibilities for improvement of the firing procedure are then considered, with emphasis on the composition variable and the use of additives. Results on the sintering and hot-pressing of zirconia with different levels of stabilizer content and different trace additives are reported and used to suggest mechanisms for the influence of the compositional differences; the significance of composition control for the preparation of dense ceramics is considered and set in the context of the powder development work.

With the recognition of the great potential offered by the mechanical and electrical properties of zirconia, an increased level of attention has been given to the fabrication of materials which can fully exploit these properties. This attention has been concerned not only with methods which allow the preparation of homogeneous, dense microstructures free of second phases at the boundaries but also with methods that are attractive in terms of product reproducibility and cost. The issue of reproducibility is seen as a key factor in ensuring the successful application of the material since a number of the important properties, such as strength, are particularly vulnerable to isolated faults in the material, and the consequences of accidents or errors during processing are accordingly severe.

The two main approaches adopted for the preparation of reliable ceramics are, first, to develop powder-processing routes which are capable of leading to powder compacts with homogeneous structures and, second, to develop firing procedures which convert these compacts to dense products. Both aspects are briefly considered in the following sections.

Powder Processing

Considerable attention has been given to powder preparation and characterization,[1-5] the ideal objective[6] being fine powders (typically <0.5 μm), which are free from agglomeration, contain a narrow distribution of particle sizes, comprise spherical-shaped particles, and have closely controlled composition. Generally, processing is seen as an optimization, with the attainment of this ideal form not only being technologically difficult but also expensive. The merits of refined powders are clear: With standard-grade zirconias, firing times of several hours at temperatures above 1700°C are not unusual; with active, deagglomerated powders,[7] densities above 99% of the theoretical value have been obtained at 1100°C after 1 h. The problem lies in bringing the refined powders to economic and flexible use.

Techniques for the preparation of active powders have been widely studied, and progress from the point of view of zirconia has been mentioned in an earlier review.[8] The feature that has characterized more recent work is the wide recognition of the problems posed by the formation of powder agglomerates, i.e., the bonding together of the fine powder crystallites into larger porous particles which are strong enough to survive the pressing and forming operations. Since these operations cause density variation and faults in the finished product either because of initial density differences between the agglomerate and the surrounding powder or because of differences in the densification rate between the agglomerate[9] and its surroundings, much effort has been devoted to the removal of the agglomerates by such techniques as milling[10] or sedimentation,[7,10] the reduction of agglomerate strength by control of the conditions of powder formation[11] and beneficiation,[12] or the avoidance of agglomerate formation in the first place.[13,14] In this last connection, the use[15] of controlled nucleation and growth to achieve TiO_2 powders close to the ideal configuration and the subsequent formation of an ordered arrangement of these powders by modification of the powder/fluid suspension characteristics is worthy of note as representing the degree to which the green structure can be controlled. The papers concerned with processing in these proceedings provide a full indication of the range of methods that is available and of the progress that is being made.

Firing Procedures

Recent work concerned with the firing of zirconia materials can be discussed in terms of solid-state sintering, liquid-phase sintering, and some special methods.

With regard to solid-state sintering, a significant development has been the clear demonstration[7] of the importance of agglomerates in terms of their effect on fired densities. The dependence of the densities on agglomerate size has been shown, and the value of using deagglomerated powders underlined. For such powders, the anticipated benefit of very fine (≈ 12 nm) powders has been realized in the very low sintering temperatures noted above.

The mechanism of densification has been explored in work[16] on the hot-pressing kinetics of ZrO_2-CaO, ZrO_2-Y_2O_3, and CeO_2-Y_2O_3 powders. A linear dependence of the densification rate on the applied stress coupled with the approximate agreement of the measured diffusion coefficient with tracer diffusion values has suggested that the controlling process in densification is cation lattice diffusion. The finding that the densification rate shows a maximum with increasing stabilizer content (Fig. 1 shows data for yttria-doped ceria in which no phase boundary exists to complicate interpretation) suggests, by correlation with the similar maximum in ionic conductivity data, that the cation defect responsible for diffusion is the interstitial. In this way, the controlling conducting species, $V_{\ddot{O}}$, and the controlling densification species have the same sign of effective charge and hence the same qualitative dependence on the stabilizer content.

The mechanism responsible for grain growth or coarsening in zirconia is not known. For the early stages of sintering, where the choice lies between surface diffusion or vapor-phase transport, the relatively low volatility of zirconia suggests the significance of surface diffusion. In this connection, it has been found[17] by grain-boundary grooving studies on zirconia bicrystals that magnesia acts to sup-

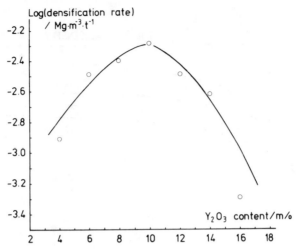

Fig. 1. Densification rate of CeO_2 during hot-pressing (1400°C, 20 MPa) rises to a maximum as the concentration of Y_2O_3 additive is increased.

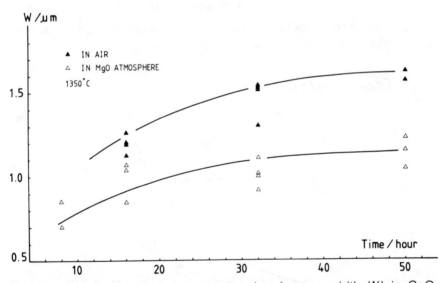

Fig. 2. Extent of grain-boundary grooving (groove width W) in CaO-stabilized ZrO_2 reduced in an MgO-containing atmosphere. The observed kinetics are consistent with control by surface diffusion, the coefficient being reduced by a factor of 3.

press the rate of groove formation (Fig. 2). Since the attainment of high density is most readily achieved under conditions where the ratio of densification rate to coarsening rate is enhanced,[18] this helps to explain the finding[19] that magnesia is helpful as an additive for the sintering of zirconia (Fig. 3), the additive yielding a smaller grain size at a given density (Figs. 4 and 5) and allowing a limited but significant improvement in final fired densities.

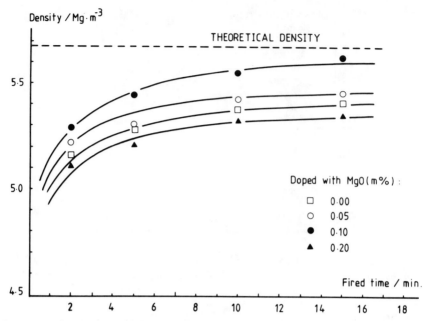

Fig. 3. Influence of MgO additions on the fired densities of CaO-stabilized ZrO_2 samples fast-fired at 1700°C.

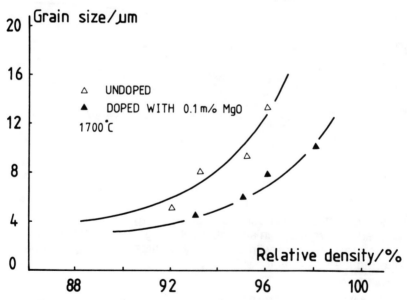

Fig. 4. Tendency for grain growth during densification reduced by the presence of MgO additive in CaO-stabilized ZrO_2.

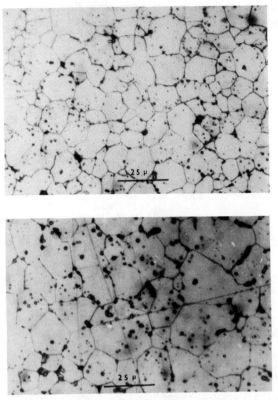

Fig. 5. Microstructures of CaO-stabilized ZrO_2 fast-fired (1700°C; 10 and 15 min, respectively) to 96% of theoretical density with and without MgO additive.

With regard to liquid-phase sintering, work has continued both in the search for sintering aids, e.g., the use of borates,[20] and in the search for ways to minimize the detrimental effects of boundary phases on properties. In this last respect, the finding[21] that alumina additions can act as scavengers for residual silicate phases has been a striking indication of the potential complexity of additive interactions; the improvement in electrical properties that can result from alumina additions is significant for materials prepared from standard-grade powders.

In terms of more specialized techniques, hot-pressing continues to be used as a convenient procedure for making test materials. Fast firing, or the use of short, high-temperature firing cycles, appears to be potentially beneficial in view of the reported activation energies for densification[22] and grain growth[23] ($E_d > E_g$). As seen in Fig. 3, short firings are indeed practicable at 1700°C for calcia-stabilized material.

Conclusion

Of late, much work on powder preparation has been done to demonstrate the degree of improvement in powder quality that is available. There seems little doubt that, with closer identification of the applications and markets that lie open to zirconia, the trend to use progressively better controlled and characterized powders that has occurred for other technical ceramics will take place for zirconia also. At the same time, it appears that distinct benefits are also available from the use of additives, and the opportunity for optimization in terms of overall economic production that is provided by the combination of the two approaches is a promising avenue for development.

References

[1] Hayne Palmour III, R. F. Davis, and T. M. Hare, Processing of Crystalline Ceramics. Material Science Research, Vol. 11, Plenum, New York, 1978.
[2] F. F. Y. Wang, "Ceramic Fabrication Processes," *Treatise Mater. Sci. Tech.,* **9** (1976).
[3] G. Y. Onoda, Jr. and L. L. Hench; Ceramic Processing before Firing. Wiley & Sons, New York, 1978.
[4] D. W. Johnson, Jr., *Am. Ceram. Soc. Bull.,* **60** [2] 221–24, 243 (1981).
[5] K. S. Mazdiyasni, *Ceramurgia Int.,* **8** [2] 42–56 (1982).
[6] H. K. Bowen, *Mater. Sci. Eng.,* **44** [1] 1–56 (1980).
[7] W. H. Rhodes, *J. Am. Ceram. Soc.,* **64** [1] 19–22 (1981).
[8] R. J. Brook; pp. 272–85 in Advances in Ceramics, Vol. 3. Edited by A. H. Heuer and L. W. Hobbs. The American Ceramic Society, Columbus, OH, 1981.
[9] A. G. Evans, *J. Am. Ceram. Soc.,* **65** [10] 497–501 (1982).
[10] C. E. Scott and J. S. Reed, *Am. Ceram. Soc. Bull.,* **58** [6] 587–90 (1979).
[11] M. A. C. G. van de Graaf, K. Keizer, and A. J. Burggraaf; pp. 83–992 in Science of Ceramics, Vol. 10. Edited by H. Hausner. Deutsche Keramische Gesellschaft, Berlin, 1980.
[12] R. Pampuch and K. Haberko, *Keram. Z.,* **31**, 478 (1979).
[13] A. Roosen and H. Hausner; p. 773 in Ceramic Powders. Edited by P. Vincenzini. Elsevier, Amsterdam, 1983.
[14] Y. Murase and E. Kato, *J. Am. Ceram. Soc.,* **66** [3] 196–200 (1983).
[15] E. A. Barringer and H. K. Bowen, *J. Am. Ceram. Soc.,* **65** [12] C-199–C-201 (1982).
[16] Suxing Wu and R. J. Brook; to be published in *Solid State Ionics.*
[17] Suxing Wu and R. J. Brook; pp. 371–80 in Science of Ceramics, Vol. 12. Edited by P. Vincenzini. National Research Council Research Institute for Ceramics Technology, Faenza, Italy, 1984.
[18] M. F. Yan, *Mater. Sci. Eng.,* **48**, 53 (1981).
[19] Suxing Wu and R. J. Brook, *Trans. J. Br. Ceram. Soc.,* **82** [6] 200–205 (1983).
[20] R. C. Buchanan and A. Sircar, *J. Am. Ceram. Soc.,* **66** [2] C-20–C-21 (1983).
[21] E. P. Butler and J. Drennan, *J. Am. Ceram. Soc.,* **65** [10] 474–78 (1982).
[22] P. J. Jorgensen; p. 401 in Sintering and Related Phenomena. Edited by G. C. Kuczynski, N. A. Hooton, and C. F. Gibbon. Gordon & Breach, New York, (1967).
[23] T. Y. Tien and E. C. Subbarao, *J. Am. Ceram. Soc.,* **46** [10] 489–92 (1963).

Sinterability of ZrO_2 and Al_2O_3 Powders: The Role of Pore Coordination Number Distribution

F. F. Lange and B. I. Davis

Rockwell International Science Center
Thousand Oaks, CA 91360

The sinterability of two ZrO_2 powders containing 2.2 and 6.6 mol% Y_2O_3, respectively, an Al_2O_3 powder, and Al_2O_3/10 vol% ZrO_2 composite powder was investigated. Shrinkage was measured at a constant heating rate. Capillary size distributions were determined by mercury intrusion after compacts were heated to various temperatures. Isopressing eliminated larger pores with larger coordination numbers, which results in greater sinterability. Local densification, i.e., densification of multiple-particle packing units, enlarged pores between these packing units. This phenomenon of local densification initiated prior to bulk shrinkage and stopped when the shrinkage rate reached its maximum. ZrO_2 inclusions, known to inhibit grain growth, also inhibited sinterability. This observation, coupled with previous work, strongly suggests that grain growth, supported by dense packing units, helps lower the coordination number of pores and therefore is helpful in the sintering process. Large multiple-particle packing units (large agglomerates) produce unsinterable pores, i.e., pores with high coordination numbers. It was shown that colloidal/sedimentation treatments, which decrease the size of the soft and hard agglomerates, reduced sintering temperatures and increased strength.

A previous paper[1] outlined a new concept concerning the sinterability of consolidated, agglomerated powders, i.e., compacts that exhibit bulk density variations on both the macro and micro scale. Only monosized (e.g., spherical) powders, packed with a periodic arrangement, do not exhibit these density variations. Thus, nonperiodic arrangements of both mono- and polydispersed powders, spherical or otherwise, must be considered as agglomerated.

Pores are the central characters of this new concept, and their coordination number (R, defined as the number of touching particles that form the pore "surface" in the consolidated state and/or the number of surrounding grains in the partially sintered state) distribution ($V(R)$, the function that describes the volume fraction of pores with a given R), is their principal functional property. A thermodynamic concept, first introduced by Kingery and Francois[2] and modified by Cannon,[3] which shows that pores with a coordination number less than a critical value (R_c) can shrink and disappear (kinetics permitting) by volume/grain-boundary diffusion was used, as shown in Fig. 1(a). Pores with a coordination number $\geq R_c$ will only shrink to an equilibrium size[3] as shown in Fig. 1(b). The critical coordination number, R_c, depends on the dihedral angle, which depends on surface and grain-boundary energies and is therefore influenced by atomic bonding and surface chemistry. Thus, from a thermodynamic viewpoint, the volume fraction of pores

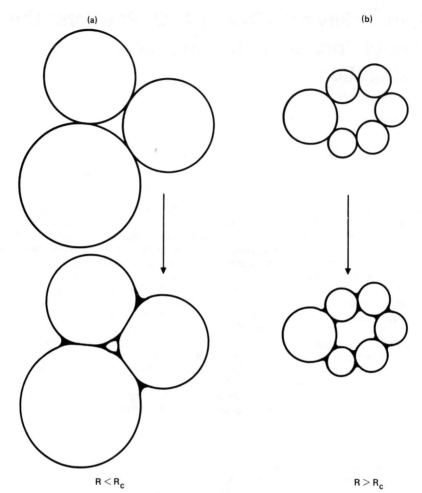

Fig. 1. Coordination number development for two pores (a) $R < R_c$ and (b) $R > R_c$.

that can disappear is given by

$$\int_4^{R_c} V(R)dR \tag{1}$$

where 4 is considered the lowest coordination number.

The previous paper[1] suggested that the pore coordination number distribution of a powder compact was dependent on how multiple-particle packing units were arranged in space. Two types of multiple-particle packing units were visualized, i.e., domains that consisted of relatively few highly coordinated particles, thus having the lowest pore coordination distribution and agglomerates that consist of packed domains. The interdomain pores have a higher coordination number distribution than the domains themselves. The agglomerates pack together to form the powder compact. The interagglomerate pores have the highest coordination

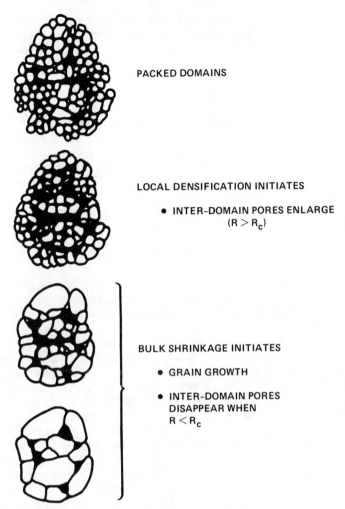

Fig. 2. Schematic for the densification of an agglomerate: local densification (or rearrangement) and grain growth.

number distribution. The pore coordination number distribution is thus the sum of the distributions within the domains, between the domains, and between the agglomerates.

The effect of consolidation forces (e.g., isopressing) and heating on the pore coordination distribution were discussed.[1] It was suggested that consolidation forces cause the multiple-particle packing units to deform to fill in the largest pores and thus increase bulk density by removing higher coordinated pores. Increasing the bulk density increases the proportion of pores with coordination numbers $\leq R_c$, thus helping to explain why compacts with a higher bulk density achieve higher end-point densities for a given sintering schedule.

The effects produced during heating to temperatures where diffusional processes promote neck formation etc. are summarized in Fig. 2. One agglomerate is shown. Domains within the agglomerate are the most highly coordinated with respect to particles (i.e., greatest number of particle contacts per volume) and contain pores with the lowest coordination number distribution. It can be shown that the free energy charge per unit volume (i.e., the driving force for shrinkage) can be expressed as

$$\frac{\delta F}{\delta V} = n\gamma dA' \tag{2}$$

where n is the number of contacts per unit volume ($n \alpha 1/R$), γ the pore surface energy per unit area, and dA' the decrease in pore surface per unit compact. Domains will therefore be the first multiple-particle packing unit to undergo shrinkage. Since each domain cannot be expected to shrink at the same rate (both n and the $V(R)$ within each domain will differ from domain to domain), the net effect is that domains will shrink on themselves, breaking some interdomain necks to enlarge and increase the coordination number of interdomain pores as shown in Fig. 2. The net effect on $V(R)$ is to eliminate pores with a low R and increase R of other pores, i.e., a process that hinders sinterability. This process of local densification can occur without bulk shrinkage. For this case, pore volume within the domains is transferred to interdomain pores.

The multiply connected, dense domains can now support grain growth. Grain growth (Fig. 2) reduces the coordination number of interdomain pores. Thus, grain growth at this stage allows interdomain pores to spontaneously disappear. The end result is a dense agglomerate.

Multiply connected agglomerates will undergo the same process as described for domains, i.e., enlargement of interagglomerate pores and grain growth to decrease the coordination number of the enlarged interagglomerate pores. The differential shrinkage between agglomerates should not be as large as that for domains since they contain many more contacts per unit volume. Bulk shrinkage is thus expected to begin with densification of agglomerates.

Thus, local densification, first of domains and then of agglomerates, increases the coordination number of the enlarged pores between the packing units. Grain growth, first supported by dense domains and then by dense agglomerates, decreases the coordination number of these enlarged pores. The effect of the effective compressive stress due to "surface tension" ($\sigma = \delta F/\delta V = \bar{n}\gamma dA'$) and applied compressive stresses (hot-pressing, isostatic hot-pressing) on reducing the coordination number of pores will not be elaborated here.

The purpose of the current paper is to review the experimental observations that appear to support the concept outlined above. These observations will be detailed for ZrO_2, Al_2O_3, and Al_2O_3/ZrO_2 powder compacts.

Powder Preparation, Consolidation, and Experimental Procedure

Two ZrO_2 powders were used; one contains 2.2 mol% Y_2O_3, which could be sintered to a single-phase, polycrystal tetragonal ZrO_2 ceramic, and one contains 6.6 mol% Y_2O_3, which produced a cubic ZrO_2 ceramic. These powders are manufactured* by heating cellulose soaked with $ZrClO_2$ and a soluble yttrium salt in air to produce the ZrO_2 solid solution and to burn off the carbon. The hard, partially sintered agglomerates of ZrO_2 solid solution crystallites are milled and washed to remove excess chlorine. Transmission electron microscopy (TEM) examination indicated that the as-received powder still contains many large, hard agglomerates, whereas the ZrO_2 crystallites were ≤ 0.1 μm.

---

*Zircar, Florida, New York.

TWO-PHASE MIXTURES

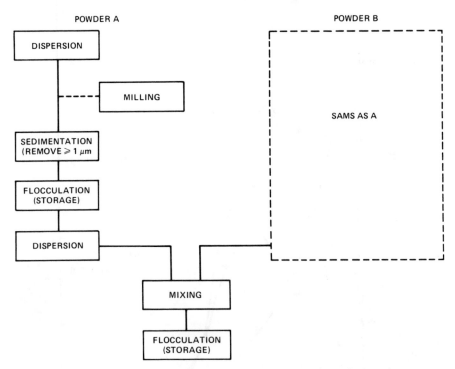

Fig. 3. Schematic for powder preparation and mixing of two phases.

The α-Al_2O_3 powder is believed to be manufactured[†] by atomizing an alkoxide in a manner described by Visca and Matijevic.[5] The particles are nearly spherical, and the powder is claimed to have a narrow size distribution with a mean of 0.59 μm. The as-received dry powder contained many large, soft agglomerates that could be broken apart with ultrasonic treatment.

Previous work[6] showed that all three powders could be dispersed in water at pH = 2.5 (using HCl) and flocced at a pH between 7 and 9 (using NH_4OH). All three powders were subjected to a colloidal treatment outlined in Fig. 3. The purpose of this colloidal treatment was to break down soft agglomerates with the surfactant and to eliminate all hard agglomerates (or mill particles) ≥ 1 μm by sedimentation. Flocculation prevents mass segregation during storage and/or consolidation. Flocculation also consolidates the powder to a higher volume fraction. Washing to remove excess salt introduced during the dispersion/flocculation steps is performed by removing the clear supernate, remixing with deionized water, and reflocing several times.

Only the ZrO_2 (+6.6 mol% Y_2O_3) powder was milled. A high-purity[‡] Al_2O_3 jar and media were used, which introduced fractured Al_2O_3 grains (i.e., the wear product) with sizes between 5 and 20 μm. These were eliminated during sedimentation.

[†]Sumitomo Chemical Co., Osaka 569, Japan (Type AKP-30).
[‡]Coors Porcelain Co., Golden, CO.

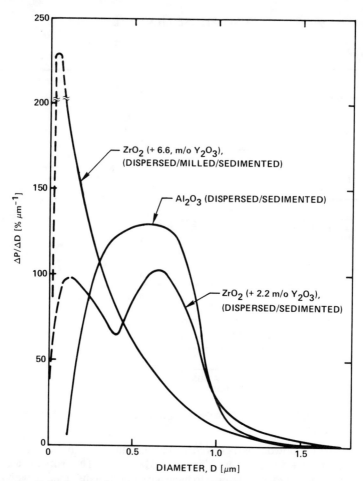

Fig. 4. Particle (or agglomerate) size distribution for three powders used in study after colloidal/sedimentation treatment.

Figure 4 illustrates the size distribution of the three powders after redispersion. Note the bimodal distribution of the small particles and the larger hard agglomerates for the unmilled ZrO_2 (+2.2 mol% Y_2O_3 powder).

Two-phase mixtures were prepared by determining the solid contents in each flocced system, redispersing, mixing the appropriate volumes with ultrasonics, and then reflocing to prevent mass segregation.

Consolidation was performed from the flocced state by filtration (slip casting). After drying, the green density of some specimens was increased by isopressing at 350 MPa.

Two types of air sintering experiments were performed. In the first, specimens were heated to 700° and then heated at a constant rate of 5°C/min to 1550° (or 1400°C for the case of ZrO_2) and furnace-cooled. Linear shrinkage was measured during these experiments with a high-temperature extensiometer.

In the second group of experiments, specimens were heated and held at the desired temperature for 30 min before furnace-cooling. These specimens were then

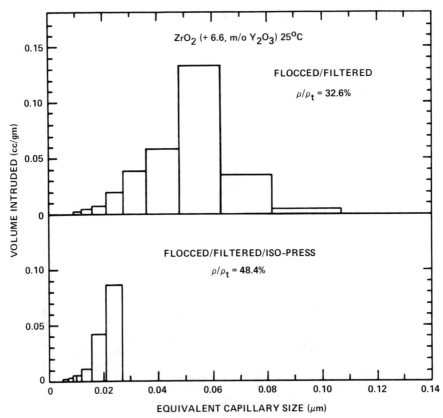

Fig. 5. Equivalent capillary size distribution (mercury intrusion) for ZrO_2 (+6.6 mol% Y_2O_3) compacts determined at 25°C.

examined with a mercury porosimeter[§] to determine the equivalent capillary size distribution as a function of temperature.

Consolidation Effects

The bulk density of the filtered ZrO_2 powders was 33% of theoretical.[¶] It could be increased to 49% of theoretical by isopressing at 350 MPa. Similarly, isopressing of the filtered/dried Al_2O_3 powder increases the bulk density from 51 to 60% of theoretical. Figure 5 illustrates the mercury intrusion data for the filtered and filtered/isopressed ZrO_2 (+6.6 mol% Y_2O_3) powder compacts. As illustrated, the capillary size distribution is larger for the lower bulk density compact than for the isopressed compact. Since the mean particle size is the same at both bulk densities, isopressing not only decreases the size of the larger pores but also reduces their coordination number. That is, the compact with the higher initial bulk density is expected to contain a higher fraction of pores with coordination numbers $\leq R_c$.

Similar results were obtained for the other two powders.

[§]Micromeritic Instrument Corp., Norcross, GA. (Auto Pore 9200)
[¶]ZrO_2 (+6.6 mol% Y_2O_3); cubic structure; ρ_t = 6.05 g/cm^3; ZrO_2 (+2.2 mol% Y_2O_3); tetragonal structure; ρ_t = 6.07 g/cm^3.

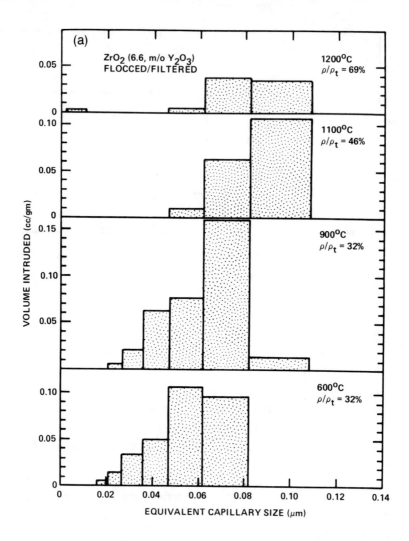

Local Densification

Local densification is a more descriptive term for what Exner[7] describes as rearrangement. As outlined above, domains will densify during the early stage of heating. Since each domain is multiply connected to one another but may not shrink at the same rate, void space between poorly bonded domains will enlarge. If neighboring domains exhibit sufficient differential shrinkage, pore enlargement, due to domain densification, can occur without bulk shrinkage. If a portion of the multiply connected domains (and then, agglomerates) exhibits the same shrinkage rate, bulk shrinkage will take place during pore enlargement. An enlarged pore is less sinterable due to its increase in coordination distribution.

Figure 6 illustrates the equivalent capillary size distribution for the filtered (Fig. 6(a)) and filtered/isopressed (Fig. 6(b)) compacts of ZrO_2 (+6.6 mol% Y_2O_3) at four temperatures. These results show (also see Fig. 5) that pore enlargement (i.e., enlargement of equivalent capillaries) precedes bulk shrinkage, viz., up to 900°C. Pore enlargement continues with the initiation of bulk shrinkage at

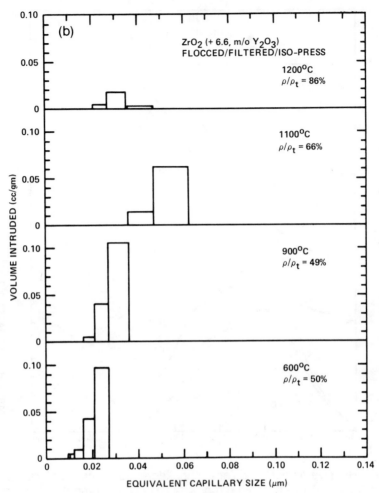

Fig. 6. Equivalent capillary size distribution (mercury intrusion) for ZrO_2 (+6.6 mol% Y_2O_3) as a function of temperature (30-min hold) for (a) filtered compact and (b) filtered/isopressed compact.

temperatures >900°C. The process of pore enlargement stops between 1100° and 1200°C, at which point other processes cause the large pores to shrink.

Figure 6 (and Fig. 5) also illustrates that pore enlargement takes place by a redistribution of pore sizes; viz., small pores disappear** and larger pores are formed. This observation is consistent with the concept of local densification, i.e., pores within densifying domains (then agglomerates) redistributing to enlarge interdomain (or agglomerate) pores. Domain (or agglomerates) densification leads to bulk shrinkage only at temperatures >900°C.

Figure 7 illustrates the constant heating rate, shrinkage results plotted as relative density vs temperature, and shrinkage rate vs temperature. It is interesting

---

**Due to plotting, Fig. 6 does not show intrusion into smaller equivalent capillaries where the intrusion values are ≤0.002 cm^3/g.

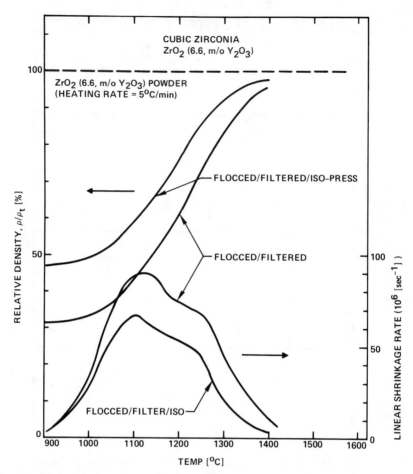

Fig. 7. Relative density and linear shrinkage rate vs temperature (heating rate = 5°C/min) for ZrO_2 (+6.6 mol% Y_2O_3) powder compacts (filtered and filtered/isopressed).

to note that the shrinkage rate increases to its maximum at ≈1100°C during the stage of local densification.

Effect of Grain Growth

As shown previously and reviewed above, grain growth, supported by dense domains and then agglomerates, occurs throughout the sintering process. It was previously hypothesized that grain growth is helpful in reducing the coordination number of pores that would be otherwise unsinterable. To test this hypothesis, several experiments have been initiated. One of these, which will be detailed elsewhere,[8] will be reviewed here.

Previous work[9] showed that ZrO_2 inclusions, introduced as a second-phase powder into Al_2O_3 powder, locate at 4-grain junctions during densification and hinder, but do not stop, growth of the Al_2O_3 grains. Dihedral angle measurements showed that the average Al_2O_3 grain-boundary energy is ≈1.5 times the

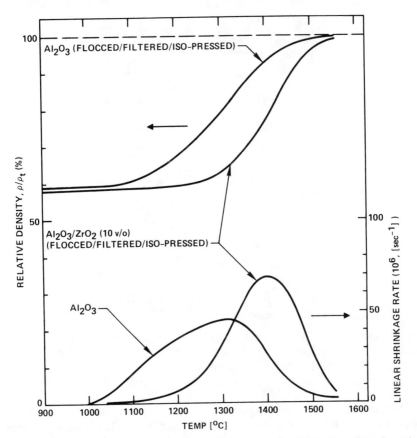

Fig. 8. Relative density and linear shrinkage rate vs temperature (heating rate = 5°C/min) for Al_2O_3 and Al_2O_3/ZrO_2 (+2.2 mol% Y_2O_3) powder compacts (both isopressed).

Al_2O_3/ZrO_2 interfacial energy. One might conclude from these relative surface energies that additions of ZrO_2 to Al_2O_3 would increase the driving force for sintering. On the other hand, if grain growth aids sintering, one might conclude that ZrO_2 additions, which hinder grain growth, would also hinder sintering.

Constant heating rate experiments were carried out for Al_2O_3 powders containing 10 vol% ZrO_2 (+2.2 mol% Y_2O_3), which were filtered and isopressed as described above. Results are compared with those for single-phase Al_2O_3 in Fig. 8. As shown, the ZrO_2 addition delayed the start of bulk shrinkage and the maximum shrinkage rate by ≈100°C. Note that, when bulk shrinkage did initiate for the composite, it was much more rapid than for the pure Al_2O_3. These results are consistent with the hypothesis that phenomena that hinder grain growth will delay sinterability.[††]

[††]Helpful grain growth should not be confused with abnormal grain growth, which entraps pores and limits end-point density.

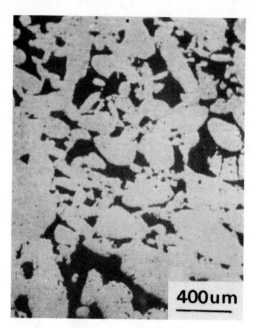

Fig. 9. Large, soft agglomerates obtained by sedimentation, packed by filtration, and sintered (heating rate = 5°C/min to 1550°C). Note large pores between dense, polycrystalline domains.

Similar composites were prepared in which only a portion of the Al_2O_3 was deagglomerated (ultrasonic treatment was not employed). The resulting sintered microstructure (sintered to only 92% of theoretical) contained large, dense regions of polycrystalline Al_2O_3 without the dispersion (i.e., Al_2O_3 agglomerates that did not incorporate the ZrO_2). These dense regions contained large Al_2O_3 grains and were surrounded by a lower-density (porous), two-phase, finer-grained matrix. This observation similarly supports the hypothesis that grain growth is helpful in decreasing the pore coordination number and, thus, sinterability.

Effect of Agglomerated Size

The size and coordination number of an interagglomerate pore are proportional to the agglomerate size. Powders produced by decomposition reactions contain many partially sintered, hard agglomerates. Powders consolidated by dry routes contain large, soft agglomerates. If these powders are not treated by the colloidal/sedimentation route (or a similar route) described above, these agglomerates persist through sintering.[10]

The extreme case where agglomerates are simply packed together without applied consolidation forces (no isopressing) was simulated for the Al_2O_3 powder described above. The large, soft agglomerates, collected after repeated sedimentation, were consolidated by filtration. After being heated to 1550°C (5°C/min; no hold time), the microstructure consisted of multiply connected, dense agglomerates and pores, as shown in Fig. 9. Since the grains within the dense agglomerates are ≈ 10 times smaller, the interagglomerate pores are virtually unsinterable (without further grain growth) due to their high coordination number. Thus, powders

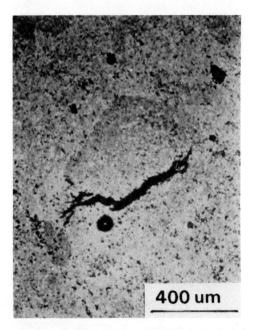

Fig. 10. Large cracklike pore produced by differential shrinkage of large Al_2O_3 agglomerate and surrounding colloidal/sedimented treated Al_2O_3 matrix (sintered 5°C/min to 1550°C).

that consist of agglomerates that are much larger than their constituent crystallites will require high sintering temperatures (despite their so-called active crystallite size) to achieve the grain growth required to reduce the coordination number of interagglomerate pores to $<R_c$. Preliminary observations currently suggest that grains must grow to the size of the agglomerate before the interagglomerate pore can disappear spontaneously.

The highly agglomerated (hard-type) ZrO_2 powders used here are a case in point. If these powders are untreated (as-received and slightly milled), densities of 95% theoretical can be achieved only by sintering at 1600°C/2 h. Treated to break up soft agglomerates and eliminate hard agglomerates, $\geqslant 1$ μm as described above, the ZrO_2 (+6.6 mol% Y_2O_3) powders can be sintered to >98% of theoretical at 1300°C/1 h. The ZrO_2 powders (2.2 mol% Y_2O_3), which contain a large fraction of hard agglomerates <1 μm (see Fig. 4), can be sintered to the same density at a slightly higher temperature.

The more general case is where the larger agglomerates make up only a small fraction of the powder compact, viz., the case where milling reduces a large fraction of the large agglomerates to their consistent crystallites (or domains) and the compact is consolidated from the colloidal state. This case was also simulated by mixing the colloidally sedimented Al_2O_3 with sedimented large Al_2O_3 agglomerates. As shown previously,[10] the differential shrinkage of the agglomerate relative to the matrix produces a cracklike void with a size proportional to the agglomerate, as shown in Fig. 10. Again, these cracklike voids are unsinterable without either an applied pressure or extensive grain growth. They not only limit the end-point density but also are fracture origins that limit potential strength.

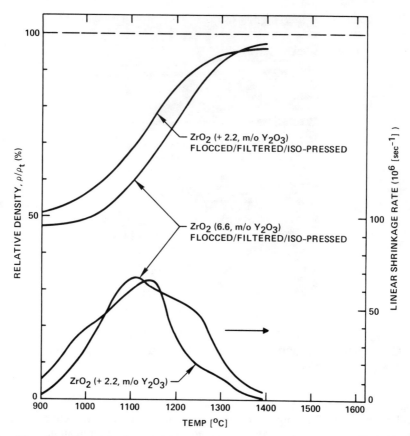

Fig. 11. Relative density and linear shrinkage rate vs temperature (heating rate = 5°C/min) for filtered/isopressed compacts of ZrO_2 (+2.2 mol% Y_2O_3) and ZrO_2 (+6.6 mol% Y_2O_3) powders.

Comparison of the two ZrO_2 powders is a case in point. The unmilled ZrO_2 (+2.2 mol% Y_2O_3) powder contained a larger fraction of hard agglomerates (≤ 1 μm) than the milled ZrO_2 (+6.6 mol% Y_2O_3) powder after the same colloidal/sedimentation treatment (see Fig. 4). A comparison of their sintering behaviors (constant heating rate of 5°C/min to 1400°C) shows (Fig. 11) that the powder containing the large fraction of hard agglomerates achieved a lower endpoint density despite the fact that they contributed to a higher shrinkage rate at lower temperatures (<1200°C).

Conclusions

(1) The thermodynamic potential of a crystalline powder compact to densify is governed by the coordination number distribution of its pore space and the critical coordination number. That is, a compact's sinterability (kinetics permitting) is determined by the arrangement of particles and multiple-particle packing units and the dihedral angle formed between a pore surface and a grain boundary.

(2) Consolidation forces (e.g., isopressing) first eliminate pores with the highest coordination numbers (least sinterable) (see Fig. 5).

(3) During neck formation, multiple-particle packing units, which have the largest number of contacts per unit volume (or lowest pore coordination numbers), densify on themselves to increase the coordination number (and size) of interpacking unit pores. This process of local densification (or rearrangement) can occur, to a limited extent, prior to bulk shrinkage, and to a greater extent, during bulk shrinkage (Fig. 6). Pore enlargement due to local densification appears to cease when the shrinkage rate exceeds its maximum value (Figs. 6 and 7).

(4) Grain growth is helpful in reducing the coordination number of those pores enlarged by local densification. When grain growth is inhibited with a dispersion of ZrO_2 particles in Al_2O_3, sinterability is also inhibited (Fig. 8).

(5) The coordination numbers (and size) of interpacking unit pores are proportional to the size of the packing unit (Figs. 9 and 10).

The directions resulting from these conclusions only add reason to something we have already conceived. Namely, the most sinterable powders are monodispersed crystallites (or amorphous particles) that are densely packed in a periodic array. Lacking this utopian situation, the conclusions direct the fabricator to powder preparation and consolidation methods (e.g., those described in powder preparation, consolidation, and experimental proceeding) of reducing sintering temperatures (viz., reducing grain-growth requirements) and reducing the size of the strength-degrading cracklike pores. These methods have enabled the present authors to reduce sintering temperatures and to achieve mean flexural strengths for transformation-toughened, pressureless-sintered ceramics of 1300 MPa (185 000 psi).

Acknowledgment

This work was supported by the Office of Naval Research under Contract No. N00014-82-C-0341.

References

[1] F. F. Lange, *J. Am. Ceram. Soc.*, **67** [2] 83-89 (1984).
[2] W. D. Kingery and B. Francois; pp. 471-98 in Sintering and Related Phenomena. Edited by G. C. Kuczynske, N. A. Hooton, and G. F. Gibbon. Gorden & Breach, New York, 1967.
[3] R. M. Cannon; unpublished work.
[4] Y. Abe, S. Horikim, K. Fijimura, and E. Ichiki, *Prog. in Sci. and Eng. of Comp. Proc.*, ICCM-IV, 1427-34 (1982).
[5] M. Visca and E. Matijevic, *J. Colloid Interface Sci.*, **68** [2] 308-19 (1979).
[6] I. A. Aksay, F. F. Lange, and B. I. Davis, *J. Am. Ceram. Soc.*, **66** [10] C-190-C-192 (1983).
[7] H. E. Exner, *Rev. Powder Metall. Phys. Ceram.*, **1** [1-4] 1-251 (1979).
[8] F. F. Lange, T. Yamaguchi, and P. E. D. Morgan; unpublished work.
[9] F. F. Lange and M. Hirlinger, *J. Am. Ceram. Soc.*, **67** [3] 164-68 (1984).
[10] F. F. Lange and M. Metcalf, *J. Am. Ceram. Soc.*, **66** [6] 398-406 (1983).

Sintering Kinetics of ZrO$_2$ Powders

A. ROOSEN AND H. HAUSNER

Technische Universität Berlin
Institut für Nichtmetallische Werkstoffe
D-1000 Berlin 12, Federal Republic of Germany

The powder and compact characteristics and the sintering behavior of two wet-chemically prepared CaO-stabilized ZrO$_2$ powders are compared with the characteristics of a mechanically mixed powder of identical composition. The wet-chemically prepared powders could be sintered to densities of 95% of theoretical at 1250°C if the coprecipitated hydroxides were submitted to a freeze-drying step; otherwise, sintering temperatures of 1600°C were necessary to obtain the same density. The distinct differences in the shrinkage-rate curves could be related to differences in the pore-size distribution in the compacted powders.

In the preparation of ceramic powders the processing parameters have a significant influence on the powder characteristics and the sintering behavior of the compact.[1]

In the case of wet-chemically prepared, sinter-active, stabilized ZrO$_2$ powders, it has been demonstrated that agglomerate formation has to be avoided in order to obtain a high density and a homogeneous microstructure in the sintered material via an optimal pore-size distribution in the compact.[2-5] Pore-free areas are formed within the agglomerates during the sintering process, and contact is lost with surrounding matter which results in the formation of large pores.[6] The dense parts are possible sources for discontinuous grain growth. This negative influence on density and microstructure requires either the destruction or elimination of the agglomerates after they have been formed[2,3,7,8] or their appearance must be suppressed from the beginning by an optimization of the precipitation conditions[9,10] or by reducing the capillary forces[11] which arise in the drying step by a treatment of the hydroxides with suitable organic solvents[10,12-16] or by freeze-drying the precipitates.[5]

The importance of the pore-size distribution for the sintering process (densification, grain growth) and the changes in pore size during the different stages of sintering is the object of several publications.[17,18] Several authors observed a shrinkage behavior, especially with ultrafine, sinter-active powders, which cannot be explained by the usual sintering mechanisms and equations.[8,13,19-22] Morgan and Yust[23,24] and Pampuch and Haberko[4] assumed that a dislocation glide was responsible for densification during the initial stages of sintering. Other authors[2-5,9,10,16,25,26] stress the importance of a homogeneous pore-size distribution in the compacted powder by a reduction of the agglomerate formation in order to reach high sintering densities. Only Carbone and Reed[27] investigated the correlation between pore-size distribution and shrinkage. They reported that the initial shrinkage was independent of the porosity, whereas the intermediate and the final-stage shrinkage was strongly influenced by the pore structure in the compact.

In this investigation, the interdependence between processing parameters, the pore structure in the compact, and the sintering characteristics has been studied for three CaO-stabilized ZrO$_2$ powders.

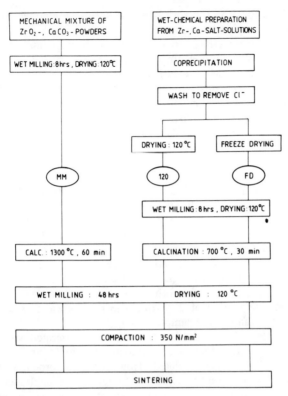

Fig. 1. Processing flow sheet.

Experimental Procedure

Two coprecipitated powders and one powder prepared from a mechanical mixture of ZrO_2 and $CaCO_3$ were used. The chemical composition of all powders was 87 mol% ZrO_2–13 mol% CaO. The processing flow sheet with experimental details is shown in Fig. 1.

The mechanical mixture (MM) was prepared from a ZrO_2 powder* (0.25% SiO_2, 0.15 TiO_2, 0.02 Fe_2O_3, 0.01 Na_2O) and a high-purity $CaCO_3$ powder (analytical grade). After drying, the powder mixture was calcined and remilled.

Precipitates were obtained from an aqueous solution of high-purity $ZrOCl_2$ and $Ca(NO_3)_2$ with ammonia and ammonium carbamate. After centrifuging and removal of all chlorine ions by several washing steps, the sample 120 was dried at 120°C in a drying oven, whereas sample FD was freeze-dried. All dried precipitates were then milled in ethanol and transformed to the oxide by calcination. The oxide powders were remilled in ethanol and compacted.

Plastic-lined ball jars with CaO/MgO-stabilized ZrO_2 balls were used for milling. The milled oxide powders were carefully dried at 120°C and granulated by passing through a 0.5-mm screen after the addition of an aqueous polyvinyl alcohol solution as a pressing aid. From these granulated powders, cylindrical samples (14 mm in diameter and 5.5 mm high) were pressed and sintered between

---

*E16 grade, Magnesium Elektron Ltd., Twickenham, England.

1100° and 1580°C for 3 h. The shrinkage and shrinkage rate were determined up to 1550°C, using samples 10 mm in diameter and 10 mm high.

The powders were characterized by phase composition (X-ray analysis), crystallite size (X-ray line broadening measurement), particle morphology (SEM), particle-size distribution (MSA centrifuge), tap density, specific surface area (N_2 adsorption), and pore-size distribution (N_2 adsorption, Hg porosimetry). The compacts were characterized by specific surface area, pore-size distribution, green density, and shrinkage characteristics (dilatometry). The sintering density (buoyancy) and the microstructure (SEM) were determined for the sintered samples.

Results

Powder Characteristics

The mechanical mixture of ZrO_2 and $CaCO_3$ was heat-treated at 1300°C. After 1 h, 94% cubic solid solution and 6% monoclinic ZrO_2 was present.

In the case of the wet-chemical powder preparation, the dried precipitate is X-ray amorphous. Calcination at 700°C results in a complete transformation into the cubic solid solution.

The dried precipitates differ greatly in appearance. After drying at 120°C, up to 20-mm large, hard lumps with glassy surfaces have been formed, whereas the FD precipitate is obtained in the form of a loose, voluminous powder with an average particle size (SEM) of 2–8 μm. The mechanical mixture has a much higher tap density than the FD sample.

The precipitates 120 and FD consist of agglomerated small crystallites (SEM). From the specific surface area of 326 and 320 m^2/g, an equivalent particle size of about 5 nm can be calculated. However, both products differ in their porosity, resulting in different mechanical strength of the agglomerates.[5] The FD powder has a pore volume which is an order of magnitude larger and a much broader pore-size distribution; therefore, the strength of the agglomerates is much lower in this case. The different morphologies of the precipitates influence the results of all following processing steps, including calcination, and they are finally responsible for the characteristics of the compact. After calcining and milling, the FD powder has the smallest particle size, although agglomerates are still present that can be observed with the SEM or that can be deduced from the powder data of Table I.

After calcining and milling, the FD powder has the same specific surface area and crystallite size as the 120 powder, but a smaller particle size and agglomeration parameter and a lower tap density. This is caused by the higher porosity of the FD powder, which in turn is related to the different drying conditions of the precipitates.

Table I. Characteristics of Calcined Powders after Milling

Powder	Tap density (% theor. dens.)	Spec. surf. area (m^2/g)	d_{50}* (nm)	d_{ads}† (nm)	d_x‡ (nm)	Aggl. parameter (d_{50}/d_{ads})
MM	29.6	2.5	1700	423	48.2	4
120	21.5	68	1900	16	12.2	127
FD	15.6	68	700	16	12.2	44

*Average grain size (MSA centrifuge). †Equivalent crystallite size (N_2 adsorption). ‡Crystallite size (X-ray line broadening measurement).

The MM powder, obtained by a completely different route, has a small agglomeration parameter, a specific surface area that is several times smaller than that of the coprecipitated powders, but shows a similar particle size distribution as powder 120 (Fig. 2). The high d_{ads} value in comparison to the d_x value is caused by the reaction and the loss of porosity during calcination at 1300°C. The MM powder consists of agglomerates composed of a few crystallites. In contrast to the agglomerates of the FD and 120 powders, they are very dense.

Compact Characteristics

The compaction behavior of powders is very important for their sintering behavior. As expected, the MM powder, which consists of relatively dense particles, can be compacted to a green density of 56.0% of theoretical, with a specific surface area of 2 m^2/g. The two coprecipitated powders (120 and FD) can be compacted to almost the same green density of 40.8 and 39.7% theoretical density, respectively, having specific surface areas of 68 and 54 m^2/g, respectively. The pore volume of the compacts in relation to the pore size is shown in Fig. 3. The FD compact has the largest pore volume, the MM compact the smallest. These results are confirmed by N_2 adsorption measurements, which show at the same time that in the MM compact there are no pores with a diameter smaller than 5–10 nm that contribute significantly to the pore volume of the 120 and FD compacts.

The pore-size distribution curve (Fig. 4) demonstrates that the MM compact has a very small intraagglomerate porosity due to its dense particles. The major part of the porosity is present as an interagglomerate or interparticle porosity, which has a maximum pore radius of 100 nm. The pore-size distribution of the 120 compact shows a maximum at a radius of 17 and one at 80 nm; the smaller one can be attributed to the intraagglomerate pores, the one at larger pore radii to the interagglomerate porosity. The FD compact has a very narrow pore-size distribution with a distinct maximum of the intraagglomerate pores. The interagglomerate pores are indicated by a maximum at 30 nm, but their peak is overlapped by the much higher peak of the intraagglomerate pores with a radius of 9 nm. N_2 adsorption measurements (Fig. 5) confirm these pore-size distributions.

Sintering Characteristics

The sintering densities after sintering in air for 3 h at 1150°, 1250°, 1300°, 1355°, and 1580°C are shown in Fig. 6. The FD sample reaches a constant density of 94.5% of theoretical at 1250°C, whereas the 120 and MM samples have density values of only 76.9 and 63.0% of theoretical at the same temperature. The densities after sintering at 1580°C are 95.6% of theoretical for the FD, 93.5 for the 120, and 94.9 for the MM sample. The curves indicate that the FD powder is very sinter-active. The ratio (SD-GD)/(100-SD) can be taken as an indication for the sintering activity (Table II).

The microstructures (fracture surfaces) of the sintered samples are quite different. In Fig. 7(A)–(C) the microstructures of the MM, 120, and FD specimens are shown after sintering at 1250°C, in Fig. 8(A)–(C) after sintering at 1355°C. In the case of the FD specimen, the grain size increases from 0.5–1.5 μm at 1250° to 3–10 μm at 1355° and to 15–25 μm at 1580°C.

The shrinkage and shrinkage rate of the oxide powder compacts are shown in Figs. 9 and 10. Shrinkage of the 120 and FD compacts starts at about 700° and that of the MM sample at 1300°C in accordance with the respective calcination temperatures. The sintering curves are identical for 120 and FD up to a shrinkage of

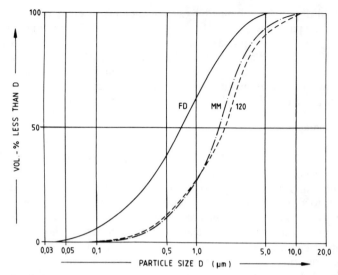

Fig. 2. Particle-size distribution of calcined powders after milling.

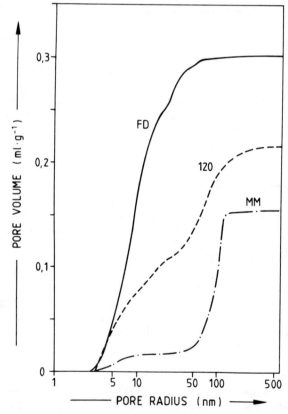

Fig. 3. Pore-size distribution of oxide powder compacts (Hg porosimetry).

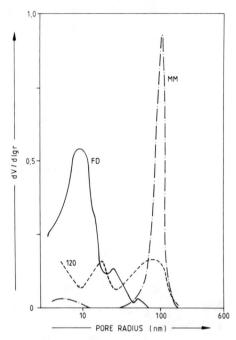

Fig. 4. Pore-size frequency distribution of oxide powder compacts (Hg porosimetry).

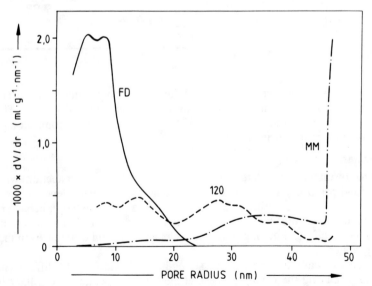

Fig. 5. Pore-size frequency distribution of oxide powder compacts (N_2 adsorption).

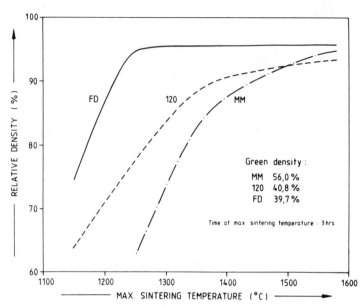

Fig. 6. Relative density of sintered compacts at different sintering temperatures (ρ_{th} = 5.67 g/cm³).

Table II. Sintering Activity of Compacts after Sintering at 1250° and 1580°C

Compact	(SD-GD)*/(100-SD) for 1250°C	(SD-GD)/(100-SD) for 1580°C
MM	0.19	7.63
120	1.56	8.11
FD	9.96	12.70

*SD: sintered density; GD: green density.

4%. Then the FD compact has a higher shrinkage with a maximum rate at 1185°C; a smaller peak in the shrinkage rate follows at a temperature of 1352°C. The 120 compact has two maxima at 1195° and 1370°, whereas the MM compact has only one maximum in the shrinkage rate at 1470°C.

Discussion

The results of the sintering experiments and their connection with the powder and compact characteristics demonstrate the interdependence between the processing parameters during powder preparation and the sintering behavior. The different dehydration of the precipitates and the wet-chemical preparation in contrast to the mechanical mixing of the powders are the main differences in the processing routes.

In the first case, the agglomerate strength is influenced by the different drying conditions. During the drying process at elevated temperatures, capillary forces cause a significant shrinkage of the gel-type precipitates, resulting in the formation of hard and strong agglomerates. If the precipitates are freeze-dried, the particles

Fig. 7. Microstructure of sintered compacts, 1250°C, 3 h (A) MM, 63.0% theoretical density; (B) 120, 76.9% theoretical density; (C) FD, 94.5% theoretical density.

are fixed in their position and their movement is avoided by the absence of a liquid phase and the resulting capillary forces during the drying process. The dried product is a loose powder with a high internal porosity; high-strength agglomerates could not form. The high internal porosity facilitates the destruction of agglomerates during milling down to dimensions smaller than 1 μm. The consolidation of these rather small, porous agglomerates in the FD powder, in contrast to the coarser and stronger agglomerates in the 120 powder, facilitates the preparation of a compacted body, in which the inter- and intraagglomerate pores are of identical

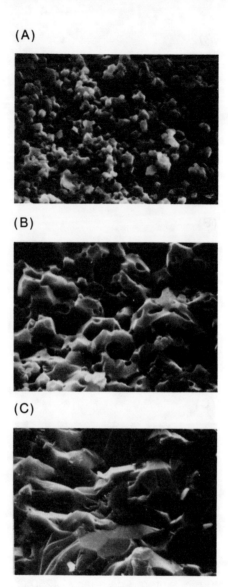

Fig. 8. Microstructure of sintered compacts, 1355°C, 3 h (A) MM, 83.5% theoretical density; (B) 120, 89.1% theoretical density; (C) FD, 95.7% theoretical density.

dimensions. This is demonstrated by the narrow pore-size distribution with a maximum at a small pore radius of 9 nm. This homogeneously distributed porosity can be eliminated at rather low temperatures around $0.5T_m$. This homogeneity in the pore-size distribution cannot be obtained with the 120 powder, resulting in a lower sintering activity.

When the sintering results obtained with the mechanically mixed powder and the wet-chemically prepared powders, especially the freeze-dried powder, are

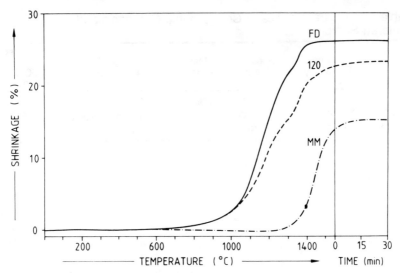

Fig. 9. Linear shrinkage of oxide powder compacts (heating rate, 5°C/min).

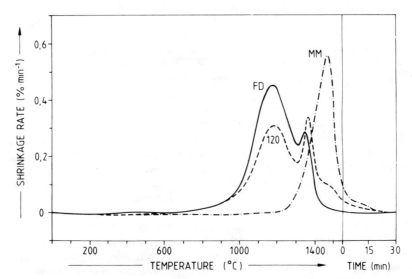

Fig. 10. Shrinkage rate of oxide powder compacts (heating rate, 5°C/min).

compared, it can be stated that the pore size in the consolidated MM compact is too large to achieve high sintering densities at low temperatures.

After three powders of different characteristics and different sintering behavior have been synthesized, the question arises which parameter of the powder or compact characteristics can be used to predict the sintering behavior. When Tables I and III are considered, it can be stated that the usual powder and compact

Table III. Characteristics of Consolidated Compacts and Sintered Samples

Powder	Green density (% theor. dens.)	Spec. surf. area (m²/g)	Max pore radius (nm)	Temp. of max shrinkage rate (°C)	Sintered density (1250°C, 3 h) (% theor. dens.)
MM	56.0	2	100	1470	63.0
120	40.8	68	17 a. 80	1195 a. 1370	76.8
FD	39.7	54	9	1185	94.5

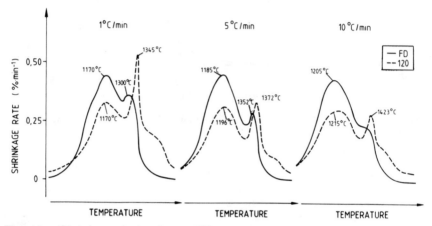

Fig. 11. Shrinkage behavior at different heating rates.

properties, e.g., particle size, specific surface area, and green density, are not very useful for this purpose, although the tap density seems to correlate powder properties and sintering characteristics in the case of coprecipitated powders.[5]

The pore-size distribution in the compact, however, seems to be an excellent indicator for the sintering behavior, which becomes evident by comparing Figs. 4 and 10. The respective curves of the pore-size distributions and the shrinkage-rate curves are very similar in their appearance. Each maximum in the pore-size distribution seems to correspond to a maximum in the shrinkage rate at a certain temperature, whereas the height of the maximum is influenced by the pore volume of the respective pore size. Each pore size seems to require a certain temperature for its elimination. Figure 11 shows that the typical shape of these shrinkage-rate curves is not influenced by the heating rate. Only the location and extension of the maxima are changed.

Figure 12 shows that the second maximum in the shrinkage-rate curves for samples 120 and FD cannot be eliminated by an extended sintering time at lower temperatures. Here the green compacts have been heated up to the temperature where the first maximum appeared. After 3 h at temperature the sample was brought to room temperature and reheated to 1500°C. Besides the fact that the first peak disappeared and the second peak remained, it can be seen by comparing Figs. 12 and 10 that the second peak of powder FD had been overlapped by the first major peak.

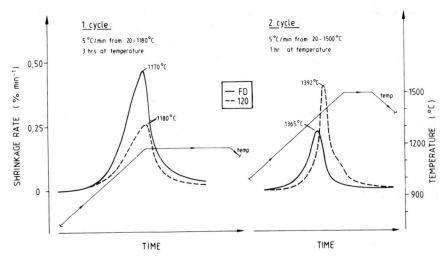

Fig. 12. Shrinkage behavior during two subsequent heating cycles.

On the basis of the results, it can be concluded that elimination of the intraagglomerate pores corresponds to the lower-temperature peak of the shrinkage rate; the higher-temperature peak is connected with the interagglomerate pore elimination. Van de Graaf et al.[10] reported too that their samples having densities of 95.7% of theoretical at 1180°C no longer contained intraagglomerate pores at 850°C. Pampuch and Haberko[4] described the fast elimination of intraagglomerate pores at low temperatures, whereas large pores needed higher temperatures and a longer time to be eliminated.

The fundamental sintering theory[28] is sufficient to explain the results. Between the negative curved surface of the neck (radius r) and the nearly flat surface of the particle, there is a difference in the vacancy concentration Δc:

$$\Delta c = \gamma a^3 c_0 / kTr \tag{1}$$

where c_0 is the concentration of vacancies under a plane surface, γ the surface energy, a^3 the atomic volume of the diffusing vacancy, k the Boltzmann constant, T the absolute temperature, and r the radius of the neck.

A flux of vacancies J diffuses away from the neck area. Under the concentration gradient Δc, the flux of vacancies J per second and per centimeter of circumferential length is given by

$$J = 4D_v \Delta c \tag{2}$$

where D_v is the diffusion coefficient for vacancies. With the expression for Δc, Eq. (1), one obtains

$$J = 4\gamma a^3 c_0 D_v / kTr \tag{3}$$

Since r, the concave radius of the neck curvature, can be considered as the convex radius of the corresponding pore, a very small pore radius increases the vacancy flux away from the neck/pore interface for T and D_v being constant. The temperature must be high enough that the diffusion flux causes a noticeable shrinkage after a short time. It must be taken into consideration that the influence of the temperature T on D_v is larger than its effect on Δc. At a constant temperature a compact containing small pores will have a higher shrinkage rate. In addition, the spatial distribution of the porosity (small pores in contrast to a few large pores and their position relative to grain boundaries) influences the length of the diffusion path and therefore controls the rate of pore elimination.[28]

Conclusions

Powders prepared by various processing routes show distinct differences in the pore-size distribution after compaction. Different maxima in the pore-size distribution of the compact are connected to the maxima in the sintering rate. The absence of hard agglomerates after freeze-drying of the hydroxides enhances the sintering activity of the powders.

Acknowledgments

This work was supported by the Deutsche Forschungsgemeinschaft (DFG) and by the Arbeitsgemeinschaft Industrieller Forschungsvereinigungen (AIF) of the Federal Republic of Germany.

References

[1] R. J. Brook; pp. 272–85 in Science and Technology of Zirconia, Advances in Ceramics, Vol. 3. Edited by A. H. Heuer and I. W. Hobbs. The American Ceramic Society, Columbus, OH, 1981.
[2] T. Vasilos and W. Rhodes; pp. 137–72 in Ultrafine-Grained Ceramics. Edited by J. J. Burke, N. L. Reed, and V. Weiss. Syracuse University, Syracuse, NY, 1970.
[3] J. S. Reed, T. Carbone, C. Scott, and S. Lukasiewic; pp. 171–80 in Processing of Crystalline Ceramics. Edited by P. H. Palmour, R. F. Davies, and T. Hare. Plenum, New York, 1977.
[4] R. Pampuch and K. Haberko; pp. 623–34 in Ceramic Powders. Edited by P. Vincenzini. Elsevier, Amsterdam, 1983.
[5] A. Roosen and H. Hausner; pp. 773–82 in Ceramic Powders. Edited by P. Vincenzini. Elsevier, Amsterdam, 1983.
[6] K. D. Reeve, *Am. Ceram. Soc. Bull.*, **42** [8] 452 (1963).
[7] C. E. Scott and J. S. Reed, *Am. Ceram. Soc. Bull.*, **58** [6] 587–90 (1979).
[8] W. H. Rhodes, *J. Am. Ceram. Soc.*, **64** [1] 19–22 (1981).
[9] M. A. C. G. van de Graaf, K. Keizer, and A. J. Burggraaf, *Sci. Ceram.*, **10**, 83–92 (1979).
[10] M. A. C. G. van de Graaf, J. H. H. ter Maat, and A. J. Burggraaf; pp. 783–94 in Ceramic Powders. Edited by P. Vincenzini. Elsevier, Amsterdam, 1983.
[11] W. D. Kingery; pp. 3–18 in Ceramic Powders. Edited by P. Vincenzini. Elsevier, Amsterdam, 1983.
[12] H. Takagi, S. Sano, and E. Ishii, *Ber. Dtsch. Keram. Ges.*, **51**, 234–35 (1974).
[13] M. Hoch and K. M. Nair, *Ceram. Int.*, **5**, 88–97 (1976).
[14] R. Y. Sheinfain and T. F. Makoskaya, *Kolloidn. Zh.*, **38**, 783–90 (1976).
[15] S. L. Dole, R. W. Scheidecker, L. E. Shiers, M. F. Berard, and O. Hunter, *Mater. Sci. Eng.*, **32**, 277–81 (1978).
[16] K. Haberko, *Ceram. Int.*, **5**, 148–54 (1979).
[17] R. J. Brook; pp. 331–64 in Treatise on Materials Science and Technology, Vol. 9. Edited by F. F. Y. Wang. Academic Press, New York, 1976.
[18] F. M. Carpay; pp. 261–75 in Ceramic Microstructures 1976. Edited by R. M. Fulrath and J. A. Pask. Wiley & Sons, New York, 1977.
[19] W. H. Rhodes and R. M. Haag; AFML-TR-70-209, Sept. 1970.
[20] W. S. Young and I. B. Cutler, *J. Am. Ceram. Soc.*, **53** [12] 659–63 (1970).
[21] P. A. Badkar, J. E. Bailey, and H. A. Barker; pp. 311–21 in Sintering and Related Phenomena. Edited by G. G. Kuczynski. Plenum, New York, 1973.
[22] M. J. Bannister, *J. Am. Ceram. Soc.*, **58** [1] 10–14 (1975).
[23] C. S. Morgan and C. S. Yust, *J. Nucl. Mater.*, **10**, 182–90 (1963).
[24] C. S. Morgan, *Mater. Sci. Res.*, **4**, 349–59 (1969).
[25] P. H. Rieth, J. S. Reed, and A. W. Naumann, *Am. Ceram. Soc. Bull.*, **55** [8] 717–21 (1976).
[26] M. A. C. G. van de Graaf, T. van Dijk, M. de Jongh, and A. Burggraaf, *Sci. Ceram.*, **9**, 75–83 (1977).
[27] T. J. Carbone and J. S. Reed, *Am. Ceram. Soc. Bull.*, **57** [8] 748–55 (1978).
[28] W. D. Kingery, H. B. Bowen, D. R. Uhlmann, Introduction to Ceramics, Ch. 10. Wiley & Sons, New York, 1976.

Sintering of a Freeze-Dried 10 mol% Y_2O_3-Stabilized Zirconia

L. Rakotoson and M. Paulus

Laboratoire d'Etude et de Synthèse des Microstructures
CNRS-ESPCI
75231 Paris, Cedex 05, France

After presenting the results of freeze drying a sulfate solution, we describe a preparation process in which we improve the freeze-drying technique by addition of a suspension of stabilized zirconia in the liquid solution before freeze-drying. This process breaks the polymeric chains, increases the green density of the compact, and decreases the sintering temperature. The mechanisms involved are discussed.

Preparation of a powder by conventional methods, such as mixing of powders, leads to chemical heterogeneities. The consequence is of course an increase in the formation time of the final compound, but most important is the Kirkendall effect occurring during the thermal homogenization and formation. The differences in the diffusion coefficient between species induce a pore migration with coalescence, which opposes rapid and complete sintering due to the low vacancy density around the large pores and the increase of the migration path of vacancies.[1,2] A statistical approach to the mixing problems shows that we can reduce the chemical heterogeneity only up to a certain value, depending on particle size and the size distribution.[3,4] The only valid way to increase the chemical homogeneity seems to be mixing on a molecular scale (liquid salt solution, gases) and then stabilizing the resulting compound. Two main processes may be considered, sol-gel and freeze-drying.

We have applied the freeze-drying technique for the preparation of a 10 mol% Y_2O_3-stabilized ZrO_2. We present here the results obtained by freeze-drying a pure solution of sulfate and a freeze-dried suspension of micrometer-sized Y_2O_3-stabilized ZrO_2 in a sulfate solution of the same cations.

Freeze-Drying Process[5,6]

We start from a sulfate solution of Y and Zr, with salt concentrations of 30, 45, 70, and 135 g/L (expressed in oxide content after thermolysis). This sulfate solution is sprayed into liquid nitrogen to stabilize the molecular homogeneity of the chemical species. After freezing, the sulfate solution looks like a very low density sponge (Fig. 1). Drying following freeze-drying lasts 15 h in vacuum at 1.3–0.13 Pa. The sulfate sponge keeps its general feature. Residual water is removed at 150°C. Finally the sulfate is transformed by thermolysis into the oxide at temperatures ranging from 800° to 1100°C. Figure 2 shows that the sponge aspect of the product is maintained.[7,8]

Characteristics of the Freeze-Dried Y_2O_3-ZrO_2 Powders

Whatever the solution concentration on freeze-drying and thermolysis parameters, the powder is chemically very homogeneous — a very good distribution

Fig. 1. Particles of sulfate solution freeze-dried in liquid nitrogen.

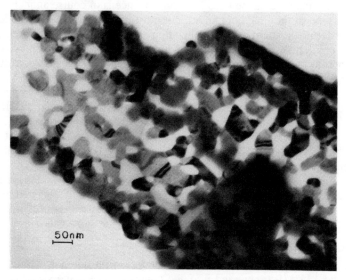

Fig. 2. Particles of freeze-dried sulfates after thermolysis at 900°C.

of Y_2O_3 is found in comparison with that observed after a careful mechanical mixing of Y_2O_3 and ZrO_2.

The grain size of the oxides prepared from the freeze-dried solution can be monitored by changing the solution concentration and the thermolysis parameters. Increase of the solution concentration from 30 to 135 g/L with thermolysis for 17 h at 900°C under oxygen decreases the surface area from 27 to 2.7 m^2/g. On the other hand, an increase in the thermolysis temperature drastically increases the grain size. Thermolysis for a lengthy time (17 h) at low temperature (900°C) seems

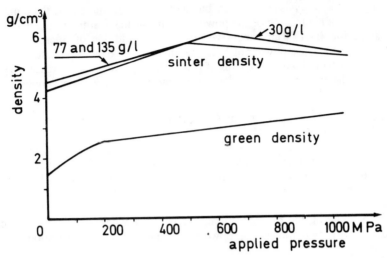

Fig. 3. Green and sintered density of freeze-dried powder vs compaction pressure. Thermolysis for 17 h at 900°C and sintering at 1400°C for 2 h.

to be the best compromise between almost complete transformation from sulfate to Y_2O_3-stabilized ZrO_2 and a small grain size (10–20 nm).

Pressing and Sintering

Figure 3 shows that the green density of the compact increases uniformly with unidirectional compaction pressure. Nevertheless, the sintered density exhibits a maximum (98% of the theoretical density) for a compaction pressure of 600 MPa. The decrease in sintered density is attributed to the formation of cracks in the green pellet over 600 MPa. They open during sintering; therefore, the macroscopic density decreases, although the local density increases.

Features of the Freeze-Drying Process

Some of the advantages of the freeze-drying process are (1) preparation of a chemically homogeneous powder with fine grains, (2) sintering at relatively low temperature with high density (98% of the theoretical density after 2 h at 1400°C), and (3) all the species in solution are found in the powder because we have no precipitation but freezing and dehydration in the solid state.

But some limits are set for this process. (1) The green density is relatively low. (2) The formation of low cohesion areas in the green pellet are observed at pressures greater than 600 MPa. (3) Grains of the powder form polymeric chains, as in the case of organic materials. This means that, due to the very low packing density of the free powder, the number of contacts between particles is reduced to 2 or 3. During thermolysis, neck formation, mainly by evaporation-condensation and the beginning of interparticle shrinkage, leads to the formation of particle chains. They are not destroyed during pressing, and they form, after a low-temperature sintering, porous macroscopic patterns due to the polymeric chains. But some parts of the same sample that show areas without polymeric patterns are almost without porosity. This observation suggests that the capability of the freeze-dried powder for sintering is limited by the polymeric chainlike agglomerates.

Three measures may be considered in order to solve this problem: (1) preparation of the powder without polymeric chains — this objective may be reached by reducing the thermolysis temperature but at the same time increasing the remaining sulfate content; (2) milling of the powder—this mechanical treatment may introduce impurities and is an expensive process; (3) addition of micrometer-sized (1–10-μm) fully dense powder of Y_2O_3-stabilized ZrO_2 to the Y-Zr solution before freezing in order to break the chains of freeze-dried powder and increase the green density of the freeze-dried powder. These measures have been adopted in the following procedures.

Freeze Drying of a Suspension of Y_2O_3-Stabilized ZrO_2 Powder in a Y-Zr Sulfate Solution

Experiments were performed with Rhone Poulenc micrometer-sized powder Y/Z9,[*] 5 μm in diameter, and a solution of Y-Zr sulfate with a concentration of 30 g/L (oxide). The freezing and freeze-drying were carried out under the same conditions as for a pure solution, and thermolysis was conducted for 17 h at 900°C in oxygen.

Depending on the ratio of micrometer-sized powder (MSP) to freeze-dried powder (FDP), two models of powder distribution and densification may be considered.

Low Micrometer-Sized Powder (MSP) Content

For this range of MSP content, we evaluated the green density on the assumption that the freeze-dried powder part (FDP) is pressed at the same green density as the pure FDP and that the MSP is fully dense without contact and well dispersed.

The theoretical green density of the compact for a low MSP content is given by the relation

$$d = \frac{m_{FDP} + m_{MSP}}{(m_{FDP}/d_{FDP}) + (m_{MSP}/d_c)} \qquad (1)$$

where d_{FDP} = density of the freeze-dried powder alone after compaction under pressure, d_{MSP} = density of the micrometer-sized powder alone after compaction under the same pressure, m_{FDP} = mass of freeze-dried powder, m_{MSP} = mass of micrometer-sized powder, and d_c = crystallographic density of Y_2O_3-stabilized ZrO_2.

In Fig. 4, the left part of the theoretical curve gives the result of this evaluation. Thus, we may expect an increase of the green density with no change in the FDP behavior during pressing. But, in addition, we expect a break of the polymeric chains.

High Micrometer-Sized Powder Content

For these high contents, we assume the MSP is pressed at the same green density as the pure MSP, but the interstitial voids are filled with FDP proportional to their content. The equation simplifies to

$$d = \frac{m_{FDP} + m_{MSP}}{m_{MSP}/d_{MSP}} \qquad (2)$$

The right part of the theoretical curve of Fig. 4 gives the expected green density.

Experimental Results

The experimental curve of the green density (Fig. 4) does not fit the theoretical one, but the shape is similar. This discrepancy between the two curves

---

[*]Rhone Poulenc Co., Monmouth Junction, NJ.

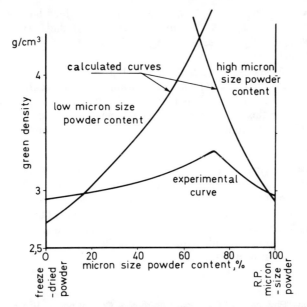

Fig. 4. Theoretical and experimental green density vs micrometer-sized powder content. Solution concentration is 30 g/L, thermolysis for 17 h at 900°C.

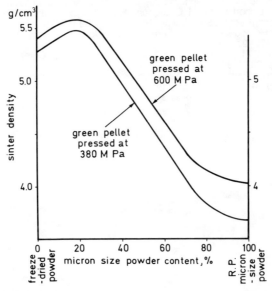

Fig. 5. Sintered density vs micrometer-sized powder content. Sintering at 1230°C for 2 h. Maximum density is 94% of theoretical. Concentration of the solution is 30 g/L, thermolysis is for 17 h at 900°C.

is due to the fact that MSP is made of agglomerates which have a much lower density than that of crystals. We may conclude that the basis of theoretical evaluation was valid, but the assumption of fully dense particles for the MSP was wrong (agglomerates).

The sintered density after 2 h at 1230°C (Fig. 5) shows an increase in density with the freeze-dried powder content. Furthermore, the curve exhibits a maximum of density for $\simeq 20\%$ of MSP. This maximum is attributed to the expected break of the polymeric chains during compaction. We get a sintered density of 94% after 2 h at 1230°C, and the remaining porosity is very fine and uniformly distributed.

Conclusion

The freeze-drying of a suspension of micrometer-sized powder of Y_2O_3-stabilized ZrO_2 in a sulfate solution with the same cation ratio appears as one way to take advantage of the sintering capability of freeze-dried powder (high chemical homogeneity and low sintering temperature) and to increase the green density and break the polymeric chains. Efficiency of this technique depends mainly on the micrometer-sized powder which must be constituted of nonagglomerated fully dense grains 1–10 μm in diameter.

References

[1] M. Paulus, "Physical and Chemical Parameters Controlling the Homogeneity of Fine Grained Powders and Sintering Materials; pp. 17–31 in Ceramics Materials Science Research, Vol. 11. Edited by Hayne Palmour, R. F. Davis, and T. M. Hare. Plenum, New York, 1978.

[2] "The Needs for Tailored Powder Properties in Ceramic Processing"; in Science of Ceramics, Vol. 11. Edited by R. Carlsson and S. Karlsson. The Swedish Ceramic Society, Stockholm, 1981.

[3] C. Lacour, "Sintering and Random Structures," *Phys. Stat. Solidi Sect. B*, **96**, 785 (1979).

[4] C. Lacour, A. Dubon, and P. M. Grangeon, "Random Aspect of the Compound Formation by Reactive Sintering," *Phys. Stat. Solidi, Sect. A*, **77**, 309 (1983).

[5] M. Paulus, "Homogeneization by Freeze-drying for Low Temperature Sintering," *Ann. Chim. (Paris)*, **1**, 187–97 (1976).

[6] C. Lacour and M. Paulus, "Lyophilisation Parameters of Ceramic Compound," *Sci. Sinter.*, **11** [3] 193–202 (1979).

[7] L. Rakotoson, "Synthesis and Sintering of Y-Stabilized Zirconia by Thermolysis of a Freeze-Dried Sulphate Liquid Solution," 3rd cycle thesis, E.S.P.C.I., 1981.

[8] L. Rakotoson, *Ann. Chim.* in press.

ZrO₂ Micropowders as Model Systems for the Study of Sintering

ROMAN PAMPUCH

Institute of Materials Science
AGH Cracow, Poland

Owing to a variable separation of pore sizes within and between agglomerates of crystallites, stabilized ZrO_2 micropowders obtained by wet-chemical methods represent an interesting model system for a study of the inhomogeneity effects in sintering of real systems. Using an approach similar to the one developed by Kiparisov and Perel'man and by Evans, the effects, on densification and final microstructure of sintered products, of the transient hydrostatic stresses which develop during sintering due to the shrinkage-rate differential between rapidly shrinking zones (within agglomerates) and slowly shrinking ones (between agglomerates) have been evaluated.

Most real powders and powder compacts are inhomogeneous with regard to what affects their behavior on sintering. With the now widely used micropowders, the inhomogeneity is due mainly to a formation of agglomerates of crystallites. The presence of agglomerates in concentrated dispersed systems is due to the marked tendency of the crystallites to form phase contacts with each other. According to the generally valid results of the percolation theory, dispersed systems should be composed mainly of clusters (agglomerates) of a finite size if the volume fraction of particles in the system is lower than a certain threshold value, p_c (Fig. 1). The bonds between crystallites within the agglomerates are of the first order, as found experimentally with Y_2O_3-stabilized ZrO_2 micropowders.[1]

There are methods which allow us to obtain initially monodisperse powders, but except for some hydrothermal methods, a calcination step is usually involved in the production of powders. This inevitably results in the formation of agglomerates. Owing to the tighter bonding between crystallites within agglomerates, the pores are here typically smaller than the ones between agglomerates. This is tantamount to an inhomogeneous distribution of voids in the compact.

In the presence of hydrostatic compressive stresses due to capillary forces, this should give rise to a difference in the driving force for sintering and, hence, to different strain rates in the two types of the microzones of the compact. Since all the microzones constitute a single body, the strain rate of the more rapidly densifying zones should be restrained by the action of the surrounding, less rapidly densifying zones, and transient stresses should develop in the compact during sintering. Taking into account the principle of conservation of equilibrium of a body, according to which the sum of all forces acting in the body must be equal to zero, it may be concluded that, in inhomogeneous compacts, stresses of two signs should occur. The microzones in which additional tensional hydrostatic stresses are developed should show a lower shrinkage rate than the neutral zones, whereas the reverse applies to zones under the action of compressional hydrostatic

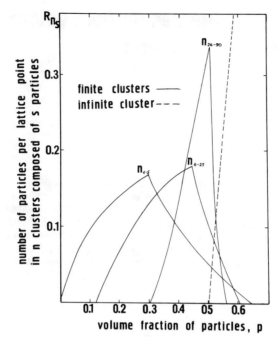

Fig. 1. Number of particles agglomerated in clusters, R_{n_s}, vs volume fraction of particles in a dispersed system, p, for three groups of cluster sizes, n_s, where s is the number of particles in the cluster, according to computer simulation for a triangular plane lattice.

stresses overlapping the sintering stress due to capillary forces. Such phenomena are of great importance for sintering of real compacts and cannot be dealt with on the basis of the usual sintering models. It is only in the last two years that the problems of sintering have been formulated in this way.[1-4] The most extensive treatment is by Evans,[4] who solved the problem of stress analysis in compacts during sintering by using the approach developed by Eshelby[5] and introducing a "transformation" strain rate for the elastic transformation strain. The former is related to the unconstrained densification rate differential between two microzones, subject to the resultant hydrostatic stress.

Evans[4] considered theoretically the case of an inner fine-grained zone surrounded by an infinite, outer coarse-grained matrix. No evaluation of direct experimental data, particularly with respect to systems having microzones of a different pore size but a uniform crystallite (grain) size, has been published so far. From this point of view, the compacts of stabilized ZrO_2 micropowders obtained by a wet-chemical method represent an interesting model system. There occur in compacts of these micropowders two more or less distinct pore populations (Figs. 2 and 3). The population of the smaller pores has been identified with pores within the agglomerates, while the population of the larger pores is identified with pores between the agglomerates.[6] Hence, a study of the behavior of each of the two populations on sintering is possible here and is the subject of the present paper.

Experimental Procedures

The powders were obtained by a method described in Ref. 6, in which precipitated gels of an Y_2O_3-MgO-stabilized ZrO_2 (with 1.83 mol% of Y_2O_3 and

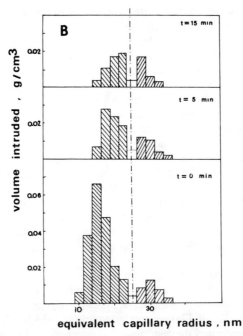

Fig. 2. Pore-size distribution in the green B compact and after different times of isothermal sintering at 1270 K. The line dividing the pores within the agglomerates (smaller pores) and between the agglomerates (larger pores) is indicated.

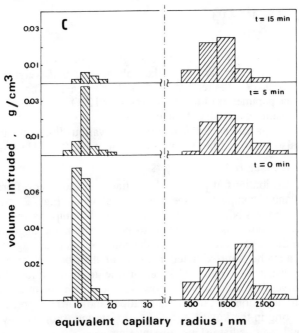

Fig. 3. Pore-size distribution in the green C compact and after different times of isothermal sintering at 1270 K.

8.55 mol% MgO) were washed with water or ethyl alcohol, dried, and isothermally calcined at 1070 K for 30 min.

Two types of compacts were obtained: (1) Compacts (called B) were formed from powders obtained via the alcohol-washing route by uniaxial cold-pressing under loads of 196 MPa with addition of 7% oleic acid as lubricant. The lubricant was removed by heat treatment at 770 K for 1.5 h. (2) Compacts (called C) were formed from powders obtained via the water-washing route by uniaxial cold-pressing under the same conditions as compacts B.

The compacts had an apparent density of $(2.264 \pm 0.02) \times 10^3 < \rho < (2.308 \pm 0.01) \times 10^3$ (kg/m^3) and dimensions of 7 by 7 by 2 mm. The compacts were isothermally sintered at 1270 ± 1 K for 30 min by rapidly inserting the samples into a furnace held at the sintering temperature. According to measurements by a Pt–PtRh thermocouple, placed directly on the sample surface, the sintering temperature was attained at the surface in a few seconds. The time necessary for attainment of this temperature in the sample center was estimated, using the method proposed in Ref. 7, to be about 2 min. In view of this result, only the measurements made after 2 min and more of isothermal sintering were evaluated.

The cumulative pore-size distribution curves and pore volume (per unit weight) were obtained by Hg porosimetry in both the green compacts and the compacts sintered isothermally for different periods of time. In view of the relative separation of the smaller and larger pore populations, which subsisted also during the first 30 min of sintering, the distribution of pore radii and the pore volumes (per unit weight) of the pores within and between the agglomerates, respectively, could be estimated from the cumulative pore-size distributions. The porosities within the agglomerates, P_i, and between the agglomerates, P_o, were calculated using the formulas

$$P_o = V_o/(V_o + V_i + V_s)$$
$$P_i = V_i/(V_i + V_s) \tag{1}$$

where V_i and V_o are the volume (per unit weight) of pores within and between the agglomerates, respectively, and V_s is the volume (per unit weight) of the solid nonporous material, i.e., the reciprocal of the X-ray density. The latter was calculated from lattice parameters of the Y-Mg-stabilized ZrO$_2$ solid solution, published in Ref. 8, by assuming a model of vacancies in the oxygen sublattice. In the green compacts and compacts sintered isothermally for various times, the mean crystallite size was also measured by X-ray broadening using the (111) reflection.

Results and Discussion

The size distributions of pores in the initial state and after different sintering times in the B and C compacts, respectively, are shown in Figs. 2 and 3. The pore populations, with sizes below 25 nm in the case of B compacts and below 22 nm in the case of C compacts, correspond to pores within the agglomerates. The populations of the larger pores constitute the pores between the agglomerates. The distributions may be approximated by normal distribution, especially when the pore-size distribution data are given in the form of differential pore-size distribution curves (not shown). This allows for the calculation of the mean values, $\bar{a}$, and standard deviations, σ, of the distributions and, furthermore, the exponents n and m appearing in the kinetic equations of sintering, proposed by Kuczynski[9] for the first and intermediate stages of sintering:

$$(\bar{a}/\bar{a}_o)^{n/(m-1)} = (P_o/P)^n = 1 + zBt$$
$$n = (3/2)x = (3/2)(y/z)[1 + 2z - (1-z)y] \quad (2)$$
$$m = (1+x)/2 \qquad y = \sigma/\bar{a}_o$$
$$B = (36\pi/P_o)^{1/2}(\gamma D\Omega/\bar{a}_o \bar{l}_o^2 kT)$$

where $\bar{a}_o$ and $\bar{a}$ are the mean values of the pore radii in the green compact and after a time interval t of sintering, respectively; P_o and P are the porosities of the green compact and after a time interval t of sintering, respectively, z is the volume fraction of pores which is eliminated and not exchanged during sintering, D the appropriate coefficient of diffusion, controlling the densification (with micropowders it is usually the coefficient of grain-boundary diffusion of cations), γ the surface energy, Ω the atomic volume, k the Boltzmann constant, T the temperature, and $\bar{l}_o$ the crystallite size in the green compact. The crystallite size $\bar{l}_o$ in the green compact and the crystallite sizes $\bar{l}$ after a given sintering time, used in further calculations, are shown in Fig. 4.

Table I shows the calculated values of n and m and Table II collects the data which were used to calculate B at a temperature of 1270 K. The values of n and m obtained for the populations of pores between the agglomerates in C compacts agree well with the ones cited by Kuczynski[9] for various oxide compacts composed of dense, relatively coarse grains, where the porosity is equivalent to the porosity between the agglomerates in the C compacts studied.

The results obtained using the Kuczynski kinetic equations represent the situation where there is no interaction between pore populations on sintering, i.e.,

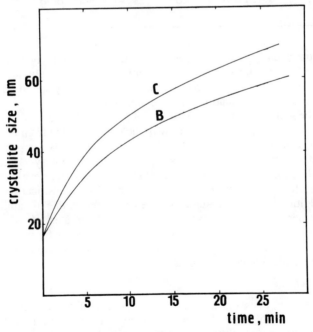

Fig. 4. Crystallite size vs time of sintering at 1270 K found with the B and C compacts, respectively.

Table I. Coefficients n and m Calculated from the Pore-Size Distribution

Type of compact	Population of pores	n		m	
		$z = 1$	$z = 0.5$	$z = 1$	$z = 0.5$
B	Within agglomerates	0.8	1.0	0.8	0.8
	Between agglomerates	0.6	0.7	0.7	0.7
C	Within agglomerates	0.9	1.1	0.8	0.9
	Between agglomerates	2.2	2.6	1.2	1.4

Table II. Data Used in Calculations of the Coefficient B

Coefficient of diffusion at 1270 K, D (m²/s)	2.6×10^{-18}	Estimated value for grain-boundary diffusion coefficient of Zr^{4+} on the basis of data from Ref. 10
Thickness of the boundary layer, δ (m)	2×10^{-10}	
Atomic volume, Ω (m³)	1.17×10^{-29}	For tetragonal zirconia
Surface energy at 1270 K, γ (J/m²)	0.8	According to Ref. 11

where the strains due to the shrinkage differential between the two microzones of different pore size (within and between agglomerates, respectively) are absent. The data obtained using these equations are indicated in Fig. 5 by dotted lines, to allow a comparison of such a situation with the actual data.

It may be seen from Fig. 5 that the actual porosity changes within the agglomerates in C compacts agree rather well with the ones expected from the Kuczynski equations. The actual porosities within the agglomerates in B compacts are, however, consistently higher than the ones expected from the Kuczynski equations. It may plausibly be assumed that these discrepancies arise from a development within the agglomerates of transient tensile stresses due to the densification-rate differential, in addition to the hydrostatic compressive stresses due to capillary forces.

The former hydrostatic stresses within the agglomerates, p^I, were estimated using equations developed by Evans[4] but modified to account for the case of a zone having pores smaller than the surrounding matrix, the crystallite size in the two zones being equal. According to Evans

$$p^I = -4\eta_o \Delta \dot{e}^T \quad (3)$$

where $\Delta \dot{e}^T = \dot{e}_i^T - \dot{e}_o^T$ is the hydrostatic component of the unconstrained strain (densification)-rate differential between the two zones, subject to the resultant hydrostatic stress, η_o is the viscosity of the surrounding matrix (zone between the agglomerates); the indices i and o stand for inner (within agglomerates) and outer (between agglomerates), respectively.

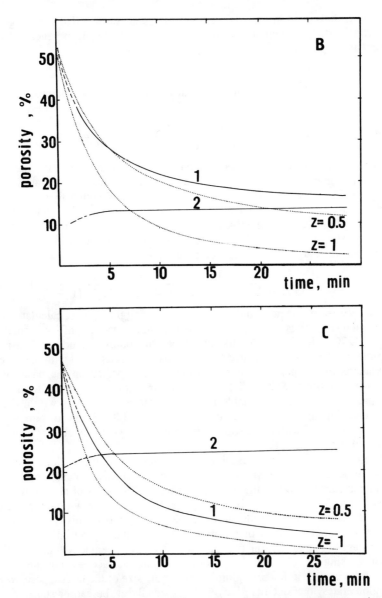

Fig. 5. Porosity (1) within and (2) between the agglomerates vs sintering time, found with the B and C compacts, respectively. Dotted lines represent porosities within the agglomerates, as expected from the Kuczynski equations at $z = 1$ and $z = 0.5$.

According to Geguzin[12]
$$\eta_j \cong kT\bar{l}_j^2 G(P_j)/D\Omega$$
$$G(P_j) \cong (1 - P_j)/(\bar{a}_j/\bar{l}_j)(1 + 2\bar{a}_j/\bar{l}_j) \qquad j = i, o \tag{4}$$

The symbols used in Eq. (3) are the same as in Eq. (2). The shrinkage rate of a porous particle system subject to stress can be described, in general, by the relation (see, e.g., Ref. 13):

$$\dot{e}_j^T = [D\Omega/kT\bar{l}_j^2 G(P_j)][p'/3 - \sigma_j^s F(P_j)] \quad (5)$$

where $\sigma_j^s = (\gamma/a_j)f(\psi)$ is the sintering stress due to capillary forces; ψ is the dihedral angle. Hence

$$\dot{e}^T = \dot{e}_i^T - \dot{e}_o^T = (D\Omega/kT\bar{l}^2)\left[\frac{p'/3 - \sigma_i^s F(P_i)}{G(P_i)} + \frac{\sigma_o^s F(P_o)}{G(P_o)}\right] \quad (6)$$

In deriving Eq. (6) it has been taken into account that, in the outer zone (between agglomerates), the stresses related to p' are exclusively deviatoric at all locations and thus cannot contribute to densification, although they may cause growth of cavities in this zone (see Ref. 4). Assuming that, with identical crystallite sizes, $f(\psi)$ is equal in the two zones, and taking into account average values of $\bar{a}$ and $\bar{l}$ only, as well as $F(P_i) = F(P_o) = 1$, one obtains after insertion of $\Delta \dot{e}^T$ from Eq. (3) into Eq. (6) and after rearranging:

$$\bar{p}'/\sigma_i^s = 12\bar{a}_i\left[\frac{(1 + 2\bar{a}_i/\bar{l})(1 - P_o) - (1 + 2\bar{a}_o/\bar{l})(1 - P_i)}{3\bar{a}_o(1 + 2\bar{a}_o/\bar{l})(1 - P_i) + 4\bar{a}_i(1 + 2\bar{a}_i/\bar{l})(1 - P_o)}\right] \quad (7)$$

The average $\bar{p}'/\sigma_i^s$ ratio in the zone within the agglomerates, calculated from experimental data (Figs. 2–5) using Eq. (7), is shown vs time of isothermal sintering at 1270 K in Fig. 6, for the two types of compacts studied. The rather unimportant p'/σ_i^s ratios calculated for C compacts correlate well with the absence of deviations of the porosity changes from those expected on the basis of the Kuczynski equations. Substantial positive p'/σ_i^s ratios have been found with B compacts, indicating appreciable hydrostatic tensional stresses within the agglomerates. This allows a rational explanation of the lower rate of porosity decrease, as compared with that expected from the Kuczynski equations, which has been found with B compacts.

The changes in the size distribution of pores within the agglomerates with sintering time (Figs. 2 and 3) are also consistent with this explanation. A rather uniform decrease of the volume of pores within agglomerates of all size ranges is observed with C compacts. The B compacts show, for the pores within the agglomerates, an increasing ratio of the larger to the smaller pores. This is equivalent to the frequently found growth of the mean pore size during the initial and intermediate stages of sintering, the reasons for which have been not explained. If one calculates, using Eq. (7), the p'/σ_i^s ratio, by taking into account not the average values of the pores within the agglomerates but the size of the largest pores within the agglomerates, one obtains $p'/\sigma_i^s \approx 1$, especially at the earlier stages of sintering. The transient hydrostatic tensional stresses should thus be equal, if not larger than, the sintering stress with this size range of pores within the agglomerates, especially at the early stages of sintering.

Implications and Conclusions

The present study of the behavior of two types of compacts has shown the following:

In compacts where there occur two pore populations between and within the agglomerates, respectively, which are widely separated in size, only unimportant hydrostatic stresses due to the densification-rate differential develop within the agglomerates during sintering. Therefore, shrinkage of the smaller pores within

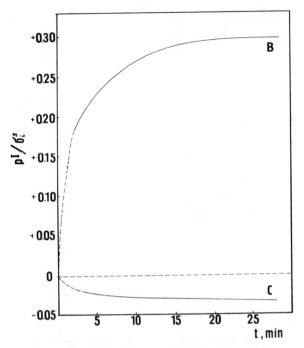

Fig. 6. Ratio of the average hydrostatic stress, p', to the average sintering stress, σ_i^s, vs sintering time, calculated for the zone within the agglomerates in the B and C compacts, respectively.

the agglomerates is unconstrained and they shrink more rapidly than the pores between the agglomerates. The porosity changes of the two populations show a divergence which increases with the sintering time (Fig. 5(C)). This should give rise to a formation of dense zones, originating from the agglomerates, surrounded by a highly porous matrix, in the sintered product. Such inhomogeneous sintering products have been found, indeed, in case of the C compacts studied.

In compacts where there occur two pore populations within and between the agglomerates, but where the mean size of the two populations is not very different, substantial tensional hydrostatic stresses develop within the agglomerates due to the densification-rate differential. Hence, the shrinkage rate of the pores within the agglomerates is decreased. This gives rise to a convergence of this porosity with the generally more slowly decreasing porosity between the agglomerates, such as observed with B compacts (Fig. 5(B)). The same applies to the mean size of the two pore populations (Fig. 2). Due to the more homogeneous pore substructure thus formed, a more homogeneous microstructure of the sintered product should be obtained. Such a behavior has been observed with the B compacts studied and also in some previous works[6,14,15] where stabilized ZrO_2 compacts of similar characteristics have been sintered.

The discussion indicates that, with systems containing inherent inhomogeneities, such as powder compacts, it is necessary that, during the earlier operations of the production of sintered bodies (e.g., during cold-pressing), adequate inhomogeneities are produced which can eventually be compensated by the inhomogeneous shrinkage during sintering. Some of the characteristics of com-

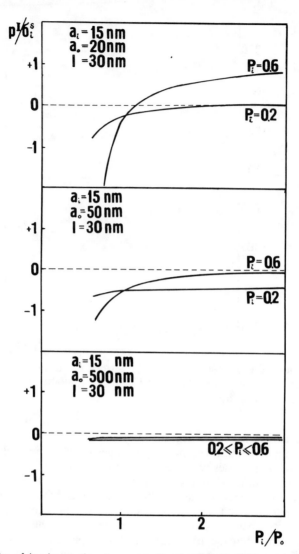

Fig. 7. Ratio of hydrostatic stresses due to the densification-rate differential, p', to the sintering stress, σ_i^s, within the agglomerates or microzones constituted by small pores vs the P_i/P_o ratio in powder compacts. Data calculated from Eq. (7) for different values of a_j and P_j ($j = i, o$) at a constant crystallite size $l = 30$ nm.

pacts which should favor the development of transient hydrostatic tensile stresses within the agglomerates or, generally, in microzones composed of smaller pores, are indicated by Fig. 7, providing a broader applicability of the concepts presented here.

If this is neglected, inhomogeneities will not be eliminated and can even be increased during sintering. The subject of a controlled microstructure of green compacts, with reference to stabilized ZrO_2 powder compacts, was treated in detail previously.[1]

Acknowledgment

The experimental data evaluated in the present paper were obtained by Miss Podwórna, under supervision of Dr. K. Haberko, in the author's laboratory.

References

[1] R. Pampuch and K. Haberko; in Ceramic Powders. Edited by P. Vincenzini. Elsevier Science Publishers, Amsterdam, 1983.
[2] F. F. Lange; in Ceramic Powders. Edited by P. Vincenzini. Elsevier Science Publishers, Amsterdam, 1983.
[3] S. S. Kiparisov and V. E. Perel'man; p. 340 in Advances in Sintering. Edited by D. Kolar and M. M. Ristic. Elsevier Science Publishers, Amsterdam, 1982.
[4] A. G. Evans, *J. Am. Ceram. Soc.*, **65** [10] 497–501 (1982).
[5] J. D. Eshelby, *Proc. R. Soc. London, Ser. A*, **241**, 376 (1957).
[6] K. Haberko, *Ceramurgia Int.*, **5** 148 (1979).
[7] V. Dauknys, K. Kazakevicius, G. Pranckevicius, and V. Jurenas; p. 20 ff in Investigations into Thermal Stress Resistance of Refractory Ceramics (in Russian). Edited by Mintis. Vilnius, 1971.
[8] R. V. Wilhelm and D. S. Eddy, *Am. Ceram. Soc. Bull.*, **56** [5] 509–12 (1977).
[9] G. C. Kuczynski; p. 325 ff in Sintering and Catalysis. Edited by G. C. Kuczynski. Plenum, New York, 1975.
[10] W. H. Rhodes and R. E. Carter, *J. Am. Ceram. Soc.*, **49** [5] 244–49 (1966).
[11] R. Pampuch, *Silic. Ind.*, **23**, 119, 191 (1958).
[12] Ya. E. Geguzin; p. 262 ff in Physics of Sintering (in Russian). Izd. Nauka, Moscow, 1967.
[13] H. C. Hsueh and A. G. Evans, *Acta Metall.*, **29**, 1907 (1981).
[14] R. Pampuch and K. Haberko, *Keram. Ztschr.*, **31**, 478 (1979).
[15] M. A. G. C. van de Graaf, K. Keizer, and A. J. Burggraaf, *Dtsch. Keram. Ges.*, **10**, 83 (1980).

Wet-Chemical Preparation of Zirconia Powders: Their Microstructure and Behavior

M. A. C. G. van de Graaf and A. J. Burggraaf

Twente University of Technology
Department of Chemical Engineering
Laboratory of Inorganic Chemistry and Materials Science
7500 AE Enschede, The Netherlands

Ultrafine homogeneous substituted zirconia powders are of importance for the preparation of ceramics for electronic devices. Such powders not only enable the production of dense ceramics at moderate temperatures but also improve thermomechanical behavior because of the resulting small grain sizes. Many methods are available for the chemical preparation of powders. However, the application of these methods to highly refractory compounds, such as zirconia, results in powder morphologies that frequently give poor sintering behavior. This is commonly caused by the inadequate control of agglomerate structures. This paper reviews the results of a number of modern synthesis methods applied to the preparation of zirconia, with emphasis on agglomeration control and microstructural development.

The employment of wet-chemical methods for the preparation of ceramic powders especially aims at the achievement of a high degree of homogeneity and improved sinterability. A number of synthesis methods that were developed in the past decade generally enable simple preparation of powders of nearly any composition, with high purity, close control of composition, and good compositional homogeneity. From the authors' experience, however, many of the methods produce powders exhibiting relatively poor sintering behavior, especially for highly refractory materials, such as ZrO_2. In many cases, microstructural inhomogeneities in powder compacts, which originate from inadequate control of the aggregates and agglomerates in the starting powders, are the cause of this poor sintering behavior.[1,2]

At first, little attention was paid in literature to the occurrence of agglomerate structures in ceramic powders, and the term "fine grained" was commonly related to crystallite size rather than to the existing crystallite clusters. Agglomerate structures present in these powders, however, have been shown to play a vital role in the densification processes during compaction and sintering. The recent literature clearly accentuates this fact.[1-10]

During synthesis, agglomerates are formed in the early stages of the preparation process and are transformed to strong agglomerates during thermal treatments like drying, pyrolysis, and calcination. Agglomerates which do not completely collapse during compaction give rise to the formation of pores which are considerably larger than crystallites. As a consequence of this deviation from an ideally packed structure[11] the sintering temperature must be increased to achieve sufficiently dense materials, and only relatively coarse-grained ceramics can be produced.

The availability of ultrafine, highly reactive powders is of interest with regard to the production of ceramic materials with enhanced mechanical strength,[12] the formation of ceramic films, or the achievement of ZrO_2-containing ceramics with special ion-conducting properties.[13]

Several wet-chemical preparation methods, which were published in the literature, have been applied to the preparation of stabilized ZrO_2 by the present authors. In this paper the experimental results will be reviewed, with emphasis on the effect of resulting agglomerate structures on compaction and sintering behavior.

Synthesis Methods Investigated

Hot-Kerosene Synthesis

This preparation method, which is basically a spray-drying process, was originally developed by Reynen et al.[14] and applied by them for the production of $CaO \cdot ZrO_2$ and a number of other compositions. We applied this method for the preparation of 17 at.% yttria-stabilized ZrO_2 (to be referred to as "ZY17").

An aqueous solution of $Zr(SO_4)_2 \cdot 4H_2O$ and $Y(NO_3)_3$ was emulsified in kerosene, with "SPAN 80"* as an emulsifier, by means of vibromixing. This emulsion was slowly dripped into hot (170°C) kerosene (boiling range, 210°–220°C) contained in a stirred vessel. After complete removal of water, the product was filtered, washed, dried, and finally calcined at 600°C for 4 h in oxygen. The resultant powder consists of hollow spherelike agglomerates, having diameters in the range 2–20 μm (Fig. 1). The tap density[†] of this powder was measured to be 18% of theoretical. After calcination, the specific surface area was 4 $m^2 \cdot g^{-1}$.

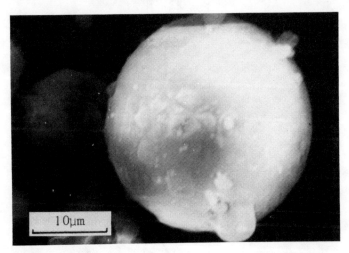

Fig. 1. Scanning electron micrograph of powder morphology resulting from hot-kerosene method.

---

*ICI Americas, Inc., Wilmington, DE.
†Tap-density measurement methods are not standardized. Therefore, tap-density values mentioned ought to be compared merely within the scope of this paper.

Citrate Synthesis

In concept, this preparation method was developed by Marcilly and co-workers.[15,16] In our laboratory, the synthesis method was applied for the preparation of lanthanum-doped lead zirconate titanate (PLZT) and substituted zirconia, with the evaporation and thermolysis steps modified to obtain powders with better homogeneity and improved sintering behavior. Experimental data concerning the preparation of gadolinia-stabilized zirconia were reported in a previous paper[11] and will be subject to a more detailed discussion in a future publication.[17]

Essentially, this synthesis method proceeds as follows: A zirconyl nitrate solution is prepared by dissolving $ZrOCl_2 \cdot 8H_2O$ in nitric acid. The dopant (e.g., Y or Gd) is next added in the form of the oxide. The solids are subsequently precipitated with ammonia, washed with water, filtered, and dissolved in nitric acid again. This procedure is repeated until the product is completely free of Cl^-.

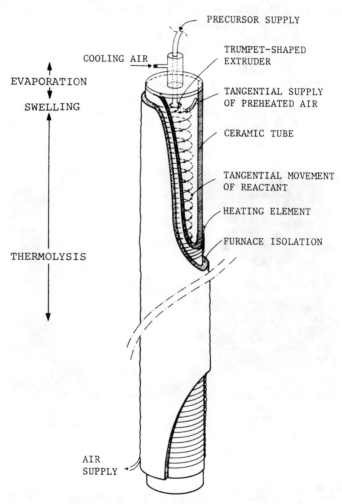

Fig. 2. Three-zone thermolyzer.

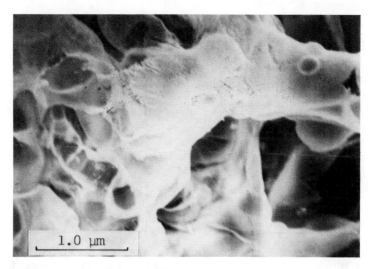

Fig. 3. Scanning electron micrograph of two-dimensional tissuelike agglomerates obtained by citrate synthesis.

The resulting product is dissolved in a large excess of HNO_3, and then ammonia is added until a value of $6 \leq pH \leq 7.5$ is achieved. Finally, citric acid is added in the ratio of 2 moles of citric acid per mole of metal to form the organometallic complex. After partial dehydration, a highly viscous system is obtained, which is then continuously processed in a specially developed vertical thermolyzer (cf., Fig. 2). This apparatus basically consists of a three-zone vertical furnace, with a trumpet-shaped extruder at the top. In the top section of the furnace, continued evaporation of liquid takes place, whereas the evolution of gases due to the decomposition of NH_4NO_3 causes substantial swelling of the viscous precursor mass. Because of this increase in volume of the precursor, the swelling mass transports itself along the conical bottom side of the extruder. The swelling is followed by ignition of the reactant, after which continued exothermic decomposition proceeds. When the stabilized zirconia powders prepared according to the citrate synthesis are calcined in air at 650°C, they show a loosely packed, tissuelike morphology (Fig. 3) with a surface area of 58 $m^2 \cdot g^{-1}$ and a tap density of 5% of theoretical.

Sol-Gel Microsphere Synthesis

The sol-gel process, originally applied for the production of nuclear fuels, was modified for the preparation of stabilized zirconia. Extended experimental data will not be given here, because they were previously reported in detail elsewhere.[7,8,11] (Fig. 4.)

The most characteristic feature of this type of sol-gel powders is the perfectly spherical shape of the powder agglomerates. When these highly dense ($\rho = 72\%$) microspheres, having diameters between 1 and 20 μm (cf. Figs. 5 and 6), are calcined at 650°C, they are found to be built up of very small and regularly packed crystallites, exhibiting a narrow crystallite size distribution ($\bar{d}_{cr} = 80$ Å) and small pores ($\bar{d}_p = 40$ Å).[7] These microspheres show extremely good *internal* sintering behavior, whereas mutual sintering after compaction of these spheres is very poor.

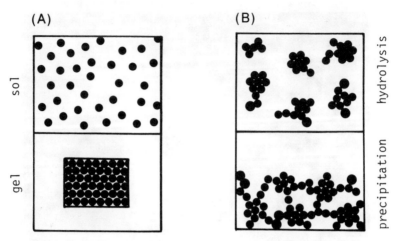

Fig. 4. Influence of gel formation principle on resulting structure: forced removal of liquid phase from a stable sol (A) and gel formation from an instable state (precipitation) (B).

Fig. 5. Scanning electron micrograph of stabilized zirconia sol-gel microspheres.

Peroxide Synthesis

This method was reported by Murata et al.[18] and used by them to prepare PLZT. Essentially, the peroxide synthesis implies coprecipitation from a salt solution by adding it dropwise to ammonia. Besides the stoichiometric amount of metal salts, the solution contains hydrogen peroxide. The authors state that, during drying, the gelatinous precipitate becomes porous as a consequence of retained peroxide decomposition. We applied this synthesis to the preparation of ZY17 in order to determine the resulting powder morphology. Briefly, the synthesis was performed as follows: An aqueous solution, containing the desired concentration

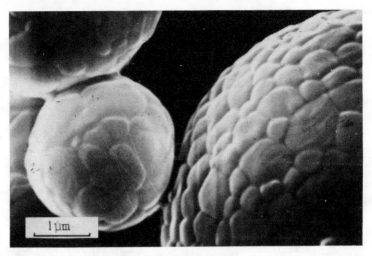

Fig. 6. Scanning electron micrograph of stabilized zirconia sol-gel microspheres; surface thermally etched at 1300°C.

Fig. 7. Scanning electron micrograph of dense Y_2O_3-ZrO_2 lumps resulting from peroxide method.

of metal ions, was prepared by dissolving $Y(NO_3)_3$ and $ZrOCl_2$. The solution was heated to 40°C, hydrogen peroxide was added in a 1.1 : 1.0 H_2O_2-metal molar ratio, and the clear solution was dripped into a stirred bath of ammonia. Temperature was kept constant at 40 ± 2°C and pH between 8.5 and 9.5 by means of simultaneous addition of diluted ammonia. The precipitate was filtered, rinsed with water, dried, and finally calcined at 600°C for 2 h.

The electron micrograph in Fig. 7 shows the resulting powder morphology. Although this picture gives an impression of dense lumps of material, Brunauer–Emmett–Teller (BET) measurement reveals a considerable specific

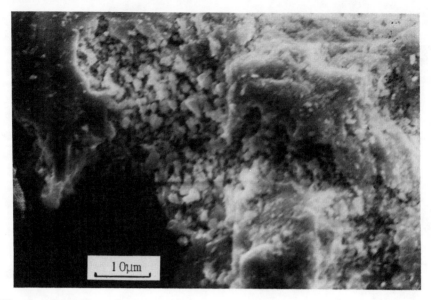

Fig. 8. Scanning electron micrograph of stabilized zirconia powder morphology obtained by the acetone toluene method.

surface area of 82 $m^2 \cdot g^{-1}$. Tap density was estimated to be as high as 38% of theoretical density.

Acetone-Toluene Synthesis

Dole et al.[19] reported this preparation route to be an attractive method for the synthesis of stabilized zirconia. Because of the use of both solvents as washing liquids, this method will be referred to as "acetone-toluene synthesis." The authors stated that the sinterability of stabilized zirconia, precipitated from aqueous salt solutions, is improved when the water in the hydrous precipitate is replaced by acetone, prior to drying. However, they report that rather high sintering temperatures of 1800°–2100°C are needed for the achievement of >95% density. We performed this synthesis for the preparation of ZY17. A hydrous gel was formed by ammonia-assisted precipitation from a mixed metal solution in HNO_3. The gelatinous product was washed with water to remove the ammonium salts formed, rinsed with acetone to remove a substantial part of the water, and subsequently washed with toluene to further remove free water. This was filtered, and then a final acetone washing step removed remaining traces of toluene. The resulting material was dried, crushed, and calcined at 650°C. The as-calcined powder reveals the morphology shown in Fig. 8. Specific surface area and tap density were measured to be 16 $m^2 \cdot g^{-1}$ and 32%, respectively.

Alkoxide Synthesis

This preparation method, which was at first described by Mazdiyasni et al.[20] and evaluated by Hoch and Nair,[21] produces powders with extremely good sintering qualities. In recent papers,[2,10] an extensive study was presented on the microstructural development of ZY17 ceramics, during the various stages of compaction and sintering. The vital role of agglomerates and aggregates, present in the powders, during these processes was illustrated in these papers. It was shown that

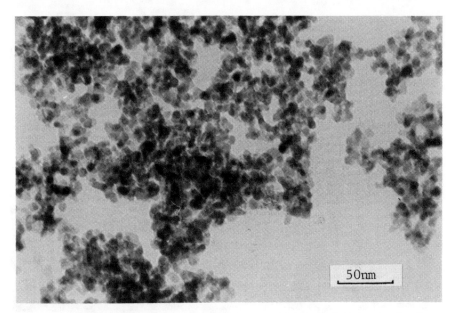

Fig. 9. Transmission electron micrograph of alkoxide-derived, ultrafine, weakly agglomerated zirconia powders, T_{calc} = 650°C.

the total removal of free water from the accessible surface of the precipitated gels, prior to drying, is essential to obtain ultrafine, weakly agglomerated powders. For that purpose, hydrolysis and washing procedures were accomplished in a specially developed reaction vessel using a dispersion turbine and (during washing) a high-energy disk turbine. After calcination the resultant powder contains extremely weak agglomerates (agglomerate density = 23%) and shows a high specific surface area of 90 m$^2 \cdot$g^{-1}. Tap density was measured to be 12%. Figure 9 shows the resultant powder morphology. As will be shown later, these powders exhibit improved sintering behavior, leading to >95% density at temperatures below 1200°C.

Gel routes for the preparation of ceramics may be divided into two groups of methods, viz., gel preparation starting from colloidal sols and gel-precipitation techniques. These methods produce essentially different gel structures.

Starting from stable sols, gelation may be accomplished by (partial) removal of the dispersing liquid. During removal of the liquid, the colloidal particles are forced to approach each other and, because of their similar surface charge, finally pack to a very regular and dense structure (cf. Fig. 4(A)). The dimensions of ceramics prepared according to this so-called monolithic process are restricted to relatively small sizes: microspheres up to few millimeters[11] or thin layers with thicknesses commonly <20 μm.[22] With larger bodies, cracking problems arise during drying of the gel monolite.

For the preparation of larger ceramics, highly reactive ceramic powders may be produced by means of gel-precipitation techniques, such as the alkoxide synthesis. Here, a gelatinous precipitate is formed by hydrolysis of mixed metal organic (or salt) solutions. The gel structure and final oxidic morphology obtained in this way deviate significantly from those resulting from the monolithic process:

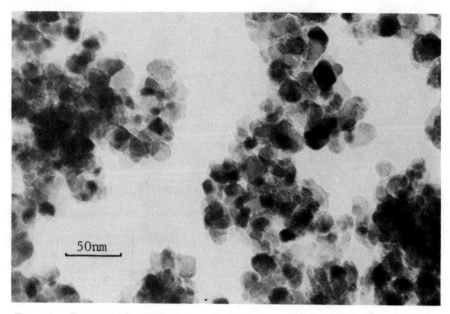

Fig. 10. Transmission electron micrograph of alkoxide-derived zirconia powders. After heat treatment at 850°C, the powders more clearly show the existence of small crystallite aggregates.

During hydrolysis, colloidal particles are formed which do not possess a surface charge in the neutral or basic ambient medium. As a result, the colloids exhibit coalescence, whereas continuing flocculation causes the small particle clusters to precipitate. Hydrous precipitates, so formed, consist of closely packed small areas[2,10] which are cross-linked by crystallite chains (see Fig. 4(B)). On calcining, this structure is consolidated and represents the internal microstructure of the agglomerates observed in this type of powder. The existence of very small closely packed aggregates in these powders (remnants of the original colloid clusters) may be detected more clearly after grain growth during calcining at 850°C (cf. Fig. 10). In the authors' opinion, the average size of the small densely packed areas in the gel (the later aggregates) and the extent of cross-linking between them, later making up the internal agglomerate microstructure, determine agglomerate strength and thus compaction and sintering behavior of the powder involved.

More recently, a growing number of workers discerned the influence of the aging medium of gelatinous zirconia precipitates on the sintering behavior of the resulting oxidic powders.[1,2,5,10,19,21] In some cases variations in sintering behavior of powder batches are ascribed to differences in agglomerate size.[9,21] In the authors' opinion, however, this is an inaccurate explanation: The occasionally reported agglomerate *size* data likely are merely values being affected by the measuring method. In fact, these agglomerate size values are related to agglomerate *strength* rather than to actual agglomerate *dimensions* (stronger agglomerates produce higher apparent agglomerate size values).

In the literature, no consistent information is available about the way in which the replacement of water by organic liquids affects internal gel structures. As was shown by Rijnten[23] and Gimblett et al.,[24] aging in water of precipitated hydrous zirconia gels causes progressive polymerization and densification of the gel structure. Furthermore, some other workers[25-27] showed that the solvent, in which gels are aged, influences pore size as well as pore volume of the dried product. In addition, higher surface tensions of a liquid present in a lyogel give rise to higher capillary forces during drying and, consequently, lead to a higher degree of pore collapse. Finally, Rijnten suggested that water might be a sintering aid for zirconia in the low-temperature region (<600°C), which promotes densification of the gel structure and thus strengthening of agglomerates during calcining.

In spite of the above uncertainties, experimental results, to be discussed later, convincingly demonstrate the drastic influence of water removal prior to drying and calcining on the resulting powder characteristics.

Chloride Synthesis

Though alkoxide-derived zirconia powders are very attractive because of their high sinter reactivity, synthesis is time-consuming owing to the rather difficult preparations of metal alkoxide precursors. In 1979, Haberko[5] reported the preparation of ZY12 from precipitated gels, starting from $ZrCl_4$ and YCl_3. Both chlorides are commercially available or may be prepared easily. Haberko extensively discussed the influence of alcohol washing on gel structure and the resulting compaction characteristics and microstructure. He also established that alcohol washing, prior to drying and calcining, significantly decreases agglomerate density (and strength) and improves sintering behavior.

In our laboratory, the synthesis of ZY17 starting from chlorides was performed according to the data given by Haberko,[5] but the circumstances employed with the alkoxide synthesis were also applied. The resulting powder showed the following properties: specific surface area, 123 $m^2 \cdot g^{-1}$; tap density, 10% (measured after intensively vibrating the powder for 5 min in a tube); agglomerate density, 19.0%. After isostatical compaction at 400 MPa the pellet reveals an overall relative density of 44% and contains no pores larger than 80 Å. Sintered densities of 95 and 98% are achieved at temperatures as low as 1180° and 1220°C, respectively. From the preceding, it appears that the excellent powder properties as reported by Haberko may even be improved to some extent by applying an adequate gel treatment procedure.

In Table I, a survey is presented of powder characteristics summarized in this section. In the next section, compaction and sintering behavior and microstructural

Table I Survey of Powder Characteristics

Method	Calcining temp (°C)	Agglomerate size (μm)	Relative agglomerate density (% theor.)	Agglomerate compression strength (MPa)	Tap density (% theor.)	Surface area ($m^2 \cdot g^{-1}$)
Hot kerosene	600	2–20		≥100	18	4
Citrate	650				5	58
Sol-gel microsphere	650	1–20	72	>400	26	32
Peroxide	600	≤50	80	>400	38	82
Acetone-toluene	650	≤100	58		32	16
Alkoxide	650		23	30	12	90
Chloride	650		19	30	10	123

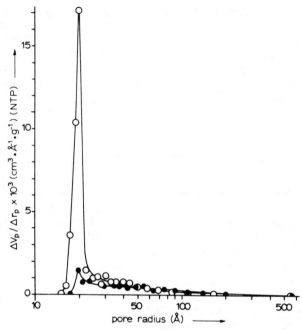

Fig. 11. Development of the internal microstructure of stabilized zirconia sol-gel microspheres: pore-size distribution curves of microspheres heat treated at 650° (○) and 950°C (●).

development during these processes will be reviewed and compared for the different powders concerned.

Evaluation of Compaction and Sintering Behavior

The powders under discussion and their behavior during ceramic processing have been studied by means of scanning electron microscopy (SEM), transmission electron microscopy (TEM), and BET measurement, gas desorption, Hg porosimetry, and X-ray methods. Part of these studies have been published elsewhere.[1,2,7,8,10,11] In this section a compilation of these results will be summarized, on the basis of a condensed review.

In previous work,[11] we investigated the *internal* microstructure and densification during sintering of zirconia sol-gel microspheres. In this paper it was shown that the internal microsphere structure represents an almost ideal homogeneous packing of small crystallites. The initial microsphere structure shows a narrow crystallite-size distribution and a uniform porosity distribution (cf. Fig. 11), with no pores large in comparison with crystallite sizes. During sintering, densification and grain growth occur, but narrow grain size and pore-size distributions, as well as an acceptable pore-to-grain-size ratio, are maintained. As a consequence, zirconia sol-gel microspheres show excellent internal densification (Fig. 12) during sintering: 95% relative density is reached at an uncommonly low temperature of 1050°C. As is illustrated in Fig. 12, mutual sintering between compacted zirconia microspheres, however, is extremely poor and does not lead to the achievement of useful dense ceramics.

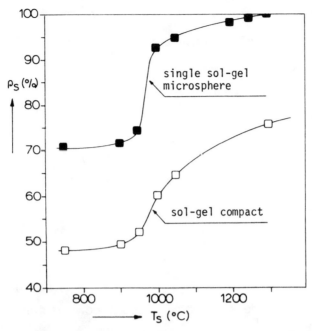

Fig. 12. A single zirconia microsphere shows ideal densification behavior, whereas a microsphere compact exhibits poor sintering.

From this study, a number of criteria were defined[11] which must be satisfied to obtain highly sinter-reactive powder compacts, the main conclusion being that a uniform distribution of small grains and pores throughout the entire powder compact is required. This implies that the occurrence of strong agglomerates in ultrafine powders should be avoided. As will be shown later, densely packed (and consequently strong) agglomerates persist during the compaction cycle and during sintering and give rise to preferential intraagglomerate sintering in early stages of the sintering process. This results in a poor overall sintering behavior at higher temperatures, the formation of large pores, and (probably) exaggerated grain growth.

Compaction Results

Examination of the densification of powder compacts at increasing isostatic pressures provides a suitable and simple method for the qualitative investigation of powder morphologies.[1,2,5,28] When compact density is plotted vs logarithm of compaction pressure, sudden changes in the compaction mechanism cause the occurrence of breakpoints in the density vs pressure curve. Powders containing distinct agglomerates, with sufficiently low strength, typically show compaction curves exhibiting two intersecting linear portions.

Figure 13 shows compaction curves of powder compacts formed of three zirconia powders. The lower curve shows the phenomenon mentioned above. The alkoxide powder considered is built up of very low-density ($\rho_{rel} = 23\%$) agglomerates.[2] In the low-pressure range, densification proceeds by fragmentation and rearrangement of these agglomerates *without* the occurrence of *compression* of their internal structure. In the high-pressure area (>30 MPa) densification con-

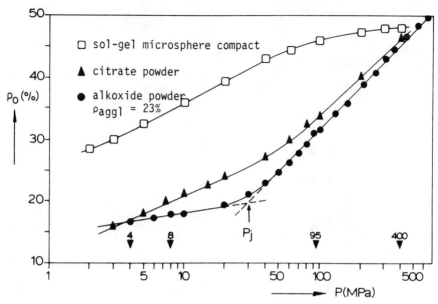

Fig. 13. Pressure-density graphs of zirconia powders prepared according to three methods.

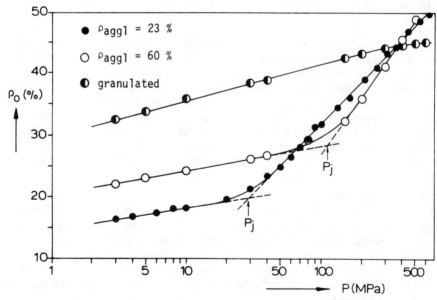

Fig. 14. Influence of alkoxide agglomerate density on the shape of the resulting pressure-density curves.

tinues by compression and, consequently, densification of the internal agglomerate microstructure. At the point of intersection (joining pressure $P_j = 30$ MPa), the entire compact has taken the agglomerate structure; at pressures exceeding P_j,

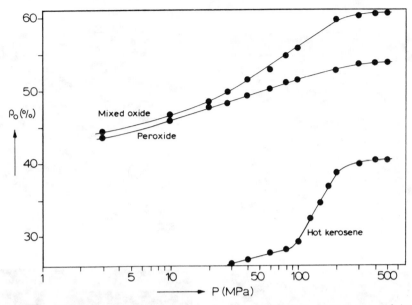

Fig. 15. Densification on increasing compaction pressure of hot-kerosene- and peroxide-derived zirconia compared with a mixed oxide powder.

agglomerate boundaries are no longer detectable. The value of P_j is a measure for agglomerate compression strength. In Fig. 16 the compaction sequence of this alkoxide powder is outlined schematically. The four stages of microstructure development shown correspond with compaction pressures as indicated in the lower side of Fig. 13.

Figure 13 also shows the compaction curves of citrate and sol-gel microsphere zirconia compacts. Neither curve reveals distinct bending points. In the case of the citrate material, this is due to the absence of distinct agglomerates with defined microstructures, while sol-gel microspheres only rearrange but can neither be compressed nor fragmentized, because of their high strength.[7,8]

Compression strength of zirconia alkoxide agglomerates depends greatly on their internal density. In a previous paper,[1] it was shown that alkoxide-derived gadolinia-substituted zirconia powders, consisting of high-density agglomerates (pore volume $V_p = 0.07$ cm$^3 \cdot$g^{-1} and relative agglomerate density = 60%), show a different compaction behavior. Figure 14 illustrates this in comparison with the alkoxide powder discussed above. As can be seen from this figure, because of lower agglomerate porosity, the compact with $\rho = 60\%$ agglomerates exhibits a densification sequence initially starting with higher densities, whereas the P_j value of 110 MPa is an indication of the presence of fairly strong agglomerates. After compaction at 400 MPa, a bimodal pore-size distribution is formed, as a result of incomplete breakdown of the agglomerates.

An even more dramatic influence of dense agglomerate structures was shown in the same paper[1]: powders which were granulated by precompaction and subsequent crushing (in order to ease compaction) appear to consist of extremely strong agglomerates. As is shown in Fig. 14, alkoxide powders granulated in this way reveal a compaction curve without a bending point as a result of the presence

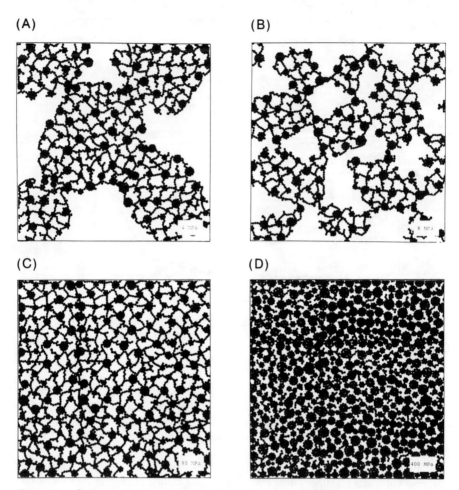

Fig. 16. Schematic representation of the different stages of microstructural development during compaction of low agglomerate density (ρ_{rel} = 23%) alkoxide powders.

of these high-strength agglomerates. These agglomerates are thought to be formed in the following way: During the precompaction step (400 MPa), a highly densified compact structure forms, as mentioned before. On subsequent crushing, powder fragments having the same highly dense internal structure are formed. During the second compaction step, only rearrangement and partial fragmentation of these powder fragments takes place, without the occurrence of compression. Figure 17 schematically illustrates the compaction sequence of these granulated powders.

Figure 15 reveals the compaction curves of ZY17 powders as prepared by the peroxide and the hot-kerosene methods. Hot-kerosine-derived powders, as mentioned before, consist of hollow spherical agglomerates. The hollow shape of these agglomerates explains the low starting density of their compaction curve. The sudden increase in the densification rate at pressures ≥100 MPa is caused by the

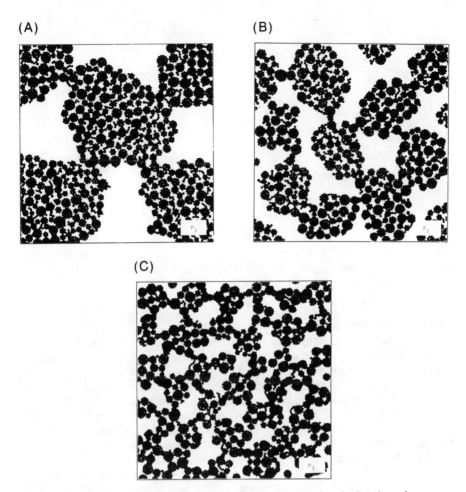

Fig. 17. Schematic representation of the compaction behavior of granulated alkoxide powders.

collapse of these hollow spheres in the pressure range $100 \leq P \leq 200$ MPa. At pressures exceeding 200 MPa, however, density hardly increases as a result of the high strength (density) of the sphere fragments. This results in a relatively porous green structure, containing very large pores.

Peroxide-derived powders show a compaction curve similar to that of mixed oxide powders (cf., Fig. 15). For comparison, a typical compaction curve of a mixed powder (prepared by wet-milling and solid-state reaction of ZrO_2 and Y_2O_3) is inserted in this graph. As was reported in the previous chapter, peroxide-derived zirconia exhibits a rather high apparent specific surface area of 82 $m^2 \cdot g^{-1}$. This high surface, however, is caused by micropores present in the extremely dense (80%) agglomerates. As a result, these powders produce a compaction curve without a bending point and resemble mixed oxide powders, exhibiting relatively high density green compacts with rather large pores.

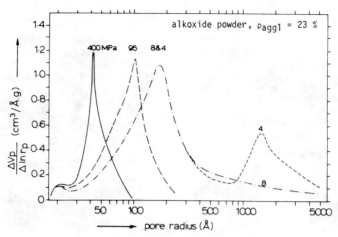

Fig. 18. Pore-size distribution curves showing microstructural development during compaction of low-density alkoxide powders. Pressures correspond with values indicated in Fig. 13.

Microstructural Development

Gas-desorption techniques and density measurements were applied to study the evolution of microstructures during the compaction process. Figure 18 shows the pore-size distribution curves of the ZY alkoxide compact at four stages of the compaction sequence, as was indicated in Fig. 13. As was reported before,[2] sinter-reactive alkoxide powders are built up of three different microstructural elements, namely, primary *crystallites* ($\bar{d}_{Cr} \approx 80$ Å), relatively dense and small *aggregates* ($\bar{d}_{aggr} \sim 200$ Å, $\rho \approx 55\%$), which consist of bound crystallites, and weak, highly porous agglomerates ($\rho_{aggl} = 23\%$) which are formed by mutual clustering of a large number of aggregates. Figure 16 schematically represents the four stages of compact densification, connected with the pore-size distribution curves of Fig. 18:

In the first stages of compression ($<P_j$), in addition to relatively small ($\bar{d}_p = 180$ Å) intraagglomerate pores, a second class of larger ($\bar{d}_p = 1500$ Å) interagglomerate pores is present; on increasing the pressure, the volume as well as the size of the larger pores diminish.

At pressures exceeding bending-point pressure, interagglomerate pores are no longer present; agglomerate boundaries have disappeared. On increasing compaction pressure, intraagglomerate pore dimensions continuously decrease as a result of microstructure compression. Finally, at a pressure of 400 MPa, a rather narrow distribution results, consisting of interaggregate pores ($\bar{d}_p = 86$ Å; $d_p(\max) < 200$ Å) and a small shoulder, originating from the internal aggregate structure (intraaggregate pores: $\bar{d}_p \approx 40$ Å). As mentioned before, the extremely weak agglomerate structure of these powders is due to the small dimensions as well as the moderate degree of interconnection of these aggregates.

For comparison, Fig. 19 illustrates a bimodal pore-size distribution, resulting after compaction at 400 MPa of a dense-agglomerate alkoxide powder.[1]

Stabilized zirconia powders, prepared according to the citrate synthesis, show a tissuelike morphology. These powders show acceptable sintering reactivity in spite of the much larger pores (40 Å $< d_p <$ 1 μm) present in the compact. This is ascribed[2,8,11] to the uniform distribution of the small radii of curvature, present

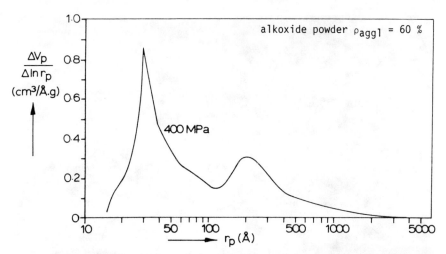

Fig. 19. Bimodal pore-size distribution resulting after compaction at 400 MPa of high-density alkoxide powders, showing relatively coarse interagglomerate pores in addition to intraagglomerate pores.

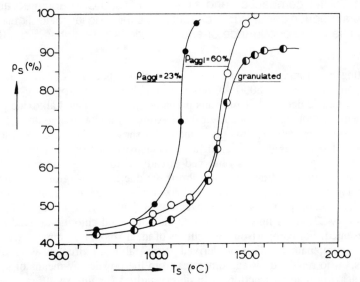

Fig. 20. Comparison of sintering behavior of different alkoxide-derived zirconias.

in the two-dimensional crystal agglomerates through the powder compact, and the hindered (three-dimensional) crystallite growth.

Sintering Behavior

Figure 20 shows the sintering behavior of zirconia alkoxide powders discussed in the preceding section. As can be seen from this figure, the low-density agglomerate powder reaches >95% density below 1200°C, whereas the powder

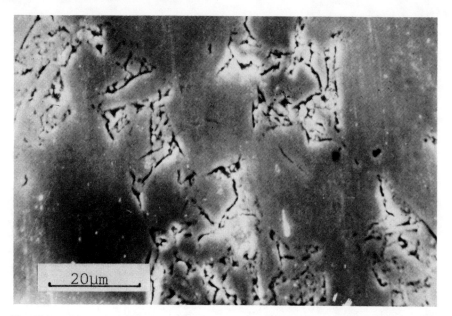

Fig. 21. After compaction and sintering at 1300°C, granulated alkoxide powders produce ceramic microstructures exhibiting large, lenticular voids giving rise to the occurrence of end-point density (see also Fig. 20).

consisting of stronger agglomerates ($\rho_{aggl} = 60\%$), because of the significantly larger pores ($d_p(\text{max}) = 6000$ Å), requires notably higher temperatures in order to reach acceptable density.[1,2] Compacts prepared from granulated alkoxide powders show densification up to about 90%, the maximum value which could be achieved in this case. The cause of this phenomenon is shown in Fig. 21: because of the presence of dense agglomerates in a lower-density matrix, preferential growth within these agglomerates is observed already at low temperatures. As a consequence of significant agglomerate shrinkage due to intraagglomerate sintering, they pull away from the surrounding matrix, leaving large lenticular voids behind. This phenomenon prevents the achievement of density values over 90%.

In Fig. 22, sintering curves of both alkoxide- and chloride-derived powders are presented. For comparison, two curves of ideal "model" compacts are inserted in this figure, namely alkoxide-derived yttria-stabilized zirconia as reported by Rhodes,[9] who removed all agglomerates from the powder by means of sedimentation, leaving a small fraction of basic crystallites (or ultrasmall aggregates) in suspension, which was subsequently centrifugally cast, and a single sol-gel zirconia microsphere.[11]

Figure 22 shows a much higher initial density for both model systems, as a consequence of the high starting level. Nevertheless, in the temperature region around 1200°C, both alkoxide- and chloride-powder compacts approach ideal density. Furthermore, this figure shows no large differences between the densification of chloride- and alkoxide-derived powders, although they were prepared from different precursors. This supports the opinion that the way in which precipitated gels are processed, rather than the nature of precursors used, has a dominant influence on the resulting powder properties. Both chloride and alkoxide compacts

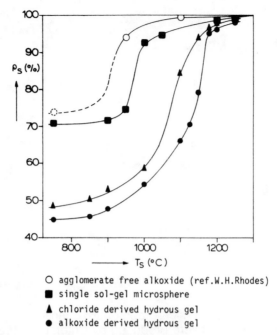

○ agglomerate free alkoxide (ref.W.H.Rhodes)
■ single sol-gel microsphere
▲ chloride derived hydrous gel
● alkoxide derived hydrous gel

Fig. 22. Densification during sintering of stabilized zirconia prepared according to the chloride and alkoxide synthesis, respectively. Both upper curves represent the sintering behavior of rather ideally packed systems.

produce >95% dense ceramics at temperatures below 1200°C (see Fig. 23). As was reported in a recent paper,[12] the small grain size ($\ll 1.0$ μm) of these ceramics leads to an improvement in mechanical behavior.

Sintering behavior of the more conventional, wet-chemically prepared powders, as discussed in this paper, is surveyed in Fig. 24. From this graph it can be seen that citrate powders exhibit a fairly good sintering behavior, whereas all other wet-chemically prepared powders fail in the achievement of sufficient ceramic density, as a result of the presence of dense agglomerates. From Fig. 24, it may be concluded that even mixed oxide samples, in this respect, have preference over wet-chemically prepared powders, in the event of inadequate agglomeration control.

Summary

Several modern wet-chemical-preparation methods have been applied to the preparation of stabilized zirconia powders. The usefulness of these synthesis methods with respect to resulting powder morphology, compaction behavior, and sinterability has been evaluated. It was shown that synthesis methods investigated frequently produce strong powder agglomerates, giving rise to inhomogeneous compact microstructures and, consequently, poor sintering behavior. Only gel route methods, applying adequate gel-processing procedures, enable the preparation of weakly agglomerated, highly sinter-reactive powders. These powders show excellent compaction characteristics and enable the achievement of >95% dense zirconia ceramics, at sintering temperatures as low as 1180°C, in combination with average grain sizes of 0.3 μm.

Fig. 23. Scanning electron micrograph of 96% dense, 0.3-μm grain-size zirconia ceramic microstructure produced by sintering at 1180°C of reactive, low-density stabilized zirconia.

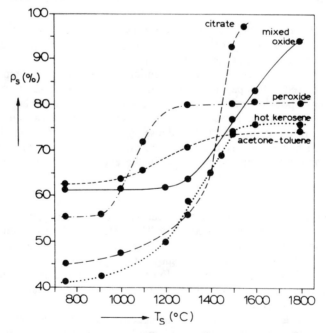

Fig. 24. Survey of sintering behavior of stabilized zirconia powders, prepared according to different methods.

References

[1] M. A. C. G. van de Graaf, K. Keizer, and A. J. Burggraaf; pp. 83–92 in Science of Ceramics, Vol. 10. Edited by H. Hausner. Deutsche Keramische Gesellschaft, Berchtesgaden, Federal Republic of Germany, 1980.

[2] M. A. C. G. van de Graaf, J. H. H. ter Maat, and A. J. Burggraaf; pp. 783–94 in Ceramic Powders. Edited by P. Vincencini. Elsevier, Amsterdam, 1983.

[3] P. H. Rieth and J. S. Reed, *Am. Ceram. Soc. Bull.*, **55** [8] 717–27 (1976).

[4] J. S. Reed, T. Carbone, C. Scott, and S. Lukasiewicz, *Mater. Sci. Res.*, **11** [3] 171–80 (1978).

[5] K. Haberko, *Ceramurgia Int.*, **5** [4] 148–54 (1979).

[6] J. van der Zwan and C. A. M. Siskens; pp. 159–68 in Science of Ceramics, Vol. 10. Edited by H. Hausner. Deutsche Keramische Gesellschaft, Berchtesgaden, Federal Republic of Germany, 1980.

[7] M. A. C. G. van de Graaf, *Klei/Glas/Keram.*, **2** [10] 300–305 (1981).

[8] M. A. C. G. van de Graaf, *Klei/Glas/Keram.*, **3** [1] 8–13 (1982).

[9] W. H. Rhodes, *J. Am. Ceram. Soc.*, **64** [1] 19–22 (1981).

[10] M. A. C. G. van de Graaf, J. H. H. ter Maat, and A. J. Burggraaf, *J. Mater. Sci.*, in press.

[11] M. A. C. G. van de Graaf, T. van Dijk, M. A. de Jongh, and A. J. Burggraaf; pp. 75–83 in Science of Ceramics, Vol. 9. Edited by K. J. de Vries. Dutch Ceramic Society, Noordwijkerhout, The Netherlands, 1977.

[12] A. J. A. Winnubst, K. Keizer, and A. J. Burggraaf, *J. Mater. Sci.*, **18**, 1958–66 (1983).

[13] M. J. Verkerk, B. J. Middelhuis, and A. J. Burggraaf, *Solid State Ion.*, **6**, 159–70 (1982).

[14] P. Reynen, H. Bastius, M. Faizullah, and H. v. Kamptz, *Ber. Dtsch. Keram. Ges.*, **54** [3] 63–68 (1977).

[15] C. Marcilly, P. Courty, and B. Delmon, *J. Am. Ceram. Soc.*, **53** [1] 56–57 (1970).

[16] C. Marcilly; Ph.D. Thesis, University of Grenoble, Grenoble, France, 1968.

[17] M. A. C. G. van de Graaf and A. J. Burggraaf, unpublished work.

[18] M. Murata, K. Wakino, K. Tanaka, and Y. Hamakawa, *Mater. Res. Bull.*, **11** [3] 323–38 (1976).

[19] S. L. Dole, R. W. Scheidecker, L. E. Shiers, M. F. Berard, and O. Hunter, Jr., *Mater. Sci. Eng.*, **32**, 277–81 (1978).

[20] K. S. Mazdiyasni, C. T. Lynch, and J. S. Smith, II, *J. Am. Ceram. Soc.*, **50** [10] 532–37 (1967).

[21] M. Hoch and K. M. Nair, *Ceramurgia Int.*, **2** [2] 88–97 (1976).

[22] A. F. M. Leenaars, K. Keizer, and A. J. Burggraaf, *J. Mater. Sci.*, **19**, 1077–88 (1984).

[23] H. Th. Rijnten, pp. 79–85 in Zirconia; Ph.D. Thesis, Delft University of Technology, Delft, The Netherlands, 1971.

[24] F. G. R. Gimblett, A. A. Rahman, and K. S. W. Sing, *J. Coll. Interface Sci.*, **84** [2] 337–45 (1981).

[25] R. Yu. Sheinfain and T. F. Makovskaya, *Koll. Zh.*, **38** [4] 816–818 (1976).

[26] O. P. Stas', R. Yu. Sheinfain, and I. É. Neimark, *Koll. Zh.*, **39** [2] 393–96 (1977).

[27] M. L. Veiga, M. Vallet, and A. Jerez, *Ann. Chim. (Paris)*, [6] 341–44 (1981).

[28] D. E. Niesz, R. B. Bennett, and M. J. Snyder, *Am. Ceram. Soc. Bull.*, **51** [9] 677–80 (1972).

Preparation of Y$_2$O$_3$-Stabilized Tetragonal ZrO$_2$ Polycrystals (Y-TZP) from Different Powders

Helmut Schubert, Nils Claussen, and Manfred Rühle

Max-Planck-Institut für Metallforschung
Institut für Werkstoffwissenschaften
Stuttgart, Federal Republic of Germany

Y-TZP samples were prepared by sintering and isostatically hot-pressing of coprecipitated (CP; 2.2 mol% Y$_2$O$_3$) and spray-reacted and attrition-milled powder (SR; 2.0 mol% Y$_2$O$_3$). The glassy phase which formed in the sintered TZP made from SR powders scavenged a large quantity of Y$_2$O$_3$ from the grains, making them highly susceptible to stress-induced transformation. Isostatic hot-pressing drastically reduced this glassy grain-boundary phase. The Y distribution within the grains of the essentially glass-free TZP, made from CP powders, was more homogeneous than in the SR material.

A major problem encountered in the preparation of dense, Y$_2$O$_3$-stabilized tetragonal (t)-ZrO$_2$ polycrystals (Y-TZP)[1-5] is the attainment of a homogeneous distribution of Y$_2$O$_3$ within the ZrO$_2$ grains. Experiments have shown that, at the usual sintering temperatures ($\approx$1400°C), it is especially difficult to achieve a uniform distribution by reaction-sintering of mechanically mixed oxide powders, at least at low Y$_2$O$_3$ concentrations (<4 wt%). Obviously, the diffusion of Y$_2$O$_3$ into ZrO$_2$ is very slow at these temperatures. Higher temperatures, however, required for the formation of a uniform solid solution, result in rapid grain growth which, in the absence of sufficient stabilizer content, leads to the transformation of the larger grains to monoclinic (m) symmetry on cooling. Therefore, wet-chemically produced powders appear to be a better route for achieving low sintering temperatures at low Y$_2$O$_3$ levels.

The present paper compares the powder characteristics and sintering behavior of a coprecipitated (CP) and a spray-reacted (SR) powder. The CP powder was essentially SiO$_2$-free, whereas the SR powder was contaminated with SiO$_2$ and Al$_2$O$_3$ from the attrition-milling process. The influence of the resulting glassy grain-boundary phase on the Y$_2$O$_3$ distribution and transformability of sintered and isostatically hot-pressed TZP samples is reported.

Experimental Procedure

Aqueous solutions of zirconium oxychloride and yttrium nitrate were spray-reacted at 900°C, i.e., spray-dried and calcined in a single-step process,* leading to the evaporation and decomposition of the solution (SR powder). To destroy the

---

*Processed at Dornier System, Friedrichshafen, Federal Republic of Germany.

hard spherical particles, this powder, with 3.6 wt% (2.0 mol%) Y_2O_3, was attrition-milled for 5 h with Al_2O_3 balls containing SiO_2. Powder contamination, by wear of the milling media, was 0.6 wt% SiO_2 and 4 wt% Al_2O_3. A co-precipitated powder[†] with 4.0 wt% (2.2 mol%) Y_2O_3 was used for comparison (CP powder).

All the powders were isostatically pressed at 630 MPa and sintered in air at $T = 1400°$ to 1500°C for various times. A further sample of the SR powder was sintered to closed porosity (1410°C/15 min) and then isostatically hot-pressed (1450°C/15 min) in Ar at 200 MPa.

The powders and fabricated samples were analyzed by X-ray diffraction (XRD), scanning electron microscopy (SEM), and transmission electron microscopy (TEM). The chemical composition of the ZrO_2 grains and the grain-boundary phase were measured by scanning transmission electron microscopy (STEM). The transformability of the t phase was investigated by XRD analysis of as-sintered and ground surfaces.[6]

Results and Discussion

Powder Morphology

The SR powder consisted of large agglomerates made up of hard, hollow spheres, a morphology which is obviously a characteristic of the SR process (Fig. 1). The t reflections of the as-received SR powder were rather broad (Fig. 2(a)). However, after 3-months aging in air at room temperature, the t reflections sharpened and a small m content ($\approx$5%) appeared (Fig. 2(b)). The dense sphere sections were then heavily microcracked (Fig. 3) because the crystallites with a lower Y_2O_3 content transformed to m symmetry. This type of degradation may have the same cause as that described in Refs. 3, 4, and 7–9.

Fig. 1. Scanning electron micrograph of spray-reacted (SR) powder with hollow-sphere geometry.

[†]Zircar Products Inc., Florida, NY.

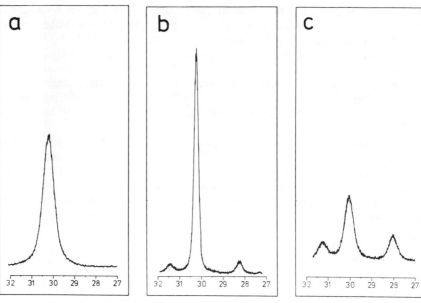

Fig. 2. X-ray diffraction pattern of (a) SR powder (as-received), (b) SR powder after 3-month aging in air at room temperature, and (c) coprecipitated (CP) powder (as-received).

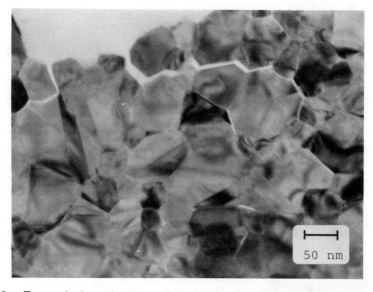

Fig. 3. Transmission electron micrograph of sectioned SR powder particle (3 months in air at room temperature). Microcracks are marked by arrows. Some t-ZrO_2 particles have transformed to m symmetry.

Fig. 4. Scanning electron micrograph of CP powder.

The CP powder was strongly agglomerated, with a large variation in agglomerate size (Fig. 4). The average crystallite size was ≈40 nm. Both t ($\approx$60%) and m phases were present (Fig. 2(c)). The XRD peaks were also broad, which may be due to an inhomogeneous Y_2O_3 distribution and to the small crystallite size.

Both the CP and SR powders retained their t symmetry after being annealed in the t single-phase field. However, the solution-annealing time at 1300°C for obtaining single-phase t powder was much shorter for the SR powder, possibly because of the closer grain-boundary-like contacts within the SR particles. The difference in the m content (Fig. 2(b) and (c)) indicates that the Y distribution in the SR powder was more homogeneous than in the CP powder.

Compaction

The as-received SR powder could be isostatically pressed only to 44% theoretical density (TD). The hollow spheres were too resistant to the applied pressure, resulting in a closed porosity which could not be eliminated during sintering. A green density of 60% TD was achieved when the SR powder was attrition-milled. Hence, only the milled SR powder was used in our experiments.

The green density of the CP powder was ≈57% TD. Attrition milling would probably have improved the green density; however, we wished to avoid any wear contamination in order to compare SiO_2-free TZP with a typical TZP[5] containing amorphous grain boundaries (i.e., the SR material). It also gave us the opportunity to study the consolidation behavior of the powders from the same green-density state.

Sintering

Shrinkage of the SR samples started at 1080° ± 40°C and was independent of the heating rate to this temperature. The final densities were 6.00 g/cm³ (98.8% TD) and 6.04 g/cm³ (99.5% TD) after 1-h hold at 1400° and 1450°C, respectively.

Fig. 5. (A) Scanning electron micrograph of thermally etched SR sample. Dark grains are Al_2O_3. (B) Scanning electron micrograph of thermally etched CP sample.

The CP powder compacts started to shrink at 1150° and had to be sintered at 1500°C for 2 h to reach a density of 6.00 g/cm³. It appeared that the superior sintering behavior of the SR powder was due to the milling contamination, allowing a liquid-phase sintering process to consolidate the powders.

Microstructure

Etched surfaces of the sintered samples are shown in the SEM pictures of Figs. 5(A) and 5(B). The mean grain size is ≈0.4 and ≈0.7 μm in the sintered SR and CP samples, respectively. The remaining porosity in the SR samples is very low when compared to that of the CP samples. The effect of thermal etching (1370°C/10 min) on the SR microsections was very marked, with all grains being rounded as a result of severe grooving.

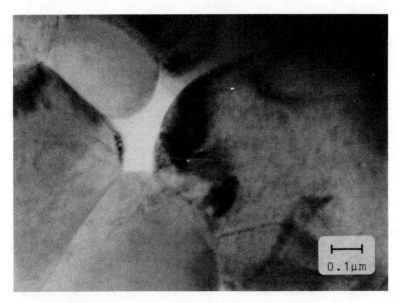

Fig. 6. Transmission electron micrograph (bright field) of SR material. Triple points are filled with glassy phase. Amorphous grain-boundary phase is ≈10 nm thick.

Transmission electron microscopy examination of the SR samples clearly revealed the presence of an amorphous phase at the triple points (Fig. 6). This phase was ≈10 nm thick and was continuous over all grain boundaries. Its composition was high in Al_2O_3 and SiO_2, with a Y_2O_3 content of 30–50 wt%. As a result of beam overlap onto neighboring grains, it was difficult to measure the exact concentrations of the boundaries. Some Al_2O_3 was also present as large, single grains (dark grains in Fig. 5(A)). The ZrO_2 content in the grain boundaries was relatively low, indicating low solubility; thus, the grain-boundary phase may act as a grain-growth inhibitor.

At an early stage in the sintering process, the chemical composition of the liquid phase should be approximately that of the milling-wear debris (85 wt% Al_2O_3 and 15 wt% SiO_2). However, the extremely high Y_2O_3 concentration in the boundary film of the final samples indicates rapid diffusion of Y_2O_3 from the ZrO_2 grains into the grain boundaries. According to the Al_2O_3-Y_2O_3-SiO_2 ternary phase diagram, the equilibrium glassy phase has a high Y_2O_3 content.

Isostatic Hot-Pressing

Only milled SR samples were isostatically hot-pressed. The glassy phase already present allowed full densification (6.07 g/cm^3) after 15 min at 1450°C (200 MPa). The samples were slightly reduced and gray. Analysis by TEM revealed a high dislocation density within the grains (Fig. 7) resulting from plastic deformation during the isostatic hot-pressing process. The grain size (≈0.4 μm) of the isostatically hot-pressed material was comparable to that of the sintered material. However, most triple points were very sharp, i.e., the amount of amorphous phase was drastically reduced, with a thickness of only 1.5 nm between the grain boundaries. It is unclear why and how the glassy phase, which is present in

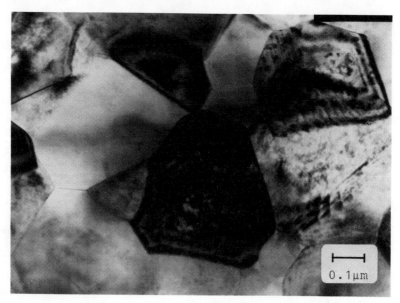

Fig. 7. Transmission electron micrograph (bright field) of isostatically hot-pressed SR material. Triple points are usually sharp, with a glassy, 1.5 nm thick grain-boundary phase. Dislocations are visible within grains.

the sintered samples, disappeared to such an extent in the isostatically hot-pressed material.

Transformability

Surface grinding of the isostatically hot-pressed SR material caused strong peak broadening of the reflections, but no transformation to m symmetry. After a reoxidation process (1100°C, 12 min) a small m content ($\approx 5\%$) appeared on ground surfaces. This is in strong contrast to the sintered SR samples where $\approx 10\%$ of the t material in the near-surface region transformed on grinding. Since the amount of glassy phase was much smaller in the isostatically hot-pressed samples, the Y_2O_3 dissolution from the grains into the grain-boundary phase was considerably lower and, as a result, the grains were more stable.

Surface grinding of sintered CP samples also transformed $\approx 10\%$ of the t-ZrO_2 grains. Due to the uniform Y distribution at a higher overall content, these samples should be more stable than the sintered SR material. However, the larger grain size (0.7 μm) is obviously the origin of a comparable transformability.

Conclusions

(1) Spray-reacted (SR) ZrO_2 powders are composed of hard-shelled, hollow spheres which must be attrition-milled to achieve sinterability.

(2) The glassy grain-boundary phase (≈ 10 nm thick), consisting principally of SiO_2 and Al_2O_3 from the milling-media wear debris, takes from 30 to 50 wt% Y_2O_3 in solution from the grains; this action results in a depletion in Y of the t-ZrO_2 grains toward the grain boundaries and, hence, the material becomes highly susceptible to stress-induced transformation.

(3) An essentially glass-free material obtained from a coprecipitated powder has a homogeneous Y distribution across the grains.

(4) Short-term isostatic hot-pressing of presintered SR material drastically reduces the amount of glassy grain-boundary phase, from ≈ 10 to ≈ 1.5 nm.

Acknowledgment

The authors thank G. Petzow and R. P. Ingel for helpful discussions.

References

[1] T. K. Gupta, J. H. Bechtold, R. C. Kuzucki, L. H. Cadoff, and B. R. Rossing, "Stabilization of Tetragonal Phase in Polycrystalline Zirconia," *J. Mater. Sci.*, **12**, 2421–26 (1977).

[2] F. F. Lange, "Stress-Induced Martensitic Reactions: II. Experiments in the ZrO_2-Y_2O_3 System," ONR Tech. Rept. No. 3, July 1978.

[3] M. Matsui, T. Soma, T. Otagini, and I. Oda, "Effect of Microstructure on the Durability of Y-PSZ"; for abstract see *Am. Ceram. Soc. Bull.*, **60** [3] 382 (1981).

[4] K. Kobayashi and T. Masaki, "High-Strength and High-Toughness Zirconia" (in Jap.), *J. Jpn. Ceram. Soc.*, **17** [6] 427–33 (1982).

[5] M. Rühle, N. Claussen, and A. H. Heuer, "Microstructural Studies of Tetragonal Zirconia Polycrystals (TZP)"; this volume, pp. 352–70.

[6] T. Kosmac, R. Wagner, and N. Claussen, "X-Ray Determination of Transformation Depths in Ceramics Containing Tetragonal ZrO_2," *J. Am. Ceram. Soc.*, **64** [4] C-72–C-73 (1981).

[7] M. Matsui, T. Soma, and I. Oda, "Effect of Microstructure on Strength of Y-TZP Components"; this volume, pp. 371–81.

[8] K. Tsukuma, Y. Kubota, and T. Tsukidate, "The Thermal and Mechanical Properties of Y-PSZ"; this volume, pp. 382–90.

[9] K. Nakajima, K. Kobayashi, and M. Murata, "Phase Stability of Y-PSZ in Aqueous Solutions"; this volume, pp. 399–407.

Preparation of Ca-Stabilized ZrO_2 Micropowders by a Hydrothermal Method

K. Haberko and W. Pyda

Institute of Materials Science
Academy of Mining and Metallurgy
Cracow, Poland

Coprecipitation was used to obtain homogeneous mixtures in the system $CaO-ZrO_2$ for CaO concentration of 13 mol%. The primary phase crystallizing on hydrothermal treatment from the coprecipitate is the cubic (c) solid solution. After prolonged treatment at 260°C the c phase decomposes into two phases, one with c symmetry and the other with monoclinic (m) symmetry. At 220°C, only the c phase has been observed within the applied time periods of the hydrothermal treatment (4 and 8 h). The single-phase powder sinters better than the two-phase one. The hydrothermally processed powder is composed of very soft agglomerates that collapse under moderate pressures, resulting in compacts of extremely uniform microstructures. Such compacts, sintered at temperatures as low as 1300°C, have shown densities >98% of theoretical.

Mechanical properties of agglomerates may be singled out as one of the most decisive physical characteristics of the powders that determine their compacting and, thus, sintering ability. A relation between microstructural properties and the strength of powder agglomerates, derived by Rumpf,[1] indicates that the strength of agglomerates at a given crystallite size can be decreased by increasing agglomerate porosity and/or decreasing the strength of contact points between crystallites. This problem has been studied in the case of ZrO_2 solid-solution powders.

ZrO_2 solid solution can be synthetized by calcination of a coprecipitated gel at temperatures of 500°–1000°C. Under these conditions, strong bonds of the first order (intercrystalline boundaries) are found within agglomerates.[2-5] Nevertheless, mechanical strength of the agglomerates can be maintained at a reasonably low level by subjecting the coprecipitates to the precalcination treatment with a suitable organic solvent[3-8] or by freeze drying.[9] Both operations result in powders whose agglomerates collapse under moderate pressures due to their high porosity. In this way, compacts of uniform pore-size distribution are obtained.

On the other hand, ZrO_2 solid solutions crystallize under hydrothermal conditions at much lower temperatures than indicated above. Temperatures of 200° to 250° and 190°C have been reported in the systems $CaO-ZrO_2$[10] and Y_2O_3,[11] respectively. The data presented in Ref. 11 show a large discrepancy between the surface area equivalent particle size, 5.7 nm, and the particle size observed under the transmission electron microscope, 150 nm, indicating that the powders are composed of porous agglomerates. No direct information is available on the nature of bonds within the agglomerates.

Japanese investigations on pure ZrO_2 submicrometer crystallites[12-15] have shown, however, that under these conditions, "single-domain," strain-free particles crystallize. This suggests that, under hydrothermal conditions, formation of strong intercrystalline bonds is limited or even excluded.

Experimental Procedure

All reagents used in the present work were of analytical grade. Coprecipitation from water solution of zirconyl chloride and calcium chloride of concentrations of 0.502 and 0.075 mol/L, respectively, was used to obtain a homogeneous gel of 13 mol% CaO. Sodium hydroxide was selected as the satisfactory precipitation reagent. Its total amount was kept constant in each run and corresponded to 1.1 times that given by the stoichiometry of $Ca(OH)_2$ and $Zr(OH)_4$. Experiments performed at two NaOH concentration levels, 0.1 and 1.0 mol/L, and at two temperatures, 30° and 95°C, showed that the maximum proportion of Ca in the filtrate is ≤0.5% of the amount introduced into the system.

A 300-cm^3 stainless steel laboratory autoclave electrically heated with an accuracy of ±1°C was used in our experiments. The autoclave was equipped with a Teflon* vessel into which coprecipitated gel separated from its mother liquor by vacuum filtering, but not dried, was inserted.

Powders prepared by the hydrothermal treatment were washed with distilled water to remove chloride and sodium ions. Chemical analysis of the filtrate for Ca showed that its concentration in each run was below the level of detection of atomic absorption spectrophotometry (AAS). Standard X-ray diffractometry was used to identify crystalline phases. Crystallite sizes (D_{hkl}) were obtained from the Scherrer formula, in which the half-width corrected for $\alpha_1\alpha_2$ overlap by the Rachinger method, and for instrumental broadening on the basis of the Cauchy-Gaussian approximation,[16] were applied. Transmission electron microscopy and scanning electron microscopy were used to examine powders and sintered samples. Pore-size distribution in pressed samples was studied by mercury porosimetry. The specific surface area was determined by BET.

Using a generally accepted oxygen vacancy model and lattice parameters of CaO-stabilized ZrO_2 taken from the literature,[17] the X-ray density of the 13 mol% CaO solid solution was calculated as 5.607 g/cm^3. Sintering characteristics were obtained from high-temperature dilatometry and density measurements (by hydrostatic weighing) of samples fired in air.

Results and Discussion

Phase Composition

Coprecipitated gels of 13 mol% CaO were treated hydrothermally at 220° and 260°C for 4 and 8 h. Samples subjected to the lower-temperature treatment contained the c phase only, whereas, in those treated at higher temperatures, both m and c phases were found (Fig. 1). The amount of m phase in samples treated at 260°C for 4 and 8 h was 4 and 18%, respectively, as determined by the method described in Ref. 18.

The c solid solution is the primary phase crystallizing under hydrothermal conditions from the coprecipitated gel. During the prolonged hydrothermal treatment, it decomposes into two solid solutions of different CaO concentrations: a low-CaO m phase and a high-CaO c phase, as was also observed by Stubican and Ray.[19] In the case of such samples heated at 10°C/min and quenched rapidly from

---

*Teflon is the tradename for tetrafluroethene of E. I. du Pont de Nemours & Co., Wilmington, DE.

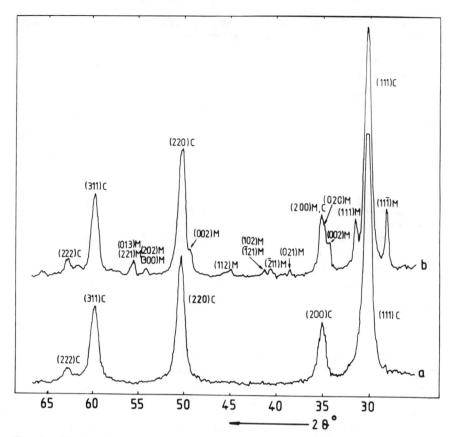

Fig. 1. X-ray diffraction patterns of the 13 mol% CaO-ZrO$_2$ system hydrothermally treated at (a) 220°C for 4 h and (b) at 260°C for 8 h. M, monoclinic solid solution; C, cubic solid solution.

preselected temperatures to room temperature, the m solid solution disappears at a temperature as low as 1000°C. It undoubtedly occurs because of the chemical homogenization of the system.

Figure 2 shows dilatometric curves of the single-phase and two-phase compacts. The two-phase sample shows the smaller shrinkage and rate of shrinkage. This is especially visible in the temperature range from about 970° to about 1160°C. The same samples fired at 1150°C show higher density for the single-phase powder (73.1 ± 0.1%[†]) than the two-phase powder (61.5 ± 0.4%). Further experiments were performed with the single-phase material.

Powder Characteristics

Conditions of the hydrothermal treatment and the resultant powder characteristics are given in Table I. For comparison, some properties of a powder of the

---

[†] ± denotes confidence interval at 0.95 confidence level.

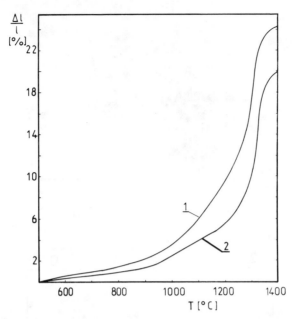

Fig. 2. Linear shrinkage vs temperature. Rate of temperature increase is 10°C/min. Powders hydrothermally treated at (1) 220°C for 4 h, and (2) 260°C for 8 h. Samples compacted under 196 MPa.

Table I. Preparation Conditions and Properties of Ca-SZ Powders

	Hydrothermal treatment	Calcination*
Temperature (°C)	235	600
Time (h)	4	0.5
CaO concentration (mol%)	13	13
Phase composition (% cubic)	100	100
Specific surface area (m²/g)	79.8	101.6
BET particle size, d_{BET} (nm)	13.4	10.5
Crystallite size, D_{111} (nm)	11.4	10.5
D_{311} (nm)	12.8	10.9
Intraagglomerate porosity (%)	56	59

*The coprecipitate before calcination was treated with C_2H_5OH; this operation doubles the intraagglomerate pore volume.

same chemical and phase composition, but obtained by calcination of the coprecipitated gel, are also shown.

Transmission electron micrographs of five different fields were taken for studying the crystallite-size distribution of the powder. One of these micrographs is shown in Fig. 3. The crystallite-size distribution is demonstrated in Fig. 4 in terms of the Rosin-Rammler equation[20]:

$$\ln\left[\ln 1/W(d)\right] = \ln b + n \ln d \tag{1}$$

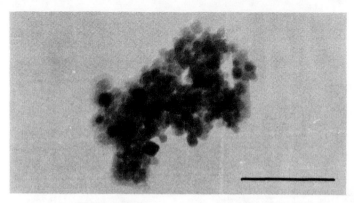

Fig. 3. Transmission electron micrograph of 13 mol% CaO-ZrO$_2$ powder. Preparation conditions indicated in Table I. Black bar denotes 100 nm.

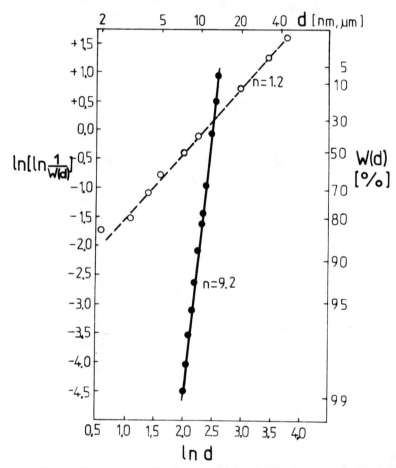

Fig. 4. Crystallite-size distribution of hydrothermally prepared powder (solid line) and the ball-milled (glass) powder (dashed line). Grain size of the ball-milled powder is expressed in μm and that of the hydrothermally crystallized ZrO$_2$ powder in nm.

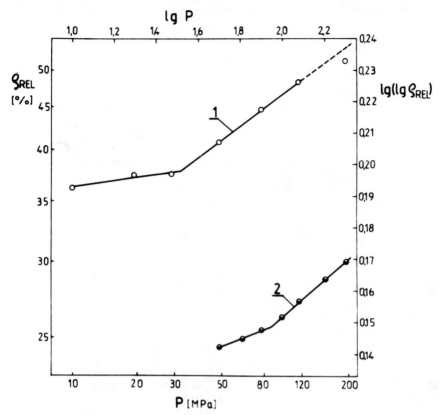

Fig. 5. Relative density, ρ_{rel}, vs compaction pressure, P. Each point is a mean of 4–6 measurements. Powder 1 obtained under hydrothermal conditions and powder 2 by calcination of a coprecipitated gel. Preparation conditions indicated in Table I.

where $W(d)$ is weight fraction of particles of sizes equal to or larger than d, and b and n are constants. For comparison, the grain-size distribution of a ball-milled (glass) powder is also shown in Fig. 4.

All the methods of particle-size determination applied in the present work yield approximately the same value of the particle size (see Table I). Since the BET particle size, d_{BET}, is calculated with the assumption of spherical or cubic shape, the coincidence between the crystallite size, obtained from the X-ray line broadening and the BET particle size, indicates that the powder is composed of isometric crystallites. This conclusion is corroborated by the electron micrograph (Fig. 3), which also offers direct evidence of the agglomerated microstructure of the powder. Simultaneously, on the basis of agreement between D_{hkl} and d_{BET}, the intercrystalline contact area is negligible in comparison to the total surface area.

The crystallite-size distribution of the powder is narrow. This is apparent when comparison is made to the mechanically ground powder, where a much smaller value of n in the Rosin-Rammler equation is found (see Fig. 4).

Effect of Pressure

Samples, 7 by 7 by 2 mm, were uniaxially pressed with 6 wt% water added as a lubricant. Figure 5 shows the compaction diagram of the hydrothermally

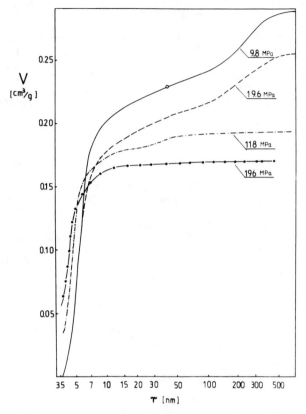

Fig. 6. Cumulative pore-size distribution curves determined by mercury porosimetry in compacts of hydrothermally prepared powder. Compaction pressures indicated.

crystallized and the calcined powder, plotted in the coordinate system $\log(\log \rho_{rel})$ vs $\log P$,[21] where ρ_{rel} is the relative density and P the applied pressure. Figure 6 shows the cumulative pore-size distribution curves in compacts of the hydrothermally prepared powder. It has been assumed that the point of inflection, marked in the plot corresponding to the lowest applied pressure (9.8 MPa), divides the total porosity into intraagglomerate, i.e., small, and interagglomerate, i.e., large, pores. On this basis, the intraagglomerate pore volume could be determined (see Table I).

As is usual in the case of agglomerated powders, two straight lines are observed in the compaction diagram (Fig. 5). It is now well recognized that the point of their intersection (the breaking point) corresponds to the pressure under which a collapse of agglomerates occurs (cf. Refs. 3 and 22). In the hydrothermally crystallized powder, this interpretation is substantiated by the fact that the pore-size distribution within agglomerates is not changed if pressures lower than those of the breaking point are applied (see Fig. 6, pressures 9.8 and 19.6 MPa). Higher pressures shift sizes of the intraagglomerate pores toward smaller radii. Such phenomena have been observed in calcined powders and ascribed to deformation of the agglomerates.[5]

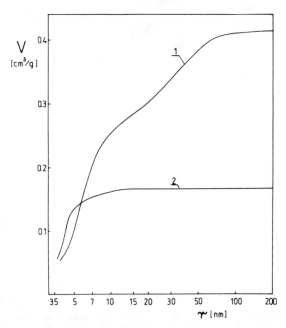

Fig. 7. Cumulative pore-size distribution curves in compacts of hydrothermally prepared (plot 2) and in the calcined powder (plot 1). Compaction pressure 196 MPa.

The breaking-point pressures of the hydrothermally processed and the calcined powders are 31.8 and 86.4 MPa, respectively. Since both powders show practically the same intraagglomerate porosity and nearly the same crystallite size (Table I), the different values of the breaking-point pressures should be related to the weaker intercrystalline bonds in the hydrothermally processed powder. Hence, deformation of these agglomerates seems to occur much more easily. Under high enough pressures, the size of the interagglomerate pores becomes identical with that of the intraagglomerate pores. This is not observed in the calcined powders (see Fig. 7). The described phenomenon may influence the compaction characteristics of the hydrothermally processed powder and plausibly explains its lower densification under the higher applied pressure (196 MPa), $51.3 \pm 0.6\%$, than predicted by the compaction diagram in Fig. 5, $53.2 \pm 0.6\%$.

Extensive deformability of the agglomerates of the hydrothermally prepared powder is probably responsible for the generally higher densities obtained with the compacts of this powder (Fig. 5).

Sintering

Compacts of the hydrothermally processed powder were sintered at 1300°C for 2 h, using a heating rate of 5°C/min. As is shown in Fig. 8, densification of the fired body is greatly affected by the compaction pressure. This is also visible in Fig. 9, which shows microstructures of the bodies compacted under 9.8 and 196 MPa. Low compaction pressures result in typically agglomerated microstructures. High compaction pressure leads to the body of $98.3 \pm 0.6\%$ density. In the case of powders prepared by classical mechanical grinding, such high densities can be obtained only at firing temperatures higher by at least 550°C, with the resultant

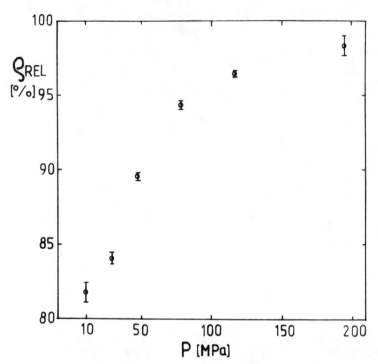

Fig. 8. Effect of compaction pressure on density of the body fired at 1300°C for 2 h. Confidence intervals (0.95) indicated.

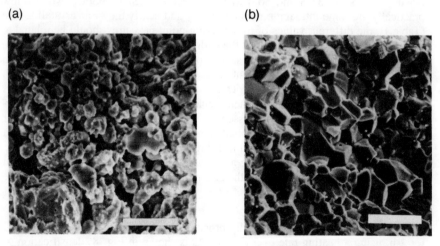

Fig. 9. Microstructures of samples sintered at 1300°C for 2 h. Compaction pressures (a) 9.8 MPa, (b) 196 MPa. White bars denote 5 μm.

grain size about 10 times greater. Comparison should be also made to the calcined micropowders whose properties are listed in Table I. They need firing temperatures higher by about 100°C to obtain densities approaching 98%.[4]

Summary

Hydrothermal treatment of the coprecipitated gel causes crystallization of the CaO-stabilized ZrO_2 in the form of a very fine powder of narrow crystallite-size distribution. The powder is composed of very soft agglomerates that easily deform under moderate pressures, resulting in compacts of very uniform microstructure. Such compacts sinter to high densities at low temperatures.

References

[1] H. Rumpf, "Principles and Methods of Granulation," *Chem. Ing. Technol.*, **30** [3] 144–58 (1958).
[2] K. Haberko, A. Cieśla, and A. Proń, "Sintering Behaviour of Yttria-Stabilized Zirconia Powders Prepared from Gels," *Ceramurgia Int.*, **1** [3] 111–16 (1975).
[3] K. Haberko, "Characteristics and Sintering Behaviour of Zirconia Ultrafine Powders," *Ceramurgia Int.*, **5** [4] 148–54 (1979).
[4] R. Pampuch and K. Haberko, "Characteristics and Fusibility of ZrO_2 Micro Powders," *Keram. Z.*, **31** [8] 478–80 (1979).
[5] R. Pampuch and K. Haberko, "Agglomeration in Ceramics"; pp. 623–34 in Ceramic Powders. Edited by P. Vincenzini. Elsevier, Amsterdam, 1983.
[6] M. Hoch and K. M. Nair, "Densification Characteristics of Ultrafine Powders," *Ceramurgia Int.*, **2** [2] 88–97 (1976).
[7] K. Haberko and R. Pampuch, "Influence of Yttria Content on Phase Composition and Mechanical Properties of Y-PSZ," *Ceramurgia Int.*, **9** [1] 8–12 (1983).
[8] K. S. Mazdiyasni, "Powder Synthesis from Metal-Organic Precursors," *Ceramurgia Int.*, **8** [2] 42–56 (1983).
[9] A. Roosen and H. Hausner; pp. 773–82 in Ceramic Powders. Edited by P. Vincenzini. Elsevier, Amsterdam, 1983.
[10] W. G. Tchukhlancev and J. M. Galkin, "On Preparation of Cubic Zirconia under Hydrothermal Conditions," *Zh. Neorg. Khim.*, **14** [2] 311–13 (1969).
[11] A. R. Burkin, H. Saricimen, and C. H. Steel, "Preparation of Yttria Stabilized Zirconia Powders by High Temperature Hydrolysis," *Trans. J. Br. Ceram. Soc.*, **79**, 105–108 (1980).
[12] T. Mitsuhashi, M. Ichihara, and U. Tatsuke, "Characterization and Stabilization of Metastable Tetragonal Zirconia," *J. Am. Ceram. Soc.*, **57** [2] 97–101 (1974).
[13] H. Nishizawa, N. Yamasaki, K. Matsuoka, and H. Mitsushio, "Crystallization and Transformation of Zirconia Under Hydrothermal Conditions," *J. Am. Ceram. Soc.*, **65** [7] 343–46 (1982).
[14] E. Tani, M. Yoshimura, and S. Somiya, "Hydrothermal Preparation of Ultrafine Monoclinic ZrO_2 Powders," *J. Am. Ceram. Soc.*, **64** [12] C-181 (1981).
[15] E. Tani, M. Yoshimura, and S. Somiya, "Formation of Ultrafine Tetragonal ZrO_2 Powder under Hydrothermal Conditions," *J. Am. Ceram. Soc.*, **66** [1] 11–14 (1983).
[16] H. P. Klug and L. E. Alexander; X-Ray Diffraction Procedures, Ch. 9. Wiley & Sons, New York, 1954.
[17] T. H. Etsel and S. N. Flangas, "The Electrical Properties of Oxide Electrolytes," *Chem. Rev.*, **70** [3] 339–76 (1970).
[18] R. C. Garvie and P. S. Nicholson, "Phase Analysis in Zirconia Systems," *J. Am. Ceram. Soc.*, **55** [6] 303–305 (1972).
[19] V. S. Stubican and S. P. Ray, "Phase Equilibria and Ordering in the System $CaO-ZrO_2$," *J. Am. Ceram. Soc.*, **60** [11–12] 534–37 (1977).
[20] T. Allen; Particle Size Measurements. Chapman and Hall Ltd., London, 1968.
[21] S. Gasiorek; "Shrinkage as a Result of the Density and Mass Changes During Sintering Process"; pp. 311–19 in *Science of Ceramics*, Vol. 10. Edited by H. Hausner. Deutsche Keramische Gesellschaft, Berlin, 1980.
[22] G. L. Messing, C. J. Markhoff, and L. G. McCoy, "Characterization of Ceramic Powder Compaction," *Am. Ceram. Soc. Bull.*, **61** [8] 857–60 (1982).

Growth and Coarsening of Pure and Doped ZrO_2: A Study of Microstructure by Small-Angle Neutron Scattering

A. F. Wright and S. Nunn

Institut Laue-Langeuin
38042 Grenoble Cedex, France

N. H. Brett

University of Sheffield
Department of Ceramics, Glasses and Polymers
Sheffield, United Kingdom

When precipitated zirconium hydroxide is heated above 300°C, an amorphous powder is formed which slowly crystallizes to tetragonal ZrO_2. A high density of very small crystallites forms, which can be detected by small-angle scattering when they exceed ≈ 2 nm in diameter. We have studied the growth and coarsening of these crystallites in pure and MgO-doped ZrO_2 from 300° to 800°C. The small-angle neutron scattering (SANS) pattern is dominated by a strong interparticle interference peak which sharpens as the coarsening proceeds. The data can be simulated by a hard-sphere progressive potential model in which the particle size and spatial distribution are refined independently. The kinetics of precipitation growth and scattering intensity confirm a pure Ostwald ripening process with a $t^{1/4}$ growth law corresponding to grain-boundary diffusion. The activation energy for coarsening calculated from the temperature dependence of the reaction is calculated to be 59 kJ·mol^{-1} for pure ZrO_2 and 114 kJ·mol^{-1} for 3% MgO-doped ZrO_2 over this temperature range, where the particle diameters range from 4 to 20 nm.

The sol-gel route to fine ceramics and inorganic glasses has been extensively developed in recent years as an alternative to melting in the preparation of highly refractory homogeneous or metastable materials.[1] When heated to moderate temperatures, the gel transforms to an amorphous anhydride which can then crystallize or, for some materials, compact to form a glass. The transformation of $Zr(OH)_4$ gel through the dehydration stage to the crystalline phase has been studied by diffraction methods.[2] The primary phase produced is metastable t-ZrO_2 which then transforms to the monoclinic (m) phase near 900°C. We report here studies by small-angle neutron scattering (SANS) on the crystallization of the tetragonal phase from the dehydrated gel, with special reference to the kinetics of the evolution of grain size and the modification of the kinetics by MgO.

Sample Preparation

Four samples of ZrO_2 gel were investigated; samples A and B were prepared by a precipitation route employing NaOH as the precipitating agent and contained, respectively, 0 and 3.26 mol% MgO (referred to as the calcined gel composition);

two further samples, C and D, were prepared by a similar route, but employing NH$_4$OH as the precipitating agent, and contained 0 and 16.4 mol% MgO, respectively. From each of these preparations, samples were calcined in air at 200°, 300°, 400°, and 500°C for 2 h (and in some cases 17 h) and pressed into disks approximately 10 mm in diameter and 2-mm thick. The disks were stored under desiccant until exposure to the beam. Disks were also pressed from the uncalcined gel samples for kinetic measurements, which were stored under an atmosphere of D$_2$O prior to SANS analysis.

Data Collection

The SANS curves were obtained on the instrument D17 at the Institut Laue Langevin, Grenoble,[3] in two measurement sessions. The first data sets were obtained from samples of compacted gel which had been heat-treated at temperatures up to 500°C for periods of either 2 or 17 h. An estimate of the background was obtained from the samples heated to 200°C to eliminate the combined water, a strong source of incoherent neutron scattering. Typical curves for a series of such samples have the form shown in Fig. 1 where a maximum in the scattering curve grows in intensity and is displaced toward a smaller scattering vector Q with increasingly severe heat treatment, either in temperature or in time. Q is defined as $2\pi \sin(2\theta/\lambda)$ where 2θ is the scattering angle. In a second series of measurements, the heat treatments were carried out in a small furnace mounted on the instrument. This setup gave a better estimate of the background scattering before

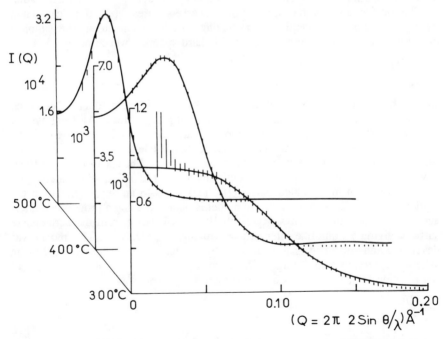

Fig. 1. Neutron small-angle scattering curves from pure ZrO$_2$ gel heat-treated for 2 h at 300°, 400°, and 500°C. The solid curves are simulations based on the extended hard-sphere model.

crystallization and enabled the kinetics of crystallization to be studied with 30-min time resolution as well as allowing temperature changes on the same sample. Kinetic measurements were made on samples A and B only due to limitations of instrument time.

Data Analysis

Precipitation processes in solid-state systems often give rise to small-angle scattering curves showing a single maximum, which is an indication of short-range order due to interactions among neighboring particles. These interactions may be related to concentration gradients around precipitates as in the case of crystallization of certain glasses,[4,5] or they may be of steric origin as proposed in the case of silica gel[6] and in the present study. Whatever the origin of the interaction, the scattering curve $I(Q)$ is made up of two functions $P(Q)$ and $S(Q)$. $P(Q)$ is the particle-form factor, directly related to the particle dimensions, and $S(Q)$ is the interparticle interference function related to the spatial arrangement of the particles. In the case of monodisperse spherical particles, we can write $I(Q) = kP(Q) \cdot S(Q)$, but it is not possible to directly separate these two components in order to obtain accurate particle-size data. Within the limits of the Guinier approximation ($QR_g < 1.2$ and $S(Q) = 1$), we can write $I(Q) = I(0) \exp(-Q^2 R_g^2/3)$ from which a particle dimension can be obtained. For spherical particles of radius R_s, $R_g = (3/5)^{1/2} R_s$, but where a strong maximum is observed, the additional modulation in $S(Q)$ leads to overestimation of R_s by up to 10%.

It is possible to simulate the small-angle scattering data by calculating $S(Q)$ for a given spatial arrangement of the particles expressed as $g(r)$, the particle radial distribution function. In the case of monodisperse spheres, $g(r)$ can be calculated from pair interaction potentials $U(x)$ having variable parameters. The simplest model is that of random noninterpenetrable hard spheres, whose potential is

$$U(x) = \infty \quad 0 < x < 1$$
$$U(x) = 0 \quad x > 1 \tag{1}$$

where x has dimensions of the particle radius R_s. This simple model has been used to explain, qualitatively, nucleation and growth of precipitates in Al-Zn alloys.[7] It fits the present data quite well when modified to an extended hard-sphere model having an exclusion zone of radius x' around each particle such that $U(x)$ is infinite for $x < x'$ and zero for $x > x'$; x' exceeds x by 20–50%. The discontinuity in $U(x)$, however, creates an unphysical discontinuity in $g(r)$ at $r/2R_s = x'$, giving too much modulation in the calculated $S(Q)$.[8] This leads to an underestimate in the particle dimensions. A more realistic model is obtained using a progressive interaction potential which decays exponentially with distance from the hard-sphere surface, giving a liquidlike interparticle structure at $r/2R_s > 1$. This model was developed and solved in analytical form by Hayter and Penfold using the mean spherical approximation[9,10] for application to charged micelle solutions. The potential function for $U(x)$ in units of $k_B T$ is then

$$U(x) = \infty \quad \text{for } x < 1$$
$$U(x) = \gamma e^{-kx}/x \quad \text{for } x > 1 \tag{2}$$

where x has dimensions of the particle radius R_s, γe^{-k} is the contact potential, and k is the inverse screening length R_s/λ_0.

At high values of γe^{-k} and k, it reduces to the simple hard-sphere model. The model has been applied successfully to scattering curves of crystallization and

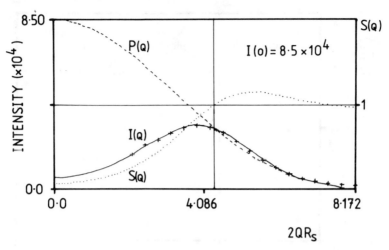

Fig. 2. Example of the progressive potential model data analysis for sample A heat-treated at 500°C for 2 h. $I(Q)$ is decomposed into a particle function $P(Q)$ and an interparticle interference function $S(Q)$. Model parameters: $R_s = 12.0$ nm, $\gamma_e^{-k} = 2.27$ $(k_B T)$, and $k = 1.5$. Volume fraction is 0.15.

phase separation in glasses.[11] The interaction parameters in these cases are related to zones of depleted solute which develop around each growing particle, thus reducing the probability of a neighboring particle existing within a given distance.

An example of the progressive potential simulation of the ZrO_2 data is shown in Fig. 2 for gel A heated to 500°C for 2 h. Also shown are the components of the calculated $I(Q)$, the particle-form factor $P(Q)$ for spheres of 12-nm diameter (refined), and the interparticle interference function $S(Q)$. The estimated error in the particle diameter is 5%.

The reduced radial distribution function $g(r/2R_s)$, where $2R_s$ is the particle diameter (Fig. 3), deviates from that for random packing of hard spheres in two ways. The probability varies continuously from zero at $r/2R_s$ equal to or slightly less than 1, rising to a broad weak maximum at $r/2R_s \approx 1.5$, whereas for hard spheres $g(r/2R_s)$ rises discontinuously to a sharp peak at $r/2R_s = 1.00$, equivalent to maximum coordination at contact. This deviation is most likely due to irregularly shaped particles with well-developed crystal faces or to anisotropy. The single-contact distance is then broadened, to range from slightly below the mean diameter (faces in contact) to considerably above this value (edges and corners in contact). It is also likely that certain closely spaced particles are not in direct contact but are rigidly bonded to a third one.

Results

The static measurement results are given in Table I. Note that the particle sizes from Guinier plots are ≈10% higher than the progressive potential simulation but that the hard-sphere model gives even more erroneous results, about 20% too low.

There is no significant difference between samples A and C, which differ only in the agent used for precipitation. The gels containing MgO are, however, some-

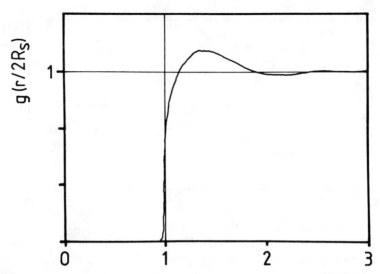

Fig. 3. The radial distribution function for crystallized particles obtained in the simulation shown in Fig. 2. Note the clustering effect when the diameter is about 1.5 $[g(r/2R_s) > 1.0]$.

Table I. Particle Diameters (nm) for Heat-Treated ZrO_2 According to Three Models

T (°C)	Period (h)	Guinier	Extended hard-sphere model		Progressive potential model
			Particle	Exclusion zone	
Sample type A					
300	2	4.88	4.41	5.5	
300	17	6.95	5.87	7.6	6.47
400	2	8.83	7.06	9.2	7.62
400	17	11.41	8.89	12.1	9.37
500	2	13.20	10.28	15.4	12.00
500	16	17.10	12.67	19.5	15.06
Sample type B					
300	2	4.18			
400	2	7.18			
500	2	12.19			11.22
Sample type C					
300	2	3.85			
400	2	8.41			7.03
500	2	13.74			12.47
Sample type D					
300	2	3.36			
400	2	4.88			4.46
500	2	8.11			7.14

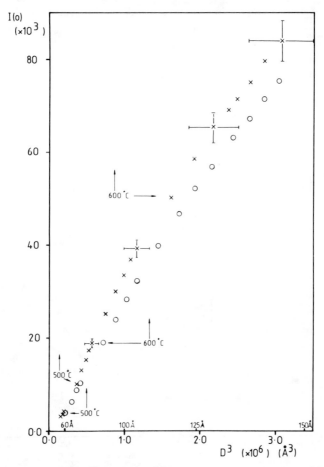

Fig. 4. Correlation between $I(0)$, from the extrapolated value of $P(Q \to 0)$, and the cube of the particle diameter. $I(0) = kD^3$ is consistent with a pure Ostwald ripening mechanism at constant volume fraction: X, sample A, O, sample B.

what stabilized against creation and growth of t-ZrO$_2$. The growth is delayed by more than 100°C in the case of sample D.

In presenting the results for the kinetic measurements, we use the relative intensities of the spectra, as well as the particle size, to obtain information on the relative particle number density during growth. The most appropriate intensity parameter is $I(0)$ obtained from $P(Q)$ extrapolated to $Q = 0$, scaled to the measured intensity [$I(0) = P(0)$ where $S(0) = 1$]. It is proportional to the square of the particle volume V_p and the number of particles N_p.

$$I(0) = KN_p V_p^2 \rho^2 \qquad (3)$$

where ρ is the contrast term for neutrons, which in the present case is invariant. We can therefore make an internal check of the nature of the variation of N_p from the variation of $I(0)$ as a function of particle size. We find experimentally that, for

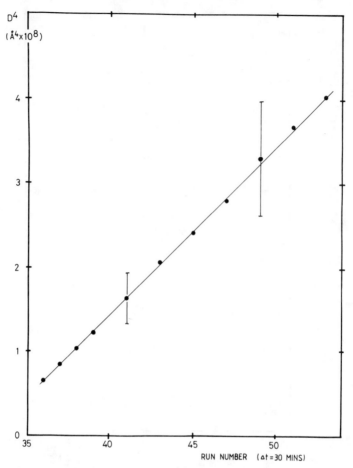

Fig. 5. Kinetics of particle growth for sample B at 600°C. $D^4 = kt$ corresponds to grain-boundary-controlled diffusion.

samples A and B, $I(0)$ varies almost linearly with the cube of the particle size over a large range (Fig. 4). This result is consistent with a pure Ostwald ripening process where the particle volume fraction $N_p V_p / V_s$ remains constant as the particle size increases. We can then write

$$I(0) = K' V_p \rho^2 \qquad (4)$$

where $K' = K N_p V_p$ and $V_p = (4/3)\pi r^3$ is determined by simulation from the shape of $I(Q)$ independent of its intensity. The difference in slopes of the two samples is not understood.

According to the L.S.W. theory of Ostwald ripening,[12] we would expect particle growth to occur by diffusion across points of contact of the crystallites, in which case the kinetic growth law would be similar to that for grain-boundary diffusion, $D^4 = kt$ for large values of t. Figure 5 shows the ripening of sample B at 600°C, consistent with this growth law. Similar data were obtained at four temperatures (sample A, 300°–600°C; B, 500°–800°C), for each of the samples A

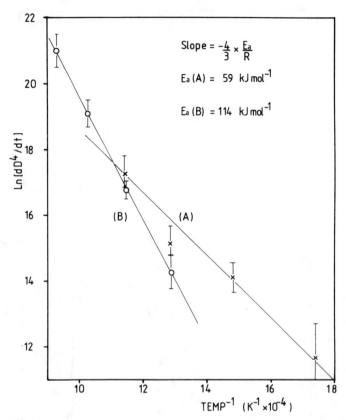

Fig. 6. Arrhenius plots used for the calculation of activation energy reaction constant data are taken from plots similar to those of Fig. 5. The slope is equal to $-(4/3)E_a/R$.

and B. We observe a sharp temperature dependence of the slope of the ripening processes which we plot in Fig. 6 in Arrhenius coordinates. The results are sufficiently linear to extract an activation energy for ripening of t-ZrO$_2$ in both cases. For sample A, we obtain 59 kJ·mol^{-1}, and for sample B it is 114 kJ·mol^{-1}. This is considerably less than the published value of 338 kJ·mol^{-1} for the activation energy of precipitation growth at 1700°C on the basis of strength data,[13] but this process may not be relevant to that described here.

Discussion

Although the Ostwald ripening process creates a broad particle-size distribution, the analysis methods used in this study are limited to the determination of a unique particle size. We might therefore be surprised that the simulations provide such good fits to the data. The answer we believe lies in the shape of the size distribution, coupled with the sensitivity of scattering processes to particle size at low Q. Jain and Hughes[12] calculated the size distribution for the grain-boundary diffusion process; a principal feature is a sharp cutoff at sizes just above the critical size R_s. At low Q, however, the intensity of scattering is dominated by the larger particles so that the sensitivity at $Q = 0$ is the V_p^2 product of the number distribu-

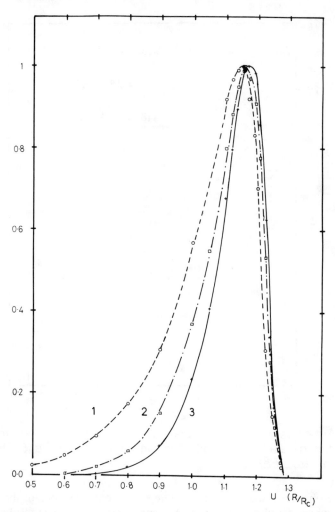

Fig. 7. Distribution functions for grain-boundary-controlled Ostwald ripening in terms of the critical radius R_c: (1) Particle number distribution, $N_p(U)$, $U = R_s/R_c$; (2) particle volume distribution, $U^3 N_p(U)$; (3) sensitivity of the particle distribution to scattering at $Q = 0$, $U^6 N_p(U)$.

tion. We show in Fig. 7 the number, volume, and intensity ($Q = 0$) distributions expanded from Ref. 12 and normalized to unity at the maximum. The neutron intensity function $N_p V_p^2(R_s)$ is much narrower than the number distribution and is displaced toward the larger particles. The narrowness of $N_p V_p^2(R_s)$ and hence the satisfactory simulations are therefore principally due to the sharp cutoff just above R_c. The method is probably a fairly sensitive indicator of R_c, whereas the smaller particles remain undetected. Any process which substantially modifies the shape of the number distribution will evidently distort the size information obtained. Such distortions could occur with changing temperatures or time scales and be the cause of deviations from ideality in the reported data. The discrepancy between samples A and B in Fig. 4 may be of this origin due to the presence of MgO.

The conclusion that the present transformation is a pure ripening process within our range of measurement is an indication of the instability of amorphous ZrO_2 once the stabilizing water is removed. The initial process is the complete transformation to metastable ZrO_2 on a scale of tens of angstroms. This high-density (homogeneous?) nucleation is followed directly by ripening with the exclusion of a normal growth process. The ripening is driven by the excess surface free-energy contribution of the smaller particles to the total free energy. Given the small particle sizes involved, this excess surface free energy must be considerable but will decrease as the size increases. We should not therefore be unduly surprised that the activation energies for ripening obtained in this work where $r_c < 10$ nm (100 Å) are considerably lower than those published elsewhere[13] for larger particle sizes. In addition, the present material is a low-density, friable, almost powdery material, for which the mass-transport processes could be quite different from those of a compact ceramic.

Conclusions

Small-angle neutron scattering is one of very few methods capable of following the kinetics of ripening processes in bulk ceramics. The present work demonstrates the validity of the method for ZrO_2 precipitation in dried amorphous ZrO_2 gel over the range 300°–700°C. The scattering curves can be analyzed using simulation methods which give insight into the microstructure and the particle size. ZrO_2 gel undergoes a pure ripening process above 300°C and follows a $t^{1/4}$ law, consistent with an interfacial diffusion process. ZrO_2 doped with 3% MgO shows substantially reduced kinetics. The observed activation energy for pure ZrO_2 in the 10-nm size range is 59 kJ·mol^{-1}, whereas for 3% doped MgO we observe 114 kJ·mol^{-1}. These figures would be expected to increase as the ripening proceeds.

References

[1]"Glasses and Glass Ceramics from Gels," *J. Non-Cryst. Solids*, **48** [1] March 1982.
[2]I. F. Guilliatt and N. H. Brett, *J. Mater. Sci.*, **9**, 2067–74 (1974).
[3]Neutron beam facilities available for users, Institut Laue-Langévin, Grenoble, France, Jan. 1981.
[4]A. F. Wright, J. Talbot, and B. E. Fender, *Nature (London)*, **277**, 366–68 (1979).
[5]J. Zarzycki, *J. Appl. Crystallogr.*, **7**, 200–207 (1974).
[6]J. B. Hayter and A. F. Wright; unpublished work.
[7]G. Laslaz, G. Kostorz, M. Roth, P. Guyot, and R. J. Stewart, *Phys. Status Solidi*, **41**, 577–83 (1977).
[8]A. F. Wright, "Neutron Scattering 1981"; pp. 359–67 in AIP Conference Proceedings. Edited by J. Faber. American Institute of Physics, New York, 1982.
[9]J. B. Hayter and J. Penfold, *Mol. Phys.*, **42**,109–18 (1981); *J. Chem. Soc., Faraday Trans.*, **77**, 1851–63 (1981).
[10]The following programs are available from I.L.L., J. B. Hayter and J. R. Hansen, "The Structure factor of charged colloidal dispersions at any density," ILL Report 82 HA 14T, Institut Laue-Langevin, Grenoble, France, 1982. J. B. Hayter, "Fast Bi-directional transforms between $f(r)$ and $S(Q)$," ILL Report 79 HA 48S, Institut Laue-Langevin, Grenoble, France, 1979.
[11]A. F. Wright, P. W. McMillan, and N. H. Brett; pp. 569–81 in The Structure of Non-Crystalline Materials. Edited by P. H. Gaskell. Taylor and Francis, London, 1982.
[12]S. C. Jain and A. E. Hughes, *J. Mater. Sci.*, **13**, 1611–31 (1978).
[13]R. T. Pascoe, R. H. J. Hannink, and R. C. Garvie; pp. 447–54 in Science of Ceramics, Vol. 9. Edited by K. J. de Vries. The Nederlandse Keramische Vereniging, Amsterdam, 1977.

Al_2O_3-ZrO_2 Ceramics Prepared from CVD Powders

SABURO HORI,* MASAHIRO YOSHIMURA, AND SHIGEYUKI SŌMIYA

Tokyo Institute of Technology
Laboratory for Hydrothermal Syntheses
Research Laboratory of Engineering Materials
Midori-Ku, Yokohama 227, Japan

RYOICHI TAKAHASHI

Kureha Chemical Industry Co., Ltd.
Chuo-Ku, Tokyo 103, Japan

The densification and tetragonal (t)-ZrO_2 content of sintered Al_2O_3-ZrO_2 ceramics were studied by starting from CVD ZrO_2 powder mixed with various Al_2O_3 powders, including CVD Al_2O_3, or from CV codeposited Al_2O_3-ZrO_2 powder. Strong dependence of t-ZrO_2 content on sintering temperature and on weight percent ZrO_2 in Al_2O_3-ZrO_2 ceramics suggested the importance of the ZrO_2 dispersion and grain-growth control for achieving higher t-ZrO_2 content. Comparisons of t-ZrO_2 contents measured on sintered and fractured surfaces indicated that the transformation-zone thicknesses on fracture were very large (4.5–7.8 μm) with specimens prepared from CVD Al_2O_3 + CVD ZrO_2 or from CV codeposited Al_2O_3-ZrO_2 probably due to extensive microcrack formation. The uniform dispersion of ZrO_2 particles in CV codeposited Al_2O_3-ZrO_2 (16 wt%) powder resulted in excellent ceramic microstructure, with small Al_2O_3 ($\approx$1 μm) and ZrO_2 (<0.6 μm) grains and high t-ZrO_2 content ($\approx$90%) only when the powder was deagglomerated by a surfactant and improved in sinterability.

It was proved first with partially stabilized ZrO_2[1] and later with other ZrO_2-containing ceramics[2] that small tetragonal (t)-ZrO_2 particles constrained in ceramic matrices increase the fracture toughness by transforming to monoclinic (m) symmetry along propagating cracks. As the dispersed ZrO_2 particles must be smaller than a critical diameter, typically 0.5–1 μm, to retain tetragonal symmetry,[3] it is important, and sometimes requires special techniques, to achieve the necessary dispersion of ZrO_2 particles into ceramic matrices. Such necessary dispersions are usually attained either by starting from very fine powders[4] or by starting from mixtures of chemicals, as in the sol–gel technique.[5]

Al_2O_3-ZrO_2 ceramics, with ZrO_2 appearing in tetragonal symmetry as a secondary dispersed phase, is one of the ZrO_2-toughened ceramics and has attracted much technological interest in recent years. The microstructure of Al_2O_3-ZrO_2 ceramics is very much dependent on the fabrication route.[6] By starting from submicrometer powders of Al_2O_3 and ZrO_2 and mixing them thoroughly,[4] the Al_2O_3-ZrO_2 ceramic was obtained, consisting of a microstructure with small Al_2O_3

---

*On leave from Kureha Chemical Industry Co., Ltd.

grains and smaller intergranular ZrO_2 particles. On the other hand, the Al_2O_3-ZrO_2 ceramic obtained by the sol-gel technique[5] consists of a microstructure with relatively large Al_2O_3 grains and most ZrO_2 as intragranular particles. Densifications of these Al_2O_3-ZrO_2 ceramics were achieved mostly by hot-pressing. However, a few reports have been made on the pressureless sintering of Al_2O_3-ZrO_2 ceramics.[2b,7]

Ceramic powders prepared by the chemical vapor deposition (CVD) technique have not been widely used as starting powders for ceramic materials despite their ultrafine particle sizes, because of difficulties such as agglomeration or low-compact densities. If these existing difficulties are overcome by better processing techniques, the CVD powders will show good sinterability, resulting in better microstructures.

The aims of this study were to investigate the effects and possible advantages of starting from CVD powders in fabricating Al_2O_3-ZrO_2 ceramics and to study the densification and the tetragonal-ZrO_2 contents of sintered Al_2O_3-ZrO_2 ceramics.

Experimental Procedures

Starting Powders

Ultrafine ZrO_2 powder (hereafter called CVD ZrO_2) was prepared in the chemical vapor deposition reaction by feeding $ZrCl_4$ vapor into an H_2/O_2 flame. Its average particle size ranged from 40 to 50 nm, and the crystalline phase was a mixture of monoclinic and tetragonal. The CVD Al_2O_3 powder was commercially available[†] (hereafter called Al_2O_3-I); its average particle size was 20 nm and the crystalline phase appeared to be γ-Al_2O_3. For comparison, two other commercial Al_2O_3 powders were obtained. Both were α-Al_2O_3, with almost the same average particle size (0.7–0.8 μm) but different size distributions (hereafter the Al_2O_3 powder with a broader distribution called Al_2O_3-II[‡] and the other with a narrower distribution called Al_2O_3-III[§]).

The Al_2O_3-ZrO_2 codeposited powder was obtained in the CVD reaction by feeding a vapor mixture of $AlCl_3$ and $ZrCl_4$ into an H_2/O_2 flame (hereafter called CV codeposited Al_2O_3-ZrO_2). Its average particle size ranged from 40 to 50 nm. The crystalline phase of Al_2O_3 appeared as δ phase, and the ZrO_2 phase was 100% tetragonal. According to observations by transmission electron microscopy (TEM) and the broadness of XRD peaks, it was assumed that the ZrO_2 particles in CV codeposited Al_2O_3-ZrO_2 powder were approximately 8 nm in average size and contained within larger Al_2O_3 particles.[8]

Processing and Measurements

The CVD ZrO_2 powder was washed several times in water to remove Cl^- before it was mixed with commercial Al_2O_3 powder (Al_2O_3-I, II, or III) by ball-milling in ethanol. The mixture was then dried and isostatically pressed at 196 MPa into small pellets.

The CV codeposited Al_2O_3-ZrO_2 powder was also washed in water, dried, ball-milled in ethanol to break agglomerates, dried again, and isostatically pressed into pellets. These pellets were sintered in air at 1000°–1600°C for 1 h. The densities of the sintered pellets were determined by the Archimedes method or by weight/dimension measurements. In the density calculation, the theoretical densi-

---

[†]Nippon Aerosil Co. (licensed from Degussa).
[‡]SA-1, Iwatani Chemical Industry Co., Ltd. Tokyo, Japan.
[§]Sholite M-08, Showa Denko KK, Tokyo, Japan.

ties of Al_2O_3 m-ZrO_2, and t-ZrO_2 were assumed as 3.987, 5.840, and 6.097 g/cm³, respectively. Phase identifications by XRD were performed on three kinds of surfaces of each sintered specimen—sintered, polished (with 0.5-μm diamond paste), and fractured surfaces. The tetragonal-ZrO_2 contents in volume percent of total ZrO_2 were determined using the integrated intensities of tetragonal (111) peak and monoclinic (111) and (11$\bar{1}$) peaks.[9] The microstructures were observed by scanning electron microscopy (SEM) on polished and thermally etched surfaces and on fractured surfaces. The average Al_2O_3 grain sizes were determined by a simple linear intercept method.

Results and Discussion

Densification Behavior

The densities of Al_2O_3-ZrO_2 green bodies appeared to depend on the starting powders. With Al_2O_3-I, the green density was low, because the density of γ-Al_2O_3 is much lower than that of α-Al_2O_3. Figure 1 indicates that the densification started at lower temperatures with Al_2O_3-I + CVD ZrO_2 and also with CV codeposited Al_2O_3-ZrO_2 than with Al_2O_3-II + CVD ZrO_2. This may be due to the finer particle sizes but is more likely due to the transformation of Al_2O_3 from γ or δ to α phase, which is supposed to aid the densification. The XRD and DTA analyses on the specimens of Al_2O_3-I + CVD ZrO_2 revealed that the transformation of Al_2O_3-I from γ to α phase started at $\approx$1100° and finished at $\approx$1200°C. In CV codeposited Al_2O_3-ZrO_2 specimens, the transformation of Al_2O_3 from δ to α phase occurred

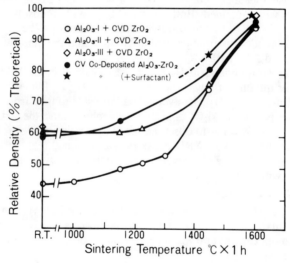

Fig. 1. Densification behavior of Al_2O_3-ZrO_2 ceramics prepared from various Al_2O_3 powders + CVD ZrO_2 (○, △, ◇) and from CV codeposited Al_2O_3-ZrO_2 powder (●). Note that the densification started at lower temperatures with Al_2O_3-I (CVD powder) + CVD ZrO_2 and CV codeposited Al_2O_3-ZrO_2 powder. Despite the good densification at low temperatures, the specimens prepared from CV codeposited Al_2O_3-ZrO_2 powder did not achieve high density at 1600°C due to the powder agglomeration. Addition of a surfactant during powder pretreatment deagglomerated the powder, resulting in better sinterability (★).

at 1200°–1250°C. The transformation of Al_2O_3 in CV codeposited Al_2O_3-ZrO_2 appeared to be retarded by the ZrO_2 particles included within Al_2O_3.

When sintered at 1600°C for 1 h, all the specimens achieved bulk densities higher than 94% of theoretical. The highest relative density (99.0%) was achieved by the pellets prepared from Al_2O_3-III + CVD ZrO_2. The specimens prepared from CV codeposited Al_2O_3-ZrO_2 showed good densification at lower temperatures but were not sintered to higher density than 97.5% of theoretical at 1600°C. It was revealed later that the specimens could be densified to 99% of theoretical when the CV codeposited Al_2O_3-ZrO_2 powder was deagglomerated by a surfactant.

Microstructural Observations

The microstructures of Al_2O_3-ZrO_2 ceramics (11.5 wt% ZrO_2) sintered at 1600°C for 1 h from various Al_2O_3s plus CVD ZrO_2 are compared in Fig. 2. The average Al_2O_3 grain sizes for Al_2O_3-I + CVD ZrO_2, Al_2O_3-II + CVD ZrO_2 and Al_2O_3-III + CVD ZrO_2 were 1.9, 3.4, and 1.5 μm, respectively. The Al_2O_3-ZrO_2 ceramics from Al_2O_3-III + CVD ZrO_2 showed the narrowest size distribution of Al_2O_3 grains, probably because of the very narrow size distribution of the starting Al_2O_3-III powder. Most of the ZrO_2 particles were located at the Al_2O_3 grain boundaries, but a few small ZrO_2 particles were trapped within Al_2O_3 grains. Large ZrO_2 particles (larger than a few micrometers) and large Al_2O_3 grains (10–20 μm) were sometimes observed and were supposed to originate from large agglomerates or insufficient mixing.

The microstructures of Al_2O_3-ZrO_2 ceramics prepared from CV codeposited Al_2O_3-ZrO_2 powder are shown in Fig. 3, one with 9.1 wt% ZrO_2 and another with 16.7 wt% ZrO_2. The average Al_2O_3 grain sizes were relatively large, 2.7 μm with 9.1 wt% ZrO_2 and 3.1 μm with 16.7 wt% ZrO_2. The remaining pores were also large and arc-shaped. These microstructures, together with the densification behavior, suggested that the CV codeposited powder was agglomerated and that the sintering did not proceed uniformly.

With larger weight percent of ZrO_2, significant grain growth of ZrO_2 occurred, as shown in Fig. 3(B). The large ZrO_2 particles transformed to monoclinic symmetry during cooling, and some more transformed while polishing and probably made microcracks around them. Some of these transformed ZrO_2 particles came out of the matrix while the specimen was polished and left holes in the matrix.

Tetragonal ZrO_2 Contents of Sintered, Polished, and Fractured Surfaces

Tetragonal ZrO_2 particles transformed to monoclinic symmetry under stresses caused by fracturing or even by polishing. The t-ZrO_2 content measured on the sintered surface of a specimen is always marginally larger than that measured on the subsequently polished surface, because a small amount of t-ZrO_2 in the surface zone transforms to monoclinic symmetry during polishing. On the other hand, large stresses introduced by fracture cause almost all the t-ZrO_2 particles near the crack surface to transform to monoclinic symmetry. The thickness of the transformation zone, within which all the ZrO_2 particles are assumed to transform, can be calculated from the difference between the t-ZrO_2 content in the bulk material and that measured on the fractured surface.[10] The t-ZrO_2 content in the bulk material is assumed equal to the t-ZrO_2 content measured on the sintered surface.

Comparisons of t-ZrO_2 contents measured on sintered surfaces with those on polished surfaces by XRD measurements are shown in Fig. 4, and comparisons of t-ZrO_2 contents between sintered surfaces and fractured surfaces are shown in

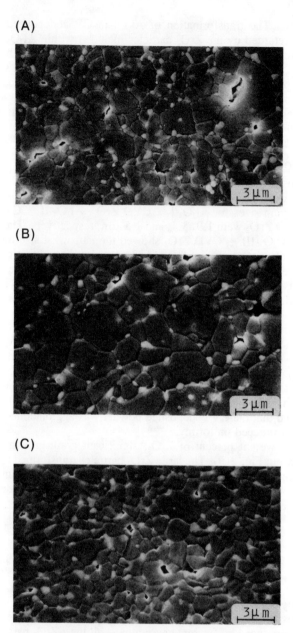

Fig. 2. Scanning electron micrographs (secondary electron mode) of Al_2O_3-ZrO_2 ceramics sintered at 1600°C for 1 h from (A) Al_2O_3-I + CVD ZrO_2 (11.7 wt%), (B) Al_2O_3-II + CVD ZrO_2 (11.3 wt%), and (C) Al_2O_3-III + CVD ZrO_2 (11.4 wt%). Al_2O_3 grain sizes and distributions were dependent on characteristics of starting Al_2O_3 powders. Specimens were polished and thermally etched before observation.

(A)

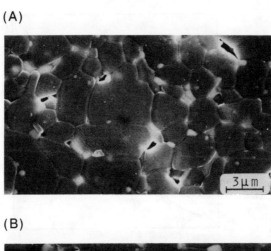

(B)

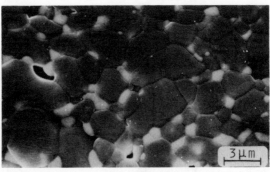

Fig. 3. Scanning electron micrographs (secondary electron mode) of Al_2O_3-ZrO_2 ceramics sintered at 1600°C for 1 h from CV codeposited Al_2O_3-ZrO_2 powder: (A) ZrO_2, 9.1 wt%; (B) ZrO_2, 16.7 wt%. With larger weight percent of ZrO_2, intergranular ZrO_2 particles coalesced to larger sizes.

Fig. 5. The measurements on sintered surfaces were not very different from the corresponding measurements on polished surfaces. But when the density was low, as with the specimen sintered at 1450°C, or when the matrix was very sensitive and easily microcracked by polishing, the differences of t-ZrO_2 contents measured on polished surfaces from those measured on sintered surfaces were relatively large. This can be explained by the degree of constraint rendered on ZrO_2 particles.

The t-ZrO_2 contents measured on fractured surfaces showed a significant drop from those measured on sintered surfaces, as shown in Fig. 5. But the data with Al_2O_3-II and -III showed smaller changes than with Al_2O_3-I or CV codeposited Al_2O_3-ZrO_2. The transformation-zone thicknesses were calculated using the equation equivalent to the one proposed by Kosmač et al.[10]

$$a = \left(\frac{\ln \left[t_{bulk} - t_{trans})/(t_{measd} - t_{trans}) \right]}{2\mu} \right) \sin \theta \quad (1)$$

where t_{bulk} is the content of t-ZrO_2 in bulk material, t_{trans} the content of t-ZrO_2 in the transformation zone (usually assumed to be 0), t_{measd} the content of t-ZrO_2

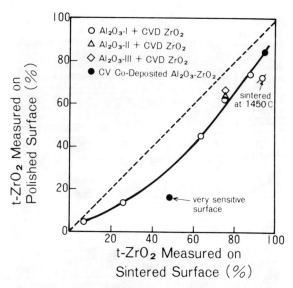

Fig. 4. Comparisons of t-ZrO_2 contents measured on sintered surface vs polished surface by XRD (CuKα) technique. t-ZrO_2 content was always a little larger on sintered surface than on polished surface because polishing induced transformation. A relatively large difference was observed when ZrO_2 particles were not strongly constrained by the matrix.

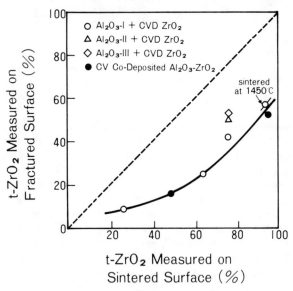

Fig. 5. Comparisons of t-ZrO_2 contents measured on sintered surface vs fractured surface. A large difference was observed especially with Al_2O_3-I (CVD powder) + CVD ZrO_2 or CV codeposited Al_2O_3-ZrO_2, suggesting a larger transformation-zone thickness probably due to extensive microcrack formation on fracture.

Table I. Content of Tetragonal ZrO_2 on Sintered and Fractured Surfaces and Calculated Transformation-Zone Thickness

Starting powder	ZrO_2 (wt%)	Tetragonal (%) Sintered surface	Tetragonal (%) Fractured surface	Absorption coefficient* (μm^{-1})	Transformation-zone thickness (μm)
Al_2O_3-I + CVD ZrO_2	11.7	76	42	0.0167	4.5
Al_2O_3-I + CVD ZrO_2	13.2	63	25	0.0174	6.8
Al_2O_3-I + CVD ZrO_2	20.0	25	9	0.0197	7.0
Al_2O_3-II + CVD ZrO_2	11.3	76	51	0.0165	3.0
Al_2O_3-III + CVD ZrO_2	11.4	75	53	0.0166	2.8
CV codeposited Al_2O_3-ZrO_2	9.1	95	53	0.0158	4.8
CV codeposited Al_2O_3-ZrO_2	16.7	48	16	0.0185	7.8

*Absorption coefficients (with $CuK\alpha$) were calculated using the data in International Tables for X-Ray Crystallography, Vol. IV.

measured on the fractured surface, μ the X-ray absorption coefficient of bulk material, and θ the diffraction angle.

The calculation results are given in Table I. The transformation-zone thicknesses ranged from 2.8 to 7.8 μm. The thicknesses were very large with specimens prepared from Al_2O_3-I (CVD Al_2O_3) + CVD ZrO_2 or from CV codeposited Al_2O_3-ZrO_2. Scanning electron microscopy observations of the fractured surfaces suggested that extensive microcrack formation and crack branching occurred in these specimens.

Change of t-ZrO_2 Content During Sintering

The Al_2O_3-ZrO_2 ceramics were sintered at various temperatures for 1 h, cooled to room temperature, and then the t-ZrO_2 contents were measured. The results shown in Fig. 6 indicate several interesting facts.

When the compacts were sintered slightly above 1170°C (equilibrium transformation temperature of ZrO_2 from monoclinic to tetragonal symmetry), the ZrO_2 particles transformed to tetragonal symmetry, and most of them retained tetragonal symmetry when cooled to room temperature unless excessive ZrO_2 grain growth occurred.

When the compacts were sintered at higher temperatures (1450°–1600°C), ZrO_2 grains were considerably larger, and those above the critical diameter transformed to monoclinic symmetry during cooling to room temperature. Even though starting from 100% tetragonal ZrO_2, compacts prepared from CV codeposited Al_2O_3-ZrO_2 powder showed the same decreasing trend of t-ZrO_2 content when sintered above 1450°C.

Dependence of t-ZrO_2 content on the weight percent of ZrO_2 is shown in Fig. 7. Higher sintering temperature and larger weight percent of ZrO_2 caused excessive ZrO_2 grain growth, resulting in lower t-ZrO_2 content. Tetragonal ZrO_2 contents of the specimens sintered at 1600°C for 1 h did not depend on the starting powder but depended only on the weight percent of ZrO_2. The specimens prepared from CV codeposited Al_2O_3-ZrO_2 showed almost the same t-ZrO_2 content in spite of better ZrO_2 dispersion in the starting material, because ZrO_2 grain growth occurred to the same extent as with the specimens prepared from various Al_2O_3 mixed with CVD ZrO_2.

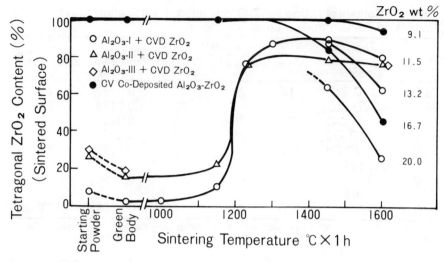

Fig. 6. Tetragonal ZrO_2 content as a function of sintering temperature. Some portion of t-ZrO_2 particles in the mixture of Al_2O_3 (I, II, or III) and CVD ZrO_2 transformed to monoclinic symmetry when the powder mixture was only compacted. Also there was a sudden increase of t-ZrO_2 content between 1150° and 1225°C. When sintered above 1450°C, a sharp decrease in t-ZrO_2 content occurred due to excessive ZrO_2 grain growth, especially with specimens of large ZrO_2 weight percent.

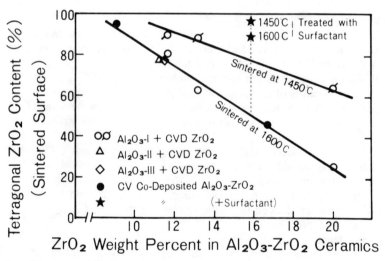

Fig. 7. Tetragonal ZrO_2 content as a function of ZrO_2 weight percent in Al_2O_3-ZrO_2 ceramics. Tetragonal ZrO_2 content was strongly dependent on the weight percent of ZrO_2 and sintering temperature. Even though starting from 100% t-ZrO_2, specimens prepared from CV codeposited powder did not give any higher t-ZrO_2 content in the first series of experiments (●). But the t-ZrO_2 content was much improved by addition of a surfactant during powder pretreatments (★).

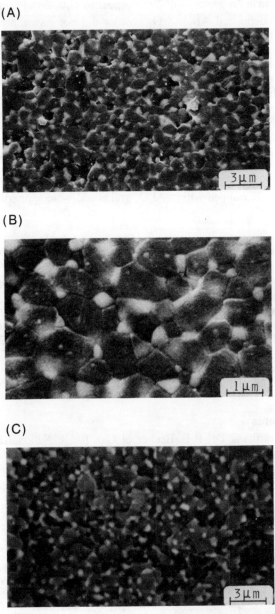

Fig. 8. Microstructure of Al_2O_3-ZrO_2 ceramics sintered at 1600°C for 1 h from CV codeposited Al_2O_3-ZrO_2 (16 wt%) powder with surfactant pretreatment: (A, B) polished and thermally etched surface at two different magnifications. (SEM, secondary electron mode); (C) fractured surface (SEM, backscattered electron mode). Average Al_2O_3 grain size was 1.1 μm by intercept method. Intergranular ZrO_2 particles were 0.3 to 0.6 μm large, whereas intragranular ZrO_2 particles were about 0.1 μm large. The uniformity of ZrO_2 dispersion in the starting powder resulted in excellent ZrO_2 dispersion in the sintered specimen as well.

Effect of Surfactant in Pretreatment of CV Codeposited Al_2O_3-ZrO_2 Powder

The CV codeposited Al_2O_3-ZrO_2 powder did not give the expected better dispersion of ZrO_2 in sintered specimens probably because of agglomeration during powder pretreatment. To prevent agglomeration, 1 wt% of a nonionic surfactant[¶] was added to the powder during ball-milling, while the other processing parameters were kept the same as described before. Thus, prepared powder with 16 wt% ZrO_2 showed excellent sinterability, as indicated in the densification curve in Fig. 1. The bulk density reached 99.0% of theoretical by sintering at 1600°C for 1 h. Along with the improved densification, the t-ZrO_2 content also improved, as shown in Fig. 7. The microstructure of this specimen is shown in Fig. 8, indicating that grain growth of Al_2O_3 and ZrO_2 was suppressed. The average Al_2O_3 grain size was 1.1 μm by linear intercept method. Most of the ZrO_2 particles were located at Al_2O_3 grain boundaries, whose particle sizes ranged from 0.3 to 0.6 μm. Fewer ZrO_2 particles were trapped within Al_2O_3 grains and were approximately 0.1 μm in size.

Summary

(1) Densification and t-ZrO_2 content of sintered Al_2O_3-ZrO_2 ceramics were studied.

(2) Even with very fine ZrO_2 starting powders, higher sintering temperatures and larger weight percent of ZrO_2 caused excessive ZrO_2 grain growth, which resulted in lower t-ZrO_2 content measured at room temperature.

(3) Transformation-zone thickness calculated from t-ZrO_2 contents measured on sintered and fractured surfaces appeared very large (4.5–7.8 μm) when Al_2O_3-ZrO_2 ceramics were prepared from CVD powders.

(4) With an improved pretreatment, CV codeposited Al_2O_3-ZrO_2 powder was proved to be an excellent starting powder for Al_2O_3-ZrO_2 ceramics.

Acknowledgments

The authors thank Dr. P. F. Becher of Oak Ridge National Laboratory for useful suggestions.

References

[1] R. C. Garvie, R. H. Hannink, and R. T. Pascoe, "Ceramic Steel?," *Nature (London)*, **258** [5537] 703–704 (1975).

[2] (a) N. Claussen, "Fracture Toughness of Al_2O_3 with an Unstabilized ZrO_2 Dispersed Phase," *J. Am. Ceram. Soc.*, **59** [1–2] 49–51 (1976). (b) N. Claussen, "Stress-Induced Transformation of Tetragonal ZrO_2 Particles in Ceramic Matrices," *J. Am. Ceram. Soc.*, **61** [1–2] 85–86 (1978).

[3] (a) F. F. Lange and D. J. Green, "Effect of Inclusion Size on the Retention of Tetragonal ZrO_2: Theory and Experiments"; pp. 217–25 in Advances in Ceramics, Vol. 3. Edited by A. H. Heuer and L. W. Hobbs. The American Ceramic Society, Columbus, OH, 1981. (b) F. F. Lange, "Transformation Toughening; Part 1. Size Effects Associated with the Thermodynamics of Constrained Transformations," *J. Mater. Sci.*, **17** [1] 225–34 (1982).

[4] (a) F. F. Lange, "Transformation Toughening; Part 4. Fabrication, Fracture Toughness and Strength of Al_2O_3-ZrO_2 Composites," *J. Mater. Sci.*, **17** [1] 247–54 (1982). (b) D. J. Green, "Critical Microstructures for Microcracking in Al_2O_3-ZrO_2 Composites," *J. Am. Ceram. Soc.*, **65** [12] 610–14 (1982).

[5] P. F. Becher, "Transient Thermal Stress Behavior in ZrO_2-Toughened Al_2O_3," *J. Am. Ceram. Soc.*, **64** [1] 37–39 (1981).

[¶] Yukanol NCS, Tetsuno Yuka KK, Tokyo, Japan.

[6]A. H. Heuer, N. Claussen, W. M. Kriven, and M. Rühle, "Stability of Tetragonal ZrO_2 Particles in Ceramic Matrices," *J. Am. Ceram. Soc.*, **65** [12] 642–50 (1982).

[7]T. Kosmač, J. S. Wallace, and N. Claussen, "Influence of MgO Additions on the Microstructure and Mechanical Properties of Al_2O_3-ZrO_2 Composites," *J. Am. Ceram. Soc.*, **65** [5] C-66–C-67 (1982).

[8]S. Hori, "ZrO_2 and ZrO_2-Containing Ultrafine Powders by Chemical Vapor Deposition Method"; pp. 21–28 in Zirconia Ceramics 1. Edited by S. Sōmiya. Uchida Rokakuho, Tokyo, 1983.

[9]R. C. Garvie and P. S. Nicholson, "Phase Analysis in Zirconia Systems," *J. Am. Ceram. Soc.*, **55** [6] 303–305 (1972).

[10]T. Kosmač, R. Wagner, and N. Claussen, "X-Ray Determination of Transformation Depths in Ceramics Containing Tetragonal ZrO_2," *J. Am. Ceram. Soc.*, **64** [4] C-72–C-73 (1981).

Preparation of Mixed Fine Al_2O_3-HfO_2 Powders by Hydrothermal Oxidation

HIDEO TORAYA,* MASAHIRO YOSHIMURA, AND SHIGEYUKI SŌMIYA

Tokyo Institute of Technology
Laboratory for Hydrothermal Syntheses
Research Laboratory of Engineering Materials
Yokohama 227, Japan

Mixed Al_2O_3-HfO_2 powders were prepared by hydrothermal oxidation of the intermetallic compound $HfAl_3$ and the mixture $HfAl_3$-4Al at 400° to 700°C under 100 MPa. Monoclinic HfO_2 with particle sizes of 10–20 nm was finely dispersed among the α-Al_2O_3 with sizes ranging from nanometers to micrometers. The temperature-time dependence of the reactions of Al, $HfAl_3$, and $HfAl_3$-4Al with H_2O was examined. The hydrothermal oxidation mechanism of the compound $HfAl_3$ was different from those of Hf and Al metals. The mixture with fine precipitation of $HfAl_3$ among the Al matrix is supposed to be effective for the preparation of mixed Al_2O_3-HfO_2 powders.

In hydrothermal oxidation, metals are pulverized and oxidized by reaction with high-temperature, high-pressure water. The oxidation proceeds at relatively lower temperatures (400°–700°C), and the formed oxide powders, such as those of ZrO_2[1,2] or HfO_2,[3,4] have very small particle sizes (25–35 nm) and narrow ranges of size distribution. Recently, $ZrAl_3$ and Zr_5Al_3 alloys were used as starting materials to prepare mixed Al_2O_3-ZrO_2 powders by hydrothermal oxidation.[5,6] Since Zr and Al atoms are distributed on their respective crystallographic positions of these intermetallic compounds, the experiment was expected to result in the fine dispersion of formed oxides, Al_2O_3 and ZrO_2. In that study, well-mixed powders of α-Al_2O_3 and monoclinic and tetragonal ZrO_2 were obtained; the particle sizes of the respective oxides were less than those obtained separately by the hydrothermal oxidation of Al and Zr metals.[5,6]

Mixed Al_2O_3-HfO_2 powders are potentially useful as ceramic materials in a higher temperature range, compared to Al_2O_3-ZrO_2 powders, due to a higher transition temperature of monoclinic HfO_2 ⇌ tetragonal HfO_2.[7] The present study reports the hydrothermal oxidation of Al-Hf alloys. The Al metal, an intermetallic compound on the Al-rich side, $HfAl_3$, and a mixture, $HfAl_3$-4Al, with an intermediate composition between the Al and $HfAl_3$ were selected as starting materials. The hydrothermal oxidation of Al metal powders has already been reported,[8] whereas metal chips were used in the present study. In the hydrothermal oxidation of Hf metals, the structural change associated with the formation of hafnium hydride plays an important role, inducing microcracks and bringing about the succeeding pulverization and oxidation.[4,9] The formation of zirconium hydride was

---

*Now with the Nagoya Institute of Technology, Ceramic Engineering Research Laboratory, 10-6-29, Asahigaoka, Tajimi, 507, Japan.

observed in the hydrothermal oxidation of Zr metal[2] as in the case of Hf, whereas no formation of hydride was reported in the case of $ZrAl_3$ and Zr_5Al_3.[5,6] The hydrothermal oxidation mechanism of the compounds $HfAl_3$ and $ZrAl_3$ were, therefore, expected to be different from those of Hf and Zr metals. The pulverization-oxidation mechanism of the compound $HfAl_3$ and the effect of precipitated $HfAl_3$ on the hydrothermal oxidation of the mixture $HfAl_3$-4Al are the points of interest in the present study.

Experimental Procedure

Preparation of Starting Materials

The Al metal ingot (99.999%)[†] was rolled into ribbons 0.7-mm thick, and then cut into chips. The Al and Hf (containing 3.9% Zr)[‡] metals were weighed in a 1:3 Hf-Al atomic ratio for $HfAl_3$ and 1:7 for $HfAl_3$-4Al. These metals were fused by the radiation of a tungsten arc in a vacuum arc furnace.[§] The fused alloys were solidified quickly by cutting off the arc, and they were remelted more than five times to improve the homogeneity. These alloys were crushed into fragments, washed by acetone and alcohol in an ultrasonic washer, and dried in air. Microscopic observation of the polished surface of the mixture $HfAl_3$-4Al by reflected light showed that $HfAl_3$ crystals with acicular form (about 50 μm in length) were precipitated among the Al matrix. The $HfAl_3$ was identified to be the $ZrAl_3$ type by X-ray powder diffraction. The $HfAl_3$ sample included a small amount of $HfAl_2$.

Hydrothermal Oxidation

Closed- and open-system runs were performed to change the atmospheres for each of three kinds of starting materials. For the closed-system runs, Al metal chips (or crushed fragments of $HfAl_3$ or $HfAl_3$-4Al) (about 0.7 mm in length) and $\approx$35 mg of redistilled water (3:2 H_2O/Al molar ratio for Al, 13:2 H_2O/$HfAl_3$ for $HfAl_3$, and 25:2 H_2O/$HfAl_3$-4Al for $HfAl_3$-4Al) were sealed by electric arc welding in a 2.7-mm ID Pt capsule (0.15-mm thick and 35-mm long). For the open-system runs, a capsule with the same dimensions and contents as the closed ones was prepared without sealing the top, leaving an opening of 4 by 0.7 mm^2. These capsules were heated in a test-tube-type pressure vessel.

For the six systems, the temperature dependence of the reactions was first examined by varying the temperature for heat treatment from 400° to 700°C for 3 h. Next, the time dependence was examined by varying the time for heat treatment from 0 to 120 h at 500°C. The runs at 400° for 120 h and at 600°C for 30 h were also examined. The heating rate was 16°C/min, and the pressure was kept at 100 MPa in all runs. The temperature was measured by a platinel thermocouple and the pressure by a Bourdon gauge. They were held within ±10°C and ±1 MPa during the run. After the heat treatment, the vessel was quenched in water.

X-ray Diffraction and Microscopic Analyses

Products in the capsule were identified by X-ray powder diffractometry. For quantitative analyses, X-ray powder diffraction intensity data were collected with CuKα radiation monochromated with a graphite monochromator by using the step-scan technique (step width = 0.02° in 2θ and fixed time of 10 or 20 s).[¶] The

---

[†]Osaka Asahi Metal Kōjō Co., Ltd., Tokyo, Japan.
[‡]Nippon Mining Co., Ltd., Tokyo, Japan.
[§]Daia Sinku Giken Co., Ltd., Tokyo, Japan.
[¶]RU-200, Rigaku Denki Co., Ltd., Tokyo, Japan.

Table I. Products Obtained by Heat Treatment at Various Temperatures under 100 MPa for 3 h.

Product	Closed system				Open system			
	400°C	500°C	600°C	700°C	400°C	500°C	600°C	700°C
Al-H_2O								
Al	m*	m	w	vw	vw			
γ-AlOOH	m				s			
χ-Al_2O_3	w	vw			vw			
κ_1-Al_2O_3		w	w			w	w	
α-Al_2O_3	vw	m	s	vs	vw	s	vs	vs
$HfAl_3$-H_2O								
$HfAl_3$	s	m	w	vw	s	w	vw	
γ-AlOOH	w				w			
Unidentified	w	vw			w			
α-Al_2O_3		m	s	vs		s	vs	vs
HfO_2	w	m	s	vs	w	s	vs	vs
$HfAl_3$-4Al-H_2O								
$HfAl_3$	s	s	w	vw	m	w	vw	
γ-AlOOH	s	vw			s			
κ_1-Al_2O_3		m	m					
α-Al_2O_3		m	m	vs	vw	vs	vs	vs
HfO_2	w	m	s	vs	s	s	vs	vs

*vw, very weak; w, weak; m, medium; s, strong; vs, very strong.

relative amounts of products (Al, $HfAl_3$, α-Al_2O_3, and HfO_2) were determined from observed integrated intensity ratios of relevant reflections of each component compared with the calibration curves previously obtained for the mixtures of known amounts. Crystallite sizes were determined from the integral breadths of true diffraction profiles obtained by the deconvolution[10] of observed line profiles with standard ones. Line broadening was assumed to be due only to the size effect of the spherical crystallite.[11]

The powders obtained by hydrothermal treatment were examined by scanning** and transmission†† electron microscopy.

Results and Discussion

Reactions between the Hf-Al and H_2O

Table I gives the products obtained by the heat treatment at various temperatures under 100 MPa for 3 h. Except for the formation of χ-Al_2O_3 at 400° and 500°C, the products in the system Al-H_2O generally coincide with those obtained from the corresponding runs using Al metal powders as starting material.[8] Boehmite (γ-AlOOH) was formed in all systems at 400°–500°C, and an unidentified phase, which seemed to be δ-Al_2O_3, was found in the system $HfAl_3$-H_2O at 400°–500°C. κ_1-Al_2O_3[12] was formed in both closed- and open-system runs in Al-H_2O at 500°–600°C but only in the closed-system runs in $HfAl_3$-4Al-H_2O.

**JSM T-200, JEOL Co., Ltd., Tokyo, Japan.
††JEM-200CX, JEOL Co., Ltd., Tokyo, Japan.

These intermediate products were changed into α-Al_2O_3 with increasing temperature. In the hydrothermal oxidation of $ZrAl_3$ and Zr_5Al_3,[5,6] both tetragonal and monoclinic ZrO_2 were formed, whereas only monoclinic HfO_2 was formed in the present experiment. The formation of hafnium hydride was not observed in all runs.

In the system Al-H_2O, the amounts of products with time at 500°C under 100 MPa varied in the same manner in both closed and open systems but at a much faster rate in the latter. The Al metal vanished within 1 h in the open system, while its trace amount remained even after 120 h in the closed system. γ-AlOOH and χ-Al_2O_3, which were believed to be formed while raising the temperature to 500°C, disappeared in the early stage of reaction (0–1 h). Instead of these products, κ_1-Al_2O_3 and α-Al_2O_3 appeared after the 1-h run. κ_1-Al_2O_3 first increased and then decreased gradually, while α-Al_2O_3 increased monotonically. A small amount of κ_1-Al_2O_3 persisted even after 120 h in the closed system, while it disappeared after 30 h in the open one. κ_1-Al_2O_3 is transformed into stable α-Al_2O_3 by prolonging the reaction time or raising the temperature or pressure.[12] The high concentration of H_2O will also promote the transformation of κ_1 to α. The cause of the large difference in the reaction rates between the closed and open systems in Al-H_2O is attributed primarily to the difference in the concentration of H_2O in the respective systems.

In contrast to the system Al-H_2O, the concentration of H_2O had little effect on the reaction rates in the system $HfAl_3$-H_2O, where a trace amount of $HfAl_3$ remained after 120-h runs in both closed and open systems (Fig. 1(A)). Since the present $HfAl_3$ sample contained no Al metals, when examined by X-ray powder diffraction, γ-AlOOH would be formed by the reaction of Al in $HfAl_3$ with H_2O when the temperature is increased. Small amounts of γ-AlOOH and an unidentified phase appeared only in the initial stage of reaction. The remaining $HfAl_3$, therefore, would be changed into HfO_2 and α-Al_2O_3 without formation of γ-AlOOH.

The reactions in the system $HfAl_3$-4Al-H_2O were the most complex of all the reactions in two systems Al-H_2O and $HfAl_3$-H_2O (Fig. 1(B)). As can be seen from the rapid decrease of Al in both closed and open systems, the reaction of the Al with H_2O proceeded at a much faster rate than the reaction of $HfAl_3$. The reaction was completed very rapidly in the open system, while it proceeded more slowly in the closed systems. Furthermore, the ratio of κ_1-Al_2O_3 to α-Al_2O_3 is very large in the closed system. The κ_1-Al_2O_3 was formed in the systems Al-H_2O and $HfAl_3$-4Al-H_2O but could not be observed in the system $HfAl_3$-H_2O with no Al metals.

Pulverization Mechanism

In the system Al-H_2O, the formed aluminum oxides were well-dispersed, white, fine powders. In the system $HfAl_3$-H_2O, the products retained the original bulk shape almost completely, even after the run for 120 h at 500°C; the bulks were hard to grind, in spite of the fact that most of the $HfAl_3$ was changed into HfO_2 and α-Al_2O_3 (Fig. 1(A)). Well-developed crystals of α-Al_2O_3 were observed on the surface exposed to water in the open system (Fig. 2(A)). In the interior, individual crystals were tightly agglomerated, as observed on the fractured surface (Fig. 2(B)). In the composite system $HfAl_3$-4Al-H_2O, the products were loosely gathered.

Figure 3 shows some transmission electron micrographs and energy dispersive spectra (EDS) of mixed Al_2O_3-HfO_2 powders formed in the system $HfAl_3$-H_2O. Very small particles (10–20 nm (Fig. 3(B)) adhered to the surface

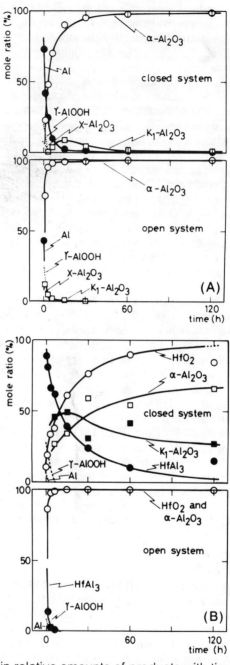

Fig. 1. Variation in relative amounts of products with time at 500°C under 100 MPa in the systems (A) $HfAl_3$-H_2O and (B) $HfAl_3$-4Al-H_2O.

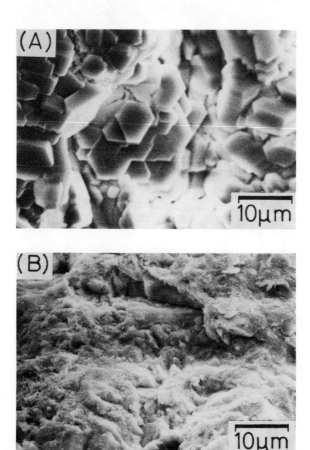

Fig. 2. Scanning electron micrographs of products obtained in the open system run at 500°C for 30 h in HfAl$_3$-H$_2$O: (A) The surface exposed to water and (B) the fractured surface.

of large crystallites (Fig. 3(A)); isolated particles were revealed to be HfO$_2$ by EDS (Fig. 3(C)). Large platelike crystals (1–10 μm long), which appeared in both HfAl$_3$-H$_2$O and HfAl$_3$-4Al-H$_2$O, were confirmed to be α-Al$_2$O$_3$ (Fig. 3(D)). The EDS spectra obtained by scanning the region, as shown in Fig. 3(E), gave both AlKα and HfKα lines. Therefore, some α-Al$_2$O$_3$ crystals would have a size compatible with that of HfO$_2$. The average crystallite size of α-Al$_2$O$_3$ was estimated to be ≫150 nm in the systems Al-H$_2$O and HfAl$_3$-4Al-H$_2$O by XRD but in the range of 60–90 nm in HfAl$_3$-H$_2$O. The range of size distribution is, however, very wide as has been observed by TEM.

Figure 4 shows the variation of crystallite size of HfO$_2$ with time at 500°C, together with values for the runs at higher temperatures. The crystallite sizes of HfO$_2$ determined by the X-ray technique were in good agreement with those measured by TEM, having a narrow range of size distribution. In the system HfAl$_3$-H$_2$O, the presence of excess water has no apparent effect on the crystallite size, showing the same variation for both closed and open systems. On the other

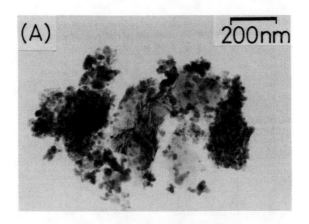

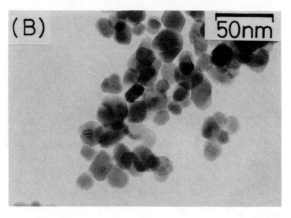

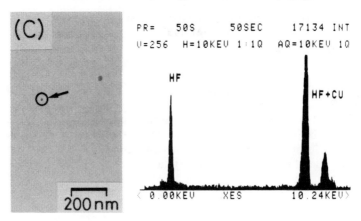

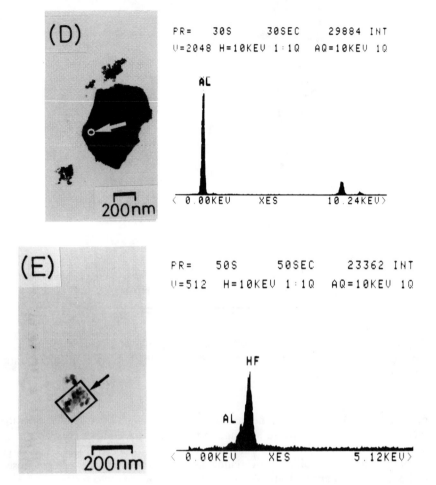

Fig. 3. (A) and (B) Transmission electron micrographs of the products at 500°C for 120 h in the system $HfAl_3$-H_2O. (C), (D), and (E) EDS spectra obtained by scanning the regions indicated for the products at 600°C for 3 h in the same system as above.

hand, its effect is significant in the system $HfAl_3$-4Al-H_2O, where the variation is similar to that in the system Hf-H_2O.[4,9]

In the system $HfAl_3$-H_2O, hafnium hydride was not formed, so the pulverization mechanism differs from that in Hf-H_2O.[4,9] The Hf and Al atoms react with H_2O mainly on the bulk surface. Thus the diffusion of Hf and Al atoms through $HfAl_3$ grains will primarily control the rate of the reaction on the limited area of interface between the alloys and H_2O. The very small crystallite size of HfO_2 will be due to the interference between the Hf and Al atoms.[5,6] Furthermore, the agglomeration of formed oxides (Fig. 2) will prevent the penetration of H_2O into the interior of the bulk sample, so that the presence of excess water in the open system has little effect. The same variation of crystallite size and reaction rate, therefore, results in both closed and open systems. On the other hand, in the

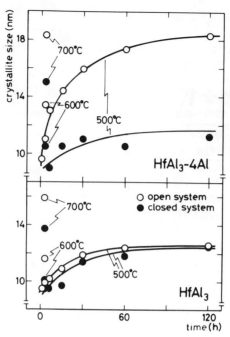

Fig. 4. Variation of crystallite size for monoclinic HfO_2 with time at temperatures indicated in the systems $HfAl_3$-$4Al$-H_2O (upper diagram) and $HfAl_3$-H_2O (lower diagram).

mixture $HfAl_3$-$4Al$, the Al in the matrix is oxidized much more rapidly than the precipitated $HfAl_3$ (Fig. 1(B)), leaving the finely dispersed $HfAl_3$ crystals among the aluminum oxide powders. The bulk, therefore, becomes loose. The large differences in the crystallite size variation and the reaction rate between the closed and open systems arises from the easy penetration of H_2O into the bulk.

Summary

Al, $HfAl_3$, and $HfAl_3$-$4Al$ were hydrothermally oxidized at 400°–700°C under 100 MPa. Mixed Al_2O_3-HfO_2 powders were formed from the $HfAl_3$ and $HfAl_3$-$4Al$, where monoclinic HfO_2 with particle sizes of 10–20 nm was finely dispersed among the α-Al_2O_3, with sizes ranging from several nanometers to micrometers. The hydrothermal oxidation mechanism of $HfAl_3$ is different from those of Hf and Al metals. Furthermore, the Al metal has a much faster reaction rate than $HfAl_3$. The formed oxide powders were not perfectly homogeneous because of the wide size distribution of α-Al_2O_3. To suppress the growth of α-Al_2O_3 crystallites, it will be necessary to increase the reaction rate of $HfAl_3$ and change the Al into α-Al_2O_3, without the formation of intermediates, such as γ-AlOOH and κ_1-Al_2O_3. The starting materials of the mixture, with finely precipitated $HfAl_3$ among the Al matrix, are considered to be effective for the preparation of mixed Al_2O_3-HfO_2 powders by hydrothermal oxidation.

Acknowledgments

The authors express their thanks to Professor T. Suzuki and Mr. Y. Ōya of Tokyo Institute of Technology for the preparation of alloy samples. They also thank JEOL Co., Ltd., for use of the transmission electron microscope.

References

[1] M. Yoshimura and S. Sōmiya, "Fabrication of Dense, Non Stabilized ZrO_2 Ceramics by Hydrothermal Reaction Sintering," *Am. Ceram. Soc. Bull.*, **59** [2] 246 (1980).

[2] M. Yoshimura, S. Kikugawa, and S. Sōmiya, "Preparation of Zirconia Fine Powders by the Reactions between Zirconium Metal and High Temperature-High Pressure Solutions"; in High Pressure in Research and Industry. Edited by C.-M. Backman et al. Arkitektkopia ISBN, Uppsala, Sweden, 1982.

[3] H. Toraya, M. Yoshimura, and S. Sōmiya, "Preparation of Fine Monoclinic Hafnia Powders by Hydrothermal Oxidation," *J. Am. Ceram. Soc.*, **65** [5] C-72 (1982).

[4] H. Toraya, M. Yoshimura, and S. Sōmiya, "Hydrothermal Oxidation of Hf Metal Chips in the Preparation of Monoclinic HfO_2 Powders," *J. Am. Ceram. Soc.*, **66** [2] 148–50 (1983).

[5] S. Sōmiya, M. Yoshimura, and S. Kikugawa, "Preparation of Zirconia-Alumina Fine Powders by Hydrothermal Oxidation of Zr-Al Alloys," Proceedings of the 19th University Conference on Ceramic Science, North Carolina University, Nov. 8–10, 1982.

[6] M. Yoshimura, S. Kikugawa, and S. Sōmiya, "Alumina-Zirconia Fine Powders Prepared by Hydrothermal Oxidation" (in Jap.), *Yogyo-Kyokai-Shi*, **91** [4] 182–88 (1983).

[7] C. T. Lynch; pp 193–216 in Refractory Materials: High Temperature Oxides, Pt. II, Vol. 5. Edited by A. M. Alper. Academic Press, New York, 1970.

[8] M. Yoshimura, S. Kikugawa, and S. Sōmiya, "Preparation of Alpha-Alumina Fine Powders by Hydrothermal Oxidation Method," *J. Jpn. Soc. Powder Powder Metall.*, **30** [5] 207–10 (1983).

[9] H. Toraya, M. Yoshimura, and S. Sōmiya, "Reaction Kinetics in the Hydrothermal Oxidation of Hf," *J. Am. Ceram. Soc.*, **66** [11] 818–22 (1983).

[10] H. Toraya, M. Yoshimura, and S. Sōmiya, "A Computer Program for the Deconvolution of X-ray Diffraction Profiles with the Composite of Pearson Type VII Functions," *J. Appl. Crystallogr.*, **16** [6] 653–57 (1983).

[11] A. J. C. Wilson; pp 37–54 in X-ray Optics, 2d ed. Methuen, London, 1962.

[12] M. Torkar, "Untersuchungen uber Aluminiumhydroxyde und -oxide, 4. Mitt," *Mh. Chem.*, **91**, 658–68 (1960).

Applications of Rapid Solidification Theory and Practice to Al_2O_3-ZrO_2 Ceramics

GRETCHEN KALONJI, JOANNA MCKITTRICK, AND L. W. HOBBS

Massachusetts Institute of Technology
Cambridge, MA 02139

Materials from the system Al_2O_3-ZrO_2 were rapidly solidified using two techniques: piston and anvil splatting of laser-melted droplets and plasma-spraying. Two compositions were investigated: the eutectic at 42 wt% ZrO_2 and an off-eutectic alloy of 25 wt% ZrO_2. Microstructures were characterized using transmission electron microscopy (TEM), scanning transmission electron microscopy (STEM), and X-ray diffraction (XRD) techniques. A wide variety of structures were attained, including amorphous phases at both compositions, lamellar eutectics with spacings as fine as 15 nm (150 Å), and various dendritic growth morphologies. Microstructural evolution in this system as a function of solidification rate and composition are discussed in terms of eutectic solidification theory.

Rapid solidification is a processing technique that can allow one to select, in a controlled manner, from a wide variety of metastable phases and phase assemblages that are sometimes inaccessible by other routes. The ability to use this technique meaningfully, however, requires a knowledge of the thermodynamics and kinetics of processes driven far from equilibrium. While the possibilities the technique offers for microstructural development are wide, they are not infinite; they are strictly limited by thermodynamics. It is the purpose of this paper to describe how those limitations may be understood to constrain structural development in rapid solidification of a model ceramic eutectic system, Al_2O_3-ZrO_2, and to compare theoretical expectations with experimental results.

Theory

Consider first the characteristics of the Al_2O_3-ZrO_2 phase diagram. It is a simple eutectic with very limited solubility in both end-members.[1] For rapid solidification theory, a most important phase diagram feature is the so-called T_0 curve, which is the locus of equal Gibbs free energy of the liquid and solid phases. The T_0 curve also delineates the maximum composition of the solid phase that may form from a liquid of arbitrary composition at that temperature. For systems, such as Al_2O_3-ZrO_2, with low solubilities it can be shown from the underlying Gibbs free energy curves that the T_0 curves for both solid phases must closely hug the solidi. This is illustrated schematically in Fig. 1(c), taken from Ref. 2. The significance of the T_0 curve in defining constraints on the attainable microstructures is the following: When a liquid of a given composition is taken below its T_0 temperature, a diffusionless or massive liquid-solid transition becomes possible.

Above the T_0 curve, crystallization cannot occur without concomitant diffusion. The implication for microstructural development in systems having different T_0-curve geometries, then, is quite profound. Three prototypical examples are

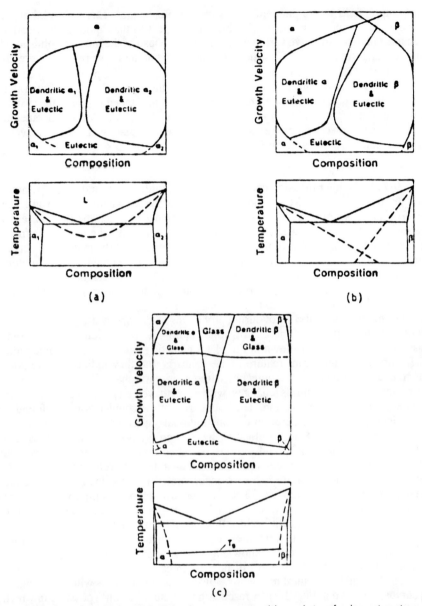

Fig. 1. Schematic growth velocity vs composition plots of microstructures expected for growth into a positive temperature gradient for simple eutectic systems (from Ref. 2). (Used by permission.)

shown in Fig. 1. In Figs. 1(*a*) and (*b*), respectively, the T_0 curves either intersect or are continuous, yielding large regions in temperature-composition space where appropriate processing can allow one to quench in a single-phase, microsegregation-free solid with greatly enhanced, and in fact tailorable, solid solubilities. For eutectic systems with plunging T_0 curves, however, for almost the entire composition field such a phenomenon is impossible, as there exists no solid phase of the same composition which is stable with respect to the liquid. Solidification then must involve diffusive sorting of components ahead of the moving solid-liquid interface and the creation of a solid-solid interfacial area. These processes greatly restrict the speed at which crystallization may occur. For the plunging T_0 curves case, then, if one is able to withdraw heat from the material at a rate such that the isothermal velocity is greater than the maximum interfacial velocity, the system will have no alternative but to form a glass. Using well-known relations for eutectic growth,[3] one can write an expression for the maximum velocity at which a eutectic structure can grow for a given interfacial temperature as

$$V_c = D(T)(\Delta T)^2/4A_1A_2 \tag{1}$$

Here $D(T)$ is the interdiffusion coefficient in the liquid, and A_1 and A_2 are parameters approximated by $m\Delta C/8$ and $2\gamma/\Delta S$, respectively, where m is an average positive liquidus slope, ΔC the composition difference between the two phases, γ the interfacial energy between the two solid phases, and ΔS the average entropy of fusion.[4] Although parameters in this expression are clearly material-specific, eutectic solidification velocities are not expected to exceed approximately 10 cm/s. This is to be contrasted with the case in which the eutectic solidification velocity map lies below the T curve, partitionless solidification is possible, and interface velocities can reach many meters per second.

For microstructural control during processing, then, a very useful type of map gives morphology as a function of solid-liquid interfacial velocity and composition. These maps are shown in Fig. 1 for eutectic systems with the three types of T_0 curves. A significant feature of the maps is that they indicate the regions where coupled growth of the eutectic, without primary phase formation, occurs. This range of compositions changes with growth velocity and need not even include the eutectic composition. The boundaries of the coupled zone are found at low velocities from consideration of constitutional supercooling. The boundaries of the coupled zone at high velocities have been analyzed using Mullins and Sekerka stability analysis[5-7] and various so-called competitive growth theories, which compare dendrite tip temperatures to attain a dendrite velocity with eutectic velocities at that temperature.[8-10] Also incorporated into the maps of Fig. 1 are the regions of "absolute stability." These are growth regions, at high velocity, into positive temperature gradients, for which the plane front interface of a single-phase solid becomes morphologically stable. Because it is surface tension that stabilizes the interface at high growth velocities, this phenomenon is also called capillarity stabilization. Clearly, a detailed knowledge of the bounds of the absolute stability region is important for the control of microsegregation.

The last microstructural region to be discussed, for eutectics with the plunging T_0 curves, is shown in Fig. 1(c). A fascinating structure becomes possible in which dendrites of a primary phase are surrounded by glass.

It should be kept in mind that the details of the morphology/interfacial velocity/composition maps discussed above may hold true only if the temperature gradient in the liquid is positive. Thus, the insight gained from these schematic drawings is rightfully applied to materials processed by a technique that meets that criterion. Rapid solidification processing techniques which are characterized by growth into positive temperature gradients may include the substrate quenching methods, such as melt-spinning, laser-glazing, piston and anvil technique, and others, such as vertical Bridgman. When, as in most atomization methods, the growth occurs into a negative temperature gradient, morphological instabilities due to heat flow become important.

Experimental Procedures

Materials from the system Al_2O_3-ZrO_2 were rapidly solidified by two techniques. One technique employs a 1500-W CO_2 laser to melt droplets on the tip of sintered feed rods of the desired composition. The droplets, which fall under the influence of gravity, are detected by an optical device and rapidly quenched between copper platens in a piston- and anvil-type apparatus. The other technique for solidification uses a plasma torch. Powders fed into the unit are collected, while in the molten state, on various substrates. In both cases, the resulting rapidly solidified materials were examined by X-ray diffraction (XRD), transmission electron microscopy (TEM), and scanning transmission electron microscopy (STEM). Two compositions were studied: the eutectic composition which contains 42 wt% ZrO_2 and an off-eutectic alloy of 25 wt% ZrO_2.

Results and Discussion

Our purpose is to understand and control microstructural evolution in this system through the scientific principles elucidated earlier in this paper. In particular, we wish to proceed to create morphology/interfacial velocity/composition maps for this system. The quenching techniques employed in this research, while useful for initial studies, leave a good deal to be desired; systematic control of solidification rate is not their strong feature. They do, however, satisfy one important criterion. They are quenching methods in which heat is withdrawn through the substrate, meaning that the temperature gradient is positive in the liquid, facilitating the comparison with theoretical expectations. They also offer the experimental convenience of crucibleless melting.

Laser-Splatting

We first describe the results on the laser-splatted materials. The solidification rates seen by the laser-splatted materials vary greatly within each splat. Evidence for this fact exists in the wide variation of eutectic spacings observed under TEM. The first material to hit the platens may solidify very rapidly. Subsequently, the remaining liquid may be sheared over and experience slower rates. Isolated nucleation events may take place in the liquid droplet on the way down. When this occurs, the local solidification conditions differ radically from those described previously. Release of latent heat of fusion from the particles into the surrounding liquid makes the temperature gradient in the liquid negative rather than positive and can lead to the formation of dendrites even at the eutectic composition. Primarily, in the laser-splatting technique, heat is withdrawn through the platens, growth occurring perpendicular to them, and the temperature gradient in the liquid must be positive.

For the eutectic composition materials that have been laser-splatted, we observe the following microstructures. The most common structure consists of fine lamellae between t-ZrO_2 and Al_2O_3. The lamellar spacings in our materials vary from 150 to $\approx 10^3$ Å $[(1 \cdot 5-10) \times 10^{-8}$ m]. By making use of the relation $\lambda^2 V =$ constant, where λ is the lamellar spacing and V the growth velocity, and obtaining the constant from previous work on more slowly directionally solidified Al_2O_3-ZrO_2 eutectics,[11] we can estimate the solid-liquid interface velocity corresponding to the finest of these spacings to be approximately 5 cm/s. We note that this provides only a rough estimate at these high speeds. Some examples of the

(a)

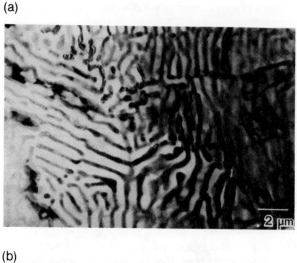

(b)

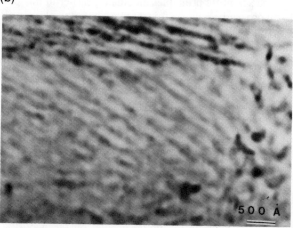

Fig. 2. Lamellar growth morphologies and laser-splatted Al_2O_3-ZrO_2 of eutectic composition.

microstructures of the laser-splatted eutectic compositions are shown in Figs. 2(a) and (b). In addition, in one of our samples, we observed formation of Al_2O_3 dendrites even at the eutectic composition. This microstructure is illustrated in Fig. 3. One explanation of this phenomenon could be that we have an asymmetric coupled zone, shifted toward ZrO_2 at higher growth velocities. This is extremely unlikely, however, due to the large range of finer spacing, and hence, the velocity regions that we have observed. We believe that this was an isolated case in which nucleation occurred in the melt on the way down, growth thus occurring into a negative temperature gradient.

The off-eutectic (25 wt% ZrO_2) material exhibits a great diversity of microstructures. One type, which we call mode 1 growth, consists of α-Al_2O_3 dendrites surrounded by a two-phase interdendritic eutectic of t-ZrO_2 and Al_2O_3. This microstructure is illustrated in Figs. 4(a) and (b). Another growth morphology, which

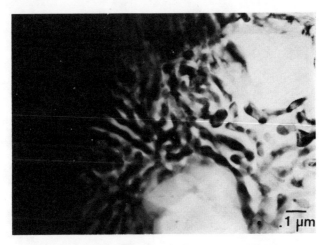

Fig. 3. Dendrite formation at the eutectic composition.

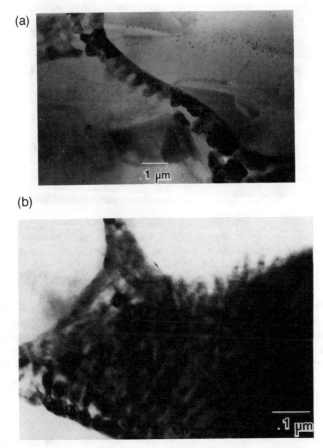

Fig. 4. Laser-splatted material of 25 wt% ZrO_2 composition. α-Al_2O_3 dendrites are surrounded by a two-phase interdendritic eutectic.

(a)

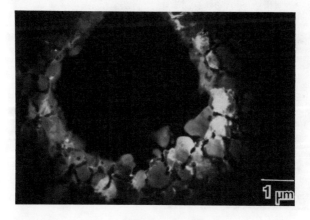

(b)

Fig. 5. (a) Laser-splatted material of composition 25 wt% ZrO_2, exhibiting mode II growth morphology; α-Al_2O_3 dendrites surrounded by t-ZrO_2. (b) Dark-field micrograph using a strong ZrO_2 reflection. Note precipitation of ZrO_2 in Al_2O_3 dendrites.

we call mode 2, consists of α-Al_2O_3 dendrites surrounded by "halos" of t-ZrO_2. Figure 5(a) is a low-magnification micrograph of this structure. Figure 5(b) is a dark-field micrograph taken with one of the strongly diffracting ZrO_2 spots. In both of the above growth types, we observe precipitates, which are as small as 5 nm (50 Å) in both phases. We presume we have exceeded the solid solubilities in both phases and are observing a postsolidification reaction. Chemical analysis in the STEM confirmed that a ZrO_2-rich phase is precipitating from the Al_2O_3. Similarly, an Al_2O_3-rich phase is precipitating out of the ZrO_2 regions. These precipitates can be seen in Figs. 5(a) and 6.

Two additional microstructural regions were observed in the off-eutectic alloys. In one, we observe a region where the growth mode has become lamellar. Presumably, we have locally exceeded the interfacial velocity necessary to reenter

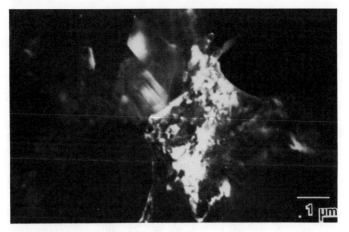

Fig. 6. Dark-field micrograph using a strong Al_2O_3 reflection. Note fine-scale Al_2O_3 precipitation in ZrO_2 areas in the upper left corner.

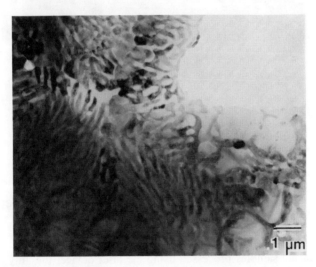

Fig. 7. Region of transition from dendrite + eutectic to eutectic growth morphology in the laser-splatted 25 wt% ZrO_2 alloy.

the coupled zone. A region in which the transition from mode 1 growth to lamellar growth appears is shown in Fig. 7.

In other regions of samples of both compositions, we observe amorphous phases. Figure 8 shows a dark-field micrograph and an electron-diffraction pattern from such a region. Adjacent the amorphous regions are regions in which very fine crystallites, some of them finer than 5 nm (50 Å), are embedded in a glassy phase. Figure 9 shows a micrograph and diffraction pattern taken from this region. More extensive high-resolution electron microscopy will be necessary to determine if this region is a product of postsolidification partial devitrification or if it is a material that solidified in the dendrites + glass region of the processing map.

Fig. 8. Dark-field micrograph and electron-diffraction pattern from an amorphous region in the 25 wt% ZrO_2 laser-splatted material.

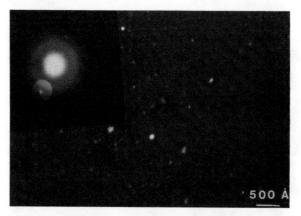

Fig. 9. Dark-field micrograph and diffraction pattern from a region containing glass and fine crystallites. Material is laser-splatted 25 wt% ZrO_2.

For all of the samples processed using the laser device, the only crystalline phases for which we have evidence are α-Al_2O_3, t-ZrO_2, and m-ZrO_2. This is to be contrasted with the results of previous workers,[12] who rapidly solidified Al_2O_3-ZrO_2 ceramics of the eutectic composition using atomization techniques, and found ε-Al_2O_3 for some of their more rapidly quenched particles. It should be pointed out that, because of the diversity in nucleation and growth kinetics of the various thermodynamically possible solid phases from the liquid, it is quite plausible that these fundamentally different techniques should result in different phase selection.

Plasma-Spraying

We used the plasma-spraying technique on the eutectic composition only. Samples examined in the TEM appear amorphous. This is in agreement with the results of previous workers who did plasma-spraying in this system.[13]

Conclusions

Al_2O_3-ZrO_2 ceramics rapidly solidified using a laser with an attached piston and anvil device exhibit a wide variety of microstructures, depending on local solidification conditions. These microstructures may be understood in terms of the theory of rapid solidification of eutectic systems. It would be desirable, however, to have experimental methods which enable one to measure and control the solid-liquid interface velocity, in order to have a more quantitative understanding of the system. For this reason, we are currently working on developing melt-spinning methods for our materials.

We believe that the type of detailed knowledge embodied in the morphology/interfacial velocity/composition maps is essential if one is to select a structure, desirable for a particular physical characteristic, from the wide range of thermodynamically possible events.

Acknowledgments

Many thanks are due to W. J. Boettinger for helpful discussions and for permission to reproduce the figure from his paper. We are grateful to Dr. J. Haggerty for the use of his laser and Dr. Jean-Marc Lihrmann for help in sample preparation. We thank Professor M. C. Flemings for donating the piston and anvil device and for stimulating comments. We thank Thao Nguyen, Yet Ming-Chiang, and Howard Sawhill for microscopy help. This work was financially supported by a grant from The Norton Company.

References

[1] A. M. Alper; p. 339 in Science of Ceramics, Vol. 3. Edited by G. H. Stewart. Academic Press, London, 1967.
[2] William J. Boettinger; in Rapidly Solidified Amorphous and Crystalline Alloys. Edited by B. H. Kear and B. C. Giessen. Elsevier, New York, 1982.
[3] K. A. Jackson and J. D. Hunt, Trans. Metall. Soc. AIME, 245, 1129–42 (1966).
[4] W. J. Boettinger, F. S. Biancaniello, G. Kalonji, and J. W. Cahn; pp. 50–55 in Rapid Solidification Processing: Principles and Technologies II. Edited by R. Mehrabian, B. H. Kear, and M. Cohen. Claitor's, Baton Rouge, LA, 1980.
[5] D. T. J. Hurle and E. Jakeman, J. Cryst. Growth, 11, 141 (1968).
[6] H. E. Cline, Trans. Metall. Soc. AIME, 242, 1613–18 (1968).
[7] S. Strasler and W. R. Scheider, Phys. Condens. Matter, 17, 153–78 (1974).
[8] K. A. Jackson, Trans. Metall. Soc. AIME, 242, 1275–79 (1968).
[9] M. H. Burden and J. D. Hunt, J. Cryst. Growth, 22, 328–30 (1974).
[10] W. Kurz and D. J. Fisher, Acta Metall., 29, 11–20 (1981).
[11] V. S. Stubican and R. C. Bradt, Ann. Rev. Mater. Sci., 11, 267–97 (1981).
[12] N. Claussen, G. Lindemann, and G. Petzow; 5th CIMTEC, Lignano-Sabbiadoro, Italy, 1982.
[13] Von A. Krauth and H. M. Meyer, Ber. Dtsch. Keram. Ges., 42, 61–67 (1965).

Synthesis and Sintering of ZrO₂ Fibers

I. N. Yermolenko, T. M. Ulyanova, P. A. Vityaz, and I. L. Fyodorova

Academy of Sciences of Byelorussian SSR
Byelorussian Powder Metallurgy Association
Minsk, USSR

Formation of ZrO_2 fibers obtained as the result of heat-treating salt-bearing polymers, as well as sintering of such fibers, were studied. Physical-chemical conversions of the materials were analyzed by structural and by spectroscopic and thermal methods. The possibility of obtaining ZrO_2 fibers of high porosity, possessing increased reactivity and predetermined composition and properties, is shown. Synthesized fibers of ZrO_2 are fit for use in high-temperature materials.

Progress in high-temperature technology demands creation of structural refractory ceramic materials, the development and study of which are presented in a large number of papers.[1-4] Together with refractoriness, these materials must meet a number of other requirements; namely, the materials must be able to withstand large temperature differences and possess thermal insulating properties and low-volume mass; all these properties vary with applications. Fibrous structural materials possess these properties. At present a large number of refractory fibers are known: carbon, boron, nitride, oxides, and others.[5-7] The properties of ceramic fibers depend both on their compositions and on the method of manufacture. Synthesis of inorganic fibers is accompanied by phase and structural transformations, which make it possible to control the process and obtain materials with predetermined properties. The purpose of the present research was to study the process of forming ZrO_2 fibers obtained by heat treatment of salt-bearing polymer fibers and the process of their sintering, as well as the effect of additives on the structure and properties of the synthesized fibers of ZrO_2.

Experimental Procedures

ZrO_2 fibers were obtained by oxidizing hydrated cellulose fibers impregnated with Zr salts. The initial polymer fibers were preimpregnated in 2–2.5 M solutions of zirconium oxychloride or zirconium oxynitrate for 3–6 h at 20°–25°C. The impregnated materials were dried in air and then oxidized; the temperature was increased during oxidation at a rate of 5°/min in the temperature range 20°–1000°C. To study the process of sintering, fiber samples were also annealed in air up to 1600°C.

At different stages of synthesis and sintering of ZrO_2 fibers the structure and properties of the materials were studied by thermogravimetric, elemental, infrared spectroscopic, X-ray, adsorption, and electron microscopy methods of analysis.

Results and Discussion

Thermograms of the initial fibers of hydrated cellulose and fibers impregnated in solutions of zirconium oxychloride and zirconium oxynitrate are shown in Fig. 1. While the hydrated cellulose fibers are heated to 200°C, sorbed and bound

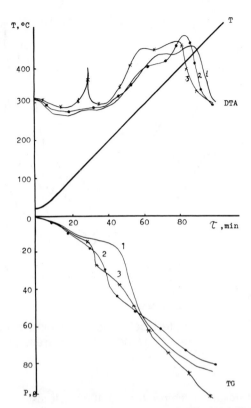

Fig. 1. DTGA curves: (1) hydrated cellulose (HC); (2) HC impregnated by zirconium chloride; (3) HC impregnated by zirconium nitrate.

waters of the added salts and polymer are expelled, accompanied by reduction of the sample mass by 15%. Further heating causes carbonization of the polymer and subsequent oxidation of formed carbon to carbon dioxide. The process is completed by 450°–500°C; the greatest rate of mass loss is observed in the temperature range 300°–400°C. Presence of the added Zr salts in the initial polymer fibers accelerates the process of cellulose dehydration and shifts this process to lower temperatures. Besides, during thermal dissociation of zirconium nitrate, separation of nitrogen oxides takes place, which promotes cellulose oxidation at lower temperatures. This is confirmed by an exotherm on the DTA curve at 155°C and a high rate of mass loss in the respective part of the TG curve. Full oxidation of hydrated cellulose containing zirconium nitrates is completed at temperatures lower than those for fibers containing chlorides. Analysis carried out for determining C and Zr contents during the oxidation process showed consistency with thermogravimetric curves. Further heating from 450° to 600°C causes a perceptible reduction in the mass loss rate; the process of carbon burning out is completed, and ZrO_2 with a fiberlike structure is formed.

Various crystallization processes occur in the material in the process of oxidation of salt-bearing organic fibers. Absorption bands in the range of 250–500 and 700–750 cm^{-1}, characteristic of ZrO_2, occur in the infrared spectra of the polymer samples containing $ZrOCl_2$ after heat treatment at 400°C (Fig. 2). In-

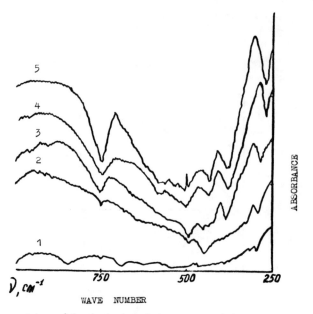

Fig. 2. Ir spectra of hydrated cellulose containing zirconium chloride, annealed in the air at the following temperatures: (1) 100°, (2) 200°, (3) 600°, (4) 800°, and (5) 1000°C.

creasing the heat-treatment temperature to 1000°C leads to improved ordering of the ZrO_2 crystal lattice, as is evident by the appearance of sharply defined absorption bands in the infrared spectra.

X-ray analysis of samples oxidized at various temperatures showed that the tetragonal (t)-ZrO_2 appeared first in fibers after heat treatment at 400°–500°C. Monoclinic (m)-ZrO_2 is normally found at low temperatures. The appearance of the t modification is probably accounted for by the crystal structure of the initial ZrO_2 salts which are tetragonal. At 800°C, t-ZrO_2 changes into m-ZrO_2 which is preserved up to 1200°C. Addition of Y_2O_3 up to 10 mol% at the impregnation stage partially stabilizes t-ZrO_2. At Y_2O_3 concentrations above 15 mol%, c-ZrO_2 fibers are formed. The data of X-ray phase analysis made it possible to calculate parameters of the crystal lattice of both ZrO_2 modifications (Table I). The results show that the parameter values of the crystal structure of both nonmodified and partly modified ZrO_2 differ from those presented in other papers.[8] In fibers containing 15 mol% Y_2O_3, the degree of tetragonality of the lattice cell approaches 1 and the elementary cell approaches the cubic lattice. Such deviations are probably related to a homogeneous solute dispersion of the refractory compound in the synthesized oxide fibers.

The homogeneity of the oxide in the material is characteristic of fibrous materials synthesized in this way.[9] While the organic component of salt-bearing fibers of hydrated cellulose is being oxidized, a substantial gas evolution occurs, resulting in formation of a system of pores and channels. Fibers synthesized at 550°–600°C have large values of specific surface (above 150 m^2/g), the majority of pores having a radius of 2–2.2 nm. With the sintering temperature growing up

Table I. Parameters of ZrO$_2$ Fiber Crystal Structure

Oxide composition	Temperature (°C)	Lattice parameters (Å)				Phase composition[§]
		a	b	c	c/a	
ZrO$_2$ (Cl$^-$)	400	5.12	5.12	5.18	1.017	tetrag
	900	5.10	5.10	5.21	1.022	tetrag + monocl
	1000	5.02	5.29	5.32	*	monocl + tetrag
	1200	5.14	5.16	5.31	†	monocl
ZrO$_2$ + 10 mol% Y$_2$O$_3$ (Cl$^-$)	400	5.06	5.06	5.18	1.021	tetrag
	600	5.06	5.06	5.23	1.034	tetrag + monocl
	1000	5.15	5.15	5.17	1.004	tetrag + monocl
ZrO$_2$ + 10 mol% Y$_2$O$_3$ (NO$_3^-$)	700	5.15	5.15	5.16	1.002	tetrag
	800	5.12	5.12	5.19	1.013	tetrag + monocl
	1100	5.12	5.19	5.33	‡	monocl + tetrag
	1200	5.14	5.14	5.15	1.002	tetrag + monocl
ZrO$_2$ + 15 mol% Y$_2$O$_3$ (Cl$^-$)	600	5.12	5.12	5.15	1.006	tetrag
	1000	5.12	5.12	5.18	1.012	tetrag
	1200	5.14	5.14	5.14	1.000	tetrag

*$\alpha = 98.67$.
†$\alpha = 100.00$.
‡$\alpha = 99.15$.
[§]Abbreviations: tetrag = tetragonal; monocl = monoclinic.

to 1600°C, the specific surface value decreases and the average pore size increases up to 5–7 nm.

Figures 3 and 4 show the nature of changes in specific surface and pore distribution due to sintering temperature for ZrO$_2$ fibers. These changes in parameters of the porous structure are, to a considerable extent, caused by both the anion type of the zirconium salt and the fibrous oxide composition. As the DTG data showed, the effect of the anion of the initial salt manifests itself in the oxidation process of the polymer and consequently in the process of forming the porous structure of ZrO$_2$ fibers and their subsequent sintering. The specific surface value for fibers obtained from zirconium nitrate at 550°C was 30% greater than that for ZrO$_2$ fibers obtained from zirconium chloride. Therefore, in the first case the fibers were of increased reactivity and the sintering rate was greater. This is confirmed by the pore-distribution data for samples annealed at temperatures up to 1000°C, the average pore radius of the samples being 1.5 times greater.

Addition of Y$_2$O$_3$ results in moderation of particle growth in ZrO$_2$, slowing of the sintering rate, and formation of macropores in the material. An increased solute results in an increase in the effect of the additive on the process mentioned above.

Study of the sintering kinetics for fibrous ZrO$_2$ of various compositions made it possible to calculate the sintering activation energy of the dispersed refractory oxide in the fiber. The activation energy of the sintering process changes from 85 kJ/mol for ZrO$_2$ to 92 kJ/mol for ZrO$_2$ containing 15 mol% Y$_2$O$_3$.

The initial fiber of hydrated cellulose has a rib surface; this form is preserved during oxidation. At high oxidation rates, gaseous products that evolve may lead to a mechanical destruction of the initial polymer, which causes a sharp lowering of quality of fibers synthesized. Figure 5 shows that the fiber diameter decreases considerably with oxidation, due to the removal of the organic component and thermal dissociation of salts present in the fibers. Shrinkage of a fiber oxidized at

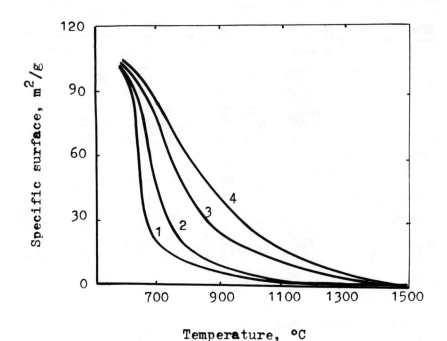

Fig. 3. Specific surface vs annealing temperature: (1) ZrO_2 + 10 mol% Y_2O_3 (NO_3^-); (2) ZrO_2 (Cl^-); (3) ZrO_2 + 10 mol% Y_2O_3 (Cl^-); (4) ZrO_2 + 15 mol% Y_2O_3 (Cl^-).

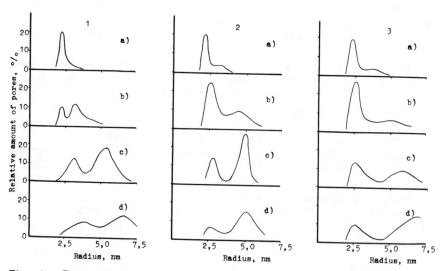

Fig. 4. Pore distribution in ZrO_2 fibers: (1) ZrO_2 (Cl^-); (2) ZrO_2 + 10 mol% Y_2O_3 (Cl^-); (3) ZrO_2 + 10 mol% Y_2O_3 (NO_3^-), annealed in air at temperatures of (a) 600°, (b) 800°, (c) 1000°, and (d) 1200°C.

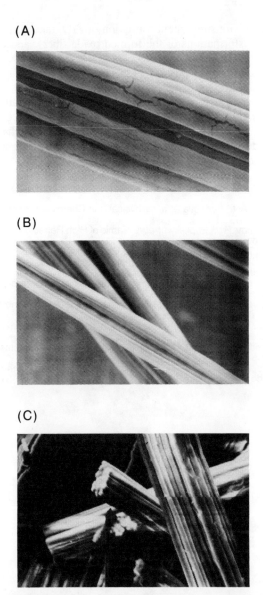

Fig. 5. Microstructure of ZrO_2 fibers annealed at temperatures of (A) 80°C (rapid heating), (B) 600°C, and (C) 1000°C (slow heating).

temperatures up to 1000°C is 60–65% in length and 70–75% in diameter. Shrinkage irregularity, both in the longitudinal and in the cross-sectional directions, is accounted for by the initial fiber anisotropy. The final diameter of the oxidic fibers is 7–9 μm.

Conclusions

The process of forming oxidic fibers is found to involve two stages. At the first stage, with heat treatment up to 600°C, thermal transformations of the polymer and salt dissociation with subsequent formation of ZrO_2 take place. At the second stage, structural transformations of the formed oxide fibers, and changes in their physical–chemical parameters take place.

Synthesized oxide fibers possess high porosity, high surface area, and increased reactivity due to the dispersed condition of ZrO_2 in the material. Homogeneity of the refractory fibrous oxide can be controlled by adding stabilizing oxides and by changing the anion type of the impregnating zirconium salt.

The synthesis of ZrO_2 by thermolysis of salt-bearing organic fibers in air makes it possible to produce high-porosity materials possessing a predetermined composition and good properties.

References

[1] G. V. Samsonov, A. L. Borisova et al.; Physical and Chemical Properties of the Oxides; Metallurgia, Moscow, 1977: 472 pp.
[2] D. N. Poluboyarinov, R. Y. Popilsky et al; Ceramic of High Refractory Oxides. Metallurgia, Moscow, 1977: 304 pp.
[3] W. D. Kingery; Kinetics of High-Temperature Processes. Wiley & Sons, New York, 1959.
[4] N. Claussen, "Stress-Induced Transformation of Tetragonal ZrO_2 Particles in Ceramic Materials," J. Am. Ceram. Soc., **61** [1] 83–86 (1978).
[5] A. A. Konkin, G. I. Kudryavcev, A. M. Schetinin et al.; Thermo-, Refractory and Uncombustible Fibres. Chimia, Moscow, 1978; 42 pp.
[6] T. Y. Kosolapova; pp. 16–20 in Fibrous and Disperse-Armed Composite Materials. Nauka, Moscow, 1976.
[7] A. Levitt and C. Wolf; pp. 220–44 in Army Materials and Mechanism Research Center, Watertown, MA. Edited by A. Levitt. Wiley & Sons, New York, 1970.
[8] R. C. Garvie and P. S. Nicholson, "Phase Analysis in Zirconia Systems," J. Am. Ceram. Soc., **55** [6] 303–305 (1972).
[9] I. N. Ermolenko, I. L. Fyodorova, T. M. Ulyanova, and P. A. Vityaz, "Thermal and Structure Conversions of Oxide Fibres during Synthesis," DAN BSSR, **26** [7] 628–31 (1982).

Epilogue

R. J. Brook
Department of Ceramics
University of Leeds
Leeds LS2 9JT, U.K.

There are disadvantages in attempting to summarize the progress achieved at a particular meeting such as Zirconia '83; the problem is that a summary tends either to be limited to a set of bland and general phrases or it risks getting too close to the arguments, with the consequent dangers of displaying partisanship or credulity. Nonetheless, Zirconia '83 has been an important stage in the development of the subject and it is perhaps worthwhile to pick out some of the more striking aspects of this development. The following paragraphs accordingly summarize one person's view of the changes in the science and technology of zirconia that have occurred since Zirconia '80 and that have been reflected by the two meetings.

The organizers have made a helpful breakdown of the subject into six topics; the common features between these topics allow them to be treated in three groups.

Processing and Structural Applications

One great difference between the two conferences lies in the fact that the two topics of processing and structural applications which could almost be omitted from the first meeting now find themselves promoted onto the program as the opening sections. It is an attractive feature of the subject that a rapid interplay between the academic and industrial communities has been produced by the great potential of zirconia materials; this interplay quite naturally has been very much concerned with fabrication and applications.

On the processing side, the papers at the meeting have given the most attention to powders and powder preparation; a second important topic has been the study and control of liquid phases or boundary phases in zirconia microstructures. In both groups, substantial progress has been made, and the quality of ceramic on display in the meeting exhibition has been a most encouraging indicator of the quality of response which the ceramics community can bring to an identified objective. Processing targets become more difficult (polycrystalline, 0.03 μm grain size materials have been mentioned as objectives) and the processing sector will no doubt continue to require this higher level of attention.

The discussions of structural applications have also been much changed from those of Zirconia '80, an important development being that, with the availability of a full range of commercial zirconia ceramics, we now have some feedback from markets and the chance of focusing more clearly on the potential applications. The existence of several sources for tetragonal zirconia polycrystals, for example, affords an excellent opportunity for the identification of real as opposed to possible applications. At the same time, as the papers at the meeting show, this section is still a very volatile and provisional one: the overselling of Si_3N_4 is a widely recognized lesson in the difficulties of developing ceramics for structural applications and much energy, persistence, and imagination will be needed to bring zirconia to all the markets suggested by its properties.

Physical Chemistry and Nonstructural Applications

Work on these topics (phase equilibria, point defects, electrolyte applications of zirconia) had already reached maturity by the time of Zirconia '80, and the interim has really been a matter of further steady development. The consequence of this more restrained rate of advance as compared with the other sectors has been that the whole group of six themes are on a much more equal footing now than was the case three years ago. This trend seems likely to continue.

Phase Transformations and Transformation-Toughening

These topics, while well represented at Zirconia '80, have the common factor that both have been the scene of notable advances in the intervening years. In respect to phase transformations, the field has been much expanded by the realization of the wide range of possible transformations that can occur in these systems, particularly when lower-temperature changes are taken into account. An additional feature is the complexity of the transformation mechanisms that is now foreseen; as indicated by the surface energy/nucleation debate that still centers around the explanations for the size effect in the tetragonal/monoclinic transformation in zirconia, this complexity can give great stimulus to work and can lead to results that are of interest well beyond the immediate concerns of zirconia. Since the debate is active and ably defended on both sides, further comment here would be out of place; it is worth mentioning, however, that an earlier (and equally strenuously defended) debate — on the validity of microcracking as a mechanism — has also been a most fruitful guide to work of significance. (An interesting conclusion from the earlier debate is perhaps also worth bearing in mind, namely that the eventual truth can lie closer to the middle ground than participants have at first been inclined to believe!)

The topic of transformation-toughening has made impressive progress in the time since Zirconia '80: the recognition of multiple toughening mechanisms, the realization of the wide range of microstructural designs to exploit these mechanisms, the availability of successful processing methods, and the elegance of the microstructural characterization that has been employed have made this a most exciting and convincing topic. As consideration of the papers at the meeting shows, this is an area undergoing very rapid advance.

General Comments

Even the most dispassionate observer would, I think, concede that there has been something special about the topic and mood of Zirconia '83. In part this has arisen from the mix of industrial and academic participants; the extent of collaboration, sometimes active but never less than at the level where each pays close attention to what the other is saying or doing, has been a clear stimulus to the rapid progress. In part it has arisen quite simply from the quality of the work; in terms of the criteria that have been proposed to assess scientific progress, the subject has excellent qualifications. In intrinsic terms, the field is ripe for study (the central and necessary role played by the electron microscope is a persuasive element in this respect), and the appropriate expertise is evidenced by the work presented at the meeting; in extrinsic terms, the property improvements are dramatic and the insights into the transformation behavior have interest beyond the immediate zirconia community (the diversity of background of those attending is striking in this regard). So far, zirconia has amply repaid the attention it has received.

At the time of Zirconia '80, the concept of a conference on zirconia looked premature, if not artificial. The conference in Cleveland had the major attribute that it did start discussions between the mechanical and electrical sides of the topic and between the academic and industrial groups interested in the material. By the time of Zirconia '83, the concept of such a conference has become a natural one; indeed, the balance that lies across the subject, the variety of potential or achieved applications, and the opportunities for good science are such that other branches of ceramics may begin to see these meetings as something of a model! May we hope that Zirconia '86 can build on the achievement of Zirconias '80 and '83 without losing the spontaneity and excitement that are associated with a subject that is still confidently seeking its way.

Author Index

Abelard, P. See Baumard, J. F.
Allen, R. V. The Reaction-Bonded Zirconia Oxygen Sensor: An Application for Solid-State Metal-Ceramic Reaction-Bonding, 537
Andersson, C. A. Diffusionless Transformations in Zirconia Alloys, 78
Anthony, A. M. Collaborative Study on ZrO_2 Oxygen Gauges, 627
Arora, A. See Clarke, D. R.
Badwal, S. P. S. Low-Temperature Behavior of ZrO_2 Oxygen Sensors, 598
Bannister, M. J. See Badwal, S. P. S.
Bauer, G. See Vizethum, F.
Baumard, J. F. Defect Structure and Transport Properties of ZrO_2-Based Solid Electrolytes, 555
Baumard, J. F. See Anthony, A. M.
Bender, B. A. See Ingel, R. P.
Binner, J. G. P. Improvement in the Toughness of β''-Alumina by Incorporation of Unstabilized Zirconia Particles, 428
Bischoff, E. See Kriven, W. M.
Boch, P. Plasma-Sprayed Zirconia Coatings, 488
Bohrer, P. See Krafthefer, B.
Bonanos, N. See Butler, E. P.
Bonne, U. See Krafthefer, B.
Borbidge, W. E. See Allen, R. V.
Brandon, D. G. See Chaim, R.
Brett, N. H. See Wright, A. F.
Brook, R. J. See Wu, S.
Buresch, F. E. Microcrack Extension in Microcracked Dispersion-Toughened Ceramics, 306
Burggraaf, A. J. See van de Graaf, M. A. C. G.
Burns, S. J. See Seyler, R. J.
Butler, E. P. Microstructural-Electrical Property Relationships in High-Conductivity Zirconias, 572
Buykx, W. J. Thermal Diffusivity of Zirconia Partially and Fully Stabilized with Magnesia, 518
Caneiro, A. See Fouletier, J.
Chaim, R. Short-Range Order Phenomena in ZrO_2 Solid Solutions, 86
Chen, I. -W. Martensitic Transformations in ZrO_2 and HfO_2—An Assessment of Small-Particle Experiments with Metal and Ceramic Matrices, 33
Chiao, Y.-H. See Chen, I. -W.
Choudhry, M. A. Theory of Twinning and Transformation Modes in ZrO_2, 46
Clarke, D. R. Acoustic Emission Characterization of the Tetragonal-Monoclinic Phase Transformation in Zirconia, 54
Claussen, N. Microstructural Design of Zirconia-Toughened Ceramics (ZTC), 325
Claussen, N. See Rühle, M.
Claussen, N. See Schubert, H.
Claussen, N. See Wallace, J. S.
Corish, J. See Anthony, A. M.
Corman, G. S. See Stubican, V. S.
Crocker, A. G. See Choudhry, M. A.
Davis, B. I. See Lange, F. F.
De Aza, S. See Pena, P.
Drennan, J. See Butler, E. P.
Dworak, U. ZrO_2 Ceramics for Internal Combustion Engines, 480
Erdle, E. Preparation and Operation of Zirconia High-Temperature Electrolysis Cells for Hydrogen Production, 685
Esper, F. J. Mechanical, Thermal, and Electrical Properties in the System of Stabilized ZrO_2 $(Y_2O_3)/\alpha-Al_2O_3$, 528
Esper, F. J. See Erdle, E.
Evans, A. G. Toughening Mechanisms in Zirconia Alloys, 193
Faber, K. T. Microcracking Contributions to the Toughness of ZrO_2-Based Ceramics, 293
Farmer, S. C. Diffusional Decomposition of c-ZrO_2 in Mg-PSZ, 152

Fauchais, P. See Boch, P.
Fingerle, D. See Dworak, U.
Fouletier, J. Accurate Monitoring of Low Oxygen Activity in Gases with Conventional Oxygen Gauges and Pumps, 618
Friese, K. H. See Erdle, E.
Friese, K. H. See Esper, F. J.
Fukuura, I. See Watanabe, W.
Fyodorova, I. L. See Yermolenko, I. N.
Garrett, W. G. See Badwal, S. P. S.
Garvie, R. C. Structural Applications of ZrO_2-Bearing Materials, 465
Geier, H. See Esper, F. J.
Goge, M. Low-Temperature Properties of Samaria-Stabilized Zirconia, 585
Gouet, M. See Goge, M.
Green, D. J. Residual Surface Stresses in Al_2O_3-ZrO_2 Composites, 240
Greggi, J. See Andersson, C. A.
Gupta, T. K. See Andersson, C. A.
Haberko, K. Preparation of Ca-Stabilized ZrO_2 Micropowders by a Hydrothermal Method, 774
Hangas, J. Ordered Compounds in the System CaO-ZrO_2, 107
Hannink, R. H. J. See Rossell, H. J.
Hannink, R. H. J. See Swain, M. V.
Hausner, H. See Roosen, A.
Hellmann, J. R. See Stubican, V. S.
Heuer, A. H. Phase Transformations in ZrO_2-Containing Ceramics: I, The Instability of c-ZrO_2 and the Resulting Diffusion-Controlled Reactions, 1
Heuer, A. H. See Hangas, J.
Heuer, A. H. See Farmer, S. C.
Heuer, A. H. See Kibbel, B. W.
Heuer, A. H. See Lanteri, V.
Heuer, A. H. See Rühle, M.
Heuer, A. H. See Schoenlein, L. H.
Hobbs, L. W. See Kalonji, G.
Hori, S. Al_2O_3-ZrO, Ceramics Prepared from CVD Powders, 794
Ichise, I. See Iwase, M.
Iio, S. See Watanabe, W.
Ingel, R. P. Physical, Microstructural, and Thermomechanical Properties of ZrO_2 Single Crystals, 408
Iwase, M. Mixed Ionic and Electronic Conduction in Zirconia and Its Application in Metallurgy, 646
Jacob, K. T. See Iwase, M.
James, M. R. See Green, D. J.
Janke, D. Oxygen Sensing in Iron- and Steelmaking, 636
Kalonji, G. Applications of Rapid Solidification Theory and Practice to Al_2O_3-ZrO_2 Ceramics, 816
Kawata, Y. See Shindo, I.
Keller, R. J. See Suhr, D. S.
Kibbel, B. W. Ripening of Inter- and Intragranular ZrO_2 Particles in ZrO_2-Toughened Al_2O_3, 415
Kobayashi, K. See Nakajima, K.
Koch, A. See Erdle, E.
Kolar, D. See Kosmac, T.
Kosmac, T. Diffusion Processes and Solid-State Reactions in the Systems Al_2O_3-ZrO_2 (Stabilizing Oxide)(Y_2O_3, CaO, MgO), 546
Kosuda, K. See Shindo, I.
Krafthefer, B. Life and Performance of ZrO_2-Based Oxygen Sensors, 607
Kraus, B. See Rühle, M.
Kreher, W. See Pompe, W.
Kriven, W. M. Anomalous Thermal Expansion in Al_2O_3-15 Vol% $(Zr_{0.5}Hf_{0.5})O_2$, 425
Kriven, W. M. The Transformation Mechanism of Spherical Zirconia Particles in Alumina, 64
Krohn, U. See Dworak, U.
Kubota, Y. See Tsukuma, K.
Lange, F. F. Sinterability of ZrO_2 and Al_2O_3 Powders: The Role of Pore Coordination Number Distribution, 699

Lange, F. F. See Green, D. J.
Lanteri, V. Tetragonal Phase in the System ZrO_2-Y_2O_3, 118
Lee, S. See Seyler, R. J.
Letisse, G. See Goge, M.
Lewis, D. See Ingel, R. P.
Lombard, D. See Boch, P.
Mader, W. See Schmauder, S.
Matsui, M. Effect of Microstructure on the Strength of Y-TZP Components, 371
Mazerolles, L. See Michel, D.
McKittrick, J. See Kalonji, G.
Meonkhaus, P. See Krafthefer, B.
Michel, D. Polydomain Crystals of Single-Phase Tetragonal ZrO_2: Structure, Microstructure, and Fracture Toughness, 131
Michel, D. Relationship Between Morphology and Structure for Stabilized Zirconia Crystals, 455
Mitchell, T. E. See Farmer, S. C.
Mitchell, T. E. See Hangas, J.
Mitchell, T. E. See Lanteri, V.
Mitchell, T. E. See Suhr, D. S.
Müller, I. Size Effect on Transformation Temperature of Zirconia Powders and Inclusions, 443
Müller, W. See Müller, I.
Murata, Y. See Nakajima, K.
Näfe, H. See Reetz, T.
Nakajima, K. Phase Stability of Y-PSZ in Aqueous Solutions, 399
Neidrach, L. W. Application of Zirconia Membranes as High-Temperature pH Sensors, 672
Nunn, S. See Wright, A. F.
Oda, I. See Matsui, M.
Opalinski, H. See Dworak, U.
Pampuch, R. ZrO_2 Micropowders as Model Systems for the Study of Sintering, 733
Paulus, M. See Rakotoson, L.
Pena, P. Compatibility Relationships of Al_2O_3 and ZrO_2 in the System ZrO_2-Al_2O_3-SiO_2-CaO, 174
Perez y Jorba, M. See Michel, D.
Pertl, L. See Krafthefer, B.
Petzow, G. See Wallace, J. S.
Pompe, W. Theoretical Approach to Energy-Dissipative Mechanisms in Zirconia and Other Ceramics, 283
Pyda, W. See Haberko, K.
Rakotoson, L. Sintering of a Freeze-Dried 10 Mol% Y_2O_3-Stabilized Zirconia, 727
Reetz, T. Influence of Impurities in Solid Electrolytes on the Voltage Response of Solid Electrolyte Galvanic Cells, 591
Rettig, D. See Reetz, T.
Rice, R. W. See Ingel, R. P.
Rogeaux, B. See Boch, P.
Roosen, A. Sintering Kinetics of ZrO_2 Powders, 714
Rossell, H. J. The Phase $Mg_2Zr_5O_{12}$ in MgO Partially Stabilized Zirconia, 139
Ruh, R. Properties of Metal-Modified and Nonstoichiometric ZrO_2, 544
Rühle, M. In-Situ Observations of Stress-Induced Phase Transformations in ZrO_2-Containing Ceramics, 256
Rühle, M. Microstructural Studies of Y_2O_3-Containing Tetragonal ZrO_2 Polycrystals (Y-TZP), 352
Rühle, M. Phase Transformations in ZrO_2-Containing Ceramics: II, The Martensitic Reaction in t-ZrO_2, 14
Rühle, M. See Heuer, A. H.
Rühle, M. See Schmauder, S.
Rühle, M. See Schoenlein, L. H.

Rühle, M. See Schubert, H.
Schäfer, W. See Erdle, E.
Schmauder, S. Calculations of Strain Distributions in and around ZrO_2 Inclusions, 251
Schoenlein, L. H. In-Situ Straining Experiments of Mg-PSZ Single Crystals, 275
Schubert, H. Preparation of Y_2O_3-Stabilized Tetragonal Polycrystals (Y-TZP) from Different Powders, 766
Senft, G. See Stubican, V. S.
Seyler, R. J. A Thermodynamic Approach to Fracture Toughness in PSZ, 213
Shindo, I. Phase Relations in the Ternary System ZrO_2-Al_2O_3-SiO_2 by the Slow-Cooling Float-Zone Method, 181
Siebert, E. See Fouletier, J.
Slotwinski, R. K. See Butler, E. P.
Some, T. See Matsui, M.
Sōmiya, S. See Hori, S.
Sōmiya, S. See Toraya, H.
Steele, B. C. H. See Butler, E. P.
Stevens, R. See Binner, J. G. P.
Strecker, A. See Rühle, M.
Stubican, V. S. Phase Relationships in Some ZrO_2 Systems, 96
Suhr, D. S. Microstructure and Durability of Zirconia Thermal Barrier Coatings, 503
Suzuki, T. See Shindo, I.
Swain, M. V. R-Curve Behavior in Zirconia Ceramics, 225
Swain, M. V. See Buykx, W. J.
Takahashi, R. See Hori, S.
Takekawa, S. See Shindo, I.
Tan, S. R. See Binner, J. G. P.
Thomas, G. See van Tendeloo, G.
Tomandl, G. See Vizethum, F.
Toraya, H. Preparation of Mixed Fine Al_2O_3-HfO_2 Powders by Hydrothermal Oxidation, 806
Trontelj, M. See Kosmac, T.
Tsukidate, T. See Tsukuma, K.
Tsukuma, K. Thermal and Mechanical Properties of Y_2O_3-Stabilized Tetragonal Zirconia Polycrystals, 382
Ulmer, G. C. ZrO_2, Oxygen and Hydrogen Sensors: A Geologic Perspective, 660
Ulyanova, T. M. See Yermolenko, I. N.
van de Graaf, M. A. C. G. Wet-Chemical Preparation of Zirconia Powders: Their Microstructure and Behavior, 744
van Tendeloo, G. High-Resolution Microscopy Investigation of the System ZrO_2-ZrN, 164
Vardelle, M. See Boch, P.
Vityaz, P. A. See Yermolenko, I. N.
Vizethum, F. Computer-Controlled Adjustment of Oxygen Partial Pressure, 631
Waidlich, D. See Rühle, M.
Wallace, J. S. Microstructure and Property Development of In Situ-Reacted Mullite-ZrO_2 Composites, 436
Watanabe, W. Aging Behavior of Y-TZP, 391
Whelan, P. T. See Allen, R. V.
Wright, A. F. Growth and Coarsening of Pure and Doped ZrO_2: A Study of Microstructure by Small-Angle Neutron Scattering, 784
Wu, S. Processing Techniques for ZrO_2 Ceramics, 693
Yermolenko, I. N. Synthesis and Sintering of ZrO_2 Fibers, 826
Yoshimura, M. See Hori, S.
Yoshimura, M. See Toraya, H.
Zook, D. See Krafthefer, B.

Subject Index

Acoustic emission characterization of the tetragonal-monoclinic phase transformation in zirconia, 54
Aging behavior of Y-TZP, 391
Alloys, ZrO_2, diffusionless transformations in, 78
  toughening mechanisms in, 193
Alumina and ZrO_2, compatibility relationship of in the system ZrO_2-Al_2O_3-SiO_2-CaO, 174
  powders, sinterability of: role of pore coordination number distribution, 699
  β''-, improvement in the toughness of by incorporation of unstabilized zirconia particles, 428
  -HfO_2 powders, mixed, fine, preparation of by hydrothermal oxidation, 806
  transformation mechanism of zirconia particles in, 64
  -ZrO_2 ceramics, applications of rapid solidification theory and practice to, 816
    prepared from CVD powders, 794
    composites, residual surface stresses in, 240
    (stabilizing oxide) (Y_2O_3, CaO, MgO), diffusion processes and solid-state reactions in, 546
    tetragonal phase in the system, 118
    -toughened, ripening of inter- and intragranular ZrO_2 particles in, 415
Behavior, low-temperature, of ZrO_2 oxygen sensors, 598
Calcia-ZrO_2, ordered compounds in the system, 107
Calcium-stabilized zirconia micropowders by a hydrothermal method, 774
Calculations of strain distributions in and around ZrO_2 inclusions, 251
Characterization, acoustic emission, of the tetragonal-monoclinic phase transformation in zirconia, 54
Cells, electrolysis, zirconia high-temperature, for hydrogen production, preparation and operation of, 685
  solid electrolyte galvanic, influence of impurities in solid electrolytes on the voltage response of, 591
Ceramics, Al_2O_3-ZrO_2, applications of rapid solidification theory and practice to, 816
  prepared from CVD powders, 794
  and metal matrices, an assessment of small-particle experiments with—martensitic transformations in ZrO_2 and HrO_2, 33
  microcracked dispersion-toughened, macrocrack extension in, 306
  ZrO_2-based, microcracking contributions to the toughness of, 293
  ZrO_2-containing, in situ observations of stress-induced phase transformations in, 256
  R curve behavior in, 225
  phase transformations in—the instability of c-ZrO_2 and the resulting diffusion-controlled reactions, 1
  processing techniques for, 693
  -toughened (TZC), microstructural design of, 325
Coarsening and growth of pure and doped ZrO_2: a study of microstructure by small-angle neutron scattering, 784
Coatings, plasma-sprayed zirconia, 488
  thermal barrier, zirconia, microstructure and durability of, 503
Combustion, internal, engines, ZrO_2 ceramics for, 480
Compatibility relationship of Al_2O_3 and ZrO_2 in the system ZrO_2-Al_2O_3-SiO_2-CaO, 174
Components, Y-TZP, effect of microstructure on strength of, 371
Composites, Al_2O_3-ZrO_2, residual surface stresses in, 240
  in situ-reacted mullite-ZrO_2, microstructure and property development of, 436
Compounds, ordered, in the system CaO-ZrO_2, 107
Computer-controlled adjustment of oxygen partial pressure, 631
Conduction, mixed ionic and electronic, in zirconia and its application in metallurgy, 646
Crystals, polydomain, of single-phase tetragonal ZrO_2: structure, microstructure, and fracture toughness, 131
  single, of Mg-PSZ, in situ straining experiments of, 275
  ZrO_2, physical, microstructural, and thermomechanical properties of, 408
  stabilized zirconia, relation between morphology and structure for, 455
Decomposition, diffusional, of c-ZrO_2 in Mg-PSZ, 152
Defect structure and transport properties of ZrO_2-based solid electrolytes, 555
Design, microstructural, of zirconia-toughened ceramics (ZTC), 325
Diffusion-controlled reactions and the instability of c-ZrO_2—phase transformations in ZrO_2-containing ceramics, 1
  processes and solid-state reactions in the systems Al_2O_3-ZrO_2 (stabilizing oxide) (Y_2O_3, CaO, MgO), 546
Diffusivity thermal, of zirconia partially and fully stabilized with magnesia, 518
Dispersion-toughened ceramics, microcracked, macrocrack extension in, 306
Durability and microstructure of zirconia thermal barrier coating, 503
Electrolysis cells, zirconia high-temperature for hydrogen production, preparation and operation of, 685
Electrolytes, solid, influence of impurities in on the voltage response of solid electrolyte galvanic cells, 591
  ZrO_2-based, defect structure and transport properties of, 555
Emission, acoustic, characterization of the tetragonal-monoclinic phase transformation in zirconia, 54
Energy-dissipative mechanisms in zirconia and other ceramics, a theoretical approach to, 283
Engines, internal combustion, ZrO_2 ceramics for, 480
Expansion, anomalous, thermal, in Al_2O_3-15 vol% $(Zr_{0.5}Hf_{0.5})O_2$, 425
Fibers, ZrO_2, synthesis and sintering of, 826
Float-zone, slow-cooling, method, phase relations in the ternary system ZrO_2-Al_2O_3-SiO_2 by, 181
Fracture toughness in PSZ, a thermodynamic approach to, 213
Freeze-dried, 10 mol% Y_2O_3 stabilized zirconia, sintering of, 727
Gases, accurate monitoring of low oxygen activity in with conventional oxygen gauges and pumps, 618
Gauges and pumps, conventional oxygen, accurate monitoring of low oxygen activity in gases with, 618
  ZrO_2 oxygen, collaborative study on, 627
Growth and coarsening of pure and doped ZrO_2: a study of microstructure by small-angle neutron scattering, 784
Hafnia-Al_2O_3 powders, mixed, fine preparation of by hydrothermal oxidation, 806
  and ZrO_2, martensitic transformations in—an assessment of small-particle experiments with metal and ceramic matrices, 33
Hydrogen and oxygen ZrO_2 sensors: a geologic perspective, 660
  production, preparation, and operation of zirconia high-temperature electrolysis cells for, 685
Impurities, influence of in solid electrolytes on the voltage response of solid electrolyte galvanic cells, 591

Inclusions, ZrO_2, calculations of strain distributions in and around, 251
Instability of $c\text{-}ZrO_2$ and the resulting diffusion-controlled reactions—phase transformations in ZrO_2-containing ceramics, 1
Iron- and steelmaking, oxygen sensing in, 636
Kinetics, sintering, of ZrO_2 powders, 714
Life and performance of ZrO_2-based oxygen sensors, 607
Macrocracks, extension in microcracked dispersion-toughened ceramics, 306
Magnesia, thermal diffusivity of zirconia partially and fully stabilized with, 518
Martensitic reaction, in $t\text{-}ZrO_2$, 14
  transformations in ZrO_2 and HfO_2—an assessment of small-particle experiments with metal and ceramic matrices, 33
Materials, ZrO_2-bearing, structural applications of, 465
Matrices, metal and ceramic, an assessment of small-particle experiments with—martensitic transformations in ZrO_2 and HfO_2, 33
Mechanisms, energy-dissipative, in zirconia and other ceramics, a theoretical approach to, 283
  toughening, in zirconia alloys, 193
Membranes, zirconia, application of as high-temperature pH sensors, 672
Metal and ceramic matrices, an assessment of small-particle experiments with—martensitic transformations in ZrO_2 and HfO_2, 33
  -modified and nonstoichiometric zirconia, properties of, 544
Metallurgy, mixed ionic and electronic conduction in zirconia and its application in, 646
Microcracking contributions to the toughness of ZrO_2-based ceramics, 293
  dispersion-toughened ceramics, macrocrack extension in, 306
Micropowders, Ca-stabilized zirconia, preparation of by a hydrothermal method, 774
  zirconia, as model systems for the study of sintering, 733
Microscopy, high-resolution, investigation of the system $ZrO_2\text{-}ZrN$, 164
Microstructure and durability of zirconia thermal barrier coating, 503
  and fracture toughness, of polydomain crystals of single-phase tetragonal ZrO_2, 131
  property development of *in situ*-reacted mullite-ZrO_2 composites, 436
  effect of on the strength of Y-TZP components, 371
  study of by small-angle neutron scattering: growth and coarsening of pure and doped ZrO_2, 784
Modes, transformation, and twinning, theory of in zirconia, 46
Monitoring, accurate, of low oxygen activity in gases with conventional oxygen gauges and pumps, 618
Morphology and structure, relation between for stabilized zirconia crystals, 455
Mullite-ZrO_2 composites, *in situ*-reacted, microstructure and property development of, 436
Neutron scattering, small-angle, a study of microstructure by: growth and coarsening of pure and doped ZrO_2, 784
Oxidation, hydrothermal, preparation of mixed, fine $Al_2O_3\text{-}HfO_2$ powders by, 806
Oxides, stabilizing (Y_2O_3, CaO, MgO) in the system $Al_2O_3\text{-}ZrO_2$, diffusion processes and solid-state reactions in, 546
Oxygen activity, low, accurate monitoring of in gases with conventional oxygen gauges and pumps, 618
  and hydrogen ZrO_2 sensors: a geologic perspective, 660
  gauges and pumps, conventional, accurate monitoring of low oxygen activity in gases with, 618

ZrO_2, collaborative study on, 627
  partial pressure, computer-controlled adjustment of, 631
  sensing in iron- and steelmaking, 636
  sensor, reaction-bonded zirconia: an application for solid-state metal-ceramic reaction-bonding, 537
  ZrO_2-based, life and performance of, 607
  ZrO_2, low-temperature behavior of, 598
Particles, inter- and intragranular ZrO_2, ripening of in ZrO_2-toughened Al_2O_3, 415
  small, an assessment of experiments with metal and ceramic matrices—martensitic transformations in ZrO_2 and HfO_2, 33
  unstabilized zirconia, improvement in the toughness of β''-alumina by the incorporation of, 428
  zirconia, in alumina, transformation mechanism of, 64
Phase $Mg_2Zr_5O_{12}$ in MgO partially stabilized zirconia, 139
  relations in the ternary system $ZrO_2\text{-}Al_2O_3\text{-}SiO_2$ by the slow-cooling, float-zone method, 181
  in some ZrO_2 systems, 96
  stability of Y-PSZ in aqueous solutions, 399
  transformations in ZrO_2-containing ceramics—the instability of $c\text{-}ZrO_2$ and the resulting diffusion-controlled reactions, 1
  martensitic reaction in $t\text{-}ZrO_2$, 14
  stress-induced, in ZrO_2-containing ceramics, *in situ* observations of, 256
  tetragonal, in the system $ZrO_2\text{-}Y_2O_3$, 118
PSZ, a thermodynamic approach to fracture toughness in, 213
Phenomena, short-range order, in ZrO_2 solid solutions, 86
Plasma-sprayed zirconia coatings, 488
Polycrystals, Y_2O_3-containing tetragonal ZrO_2 (Y-TZP), microstructural studies of, 352
  stabilized tetragonal ZrO_2 (Y-TZP), preparation of from different powders, 766
  zirconia, Y_2O_3-stabilized tetragonal, thermal and mechanical properties of, 382
Polydomain crystals of single-phase tetragonal ZrO_2: structure, microstructure, and fracture toughness, 131
Powders, CVD, $Al_2O_3\text{-}ZrO_2$ ceramics prepared from, 794
  different, preparation of Y_2O_3-stabilized ZrO_2 polycrystals (Y-TZP) from, 766
  mixed fine $Al_2O_3\text{-}HfO_2$, preparation of by hydrothermal oxidation, 806
  ZrO_2 and Al_2O_3, sinterability of: role of pore coordination number distribution, 699
  sintering kinetics of, 714
  wet-chemical preparation of: their microstructure and behavior, 744
Preparation and operation of zirconia high-temperature electrolysis cells for hydrogen production, 685
Pressure, partial, oxygen, computer-controlled adjustment of, 631
Processing techniques for ZrO_2 ceramics, 693
Properties, low-temperature, of samaria-stabilized zirconia, 585
  mechanical, thermal, and electrical, in the system of stabilized $ZrO_2(Y_2O_3)/\alpha\text{-}Al_2O_3$, 528
  of metal-modified and nonstoichiometric zirconia, 544
  physical, microstructural, and thermomechanical, of ZrO_2 single crystals, 408
  thermal and mechanical, of Y_2O_3-stabilized tetragonal zirconia polycrystals, 382
  development and microstructure of *in situ*-reacted mullite-ZrO_2 composites, 436
PSZ-Mg single crystals, *in situ* straining experiments of, 275
Pumps and gauges, conventional oxygen, accurate monitoring of low oxygen activity in gases with, 618

R curve behavior in zirconia ceramics, 225
Reaction-bonded zirconia oxygen sensor: an application for solid-state metal-ceramic reaction-bonding, 537
—bonding, solid-state metal-ceramic, an application for the reaction-bonded zirconia oxygen sensor, 537
Reactions, solid-state, and diffusion processes in the systems Al_2O_3-ZrO_2 (stabilizing oxide) (Y_2O_3, CaO, MgO), 546
Relationships, compatibility of Al_2O_3 and ZrO_2 in the system ZrO_2-Al_2O_3-SiO_2-CaO, 174
phase, in some ZrO_2 systems, 96
Ripening of inter- and intragranular ZrO_2 particles in ZrO_2-toughened Al_2O_3, 415
Samaria-stabilized zirconia, low-temperature properties of, 585
Sensors, oxygen, ZrO_2, low-temperature behavior of, 598
pH, high-temperature, application of zirconia membranes as, 672
reaction-bonded zirconia oxygen: an application for solid-state metal-ceramic reaction-bonding, 537
ZrO_2-based oxygen, life and performance of, 607
oxygen and hydrogen: a geologic perspective, 660
Sintering and synthesis of ZrO_2 fibers, 826
kinetics of ZrO_2 powders, 714
of a freeze-dried 10 mol% Y_2O_3-stabilized zirconia, 727
zirconia micropowders as model systems for the study of, 733
Size effect on transformation temperature of zirconia powders and inclusions, 443
Solidification, rapid, theory and practice, applications of to Al_2O_3-ZrO_2 ceramics, 816
Solutions, aqueous, phase stability of Y-PSZ in, 399
solid, ZrO_2, short-range order phenomena in, 86
Stability, phase, of Y-PSZ in aqueous solutions, 399
Steelmaking, and iron, oxygen sensing in, 636
Strain distributions, calculations of, in and around ZrO_2 inclusions, 251
Straining experiments, *in situ*, of Mg-PSZ single crystals, 275
Strength of Y-TZP components, effect of microstructure on, 371
Stress-induced phase transformations in ZrO_2-containing ceramics, *in situ* observations of, 256
surface, residual, in Al_2O_3-ZrO_2 composites, 240
Structure and morphology, relation between for stabilized zirconia crystals, 455
applications of ZrO_2-bearing materials, 465
defect, and transport properties of ZrO_2-based solid electrolytes, 555
microstructure, and fracture toughness, of polydomain crystals of single-phase tetragonal ZrO_2, 131
Surface stresses, residual, in Al_2O_3-ZrO_2 composites, 240
Synthesis and sintering of ZrO_2 fibers, 826
Systems CaO-ZrO_2, ordered compounds in, 107
model, for the study of sintering, zirconia micropowders, 733
ZrO_2, phase relationships in, 96
ZrO_2-Al_2O_3-SiO_2, ternary, phase relations in, by the slow-cooling, float-zone method, 181
ZrO_2-Al_2O_3-SiO_2-CaO, compatibility relationship of Al_2O_3 and ZrO_2 in, 174
ZrO_2-Y_2O_3, tetragonal phase in, 118
ZrO_2-ZrN, high-resolution investigation of, 164
Temperature transformation, of zirconia powders and inclusions, size effect on, 443
Thermodynamics approach to fracture toughness in PSZ, 213
Toughening mechanisms in zirconia alloys, 193
Toughness, fracture, in PSZ, a thermodynamic approach to, 213
structure, and microstructure, of polydomain crystals of single-phase tetragonal ZrO_2, 131
of β''-alumina, improvement in by incorporation of unstabilized zirconia particles, 428
of ZrO_2-based ceramics, microcracking contributions to, 293
Transformation, diffusionless, in ZrO_2 alloys, 78
martensitic, in ZrO_2 and HfO_2—an assessment of small-particle experiments with metal and ceramic matrices, 33
mechanism of spherical zirconia particles in alumina, 64
modes, and twinning, in zirconia, theory of, 46
phase, in ZrO_2-containing ceramics—the instability of c-ZrO_2 and the resulting diffusion-controlled reactions, 1
martensitic reaction in t-ZrO_2, 14
stress-induced, in ZrO_2-containing ceramics, *in situ* observations of, 256
tetragonal-monoclinic, acoustic emission characterization in zirconia, 54
temperature of zirconia powders and inclusions, size effect on, 443
Twinning and transformation modes in zirconia, theory of, 46
Voltage response of solid electrolyte galvanic cells, influence of impurities in solid electrolytes on, 591
Yttria-TZP, aging behavior of, 391
—containing tetragonal ZrO_2 polycrystals (Y-TZP), 352
-PSZ, phase stability of in aqueous solutions, 399
-stabilized tetragonal ZrO_2 polycrystals (Y-TZP) from different powders, 766
thermal and mechanical properties of, 382
Zirconia, acoustic emission characterization of the tetragonal-monoclinic phase transformation in, 54
-Al_2O_3 ceramics, applications of rapid solidification theory and practice to, 816
prepared from CVD powders, 794
-Al_2O_3 composites, residual surface stresses in, 240
Al_2O_3-, (stabilizing oxide) (Y_2O_3, CaO, MgO), diffusion processes and solid-state reactions in, 546
alloys, toughening mechanisms in, 193
and Al_2O_3, compatibility relationship of in the system ZrO_2-Al_2O_3-SiO_2-CaO, 174
and Al_2O_3 powders, sinterability of: role of pore coordination number distribution, 699
and HfO_2, martensitic transformation in—an assessment of small-particle experiments with metal and ceramic matrices, 33
and other ceramics, theoretical approach to energy-dissipative mechanisms in, 283
-based ceramics, microcracking contributions to the toughness of, 293
oxygen sensors, life and performance of, 607
solid electrolytes, defect structure and transport properties of, 555
-bearing materials, structural applications of, 465
c-, instability of and the resulting diffusion-controlled reactions—phase transformations in ZrO_2-containing ceramics, 1
in Mg-PSZ, diffusional decomposition of, 152
CaO-, ordered compounds in the system, 107
ceramics, processing techniques for, 693
R curve behavior in, 225
coatings, plasma-sprayed, 488
-containing ceramics, *in situ* observations of stress-induced phase transformation in, 256
phase transformations in—the instability of c-ZrO_2 and the resulting diffusion-controlled reactions, 1
crystals, stabilized, relation between morphology and structure for, 455

841

fibers, synthesis and sintering of, 826
freeze-dried 10 mol% Y_2O_3 stabilized, sintering of, 727
high-conductivity, microstructural-electrical property relationships in, 572
high-temperature electrolysis cells for hydrogen production, preparation and operation of, 685
inclusions, calculations of strain distributions in and around, 251
membranes as high-temperature pH sensors, application of, 672
metal-modified and nonstoichiometric, properties of, 544
micropowders as model systems for the study of sintering, 733
mixed ionic and electronic conduction in, and its application in metallurgy, 646
-mullite composites, *in situ* reacted, microstructure and property development of, 436
oxygen and hydrogen sensors: a geologic perspective, 660
  gauges, collaborative study on, 627
  sensors, low-temperature behavior of, 598
  reaction-bonded, an application for solid-state metal-ceramic reaction-bonding in, 537
partially and fully stabilized with magnesia, thermal diffusivity of, 518
particles, inter- and intragranular, in ZrO_2-toughened Al_2O_3, ripening of, 415
unstabilized, improvement in the toughness of β''-alumina by the incorporation of, 428
polycrystals, Y_2O_3-containing tetragonal (Y-TZP), microstructural studies of, 352

Y_2O_3-stabilized tetragonal, thermal and mechanical properties of, 382
powders and inclusions, size effect on transformation temperature of, 443
  Ca-stabilized, preparation of by a hydrothermal method, 774
  sintering kinetics of, 714
  wet-chemical preparation of: their microstructure and behavior, 744
pure and doped, growth and coarsening of: a study of microstructure by small-angle neutron scattering, 784
samaria-stabilized, low-temperature properties of, 585
single crystals, physical, microstructural, and thermomechanical properties of, 408
single-phase tetragonal, polydomain crystals of: structure, microstructure, and fracture toughness, 131
solid solutions, short-range order phenomena in, 86
systems, phase relationships in, 96
theoretical approach to, 283
theory of twinning and transformation modes in, 46
-toughened Al_2O_3, ripening of inter- and intragranular ZrO_2 particles in, 415
  ceramics, microstructural design of, 325
  $(Y_2O_3)/\alpha$-Al_2O_3, stabilized, mechanical, thermal, and electrical properties of, 528
Y_2O_3-stabilized, tetragonal polycrystals (Y-TZP), preparation of from different powders, 766
-Y_2O_3, tetragonal phase in the system, 118